AA001079

2013 9th Conference on Ph.D. Research in Microelectronics and Electronics

(PRIME 2013)

Villach, Austria
24 – 27 June 2013

IEEE Catalog Number: CFP13622-POD
ISBN: 978-1-4673-4579-8

Copyright © 2013 by the Institute of Electrical and Electronic Engineers, Inc
All Rights Reserved

Copyright and Reprint Permissions: Abstracting is permitted with credit to the source. Libraries are permitted to photocopy beyond the limit of U.S. copyright law for private use of patrons those articles in this volume that carry a code at the bottom of the first page, provided the per-copy fee indicated in the code is paid through Copyright Clearance Center, 222 Rosewood Drive, Danvers, MA 01923.

For other copying, reprint or republication permission, write to IEEE Copyrights Manager, IEEE Service Center, 445 Hoes Lane, Piscataway, NJ 08854. All rights reserved.

***This publication is a representation of what appears in the IEEE Digital Libraries. Some format issues inherent in the e-media version may also appear in this print version.*

IEEE Catalog Number: CFP13622-POD
ISBN 13:   978-1-4673-4579-8

Additional Copies of This Publication Are Available From:

Curran Associates, Inc
57 Morehouse Lane
Red Hook, NY 12571 USA
Phone: (845) 758-0400
Fax: (845) 758-2633
E-mail: curran@proceedings.com
Web: www.proceedings.com

2013 9th Conference on Ph.D. Research in Microelectronics and Electronics (PRIME 2013)

Villach, Austria
24-27 June 2013

IEEE Catalog Number: CFP13622-POD
ISBN: 978-1-46734-579-8

PRIME 2013, Villach, Austria

Table of Contents

Welcome xiii

Committees xv

Company Fair xix

Side Events xxi

Plenary Papers

Infineon Technologies: from Product to System and the Human Resources Excellence as Key Competitive Advantage
S. Herlitschka
Infineon Technologies Austria 1

Analog Design Trends in Communication Systems
J. Hauptmann, D. Giotta, U. Gaier
Lantiq 3

Challenges of ASIC Development for Sensor Nodes in Automotive and Consumer Applications
D. Droste
Robert Bosch 7

Innovative RFDAC Concepts for Digital Multi-Mode Transmitter in Cellular Applications
M. Fulde, F. Kuttner, E. Thaller, D. Ponton, V. Kampus, S. Gruenberger, H. Habibovic, A. Santner, C. Duller, C. Krassnitzer, G. Babin, G. Knoblinger
Intel 9

FPGA-Based Silicon Innovation Exploiting "More than Moore" Technology
P. Quinn
Xilinx Ireland 11

Organic Electronics: Material Aspects, Devices and Microelectronic Applications
B. Stadlober
Joanneum Research 13

Smart Power: From Problem Statement to System Solution
D. Draxelmayr, K. Norling, C. Lindholm
Infineon Technologies Austria 19

Conquering Variability in Mixed-signal ICs
C. Wegener, M. von Staudt
Dialog Semiconductor 23

iii

Regular Papers

M1A – Analog Techniques 1
Monday 24/06/2013, 09:40–10:40, Gottfried von Einem Saal

M1A1 – CMOS 8-Channel Frequency Division Multiplexer for 9.4T Magnetic Resonance Imaging
M. Jouda, O. Gruschke, J. Korvink
University of Freiburg
25

M1A2 – Comparison of High-voltage Linear Transmitter Topologies for Ultrasound CMUT Applications
S. Khandelwal, T. Ytterdal
Norwegian University of Science and Technology
29

M1A3 – Bluetooth Transceiver Modeling Using SystemC-AMS
F. Li[1], R. Butaud[2], E. Dekneuvel[1], G. Jacquemod[1]
[1]University of Nice, [2]RivieraWaves
33

M2A – Oversampled ADCs
Monday 24/06/2013, 11:30–12:30, Gottfried von Einem Saal

M2A1 – A Very-Low OSR 90-nm 1-MS/s Incremental $\Sigma\Delta$ ADC
D. Cavallo[1], M. De Matteis[1], M. Ronchi[2], G. Leggeri[3], A. Baschirotto[1]
[1]University of Milano-Bicocca, [2]STMicroelectronics, [3]Pegasus Micro Design
37

M2A2 – Variable-Step 12-bit ADC Based on Counter Ramp Recycling Architecture Suitable for CMOS Imagers with Column-Parallel Readout
T. Hassan, M. Strobel, H. Richter, J. Burghartz
Institute for Microelectronics Stuttgart
41

M2A3 – An Automatic Calibration Circuit for 12-Bit Single-Ramp A-to-D Converter in LHC Environments
T. Vergine[1, 1], M. De Matteis[2], L. Rota[2], A. Marchioro[3], A. Baschirotto[2]
[1]University of Pavia, [2]University of Milano-Bicocca, [3]CERN
45

M3A – Sigma-Delta Modulators
Monday 24/06/2013, 14:30–15:30, Gottfried von Einem Saal

M3A1 – A New Decoding Solution for the Asynchronous Sigma-Delta Modulator
W. Chen, C. Papavassiliou
Imperial College
49

M3A2 – A Low-Power, Continuous-Time Sigma-Delta Modulator for MEMS Microphones
C. De Berti[1], P. Malcovati[1], L. Crespi[2], A. Baschirotto[3]
[1]University of Pavia, [2]Conexant Systems, [3]University of Milano-Bicocca
53

M3A3 – System-Level Power Optimization for a $\Sigma\Delta$ D/A Converter for Hearing-Aid Application
P. Pracný, I. Jørgensen, E. Bruun
Technical University of Denmark
57

M4A – Amplifiers
Monday 24/06/2013, 15:50–17:10, Gottfried von Einem Saal

M4A1 – CMOS Current Amplifier for AFM Impedance Sensing on Chip with ZeptoFarad Resolution
D. Bianchi, M. Carminati, G. Ferrari, M. Sampietro
Polytechnic of Milano
61

M4A2 – A Wide Input Range Instrumentation Amplifier for Impedance Spectroscopy Applications
F. Del Cesta[1], A. Longhitano[1], P. Bruschi[1], R. Simmarano[2]
[1]University of Pisa, [2]Sensichips
65

M4A3 – A 36-μW Rail-to-Rail-Input Chopper Stabilized Amplifier Using Correlated Double Sampling
A. Pipino[1], M. De Blasi[2, 1], M. De Matteis[2], A. Fornasari[3], A. Baschirotto[2]
[1]University of Salento, [2]University of Milano-Bicocca, [3]Texas Instruments
69

M4A4 – Integrated Class-D Audio Amplifier Virtual Test for Output EMI Filter Performance
R. Mrad[1, 2], F. Morel[1], G. Pillonnet[1], C. Vollaire[1], A. Nagari[2]
[1]University of Lyon, [2]ST-Ericsson
73

M1B – Mixers and Oscillators
Monday 24/06/2013, 09:40–10:40, Room Draublick

M1B1 – Design of Mixers for a 130-GHz Transceiver in 28-nm CMOS
D. Parveg, M. Varonen, M. Kärkkäinen, K. Halonen
Aalto University
77

M1B2 – A 12-dBm IIP3 Reconfigurable Mixer for High/Low Band IR-UWB Receivers
M. Pasca[1], V. Chironi[1], S. D'Amico[1], M. De Matteis[2], A. Baschirotto[2]
[1]University of Salento, [2]University of Milano-Bicocca
81

M1B3 – Phase Noise Comparative Analysis of LC Oscillators in 28-nm CMOS through Impulse Sensitivity Function
I. Chlis, D. Pepe, D. Zito
Tyndall National Institute
85

M2B – RF Techniques 1
Monday 24/06/2013, 11:30–12:30, Room Draublick

M2B1 – Investigation of Two-Point Modulation to Increase the GFSK Data Rate of PLL-Based Wireless Transceivers of Wireless Sensor Nodes
H. Unterassinger[1], M. Flatscher[2], T. Gschier[2], W. Pribyl[1]
[1]TU Graz, [2]Infineon Technologies Austria
89

M2B2 – Transient Analysis Applying S-Parameters as Operators
H. Nuszkowski, R. Wolf, F. Ellinger, G. Fettweis
TU Dresden
93

M2B3 – Design of a W-Band 2-Bit Differential CMOS Phase Shifter
A. Vahdati, M. Varonen, M. Kärkkäinen, K. Halonen
Aalto University
97

M3B – Signal Processing 1
Monday 24/06/2013, 14:30–15:30, Room Draublick

M3B1 – Impact of PA Class on Reconstruction Filters Sizing for a WCDMA Base Station LINC Transmitter
N. Prou[1, 1], F. Rivière[1], C. Berland[2]
[1]Crismat/LaMIPS, [2]University of Paris-Est
101

M3B2 – Scalable Hybrid CORDIC-LUT Architectures for CG-FFT Processors
A. Congiu[1], A. Picciau[1], M. Barbaro[1], E. Bodano[2]
[1]University of Cagliari, [2]Infineon Technologies
105

Table of Contents   *PRIME 2013, Villach, Austria*

M3B3 – Off-line BIST in Watt Hour Meters
R. Ribnikar[1], D. Strle[2]
[1]Iskraemeco, [2]University of Ljubljana   109

M4B – MEMS
Monday 24/06/2013, 15:50–17:10, Gottfried von Einem Saal

M4B1 – A MEMS BPSK Demodulator
J. Scerri, I. Grech, E. Gatt, O. Casha
University of Malta   113

M4B2 – A MEMS-Based Oscillator with Synchronous Amplitude Control
J. Gronicz[1], N. Chekurov[2], L. Aaltonen[3], K. Halonen[1]
[1]Aalto University, [2]KTH Royal Institute of Technology, [3]Murata Electronics   117

M4B3 – Fully Electrical Test Procedure for Inertial MEMS Characterization at Wafer-Level
A. Sisto[1], O. Schwarzelbach[2], L. Fanucci[1]
[1]University of Pisa, [2]Fraunhofer Institute for Silicon Technology   121

M4B4 – Capacitive Out of Plane Large Stroke MEMS Structure
C. Glacer[1], A. Dehé[2], D. Tumpold[3], R. Laur[1]
[1]University of Bremen, [2]Infineon Technologies, [3]TU Vienna   125

M1C – FPGA
Monday 24/06/2013, 09:40–10:40, Room Drau 3

M1C1 – Graph Coverage: an FPGA-Targeted Implementation
A. Cinti, A. Rizzi
University of Roma "La Sapienza"   129

M1C2 – Design-Space Exploration of an eFPGA Soft-Core Based on Multi-Stages Switching Networks
M. Cuppini[1], E. Franchi Scarselli[1], C. Mucci[2]
[1]University of Bologna, [2]STMicroelectronics   133

M1C3 – Direct Digital Frequency Synthesizers Implemented on High-End FPGA Devices
M. Genovese, E. Napoli
University of Napoli "Federico II"   137

M2C – Digital Techniques 1
Monday 24/06/2013, 11:30–12:30, Room Drau 3

M2C1 – A Modified Truncation Scheme for Recursive Multipliers
K. Sashank, S. Ahmed, A. Singh, M. Srinivas
Birla Institute of Technology and Science Pilani   N/A

M2C2 – Random Interleaved Pipeline Countermeasure against Power Analysis Attacks
R. Menicocci[1], A. Trifiletti[2], F. Trotta[2]
[1]Fondazione Ugo Bordoni, [2]University of Roma "La Sapienza"   145

M2C3 – Fast Constant Time Memory Allocator for Inter Task Communication in Ultra Low Energy Embedded Systems
G. Rebel[1], F. Estevez[1], I. Schulz[2], P. Glösekötter[1]
[1]University of Applied Sciences Munster, [2]TU Dortmund   149

PRIME 2013, Villach, Austria   *Table of Contents*

M3C – Digital Circuits
Monday 24/06/2013, 14:30–15:30, Room Drau 3

M3C1 – Novel Field Programmable Embryonic Cell for Adder and Multiplier
G. Malhotra, J. Becker, M. Ortmanns
University of Ulm   153

M3C2 – Design and Applications of Magnetic Tunnel Junction Based Logic Circuits
H. Mahmoudi, T. Windbacher, V. Sverdlov, S. Selberherr
TU Vienna   157

M3C3 – Optimal Deployment of Shadowed Registers in Systems with Serial Clock Distribution
M. Trifković, D. Raič, D. Strle
University of Ljubljana   161

M4C – CAD
Monday 24/06/2013, 15:50–17:10, Room Drau 3

M4C1 – Monte Carlo Based Post-Silicon Verification Considering Automotive Application Variances
M. Harrant[1], T. Nirmaier[1], J. Kirscher[1], C. Grimm[2], G. Pelz[1]
[1]Infineon Technologies, [2]TU Kaiserslautern   165

M4C2 – tLIFTING: an Open-Source Multi-Level Fault Simulator for Ionizing Effects
F. Lu, G. Di Natale, M. Flottes, B. Rouzeyre
LIRMM   169

M4C3 – CIRSIUM: A Circuit Simulator in MATLAB with Object Oriented Design
A. Mahmutoglu, A. Demir
Koc University   173

M4C4 – Circuit Simulation Using State Space Equations
K. Lam, M. Zwolinski
University of Southampton   177

T1A – Analog Techniques 2
Tuesday 25/06/2013, 09:40–10:40, Gottfried von Einem Saal

T1A1 – Analog Performance of PD-SOI MOSFETs at High Temperatures Using Reverse Body Bias
A. Schmidt, H. Kappert, R. Kokozinski
Fraunhofer Institute for Microelectronic Circuits and Systems   181

T1A2 – Technique for Reducing On-Resistance of High-Voltage Drivers Based on Stacked Standard CMOS
S. Pashmineh[1], H. Xu[2], D. Killat[1]
[1]TU Cottbus, [2]University of Ulm   185

T1A3 – Diode Detector with Voltage Gain
R. Wolf, F. Ellinger
TU Dresden   189

T2A – ADCs and DACs
Tuesday 25/06/2013, 11:30–12:30, Gottfried von Einem Saal

T2A1 – An Analog 1:16 Demultiplexer for Time-Interleaved A/D-Converters with a Sampling Rate of up to 64 GS/s
F. Lang, J. Gerigk, D. Ferenci, M. Grözing, M. Berroth
University of Stuttgart   193

vii

T2A2 – A Continuous-Time Switched-Capacitor DAC with Offset and Flicker Noise Cancellation
A. Longhitano[1], F. Del Cesta[1], P. Bruschi[1], R. Simmarano[2]
[1]University of Pisa, [2]Sensichips
197

T2A3 – A 10-Bit 100-MS/s Low-Power Pipeline ADC for High Energy Physics Experiments
A. Donno[1], S. D'Amico[1], M. De Matteis[2], A. Baschirotto[2]
[1]University of Salento, [2]University of Milano-Bicocca
201

T3A – Wireless Sensor Nodes
Tuesday 25/06/2013, 14:00–15:20, Room Draublick

T3A1 – An Improved Ultra-Low-Power Wireless Sensor-Station Supplied by a Photovoltaic Energy Harvester
A. Lazzarini Barnabei, E. Dallago, P. Malcovati, A. Liberale
University of Pavia
205

T3A2 – Wireless Energy and Data Transfer for Neural Recording and Stimulation Applications
G. Yilmaz, C. Dehollain
EPFL
209

T3A3 – Design of a Passive UHF RFID Tag for Capacitive Sensor Applications
K. Kapucu, J. Merino Panadés, C. Dehollain
EPFL
213

T3A4 – Full-Duplex Communication and Remote Powering Implementation of an Electronic Knee Implant
O. Atasoy, C. Dehollain
EPFL
217

T1B – Transmitters and Receivers
Tuesday 25/06/2013, 09:40–10:40, Room Draublick

T1B1 – Efficiency Enhancement of Burst Mode Transmitters by RF Energy Recovery
D. Seebacher[1], W. Bösch[1], P. Singerl[2], C. Schuberth[2]
[1]TU Graz, [2]Infineon Technologies Austria
221

T1B2 – A 23-mW 4.5/8-GHz IR-UWB Transmitter in 65-nm TSMC CMOS Technology
A. Donno[1], S. D'Amico[1], M. De Matteis[2], A. Baschirotto[2]
[1]University of Salento, [2]University of Milano-Bicocca
225

T1B3 – System Design for an Experimental Cognitive Transceiver
A. Ashok, I. Subbiah, S. Heinen
RWTH Aachen University
229

T2B – Digital Systems
Tuesday 25/06/2013, 11:30–12:30, Room Draublick

T2B1 – High Speed Interface for Digital Centric Transmitters
B. Mohr[1], J. Mueller[1], Y. Zhang[1], R. Leys[2], S. Schenk[2], U. Bruening[2], S. Heinen[1]
[1]RWTH Aachen University, [2]University of Heidelberg
233

T2B2 – Design of a Reconfigurable Multi-Core Architecture for Streaming Applications with a Case Study on Performance Evaluation of FIR-Filters
L. Ghazanfari, R. Airoldi, J. Nurmi, T. Ahonen
Tampere University of Technology
237

PRIME 2013, Villach, Austria　　　　　　　　　　　　　　　　　　　　*Table of Contents*

T3B – Modeling
Tuesday 25/06/2013, 14:00–15:20, Room Draublick

T3B1 – On the Calculation of Hall factors for the Characterization of Electronic Devices
V. Uhnevionak[1], A. Burenkov[1], P. Pichler[2, 1]
[1]Fraunhofer Institute for Integrated Systems and Device Technology, [2]University of Erlangen-Nuremberg　　241

T3B2 – Analysis and Modeling of Minority Carrier Injection in Deep-Trench Based BCD Technologies
M. Kollmitzer[1], M. Olbrich[2], E. Barke[2]
[1]Infineon Technologies Austria, [2]Leibniz University of Hannover　　245

T3B3 – S-Parameter Models for Transient Simulation in Verilog-A
T. Maier[1], D. Droste[1], M. Siegel[2]
[1]Robert Bosch, [2]TU Karlsruher　　249

T3B4 – Extended Model for Platinum Diffusion in Silicon
E. Badr[1], P. Pichler[2, 1], G. Schmidt[3]
[1]Fraunhofer Institute for Integrated Systems and Device Technology, [2]University of Erlangen-Nuremberg, [3]Infineon Technologies Austria　　253

W1A – High-Frequency Analog Circuits
Wednesday 26/06/2013, 09:40–10:40, Gottfried von Einem Saal

W1A1 – A SiGe Wideband VCO and Divider MMIC with Low Gain Variation for Multi-band Systems at 2.4 and 5.8 GHz
N. Joram, R. Wolf, F. Ellinger
TU Dresden　　257

W1A2 – A CMOS 28-nm 880-MHz Fourth-Order Low-Pass Active-RC Filter for 60-GHz Transceivers
A. Pezzotta[1], M. De Matteis[1], A. Baschirotto[1], S. D'Amico[2]
[1]University of Milano-Bicocca, [2]University of Salento　　261

W1A3 – An Inductorless 34 GBit/s Half-Rate 4:1 Multiplexer in 0.25-μm SiGe Technology
M. Khafaji[1], C. Carta[1], E. Sobotta[1], D. Micusik[2], F. Ellinger[1]
[1]TU Dresden, [2]IHP Microelectronics　　265

W2A – Power Management
Wednesday 26/06/2013, 11:30–12:30, Gottfried von Einem Saal

W2A1 – A Nano-Power Power Management IC for Piezoelectric Energy Harvesting Applications
M. Dini, M. Filippi, M. Tartagni, A. Romani
University of Bologna　　269

W2A2 – Portable Battery Charging Circuits for Enhanced Magnetic Resonance Wireless Power Transfer (WPT) System
H. Choi, E. Ahn, S. Park, J. Choi
Kyungpook National University　　273

W2A3 – Monolithic Power Management Front-End with High-Voltage Dense Energy Storage for Wireless Powering
H. Meyvaert, A. Crouwels, S. Indevuyst, M. Steyaert
KU Leuven　　277

W3A – Power Electronics
Wednesday 26/06/2013, 14:30–15:30, Gottfried von Einem Saal

W3A1 – Application of Bayesian Networks to Predict SMART Power Semiconductor Lifetime
K. Plankensteiner[1, 2], O. Bluder[1], J. Pilz[2]
[1]KAI Kompetenzzentrum Automobil- und Industrie-Elektronik, [2]Alpen-Adria University of Klagenfurt 281

W3A2 – Optimal Design of Experiments for Semiconductor Lifetime Data
A. Zernig[1, 2], O. Bluder[1], G. Spöck[2]
[1]KAI Kompetenzzentrum Automobil- und Industrie-Elektronik, [2]Alpen-Adria University of Klagenfurt 285

W3A3 – A Metric Driven Verification and Validation Approach for Smart Power Devices
O. Melnychenko[1], H. Kreuter[2]
[1]TU Vienna, [2]Infineon Technologies Austria 289

W4A – Advanced Devices
Wednesday 26/06/2013, 15:50–16:50, Gottfried von Einem Saal

W4A1 – Modeling of an Electrostatically Actuated Micro-Electro-Mechanical (MEMS) Speaker System
D. Tumpold[1], M. Kaltenbacher[1], C. Glacer[2], A. Dehé[2], M. Nawaz[2]
[1]TU Vienna, [2]Infineon Technologies 293

W4A2 – A Promising Technology of Schottky Diode Based on 4H-SiC for High Temperature Application
R. Pascu, F. Craciunoiu, M. Kusko
National Institute for Research and Development in Microtechnologies 297

W4A3 – Comparative Analysis of Methods for Computing Pole Angle Offsets in Magnetic Pole Wheels Using ABS Sensor
M. Adnan[1], D. Hammerschmidt[2]
[1]Alpen-Adria University of Klagenfurt, [2]Infineon Technologies Austria 301

W1B – RF Techniques 2
Wednesday 26/06/2013, 09:40–10:40, Room Draublick

W1B1 – A Wideband Planar Microstrip to Coplanar Stripline Transition (Balun) at 35 GHz
J. Leufker, A. Strobel, C. Carta, F. Ellinger
TU Dresden 305

W1B2 – Cascode Class E Power Amplifier in 180/350-nm CMOS for EER System
I. Rumyancev[1], A. Korotkov[1], J. Hauer[2]
[1]Saint Petersburg State Polytechnical University, [2]Fraunhofer Institute for Integrated Circuits 309

W1B3 – A SiGe LTE Power Amplifier with Capacitive Tuning for Size-Reduction of Biasing Inductor
J. Zhao, R. Wolf, F. Ellinger
TU Dresden 313

W2B – Sensors
Wednesday 26/06/2013, 11:30–12:30, Room Draublick

W2B1 – An Ultra-Thin CMOS in-Plane Stress Sensor
Y. Mahsereci, N. Wacker, H. Richter, J. Burghartz
Institute for Microelectronics Stuttgart 317

W2B2 – A Low-Power Sensor Interface Circuit for Remotely Powered Implants
X. Liu, C. Dehollain
EPFL 321

W2B3 – Low-Noise Low-Offset Current-Mode Hall Sensors
H. Heidari, U. Gatti, E. Bonizzoni, F. Maloberti
University of Pavia   325

W3B – Signal Processing 2
Wednesday 26/06/2013, 14:30–15:30, Room Draublick

W3B1 – Dynamic Programming Based Grouping Method for RO-PUFs
G. Kömürcü[1], G. Dundar[2], A. Pusane[2]
[1]National Research Institute of Electronics and Cryptology Tübítak, [2]Bogazici University   329

W3B2 – Power-Aware Architectural Exploration of the CORDIC Algorithm
J. Manica, R. Passerone, L. Rizzon
University of Trento   333

W3B3 – Zoom FFT for Precise Spectrum Calculation in FMCW Radar Using FPGA
B. Al-Qudsi, N. Joram, A. Strobel, F. Ellinger
TU Dresden   337

W4B – Digital Techniques 2
Wednesday 26/06/2013, 15:50–16:50, Room Draublick

W4B1 – A JTAG Based 3D DfT Architecture Using Automatic Die Detection
Y. Fkih[1, 2], P. Vivet[2], B. Rouzeyre[1], M. Flottes[1], G. Di Natale[1]
[1]LIRMM, [2]CEA   341

W4B2 – Simulated Power Analysis Attacks on a DDPL Crypto-Core without Routing Constraints
S. Bongiovanni, G. Scotti, A. Trifiletti
University of Roma "La Sapienza"   345

W4B3 – An Improved Instruction-Level Energy Model for RISC Microprocessors
W. Wang, M. Zwolinski
University of Southampton   349

Author Index

PRIME 2013, Villach, Austria

Welcome

It is our great pleasure to welcome you to the 9th Conference on Ph. D. Research in Microelectronics and Electronics (PRIME 2013). This year this unique event specifically dedicated to Ph. D. students in Microelectronics and Electronics will be held in Villach, Austria, from June 24th to 27th, 2013.

The participation of researchers, professors and companies working in the field of microelectronics and electronics has been fostered at PRIME 2013 in aspiration to provide useful feedback onto the research results presented by Ph. D. students. A top-quality Technical Program Committee of world-class experts from academia and industry has organized a varied and extraordinarily appealing scientific program. The selected papers cover all aspects of Microelectronics and Electronics, ranging from digital to analog as well as from DC to RF.

All of the presented papers at PRIME 2013 will be published in the IEEExplore database, which provides a large audience and excellent dissemination of the event. Moreover, the best papers will be honored by the traditional PRIME Leaf Certificates, which will be awarded on the closing session on Wednesday June 26th in the afternoon.

The aim of PRIME 2013 is to provide an opportunity for Ph. D. students to present their research activity and contact other people in the research community. During each of the three days of the Conference (from June 24th to 26th, 2013) two time slots will be dedicated to a Company Fair, for allowing Ph. D. students to get in touch with companies and other institutions. Moreover, on June 25th, 2013 there will be a Podium Discussion dedicated to the perspectives that microelectronics companies offer to Ph. D. graduates and on June 27th, 2013 there will be a "EU Research Day", followed by a short course on "Power IC and Discrete Technologies", as well as a visit to Infineon Technologies Austria.

We look forward to welcoming you in Villach. We are sure that the technical contents of PRIME 2013 and the wonderful surroundings of the conference site will result in an exciting "cocktail" you will enjoy.

Alberto Gola and Andrea Baschirotto
Conference General Co-Chairmen

PRIME 2013, Villach, Austria

Committees

General Chairmen

- Andrea Baschirotto, University of Milano-Bicocca, Italy
- Alberto Gola, Dialog Semiconductor, Italy

Technical Program Chairman

- Piero Malcovati, University of Pavia, Italy

Secretariat

- Marcello De Matteis, University of Milano-Bicocca, Italy

Steering Committee

- Franco Maloberti, University of Pavia, Italy
- Andrea Baschirotto, University of Milano-Bicocca, Italy
- Catherine Dehollain, EPFL, Switzerland
- Alberto Gola, Dialog Semiconductor, Italy
- Frank Henkel, IMST GmbH, Germany
- Peter Kennedy, University College Cork, Ireland
- Wolfgang Pribyl, Graz University of Technology, Austria

Committees   *PRIME 2013, Villach, Austria*

Technical Program Committee

- Eduard Alarcón, UPC, Spain
- Federico Alimenti, University of Perugia, Italy
- Aytac Atac, RWTH Aachen University, Germany
- Bach Elmar, Infineon, Austria
- Diego Barrettino, SUPSI, Switzerland
- Didier Belot, STMicroelectronics, France
- Roc Berenguer, CEIT, Spain
- Edoardo Bonizzoni, University of Pavia, Italy
- Ralf Brederlow, Texas Instruments, Germany
- Klaas Bult, Broadcom, Netherlands
- Alessandro Cabrini, University of Pavia, Italy
- Massimo Conti, Università Politecnica delle Marche, Italy
- Jan Craninckx, IMEC, Belgium
- Stefano D'Amico, University of Salento, Italy
- Gian-Franco Dalla Betta, University of Trento, Italy
- Marcello De Matteis, University of Milano-Bicocca, Italy
- Carl James Debono, University of Malta, Malta
- Catherine Dehollain, EPFL, Switzerland
- Manuel Delgado-Restituto, IMSE-CNM, Spain
- Günhan Dundar, Bogazici University, Turkey
- Ahmed Elwakil, University of Sharjah, United Arab Emirates
- Luca Fanucci, University of Pisa, Italy
- Vittorio Ferrari, University of Brescia, Italy
- Giuseppe Ferri, University of L'Aquila, Italy
- Tzeno Galchev, University of Michigan, United States
- Massimo Gottardi, Fondazione Bruno Kessler, Italy
- Marco Grassi, University of Pavia, Italy
- Stefan Heinen, RWTH Aachen University, Germany
- Robert Henderson, University of Edinburgh, United Kingdom
- Frank Henkel, IMST, Germany
- David Johns, University of Toronto, Canada
- Peter Kennedy, University College Cork, Ireland
- Juha Kostamovaara, University of Oulu, Finland
- Paolo Madoglio, Intel, United States
- Franco Maloberti, University of Pavia, Italy
- Rui Paulo Martins, University of Macau, Macau
- Nicola Massari, Fondazione Bruno Kessler, Italy
- Peter Mole, Intersil, United Kingdom
- Angelo Nagari, ST-Ericsson, France
- Tobias Noll, RWTH Aachen University, Germany
- Pierluigi Nuzzo, University of California at Berkeley, California
- Tom O'Dwyer, Analog Devices, Ireland
- Gaetano Palumbo, University of Catania, Italy
- Daniele Passeri, University of Perugia, Italy
- Roberto Passerone, University of Trento, Italy
- Michiel Pertijs, TU Delft, Netherlands
- Wolfgang Pribyl, TU Graz, Austria
- Rüdiger Quay, Fraunhofer IAF, Germany
- Patrick Quinn, Xilinx Dublin, Ireland
- Valerio Re, University of Bergamo, Italy

PRIME 2013, Villach, Austria   *Committees*

- Patrick Reynaert, KU Leuven, Belgium
- Angel Rodriguez-Vazquez, AnaFocus , Spain
- Stefan Rusu, Intel, United States
- Gilles Sicard, TIMA-CMP, France
- Pietro Siciliano, CNR-IMM Lecce, Italy
- David Stoppa, Fondazione Bruno Kessler, Italy
- Himanshu Thapliyal, Qualcomm, United States
- Roland Thewes, TU Berlin, Germany
- Guido Torelli, University of Pavia, Italy
- Alberto Tosi, Politecnico di Milano, Italy
- Maurizio Valle, University of Genova, Italy
- Udo Weimar, University of Tuebingen, Germany
- Zhipeng Ye, ASTC, Australia

PRIME 2013, Villach, Austria

Company Fair

Every day from June 24th to 26th, 2013 in the morning and in the afternoon there will be time slots dedicated to establish new contacts between Ph. D. students and institutions (research centers, universities, and companies) for the creation of future joint activity (like, for instance, fellowships, internships, or also employment). The following institution confirmed their presence:

- Bosch
- Dialog Semiconductor
- Infineon
- Intel
- Joanneum Research
- Lantiq
- Xilinx

Company Fair

PRIME 2013, Villach, Austria

PRIME 2013, Villach, Austria

Side Events

Podium Discussion (Tuesday, June 25th, 2013)

Infineon Austria CEO and CTO as well as other high level executives, managers and former Ph. D. students will discuss which perspectives microelectronics companies offer to Ph. D. graduates.

EU Research Day (Thursday, June 27th, 2013)

An overview on EU Research funding programs with particular focus on funding opportunities for microelectronics and training as well as exchange programs.

Short Course (Thursday, June 27th, 2013)

Power IC and Discrete Technologies

Martin Henning Vielemeyer
Infineon Technologies Austria

During the last decades MOSFETs have been continuously improved enabling them to switch higher powers without overheating. Devices with a few square millimeters can handle currents up to the triple digit range. At the same time they are switching with many hundred kilohertz. It is crucial to work at very high efficiency. In addition to the energy savings it avoids thermal destruction of the devices. For this it is needed to optimize the parameters causing most losses in application. Some can be improved by understanding the MOSFET and the physics behind it. For others there are trade-offs, on which the target must be carefully chosen.

PowerMOSFET:

- Short overview of power devices
- History of PowerMOSFET (from lateral to trench)
- Physics of a trench MOSFET
- TrenchMOSFET parameters and parasitic
- What's the influence (in the application) of the parameters and the parasitic and how can they be influence

Visit of Infineon Technologies Austria (Thursday, June 27th, 2013)

Infineon Technologies: from Product to System and the Human Resources Excellence as Key Competitive Advantage

Sabine Herlitschka

Infineon Technologies Austria, Villach, Austria

Major global challenges of the 21st century are drivers for the industry and especially for the semiconductor industry. Microelectronics and semiconductors are key enabling technologies and will contribute significantly to increasing Innovation and Productivity. Infineon provides semiconductors and system solutions, focusing on three central needs of modern society: Energy Efficiency, Mobility and Security.

Tight Customer Relationships are based on system know-how and application understanding. People in our company are the key to our success. Their experiences, knowledge and passion drive us to new levels of excellence. Thus, people excellence is a core element of our strategy. Our success depends on creating an attractive and stimulating working environment for our employees, who represent "living" diversity with staff coming from 55 nations.

As one of the most innovative companies, Infineon has set up a corporate PhD network. For our PhD's we offer best conditions for up-to-date research & development and an attractive multi-disciplinary educational opportunity at an international workplace.

The presentation will provide an overview on current technology developments in Infineon's focus areas and PhDs' opportunities to become part of a successful global company.

Analog design trends in communication systems

Joerg Hauptmann, Dario Giotta, Ulrich Gaier
Lantiq A GmbH
Austria

Abstract—Modern high-speed silicon technologies in the sub-micron region enable today very efficient and cost effective digital communication systems. Highly integrated system-on-chip (SoC) solutions cover satisfying the ever-increasing demand for broadband communication from home. High data transfer rates with excellent quality are achieved through the use of Wi-Fi, DSL, Ethernet and optical transport. The presentation will deal with the most important design aspects and system trends that occur in the development of RF- and MS-front-end modules for SoC integration. New trends in wireless, as well as requirements for individual MS / RF blocks are considered with respect to high integration and cost reduction.

I. INTRODUCTION

Analog telephony was invented in last quarter of the 19th century. Since its introduction, there have been plenty of installments and innovative development steps, and today analog telephony has more than 1 billion telephone lines (twisted pairs) installed worldwide for use with so-called Plain Old Telephone Services (POTS). However, highly sophisticated modem development was necessary to fully exploit the potential of the copper wire connections. Digital Subscriber Line (DSL) was introduced as a very promising new technology for wired communications over twisted pairs with data rates in the 100Mbit range.

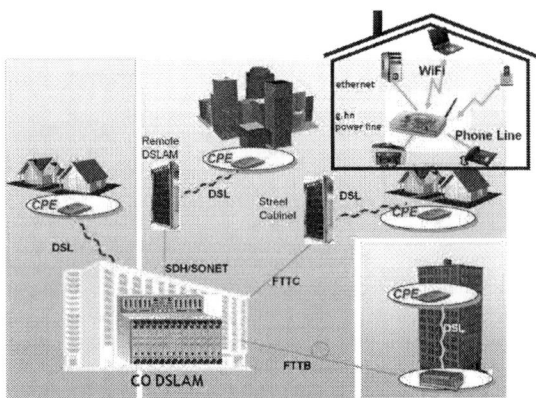

Fig. 1. End to End network solution

Figure 1 shows the end to end network solution, targeted by wire line communication. DSL uses the twisted pair copper cables of the POT's as transport medium and can directly connect the traditional Central Office with end users (so called customer premises or CPE). Nowadays fiber is often used to connect the street cabinets or directly the buildings, while DSL is bridging the so called ,last mile' to the house.

Inside the house Ethernet, Wi-Fi, phone lines and power lines are used for further data distribution. Thus all these different communication systems are integrated together into a modern home gateway. The following chapters will describe the architectures used for communication systems, the integration strategy both on central office line cards and in home gateways in order to achieve the cost, power and performance targets and the challenges on the different building blocks needed.

II. ARCHITECTURE OF COMMUNICATION SYSTEMS

A. Architectural considerations

All modern communication systems became digital, since highly sophisticated algorithms of modern modulation schemes running on very fast DSPs help to enable high quality and high data rate at a very low cost. This is feasible due to the fact that the availability of deep sub-micron technologies makes it possible to build very-high-speed digital signal processors with a lot of integrated data memory. However, in digital communication systems, the transmission media (cable or radio) is still analog. Thus building modern communication systems not only demands sophisticated algorithms running on a high-speed DSP; the Analog Frontend (AFE) circuitry and the power amplifier, which are the interface to the real cable or the radio channel, plays a key role in the successful and practical implementation of any communication system. The AFE must capture and digitize highly attenuated received signals superimposed on very strong echo signals or neighboring blockers, requiring the utilization of high-dynamic-range converters and strong analog echo cancellation [1] or filtering - all while keeping the power dissipation and system costs at a minimum

Figure 2 show the various building blocks of a wired and a wireless solution, respectively. In principle, they look very similar besides some differences due to different physical transmission media - twisted pair in case of wire line, and radio waves in case of wireless. Additional mixers are needed in the RF-Front-End of a wireless system for the up- and down-mixing of signals between the RF- and Base Band (BB) domain. Both amplifiers, one operating in the baseband and the other operating in the RF domain, have two essential things in common: They need high-voltage capabilities, as the transmitted signals have high-amplitude, and consume a huge part of the overall system power. The hybrid circuitry reduces the influence of dominating disturbing signals into the receive channel. In the wired solution, its own transmit signal is the

978-1-4673-4579-8/13 $31.00 © 2013 IEEE

Paper PL2   *PRIME 2013, Villach, Austria*

disturbing factor that has to be attenuated, while in the wireless application, foreign neighborhood signals have to be filtered out. Again both hybrids need very good disturber suppression for high-bit-rate operations.

The AFE or RF-FE convert digital signals into analog and vice versa, together with analog filtering and amplifications. Both frontends have high gain programmability in RX for amplifying small incoming signal and AD conversion with similar high resolution and bandwidth between 20M and 100MHz. The RF chip is a bit more complicated in design, as it has to mix up and down the BB signals to the RF frequency bands and needs additional calibration.

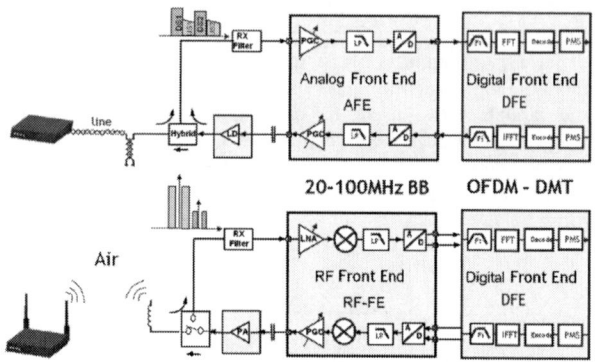

Fig. 2.   Architecture – wired versus wireless

Modern digital communication systems are using OFDM as modulation scheme and thus having multiple discrete multi-tones (DMT) for carrying the data. Digital filtering as well as FFT and IFFT for switching between time-domain and frequency domain signal processing are additionally implemented.

B. Balancing the System parameters

In analog and RF front-ends, a lot of different parameters determine the overall system performance in terms of bit rate, maximum achievable reach, PSD, out-of-band specification, and many other features specified in appropriate standards. A typical AFE for VDSL looks quite similar to those used for ADSL or SHDSL. Even wireless systems including RF have very similar block diagrams as well as system parameters. Figure 3 shows all the building blocks (yellow boxes) of a typical AFE for a wired communication system including a description of their functions from a system point-of-view, and their most important system parameters (blue boxes).

All these derived system parameters must be further break down to the various design blocks of the architecture. Each building block has a certain function and different important system parameters shown here in the blue bubbles. Very essential is as an example to achieve quite good hybrid attenuation. Transformer ratio and synthesized impedance are essential influencing this attenuation. All noise sources and nonlinearities are distributed all over the blocks, which however are not equally sensitive. All of them have to be known and a good balancing is very essential. The key building

blocks of the system with most challenges in the design are the line driver, the hybrid and the ADC.

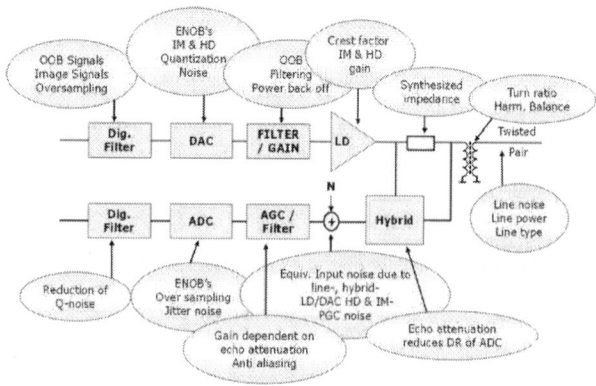

Fig. 3.   System parameters

III.   MARKET NEEDS AND SYSTEM CHALLENGES

A. Drivers in the market

The joint optimization of various system requirements is the primary challenge for the design of innovative communication solutions that always keep the most important design criteria in mind. These design criteria are valid any time and are identical for all kinds of systems:

- Minimum costs

- Minimum power dissipation

- Increasing bandwidth for higher data rates to customers

- Feature enhancement, innovation

Cost is the most important design factor in the high-volume and highly price-sensitive communication market. Although there is an average annual feature enhancement of 40%, the price also is declining by 40% year-by-year, which leads to an overall cost/feature decline by a factor of 2. This means the annual productivity gain in new designs has to be larger than 2. This can only be achieved by optimizing not only the silicon area, but also the yield, packaging cost, testing, external components, cost of system board, and last but not least, the manufacturing costs. A well-defined integration strategy is a key factor as well as the right choice of technology. The optimum chipset partitioning has to be found - separate devices, multi-chip modules (MCMs) or all integrated on one single piece of silicon. System on Chip solutions (SoC) in deep sub-micron technologies are the most cost-effective in the long run in many cases, but a limited R&D budget and the maturity of technologies used may have an important impact.

However, deep sub-micron technologies are not the only route to area benefits. Innovation in parallel is a key factor to make analog products more cost-effective. Innovative new architectures on the mixed-signal circuit level are mandatory to maximize the benefits of upcoming speed advantages from the newest technology shrinks. Innovation is not always dependent

978-1-4673-4579-8/13 $31.00 © 2013 IEEE   4

on the use of the newest technologies. One technology-independent innovation is in the field of line drivers [2] and power amplifiers, where a lot of innovative steps have been implemented in order to improve power efficiency without changing technology. Another innovation is in the field of time-continuous Sigma-Delta A/D converters for high bandwidth applications, which clearly benefits from speed improvements of deep sub-micron technologies.

Minimum power consumption is extremely important, as it has a big impact on the overall system costs as well. Integration into one package is highly depending on the power dissipation and thus defines overall package costs. System board costs are related to the achievable density. The board density is especially important at the central office, where multi-channel chip solutions are a prerequisite for enabling high-density line-cards with up to 96 DSL channels per card. Many of these cards are then placed in one rack. Obviously without minimum power dissipation, this would not be possible or would require extremely expensive cooling systems.

In the CPE and mobile arena, the power is more of an issue for achieving a small form factor and a maximum operating time rather than for reducing operating costs. However, achieving low power dissipation is still a prerequisite for a high integration level and in this way cost savings.

Besides cost and power improvements, there is in addition a strong demand for increasing the bandwidth in every new generation. A higher bandwidth allows new services such as video-on-demand, high-speed internet, and multiple voice channels with audio quality simultaneously. In wired communications, the achievable bit rate started at 28 kbit/s using 4kHz bandwidth and reaches nowadays 100-200MHz bandwidth achieving >1Gbit/sec data rates.

Last but not least, while the bandwidth is usually standardized and therefore mandatory to support, a real competitive advantage may be achieved by introducing new features having added value for the customers.

B. Integration strategy and technology scaling

A lot of introduced innovations are based on cost reduction. Using the newest deep sub-micron technologies in combination with new innovative circuitries can help a lot to minimize the silicon area and thus system costs. However, technology scaling is not always the only key success factor for cost reduction. Depending on the application, the technology nodes used and the integration level can look quite different.

On one hand, at the central office, multiple modems have to be installed on one line card [3], which leads to multiple AFEs, multiple line drivers and multiple DFEs next to each other. Vertical integration is done, where multiple channels of AFEs can be integrated together on one die and multiple channels of DFEs on another die. This has the benefit of using different optimized technologies for each function, such as analog, digital or high-voltage line driver, and in this way optimizes silicon costs depending on the function and the voltage used. For pure digital DFE chips, the latest deep sub-micron technologies with a very low supply voltage, e.g. 28 nm, can be used to achieve extremely low area, whereas in analog, 130-nm

down to 40-nm technology with suitable higher supply voltages is preferred for performance reasons. Integration of multi channel AFE's on one die shown in figure 4 combined with technology scaling can save up to a factor 5 in silicon- area per channel between a 130nm single channel solution and a 65nm multi-channel solution.

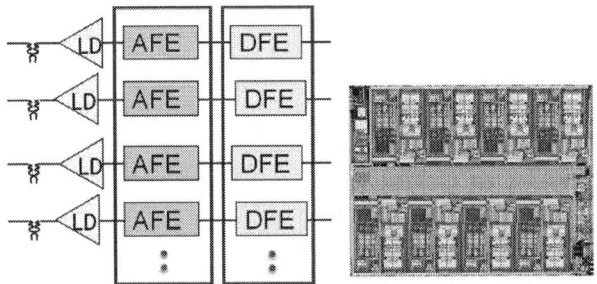

Fig. 4. Multi channel integration in CO applications

The digital part is scaling with the square of the minimum transistors' length, which results in 50% area improvement per technology node. Area reduction in the analog domain is not that easy. Matching is a key design parameter in analog design and can be only achieved by keeping the transistor area sufficiently high. Furthermore, the supply voltage scales down per technology shrink, which leads to a reduced signal swing. This means the noise floor has to be reduced further in order to maintain a certain signal-to-noise ratio leading to higher capacitor values. Nonetheless, several technology-scaling of Lantiq-products demonstrated a 30% area improvement per node step. The secret is that one further degree of freedom is available, which is the architecture of the system itself. Only by changing the architecture both on the system level and building-block level is it possible to fully exploit the high-speed capabilities of the transistors in the deep sub-micron technologies, which indeed is the most significant benefit of the newer technologies. In other words, analog focus is shifted from matching to speed.

On the customer premises side on the other hand, just one modem has to be implemented. This demands the integration of the line driver, the AFE, and the DFE on the same silicon die in order to save die and packaging costs. In addition, there is the need to integrate many other applications together into the system, such as UMTS, BT, WLAN and GPS in wireless phones or DSL, WLAN, Ethernet, USB and POTs in wired gateways. Integrating all functions on one die makes the technology choice challenging. The digital part in the DFE operates at the lowest possible voltage to save power, whereas the AFE operates at moderate supply voltages (>1V) to maintain a high signal-to-noise ratio and the power amplifier or line driver even operates at supply voltages of 3.3V up to 12 V. Figure 5 show different integration strategies, as different markets have to be served with different integration levels. Analog and digital functions may be integrated together on one die (SoC) or will be separated on silicon in one common package (multi chip module-MCM). Dedicated functions are even separated via appended chips in order to minimize the overhead in different markets and minimize design variants.

978-1-4673-4579-8/13 $31.00 © 2013 IEEE

Paper PL2   *PRIME 2013, Villach, Austria*

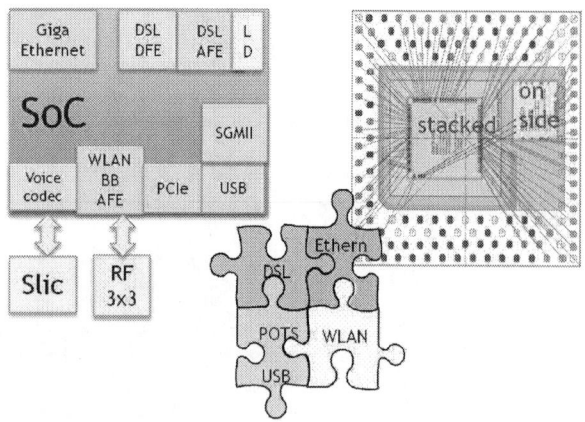

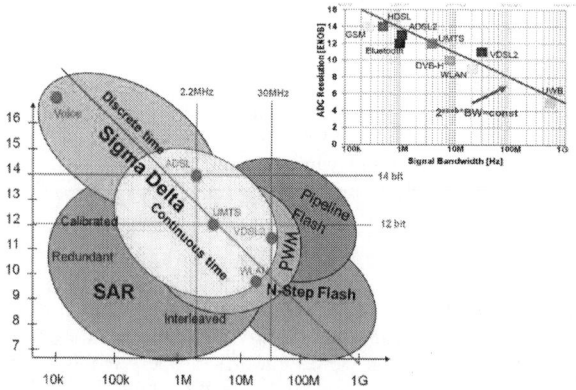

Fig. 5. SoC and SiP solutions for multi functional integration

Fig. 6. ADC mapping: resolution versus bandwidth

IV. INNOVATIVE BUILDING BLOCKS

Communication systems are always complex combinations of mixed-signal- and digital-signal processing units. The goal is to achieve very high data rates as close as possible to the available transmission channel capacity in order to save costs and to maximize communication capability over limited transmission channels.

Although modern designs in newest technologies fully exploit the potential of digital communication, the transmission media (cable or RF) is still analog. Thus the analog front-end circuitry and particularly the ADC play a key role in the successful and practical implementation of any communication system. The higher the resolution of the ADC, the more signal processing can be done digitally afterwards. Figure 6 shows the resolution and the bandwidth of the ADC needed for various communication systems. The bandwidth is increasing in modern communication systems such as VDSL and UWB, but the resolution needed is decreasing so that the product 2^{ENOB} x BW stays more-or-less constant. However, ADCs with higher bandwidths are more difficult to design and have more power dissipation and area, although the product of resolution and bandwidth is constant. This is due to the fact that higher bandwidth leads to more challenges due to slewing and speed limitations. So far, the communication systems are dominated by Sigma-Delta converters up to 30 MHz because of their feedback loop nature and thus inherent linearity advantage without the need for calibration [4]. Deep sub-micron speed capabilities and continuous-time processing even make them the most power efficient. Currently new additional innovative steps are developed in this direction in order to extend the bandwidth further, or decrease the power and area needed. Such emerging techniques are self-oscillating pulse-width modulation ADCs [5], where the pulse-width modulation (PWM) is used instead of a multi-bit quantizer in the Sigma-Delta ADC. The multi-level feedback signal is exchanged by a single level PWM signal with encoded time information. This avoids the use of a flash ADC and multi-bit feedback DAC. With these new innovative steps both the area and power could be reduced by half as is shown in figure 7.

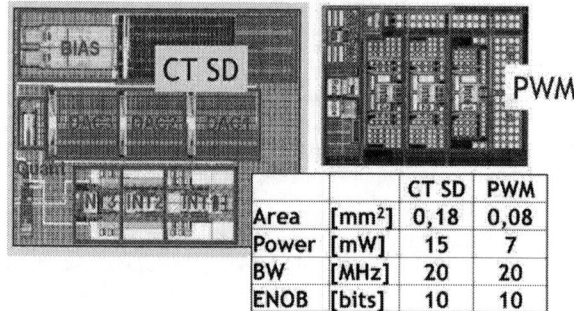

		CT SD	PWM
Area	[mm²]	0,18	0,08
Power	[mW]	15	7
BW	[MHz]	20	20
ENOB	[bits]	10	10

Fig. 7. Comparison CT SD ADC versus PWM-SD ADC

V. SUMMARY

This paper summarized the most important challenges in the analog design of communication systems. Cost pressure and increasing bandwidth demands are addressed by increased integration level using technology shrinks accompanied by innovations. It is also very essential to balance all system parameters in the optimization process in order to achieve an overall optimized solution.

REFERENCES

[1] Pécourt, J. Hauptmann, "An Analog Integrated Adaptive Balancing Hybrid for use in (A)DSL Modems", ISSCC 1999, pages 252-253

[2] T. Piessens, M. Steyaert, "SOPA: A High-Efficient Line Driver in 0.35 mm CMOS Using a Self Oscillating Power Amplifier", in ISSCC Digest of Technical Papers, pp. 306-307. February 2001.

[3] P. Pessl, R. Gaggl, J. Hohl, D. Giotta, J. Hauptmann; "A four-channel ADSL2+ analog front-end for CO applications with 75 mW per channel, built in 0.13µm CMOS"; IEEE Journal of Solid-State Circuits, Vol.39, No.12; pages 2371–2378; 2004.

[4] A.Wiesbauer, J. Hauptmann, P. Laaser; "Sigma Delta Converters in Wireline Communications"; Proceedings of 11th Workshop on Advances in Analog Circuit Design (AACD); Spa, Belgium; 2002.

[5] Enrique Prefasi, Susana Paton, Luis Hernandez, Jörg Hauptmann "A0.08mm2, 7mW Time-Encoding Oversampling Vonverter with 10bits and 20MHz BW in 65nm CMOS", Esscirc 2010

978-1-4673-4579-8/13 $31.00 © 2013 IEEE

Challenges of ASIC Development for Sensor Nodes in Automotive and Consumer Applications

Dirk Droste

Robert Bosch Automotive Electronics, Reutlingen, Germany

Contemporary development projects of ASICs for sensor-nodes in automotive and consumer industry are facing a variety of challenges due to strong requirements for robustness, performance and flexibility for automotive applications and strong requirements about power consumption and die-size for consumer applications. The ASIC development divisions of the Robert Bosch GmbH are facing these challenges in multiple ASIC projects for sensor nodes and derive suitable competitive structures of fast but sustaining development methodologies. This talk gives a brief essay about aspects of ASIC development approaches for automotive applications, covering partitioning and technology choices, frontend design for challenging requirements of performance, robustness and long-term reliability and flexible architectures for signal processing and interfacing. Also, a small essay about diverging requirements to ASIC development for automotive and consumer applications is given.

Paper PL3

PRIME 2013, Villach, Austria

978-1-4673-4579-8/13 $31.00 © 2013 IEEE

Innovative RFDAC Concepts for Digital Multi-Mode Transmitter in Cellular Applications

M. Fulde, F. Kuttner, E. Thaller, D. Ponton, V. Kampus, S. Gruenberger, H. Habibovic, A. Santner, C. Duller, C. Krassnitzer, G. Babin, G. Knoblinger

Intel

Digital polar transmitter concepts based on RF-DA-converter recently proved the potential to significantly reduce power consumption. Furthermore, external component count as well as PCB area is minimized since no TX SAW filter is required and a single multimode, multiband power amplifier can be used.

The RFDAC is a key building block for this architecture and needs to provide very high dynamic range (e.g. 17 ENOB for 3G) at high clock frequencies (e.g. 1GHz) with reasonable power efficiency in order to deal with far-off noise, spurious and repetition spectrum issues. Another key requirement for the DAC is the scalability to new CMOS nodes. A novel RFDAC concept is presented and benchmarked; the capability to fulfill the tough linearity/resolution/power specifications even in 28nm technology and below is shown.

A 15bit class-B (=single-ended) current mode DAC with distributed digital mixer has been implemented in 28nm CMOS. To minimize DNL and to meet the stringent quantization noise requirements a 10bit thermometer coded array of 1024 current sources is combined with 5 binary scaled cells. Different calibration techniques are employed to compensate for the dynamic degradation effects inherently connected to the single ended DAC structures and to achieve linearity requirements. A pre-charge mechanism prepares the current source before turning it on. To compensate coupling on the bias node a H2 dummy capacitor is used. The related challenge to match a parasitic capacitor to a dummy MOS capacitor is solved with an integrated regulation loop employing tracking ADC principle to set the correct compensation voltage. The bias voltage of the distributed mixer-cascode structure is also regulated on chip. Preliminary measurements on a 28nm test-chip show that IM3 of better than -45dBc at 6dBm output power is achievable (meeting product specs for 2G/3G/LTE). At 6dBm the RFDAC draws 7mA from 1V (cell array) and 25mA from 2V (trafo/load). The active DAC area is just 0.048mm2 (without trafo).

Furthermore a new digital data decoding and mixing scheme is presented. This approach allows splitting the current source array in two independent sub-DACs. In addition the phase of the LO signal can be shifted in the decoder which allows to use a single RFDAC for polar, signed polar and IQ modulation.

Finally an outlook to even more digitized RFDAC based transmitter concepts with integrated PA functionality is given.

FPGA based silicon innovation exploiting "More than Moore" technology

Patrick J. Quinn
Xilinx, Dublin, Ireland

Abstract—The most recent series of Xilinx FPGAs in 28nm host a wide range of the most advanced technologies. 3D-IC technology has allowed for a doubling of logic capacity in a single device, commensurate with going to the next process node, with the added advantages of reduced power and improved yield. New levels of programmable systems integration are enabled through the co-integration of digital FPGA die with dedicated analog die such as high-speed analog SerDes on a single interposer. Such heterogeneous integration delivers an industry leading 2.8Tb/s of off-chip serial connectivity.

This talk reviews what's driving Moore scaling and how it impacts FPGA technology scaling relative to ASSPs and ASICs. It goes on to cover such items as process selection, 3D-IC technology, analog features, as well as power reduction strategies and the migration steps to 20nm and 16nm.

MOORE SCALING AND ITS IMPACT ON FPGA DESIGN

The increasingly rapid expansion in global internet traffic puts increasing demands on equipment manufacturers to keep pace by continually upgrading technology [1]. Enterprise bandwidth is required to grow at a rate of almost 3x every 2 years while power levels can only increase moderately. Innovative new applications, especially those based on high quality video streaming, are driving insatiable demand for more intelligent network bandwidth (Fig. 1). Moreover, the network infrastructure is becoming more fragmented, despite efforts at industry convergence, and has to deal with a variety of media technologies (copper, optical fibre, air). Investment in internet infrastructure (only 8% annually) is not keeping pace with the rapid increase in IP traffic [2]. The consequence is that the cost per unit bandwidth must come down sharply so that future bandwidth needs can be met with the expected investment forecasts.

The way the electronics industry has ensured the continued reduction in cost per function year-in, year-out is to aggressively follow Moore's Law process scaling, where the number of transistors per unit area doubles about every 2 years, coupled with a 25% performance increase and 20% power reduction per function with scaled Vdd. Increasingly complex process technologies are needed each chip generation to ensure continued scaling with the consequence that chip costs are going up enormously. We have now reached the point where the costs to manufacture a transistor have levelled off from the 28nm node. In going from 28nm to 20nm, the extra costs are not just associated with more expensive masks but also the extra number of masks associated with double patterning. At this point, only a small cohort of ASIC and ASSP chip providers, especially those implementing very large scale SoCs, can afford to convert to beyond the 28nm node. The escalating up-front NRE costs (Fig. 2) are often prohibitive and put considerable pressure on designers to make sure their designs are "first time right" to avoid costly re-spins. FPGA technology, on the other hand, will always benefit from being in the leading edge technology node, encouraging broader adoption among end users who don't have to shell out the NRE costs for their designs and who require a fast time-to-market. FPGA technology, with the inherent benefits of programmable integration and in-the-field flexibility and scalability, can address a wide variety of different applications while the recent innovation in IO solutions allows for multi-Terabits per second data transfer rates. Xilinx judicially chose 28HPL, with high-K metal gate, which gives an optimal mix of process high-performance and low power (HPL)

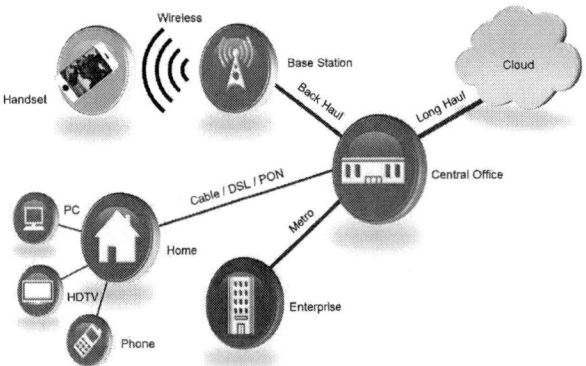

Fig. 1: Exponential growth in bandwidth driving the need for semiconductor innovation

suitable for FPGA applications.

One of the consequences of Moore's Law, allowing us to put so much extra functionality on one chip, is that the gap increases between the number of logic gates and I/O each process generation (Fig. 3). FPGA processing capability is expanding at an order of magnitude greater than the ability to get that data on and off chip. To counteract the cost and yield challenges posed by both Moore's law density increase and the associated relative reduction in off-chip bandwidth, Xilinx has successfully pioneered the use of 3D technology and is employing it in its top range products.

BEYOND MOORE SCALING

As chip die size gets larger, yield goes down exponentially and so building very large scale dice has become very difficult and expensive. Largest devices on the latest process nodes cost disproportionately more and availability is delayed due to the impact of die size on yield. With the innovative 3D-IC approach, using SSIT (stacked silicon interconnect technology) a number of smaller high yielding dice are laid side-by-side on a silicon interposer in such a way that they all function as one single integrated die. In this way, it is possible to effectively linearize what was originally an exponential curve of die cost versus die area. Regular chip metal

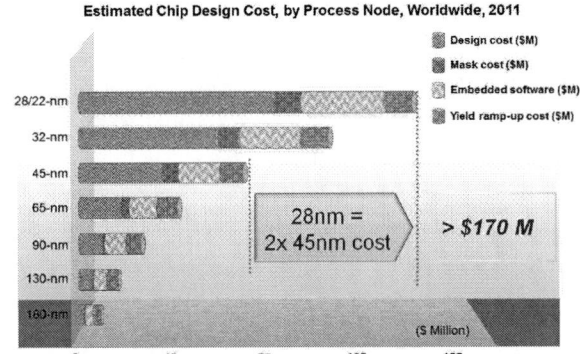

Fig. 2: Escalating NRE costs for ASIC and ASSP designs

978-1-4673-4579-8/13 $31.00 © 2013 IEEE

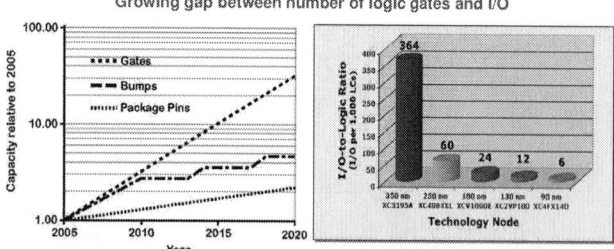

Fig. 3: Chip I/O bottleneck with a) ITRS showing 15x drop in I/O-to-logic ratio by 2020, b) I/O per 1000 logic cells in largest FPGA per node

interconnect is used to route the dice of the SSIT device. It makes possible more than 100 times the die connectivity bandwidth per Watt, at a fraction of the latency, and at much less power, than is possible using traditional I/O (either standard parallel or high-speed serial). Based on this methodology, a Virtex 2000T has been manufactured in 28HPL with double the capacity of what would have been achievable based on Moore's Law scaling alone. It contains 2 million logic cells, commensurate with >7 billion transistors, and there are up to 10,000 routing channels between individual FPGA slices in the stack.

Putting it all together, the net benefit of both Moore's Law and "More than Moore" in system power reduction through advanced integration in 28HPL can be seen in Fig. 4. The power savings in the 3D-IC implementation are principally achieved through elimination of standard I/O power between multiple devices.

The SSIT 3D-IC approach lends itself naturally to the combination of FPGA slices with slices of other technologies to create a mixed-mode SoC. The principal electronic components of logic, memory, CPU and analog are each best suited to their own optimized process type. While it is feasible to combine all such technologies in one leading edge process node, it doesn't make economic sense to do so. Indeed for analog, smaller does not necessarily mean better, since analog performance deteriorates with smaller die areas and reduced power supplies, despite the finer lithographies. The gap between current analog technology nodes and current digital nodes continues to widen each successive new CMOS generation. Most analog designs will stay parked at 40nm and above for a long time to come. Factors such as double/multi-patterning in finer processes and the analog uncertainty that it brings, as well as the lower supply voltages and expense of all the extra masks, will make it prohibitive for analog migration for quite some time.

A set of heterogeneous FPGAs has been developed at Xilinx based on SSIT [3], the largest of which (XC7VH870T) is illustrated in Fig. 5, which contains 3 FPGA slices and 2 28-Gbps serial transceiver slices. The total IO throughput rate is 2.8Tb/s from one such module and lends itself to Nx100G Ethernet ports and 400-

➤ Highest bandwidth FPGA with 2.8 Tb/s serial connectivity (~3X Monolithic)
➤ Electrically-isolated 28G transceivers for optimal signal integrity
➤ Faster time-to-market and lower BOM costs due to low-jitter heterogeneous integration

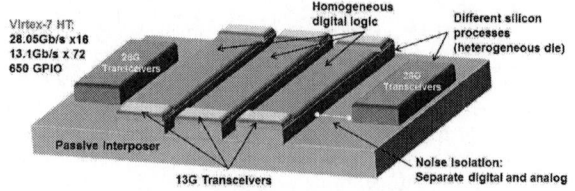

Fig. 5: Heterogeneous SSIT integration on the Virtex-7 HT

Gbps Ethernet line cards, optical transponders, and base-station and remote radio head applications. The SSIT has the added benefit of providing excellent noise isolation between the 28-Gbps analog circuits and the digital FPGA circuits, which is much superior to any monolithic offering.

With internal chip data rates scaling to Terabits per second in the latest technology nodes, SoC performance is no longer limited by just transistor performance but ultimately depends on the interconnect because of the impact it has on signal routing congestion, power dissipation and delay. To this end, the latest family of Xilinx FPGAs, the first of which has been taped out in a 20nm planar process and will have silicon back in 2013, addresses this issue in a systematic way using an UltraScale architecture (Fig. 6) [4]. Enhancements include highly optimized critical paths, which relieve routing congestion and help convey data to where it is needed much faster than has previously been possible, as well as multi-region high-performance clock networks to create low-power, very low-skew scalable clocking right across the chip. This enables maximum utilization and routeability of all the resources of the FPGA, including across SSIT slices, without loss of performance. This architecture is scalable from planar at 20nm to finfet technology at 16nm. Xilinx is cooperating with TSMC to have the first Ultrascale 16nm finfet FPGA devices in 2014.

REFERENCES

[1] D. Buss, "Technology in the internet age", *Proc. IEEE Int. Solid State Circuits Conf.*, section 1.1, 2002

[2] Scott Kipp, "Exponential bandwidth growth and cost declines", Network World, http://www.networkworld.com/news/tech/ 2012/041012-ethernet-alliance-258118.html

[3] Liam Madden et al, "Advancing high performance heterogeneous integration through die stacking", ESSCIRC, pp18-24, 2012

[4] http://www.xilinx.com/ultrascale

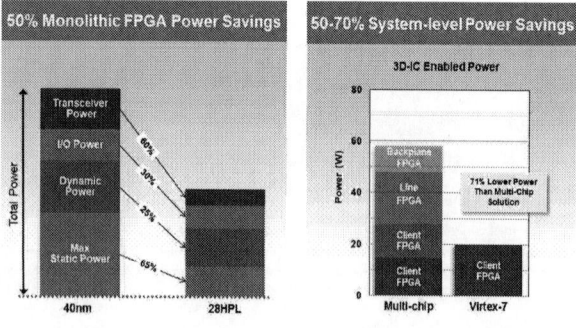

Fig. 4: Power reductions achieved thru scaling and innovation for a) 40nm to 28HPL and b) multi-chip to a single SSIT SoC

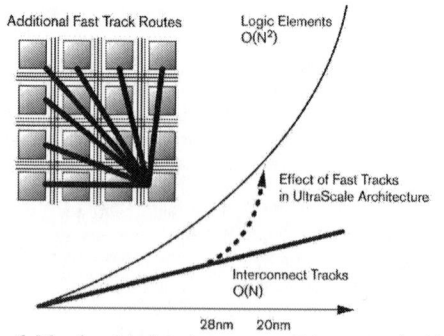

Fig. 6: Massive data flow increases on chip managed with new fast track routing architecture in 20nm and beyond

Organic Electronics: Material Aspects, Devices and Microelectronic Applications

Barbara Stadlober

JOANNEUM RESEARCH Forschungsgesellschaft mbH
Institute for Surface Technologies and Photonics
Weiz, Austria
Barbara.stadlober@joanneum.at

Abstract— **Due to their superior mechanical properties and the cost advantage associated with their processing, organic thin-film transistors can be viewed as an important technology for a broad range of applications spanning from optical displays, microelectronics to optical, chemical, and biological sensors. There is no doubt that, as the technology matures the number of applications will keep increasing with the possibility of OTFTs becoming a key technology in product innovation. Starting from a short overview of organic thin-film transistors research, this paper will concentrate on the recent developments in organic circuit design and on how the state of the art in this field can be further advanced with contributions from materials and processing research.**

Keywords—organic semiconductors, thin-film transistors, circuit design, double gate

I. INTRODUCTION

Powered by a strong market pull and owing to the large research effort with respect to materials, processing and modeling, electronics based on organic semiconductors has evolved from a curiosity for scientists to a technology that enables completely new products. Flexible electronic RFID tags manufactured from the plastic roll, lightweight displays that can be rolled-up and stowed in a fraction of the surface they occupy when used, low-cost and colorful solar cells, organic light-emitting diodes for large area and free form factor lighting elements are only some of the organic electronic products that are expected to quickly appear on the market. The advent of such innovative products is enabled by an increasingly mature material science and production technology of organic semiconductors. The "big three" of organic electronic devices are organic thin film transistors (OTFTs), organic light emitting diodes (OLEDs), and organic solar cells, however, over the last two years increasing output has been noticed for sensors and memories both in science and in product applications.

This paper reviews research developments in materials, technology and circuit design for electronics based on organic thin film transistors to learn more about upcoming research trends that will contribute to a further advancement of the state of the art. In order to determine the future challenges we will take the perspective of a circuit designer, analyzing simple state-of-the-art OTFTs circuits.

II. A SHORT OVERVIEW ON MATERIALS, PERFORMANCE AND PROCESSING OF OTFTS

A. Performance of Organic Semiconductor Materials

Over the last years the organic TFT research community has recorded a big success in the development of novel and the improvement of existing p- and n-type semiconductor materials. That resulted in devices with high charge carrier mobility values owing to the improvement of molecular ordering and stacking via crystal engineering and to better processing technology knowledge of the semiconductor materials [1]. As witnessed in many review papers [2, 3] the charge carrier mobility of polymer semiconductors has increased by more than four orders of magnitude, from values in the order of 10^{-5} cm^2/Vs before the 1990s to values around unity nowadays [3, 4]. Small-molecule organic semiconductors have dramatically improved in a similar way, from the typical mobility of 10^{-3} cm^2/Vs in the early 1990s to

values surpassing 10 cm^2/Vs today [5,6]. Processing and deposition techniques still play a significant role in determining the mobility of the semiconductor films. Depending on the degree of internal ordering and crystal packing motifs, on the presence of traps and on the properties of the semiconductor/insulator interface, the mobility of a given material can vary by more than one order of magnitude. For example, if pentacene, a very popular small molecule p-type semiconductor is processed from solution a maximum hole mobility of 1.8 cm^2/Vs was achieved for the related OTFTs [7], whereas vacuum evaporated pentacene OTFTs show up to 5.5 cm^2/Vs [8]. For OTFTs based on single crystalline pentacene (processed by hot wall epitaxy) even mobility values of 40 cm^2/Vs have been reported in interface-optimized devices [9].

Stability, i.e. shelf and operational lifetime of devices based on organic semiconductors, has always been a matter of concern and an important research domain. Very encouraging results have been obtained with p-type materials [5, 6, 10], whereas n-type materials typically are characterized by extremely poor performance stability. Resent research, though, shows interesting results also in this field. Whereas the mobility of a cyanated perylene carboxylic diimide derivative with an initial value of 0.12 cm^2/Vs decreased by more than one order of magnitude over a 400 day storage in air [11], novel organic n-type semiconductors with a much higher mobility ($1 cm^2/Vs$) were shown to be reasonably air-stable [12, 13].

B. Processing Strategies for OTFTs

One of the advantages of organic electronics is, that organic thin film transistors can be manufactured with large-area compatible techniques, using processing temperatures just above ambient and industrial high-throughput fabrication methods like roll-to-roll printing. Also, due to the very limited thermal budget, organic electronics can be produced on basically any substrate, including cheap and flexible films. The economic interest in organic electronics is generated by the fact that organic devices may be produced with tremendous throughput, at very low cost per unit area, on large surfaces, and on any substrate.

Early ground-breaking research on inkjet printing for the manufacture of OTFTs [14], has been followed by integration of different devices fabricated on separate sheets by inkjet printing, screen printing and shadow- mask evaporation, as shown in [e.g. 15-17]. Important application-related work on truly high-throughput printing techniques has been done [18] resulting in the first products (simple RFID labels) based on printed OTFT technology [19, 20]. Several other groups are also working in the field of large-throughput OTFT and organic integrated circuit (IC) fabrication based on mass printing techniques such as offset printing [21-22] and gravure printing [23]. Some are even exploring alternative techniques like transfer printing using stamps for organic circuit fabrication [24-26]. Stamp-based methods such as UV-nanoimprint lithography (UV-NIL) were further utilized for the fabrication of OTFTs with miniaturized channel lengths [27] and self-aligned electrodes with the aim to stretch their operation regime and to simplify the high-throughput fabrication of circuits [28].

III. FUTURE CHALLENGES BASED ON A CIRCUIT DESIGNER'S POINT OF VIEW

Based on the state-of-the-art some challenges for further high-impact research will be pointed out in this section for different kinds of circuits and their specific demands: digital, RF and analogue.

A. Digital circuits:

Concerning circuit design, the very first encouraging results reporting functional ICs [29] have been followed by much research work to improve the yield, robustness and complexity of digital circuits based on OTFTs. State-of-the- art digital designs contain routinely more than 1000 transistors and range from display driver circuits [30], over 128bit 13.56 MHz RFIDs [31- 32] to 8bit organic microprocessors [33].

Noise margin

Owing to their higher environmental stability most of the organic electronics circuit development work has been done on the basis of a p-type-only technology. The elementary building block of the digital gate is the inverter, a circuit that inverts an input signal. One of the major issues there is to design digital gates with enough noise margin and low-enough noise margin variability. Only under these conditions, in fact, it is possible to build digital circuits with a complexity of hundreds or more logic gates that would function properly in a real working environment.

The noise margin is a figure of merit of an inverter stage, that measures the immunity of the inverter against inevitable variations (noise), e.g. in the transistor parameters or resulting from electrical noise picked up by the circuit and adding to the input voltage. For an inverter it is defined as the side of the largest square that can be inscribed between the input–output characteristics (V_{in} vs. V_{out} , see straight line in Fig.1a,b) and the mirrored characteristics that shows V_{in} as a function of V_{out} (inverted striped line in Fig.1 a,b). In organic p-type only circuits the noise margin turns out to be strongly dependent on the threshold voltage V_{th} of the transistors used to build the inverter.

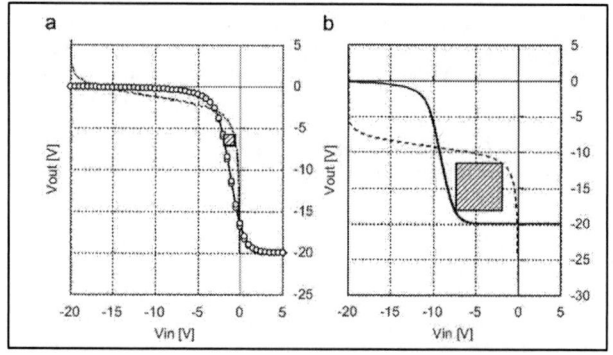

Fig. 1. a) Measured input-output characteristics of a depletion-load inverter (VDD=-20V) and determination of the noise margin of the inverter as the size of the largest square fitting between inverter transfer curve and mirrored inverter transfer curve (striped line). b) Noise margin of a depletion-load inverter with enhancement-mode driver (Vth=-8V) and depletion-mode load [After 34].

Since the threshold voltages in manufacturable p-type OTFTs usually are positive (meaning that the devices are depletion-type), the depletion-load inverter configuration with the gate of the load transistor TL being connected to its source (shown in the inset of Fig.1b) is often used - last but not least also because of its much higher gain. However, the noise margin of such an inverter with depletion-mode driver and load transistors (TL and TD) is typically not more than 1 V (Fig.1b). Using a hypothetic technology, where transistors with two different threshold voltages would be available (one with slightly positive and the other with negative Vth) the noise margin could be strongly improved. In Fig.1c for example, the driver transistor has received a Vth of -8V in simulation and the load transistor retains the small positive threshold typical of the technology employed, resulting in a noise margin of about 5V [34].

As it has been shown using the noise margin as an example, threshold voltage control is crucial to allow better performance in circuit based on p-type organic transistors. A lot of approaches for controlling the threshold voltage have been explored.

Some research attempts are directed towards a control of the Fermi level by monolayers of dipolar molecules. However, such approaches turned out to be not reproducible enough for circuit design. Moreover, tuning of the threshold voltages in depletion-load inverters by UV-illumination induced chemical doping of the p-type semiconductor pentacene was demonstrated in [35]. Although the characteristics of these inverters could be varied over a large range there is no practical use for digital circuits.

Logic gates using ambipolar transistors (semiconductors that simultaneously conduct electrons and holes), guarantee a good noise margin as the pull-up and pull-down device can be biased, respectively, in the p-type region and in the n-type region, to emulate the way complementary logic works [36]. Ambipolar transistors, however, never fully switch off, when one of the stable output voltages of the gate (corresponding to the logic high or logic low) is reached, and thus do not allow to cut the leakage path between the power supplies when the logic gate does not switch.

Another, more successful approach relies on the usage of dual-gate OTFTs incorporating two gate dielectrics and two gate electrodes. These devices have the main advantage that the threshold voltage can be set as a function of the applied second gate bias with the threshold voltage shift depending on the ratio of the capacitances of the two gate dielectrics. Based on dual-gate OTFTs it was proven that the noise margin can indeed be increased significantly [37] and that such a transistor design can form the base for the largest OTFT based circuits (microprocessor) reported so far [32].

Truly complementary OTFTs technologies, i.e. technology platforms where both n- and p-type transistors are available, have also been studied [38, 39]. The availability of complementary transistors improves the noise margin dramatically and may result in better power consumption too, as complementary gates only need power when the output switches from one stable output voltage to the other. Complementary OTFT technologies deserve, thus, the full attention of scientists active in the field.

Speed
Now we consider the dynamic performance of digital circuits, again based on the example of the simple depletion-load inverter. Fig.2 left displays a schematic diagram showing the time response of such an inverter to a variation of the incoming signal V_{in} from a logic low to a logic high value and back.
When the input voltage goes high, the driver transistor T_D is switched off and the net pull-down current is approximately equal to the current flowing for $V_{GS}=0V$ in the wide load transistor T_L. This current is small and heavily depends on the threshold voltage of the load transistor, as shown in some

measured transfer characteristics of TFTs having different threshold voltages (Fig.2 right). The time needed to pull down the output node capacitance is about inversely proportional to the current flowing at $V_{GS} = 0V$ in the load transistor, and thus again strongly depends on the threshold voltage. When the input voltage goes down, on the other end, the driver transistor switches heavily on and is capable of pulling up the capacitance connected to the output node towards the high voltage supply (Gnd) in a short time. The pull-up time depends on the output capacitance and on the pull-up current, and thus will be inversely proportional to the mobility and proportional to the square of the transistor length (assuming all transistors share the same length and that the capacitive load is only due to transistors). If one looks at the total transit time of an input pulse through a depletion-load inverter, the pull-down time being much slower than the pull-up time, the bottleneck in achieving better speed is, for this popular kind of inverters, in the threshold voltage of the load transistor (which influences the pull-down action) more than in the mobility of the semiconductor used (which has a stronger influence on the pull-up). However, if the load transistor has a distinct positive threshold voltage with a clear depletion-mode behavior it will operate in the saturation regime at high input levels and $V_{GS} = 0V$. Then the drain current and therefore the pull-down time will mainly be determined by the mobility and the channel length.

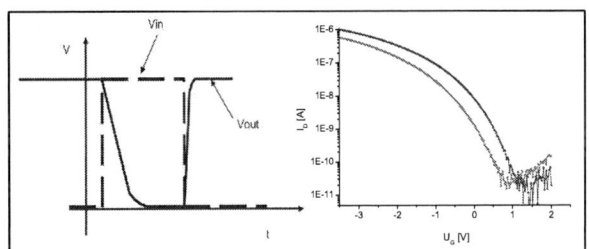

Fig. 2. (left) Schematic time response of a depletion-load inverter to an input impulse. b) Transfer characteristics of two transistors with identical layout and different, positive threshold voltages. The threshold of Transistor 1 (blue curve) is more positive than the one of Transistor 2 (red curve) thus Transistor 1 conducts a much higher current at $V_{GS} = 0V$ than Transistor 2.

The strategy to increase circuit speed by decreasing the channel length L (often referred to as downscaling) has very successfully been exploited in Si technology and it is mirrored in the Moore's law and the international technology roadmap for semiconductors (ITRS). In OTFTs, however, the miniaturization of the channel length turned out to be more complicate due to several reasons. On the one hand the materials are less robust against standard high-resolution photolithography processes and the patterning by etching. On the other hand, since the channel resistance scales with L and typical p-type OTFTs often show a non-negligible resistance of the source and drain electrodes (the so-called contact resistance that originates from the hole injection barrier due to energy level misalignment) downscaling in many cases does not result in an increase of the operation frequency but in a decrease of the current level. The latter is a consequence of the performance limitations by the contact resistance. Moreover, in order to avoid typical short channel effects the gate dielectric needs to be scaled appropriate to L, which in many cases complicates the choice of suitable organic gate dielectric materials. In a nutshell proper downscaling means not only a decrease of the channel length but also a decrease of the gate dielectric layer thickness, a minimization of the contact resistance and the avoidance of large parasitic overlap capacitance contributions from the gate layer.

By means of UV-NIL as the patterning technique for the fabrication of submicron-spaced source and drain electrodes it could have been shown that all these requirements can in principle be fulfilled, thus enabling a downscaling of the devices without adversely affecting the device characteristics [40]. Thus in submicron transistors with thin enough gate dielectrics and minimized contact resistance (resulting from a UV-ozone treatment) the drain current level scales properly with the channel length meaning that the charge carrier mobility is independent of the source-drain distance (Fig. 3 left). However, the reproducibility and stability of miniaturized devices produced on flexible substrates, their reliable processing on larger areas and their integration into circuits still need to be improved (Fig. 3 right).

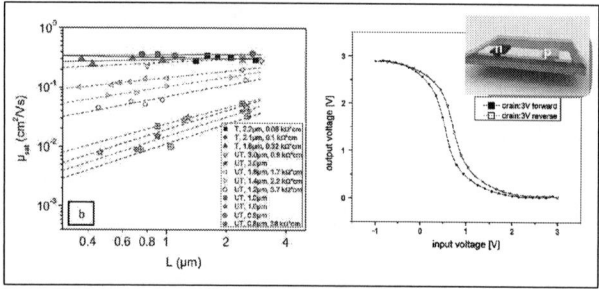

Fig. 3. (right) Channel length dependence of the charge carrier mobility in devices with treated electrodes (T, full symbols, low contact resistance) compared to devices with untreated contacts (UT, open symbols, medium to high contact resistance) (left) Complementary inverter based on NIL-patterned OTFTs fabricated on flexible substrate. Note the hysteresis in the inverter curve arising from a hysteresis in the transfer curves of the single transistors.

Speed may also be increased using complementary technologies, because in complementary logic the pull-up or pull-down actions are always performed by a fully- on device, exploiting the maximum current-driving capability for a given gate (and thus load) capacitance. In the emerging field of complementary OTFTs we can also list some interesting research challenges, mainly

- the mobility (and the thresholds) of the p- and n-type transistors needs to be balanced
- more work should be done on operational and shelf life stability
- device modeling and circuit design with complementary OTFTs should receive attention, especially in the new area of analogue circuits

B. RF circuits:

The equations are an exception to the prescribed specifications of this template. You will need to determine whether or not your equation should be typed using either the Times New Roman or the Symbol font (please no other font). Item-level radiofrequency identification (RFID) of goods, based on electronics labels (RFID tags) is one of the most interesting applications of organic and printed electronics with a huge market potential under high cost pressure. In RFID systems the reader sends electromagnetic waves that are captured by the antenna on the label: using this energy a rectifier provides the DC power needed to read an identification code stored in the label memory and the code is sent back to the reader utilizing the same radio link used to send power [31].The frequencies for the power and communication link are usually 13.56MHz or 860–960MHz according to the most common industrial standards. A non-linear circuit element is needed to rectify the electromagnetic wave captured by the antenna and transform it into a DC current, the latter being used to power the electronics on board of the RFID tag.

Due to the lack of efficient doping (so far utilized successfully only in OLEDs [41]) state-of-the-art organic electronics does not provide efficient rectifying p–n junctions, so rectifiers are normally implemented via Schottky junctions or diode-connected OTFTs. The bandwidth of these rectifiers depends on the time needed for the charge carriers to traverse the device when the incoming signal has the right polarity. Although the specific mechanisms behind the conduction are rather complicated the maximum working frequency depends on the semiconductor mobility and on the thickness of the semiconductor layer to cross. Owing to their smaller layer thickness Schottky diodes [42] are inherently faster than diode-connected transistors [43]. Thus, in RF circuits the semiconductor mobility is of high importance and it has been reported that it is already possible to rectify UHF radio waves by organic rectifiers [44].

C. Analogue circuits:

Analogue circuits are very important as they are needed to interface our physical world (via sensors, actuators, radio communication channels, etc.) to the processing performed using digital electronics. Organic analogue circuits would be needed, for instance, to interface and digitalize the signals coming from organic sensors using an integrated circuit approach. The field of analogue circuit design is still in its infancy. Analogue circuit design needs mature characterization and modeling of the transistors as well as deeper insight by the designer in device physics. Analogue design is thus more challenging and usually follows the development of transistor-level logic circuits design. The first studies were done on very basic analogue building blocks like amplifiers and differential amplifiers using p-type-only OTFTs [45]. Meanwhile also 6b D/A converters based on complementary OTFTs have been realized [46-47], but quite some work still needs to be done. The efforts of modeling and design experts will probably concentrate on this field in the near future.

REFERENCES

[1] Wang C, Dong H, Hu W, Liu Y, and Zhu D. Semiconducting π-Conjugated Systems in Field-Effect Transistors: A Material Odyssey of Organic Electronics. Chem Rev 2011;112:2208-67

[2] Klauk H. Organic circuits on flexible substrates. In: IEDM conference proceedings, 2005. p. 446–9.

[3] Zaumseil J, Sirringhaus H. Electron and ambipolar transport in organic field-effect transistors. Chem Rev 2007;107:1296–323

[4] McCulloch I, Heeney M, Bailey C, Genevicius K, MacDonald I, Shkunov M, et al. Liquid-crystalline semiconducting polymers with high charge-carrier mobility. Nat Mater 2006;5:328–33

[5] Kang MJ, Doi I, Mori H, Miyazaki E, Takimiya K, Ikeda M, and Kuwabara H. Alkylated Dinaphtho[2,3- b :2 ′ ,3 ′ - f]Thieno[3,2-b]Thiophenes(Cn -DNTTs): Organic Semiconductors for High-Performance Thin-Film Transistors. Adv Mater 2011;23:1222-1225

[6] Nakayama K, Hirose Y, Soeda J , Yoshizumi M, Uemura T, Uno M, Wanyan Li , Kang MJ, Yamagishi M, Okada Y, Miyazaki E, Nakazawa Y, Nakao A, Takimiya K, and Takeya J. Patternable Solution-Crystallized Organic Transistors with High Charge Carrier Mobility. Adv Mater 2011; 23:1626-1629

[7] Park SK, Jackson TN, Anthony JE, Mourey DA. High mobility solution processed 6,13-bis(triisopropyl-silylethynyl)pentacene organic thin film transistors. Appl Phys Lett 2007;91: 063514.

[8] Lee S, Koo B, Shin J, Lee E, Park H, Kim H. Effects of hydroxyl groups in polymeric dielectrics on organic transistor performance. Appl Phys Lett 2006;88:162109

[9] Jurchescu OD, Popinciuc Mihaita, van Wees BJ, Palstra TTM. Interface-controlled high-mobility organic transistors. Adv Mater 2007;19:688–92.

[10] Meng H, Sun F, Goldfinger MB, Jaycox GD, Li Z, Marshall WJ, et al. High-performance, stable organic thin-film field-effect transistors based on Bis-50-alkylthiophen-20-yl-2,6-anthracene semiconductors. J Am Chem Soc 2005;127:2406–7.

[11] Weitz RT, Amsharov K, Zschieschang U, Barrena Villas E, Goswami DK, Burghard M, et al. Organic n-channel transistors based on core-cyanated perylene carboxylic diimide derivatives. J Am Chem Soc 2008;130:4637–45.

[12] Kumaki D, Ando S, Shimono S, Yamashita Y, Umeda T, Tokitoa S. Significant improvement of electron mobility in organic thin-film transistors based on thiazolothiazole derivative by employing self-assembled monolayer. Appl Phys Lett 2007;90:053506.

[13] Hak Oh J, Liu S, Bao Z, Schmidt R, et al. Air-stable n-channel organic thin-film transistors with high field-effect mobility based on N,N'-bis(heptafluorobutyl)-3,4:9,10-perylenediimide. Appl Phys Lett 2007;91:212107.

[14] Sirringhaus H, Kawase T, Friend RH, Shimoda T, Inbasekaran M, Wu W, et al. High-resolution inkjet printing of all-polymer transistor circuits. Science 2005;290:2123–6.

[15] Sekitani T, Takamiya M, Noguchi Y, Nakano S, Kato Y, Sakurai T, et al. A large-area wireless power transmission sheet using printed organic transistors and plastic MEMS switches. Nat Mater 2007; 6:413–7.

[16] Noguchi Y, Sekitani T, Someya T. Organic-transistor-based flexible pressure sensors using ink-jet-printed electrodes and gate dielectric layers. Appl Phys Lett 2006;89:253507.

[17] Sekitani T and Someya T. Stretchable, Large-area Organic Electronics, Adv Mater 2010; 22: 2228-46

[18] Knobloch A, Manuelli A, Bernds A, Clemens W. Fully printed integrated circuits from solution processable polymers. J Appl Phys 2004;96(4):2286.

[19] Clemens W, Mildner W, Bergbauer B. New high volume applications with printed RFID and more. MST-News 2007;3(5):10–2.

[20] Minhun Jung, Jaeyoung Kim, Jinsoo Noh, Namsoo Lim, Chaemin Lim, Gwangyong Lee, Junseok Kim, Hwiwon Kang, Kyunghwan Jung, Ashley D. Leonard, James M. Tour, and Gyoujin Cho. All-Printed and Roll-to-Roll-Printable 13.56-MHz-Operated 1-bit RF Tag on Plastic Foils. IEEE TED 2010;57:571-80

[21] Huebler AC, Doetz F, Kempa H, Katz HE, Bartzsch M, Brandt N, et al. Ring oscillator fabricated completely by means of mass-printing technologies. Org Electron 2007;8:480–6.

[22] H. Kempa, M. Hambsch, K. Reuter, M. Stanel, G. C. Schmidt, B. Meier, and A. C. Hübler. Complementary Ring Oscillator Exclusively Prepared by Means of Gravure and Flexographic Printing. IEEE TED 2011;58:2765-69#

[23] Hongki Kang , Rungrot Kitsomboonloha , Jaewon Jang , and Vivek Subramanian. High-Performance Printed Transistors Realized Using Femtoliter Gravure-Printed Sub-10 m Metallic Nanoparticle Patterns and Highly Uniform Polymer Dielectric and Semiconductor Layers. Adv Mater 2012;24:3065-69

[24] Ofuji M, Lovinger AJ, Kloc C, Siegrist T, Maliakal AJ, Katz HE. Organic semiconductor designed for lamination transfer between polymer films. Chem Mater 2005;17:5748–53.

[25] Hines DR, Ballarotto VW, Williams ED, Shao Y, Solin SA. Transfer printing methods for the fabrication of flexible organic electronics. J Appl Phys 2007;101:024503.

[26] Lee HH, Brondjik JJ, Tassi NG, Mohapatra S, Grigas M, Jenkins P, et al. Appl Phys Lett 2007;90:233509.

[27] C. Auner, U. Palfinger, H. Gold, J. Kraxner, A. Haase, T. Haber, M. Sezen, W. Grogger, G. Jakopic, J.R. Krenn, G. Leising, and B. Stadlober. Residue-free room temperature UV-nanoimprinting of submicron organic thin film transistors. Org Electr 2009;10: 1466

[28] U. Palfinger , C. Auner , H. Gold , A. Haase , J. Kraxner , T. Haber , M. Sezen , W. Grogger , G. Domann, G. Jakopic, J. R. Krenn, and B. Stadlober, Organic Thin Film Transistors with Minimized Gate Overlaps by Self-Aligned Nanoimprinting. Adv Mater 2010;22: 5115

[29] Hart CM, de Leeuw DM, Matters M, Herwig PT, Mutsaerts CMJ, Drury CJ. Low-cost all-polymer integrated circuits. In: 24th European solid-state circuits conference proceedings 1998. p. 30–4.

[30] van Lieshout P, van Veenendaal E, Schrijnemakers L, Gelinck G, Touwslager F, Huitema E. A flexible 240x320-pixel display with integrated row drivers manufactured in organic electronics. ISSCC digest of technical papers 2005. p. 578–618

[31] Cantatore E, Geuns TCT, Gelinck GH, van Veenendaal E, Gruijthuijsen AFA, Schrijnemakers L, et al. A 13.56-MHz RFID system based on organic transponders. IEEE J Solid State Circuits 2007;42:84.

[32] K. Myny , M. J. Beenhakkers , N. A. J. van Aerle , G. H. Gelinck , J. Genoe , W. Dehaene , P. Heremans , A 128b organic RFID transponder chip, including Manchester encoding and ALOHA anti-collision protocol, operating with a data rate of 1529 b/s., IEEE Proc. Int. Solid-State Circuit Conf. (ISSCC) – Dig. Tech. Pap. 2009 , 206 .

[33] Myny K, van Veenendaal E,. Gelinck GH, Genoe J, Dehaene W, Heremans P, An 8b Organic Microprocessor on Plastic Foil, IEEE Proc. Int. Solid-State Circuit Conf. (ISSCC) – Dig. Tech. Pap. 2011, p. 322-24

[34] De Leeuw D, 2008 EMRS Spring Meeting, Straßbourg

[35] M. Marchl, M. Edler, A. Haase, A. Fian , G. Trimmel, T. Griesser, B. Stadlober, and E. Zojer, Tuning the Threshold Voltage in Organic Thin-Film Transistors by Local Channel Doping Using Photoreactive Interfacial Layers, Adv Mater 2010; 22:5361

[36] Anthopoulos TD, Setayesh S, Smits E, Colle M, Cantatore E, de Boer B, et al. Air-stable complementary-like circuits based on organic ambipolar transistors. Adv Mater 2006;18:1900–4.

[37] Spijkman M.-J., Myny K., Smits E. C. P., Heremans P., Blom P. W. M., and de Leeuw D. M., Dual-Gate Thin-Film Transistors, Integrated Circuits and Sensors, Adv. Mat. 2011; 23:3231

[38] Klauk H, Halik M, Zschieschang U, Eder F, Rohde D, Schmid G, et al. Flexible organic complementary circuits. IEEE Trans Electron Dev 2005;52(4):618.

[39] Klauk H, Zschieschang U, Pflaum J, Halik M. Ultralow-power organic complementary circuits. Nature 2007;445:745–8

[40] B. Stadlober, U. Haas, H. Gold, A. Haase, G. Jakopic, G. Leising, S. Rentenberger, E. Zojer, and N. Koch, Orders-of-Magnitude Reduction of the Contact Resistance in Short-Channel Hot Embossed Organic Thin Film Transistors by Oxidative Treatment of Au-Electrodes. Adv Funct Mater 2007; 17:2687

[41] H. Kleemann, B. Lussem, and K. Leo, Controlled formation of charge depletion zones by molecular doping in pin-diodes and its description by the Mott-Schottky relation, Jour Appl Phys 2012;101:123722

[42] Steudel S, Myny K, Arkhipov V, Deibel C, De Vusser S, Genoe J, et al. 50MHz rectifier based on an organic diode. Nat Mater 2005; 4(8):597–600

[43] Steudel S, De Vusser S, Myny K, Lenes M, Genoe J, Heremans P. Comparison of organic diode structures regarding high-frequency rectification behavior in radio-frequency identification tags. J Appl Phys 2006;99:114519

[44] Steudel S, Genoe J, Heremans P. Organic rectifiers reaching UHF frequencies. MST-News 2007;3(5):41–2.

[45] Gay N, Fischer W, Halik M, Klauk H, Zschieschang U, Schmid G. Analog signal processing with organic FETs. In: ISSCC digest of technical papers 2006. p. 1070–9

[46] Wei Xiong, Ute Zschieschang, Hagen Klauk, Boris Murmann. A 3V 6b Successive-Approximation ADC Using Complementary Organic Thin-Film Transistors on Glass. IEEE Proc. Int. Solid-State Circuit Conf. (ISSCC) – Dig. Tech. Pap. 2010, p. 134

[47] Tarek Zaki, Frederik Ante, Ute Zschieschang, Joerg Butschke,Florian Letzkus, Harald Richter, Hagen Klauk, Joachim N. Burghartz. A 3.3V 6b 100kS/s Current-Steering D/A Converter Using Organic Thin-Film Transistors on Glass. IEEE Proc. Int. Solid-State Circuit Conf. (ISSCC) – Dig. Tech. Pap. 2011, p. 324

Smart Power: From Problem Statement to System Solution

Dieter Draxelmayr, Karl Norling, Christian Lindholm

Infineon Technologies Austria AG
Villach, Austria

Abstract— **This paper demonstrates the design of a high power, high efficiency inverter based on SiC (Silicon Carbide) JFET (Junction Field Effect Transistor) switches. It starts with a problem statement, i.e. which requirements are to be fulfilled, discusses design solutions, and ends with some measurements of the complete system. The final design is able to handle more than 1000V, currents of more than 30A and achieves a peak efficiency of more than 99% in a buck configuration [1].**

Keywords— SiC, silicon carbide, power conversion, SMPS, gate driver, bootstrapping

I. INTRODUCTION

High efficiency power management is a topic which has gained significant interest during the last years. The reasons for this are manifold. They span from environmental care over the avoidance of thermal design and cooling systems to simply paying less money for the consumed energy. At the heart of such a high efficiency power management system typically sits an SMPS (Switched Mode Power Supply). This mainly consists of some inductive means, which may be simply a coil, some storage capacitors, some switches, and some circuitry to drive the switches according to some algorithm. Fig. 1 shows a typical arrangement for a simple buck converter. In the context of this paper it is of minor importance which type of SMPS at the end is chosen. We want to provide some generic thoughts and solutions for the "inverter problem": How to design a system which can by driven by some controller, will drive some inductive load, and can handle high voltages and high currents at high efficiency.

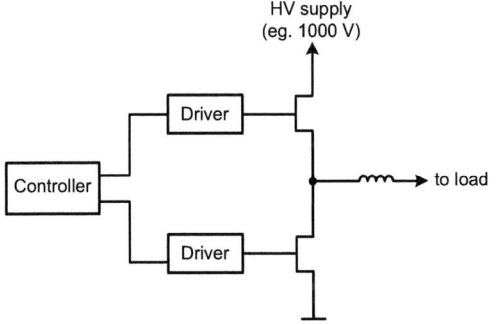

Fig. 1. Buck converter as a simple representative for a potentially more general SMPS

II. PROBLEM STATEMENT AND SOME IMPLICATIONS

We wanted to build a system which was able to handle 1000V, some 10A and reach an efficiency of 99%. Of course it should also be cheap, which means that the number of components should be small and that the switching frequencies should be high in order to reduce the size / weight / cost of passive components (inductors / capacitors). Typical applications of this could be the power supply of server farms or the power conditioning of solar power plants.

Especially the 99% efficiency puts a pretty stringent requirement on the selection of switches. Existing solutions mostly use IGBTs (Insulated Gate Bipolar Transistors) operating at frequencies between 16 and 30kHz. Another potential candidate could be a high-voltage silicon MOSFET. However, at somewhat elevated switching frequencies of, eg., 100kHz it turns out that the switching losses are simply too high to come even close to this target. Therefore alternative materials have been discussed. Devices based on SiC have been promising candidates. Of course the current manufacturing status of SiC is not as robust and not as mature as the manufacturing of silicon. Therefore we have decided to launch a series of JFET devices as first productive units [2]. However, these JFETs are normally-on devices. This means that they need an appropriate negative gate-source voltage to stop conduction. Taking a look on Fig. 1 this clearly tells that first the switches have to be turned off, which means that the switch drivers have to be powered, before the main high voltage might be turned on. It might be possible to design systems for this requirement, but generalizing this for arbitrary architectures and considering all possible failure scenarios is not an easy and foreseeable task. Therefore we have decided that we should provide an additional safety device for the SiC JFET.

III. SWITCH DESIGN

For the reasons mentioned above we have put another transistor in series to the SiC JFET [2]. This is a conventional low-voltage MOSFET transistor which is normally-off. Fig. 2 shows this arrangement. It should be noted that the MOSFET is of p-type whereas the JFET is of n-type. This means that their respective sources are connected together which makes it possible to drive them with a comparatively simple driver circuit (see next section). One could argue that an n-MOSFET could be better but this would lead to a complicated driver situation because due to the ringing between the connect point and ground, which can reach several 10V, one would have to build separated floating drivers for the two transistors.

Therefore we have decided for a combination of the high-voltage n-type SiC JFET and a low-voltage low R_{on} (on-resistance) p-type MOSFET. As will be explained later in more detail in normal operation the MOSFET is permanently on and the JFET acts as a switching device. In case that there is an insufficient driver supply voltage the MOSFET is shut down and acts as a safety guard, especially if there is no supply at all. In this situation both gate-source voltages are essentially zero. If the main high voltage power is still there the source of the JFET might go up until it reaches its pinch-off voltage. In our case this typically is 14V. On the other hand this means that the low-voltage MOSFET has only to withstand this limited voltage [1]. It is protected by the JFET from the high main power.

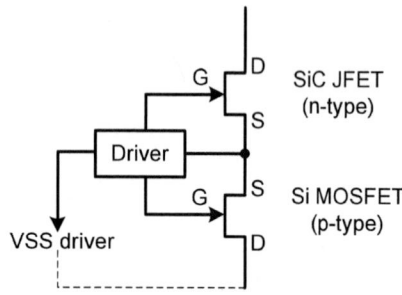

Fig. 2. Series connection of a high voltage switch and a low voltage safety guard

IV. POWERING THE SYSTEM

From Fig. 1 it can be seen that the highside driver might have an offset of 1000V to the lowside driver. This implies that there must be galvanic isolation and it also implies that one needs galvanically isolated power supplies for the respective driver circuits. The conventional approach foresees auxiliary transformers and rectifiers to individually bring power to the respective drivers. This is a pretty robust and simple scheme but it is also bulky and expensive, especially if one needs to design multi-phase systems. One can also use a bootstrapped solution which can do the job with capacitors and diodes. This is known for normally-off devices but can't be directly translated to our situation. Fig. 3 shows our approach in a simple half-bridge configuration.

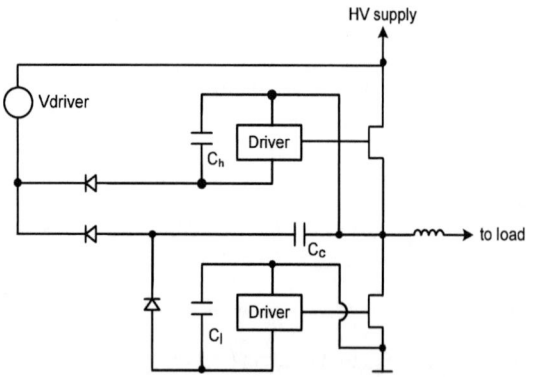

Fig. 3. Bootstrapped operation. During switching Ch is fed directly, Cl is fed indirectly via Cc with the voltage delivered by V_{driver}.

One needs only a single low-voltage power supply (V_{driver}). If one assumes that the switching nodes switches between 0 and 1000V the left-hand side of capacitor C_c will swing between (1000V- V_{driver}) and $- V_{driver}$ (neglecting diode drops). As a consequence Capacitor C_c is charged when the switching node is up and it will feed capacitor C_l each time the switching node goes low. In this way the voltage V_{driver} is transferred to the capacitor C_l (minus two diode drops). Capacitor C_h gets charged each time the switching node is in its "high" position. The main question left is how one can get power to the driver circuits if the switching node is not yet switching. In other words: how to do the start-up?

To answer this question let's reconsider Fig. 2. If there is no driver power at all, but the main power is already up then the JFET acts as cascoding device for the MOSFET. This means that the voltage drop over the MOSFET is equal to the pinch-off voltage of the JFET. In our case this is approximately 14V and this voltage can be used as a first supply for the driver chip. It is not sufficient for safe operation but it is definitely enough for powering up all the communication interfaces and all the references and biasing circuits. With this auxiliary voltage we therefore could start operation. The only constraint we have is that when turning off we also should turn off the low voltage MOSFET. This is because the voltage might still not be high enough to safely turn off the JFET. But in operating both MOSFET and JFET we can safely turn on and off the switch. This is true for high-side as well as for low-side. In this way we also can safely start operation of the switching node which, in consequence, will feed additional power to the drivers. So the startup-scenario is as follows:

- Everything off

- Wait until sufficient auxiliary power is available

- Start operation in switching both devices

- If sufficient power is available go to full operation

It should be noted that also switching both devices is functionally correct but it brings the penalty of much higher switching losses due to the operation of the silicon MOSFET. Only full operation, which is turning the MOSFET permanently on and switching only the SiC JFET, gives full performance.

V. DRIVER DESIGN

The driver usually gets its input from a controller and has to serve two power transistors (JFET + MOSFET) which can be connected to a substantially different voltage. Usually this voltage problem is solved via an opto-coupler. However, since this is yet an additional component which is not very well compatible with a standard integrated process we have chosen a different approach. Our driver contains a coreless transformer for voltage separation [3]. Basically this is some metal windings on top of some other metal windings. This forms a transformer which can transport information. At the same time the isolation capability of the two metal layers can be as high as several 1000V. The coreless transformer in our driver has been characterized up to 1700V and the data transmission has been characterized under a slew rate of up to

100V/ns. The main parts of the driver are the transmitter, the receiver, power regulation, references and control, and the two gate drivers. This is depicted in Fig. 4.

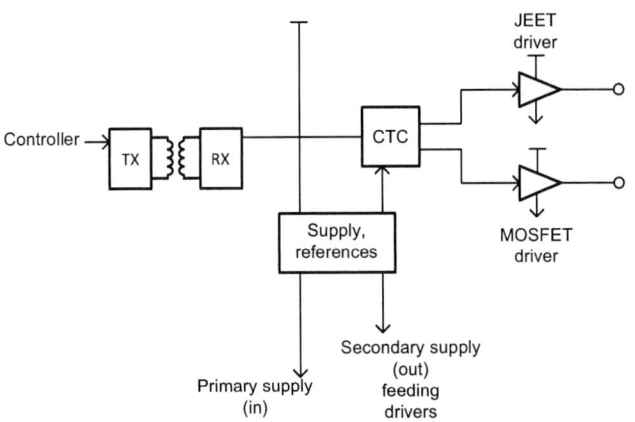

Fig. 4. Simplified overview of the driver IC. It contains two silicon dies coupled via a coreless transformer.

The power regulator takes the primary input voltage, ranging mainly from the pinchoff-voltage of the JFET up to V_{driver}. From this it generates a well-controlled secondary supply which is then used to power the actual gate drivers.

The control block takes the information of the receiver and combines it with the information of the secondary power supply. If the secondary power supply is too low it refuses operation and keeps the power transistors in off-state. If the secondary power supply is in some intermediate range it starts operation in startup-mode where both power transistors are switched. If the secondary power supply is high enough it goes to full operation. With this built-in supervision and intelligence the system can be brought up in a very simple and safe way without the need to control these different operating conditions via the microcontroller.

Fig. 5 shows a more detailed schematic of the voltage regulator. The primary supply comes in at nodes VCC2 – VEE2. In a first section (not shown) a 5V supply is built up (VDDLO_ANA – VREG). This supply is used to generate a well-controlled supply VREG which is referenced against VCC2. Basically the concept along with its regulation transistors tries to make the voltage difference VDDLO_ANA – VREG equal to the voltage drop over 5R. So the voltage drop over 5R + 14R should be 19V. If this criterion is not met transistors N2 and P2 steer the main regulation transistor P0 via N3 until the condition is met.

Finally, Fig. 6 shows a photograph of an opened driver IC.

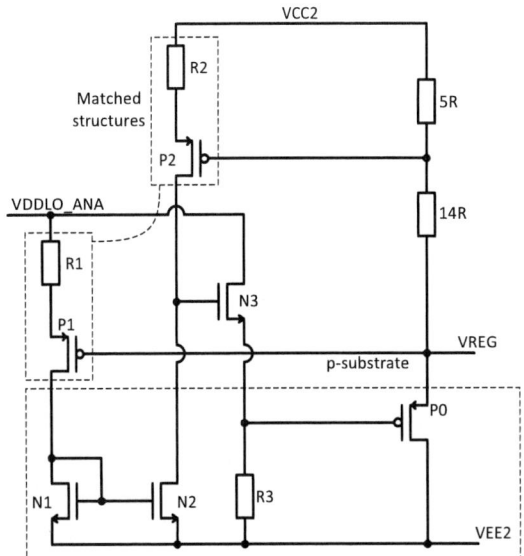

Fig. 5. Main voltage regulator. Input voltage comes from VCC2 – VEE2, output voltage is VCC2 – Vreg.

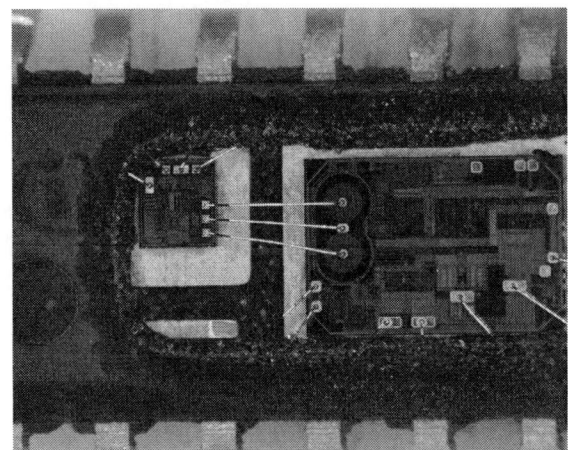

Fig. 6. Photograph of an opened driver IC.

VI. MEASUREMENT RESULTS

Fig. 7 shows a start-up sequence of our system. Ignoring the first 3 short pulses which are artifacts coming out of the used signal generator one can recognize the building up of the supplies. Once the system starts switching this is done with both power transistors. As VREG rises there comes the point where the MOSFET switch is permanently turned on. Finally one can recognize that the bootstrapped supply VEE2 is still rising while VREG already remains constant. Please note that all voltages are negative because the reference point is the source node of the power transistors.

Finally we also have taken an efficiency plot of a running buck converter system (Fig. 8). One can see the typical efficiency behavior with some decreased efficiency at light load and some decreased efficiency at high load with the maximum at the estimated typical operating point. This plot indicates that we indeed have achieved peak efficiencies above 99%.

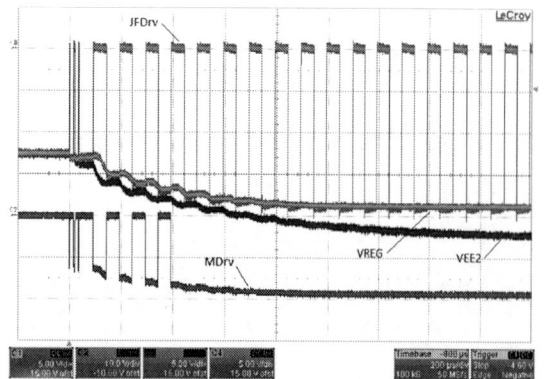

Fig. 7. Startup sequence via bootstrap mode. One can recognize the buildup of voltage and the change in operation modes as the voltages come up. JFDrv stands for the gate voltage of the JFET, MDrv stands for the gate voltage of the MOSFET (at a different offset for better visibility).

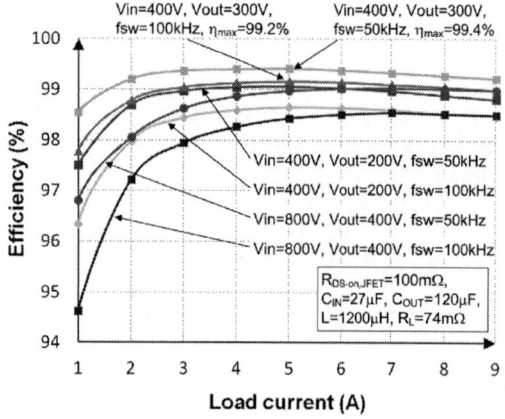

Fig. 8. Efficiency plots of a buck converter arrangement.

VII. Conclusion

In this paper we have demonstrated a high power, high efficiency system for switched mode power supplies. We have shown how some essential properties of the system finally lead to the main architectural choices. With this we have demonstrated a SiC-JFET based SMPS able to handle powers > 10kW at efficiencies > 99%.

References

[1] K. Norling, C. Lindholm, and D. Draxelmayr, "An optimized driver for SiC JFET-based switches delivering more than 99% efficiency," in IEEE Journal of Solid-State Circuits, Dec. 2012, pp 3095-3104

[2] D. Domes, C. Messelke, and P. Kanschat, "First industrialized 1200V SiC JFET module for high energy efficiency applications," in Power Conversion Intelligent Motion (PCIM) Meeting, May 2011

[3] M. Münzer, W. Ademmer, B. Strzalkowski, and K.T. Kaschani, "Insulated signal transfer in a half bridge driver IC based on coreless transformer technology," Power Electronics and Drive Systems, PEDS, vol. 1, pp 93-96, Nov. 2003

Conquering Variability in Mixed-signal ICs

Carsten Wegener and Hans Martin von Staudt

Dialog Semiconductor, Kirchheim, Germany

Shrinking feature sizes in IC manufacturing are required for implementing Moore's law of doubling device density. Higher device density enables increasing digital IC performance. For mixed-signal ICs, shrinking transistor size implies increasing variability of analog device parameters. This has prevented analog circuit designs to benefit from shrinking process geometries. At the same time, increasing (digital) performance causes increasing customer expectations also regarding analog performance parameters. Conquering the effects of increased variability is an engineering challenge. Trimming and calibration is a design solution to this challenge. However, the cost incurred during IC manufacturing can become a significant part of the overall product cost, especially, when trimming is performed as part of product test. Design-for-Test (DfT) is a key for reducing the cost of product test by providing test access mechanisms and built-in self-tests. Extending DfT to Design-for-Trimming and Calibration is part of Dialog Semiconductor's approach to design challenges for the next generation ICs. In this contribution, we show-case examples of bridging between design and test for the implementation of mixed-signal ICs used in mobile communication platforms.

CMOS 8-Channel Frequency Division Multiplexer For 9.4 T Magnetic Resonance Imaging

Mazin Jouda[1], Oliver G. Gruschke[1], Jan G. Korvink[1,2]

[1]Laboratory of Simulation, Department of Microsystems Engineering IMTEK, University of Freiburg, Germany

[2]Freiburg Institute for Advanced Studies FRIAS, University of Freiburg, Germany

Email: {mazin.jouda, oliver.gruschke, korvink}@imtek.uni-freiburg.de

Abstract—We present a CMOS 8-channel frequency division multiplexer (FDM) to interface a phased array of micro coils for 9.4 T (400 MHz Larmor precession frequency ω_0) for magnetic resonance imaging (MRI). The integrated multiplexer contributes towards a solution to achieve phased arrays with a massive number of coils without unnecessarily increasing system complexity, the size of hardware, and cost. The multiplexer is designed using commercially available $0.35 \ \mu m$ CMOS technology and consists of five major components: a low-noise amplifier (LNA), a frequency mixer, a voltage-controlled oscillator (VCO), a bandpass filter (BPF), and an adding operational amplifier. The maximum gain of a single channel is 79 dB, and the input referred noise is 1.4 nV/$\sqrt{\text{Hz}}$. The die area of the multiplexer is approximately 8 mm^2, and requires 300 mA from a 3.3 V source.

I. INTRODUCTION

The concept of a magnetic resonance (MR) phased array, where multiple coils are used to simultaneously receive the MR signals, was first introduced by Roemer *et al.* [1]. A phased array coils, instead of a single surface coil of the same area, increases the detector's signal to noise ratio (SNR), the field of view (FOV), or a combination of both.

Numerous papers have been published to investigate the benefits of increasing the number of coils of the phased array. In what has become the largest single system, Schmitt *et al.* [2] presented a 128-channel cardiac coil for highly accelerated cardiac imaging. This coil array showed a remarkable improvement in the SNR (up to 2.2-fold) and a decrease in imaging time (up to 7 fold) compared to predecessor 24-channel coils. On the other hand, an increased number of coils results in increased complexity, cost and size of the MR receiver, which becomes more challenging and is ultimately the limiting factor. This problem of increased complexity may be overcome by multiplexing the signals of the phased array.

MRI four-channel time-domain and frequency-domain multiplexing schemes were presented by Porter *et al.* [3] and He *et al.* [4] respectively. Both multiplexers were realized using discrete components. The frequency-domain multiplexer (FDM) showed better performance with less cross-talk between the channels. The number of channels for the time-domain multiplexer (TDM) was limited by the speed of the RF switch and the bandwidth of the analog to digital converter (ADC), while the number of channels in the FDM is limited by the ADC's bandwidth and input dynamic range.

In this paper we present a CMOS integrated frequency domain multiplexer to combine the signals of the phased array of micro coils in Fig. 1 [5] within a very compact footprint.

The CMOS multiplexer is a size- and power-efficient solution to interface the phased array to the scanner and replace the bulky low-noise amplifiers, and a first step towards merging a massive number of channels. The phased array in Fig. 1

Fig. 1: Phased array of several hexagonally wire bonded micro coils [5]. The phased array of micro coils with tuning and matching electronics is mounted on a PCB. Seven low-noise pre-amplifiers are connected to the phased array.

defines the following requirements which have to be met by the electronics interface:

- The frequency of the MR signals is 400 MHz.

- The amplitude of the MR signal coming from the receive coil is in the range of few micro Volts.

- The amplitude of the noise of the MR coil is in the range of 10 to 20 nV.

The FDM is implemented in a high performance commercially available $0.35 \ \mu m$ CMOS technology. As shown in Fig. 2, it consists of eight separate signal channels, where each channel has a low-noise amplifier, a frequency mixer, an oscillator, and a bandpass filter. The outputs of all channels are added up using a high speed operational amplifier.

The LNA amplifies the weak MR signal, then the MR signal (400 MHz) is down-mixed to a low intermediate frequency (IF). The oscillators generate frequencies from 390 to 355 MHz separated by 5 MHz. The BPFs are centered around these frequecies at 10, 15, 20, 25, 30, 35, 40 and 45 MHz respectively to filter out any unwanted band noise. The 5 MHz separation is a conservative choice to ensure that the signals will not overlap and to relax the requirements on the filtering.

II. THE LOW-NOISE AMPLIFIER

The low-noise amplifier is the first element in the front end of the RF receivers. The task of the LNA is to amplify the 400 MHz magnetic resonance (MR) signals whilst

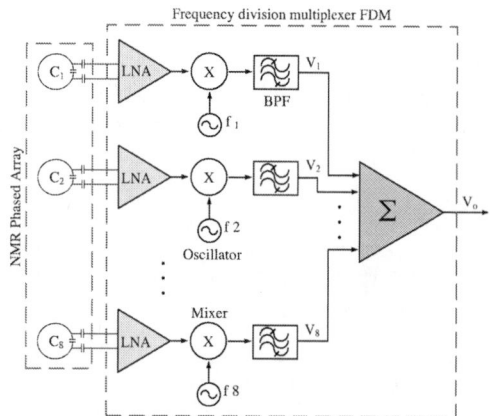

Fig. 2: Schematic of an eight-channel frequency division multiplexer FDM. Each channel consists of an LNA, a mixer, a VCO and a BPF. All signals are combined using an adding operational amplifier.

introducing as little noise as possible. Among many available circuit topologies, the common-gate common-source (CG-CS) arrangement was chosen.

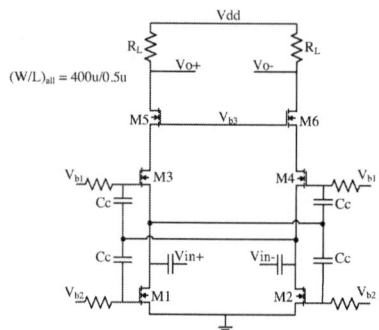

Fig. 3: Common-gate common-source (CG-CS) low-noise amplifier. Biasing circuits are not shown.

Fig. 3 shows a differential-input differential-output CG-CS LNA. When compared to the commonly used inductively-degenerated common-source (CS) low-noise amplifier, the common-gate common source topology shows better linearity, stability, and isolation performance [6]. It also uses gm-boosting technique to improve the noise figure (NF) [6]. Table I summarizes the simulation results of the CG-CS low-noise amplifier. The S-parameters are shown in Fig. 4a. It

TABLE I: Parameters of the low-noise amplifier.

Voltage Gain	28.7 dB
Noise figure (NF)	1 dB
$P_{-1\ dB}$	-10 dBm
Power consumption	16 mW

can be seen from this figure that the input and output reflection

coefficients S_{11} and S_{22} are low. The reverse isolation between output and input ports (S_{12}) is -46 dB. Fig. 4b shows the noise figure of the LNA. It can be seen that the low-noise amplifier has a reasonably low ($1 \sim 1.5$ dB) noise figure for a wide range of input frequencies.

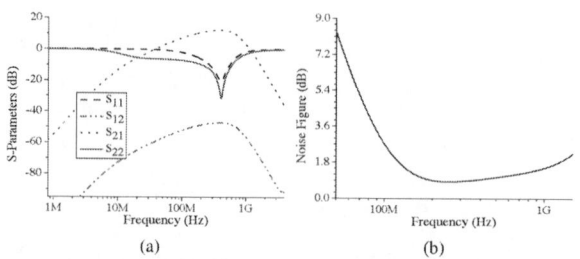

Fig. 4: LNA characteristics (a) S-parameters. The differential input of the LNA is matched to 100 Ω (b) Noise figure.

III. THE RF MIXER

The frequency mixer multiplies the radio frequency (RF) signal from the low-noise amplifier by the local oscillator (LO). The 400 MHz RF signal will thereby be shifted and scaled to two frequencies: (RF-LO) and (RF+LO). Fig. 5 shows a double balanced Gilbert cell. This mixer is chosen because of its good suppression of even distortions and its low feed-through of the local oscillator [8]. Another advantage is the good isolation of the local oscillator from feeding through the RF port, and also the good suppression of the RF and LO at the intermediate frequency (IF) port [9].

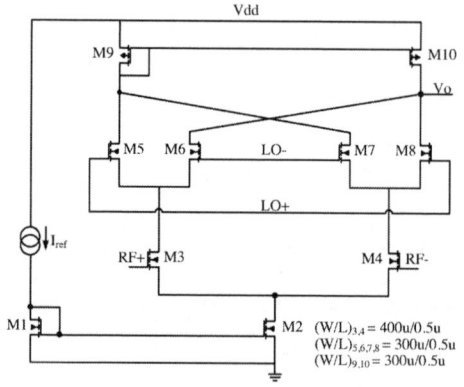

Fig. 5: Double balanced Gilbert mixer.

Table II shows the main parameters of the designed mixer. The mixer has high linearity and good local oscillator and radio frequency suppression at the IF output port.

Fig. 6 shows the conversion gain of the mixer. In Fig. 6a it can be seen that the mixer has a maximum gain of 12 dB when the input power of the local oscillator is -10 dBm, while Fig. 6b shows that the conversion gain of the mixer is nearly constant (12 dB) for a wide range of intermediate frequency IF when the input radio frequency signal and the local oscillator range from 300 MHz to 500 MHz.

TABLE II: Parameters of the double balanced mixer.

Conversion Gain	12 dB
Noise figure (NF)	4.6 dB
$S_{(RF-IF)}$	-64 dB
$S_{(IF-LO)}$	-60 dB
Third intercept point (IIP3)	40 dBm
Power consumption	16 mW

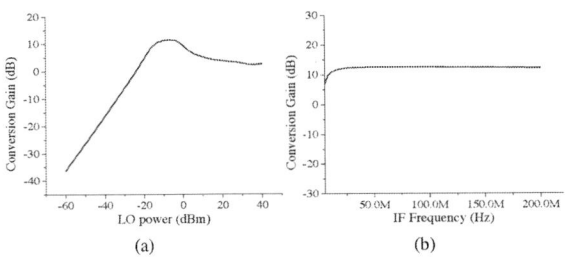

(a)   (b)

Fig. 6: Conversion gain of the mixer. (a) Conversion gain versus the local oscillator power. (b) Conversion gain versus frequency.

IV. VOLTAGE-CONTROLLED OSCILLATOR

The third component in the signal path of our CMOS frequency division multiplexer is the oscillator. Since we want to shift the MR signals to different frequencies, we need different local oscillator values and thus we need to use a voltage-controlled oscillator (VCO). Using a VCO is also beneficial to overcome the shifts in oscillator frequency due to process errors. Fig. 7a shows the schematic of a complementary cross-coupled LC voltage-controlled oscillator. The LC voltage-controlled oscillator was chosen because of its good phase-noise performance, which is critical in NMR and MRI, compared to other oscillators such as relaxation and ring oscillators [10], [11]. Table III shows the tuning range, frequency range and the phase noise of the LC-VCO. From this table it can be seen that the VCO has a wide tuning band that is sufficient to shift the MR signals to the desired frequencies, and that the phase noise is very low. The tuning and phase-locking of the oscillators is performed using external control circuitry. Using external control for the oscillators gives the ability to improve the phase-locking circuits until the desired frequency precision is achieved.

TABLE III: Parameters of the LC voltage-controlled oscillator.

Tuning range	0-2 V
Frequency range	320-400 MHz
Phase noise @(10 MHz)	-157 dBc/Hz

V. OPERATIONAL AMPLIFIER

In order to extract the desired signal after mixing, and simultaneously to suppress the out-of-band noise, noting that our signals will be located at some low intermediate frequency other than DC, we need a band-pass filter (BPF). A high speed and large gain-bandwidth product (GBW) operational amplifier (op-amp) is used to build the BPF's as well as the final adding

stage. Fig. 7b shows a two-stage operational amplifier. The

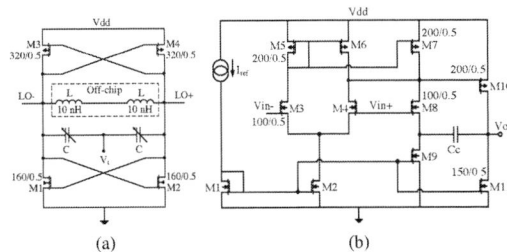

(a)   (b)

Fig. 7: (a) Complementary cross-coupled LC-VCO. (b) Two-stage operational amplifier with indirect compensation.

op-amp offers a high gain and wide output voltage swing. Moreover, it offers a high speed operation due to the indirect compensation of Cc through the common-gate transistor M8. The indirect compensation also enhances the phase margin PM. Table IV summarizes the main parameters of the designed operational amplifier.

TABLE IV: Parameters of the operational amplifier.

Open-loop gain	53 dB
Slew Rate	660 V/μs
Phase margin	65 Degrees
DC offset	<1 mV
Gain-Bandwidth product (GBW)	1.6 GHz

From this table it can be seen that the operation amplifier has a good stability for the closed-loop operation since the phase margin is 65 degrees. It also can be seen that the operational amplifier has a large GBW of 1.6 GHz which enables adding more channels up to 80 MHz with a voltage gain of 20 dB.

A. Band-pass filters BPF's

The eight bandpass filters were realized using the operational amplifier in Fig. 7b in combination with resistors and capacitors as shown in Fig. 8a. This figure shows a second-

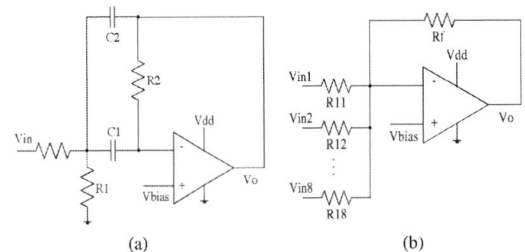

(a)   (b)

Fig. 8: Circuits realized by the high speed operational amplifier. (a) 2nd-order bandpass filter. (b) Adding op-amp.

order BPF with a center frequency f_c and a quality factor Q that can be calculated from equ. 1 and equ. 2. The order of the filter can be increased by cascading similar stages.

$$f_c = \frac{1}{2\pi\sqrt{R_1 R_2 C_1 C_2}} \qquad (1)$$

$$Q = \frac{1}{2}\sqrt{\frac{R_2}{2R_1}} \qquad (2)$$

Fig. 9 shows the normalized frequency response of the eight bandpass filters designed to extract the desired down-converted MR signal. The filters have 10, 15, 20, 25, 30, 35, 40 and 45 MHz center frequencies.

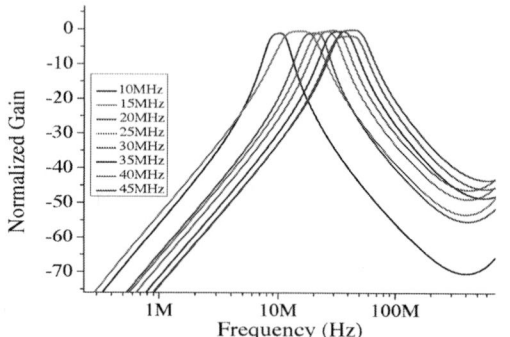

Fig. 9: Normalized gain of the eight bandpass filters. The quality factor Q ranges from 1.7 to 2.7.

B. The adding stage

The final adding stage, can be realized using an adding operational amplifier as shown in Fig. 8b. The output of this circuit will be a scaled sum of the eight input signals. The frequency response of the adding stage for a single channel with $R11 = 100\ \Omega$ and $Rf = 1\ k\Omega$ is shown in Fig. 10a. It can be seen from this figure that the op-amp offers a constant voltage gain (20 dB) over a wide band of frequencies up to 80 MHz.

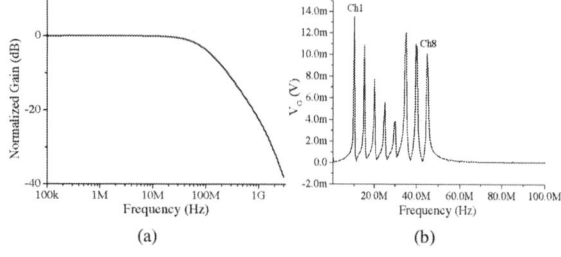

Fig. 10: (a) Frequency response of the adding operational amplifier. (b) Transient simulation of the eight-channel frequency division multiplexer.

A transient simulation of the extracted layout of the eight-channel frequency multiplexer is shown in Fig. 10b. The maximum coupling between the channels is -18 dB between channel 8 and 6, while the minimum coupling is -30 dB. The variation of the voltage gain in the channels is due to the coupling capacitors between the different stages and can be compensated by the gain resistors in the bandpass filters and the adding stage.

VI. CONCLUSION

In this paper we have demonstrated the implementation of an eight-channel CMOS frequency division multiplexer FDM to interface a nuclear magnetic resonance phased array of micro coils. The integrated multiplexer is a solution to overcome the complexity and size problems that arise when using MR phased arrays with large number of coils. It's use is not restricted to arrays of micro coils, and can be extended to macroscopic human imaging arrays. The number of channels of the frequency division multiplexer, which is mainly limited by the analog-to-digital converter (ADC) bandwidth and the channel separation, can be potentially increased to 300 channels, since current commercially-available ADCs can reach 3 GSPS or even higher.

ACKNOWLEDGMENT

MJ acknowledges financial support from the Fritz-Hüttinger foundation. OGG and JGK acknowledge support from the University of Freiburg through an operating grant.

REFERENCES

[1] Roemer, P. B., et al. *The MR phased array.* Magnetic resonance in medicine 16.2 (1990): 192-225.

[2] Schmitt, Melanie, et al. *A 128-channel receive-only cardiac coil for highly accelerated cardiac MRI at 3 Tesla.* Magnetic Resonance in Medicine 59.6 (2008): 1431-1439.

[3] Porter, Jay R., Steven M. Wright, and Nader Famili. *A fourchannel time domain multiplexer: A costeffective alternative to multiple receivers.* Magnetic resonance in medicine 32.4 (1994): 499-504.

[4] He, Wang, et al. *Four-channel magnetic resonance imaging receiver using frequency domain multiplexing.* Review of scientific instruments 78.1 (2007): 015102-015102.

[5] Gruschke, Oliver G., et al. *Lab on a chip phased-array MR multi-platform analysis system.* Lab on a Chip 12.3 (2012): 495-502.

[6] Zhuo, W., et al. *A capacitor cross-coupled common-gate low-noise amplifier.* Circuits and Systems II: Express Briefs, IEEE Transactions on 52.12 (2005): 875-879.

[7] Hoult, David I., and R. E. Richards. *The signal-to-noise ratio of the nuclear magnetic resonance experiment.* Journal of Magnetic Resonance (1969) 24.1 (1976): 71-85.

[8] Zencir, E., N. S. Dogan, and E. Arvas. *A low-power CMOS mixer for low-IF receivers.* Radio and Wireless Conference, 2002. RAWCON 2002. IEEE. IEEE, 2002.

[9] Nimmagadda, Kiran, and Gabriel M. Rebeiz. *A 1.9 GHz double-balanced subharmonic mixer for direct conversion receivers.* Radio Frequency Integrated Circuits (RFIC) Symposium, 2001. Digest of Papers. 2001 IEEE. IEEE, 2001.

[10] Berny, Axel D., Ali M. Niknejad, and Robert G. Meyer. *A 1.8-GHz LC VCO with 1.3-GHz tuning range and digital amplitude calibration.* Solid-State Circuits, IEEE Journal of 40.4 (2005): 909-917.

[11] Razavi, Behzad, and Razavi Behzad. *RF microelectronics.* Vol. 1. New Jersey: Prentice Hall, 1998.

[12] Baker, R. Jacob. *CMOS: Circuit design, layout, and simulation.* Vol. 18. Wiley-IEEE Press, 2011.

Comparison of High-voltage Linear Transmitter Topologies for Ultrasound CMUT Applications

Sourabh Khandelwal, *Student Member IEEE* and T. Ytterdal, *Senior Member IEEE*

Dept. of Electronics and Telecommunications
Norwegian University of Science and Technology
Trondheim, Norway
sourabh.khandelwal@ntnu.no and trond.ytterdal@ntnu.no

Abstract—In this paper, we present a comparison of high-voltage linear transmitter circuits for driving capacitive micro-machined ultrasound transducers (CMUTs). CMUTs are emerging transducer elements in ultrasound imaging applications. Two different circuit topologies for the high-voltage linear transmitter with a 40 V peak-to-peak output voltage are compared. We compare the two designs based on total harmonic distortion, power consumption, bandwidth and area for same voltage gain. The designs are done in AMS 0.18 μm technology, and utilize Si LDMOS device for a high-voltage output.

Keywords—Linear Transmitters, CMUTs, High-voltage Circuits

I. INTRODUCTION

Application of ultrasound in intravascular and intra-cardiac imaging systems is very popular [1, 2]. Traditionally, piezoelectric transducers were used to transmit and receive the ultrasound wave. However, capacitive micro-machined ultrasound transducers (CMUTs) are emerging as a promising alternative to piezoelectric transducers. This is due to the fact that CMUTs offer higher bandwidth and an ease of integration with frond-end electronics [3-6]. A full ultrasound imaging system employs many channels of CMUT elements. Each channel of CMUTs requires a transducer excitation and a receive circuit. Typically, separate transducer excitation (or transmit) and receive circuits are used in these systems. Integrated solutions are desirable to reduce system size, especially for ultrasound medical probes, where area is a critical parameter.

The difficulty in integrated solution arises from a large difference between the voltage levels required for transducer excitation (or transmitter circuits) and receive circuits. For the transmitter, CMUTs need a high voltage output of 40 V peak-to-peak, on the other hand, the receiver circuits operate at a maximum of 2 - 5 V. To overcome this difficulty, level shifter circuits are explored [7, 8]. The level-shifters shift the input signal to desired high-voltage level by utilizing high voltage Si LDMOS devices. However, a linear high-voltage transmitter is desirable to improve overall system performance. This is because with linear transmitter circuit, arbitrary shape input waveform can be applied to the input of transmitter and hence CMUTs. The input waveform shape can be tuned to optimize the output response and avoid undesired resonance modes of

CMUTs. This will also relax the dynamic range requirements on the receiver design. This underlines the importance of high voltage linear transmitter for CMUTs.

In this paper, we present and compare two different topologies of linear high-voltage transmitter for CMUTs. The circuits are designed in 0.18 μm CMOS, with silicon LDMOS devices as the high voltage MOSFETs. The transmitter topologies are compared based on important criteria of linearity, area, power and bandwidth for the same gain. The paper is arranged as follows. Section II is divided in two sub-sections: First a discussion on the transmitter requirements and device selection for the high voltage transmitter design is presented and then the linear transmitter topologies are shown. In section III we present simulation results for the two topologies. A comparison between the two topologies is also discussed in this section. In section IV we conclude the paper.

II. LINEAR TRANSMITTER TOPOLOGIES

A. High-voltage Transmitter Requirements and Device Considerations for Transmitter design

A linear transmitter is used to drive CMUTs. Requirements from CMUTs set the specifications about the voltage-swing, DC bias and bandwidth of the transmitter. CMUTs need a 30 V DC bias to operate. The transmitter is designed such that the DC bias to CMUTs is provided by the transmitter itself, reducing the system complexity. With change in process, voltage and temperature the DC output from transmitter can change from 30 V and an appropriate calibration technique can be used to set it at the correct value. Superimposed on the 30 V DC, a 40 V peak-to-peak signal is required by CMUTs to transmit the ultrasound wave. This means maximum voltage at the device terminals will be 50 V. A typical on-time or transmit time of T_{on} = 0.5 μs and an off- time or receive time of T_{off} = 40 μs are CMUT driven specifications in our case.

The Si LDMOS device used for amplification is chosen based on the breakdown voltage rating and gain-factor K_p. The drain-current I_d in saturation condition is given by,

$$I_d = K_p \frac{W_{eff}}{L_{eff}} \left(V_g - V_{th} \right)^2 \qquad (1)$$

where W_{eff} and L_{eff} are the effective gate-length and gate-width

Paper M1A2

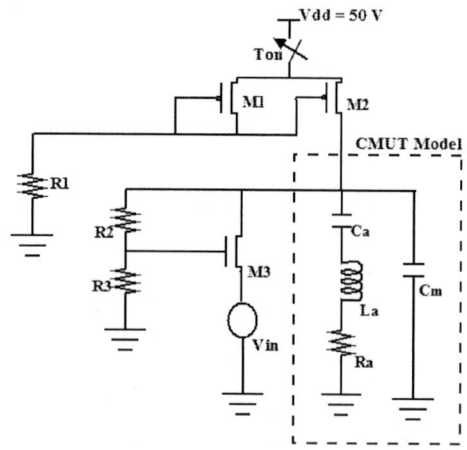

Fig. 1. Common-gate (with resistive feed-back) topology of linear transmitter. M1, M2 and M3 are high-voltage Si LDMOS devices. The switch is on during T_{on}. R1 = 800 Ω, R2 = 850 Ω and R3 = 110 Ω. The CMUT model elements are: Ra = 9 Kilo Ω, La = 46.4 µH, Ca = 2.39 pF and Cm = 3.74 pF. Widths of M1, M2 and M3 are 5 mm, 10 mm and 10 mm respectively.

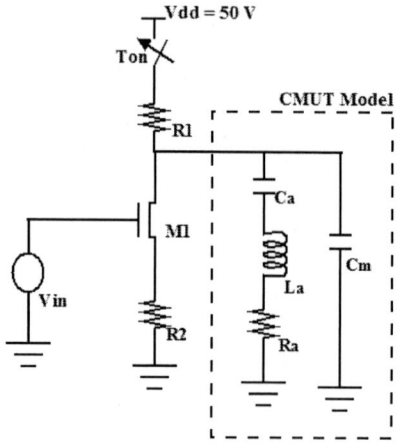

Fig. 2. Source-degenerated (SD) topology for linear transmitter design. M1 is high-voltage Si LDMOS device. R1 and R2 are chosen to optimize bandwidth, linearity and DC bias voltage for CMUT. R1 = 15 Ω and R2 = 530 Ω. M1 is 2 mm wide. The CMUT model elements are same as in Fig. 1.

respectively, V_g is the applied gate-voltage and V_{th} is the threshold voltage. A higher value of K_p indicates a higher gain from the device. Three different high-voltage MOSFET flavors in the technology with thick-, medium- and thin-gate-oxide satisfy the drain-to-source breakdown voltage requirement of 50 V. K_p for these devices increases continuously from thick to the thin gate-oxide device. However, for the thin gate-oxide device the gate-to-body breakdown voltage is found to be low and hence we choose medium-oxide device for our design. Apart from MOSFETs, resistors also need to withstand the high-voltage swing. Most of the available resistors have a terminal-to-bulk breakdown voltage limit of 50 V. However, the maximum voltage limit between two terminals of the resistor is governed by the current density in the resistor. Resistor dimensions are chosen accordingly to satisfy the current density requirements. One more important factor to

consider was linearity. We need good linearity and hence the voltage coefficient of the resistors should be low, especially because they face a large voltage swing. We choose N+ poly resistor due to its very small voltage coefficient as compared to other resistor choices.

Another important consideration in this design is related to the CMUT load. One transmitter can drive many CMUT elements and driving a large number of CMUT elements will save area. However, an increase in number of CMUT elements to be driven by one transmitter will increase the capacitive loading (see CMUT model in Fig. 1) of the transmitter. This will adversely affect the transmitter bandwidth. Hence an optimum trade-off between area saving and bandwidth degradation has to be made. Here, we design the transmitter to drive 288 individual CMUT elements.

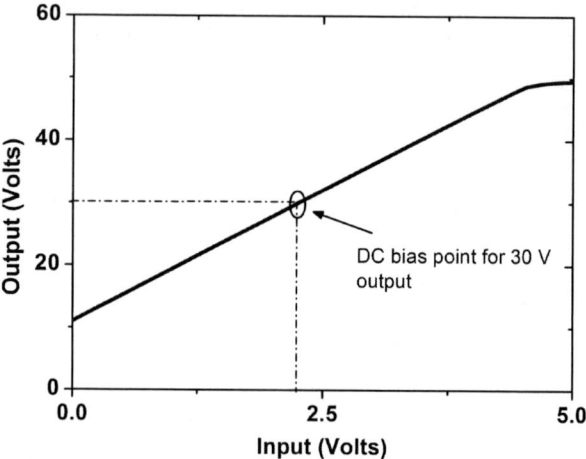

Fig. 3. Transfer curve for CG topology. The DC bias point for a 30 V output is marked in the plot.

B. Transmitter Topologies

The first topology explored for the linear transmitter design is shown in Fig. 1. This is a Common-Gate configuration with resistive feed-back. M1 and M2 are high-voltage PMOS transistors which provide the bias current for the high-voltage NMOS M3. A high-voltage switch is used to turn the circuit off during T_{off} to save power dissipation. The bias current is decided such that with no input voltage the output is 30 V, which is the DC bias requirement for CMUT. M1 and M2 are chosen to be long-channel devices with channel length L = 2 µm, so that the current mirror has a large output impedance. It is interesting to see that even with laterally diffused drain architecture, short-channel LDMOS devices are found to have considerable output conductance and this degrades the current mirror performance. M3 is chosen to have a short gate-length of L = 0.4 µm, for high gain. R2 and R3 are optimized to give best linearity and bandwidth.

The second topology explored for the design is shown in Fig. 2. It is a source-degenerated amplifier with M1 as a high voltage MOSFET. M1 is chosen to have minimum channel length of L = 0.4 µm, to achieve high gain. A high voltage switch is employed for turning the circuit off during the off-

978-1-4673-4579-8/13 $31.00 © 2013 IEEE

period. The values of the resistors R1 and R2 are chosen to provide the 30 V DC bias point for CMUT, required linearity

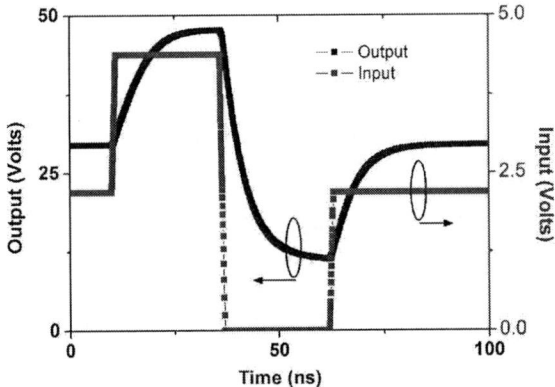

Fig. 4. Transient simulation result of CG topology of linear transmitter. Rise and fall time of input signal is 1 ns.

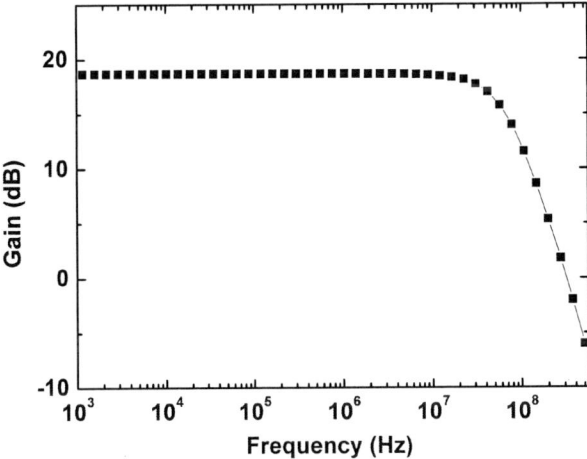

Fig. 5. AC response for CG topology of linear transmitter. A 3 dB bandwidth of 31.6 MHz is obtained. Simulations are with the CMUT load.

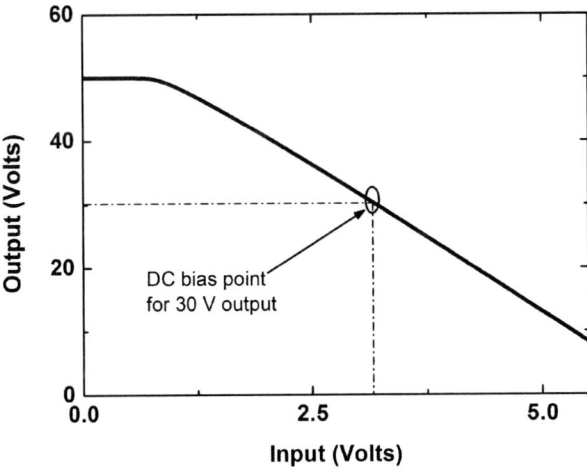

Fig. 6. Transfer curve for SD topology. The DC bias point for a 30 V output is maker in the plot.

and bandwidth. Increasing R2 will increase linearity, but, it reduces the bandwidth. R1 and R2 are thus optimized for our requirements. Simulation results of both the topologies and discussions on these are presented in the next section.

III. RESULTS AND DISCUSSIONS

Simulations for the above topologies are performed in Spectre, with AMS 0.18 μm design kit. Simulations are used to analyze the input-output transfer curve, AC response, transient response and total harmonic distortion (THD) for the two topologies. For the Common Gate (CG) topology shown in Fig. 1, the transfer curve obtained is shown in Fig. 3. The input voltage required for a 30 V DC output is obtained to be 2.2 V. The input signal will ride on this DC voltage during T_{on}. It is apparent from Fig. 3 that a reasonable linear transfer curve is obtained for the required high output voltage swing. It is also important to check the transient response of the amplifier, since CMUTs cause considerable capacitive loading. In Fig. 4, we show the transient response of this design. We applied a transient signal with 50 ns time-period and 1 ns rise and fall time. It is required that the output time constant should be below 10 ns. It is clear from Fig. 4 that the time-constant

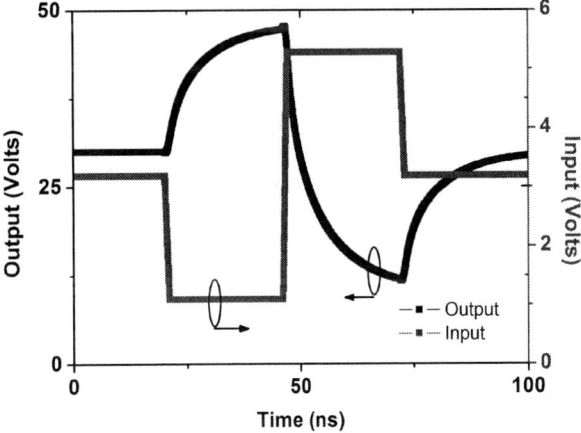

Fig. 7. Transient simulation result of CG topology of linear transmitter. Rise and fall time of input signal is 1 ns.

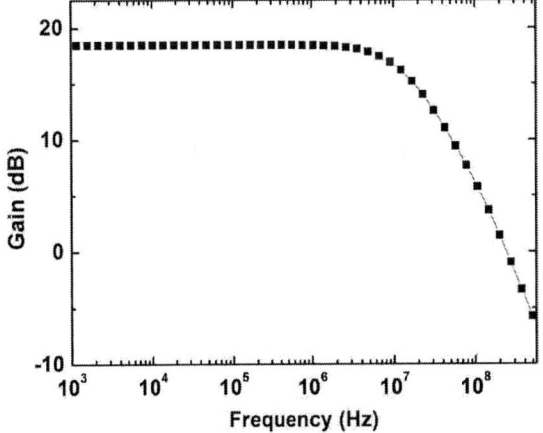

Fig. 8. AC response for SD topology of linear transmitter. A 3 dB bandwidth of 16.8 MHz is obtained. Simulations are with CMUT load.

requirement is met and with a small input voltage swing, a 40 V peak-to-peak linear output is obtained. In Fig. 5 we show AC response for this topology. A 3 dB bandwidth of 31.6 MHz is obtained.

The transfer curve for source degenerated (SD) linear transmitter is shown in Fig. 6. The DC input voltage for a 30 V output is obtained to be 3.2 V in this case. It can be seen from Fig. 7 that a reasonable linear high-voltage output is obtained with a low voltage input swing in this case also. The transient simulations for SD are shown in Fig. 7 and it shows that the topology meets the requirement of time constant. AC response for this topology is shown in Fig. 8. A 3 dB bandwidth of 16.8 MHz is obtained for this topology. Finally, we compare the two topologies on the basis of important performance parameters for this design. Table I shows the comparison between the two topologies. From Table I, it can be seen that for similar voltage-gain and almost similar linearity performance, SD topology is better in terms of area and power consumption. Area is important considering the fact that these designs can be used in-probe, while power dissipation is important since large power dissipation can cause undesirable heating in the probes. The bandwidth in case of SD is slightly poorer as compared to CG. For applications with more emphasis on bandwidth requirement CG is a better choice. However, in our case, the requirement for bandwidth is met with SD topology. Hence, overall for our requirements, SD is found to be a better topology for high voltage linear transmitter driving CMUT loads.

Table I.
Comparison of high voltage linear transmitter topologies

Parameter	Common Gate (CG)	Source Degenerated (SD)
Voltage Gain	18.8 dB	18.5 dB
Total Harmonic Distortion	-32.6 dB	-35.8 dB
Bandwidth	31.6 MHz	16.8 MHz
Average Power Consumption	43 mW	8.5 mW
Area	1.4 mm^2	0.262 mm^2

IV. CONCLUSIONS

Two high voltage linear transmitter topologies to drive CMUT loads are compared. The designs are made in AMS 0.18 μm technology with Si LDMOS as the high voltage device. A reasonable linear conversion from low voltage input to high voltage output is obtained with the two topologies. Among the two topologies: Common Gate and Source degenerated, Source degenerated is found to be better in terms of area and power consumption for similar gain and total harmonic distortion. The proposed designs and comparison can be used to develop integrated solution for ultrasound imaging applications.

ACKNOWLEDGMENT

The work was carried out with support by European Commission Grant Agreement 218255 (COMON) and the Norwegian Research Council under contract 970141669 (MUSIC).

REFERENCES

[1] Y. Wang, D. N. Stephens, and M. O'Donell, "Optimizing the beam pattern of a forward-viewing ring-annular ultrasound array for intravascular imaging," IEEE Trans. UltraSon. Ferroelectr. Freq. Control, vol. 49, no.12, pp. 1652–1664, December 2002.

[2] O. Oshiro, M. Nambu, and K. Chihara, "3D echocardiography using a 3D positioner," Proc. of International Conference of the IEEE Engineering in Medicine and Biology Society, vol. 2, pp. 783-784, 1998.

[3] M. I. Haller, and B. T. Khuri-Yakub, "A surface micromachined electrostatic ultrasonic air transducer," IEEE Trans. UltraSon. Ferroelectr. Freq. Control, vol. 43, no.1, pp. 1–6, January 1996.

[4] P. C. Eccardt, K. Niederer, T. Scheiter, and C. Hierold, "Surface micromachined ultrasound transducers in CMOS technology," in IEEE Ultrasonics Symp., 1996, pp. 959–962.

[5] I. Ladabaum, X. Jin, H. T. Soh, A. Atalar, and B. T. Khuri-Yakub, "Surface micromachined capacitive ultrasonic transducers," IEEE Trans. UltraSon. Ferroelectr. Freq. Control, vol. 45, no.3, pp. 678–690, May 1998.

[6] C. Meynier, M. Legros, G. Ferin, D. Certon, A. Nguyen-Dinh and R. Dufait, "Smart micromachined ultrasonic probe with advanced imaging performances," in European Conference and Exhibition on Integration Issues of Miniaturized Systems – MOMS, MOEMS, ICS and Electronic Components, pp 1-9, April 2008.

[7] M. Khorasani, L. V. D. Berg, P. Marshall, M. Zargham, V. Gaudet, D. Elliot, and S. Martel, "Low power static and dynamic high-voltage CMOS level shifter circuits", Proc. of IEEE International Symposium on Circuits and Systems, pp. 1946 – 1949, 2008.

[8] Y. Moghe, T. Lehmann, and T. Piessens, "Nanosecond delay floating high voltage level shifters in a 0.35 μm HV-CMOS technology", IEEE Journal of Solid State Circuits, vol. 46, no. 2, pp. 485 – 497, 2011.

Bluetooth Transceiver Modeling Using SystemC-AMS

Fangyan Li
Biomedical Implants Project Team (EPIB)
Université de Nice
Sophia Antipolis, France
Email: li.fangyan@etu.unice.fr

Rémi Butaud†, Eric Dekneuvel*, Gilles Jacquemod*
† RivieraWaves, Sophia Antipolis, France
Email: remi.butaud@rivierawaves.com
* Biomedical Implants Project Team (EPIB)
Université de Nice
Sophia Antipolis, France
Email: {eric.dekneuvel,gilles.jacquemod}
@polytech.unice.fr

Abstract—**This paper presents the high-level modeling and simulation of a GFSK RF transceiver for a Bluetooth Low Energy (BLE) system. This model, written in SystemC-AMS (SC-AMS) permits a simple integration within the existing system digital environment written in SystemC-TLM, which enables a global system simulation to obtain the system performance (e.g. Bit-Error-Rate (BER) or Packet- Error-Rate (PER)) related to the power and energy consumption.**

Keywords—BT; SystemC-AMS; RF; system design

I. INTRODUCTION

As an increasing amount of applications is developed, the demand for functional blocks to be integrated in a System-on-Chip (SoC) has rapidly grown to a big scale. Thus the system complexity and energy consumption can go up to a high level. This is the case for a Bluetooth (BT) system. Therefore, reducing the energy consumption at the local as well as at the global level becomes necessary. To do an accurate power and energy consumption analysis, the models of the analog blocks need to go down to the circuit-level to obtain their power consumption information. These models however are difficult to simulate within the digital environment of the system to obtain the system performance, like Bit-Error-Rate (BER) or Packet-Error-Rate (PER). Accurate Spice-model simulations of a complete or several BT packets takes a very long time because of the complexity of the model and the huge amount of calculation points during the simulation.

A suitable methodology is to use system-level analog models that contain the accurate power consumption information from the existing circuit-level designs, in order to achieve a global simulation of the system with the digital environment and operating software. To be compatible with the digital design and simulation environments, a high-level description tool for the analog modeling and simulation is needed [1][2].

This paper presents the system-level modeling of a GFSK transceiver which contains the RF block and the MODEM in a BT system using the TDF MoC of SC-AMS. This model permits the future integration of non-idealities and power consumption information based on the existing circuit-level designs, and simplifies the global performance related co-simulation with the digital environment.

Next section introduces the proposed methodology used to describe the BLE protocol at system level. In section 3 a high-level BLE transceiver modeling with SC-AMS and the simulation results are presented. The final section shows the conclusions and perspective of this work.

II. SYSTEMC BASED METHODOLOGY

The Bluetooth 4.0 specification introduces the Bluetooth low energy (BLE) technology which allows new Bluetooth devices to operate for months or even years on tiny, coin-cell batteries, within a short range (up to 50 metres). It facilitates new wireless device applications including health care, sports and fitness, security, and home entertainment. With its characteristics such as ultra low power consumption, fast connection times, reliability and security, BLE consumes 10-20 times less power (than Bluetooth) [3].

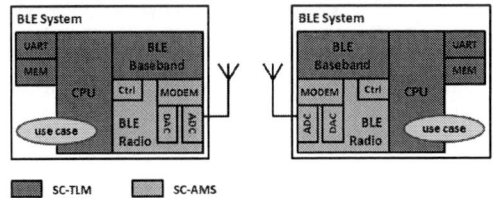

Fig. 1 Bluetooth Low Energy two-devices system.

Fig.1 presents a proposed BLE system diagram of two devices (a master scanning for devices discovering and a slave sending advertising packets). BB lies on top of the BT radio layer in the BT stack handling packets, scanning or advertising (in BLE mode) to connect Bluetooth devices in the area. Inside a BLE communication device, the Baseband (BB) typically acts at the physical layer protocol. It manages physical channels and links, and offers other services like data whitening, hop selection and BT security [4].

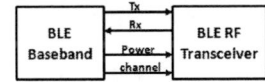

Fig. 2 BLE low level layers.

The Baseband transceiver applies a time-division duplex scheme (alternate transmit and receive), which brings out different hopping frequencies and time slots that represent the frequency and time division during the communication. BLE uses a unique packet format for both advertising and data packets. Preamble is used in the receiver to perform frequency synchronisation and symbol timing estimation. The access address shall be different for each link layer active connection. Protocol Date Unit (PDU) format depends on the nature (advertising or data) of the packet. Last, a Cyclic Redundancy Check (CRC) polynomial is calculated over the PDU.

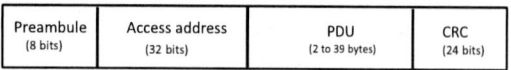

Fig. 3 Air Interface packet.

The description of the BLE protocol is achieved using the SystemC language. SystemC-AMS (SC-AMS) which is a C++ based Analog Mixed-Signal (AMS) extension library of SystemC standard is proposed as a means of efficiently modeling the AMS behavior at a high level. Users are able to control the modeling abstraction and complexity by choosing the different Models of Computation (MoC): Timed Data Flow (TDF), Linear Signal Flow (LSF) and Electrical Linear Networks (ELN) [5].

Some previous works from different domains realized using SC-AMS have shown the advantages. [6] has demonstrated that in comparison with Matlab, SC-AMS offers the same system accuracy but with a much faster simulation speed. It can achieve a simulation of a mixed signals system containing digital blocks like a microcontroller that are not available in Matlab. [7][8] compared their SC-AMS simulations with CMOS designs which proved the models accuracy from the simulation results consistency. Another work like [9] has proposed and validated the idea of including low-level power consumption into high-level model, but without considering a global system validation of the performance. In [10], an approach for a heterogeneous system with powered nodes is presented. But this work was based on a high-level transceiver model which didn't contain accurate analog performance.

The digital environment (Fig.1), which contains CPU, UART, MEM and the BLE Baseband, has been modelled with SC-TLM (SystemC-Transaction Level Model). The MODEM and RF transceiver which are presented in this paper are modelled with SC-AMS. SC-TLM models are fast to simulate, reusable and interoperable with SC-AMS models. The SC-AMS models at this first stage work are realized as the functional models which needs a future refinement with the circuit impairments and an improvement of the accurate energy consumption information. This entire platform is suitable for a fast global simulation, and allows obtaining accurate circuit-level energy consumption and the system performance at the same time. This is the meet-in-the-middle method which is based on the top-down models and bottom-up designs [11].

III. BLE TRANSCEIVER ARCHITECTURE

A high-level BLE radio model is needed to fill the missing link between the specification and the implementation in the meet-in-the-middle method. The first step work is therefore the modeling of this radio model with SC-AMS.

A. Top level functional description

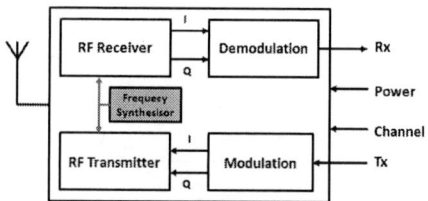

Fig. 4 BLE top level block diagram.

A BLE device operates in the unlicensed 2.4 GHz ISM (Industrial Scientific Medical) band at 2400 - 2483.5 MHz. It uses 40 RF channels which have center frequencies 2402+k*2 MHz, where k = 0, ..., 39 [4]. The modulation is Gaussian Frequency Shift Keying (GFSK) with a bandwidth-bit period product BT=0.5. Fig. 4 shows a BLE transceiver block that contains the GFSK modulator and demodulator, a Zero-IF type RF transmitter and a Low-IF type RF receiver. The most important interface between this block and the BB contains the transmitted (Tx) and received (Rx) data, the power selection signal and channel selection signal.

To simplify the first step work, the modeling begins with a GFSK transceiver which contains one transmitter and one receiver. The local oscillator frequencies are fixed at this stage. The details are presented in the next sub sections. The functional correctness of this model is our interest.

B. Modulator Data Flow

The modulator data flow is shown in Fig.5. The bits coming from the generator are '0' and '1', which represent the 2 frequency deviations " $-\Delta f$ " and " Δf ". Here an 8 bits BLE preamble of "01010101" is sent at 1 MHz. They are firstly converted to '-1' and '+1' respectively, and then pass through a Gaussian pre-modulation filter (GF) with the modulation index h = 0.5, to avoid the sharp bit edges which causes a wide RF signal bandwidth. Each bit is represented by 13 coefficients in GF, its output stream (Fig.6) is thus running at 13 MHz. This signal which represents the frequency deviations is converted to phase by an integrator, and then mapped as $\cos(\Delta\Phi)$ and $\sin(\Delta\Phi)$ with the cosine and sine look up tables.

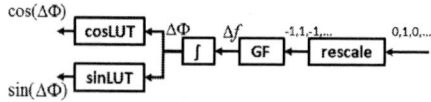

Fig. 5 Modulator data flow.

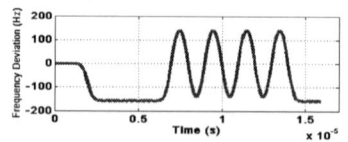

Fig. 6 GF output.

C. RF Transmitter

The modeling of a digital-to-analog conversion can be realized by repeating the digital samples (hold function). This process causes spectrum aliasing in the frequency domain, which requires the Low-Pass Filters to remove the harmonics before mixing with the local oscillators.

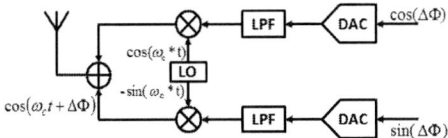

Fig. 7 RF Transmitter architecture.

The local frequency in the transmitter is fixed at 2.402 GHz which means the system communicates in the channel 0. The oscillator generates two quadrature carriers as $\cos(\omega_c t)$ and $-\sin(\omega_c t)$, which are used to generate the final RF signal sent by the antenna by multiplying with $\cos(\Delta\Phi)$ and $\sin(\Delta\Phi)$:

$$\cos(\Delta\Phi)*\cos(\omega_c t)-\sin(\Delta\Phi)*\sin(\omega_c t)=\cos(\omega_c t+\Delta\Phi) \quad (1)$$

D. RF Receiver

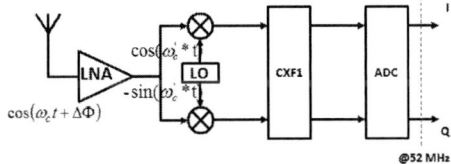

Fig. 8 RF receiver architecture.

The receiver is Low-IF type with the intermediate frequency of -1 MHz (Fig.8). The signal received is firstly amplified by the Low Noise Amplifier (LNA) which is represented simply by a gain in this design level. The LNA output is then mixed with two quadrature carriers as $\cos(\omega_c' t)$ and $-\sin(\omega_c' t)$, where ω_c' is ω_c plus 1 MHz. This operation generates a signal spectrum centered around -1 MHz. Other image or out of band signal spectrums are then filtered out by a subsequent complex filter (CXF). This filter is a Butterworth type complex band-pass filter centered at -1 MHz with the following transfer function:

$$H(j\omega)=\frac{G\omega_o^2}{\left[j(\omega-\omega_1)+j\dfrac{(\omega-\omega_1)\omega_o}{Q}+\omega_o^2 K \right]} \quad (2)$$

where the ω_o is the 1 MHz bandwidth, and ω_1 is the -1 MHz central frequency. The CXF1 output spectrum is shown in Fig.9.

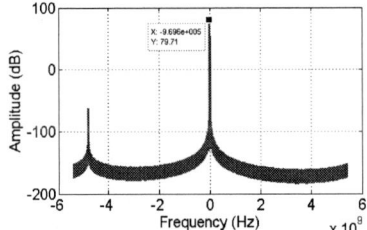

Fig. 9 CXF1 output spectrum.

The ADC is a 3rd order quadrature band pass Sigma-Delta (SD) type ADC with the sampling frequency at 52 MHz. SD type ADC is used to shape the quantization noise out of the signal bandwidth and increase the in-band signal-to-noise ratio. The three possible output levels are '+1', '-1' and '0'. The input signal (CXF output) is centered at -1 MHz and the bandwidth is 1 MHz. The ADC output spectrum is shown in Fig.10.

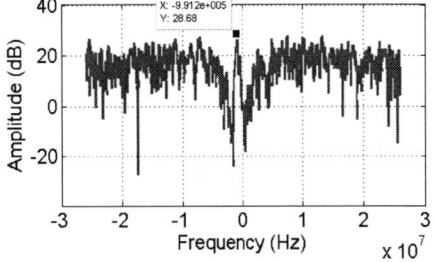

Fig. 10 ADC output spectrum.

E. Demodulation

The ADC is followed by another complex filter in the demodulation block (Fig.11) to attenuate the out of band noise and preserve the low-IF centered at -1 MHz signal. This filter is 127th order FIR type complex filter (CXF2). The output spectrum is shown in the Fig.12.

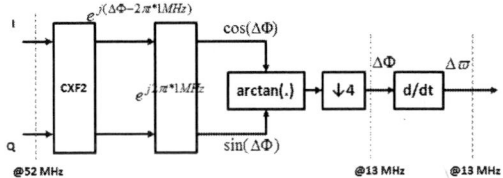

Fig. 11 Baseband demodulation architecture.

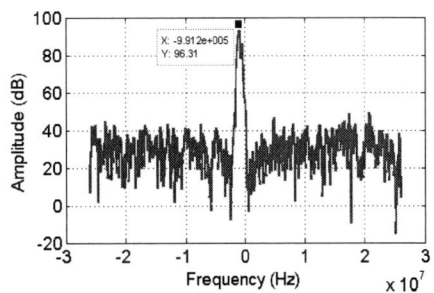

Fig. 12 CXF2 output spectrum.

The next step is to recover the real (centered on 0 Hz) signal by shifting it by 1 MHz, which can be achieved by a multiplication with the term $e^{j2\pi*1MHz}$. This behaviour generates the $\cos(\Delta\Phi)$ in the I branch and $\sin(\Delta\Phi)$ in the Q branch, which correspond to the two terms of the transmitter baseband output. To obtain the phase $\Delta\Phi$, which contains the wanted bits information, an arctangent operation is needed:

$$\Delta\Phi=\arctan\left(\frac{\sin(\Delta\Phi)}{\cos(\Delta\Phi)}\right) \quad (3)$$

The system then implements a down-sampling of 4 to reduce the frequency to 13 MHz, which is followed by a differential operation to get the angular deviations Δw (Fig.13).

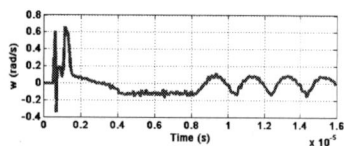

Fig. 13 Differential output.

F. Validation

The system needs a correlator to detect BT Access Code (AC) or BLE Access Address (AA). It allows the system to determine when to sample the incoming oversampled stream in order to down-sample the Rx Data stream and recover the data at 1 MHz. The correlation results in a BT/BLE system is used to extract the PDU data in a received packet by detecting its AC/AA with the synchronization word. The correlator structure is shown in the Fig.14.

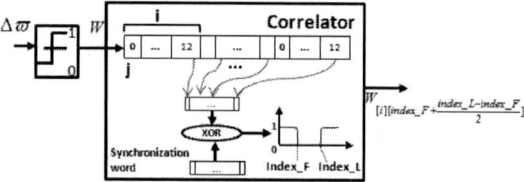

Fig. 14 Correlator.

The sign of the Differential output $\Delta\omega$ is converted into logic "0" and logic "1" and stored into a buffer. The synchronization word is compared with the incoming W oversampled data stream using the i to j + 13 bits, where i and j represent respectively the index of the group of bits every 1 μs and the index of each bit during 1 μs. The comparison (XOR) calculates the number of error found on each clock cycles. The stream is considered as matching the synchronization word when no errors are detected. During the comparison the number of errors is processed with the detection of the first instant of matching detection (indicated by index_F), and the last instant of matching detection (index_L). Those first and last index values are used to calculate the optimal sampling instant using the following formula:

$$\text{Opt\_Samp} = \text{index\_F} + \frac{(\text{index\_L} - \text{index\_F})}{2} \quad (4)$$

Once the optimal sampling instant is detected, the complete packet can be extracted with all the indexes matching the formula (5) (Preamble and AA) and (6) (PDU data).

$$\text{Packet}[i] = \text{RxStream}[\text{Opt\_Samp} - i * 13] \, (i \text{ is from 0 to 39}) \quad (5)$$

$$\text{Packet}[40 + i] = \text{RxStream}[\text{Opt\_Samp} + i * 13] \quad (6)$$
(i is from 40 to the PDU length -1)

In order to validate the model, the RF Direct Test Mode BLE AA has been used (i.e. 0x71764129), with a random PDU data length and content. The validation has been conducted over 1000 iterations, and no error has been detected so far, demonstrating the validity of the ideal model.

IV. CONCLUSION

We have presented in this paper the importance of estimating the RF energy consumption in a SoC and the necessity of analyzing it in a global system simulation including all contributors. The meet-in-the-middle method is presented to realize a high-speed simulation with circuit-level design accuracy. SC-AMS is introduced for a BLE radio modeling which is compatible with a BLE Baseband modeled with SC-TLM. This early SC-AMS model which contains the RF transceiver and the GFSK MODEM is an ideal functional model, which requires a future refinement and a completion by adding the RF circuit impairments and energy consumption. A global co-simulation of the Baseband and the BLE radio model will be applied to obtain the system performance in relationship with the power and energy consumption.

ACKNOWLEDGMENT

The authors would like to acknowledge Pr. François Pecheux who gave a lot of advices about the SC-AMS modeling, and Mr. Jean-Philippe Lambert (RivieraWaves) for his BT RF modeling advices. This work is supported by RivieraWaves and the CIM-PACA Design Platform.

REFERENCES

[1] Rob A. Rutenbar, G. Georges G E, J. Roychowdhury, "Hierarchical Modeling, Optimization, and Synthesis for System-Level Analog and RF Designs", Proceedings of the 2006 IEEE/ACM international conference on Computer-aided design, vol.95, Issue 3, pp.640-669 2007.

[2] L.Aleves.Da.Silva, "Estimation de la consummation d'énergie et du taux d'erreur bit d'un système Bluetooth LE basée sur la modélisation hiérarchique d'un amplificateur de puissance à 2.4 GHz", PhD thesis, Université de Nice-Sophia Antipolis, 2011.

[3] The Bluetooth website. [Online]. Available : http://www.bluetooth.com/Pages/low-energy.aspx.

[4] Bluetooth Core Specification v4.0.

[5] Open SystemC Initiative (OSCI), "SystemC-AMS extensions User's Guide" March 8 2010.

[6] M.Vasilevski, F.Pecheux, H.Aboushady, L.De Lamarre, "Modeling Heterogeneous Systems Using SystemC-AMS Case Study: A Wireless Sensor Network Node", International Behavioral Modeling and Simulation Conference, BMAS'07, USA, September 2007.

[7] F. Cenni, E. Simieu, S. Mir, "Macro-modeling of analog blocks for SystemC-AMS simulation: A chemical sensor case-study", 17th IFIP International Conference of VLSI (VLSI-SoC), October 2009.

[8] T. Xu, H. L. Arriës, R. Van Leuken, A. De Graaf, "A precise SystemC-AMS model for Charge Pump Phase Lock Loop with multiphase outputs", 8th International ASIC Conference, October 2009.

[9] L. Bousquet, F. Cenni, E. Simieu, "SystemC-AMS high-level modeling of linear analog blocks with power consumption information", 12th Latin American Test Workshop (LATW), March 2011.

[10] A.Leveque, F.Pecheux, M.Louerat, H.Aboushady, M.Vasilevski, "SystemC-AMS Models for Low-Power Heterogeneous Designs: Application to a WSN for the Detection of Seismic Perturbations", 23rd International Conference Architecture of Computing Systems (ARCS), Febrary 2010.

[11] L.Aleves.Da.Silva, E.Dekneuvel, A.Lewicki, B.Nicolle, G.Jacquemod, "Virtual RF system platform dedicated to heterogeneous complex SoC design", Microelectronics Journal, vol.43, no.2, pp.98-109, March 2012.

PRIME 2013, Villach, Austria

Session M2A – Oversampled ADCs

A very low OSR 90nm 1MS/s Incremental $\Sigma\Delta$ ADC

D. Cavallo[*], M. De Matteis[*†], M. Ronchi[‡], G. Leggeri[§] and A. Baschirotto[*]

[*]Department of Physics "G. Occhialini" University of Milan Bicocca. Milan, Italy
Email: see https://fisica.mib.infn.it/media/homepages/applicata/microlab/index.html
[†]Department of Innovation Engineering. University of Salento, Lecce. Italy
[‡]ST Microelectronics
[§]PEGASUS Micro Design

Abstract—A calibration free, high resolution second-order multi-channel Incremental A-to-D-Converter with multi-level quantizer is presented. The system is designed for biomedical application and combines the advantages of low oversampling ratio with SC design solution, like multi bit topology and accurate opamp design. An optimal decimation filter to minimize the weighted sum of thermal and quantization noise is used. In this paper is presented the schematic level implementation of the system in a 1.2 V 90 nm CMOS Technology and the preliminary simulation shows a 56.4 dB signal-to-noise-distortion within a 500 kHz bandwidth at a 16 MHz sample frequency.

Index Terms—Analog-to-Digital conversion, CMOS analog integrated circuit, Incremental $\Sigma\Delta$ converter, Switched-Capacitor circuit.

I. INTRODUCTION

Biomedical Instrumentation and Measurement (I&M) applications often require very high absolute accuracy and linearity, and very low offset and noise. Low power consumption is then one of the key-points in I&M applications, in order to improve battery-life and as a consequence the system portability [1].

This type of specifications are not easily satisfied with conventional A-to-D-Converters, since these do not provide accurate gain control and low offset, and require complex and power-hungry digital filters for high-accuracy performance [2] - [3]. Typically, due to the high sample rate and quasi-real time processing pipelined and SAR(Successive Approximation Register) converters are used in these kind of demanding applications. These converters need some kind of (background) calibration or error correction techniques to achieve accuracies of 12 bit or more, resulting in an increased complexity. $\Sigma\Delta$ (Sigma-Delta) modulation instead is one of the most relevant approaches for data conversion. These modulators exploit data oversampling, and so the input bandwidth is limited by both the oversampling ratio (OSR) and the maximum sampling frequency. Conventional $\Sigma\Delta$ modulators , operate continuously, and normally acquire a band-limited stream of input samples to produce the corresponding stream of output, without sample-to-sample mapping between input and output. In contrast, the incremental data converters (IDCs)operate on individual input samples. It operates for a predetermined number of clock periods and is then reset.

The properties of IDCs [3] well matched the requirements of I&M. They provide very precise conversion with accurate gain, high linearity and low offset, and the conversion time is relatively short. IDCs need only simple digital post-filters, and they can readily be multiplexed between multiple channels [4]. This paper presents the new application of incremental A-to-D-Converters as low OSR, high accuracy and high speed. The key property of these converters is that they do not rely on precisely matched analog elements to achieve high resolution, but on oversampling, noise-shaping and digital post-filtering. Thus, these converters can be integrated well into todays fine line-width CMOS technologies [5].

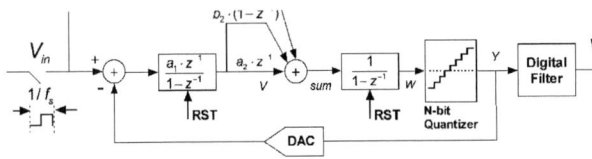

Fig. 1: Second-order $I\Sigma\Delta$ modulator model.

In biomedical applications, high resolution A-to-D-Converters are needed for the acquisition of useful signals without the need for trimming or calibration [6].

The here proposed A-to-D-Converter uses a 2nd-order multi-bit $\Sigma\Delta$ modulator topology, where each design choice is optimized for power consumption minimization while guaranteeing good dynamic range (DR). The proposed converter is targeted to achieve a number of effective bit (ENOB) of 8 at a conversion rate of 1MSample/s. This paper is organized as follows. Section II reviews the fundamentals of $I\Sigma\Delta$ modulation and provides a detailed description of the parameters focused to implement the proposed structure. Section III describes the circuital design and Section IV shows the simulation results illustrating the performance of the $I\Sigma\Delta$ architecture within the targeted operative bandwidth.

II. INCREMENTAL $\Sigma\Delta$ MODULATOR ARCHITECTURE

Incremental-$\Sigma\Delta$ converters share the same general topology with $\Sigma\Delta$ modulators, but the overall structure is reset after oversampling and processing of a specific input sample. This type of modulator can be thought as a $\Sigma\Delta$ modulator which operates permanently in transient mode since it is reset as soon as it acquires the specific input sample with a pre-arranged accuracy. This provides the needed one-to-one correspondence. Such type of mapping is essential in multiple channel applications since the data from independent sensors are generally uncorrelated.

A. Second Order $I\Sigma\Delta$ modulator

In this design , power consumption has to be minimized, without performance in terms of ENOB reduction. Any choice has been carefully considered and properly optimized. Regarding the architecture, a 2nd-order Multi-Bit (MB) Feed-Forward (FF) $\Sigma\Delta$ modulator architecture is used, as shown in Fig. 1 [7].

The 2nd-order Cascade Integrator Feed-Forward (CIFF) topology has been employed to avoid integrator clipping because that allows to obtain a reduced first integrator output swing [8] allowing a simple and more efficient opamp design [9]- [10] . In a CIFF topology, feed-forward paths are added to the loop to cancel out the input signal component at the input of the integrator. The integrator then need only process the quantization error, thus effectively reduce the voltage swings at the integrator output. The 8-level multi-bit quantizer improves system linearity and avoids parasitic idle-tones generation.

978-1-4673-4579-8/13 $31.00 © 2013 IEEE   37

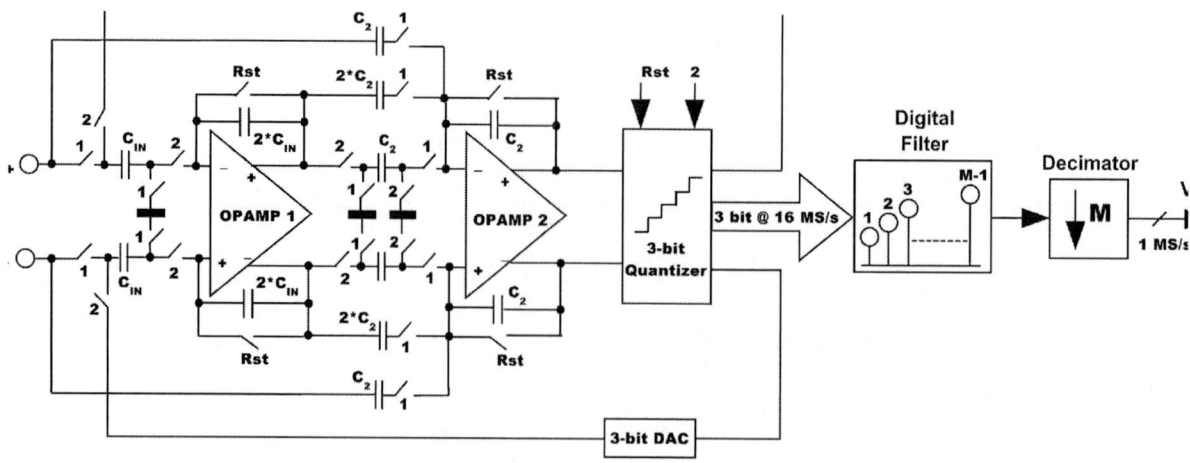

Fig. 2: Second-order $I\Sigma\Delta$ modulator schematic.

In addition, the proposed FF structure needs only one feedback DAC instead of normally used, resulting in lower design complexity, smaller area and lower power consumption.

In the here proposed structure, the FF summer is implemented as charge injected into the second integrator virtual ground [11]. This solution allows to spare an additional opamp, and so to reduce the power consumption. The architecture of Fig. 1 has been implemented with the fully-differential switched-capacitor modulator of Fig. 2.

In order to respect the transfer function precision, the input signal is sampled on the same clock phase by all input capacitors. This avoid any eventual dis-alignment (delay) between input signal sampling times in different sampling paths that introduce sampling error with consequent dynamic range loss.

Referred to block diagram to maintain a stable operation for all input signal level the modulator coefficients are set to $a_1 = 1/2$, $a_2 = 1$, $b_1 = 1$ and $b_2 = 2$ as shown in Fig. 2. The feed-forward capacitors go into the 2nd-integrator and implements the $(1 - z^{-1})$ factor of the transfer function. These paths are active only in phase 1, i.e. the phase when the second integrator is active. In the other clock phase they are disconnected to insure a stable signal to the quantizer comparators and providing the delay for the correct loop operation. Also the settling operation of the both integrators are operated in different clock phases avoiding settling response interaction between them.

An accurate system behavior analysis and optimization of block requirements related to the non-idealities of the analog part have been carried out with a dedicated Simulink model. In order to achieve the target SNR performance the developed model the requirement in terms of slew rate(SR) and closed-loop gain bandwidth(GBW) of the single integrator to meet the other non-idealities degradation is shown in Table I. These values have been used as starting point of transistor-level implementation design of operative blocks.

Parameter Value	OA1	OA2
Opamp dc-gain (A_0)	50 dB	40 dB
Opamp Unity-Gain Bandwidth (UGBW)	>35 MHz	>50 MHz
Opamp Slew Rate (SR)	>30 V/μs	>40 V/μs

TABLE I: Opamp Minimum Performance Requirements

B. Review of the fundamentals of $I\Sigma\Delta$ modulation

In Fig. 1 the block diagram of a second-order $\Sigma\Delta$ modulator is shown. This structure consists of integrators followed by a multi bit quantizer and a feedback DAC. Focusing on the simple IDC functionality, at the beginning of a new conversion cycle, all memory elements, both analog and digital must be reset. Then, a constant modulator input signal V_{in}, obtained using an input sample-and-hold (S&H) circuit, is applied to the input of the first integrator. If the loop is stable for all possible dc inputs, which can be achieved by carefully designing the loop and limiting the maximum gain of the noise transfer function (NTF), then $w[n]$ is bounded. The sampled outputs of the integrators ($v[j]$ and $w[j]$ respectively) can be calculated from the iterative equations:

$$v[j] = v[j-1] + a_1 V_{in} - a_1 V_{ref} y[j-1] \qquad (1)$$

$$sum[j] = b_1 V_{in}\delta[j-1] + a_2 v[j-1] + b_2(v[j] - v[j-1]) \qquad (2)$$

$$w[j] = w[j-1] + sum[j] \qquad (3)$$

Where a_1 and a_2 are modulation coefficients, V_{ref} is the feedback digital to analog converter (DAC) reference voltage, $y[j]$ is the output of the multi-bit quantizer , and j is the sample time index. Impose as null all signals at initial sample instant $j = 0$, which corresponds to overall structure components reset, these equations can be solved for $j = 1$ to $j = i$, to obtain the non-iterative expressions:

$$v[i] = a_1 i V_{in} - a_1 V_{ref} \sum_{j=1}^{i} y[j-1] \qquad (4)$$

$$w[i] = b_1 V_{in} + b_2 v[i] + a_2 a_1 V_{in} \sum_{j=1}^{i}(j-1) + \\ - a_1 a_2 V_{ref} \sum_{j=1}^{i}\sum_{k=1}^{j-1} y[k-1] \qquad (5)$$

After M cycles (M is the oversampling ratio) it follows that

$$w[M] = b_1 V_{in} + b_2 v[M] - a_1 a_2 \frac{M(M-1)}{2} V_{in} + \\ + a_1 a_2 V_{ref}(1\, y[M-2] + ... + (M-1)\, y[0]) \qquad (6)$$

It is evident from (6) that is possible to reconstruct the input signal V_{in} from the $M - 2$ modulator output stream bit $y[1]$ to $y[M - 2]$. Since v is bounded, due to stability constraint of the modulator, if M is large enough then $w[M]$ will be become negligible respect to the term: $a_1 a_2 V_{ref}(1 \cdot y[M - 2] + \cdots + (M - 1) \cdot y[0])$. In that case, an estimation of V_{in} can be obtained by calculating $(1 \cdot y[M - 2] + \cdots + (M - 1) \cdot y[0])$ and down-sampling of factor of M. This is accomplished in the digital domain by the decimation filter in Fig. 1. The output of the decimation filter, V_{out}, is thus:

$$V_{out} = \frac{2a_1 a_2}{2b_1 + a_1 a_2 M(M - 1)} V_{ref}(1\, y[M - 2] + \dots \\ \dots + (M - 1)\, y[0]) \quad (7)$$

and V_{out} closely approximates V_{in} when $v[M]$ and $w[M]$ are negligible.

Upon completion of the acquisition cycle of the sample, operation of the modulator continues by resetting both integrators and memory of digital filter, and repeating the same processing cycle for the next input sample.

C. Modulator order and oversampling ratio

The number of bit of the canonical multi-bit quantizer $I\Sigma\Delta$ is approximately:

$$ENOB = N + 2 \cdot log_2(M) - 2 \quad (8)$$

With N is the number of bit of quantizer and M is the oversampling ratio. A bit margin, however must be provided in the quantization noise budget to allow for the effect of the circuit non-idealities. Therefore in the Simulink-based-model the target ENOB used to evaluate the performance of the system by simulation is 9 bit that corresponds to a quantization noise level approximately of $-56\ dB$.

III. CIRCUIT DESIGN

A. Opamp design

The goals of this design is simplicity and low power consumption and so the preliminary schematic level implementation uses two stages Miller compensated opamp as integrators(Fig. 3). Since the $1/f$ is not critical due to the bandwidth of input signal, NMOS input devices are used to maximize g_m/I efficiency and in addition they operates in subthreshold. The choice of multi bit quantizer and feed-forward structure lower the 1st-opamp output swing to $\simeq 500mV_{peak}$. The 2nd-opamp implements integration and feed-forward summation. Its output swing is as large as the quantizer full scale, i.e maximum signal. Opamps frequency responses are shown in Fig. 4.

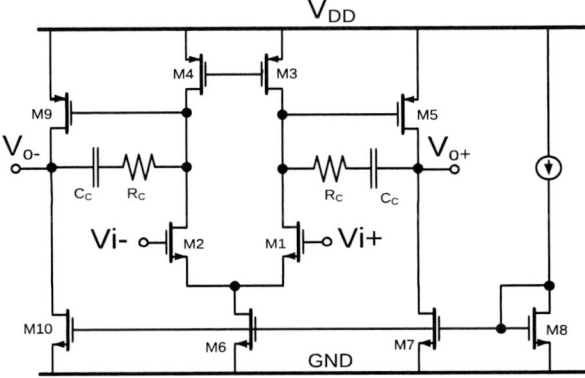

Fig. 3: Integrators two stages Miller compensated Opamp structure

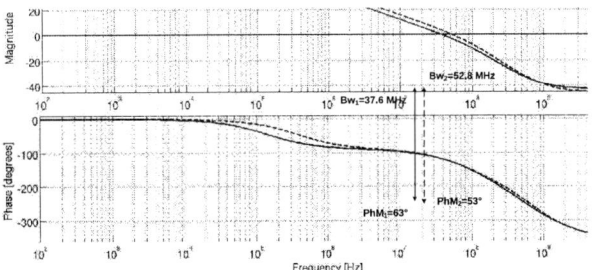

Fig. 4: Integrators Opamp frequency responses

B. Multibit quantizer design

The 3-bit quantizer uses seven fully-differential comparators, and the threshold levels are generated by a passive capacitive partition as shown in Fig. 5. This type of passive threshold generation reduces the power consumption of the overall quantizer.

The single threshold V_{thr} is realized following the relation

$$V_{thr} = (V_{R+} - V_{R-}) \cdot \frac{C_p - C_m}{C_p + C_m} \quad (9)$$

The comparators use a time continuous PMOS input pairs and a latched output stage to interface directly to the output digital filter and feedback DAC.

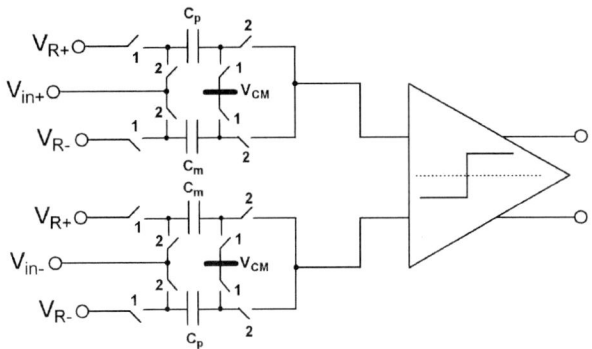

Fig. 5: Passive threshold network of ADC structure

C. Multibit feedback DAC structure

The capacitive DAC is implemented directly at the input subtractor network as shown in Fig. 6. Its operation consists of a redundant pairs of capacitors that are selectively connected to positive and negative V_{ref} following a thermometer coding scheme. This type of operation allows to have a constant load on V_{ref} for any DAC code. This scheme presents also a constant input capacitance to the 1st integrator and so the settling time being independent of DAC feedback code. The DAC capacitor array C_{IN} is composed of seven equal sub-capacitors of value $C_u = C_{IN}/7$. As shown in Fig. 6 the switches network are driven from a digital combination of the quantizer thermometric scheme bit, clock phase and reset signal i.e.

$$b_i = Ph2 \cdot b_c \cdot \overline{RST}$$
$$\bar{b_i} = Ph2 \cdot \overline{b_c} \cdot \overline{RST} \quad (10)$$
$$rst = Ph2 \cdot RST$$

where b_c is the comparator digital output bit and RST is the overall system reset signal.

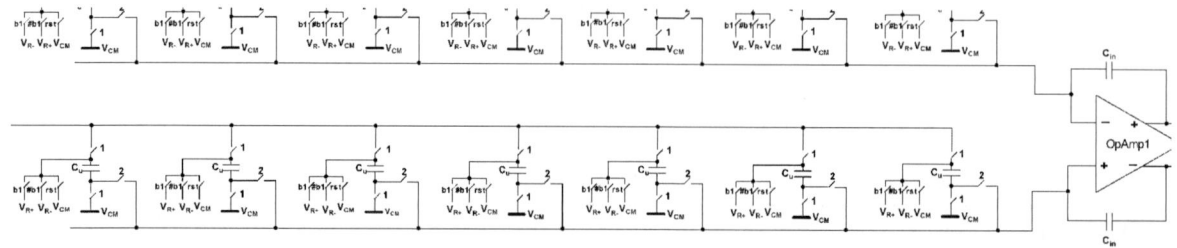

Fig. 6: 3-bit feedback DAC input network structure.

IV. SIMULATION RESULTS

The proposed $I\Sigma\Delta$ schematic is designed in a 90 nm CMOS technology. Different types of simulation has been done. Chosen the right coefficients to evaluate the stability of the loop by means of the analysis of the Signal and Noise Transfer Function(STF and NTF) of the modulator in continuous time, the dynamic range and bandwidth of the overall structure has been evaluated.

In Fig. 7 is shown the ENOB vs. Input signal amplitude and can be noted that the structure achieves a dynamic range $>60dB$. Since the signal full-scale ($1V_{pp}$) is lower than the $I\Sigma\Delta$ full scale due to the supply voltage, no saturation effect is observed and a constant noise floor for the full signal amplitude range has been observed.

Fig. 8 shows the ENOB vs. Input signal frequency for a $-10dB$-Full Scale (FS) proving the good performance of the converter over the entire bandwidth of interest.

From a single 1.2 V power supply, its total current consumption of $770\mu A$ is shared between 1st Integrator ($150\mu A$), 2st Integrator ($200\mu A$) and the multi-bit quantizer ($420\mu A$). The converter has a power consumption of $924\mu W$. Table II reports the achieved performance of the here presented converter and they are compared with other papers reported in literature.

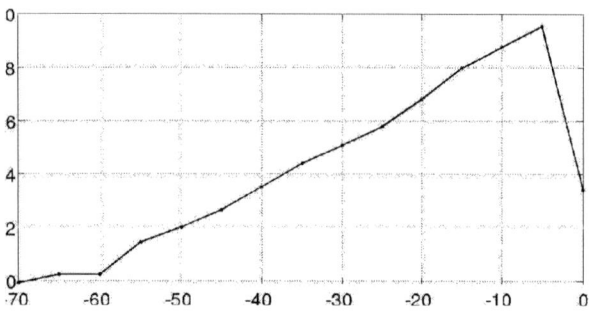

Fig. 7: Dynamic range(ENOB) vs. Input signal amplitude

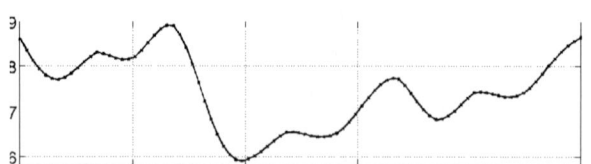

Fig. 8: Dynamic range(ENOB) vs. Input signal frequency

	This work	Garcia [12]	Agah [13]
Technology	$90nm$	$0.15\mu m$	$0.18\mu m$
SNR [dB]	56	64	90.1
Sampling Freq.[MHz]	16	n.a	45.2
Bandwidth [kHz]	500	2	500
Supply [V]	1.2	1.6	1.8
Power Consumption	$924\mu W$	$96\mu W$	$38.1mW$
P/(SNR*BW)	2.92e-12	3.02e-11	2.37e-12

TABLE II: Converter performance comparison

REFERENCES

[1] J. Márkus, P. Deval, V. Quiquempoix, J. Silva and Gabor C. Temes, *Incremental Delta-Sigma Structures for DC Measurement:an Overview*, IEEE 2006 Custom Integrated Circuit Conference(CICC).

[2] S. Hein and A. Zakhor, *Sigma-Delta Modulators: Nonlinear Decoding Algorithms and Stability Analysis.* Norwell, MA: Kluwer 1993.

[3] K. Nam, S. M. Lee, D. K. Su, and B. A. Wooley, *A low-voltage low-power sigma-delta modulator for broadband analog-to-digital conversion*, IEEE J. Solid-State Circuits, vol. 40, no. 9, pp. 1855-1864, Sep. 2005.

[4] P. Carniti, M. De Matteis, A. Giachero, C. Gotti, M. Maino, G. Pessina, *CLARO-CMOS, a very low power ASIC for fast Photon counting with pixellated photodetectors.* Volume 7. November 2012. pp: 1-24.

[5] D'Amico et al. *A 6.4 mW, 4.9 nV/Hz, 24 dBm IIP3 VGA for a multi-standard (WLAN, UMTS, and Bluetooth) receiver* Analog Integrated Circuits and Signal Processing, October 2009, Volume 61, Issue 1, pp 1-7

[6] M. De Matteis, D'Amico, S. ; Giannini, V. ; Baschirotto, A., *A 550mV 8dBm IIP3 4th order analog base band filter For WLAN receivers.*, Solid State Circuits Conference, 2007. ESSCIRC 2007. 33rd European, Page(s): 504 - 507

[7] Guinea, J. ; Sentieri, E. ; Baschirotto, A. , *An instrumentation 128dB-SNR 750μA SDM*, ESSCIRC (ESSCIRC), 2012 Proceedings of the, Page(s): 205 - 208

[8] S. R. Norsworthy, R. Schreier, and G. C. Temes, *Delta-Sigma Data Converters.* Piscataway, NJ: IEEE Press, 1997.

[9] D'Amico, S.; De Blasi, M.; De Matteis, M.; Baschirotto A., *A 255 MHz Programmable Gain Amplifier and Low-Pass Filter for Ultra Low Power Impulse-Radio UWB Receivers.*, IEEE Transactions on Circuits and Systems I: Regular Papers, Vol.59 n. 2. pp. 337-345, 2012

[10] Baschirotto, A.; Cocciolo, G.; De Matteis, M.; Giachero, A.; Gotti, C.; Maino, M.; Pessina, G., *A fast and low noise charge sensitive preamplifier in 90 nm CMOS technology*, Journal of Instrumentation, Volume 7, January 2012. pp.1-8.

[11] Hao San et al.,*Second-order Sigma-Delta modulator with novel feedforward architecture.* 50th Midwest Symposium on Circuits and Systems (MWSCAS 2007), pages 148-151.

[12] Garcia, J. ; Rodriguez, S.; Rusu, A.; *A Low-Power CT Incremental 3rd Order $\Sigma\Delta$ ADC for Biosensor Applications* , IEEE Transactions on Circuits and Systems I , Vol: 60 , Issue: 1, Page(s): 25 - 36 , Jan. 2013

[13] Agah, A., Vleugels, K., Griffin, P.B., Ronaghi, M., Plummer, J.D., Wooley, B.A. *A High-Resolution Low-Power Incremental Sigma Delta ADC With Extended Range for Biosensor Arrays.*, J. Solid-State Circuits(2010)1099-1110

[14] Quiquempoix, et al. *A low-power 22-bit incremental A-to-D-Converter*, IEEE J. Solid-State Circuits, vol. 41, no. 7, pp. 1562-1571, Jul. 2006.

978-1-4673-4579-8/13 $31.00 © 2013 IEEE

Variable-Step 12-bit ADC based on Counter Ramp Recycling Architecture suitable for CMOS Imagers with Column-Parallel Readout

Tarek M. Hassan, *Student Member, IEEE,* Markus Strobel, Harald Richter, Joachim N. Burghartz, *Fellow, IEEE*

Institute for Microelectronics Stuttgart (IMS CHIPS), Stuttgart, Germany

Abstract—A 12-bit counter ramp recycling analog-to-digital converter (ADC) is proposed, which can be configured in a single-step mode for achieving high conversion accuracy as well as in various multi-step modes for yielding high conversion speed. A unique ADC circuit realization is used for the different modes of operation, while a digital control unit is responsible for providing the necessary control signals to the ADC. Similar to common counter ramp architectures, the proposed implementation is suitable for column-parallel readout owing to its simplicity. The proposed variable-step recycling ADC is implemented in a 0.18 μm CMOS technology from UMC. Simulation results show good agreement with the expected trade-off between speed and accuracy, which is common to all conventional ADCs.

Index Terms—algorithmic, counter ramp, multi-step, column-parallel, analog-to-digital converters

I. INTRODUCTION

Column-parallel analog-to-digital converters (ADCs) are gaining increased attention in modern pixel-array implementations, as for example, in CMOS imagers. Due to their native parallel structure they can provide superb conversion speed in comparison to their single chip-level counterparts. The request for dramatically increasing pixel density with higher frame rate has posed many challenges on current column-parallel ADC architectures. Algorithmic and successive approximation ADCs are known for their high conversion speed, producing an N-bit digital word in N conversion steps. However, they primarily achieve moderate resolution with acceptable accuracy due to the accumulative nature of the offset and gain errors existing in typical implementations. Although different correction techniques are employed to challenge these non-idealities, this frequently requires substantial area, thus limiting their integration in a single column [1], [2].

On the other hand, single-step counter ramp ADCs (also known as single-slope ADCs) have proved their feasibility for column-parallel readout architectures owing to their simplicity, low power consumption, high linearity and small area. However, they suffer from low conversion speed, producing the N-bit digital word in 2^N clock cycles [3]. For modern imaging systems with high pixel count, modified architectures based on a two-step counter ramp concept have been introduced claiming to produce moderate to high speed frame rates [4], [5]. In this type of converters the analog-to-digital (A/D) conversion is carried out using two conversion steps, a coarse step for determining the most significant bits (MSBs) and a fine step for evaluating the remaining least significant bits (LSBs). However, in [4] multiple ramp generators and switch arrays are used for the fine conversion, which increases the complexity and limits the number of possible coarse bits. This problem is alleviated in [5] by storing the analog coarse level on a capacitor inside the column, enabling the ADC to complete the conversion in $2 \times 2^{N/2}$ clock cycles. Although implemented in a prototype imager with 320×240 pixels, higher pixel density will require the ramp generator to drive a huge overall column capacitance, substantially affecting the settling accuracy and speed. However, many applications require even faster conversion speed. In industrial and automotive imagers for example, it is sometimes required to scan an image with high speed and medium accuracy while in other circumstances a high accuracy is essential, affording a slightly lower frame rate. Since most conventional ADCs produce their samples in a fixed number of clock cycles, the main clock frequency has to be tuned accordingly. In fact, this complicates the overall system design, mandating careful analysis of the analog circuitry as it should still operate at the maximum possible clock frequency.

This work overcomes the above limitations through a novel variable-step 12-bit counter ramp ADC, which can be configured in a single-step mode for achieving high accuracy as well as in a multi-step mode for yielding high conversion speed. In the extreme case, it can be configured to resemble an algorithmic ADC producing the highest possible readout speed with the same circuit realization. This is achieved with the aid of a digital control unit that is responsible for providing the necessary control and clock signals to the ADC according to the desired mode of operation.

In the next section of this paper an overview of a single-step counter ramp architecture is given. It serves as the basis for the proposed variable-step counter ramp recycling ADC, which will be introduced in Section III. A prototype of the proposed ADC was implemented in 0.18 μm CMOS technology and submitted for fabrication with some simulation results presented in Section IV. Finally, a conclusion is drawn in Section V.

This work was funded by the German Federal Ministry of Education and Research, BMBF contract number 16N10370, in the CATRENE project CA301 HiDRaLoN, "High Dynamic Range Low Noise CMOS imagers".

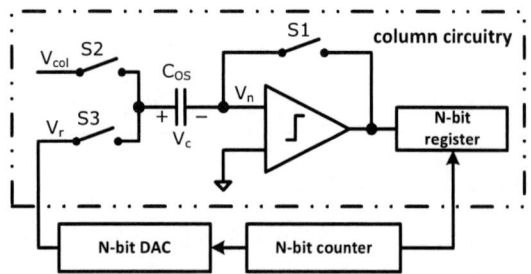

Figure 1. Block diagram of a single-step counter ramp ADC architecture.

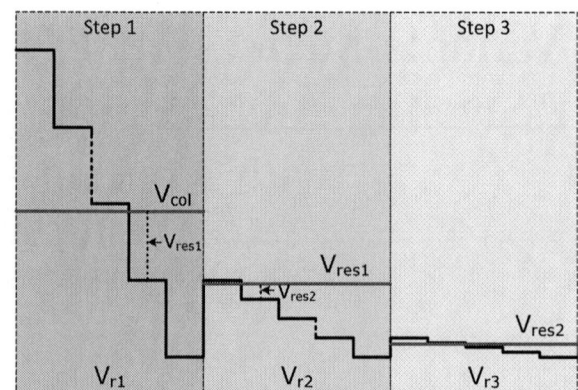

Figure 2. Multi-step counter ramp timing diagram showing voltage residues.

II. SINGLE-STEP COUNTER RAMP ARCHITECTURE

N-bit column-parallel counter ramp architectures are generally constructed using a global counter for updating a column located register and simultaneously driving an N-bit DAC for generating the digital ramp. Beside the N-bit register, each column contains a comparator, a sample-and-hold amplifier (SHA) and few switches. A possible architecture employing offset auto-zeroing is shown in Fig. 1. Here, the column signal (V_{col}) is sampled along with the comparator's internal offset on the two plates of an input capacitor (C_{OS}) by setting the comparator in unity-feedback configuration (switching on S1 and S2). At the moment the feedback loop is broken (S1 is opened) the charge stored in C_{OS} will represent a voltage difference equal to

$$V_c = V_{col} - V_{os} \qquad (1)$$

Although introducing high input impedance to the column circuitry, this architecture avoids the need for an extra SHA. In addition, opening S1 slightly before S2 ensures a signal-independent charge injected into C_{OS}, causing a constant voltage error that can be easily corrected at system level. Next, a descending digital ramp (could be also implemented in ascending order) is applied and transferred to the comparator's inverting terminal (V_n) via the floating input capacitor (S3 is switched on). Therefore, a ramp voltage containing the difference between the applied digital ramp and the previously stored voltage difference appears on the comparator's inverting terminal and is given by

$$V_n = V_r - V_c = V_r - V_{col} + V_{os} \qquad (2)$$

When the voltage difference $V_r - V_{col}$ reaches zero the comparator changes its polarity, saving the corresponding digital word of the counter inside the N-bit register.

The structure and functionality of counter ramp architectures is pretty simple, which has made them popular for column-parallel readout architectures. However, due to the distributed nature of the readout voltages along the column array the ramp generator has to go through all possible voltage levels before the conversion terminates. This makes single-step counter ramp ADCs inherently slow, lending themselves suitable for low speed applications. Yet, in order to keep the benefits of counter ramp architectures while introducing faster conversion speeds, a novel technique is introduced in this paper that enables the A/D conversion

to be split into two or more steps with the aid of a recycling architecture. As shown in Fig. 2 the A/D conversion begins with a coarse step in which the ramp moves with large voltage levels for evaluating the MSBs. The resulting voltage residue is stored and converted in the subsequent steps via smaller ramp voltages for resolving the LSBs. If each step is designed to resolve equal number of bits M, then the total conversion time can be expressed in terms of the conversion time per step as follows:

$$T_{conv} = \frac{N}{M} \times T_{step} = \frac{N}{M} \times \frac{2^M}{f_{clk}} \qquad (3)$$

Obviously, the total conversion time is proportional to the total number of steps while exponentially following M. Hence, it is basically required to reduce the number of bits per step as far as possible. However, accuracy limitations usually pose a lower bound for the value of M as will be shown later.

III. PROPOSED VARIABLE-STEP RECYCLING ADC

A. ADC Architecture and Operation

The proposed variable-step counter ramp recycling ADC architecture is shown in Fig. 3. As compared to the single-step architecture presented in Section II this ADC incorporates a bidirectional buffered SHA for storing the voltage residue after each step. The SHA has a bidirectional terminal, which acts as a buffered input port in the tracking mode. During the hold mode, the sampled voltage appears on the same port provided that the input signal is decoupled. In addition, the column register is realized using two N-bit shift registers. A parallel-in serial-out shift register (PISO) is employed for storing the M-bit counted values. After each conversion step this value is shifted into a serial-in parallel-out shift register (SIPO), forming the N-bit digital word after the final step.

The operation of the proposed ADC can be described in analogy to the previous section as follows: During the sampling phase the comparator's internal offset is sampled along with the column voltage on C_{OS}. Thereafter, S1 and S2 are opened in sequence injecting a constant charge while S3 is switched on to connect the first descending M-bit digital ramp to the floating capacitor. At the same time, the buffered SHA

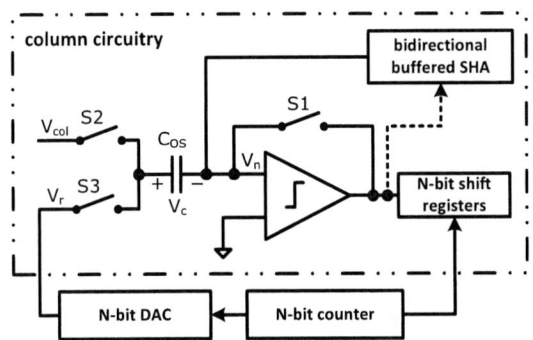

Figure 3.   Block diagram of the proposed counter ramp recycling ADC.

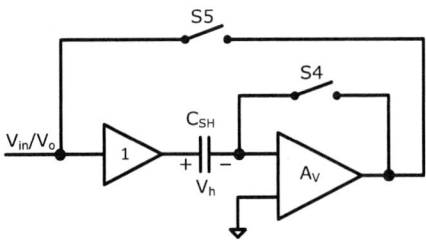

Figure 4.   Block diagram of the bidirectional buffered SHA.

is enabled in the acquisition mode, tracking the voltage level at V_n. Whenever the digital ramp falls below the previously stored column voltage the comparator strobes the input shift register to hold the M-bit counted value and simultaneously sends a signal to the SHA to sample the voltage residue given with the aid of Eq. (2) and Fig. 2 by

$$V_{n1} = V_{r1} - V_{col} + V_{os} = -V_{res1} + V_{os} \qquad (4)$$

After the first conversion step is over the SHA is switched to the hold mode placing the sampled voltage residue back on V_n, while the ramp generator provides a zero potential. This enables C_{OS} to store a charge equal to the voltage difference

$$V_{c1} = V_{res1} - V_{os} \qquad (5)$$

Interestingly, this expression is similar to Eq. (1) with the voltage residue replacing the column voltage. Hence, a second M-bit digital ramp with a full scale voltage equal to the maximum voltage residue (one LSB voltage of the first ramp) can be applied in order to convert the residue into the second M-bit code. After this step terminates the second voltage residue can be further resolved in a third step using the same recycling concept. In general, after each step is completed C_{OS} will store a charge equal to the voltage difference

$$V_{cK} = V_{resK} - V_{os} \qquad (6)$$

where K represents the step number. Obviously, the comparator's internal offset is saved along all conversion steps, ensuring that offset auto-zeroing still holds. So in practice the comparator's accuracy is insensitive to device mismatch while primarily depending on its gain and noise behavior.

In order to fulfill accuracy and speed requirements, the comparator is realized using a two-stage amplifier followed by a CMOS inverter with input offset storage. Regarding the multi-ramp generator, a global current-steering 12-bit DAC is implemented for producing the different M-bit ramps [6]. Therefore, non-linearity and gain errors arising from the ramp generator are common to all column ADCs and can be globally corrected. Besides, the ramp generator does not directly drive a capacitance for storing the voltage residue as compared to [5] thanks to the buffered SHA. Since the accuracy of the SHA will directly affect the linearity of the column ADC, an appropriate SHA topology is briefly introduced next with more details to be published in future work.

B. Bidirectional Buffered SHA

The block diagram of the bidirectional buffered SHA is shown in Fig. 4. It is composed of a unity-gain buffer, a sampling capacitor and an operational amplifier (op-amp). During the acquisition mode the op-amp is connected in unity-feedback configuration (switching on S4) while its output remains disconnected from the bidirectional port (S5 is open). The input signal is applied to the unity-gain buffer, generating a positive charge on the sample-and-hold capacitor (C_{SH}) that is proportional to the input voltage. Since the op-amp's inverting terminal is at virtual ground, the tracked voltage difference is approximately equal to the input voltage. At the moment a decision is performed by the column comparator, the unity-feedback loop is disconnected (S4 is opened) sampling the input voltage on C_{SH}, which represents the voltage residue of a certain conversion step. Moving to the hold mode, the outer feedback loop is enabled (S5 is turned on) while the input signal must be either disconnected or decoupled (as for the proposed ADC architecture). This creates a loop gain of approximately A_V (the open-loop gain of the op-amp), which ensures that the original input voltage is restored at the input of the unity-gain buffer. The accuracy of the restored voltage depends mainly on the relative error between the loop gain and A_V, which is primarily given by the gain variation of the unity-gain buffer. Concerning charge injection errors (resulting from S4) and internal offsets (due to op-amp's device mismatch) they are both divided by the loop gain. Therefore, in order to achieve a SHA with good accuracy both a linear unity-gain buffer and a high gain op-amp are desired (speed and stability are the main challenges). If that is not the case, then for the proposed recycling ADC architecture a SHA inaccuracy (arising mainly from the relative error between the loop gain and A_V) will manifest itself as a gain error in the voltage residue. In fact, this type of error can be corrected at system level. Yet, the main limiting factor is the signal-dependent charge injected from S5 into C_{OS} before starting a new conversion step. Since this type of error is depending on the sampled voltage residue, it can be considered almost constant for the last few conversion steps, and can be easily corrected. For a large number of conversion steps one should account with increased non-linear errors for the first couple of steps. Due to the accumulative nature of these errors they are usually not easy to correct, and produce integral and differential non-linearity errors (INL, DNL) in the transfer characteristics. In order to minimize these errors, S5 can be realized using minimum device dimensions.

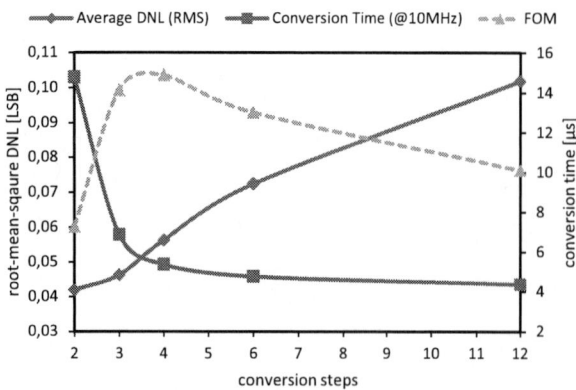

Figure 5. Speed-accuracy trade-off for the proposed variable-step ADC.

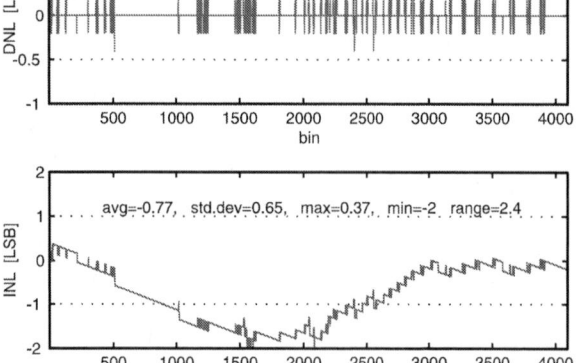

Figure 6. DNL and INL of 12-bit recycling ADC in four-step configuration.

Table I
SPECIFICATION SUMMARY OF THE PROPOSED COLUMN ADC

ADC Resolution	12-bit (defined by ramp generator)
Technology	UMC $0.18\,\mu m$ CMOS 1P4M 1.8 V/3.3 V
Power Supplies	1.8 V (digital circuitry, comparator)
	3.3 V (analog circuitry)
Dynamic Range	1.5 V
Max. Sampling Rate	$220\,kHz$ (@ $f_{clk} = 10\,MHz$)
Power Consumption	0.26 mW
Column Footprint	$6.7\,\mu m \times 500\,\mu m$ (contains $200\,\mu m$ power rails)

are shown in Fig. 6. While the DNL is having a maximum variation of $\pm0.4\,LSB$, the INL is ranging from $+0.4\,LSB$ to $-2\,LSB$. The column ADC consumes around $0.23\,mW$ at a maximum sampling frequency of $220\,kHz$. The different ADC modes of operation were tested at a constant clock frequency of $10\,MHz$. The main specifications of the proposed ADC are summarized in Table I

V. CONCLUSION

In this paper a 12-bit variable-step counter ramp recycling ADC was presented. With the aid of flexible digital control, the ADC is capable of operating in single-step as well as in different multi-step modes. The ADC was implemented in a $0.18\,\mu m$ CMOS process from UMC Co. Simulation results show good linearity for the selected four-step mode with a DNL below $\pm0.5\,LSB$. The expected trade-off between speed and accuracy was simulated, yielding an optimum value for the four-step configuration. Due to its small footprint and sampling rates up to $220\,kHz$, the proposed ADC is suitable for column-parallel readout architectures e.g. modern high speed CMOS imagers.

ACKNOWLEDGMENT

The authors thank the HiDRaLoN design team at IMS CHIPS for its cooperation. Special appreciation goes to the German Federal Ministry of Education and Research, BMBF contract number 16N10370, for funding this project.

IV. SIMULATION RESULTS

A 12-bit ADC prototype of the proposed counter ramp recycling architecture has been realized in standard $0.18\,\mu m$ CMOS technology and submitted for fabrication. Meanwhile, the ADC performance was simulated using five modes of operation, namely two, three, four, six and twelve-step (equivalent to algorithmic) configurations. The single-step mode was omitted due to expected superb linearity while requiring long simulation time.

The linearity performance of the proposed ADC was characterized based on a slow ramp histogram testing with a full scale voltage of 1.5 V. In Fig. 5 the root-mean-square DNL values for all tested modes of operation are plotted against the number of conversion steps, presenting an almost linear behavior. Concurrently, the total conversion time is drawn on a secondary axis showing a logarithmic decrease with increasing number of conversion steps as expected from Eq. (3). A figure-of-merit (FOM) representing the speed-accuracy trade-off is also illustrated (dotted curve), pointing out a maximum value for the four-step configuration. Detailed linearity results for the four-step mode of operation

REFERENCES

[1] Z. Yang and J. Van der Spiegel, "A 10-bit 8.3MS/s switched-current successive approximation ADC for column-parallel imagers," in *IEEE ISCAS*, May 2008, pp. 224–227.
[2] M. Furuta, Y. Nishikawa, T. Inoue, and S. Kawahito, "A high-speed, high-sensitivity digital CMOS image sensor with a global shutter and 12-bit column-parallel cyclic A/D converters," *IEEE J. Solid-State Circuits*, vol. 42, no. 4, pp. 766–774, Apr. 2007.
[3] T. Sugiki *et al.*, "A 60 mW 10 b CMOS image sensor with column-to-column FPN reduction," in *IEEE ISSCC Tech. Dig.*, Feb. 2000, pp. 108–109.
[4] M. Snoeij, A. J. P. Theuwissen, K. A. A. Makinwa, and J. Huijsing, "Multiple-ramp column-parallel ADC architectures for CMOS image sensors," *IEEE J. Solid-State Circuits*, vol. 42, no. 12, pp. 2968–2977, Dec. 2007.
[5] S. Lim, J. Lee, D. Kim, and G. Han, "A high-speed CMOS image sensor with column-parallel two-step single-slope ADCs," *IEEE Trans. Electron Devices*, vol. 56, no. 3, pp. 393–398, Mar. 2009.
[6] T. Zaki, T. Hassan, and C. Scherjon, "A compact 12-bit current-steering D/A converter for HDRC image sensors," in *ProRISC Workshop*, Nov. 2010, pp. 60–63.

PRIME 2013, Villach, Austria

Session M2A – Oversampled ADCs

An Automatic Calibration Circuit for 12-bits Single-Ramp A-to-D Converter in LHC Environments

T. Vergine[1][2], M. De Matteis[1], L. Rota[1], A. Marchioro[3], A. Baschirotto[1]

[1]University of Milano Bicocca – Italy. Department of Physics G. Occhialini. [2] University of Pavia – Italy.
Department of Electrical Computer and Biomedical Engineering. [3] CERN, 1211 Geneve 23, Switzerland.
tommaso.vergine, marcello.dematteis, andrea.baschirotto@unimib.it.

Abstract — A calibration circuit for single-ramp A-to-D converters is presented here. The calibration circuit allows to automatically compensate the process/mismatch and radiation effects on the A-to-D converter, improving performance and Equivalent Number of Bits. In particular, the calibration circuit is able to automatically align the ramp signal reference used for the conversion in single slope architectures A-to-D architectures, compensating slope deviations due to technological/electrical reasons.

Moreover, the calibration circuit shares the same analog circuits of the A-to-D converter, requiring only a small additional power budget and logic for the implementation. The calibration circuit has been validated, testing the overall A-to-D converter after the calibration. A 12 steps binary search is required to calibrate the A-to-D converter (about 2.5ms). This calibration circuit is able to guarantee an 11bits accuracy, in the worst case simulation corner. The technology used is a 65 nm CMOS. The clock frequency has been set to 20 MHz and the power consumption is about 400 µW.

Keywords — A-to-D, Low-Power, Radiation Hardness, High-Energy-Physics instrumentation, Automatic Calibration, Successive Approximations Register.

I. INTRODUCTION

Nowadys CMOS integrated microelectronic circuits are widely used in instrumentation for High-Energy Physic experiments [1][2][3][4]. In particular, in the LHC (Large Hadron Collider) environment, the detectors read-out channels and the additional circuits for environmental sensing/monitoring are mostly implemented in CMOS technology [5].

The LHC environment is very harsh for microelectronic circuits since the high intensity radiation present around such accelerator can damage integrated devices. Device electrical/physical parameters (like MOS threshold voltage, mobility and resistances) can significantly change due to irradiation and can lead to system-level performance degradation or even failure.

In order to optimize performance and increase life-time a practical approach in high-energy physics experiment is to externally periodically calibrate the integrated devices. In this way the electrical variation of the integrated circuits can be adequately compensated. Similar approach appears robust and reliable, but becomes critical and impracticable in presence of several integrated circuits in a limited specific area, like in LHC environment.

The 12 bit A-to-D converter object of this paper will be art of a bigger system responsible for the temperature/leakage sensing and monitoring of the detectors in the LHC CMS (Compact Muon Solenoid) central tracker.

This A-to-D converter is based on single-ramp architecture. Notice that any deviation of ramp slope, due to process, temperature, radiation, can be detrimental for overall A-to-D converter performances. For these reasons an automatic calibration circuit has been implemented, with the aim to occasionally align, in an automatic way, the ramp slope to the nominal one, compensating any ramp signal deviation due to electrical/physical parameters variations.

The calibration circuit improves significantly the overall performance of the A-to-D converter, and in addition no external calibration is then needed for A-to-D performance optimization.

An important aspect of this approach is that it does not affect the A-to-D converter overall power consumption because several A-to-D blocks are shared with the calibration circuit. It exploits the A-to-D conversion result in the array capacitor control word research. This makes the calibration circuit simpler and composed only by a few additional digital logic gates. The ramp voltage signal has been obtained charging a 12 bit grounded array capacitor with a very precise current mirror. This approach allows to control the ramp slope changing the array control word.

This paper is organized as follows. In Section II the structure and sizing of the Calibration Circuit are shown. In Section III the design of the A-to-D converter two most important blocks is reported whereas Section IV shows the A-to-D converter performance, as validation tool. At the end of the paper the conclusions are drawn.

II. CALIBRATION CIRCUIT

The calibration circuit proposed here is foreseen to work in a single-ramp A-to-D Converter [5]. The conversion method is a voltage to time conversion, comparing the analog input signal with a continuous-time ramp voltage signal. It is clear that any deviation in the ramp voltage signal slope affects the overall A-to-D converter performance. For this reason the main aim of the calibration circuit is to automatically align the ramp signal to the nominal (ideal) ramp, compensating any electrical/physical deviation, and improving the overall A-to-D conversion performance. The ramp voltage signal is obtained by charging a capacitor with a very precise reference current (I_{REF}) - see Fig. 1.

A. Calibration Algorithm

The idea of this calibration circuit is to align the ramp slope with the ideal one by adjusting the size of the integrating capacitor through a digital control word (Calibration Word). Notice that, there are two main reasons for the ramp slope deviation (and as a consequence A-to-D converter performance degradation): variation of I_{REF} (reference current), and spread of the capacitor due to process, temperature and aging.

978-1-4673-4579-8/13 $31.00 © 2013 IEEE

Paper M2A3   *PRIME 2013, Villach, Austria*

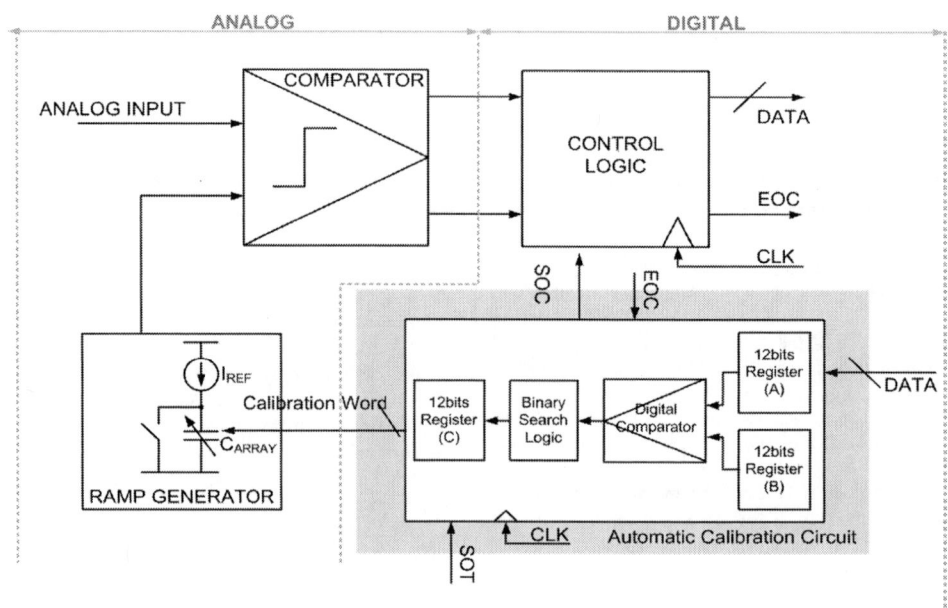

Fig. 1 – Single Slope A-to-D converter with automatic calibration circuit

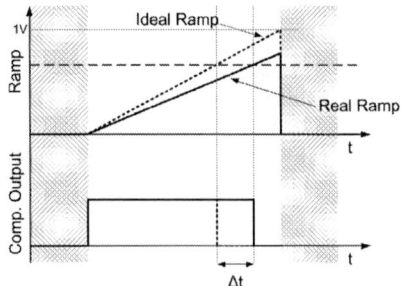

Fig. 2 - Comparator Output Duty-Cycle vs. Ramp Voltage Signal.

The main effect of both non-idealities is a deviation of the ramp slope with respect to the ideal one. The basic idea of the calibration circuits presented here is to compare the ramp signal (affected by some deviation due to I_{REF} and C_{ARRAY}) with a fixed reference voltage signal. As a consequence of a non-ideal ramp signal, the duty-cycle of the comparator output voltage will be different from the expected duty-cycle (in case of ideal ramp), as shown in Fig. 2.

Comparing the two signals duty-cycles, it is possible to evaluate how much is important the slope deviation with respect to the ideal one.

The calibration word search could be performed in several ways: starting from all zeros word, adding one bit at a time, could be the simplest way, but such an approach requires too long time. Remember that $2^{Nb}-1$ clock periods (where Nb is the number of bits in the A-to-D) are needed to perform a single conversion. For this reason the research of the right calibration word is based on a binary search. The calibration word is set, bit by bit, making a comparison between the real and ideal conversion results.

The binary search, being based on the result of A-to-D conversion, allows to share several blocks between the A-to-D converter and the binary search-based calibration circuit, saving power consumption and circuital complexity. Indeed only few additional logic gates are required to perform the calibration.

B. Calibration Circuit Implementation.

Fig. 1 shows the block diagram of the Single Slope A-to-D converter with the automatic calibration circuit. The calibration circuit is essentially composed by three 12 bit registers, a digital comparator block and some logic. The first 12 bit register (A) stores the actual conversion result, provided by the A-to-D converter. This result is compared with the ideal one, stored in the second 12 bit register (B). After the comparison, the logic sets the array capacitor control word in the third 12 bit register (C). An example of the calibration sequence is shown in Fig. 3. The algorithm starts assuming that the MSB of the (C) register is 1 and the other bits are 0. At the end of the first conversion the actual converted data is compared with the ideal one. If it is higher the MSB are set to 0, if it is lower the MSB is asserted to 1. The algorithm proceeds recursively until the last bit, the LSB, is asserted.

The circuit topology of the array capacitor is shown in Fig. 4. Its resolution has given by the equation (1) [7]:

$$\delta C = \frac{C_{max} - C_{min}}{2^N - 1} \qquad (1)$$

where $C_{max} = C_{nom}/(1-\xi)$ and $C_{min} = C_{nom}/(1+\xi)$ are the maximum and minimum array capacitance and ξ is the technological spread. This parameter (ξ) has been imposed to

978-1-4673-4579-8/13 $31.00 © 2013 IEEE   46

35%, considering the process/mismatch effects both on the array capacitor elements and on the current mirror.

In order to make negligible the minimum difference between ideal ramp and real ramp (±0.5 LSB) that the A-to-D converter can sense, the calibration procedure has been performed using a voltage signal ($V_{calibration}$) very close to the full scale (976.5625 mV = 4000 LSB). Eq. (2) shows the voltage error between the ideal and real ramp as a function of time, knowing its value in a given point. If this point is chosen very close to the full scale the voltage deviation between the ideal and the real ramp is almost equal to 0.5 LSB up to the full scale. $t_{calibration}$ is the time that the ramp needs to reach the $V_{calibration}$ voltage. As previously said, the algorithm changes the calibration word in order to match the real conversion result with the ideal result. See lower part of Fig. 3.

$$\Delta V(t) = \Delta V(t_{calibration}) \cdot \frac{t}{t_{calibration}} \qquad (2)$$

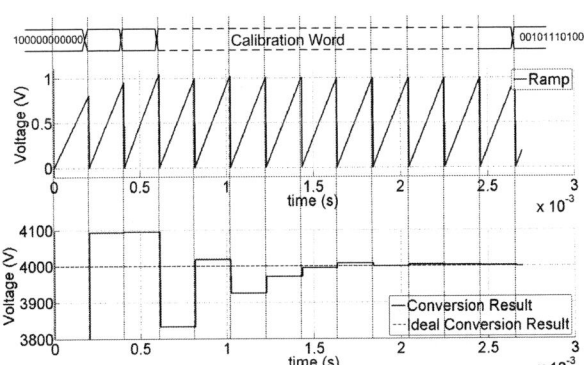

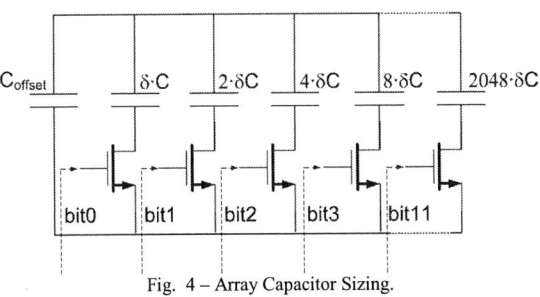

Fig. 3 – Example of Calibration.

Fig. 4 – Array Capacitor Sizing.

III. A-TO-D CONVERTER ARCHITECTURE

The A-to-D converter, as shown in Fig. 1, is composed by two sections. The analog section includes mainly the RAMP GENERATOR and the COMPARATOR blocks while the digital section includes the A-to-D CONTROL LOGIC and the calibration circuit blocks.

Since the analog input signal ranges from 0 V to 1 V the output of the RAMP GENERATOR has to start from 0 V and reach 1 V too. For this reason a grounded array capacitor is charged, with an accurate current mirror, to generate the ramp voltage signal. This implies a careful design of the RAMP GENERATOR block. The COMPARATOR is a Verilog-A block.

The A-to-D CONTROL LOGIC is responsible of the A-to-D converter overall control and timing. Every SOC (start-of-conversion) rise-edge enables an A-to-D conversion. The flag EOC points out the end of every conversion cycle.

The calibration circuit controls the array capacitor value in order to match real conversion result with the ideal one. At the SOT rising edge the automatic calibration procedure starts with 12 consecutive conversions. Each conversion is activated by its SOC signal. At the end of the 12th conversion the calibration word is set.

A. Ramp Generator

The ramp voltage signal is obtained charging a grounded array capacitor with a very accurate current mirror. Since the voltage swing at the output node of the current mirror is 1 V, its output resistance has to be carefully evaluated. A simple Matlab script of the current mirror has demonstrated that the minimum output resistance value has to be about 5 GΩ, in order to avoid performance degradations. A low voltage current mirror has been implemented [6]. A regulated coscode structure has been introduced, as reported in Fig. 5. The negative feedback tends to keep constant, when the drain voltage of transistor M4 changes from 0 V to 1 V, the drain node of transistor M3. An output resistance of 6.5 GΩ has been obtained. As it can be seen in Fig. 1, the current mirror output node voltage is periodically reset to ground. For this reason the loop stability of the low voltage current mirror has to be carefully evaluated. In addition the power consumption of the loop operational amplifier has to be minimized, to accomplish the low-power target. An operational amplifier has been designed with 55 dB of dc-gain, 2 MHz of bandwidth and a phase-margin of about 83°. Despite the very low unity gain bandwidth there aren't any negative effects on internal nodes settling time, during all the conversion phases. The low voltage current mirror performance has been also evaluated in MonteCarlo Simulations, introducing the process/mismatch effects. The maximum current deviation obtained for each corner is less than 3.5%, see Fig. 6. The ramp slope variation due to this current deviation could degrade the performance of a 12 bit A-to-D converter but the calibration circuit, acting directly on the ramp slope, is able to compensate these variations.

IV. SIMULATION RESULTS

To validate the Successive-Approximations Calibration Circuit performance a rough A-to-D converter quantization error (QE) has been evaluated. Four corners have been set (see Tab. 1).

Parameter	Values
Process	slow-slow, fast-fast
Temperature	-40°C, 120°C

Tab. 1 - Simulation Corners

This QE has been evaluated for 391 points of the input range, chosen as multiples of ¼ LSB and ¾ LSB and spaced by 10 LSB. The ideal QE, considering these input values, should be a saw-tooth waveform between ±0.25 LSB, being the distance between the input values and the corresponding ideal quantized voltage of ¼ LSB. As shown in Fig. 7 and Fig. 8 this calibration circuit has allowed to obtain a quantization error very close to the ideal one. The maximum quantization error, in the worst case, is 1 LSB as shown in Fig. 9 and Fig. 10. Some input values, multiples of ¾ LSB, have been converted with the next quantized interval. This implies a QE equal to -0.75 LSB.

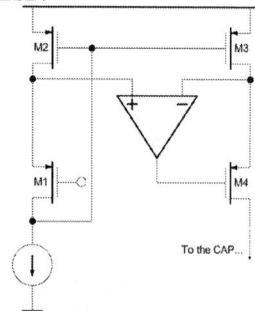

Fig. 5 - Improved Low Voltage Current Mirror.

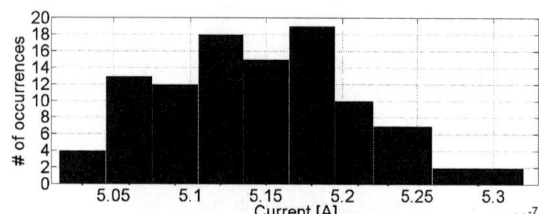

Fig. 6 - Current Mirror Process/Mismatch Deviation.

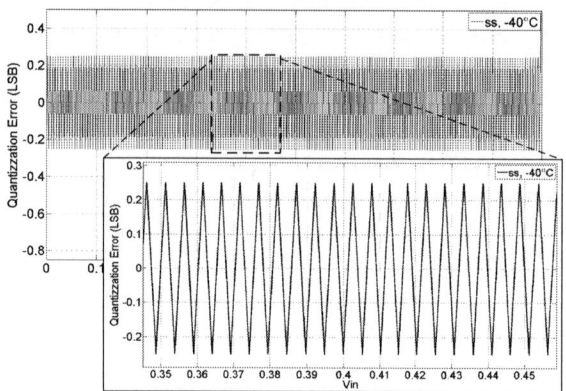

Fig. 7 – Quantization Error in ss corner, -40°C.

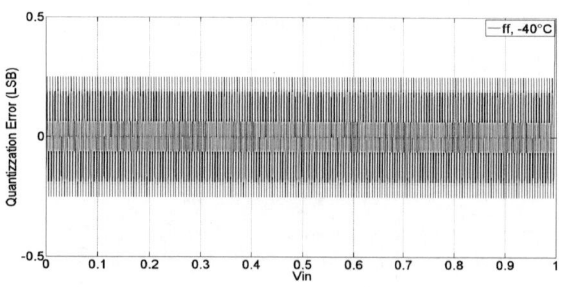

Fig. 8 – Quantization Error in ff corner, -40°C.

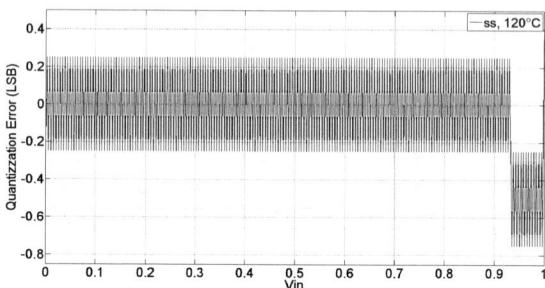

Fig. 9 – Quantization Error in ss corner, 120°C.

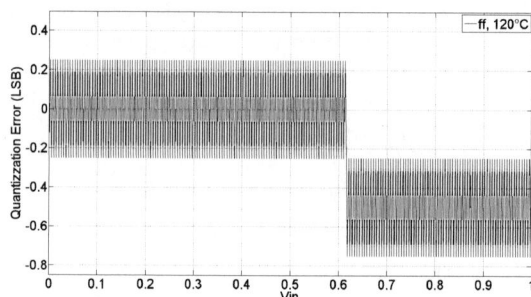

Fig. 10 - Quantization Error in ff corner, 120°C.

V. CONCLUSIONS

In this paper a Successive-Approximations Calibration Circuit for the integration capacitor on a single slope A-to-D converter has been presented. This circuit compensates the process/mismatch effects on the ramp signal slope through a 12 bit calibration procedure. An on-chip calibration circuit changes the capacitor value used to generate the ramp signal. Radiation effects that change the circuit operating point are also mitigated. The design has been realized in a 65 nm CMOS Technology. The proposed solution shares several blocks with the A-to-D converter and results in a consumption of only about 400 µW for the overall system. This solution results in an A-to-D converter maximum quantization error of 1 LSB in the worst case.

REFERENCES

[1] M. De Matteis, et al. "A 550mV 8dBm IIP3 4th order analog base band filter For WLAN receivers". Solid State Circuits Conference, 2007. ESSCIRC 2007. 33rd European. Page(s): 504 – 507.

[2] D'Amico et al. "A 6.4 mW, 4.9 nV/√Hz, 24 dBm IIP3 VGA for a multi-standard (WLAN, UMTS, and Bluetooth) receiver" Analog Integrated Circuits and Signal Processing October 2009, Volume 61, Issue 1, pp 1-7

[3] Carniti, P.; De Matteis, M.; Giachero, A.; Gotti, C.; Maino, M.; Pessina, G. "CLARO-CMOS, a very low power ASIC for fast Photon counting with pixellated photodetectors". Volume 7. November 2012. pp: 1-24.

[4] D'Amico, S.; et al. "A 255 MHz Programmable Gain Amplifier and Low-Pass Filter for Ultra Low Power Impulse-Radio UWB Receivers". Circuits and Systems I: Regular Papers, IEEE Transactions on. 2012. Vol.59 n. 2. pp. 337-345.

[5] G. Magazzù, A. Marchioro, P. Moreira, "A rad-hard 8-channel 12-bit resolution ADC for slow control applications in the LHC environment, " 7th Workshop on Electronics for LHC Experiments, Stockholm, Sweden, 10 - 14 Sep 2001, pp.338-341.

[6] J. Ramirez-Angulo, et al. " Low supply voltage high performance CMOS current mirror with low input and output voltage requirements'' IEEE Transactions on Circuits and Systems-II Express Briefs, Vol. 51, No. 3, March 2004.

[7] D'Amico, S. ; Delizia, P.; Baschirotto, A.; Azeredo-Leme, C.; Tavares, A. "A 90nm-CMOS 1.8mW 87dB-SNR 3rd order analog filter for GSM receivers". Research in Microelectronics and Electronics, 2008. PRIME 2008. Ph.D., pp: 173-176.

A New Decoding Solution for the Asynchronous Sigma Delta Modulator

Wei Chen
Department of Electrical & Electronic Engineering
Imperial College, London UK
w.chen09@imperial.ac.uk

Christos Papavassiliou
Department of Electrical & Electronic Engineering
Imperial College, London UK

Abstract—A new type of low power decoding circuit for asynchronous sigma delta modulators is presented. The circuit implements a special coarse-fine time-to-digital converter to quantize the square wave produced by asynchronous sigma delta modulators, and converts the duty cycle to a digital output. The time-to-digital converter operates asynchronously by utilizing vernier delay lines. The purpose of this circuit is to achieve a high resolution with a low frequency sampling clock, which is suitable for the ultra-low power applications. The proposed circuit is designed in 0.35um. Spectre simulations, show an 11-bit resolution is realized with 0.06LSB integral non-linearity and 0.04LSB differential non-linearity. The simulated power consumption is below16uW from a dual 0.6 V supply voltage.

Keywords— asynchronous sigma delta modulator (ASDM), time-to-digital converter (TDC), vernier delay line (VDL)

I. Introduction

A new type of sigma delta modulators originally introduced by Kikkert has recently gained [1]. This type of modulator was forgotten for many years until interest revived due to the perceived potential for high frequency operation in the absence of a fast system clock. The dynamics of asynchronous sigma delta modulators were studied in detail by Roza and Ouzounov [2-4]. The main difference between asynchronous and synchronous sigma delta modulators is that there is no sampling clock in asynchronous modulator. Asynchronous sigma delta modulators, therefore have potential for high speed and low power ADCs.

An asynchronous sigma delta modulator (Fig. 1a) includes two functional blocks: an integrator and a hysteretic comparator. The output (Fig. 1b) is a pulse width modulated square wave of period T with a pulsewidth T_1. The duty cycle α is proportional to the amplitude of the input signal (eq. (1)). Moreover, the period T of the asynchronous modulator output depends on the normalised input voltage (eq. (2)).

$$\alpha = \frac{m+1}{2} = \frac{T_1}{T_1+T_2} \qquad (1)$$

$$\frac{\omega_0}{\omega_c} = \frac{T_c}{T_1+T_2} = 1-m^2 \qquad (2)$$

In this expressions ω_0 is the output carrier frequency and ω_c is the limit cycle frequency of the asynchronous sigma delta modulator; $|m| = |v_{in}/V| < 1$ is the modulation index, namely the input voltage normalised to the dynamic range of the converter.

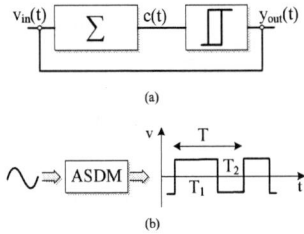

(a)

(b)

Fig. 1 (a) block diagram and (b) output waveform of the ASDM

It is interesting to observe in equations (1) and (2), that both the duty cycle and the period are input signal dependent; The asynchronous converter output is more complex that of the conversional PWM system. However, since the input signal amplitude is continuously is encoded into the time domainwithout loss of information, asynchronous sigma delta modulators can be considered as an infinite sampling frequency version of conversional synchronous sigma delta modulators. Hence, the signal-to-noise rate (SNR) can be very high even for a first-order system. This result can be extended to a slowly varying input signal, and the signal will be only corrupted by harmonic distortion. Based on the analysis in [5], the most significant distortion term is a third harmonic, the spurious-free-dynamic range (SFDR) can be approximated as:

$$SFDR \sim \frac{f_0^2}{B^2m^2} \qquad (3)$$

B is the input signal bandwidth; f_0 is the instantaneous output frequency (which is input dependent!).

In order to apply the asynchronous sigma delta modulator to data conversion, a sampler is necessary. For the conversional signal poly sampling as in Fig. 2 (a), the quantization errors are:

$$E = \frac{8}{3}f_cT_s^2 \qquad (4)$$

The signal-to-noise ratio is therefore:

$$SNR \sim OSR \cdot \sqrt{\frac{f_0}{B}} \qquad (5)$$

Where $OSR = f_s/2f_c$ is the oversampling ratio

Equation (5) suggests that, because there is no noise shaping loop applied, a high OSR is required so as to achieve a high SNR. Many solutions have been proposed for this problem. In [5], a poly-phase sampler is presented to reduce the requirement of the oversampling rate. The drawback of this method is that many phases are required, which will significantly increases the size and complexity of the multi-phase filter. Another solution is to implement time-to-digital converters to quantize the modulated output square wave, as shown in Fig. 2 (b). In [6], an irregular sampling decoder and interpolator are implemented to reduce the sampling rate. However, a very high sampling clock rate is required to trigger the irregular sampler. In [7], a time-to-digital converter with single delay chain is used to locate the rising and falling edge of the square wave. The drawback of this solution is that the input signal cannot be reconstructed directly from the output of the time-to-digital converter, and an additional algorithm must be implemented for reconstruction.

A new decoding solution for asynchronous sigma delta modulators is presently proposed; the proposed technique requires a reduced sampling frequency than previous techniques. The rest of this paper is organized as follows: In section II the system of the proposed decoding circuit is analysed. In section III the noise performance of the ASDM including decoding circuit is presented. Section IV finally presents simulations on an 11-bit decoding circuit and discusses the results.

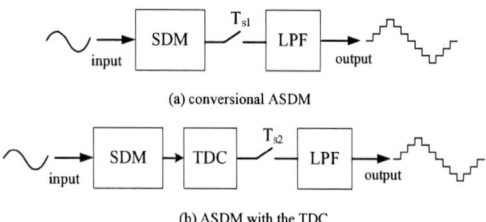

(a) conversional ASDM

(b) ASDM with the TDC

Fig. 2 system diagrams of the analogue-to-digital converter

II. The proposed decoding circuit

The normal way to determine the duty cycle of a periodic waveform is to measure both the positive and negative phases. This can be simply realized by operating two time-to-digital converters in parallel. However, this approach doubles the chip area and power dissipation. The structure we proposed can measure duty cycle with one special time-to-digital converter (Fig. 3). The proposed time-to-digital converter is a coarse-fine architecture based on vernier delay lines. In fine measurement, two vernier delay lines are used to measure the start phase and stop phase respectively. And in the coarse measurement, two counters are implemented. Fig. 4 shows the time diagram of the proposed circuit. T_1 and T_2 can be described as

$$T_1[n] = NC_1[n] \cdot T_{ref} + \left(NF_1[n] - NF_2[n]\right)\tau_{res} \quad (6)$$

$$T_2[n] = NC_2[n] \cdot T_{ref} + \left(NF_2[n] - NF_1[n+1]\right)\tau_{res} \quad (7)$$

Together with eq. (6) and (7), the period of the nth cycle is

$$
\begin{aligned}
T[n] &= T_1[n] + T_2[n] \\
&= \left(NC_1[n] + NC_2[n]\right)T_{ref} \\
&\quad + \left(NF_1[n] - NF_1[n+1]\right)\tau_{res}
\end{aligned}
\quad (8)
$$

Where $T_1[n]$ is the positive phase of the nth cycle; $T_2[n]$ the negative phase of the nth cycle; $T[n]$ is the period of the nth cycle; T_{ref} is the period of the reference clock; τ_{res} is the resolution of the time-to-digital converter

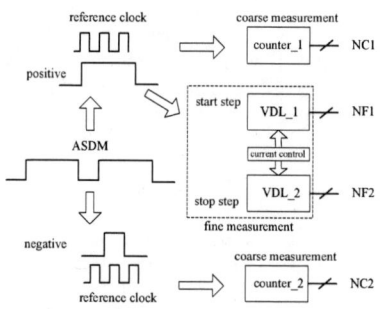

Fig. 3 system diagram of the proposed ASDM decoding circuit

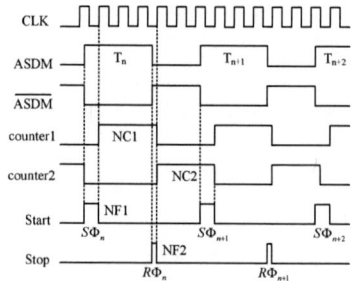

Fig. 4 time diagram of the proposed circuit

It is clear, from eq. (8), that the period can be determined without adding another time-to-digital converter; but only an additional counter is required, which has the potential to result in significant savings in chip area and power consumption. Based on eq. (1) and (2), T_1 and T_2 can be written as:

$$T_1, T_2 = \frac{T_c}{2} \cdot \frac{1}{1 \pm m} \geq \frac{T_c}{4} \quad (9)$$

The minimum value for the pulse width is $T_c/4$.

The fine measurement, the vernier delay line (Fig. 5) uses two delay buffer chains to measure delay. Assuming that the delay time τ_1 is slightly larger than τ_2, the start and stop signals represent the events whose time difference is to be measured. The time difference $t_m = t_{stop} - t_{start}$ decreases by $\tau_R = \tau_1 - \tau_2$ after each delay element of the vernier delay line. After several stages, the stop signal will catch up with start. In that case, the flip-flop at position N_x will output a "1" signal to stop the measurement. The output of the vernier delay line contains, therefore, a thermometer coded value of $N_x = t_m \bmod \tau_R$. The final measurement can be represented as:

$$N_x \tau_R < t < \left(N_x + 1\right)\tau_R \quad (10)$$

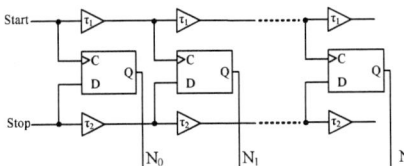

Fig. 5 the basic vernier delay line configuration

By utilizing vernier delay lines, high resolution can be achieved with a low frequency reference clock, because the resolution is no longer based on the delay time τ_1 and τ_2, but their difference. However, there is a critical issue for the vernier delay lines implemented in the conversional time-to-digital converter.

The dynamic range of an N-stage vernier delay line utilizing delay chains with τ_1 and τ_2 is

$$T_{DR} = N\tau_R \qquad (11)$$

While the maximum propagation delay of the vernier delay line is

$$T_{Max} = N\tau_1 \qquad (12)$$

It is clear to see that T_{Max} is larger than T_{DR}. This means that, the reset time of the vernier delay line is larger than the measurement time. An error will occur when a next measurement event starts to propagate in the delay line while the previous one is still propagating. P. Dudek used a read-out pipe to solve this problem [8]. The issue does not exist in the proposed time-to-digital converter, because the measurement is asynchronous. Fig. 6 shows one slice of the proposed vernier delay lines. Both D flip-flops are triggered by the rising edge of the input time interval. The function of the first D flip-flop is similar with that in basic vernier delay lines, namely to measure the location of the time interval in one period of the reference clock. The other one works like an irregular sampling gate, feeding the previously held result into the thermometer-to-binary encoder. The maximum frequency of this irregular sampling structure is the limit cycle frequency of asynchronous sigma delta modulators. As the quantization is in the time domain, it is easy to realize a high resolution without a high sampling clock.

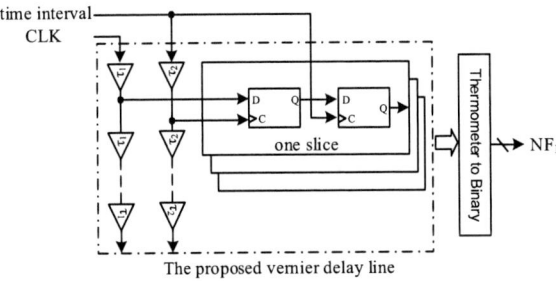

Fig.6 system diagram of the proposed vernier delay line

We rewrite eq. (1):

$$\alpha[n] = \frac{N \cdot NC_1[n] + \left(NF_1[n] - NF_2[n]\right)}{N \cdot \left(NC_1[n] + NC_2[n]\right) + \left(NF_1[n] + NF_1[n+1]\right)} \qquad (13)$$

Where N is the length of the delay chain

One drawback of this solution is that the reconstructed signal is one cycle delayed relative the original signal. This delay time can be minimized by increasing the value of f_c/B.

III. Noise performance

The SNDR of the proposed system is:

$$SNDR \sim \frac{m^2}{f_c B \tau_{res}^2} \qquad (14)$$

Increasing the modulation index will increase the SNDR. However, the equation only holds when the harmonic distortion is smaller than the noise introduced by the time-to-digital converter. For the small f_c/B and large m, the distortion in the asynchronous sigma delta modulator dominates the performance of the system, which is shown in Fig. 7.

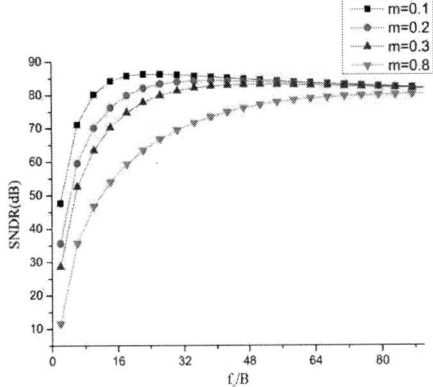

Fig. 7 relationship between f_c/B and SNDR of the whole system (B=3 kHz, $\tau_{res}=10ns$)

The main drawback of the proposed configuration is the lack of noise shaping. In order to apply noise shaping, the proposed configuration can be turned into a closed loop system by the addition of a high resolution digital-to-analog converter in the feedback path, shown in Fig. 8. However, the performance of the noise shaping is decreases with increasing modulation index, as shown in Fig. 9. Given that an ideal DAC was used to establish the feedback, in practice, the performance will be worse.

Evidently, the closed-loop system performs much better when a small signal with a wide bandwidth is chosen, which is common in communications applications. While the open loop system (without the noise shaping) performs better when a low limit cycle frequency and a large input are chosen. Hence, the open loop system is more suitable in ultra-low power applications.

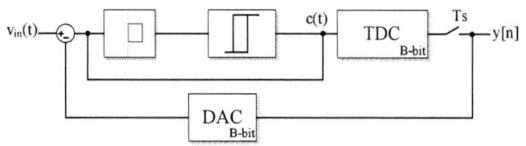

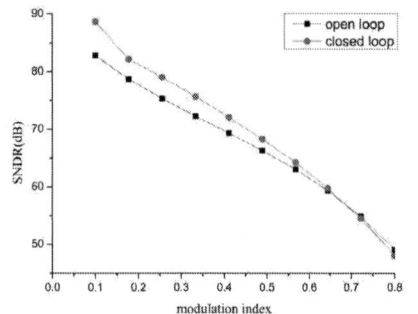

Fig. 8 ASDM with the noise shaping

Fig. 9 relationship between SNDR and modulation index (with the same sample clock)

IV. Design example

An 11-bit coarse-fine time-to-digital converter is designed to decode an ASDM with a 3 kHz signal bandwidth and a limit cycle frequency of 196 kHz. In practice, the asynchronous sigma delta modulator will overload when the modulation index m is over 80%. Hence, according to eq. (9), we can identify the range of the reference clock:

$$2^{N-1} \times \frac{4}{9} f_c < f_{clk} \le 2^N \times \frac{4}{9} f_c \qquad (15)$$

Based on eq. (15), we choose $f_{clk} = 4 f_c$. The maximum time interval for the time-to-digital converter is $10 f_{clk}$. Hence, two 4-bit counters are implemented to establish the coarse measurement. For the fine measurement, vernier delay lines with resolution of 10ns are selected. Each delay line contains 128 delay elements to realize 7-bit digital output. The proposed decoding circuit is designed in the AMS 0.35um CMOS process, and simulations were carried out on Spectre, on the Virtuoso platform of Cadence. The differential non-linearity (DNL) of the vernier delay line is within 0.04 LSB (Fig. 10), and the integral non-linearity (INL) is within 0.06 LSB (Fig. 11). The simulated power consumption is less than 16uW.

V. Conclusion

A new decoding circuit for asynchronous sigma delta modulators is presented; this structure achieves high resolution using a low sampling frequency. The principle of operation of the proposed circuit is to measure the pulse width and the period separately by implementing a special time-to-digital converter, which utilizes vernier delay lines. A design example of an 11-bit resolution converter with low sampling frequency, and less than 16uW from a dual 0.6 V voltage supply I presented.

Needless to add, perhaps, there is plenty of margin for further improvement. For example, the effects of temperature,

process variation and ambient conditions can be reduced by realisation of a delay locked loop at the expense of additionalchip area and power dissipation.

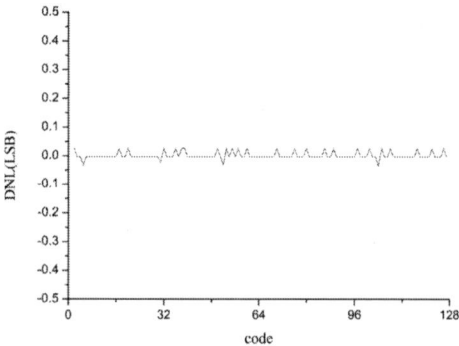

Fig. 10 DNL of the vernier delay line

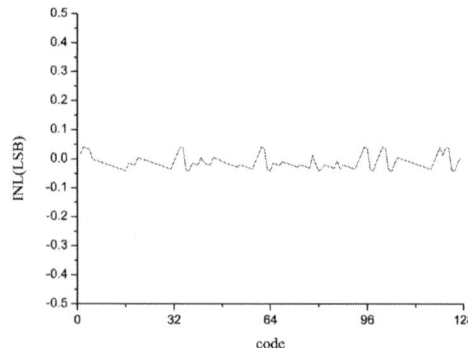

Fig. 11 INL of the vernier delay line

REFERENCE

[1] C. J. Kikkert, and D. J. Miller, "Asynchronous Delta Sigma Modulatotion," Proc. IREE, 1975, 36

[2] S. Ouzounov, E. Roza, "An 8MHz, 72 dB SFDR Asynchronous Sigma-Delta Modulator with 1.5mW power dissipation," VLSI Circuits, June 2004: 88- 91

[3] S. Ouzounov, E. Roza; Hegt, "Analysis and design of high-performance asynchronous sigma-delta Modulators with a binary quantizer," Solid-State Circuits, IEEE Journal, VOL. 41, NO.3, March 2006: 588- 596

[4] S. Ouzounov, H. Hegt, and A. van Rosermund, "Sigma-Delta Modulators Operating at a Limit Cycle," IEEE Transactions on Circuits and Systems-II: Express Briefs, VOL. 53, NO. 5, May 2006: 399-403

[5] E. Roza, "Analog-to-digital Conversion via Duty-Cycle Modulation," IEEE Transactions on Circuits and Systems-II: Analog and digital signal processing, VOL. 44, NO. 11, Nov. 1997: 907-913

[6] L.Hernandez, E. Prefasi, "Analog to digital conversion using a PulseWidth Modulator and an irregular sampling decoder," Proceedings of the13th IEEE International Conference on Electronics, Circuits, and Systems, Dec. 2006: 716 – 719

[7] J. Daniels, W. Dehaene, M. Steyaert, and A. Wiesbauer, "A/D Conversion using an Asynchronous Delta-Sigma Modulator and a Time-to-Digital Converter," IEEE International Symposium on Circuits and Systems, May 2008

[8] P. Dudek, S.Szczepanski, and J. V. Hatfield, "A High-Resolution CMOS Time-to-Digital Converter Utilizing a Vernier Delay Line," IEEE Journal of Solid-State Circuits, VOL. 35, NO. 2, Feb. 2002: 240~247

PRIME 2013, Villach, Austria *Session M3A – Sigma-Delta Modulators*

A Low-Power, Continuous-Time Sigma-Delta Modulator for MEMS Microphones

C. De Berti*, P. Malcovati*, L. Crespi' and A. Baschirotto°

*Dept. of Electrical, Computer, and Biomedical Engineering, University of Pavia, Italy
'Conexant Systems, Newport Beach, CA, USA
°Department of Physics "G. Occhialini", University of Milano Bicocca, Milano, Italy
E-Mail: claudio.deberti01@ateneopv.it, piero.malcovati@unipv.it, lorenzo.crespi@conexant.com,
andrea.baschirotto@unimib.it

Abstract—**This paper presents a continuous-time ΣΔ ADC for MEMS microphone integrated interface applications. The ΣΔ modulator features a 3rd-order loop filter realized with only two operational amplifiers to reduce power consumption. It exploits a 15-level quantizer and 3-MHz sampling frequency to minimize the impact of clock jitter. The proposed architecture can handle an excess loop delay of half period of the sampling clock. The ΣΔ modulator, simulated at transistor-level, performs a 105.5dB A-weighted dynamic range (DR) and a 100.8 dB signal-to-noise-and-distortion ratio (SNDR) at –2 dB$_{FS}$. The device operates from a single 1.8 V and consumes 200 µW.**

I. INTRODUCTION

One of the main aspects in the development of an electronic device, especially for portable applications, is the power consumption. Nowadays, the demand of low-power integrated interface circuits for miniaturized microphones to be utilized in portable devices is greatly growing. Furthermore, micro-electro-mechanical-systems (MEMS) are becoming very popular to realize such microphones with dynamic range higher than 90 dB. An emerging solution for achieving high performance and low-power consumption in microphone interface circuits is represented by Continuous-Time (CT) ΣΔ ADCs. Compared to Switched-Capacitor (SC) implementations (that was typically used in previous large DR interfaces), CT ΣΔ ADCs require lower power consumption, operate at higher sampling rate/bandwidth and feature built-in antialiasing filtering. Moreover, the use of a multibit DAC in the ΣΔ modulator structure allows the use of a low-order loop filter with a moderate out-of-band noise transfer function (NTF) gain and helps in minimizing the impact of clock jitter [1].

This paper proposes a 3rd-order multibit CT ΣΔ ADC architecture for audio applications (20 Hz-20 kHz) with a single supply voltage of 1.8 V and a differential input signal with a 1 V$_{rms}$ peak amplitude. The designed is focused on the power consumption reduction, while achieving the above quality-of-signal performance. To this task, the loop filter is realized with only two operational amplifiers (opamps). The choice of a 3rd-order loop filter and a 15-level quantizer with 3-MHz modulator sampling frequency (F$_s$) can provide more than 120-dB signal-to-quantization-noise ratio (SQNR). The following sections present the most important modulator design issues, including the excess loop delay compensation

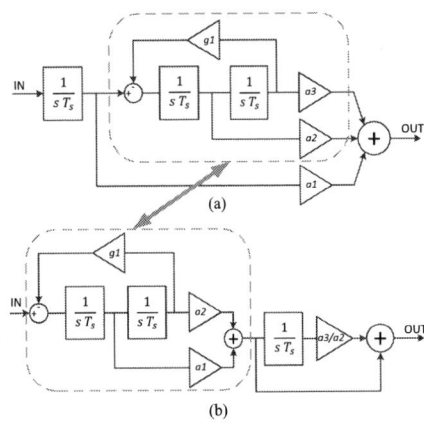

Fig. 1. (a) Traditional CIFF loop filter (b) Proposed loop filter

and the circuit implementation. Finally, the simulated results and evaluations are presented followed by some conclusions.

II. ΣΔ MODULATOR ARCHITECTURE

Typically a 3rd-order loop filter can be implemented with a cascade-of-integrators either in feedback form (CIFB) or feedforward form (CIFF). In both cases three opamps are needed. Moreover, to save power and area in CT ΣΔ modulators, the CIFF structure is more suitable since it requires fewer DACs [2].

CT ΣΔ modulators usually are designed with noise transfer functions (NTF) containing conjugate complex zeros to extend the usable bandwidth. Traditionally, these zeros are implemented with a resonance generated by a local feedback around two integrators of the loop filter, as shown in Fig. 1a.

The same NTF with complex zeros can be obtained with the proposed loop filter structure depicted in Fig. 1b. In this solution, the resonator presents a single output and, therefore, it can be implemented with a single-opamp structure to reduce power consumption [3], [4]. The resonator has the transfer function

$$H_{res}(s) = \frac{a_1 s\, T_s + a_2}{s^2\, T_s^2 + g_1} \qquad (1)$$

978-1-4673-4579-8/13 $31.00 © 2013 IEEE   53

so it has complex conjugate poles at the frequency ω_0, a zero at the frequency ω_z and a DC_{gain} given by

$$\omega_0 = \frac{\sqrt{g_1}}{T_s} \qquad \omega_z = \frac{a_2}{T_s a_1} \qquad DC_{gain} = \frac{a_2}{g_1} \qquad (2)$$

A. Implementation

A differential single-opamp resonator with cross-coupled feedback in RCX configuration [4] is suitable to achieve the required $H_{res}(s)$. The schematic is shown in Fig. 2: a notch filter made with capacitors $C1$-$C2$ and resistors $R1$-$R3$ is placed in the feedback loop of an opamp. With an ideal opamp and under the condition

$$\frac{C_2}{R_3} = \frac{C_1}{R_3} + \frac{C_3}{R_1} \qquad (3)$$

the following equations are obtained

$$\omega_0 = \frac{1}{\sqrt{R_1 R_3 C_1 C_2}} \qquad \omega_z = \frac{1}{C_2 R_3} \qquad DC_{gain} = \frac{R_1}{R_i} \qquad (4)$$

From (2) and (4) the relation between the coefficients of the loop filter and the components of the resonator circuit can be derived.

An Active-RC circuit based on differential opamp provides the final integration operation, with a frequency ω_i at 0-dB gain equal to

$$\omega_i = \frac{a_3}{a_2 T_s} = \frac{1}{R_4 C_4} \qquad (5)$$

The need of an additional summing amplifier can be avoided with the use of capacitive feedforward structures directly at the summing junction of the final integrator opamp [1]. The feedforward gain is given by the ratio $Cf/C4$ (Fig. 2).

B. Excess Loop Delay Compensation

The comparators in the ADC of a CT $\Sigma\Delta$ modulator require some time to take a decision. Therefore, the feedback DAC

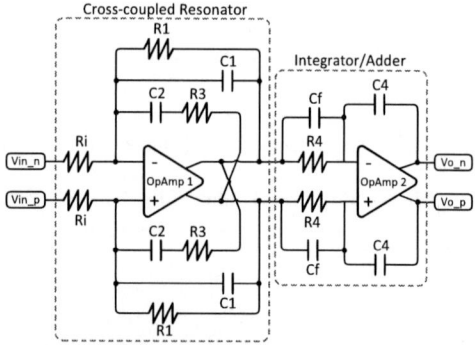

Fig. 2. Loop filter circuit schematic

must be clocked at a time τ_d delayed with respect the sampling instant of the ADC. This amount of time τ_d is called Excess-Loop-Delay (ELD). In practice, it is convenient to force the ELD to be 1/2 or 1 period of the clock (Ts), operation usually obtained by placing a D-flip-flop before the feedback DAC, in

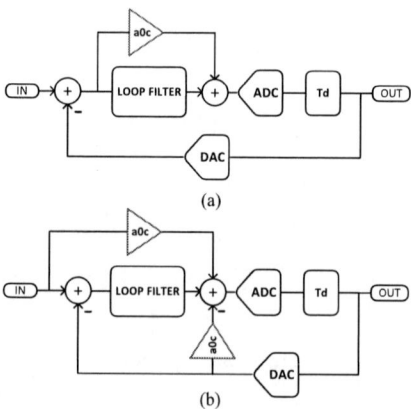

Fig. 3. (a) ELD compensation with feedforward across the loop (b) Equivalent solution scheme

order to make sure that the ELD is well known and constant, independently of process and voltage variations as well as temperature.

The ELD is a problem in CT $\Sigma\Delta$ because it degrades the performance of the loop and even a low amount of delay can result in an unstable structure. In literature several techniques for mitigating the effect of ELD have been proposed. In [5], the author introduces a feedforward path across the loop filter to restore the NTF to the one without ELD, as shown as the direct path in Fig. 3a.

According to the theory demonstrated in [5], the modified coefficients a_{1c}-a_{3c} and the gain a_{0c} of the direct feedforward path across the loop filter needed to restore the NTF are

$$a_{0c} = a_1 \cdot \tau_d + a_2 \cdot \tau_d^2/2 + a_2 \cdot \tau_d^6/6 \qquad (6)$$

$$a_{1c} = a_1 + a_2 \cdot \tau_d + a_2 \cdot \tau_d^2/2$$

$$a_{2c} = a_2 + a_3 \cdot \tau_d$$

$$a_{3c} = a_3 - a_1 g_1$$

The feedforward path across the loop filter cannot be directly implemented in the proposed circuit because the subtraction between the $\Sigma\Delta$ modulator input and the feedback DAC output is done by the resonator itself. Anyway, the same result can be accomplished with a feedforward path from the $\Sigma\Delta$ modulator input and a local feedback, both in the final adder (Fig. 3b).

The resulting circuit for the CT $\Sigma\Delta$ modulator is shown in Fig. 4. A delay equal to $Ts/2$ has been placed before the DAC and the feedforward operations for restoring the desired NTF are done by the gain ratios $Cd/C4$.

C. Noise Analisys

The performance target of the proposed CT $\Sigma\Delta$ modulator is an A-Weighted Dynamic Range larger than 100 dB. Starting from this specification, a maximum input referred noise (integrated between 20 Hz and 20 kHz) equal to 7 µV (−103 dBV) has been assumed. In the proposed CT $\Sigma\Delta$

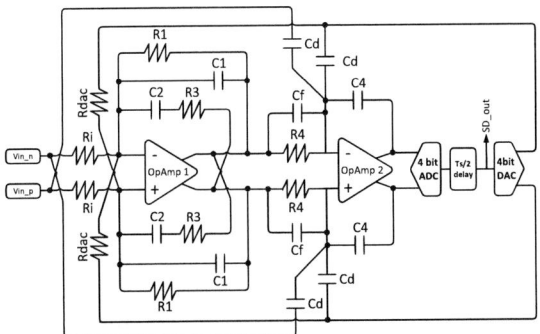

Fig. 4. ΣΔ modulator implementation with ELD

modulator the dominant noise sources referred to the input are the input resistors, the opamp of the resonator and the DAC. Since the DAC is modeled as an ideal voltage source, its noise contribution is given by the thermal noise generated from R_{dac} resistors ($R_{dac}=Ri$). Assuming that the opamp noise accounts for 30% of total noise amount, the value for each input and DAC resistor results 25 kΩ.

D. Circuit componet values

The coefficients of the loop filter were calculated to obtain a NTF with out of band gain (OBG) of 1.6 in order to have a good trade-off between stability and SQNR.

For the cross-coupled resonator, in case of a real opamp, the peak gain at ω_0 is determined by the open loop gain of the opamp at this frequency, because there is no feedback at ω_0. Therefore, the resonator Q factor increases with the open loop gain of the opamp, but decreases when the resonator DC_{gain} is increased (Fig. 5). The loop filter requires a Q greater than 1.5 for exhibiting the benefits of conjugate complex zeros in the NTF. Considering an open loop gain of the opamp at ω_0 lower than 70 dB (low power consumption), the coefficients must be scaled in order to reduce the DC_{gain} and keep the Q factor greater than 1.5. However, the drawback of scaling is the increase of the total capacitor size of the ΣΔ modulator. A good trade-off between area and Q factor is shown in Table 1.

TABLE I. COMPONENT VALUES

Resistors (Ω)					
R_i	R_{dac}	R_1	R_3	R_4	
25k	25k	19.5M	152k	4M	
Capacitors (F)					
C_1	C_2	C_4	C_f	C_d	Total Cap.
5.8p	5.9p	1.1p	1.8p	454f	28p

III. SIMULATION RESULTS

A. Behavioral Simulation

In order to verify the achievable performance with the proposed ΣΔ ADC architecture, simulations in Simulink, including most of the non-idealities (thermal noise, jitter, opamp noise) has been performed.

The clock jitter of the DAC has been modeled taking into account a non-gaussian distribution of the period Jitter (J_{per}).

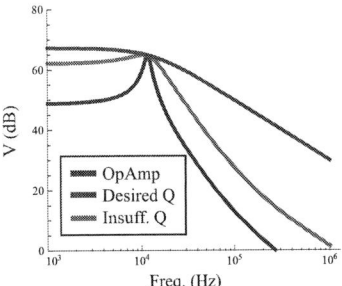

Fig. 5. Resonator Q factors limits with real opamp

Starting from defined Phase Noise profiles (Fig. 6a), "colored" distributions of J_{per} are generated, each with RMS value of 200 pS. If the power spectrum density of the clock signal is closer to the carrier, the jitter is less relevant and the performance are increased, as shown in Fig. 6b. In the proposed CT ΣΔ modulator, the clock signal has a Phase Noise profile described by the curve PN_C. Thanks to the multibit approach in the ΣΔ modulator, behavioral simulations in Simulink (Fig. 8) show that Clock Jitter noise is not dominant

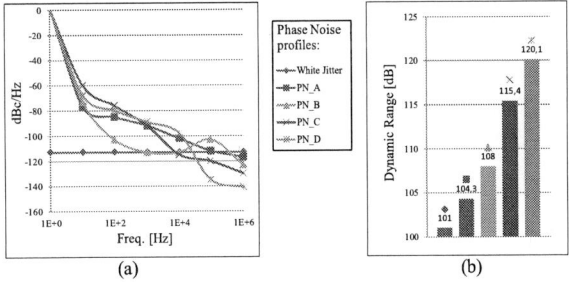

(a)   (b)

Fig. 6. (a) Phase Noise profiles (b) Corresponding Dynamic Range

on thermal noise. In fact the red line in Fig. 8 indicates the total integrated noise which at 20 kHz is around –103 dB, as expected if the thermal noise is dominant.

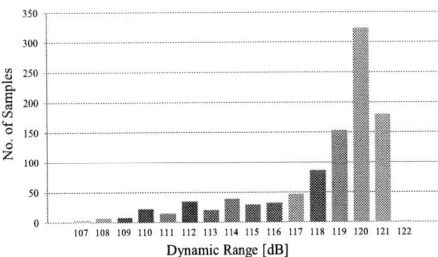

Fig. 7. Distribution of the simulated DR with ±20% capacitor variation, ±20% resistor variation and ±1% mismatch.

Resistor and capacitor variations have also been incorporated in the analysis. The modulator remains stable over ±20% resistor variation and ±20% capacitor variation, considering also ±1% mismatch for capacitors and resistors. The results of a 1000-cycles simulation are shown in Fig. 7. The OBG is inversely proportional at the RC time constant variation. As a result, the SQNR improves with a smaller RC and decrease with larger RC. Anyway, thermal noise still limits the performance of this $\Sigma\Delta$ modulator. This demonstrates that the $\Sigma\Delta$ modulator does not need an additional tuning circuit.

TABLE II.   PERFORMANCE SUMMARY

Parameters	Value
Bandwidth/Clock Rate	20 kHz/3 MHz
Signal-to-noise ratio (SNR)	101.3 dB
Signal-to-noise and distortion ratio (SNDR)	100.8 dB
Spurious-free dynamic range (SFDR)	115 dB
Dynamic-Range (DR)	105.5 dB A-w
Power supply voltage	1.8 V
Input Full Scale (FS)	± 1 V_{rms} differential
OpAmps power consumption	200 μW
Technology	180-nm CMOS

B. Transistor-level Simulation

The proposed $\Sigma\Delta$ modulator has been simulated with both the opamps at transistor-level in a 180-nm CMOS technology (1.8-V supply) and with a behavioral model for ADC and

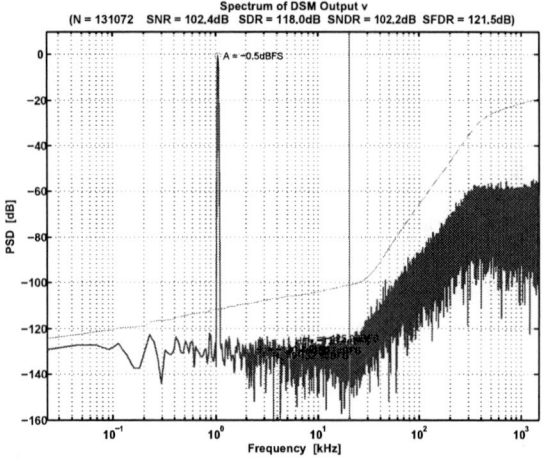

Fig. 8. Output spectrum of the proposed $\Sigma\Delta$ modulator obtained with behavioral simulations.

DAC. A two stage OTA is used in the resonator because its structure requires low output impedance in the opamp (<5kΩ). Simulations indicate that the amplifier has 77-dB gain and 50° phase margin, and consumes 130 μW. Also for the second opamp, a two stage amplifier has been used since the voltage swing at the ADC input is $\pm 1 V_{rms}$ fully-differential. Simulations indicate that the amplifier has 65-dB gain, 62° phase margin, and consumes 70 μW.

Fig. 9 shows the simulation results with an input amplitude $P_{in} = -2 dB_{FS}$ considering the noise generated from transistors and resistors. In the same condition, applying $P_{in} = -60$ dB_{FS},

105.5 dB A-Weighted Dynamic Range is achieved. Table II summarizes the simulated results of the CT $\Sigma\Delta$ ADC.

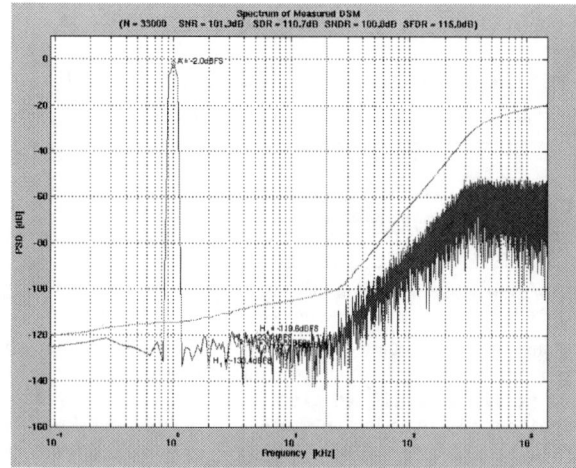

Fig. 9. Spectrum of the proposed $\Sigma\Delta$ modulator output signal, applying at the input a 1-kHz, -2.0 dB_{FS} sinusoidal signal

IV.   CONCLUSION

In this paper, a low-power CT $\Sigma\Delta$ modulator has been proposed. A third-order loop filter can be achieved with the use of only two opamps. Moreover, the NTF presents conjugate complex zeros to extend the usable bandwidth. An ELD compensation of $Ts/2$ has also been implemented.

Behavioral simulations, taking into account non-gaussian distribution of the Clock Jitter, show that the performance is mainly limited by thermal noise, thanks to the multibit approach. Simulations of the implemented architecture, with opamps at transistor-level, indicate high performance (DR = 105.5 dB A-w) with low power consumption (200 μW).

Thanks to the results achieved, the architecture will be used as building block in an interface circuit for MEMS microphones.

REFERENCES

[1] X. Jiang, J. Song, J. Chen, V. Chandrasekar, S. Galal, F. Y. L. Cheung, D. Cheung, and T. L. Brooks, "A Low-Power, High-Fidelity Stereo Audio Codec in 0.13 μm CMOS," IEEE Journal of Solid-State Circuits, vol. 47, No. 5, May. 2012.

[2] S. Pavan, N. Krishnapura, R. Pandarinathan, and P. Sankar, "A Power Optimized Continuous-Time $\Sigma\Delta$-ADC for Audio Applications," IEEE Journal of Solid-State Circuits, vol. 43, No. 2, Feb. 2008.

[3] H. Chae, J. Jeong, G. Manganaro, and M. Flynn, "A 12mW Low-Power Continuous-Time Bandpass $\Delta\Sigma$ Modulator with 58dB SNDR and 24MHz Bandwidth at 200MHz IF," ISSCC 2012, Session 8, Delta-Sigma Converters.

[4] S. Zeller, C. Muenker and R. Weigel, "X-Coupled Differential Single-Opamp Resonator for Low Power Continuous Time $\Sigma\Delta$-ADCs," Ph.D. Research in Microelectronics and Electronics (PRIME), 2011 7th Conference on.

[5] S. Pavan, "Excess Loop Delay Compensation in Continuous-Time Delta-Sigma Modulators," IEEE Transactions on Circuits an Systems II-Express Briefs, Vol. 55, No. 11, Nov. 2008.

System-Level Power Optimization for a $\Sigma\Delta$ D/A Converter for Hearing-Aid Application

Peter Pracný
Department of Electrical
Engineering
Technical University of Denmark
Kgs. Lyngby, Denmark
pp@elektro.dtu.dk

Ivan H. H. Jørgensen
Department of Electrical
Engineering
Technical University of Denmark
Kgs. Lyngby, Denmark

Erik Bruun
Department of Electrical
Engineering
Technical University of Denmark
Kgs. Lyngby, Denmark

Abstract— This paper deals with a system-level optimization of a back-end of audio signal processing chain for hearing-aids, including a sigma-delta modulator digital-to-analog converter (DAC) and a Class D power amplifier. Compared to other state-of-the-art designs dealing with sigma-delta modulator design for audio applications we take the maximum gain of the modulator noise transfer function (NTF) as a design parameter. By increasing the maximum NTF gain the cutoff frequency of modulator loop filter is increased which lowers the in-band quantization noise but also lowers the maximum stable amplitude (MSA). This work presents an optimal compromise between these. Increased maximum NTF gain combined with a multi bit quantizer in the modulator allows lower oversampling ratio (OSR) and results in considerable power savings while the audio quality is kept unchanged. The proposed optimization impacts the entire hearing-aid audio back-end system resulting in less hardware and power consumption in the interpolation filter, in the sigma-delta modulator and reduced switching rate of the Class D output stage.

Keywords—Sigma-Delta modulator; Interpolation filter; Class D; Hearing aid; low voltage, low power

I. INTRODUCTION

High audio quality, longer operation time and small device size are parameters demanded in hearing-aids today. Optimum balance between the design parameters in every part of a hearing-aid device is therefore of vital importance, making the power consumption one of the crucial parameters for the design. This is also the case of the audio signal processing path, which requires digital-to-analog conversion and power amplification at the back-end to drive the speaker (see Fig.1). As part of the digital-to-analog conversion a digital sigma-delta ($\Sigma\Delta$) modulator with Class D output stage is usually used in low-voltage low-power audio applications. This eliminates problems with device matching and reduced power efficiency experienced in case Class AB output stage is used [1, 2, 3]. The Class D output stage is usually implemented as an H-bridge (schematic in Fig.1 is simplified) and operates in switched mode. Compared to [1, 2, 3] that use Class AB power stage the Class D allows to perform all signal processing before the output filter in digital domain. Digital design provides the advantage of low-voltage low-power and cost effective implementation and scales down with integrated circuit (IC) technologies of today.

Due to the oversampling nature of the $\Sigma\Delta$ modulator an interpolation filter is needed prior to the modulator. When using a multi-bit $\Sigma\Delta$ modulator, digital pulse width modulation (DPWM) block that turns the $\Sigma\Delta$ signal into symmetrical 1 bit pulse width modulation, is needed.

This paper deals with the power optimization of the system in Fig. 1. Section II provides the design specifications for the $\Sigma\Delta$ modulator. In Section III, optimization approach is proposed. In Section IV $\Sigma\Delta$ modulator designs are compared as an example of the optimization approach. Finally, Section V concludes this work.

II. DESIGN AND FIGURE-OF-MERIT SPECIFICATIONS

A thorough discussion on hearing-aid audio back-end system specification and the $\Sigma\Delta$ modulator is provided in [4]. We assume ideal 16 bit quantization of the system input signal that has band-width (BW) of 10 kHz. This results in signal-to-quantization-noise ratio (SQNR) = 98 dB. The sampling frequency at the system input is fs_{in} = 22.05 kHz. The input signal of the back-end is then up-sampled using an

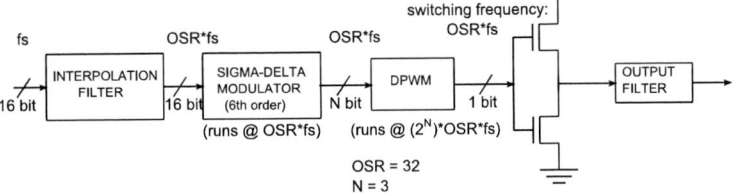

Figure 1. Simplified schematic of the back-end of audio signal processing chain: interpolation filter, $\Sigma\Delta$ modulator, Class-D output-stage and output filter.

interpolation filter [5] and passed to the ΣΔ modulator. The interpolation filter in state-of-the-art designs [1 - 3, 5 - 7] consists of multiple stages. Another requirement is the signal-to-noise-and-distortion ratio (SNDR) at the total output of the back-end of 90 dB. We designed the interpolation filter and the ΣΔ modulator to keep the quality of the audio signal at SNDR = 98 dB so that a margin of 8 dB is left for the performance reduction introduced by the output stage. MSA is also a crucial parameter, the lowest limit is -1.2 dBFS.

Note that we are dealing with a digital ΣΔ modulator in this work and we treat it as a digital filter. This allows us to judge the complexity and power savings using the FOM:

$$FOM = \sum_i b_i . OSR_i \qquad (1)$$

Where i is the number of adders in the ΣΔ modulator block, b_i is the number of bits used in individual adders and OSR_i is the oversampling used for the individual adders. In the case of the ΣΔ modulator block OSR_i is the same for all the adders. There are more precise figures of merit for sigma-delta modulators used in other works [2, 8]. However, these figures of merit can be used only after the design has been completed and possibly measured. The advantage of the figure of merit of Eq. 1 is that it allows us to compare different designs to each other early in the design process.

III. DESIGN OPTIMIZATION APPROACH

In this work we want to optimize the back-end of the audio signal processing chain in Fig. 1 [9] at system level with respect to power. With the Class D output stage being the main power consumer in the system due to the resistance in the output transistors, we aim to reduce its switching frequency. The switching frequency of the Class D stage is the same as the operating frequency of the ΣΔ modulator (see Fig. 1). Thus keeping the OSR of the ΣΔ modulator low helps to lower the power consumption of the Class D stage as well. We were not able to use the optimization approach of [9] where we trade higher modulator order for lower OSR while keeping the SQNR. With high modulator order (6th order, see Fig. 1) this increases the order even further. We tried to design a 12th order modulator with OSR=16 but experienced stability problems. To have a stable modulator with such high order it is needed to have high precision coefficients and integrator adders which results in worse modulator FOM. Such approach leads us away from optimum design. Thus the idea behind further optimization of the ΣΔ modulator and the entire back-end is to keep the modulator order, decrease the modulator OSR and increase the number of bits in its quantizer. To have lower power consumption in the Class-D output stage and have more bits in the quantizer of the ΣΔ modulator is reasonable tradeoff

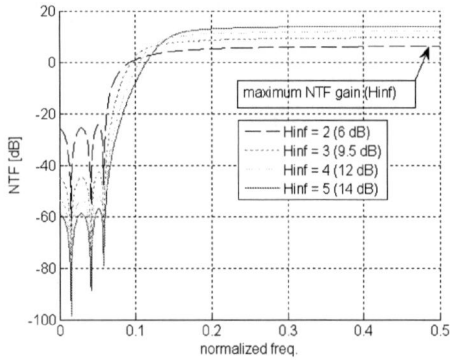

Figure 2. NTF of 6th order ΣΔ modulator with OSR = 8 and 5 bit quantizer. Maximum NTF gain H_{inf} as a parameter.

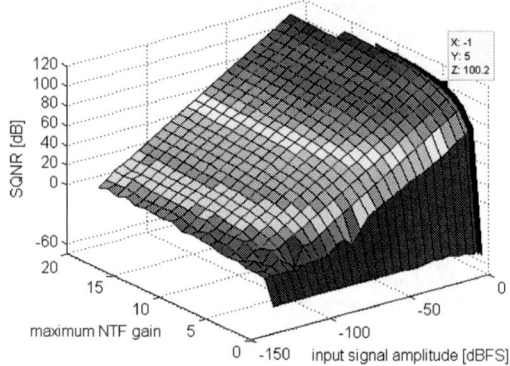

Figure 4. SQNR of the ΣΔ modulator output signal as a function of modulator input signal amplitude and max. NTF gain H_{inf}.

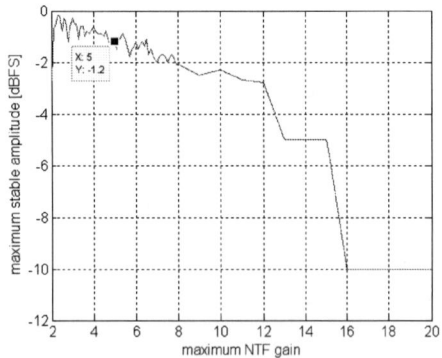

Figure 3. Maximum stable amplitude at ΣΔ modulator input as a function of max. NTF gain.

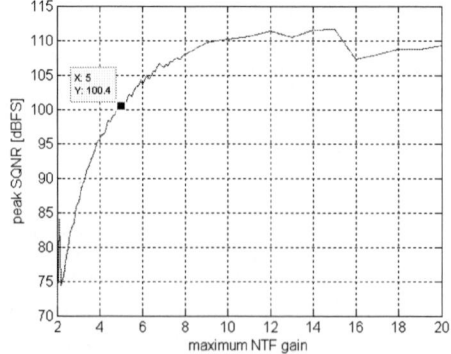

Figure 5. peak SQNR of the ΣΔ modulator output signal as a function of max. NTF gain.

since the $\Sigma\Delta$ modulator is completely digital and thus scales with technology. The same cannot be said about the Class-D output stage. In order not to increase the maximum system clock available given by the DPWM block (see Fig. 1), and at the same time decrease the OSR and keep the modulator at 6th order, combination of OSR = 8 and 5 bit quantizer is needed. However, 6th order modulator with OSR = 8 and 5 bit quantizer does not provide necessary peak SQNR = 98 dB at the output of the modulator, if maximum NTF gain H_{inf} = 1.5 is used, as recommended in [8]. As can be seen from the NTF plots in Fig.2, increase of H_{inf} above 1.5 pushes the cutoff frequency of the NTF up. This results in less in-band quantization noise and potentially gives better SQNR. At the same time increase of H_{inf} reduces the MSA which potentially gives worse SQNR (see Fig.3). These two effects contradict each other and need to be further investigated. Fig.4 and Fig.5 show that increase of H_{inf} above 5 allows us to reach peak-SQNR = 100 dB at the output of the modulator at maximum stable input amplitude (MSA) = -1.2 dBFS (Fig.3). Moreover, Fig.4 and Fig.5 show that further increase of H_{inf} reduces the in-band noise at the same rate as the MSA is reduced and results in a wide range where the SQNR is constant. Thus the highest H_{inf} is decided by the point where MSA reaches the limit of -1.2 dBFS (see Fig.3). Therefore our choice of H_{inf} = 5 is optimal for combination of $\Sigma\Delta$ modulator parameters of 6th order, OSR = 8 and 5 bit quantizer.

Performing the changes mentioned above allows us to reduce the operating frequency of the $\Sigma\Delta$ modulator and thus switching frequency of the Class D output stage by 87.5% compared to [4] and by 75% compared to the design of Fig. 1. This will result in considerable power savings. Moreover these changes will have a positive impact on the interpolation filter too as oversampling by 8 only is needed compared to oversampling by 64 in [3, 4, 6, 7] and by 32 in Fig. 1. This saves several stages in the interpolation filter operating at high frequency. Using the FOM of Eq.1 for interpolation filter of [4] and [9] we calculate FOM = 118 and FOM = 83 respectively. After the reduction of OSR down to 8 the FOM of the interpolation filter is 58. This is improvement of hardware/power saving by 49% in the interpolation filter compared to [4] and by 30% compared to [9]. With the maximum clock frequency of the DPWM block the same as in Fig. 1, and with power savings in the interpolation filter and in the Class D output stage, the only block of the back-end system that remains to be investigated to see whether or not this optimization approach is power efficient is the $\Sigma\Delta$ modulator. We discuss this in the next section.

IV. $\Sigma\Delta$ MODULATOR DESIGN AND COMPARISON

The modulator in this work is 6th order with OSR = 8, 5 bit quantizer and maximum NTF gain = 5. A model using fixed-point arithmetic was built and simulated in Matlab. The list of coefficients used for the modulator in current design can be seen in Tab. I. The FFT of the $\Sigma\Delta$ modulator fixed-point model's output signal can be seen in Fig. 6. A cascade of resonators with feedback (CRFB) $\Sigma\Delta$ modulator structure is used (see Fig. 7).

TABLE I. $\Sigma\Delta$ MODULATOR CURRENT DESIGN - COEFFICIENT LIST.

Coeff.	Value	Shift/Add	Adders
a_1	1/8	2^{-3}	0
a_2	0.1718	$2^{-3}+2^{-5}+2^{-7}+2^{-8}$	3
a_3	0.2243	$2^{-2}-2^{-5}+2^{-8}$	2
a_4	0.1604	$2^{-3}+2^{-5}+2^{-8}$	2
a_5	0.4992	2^{-1}	0
a_6	0.1203	$2^{-3}-2^{-7}+2^{-8}$	1
b_1	1/8	2^{-3}	0
c_1	1/4	2^{-2}	0
c_2	1/2	2^{-1}	0
c_3	1/2	2^{-1}	0
c_4	2	2^{1}	0
c_5	1/2	2^{-1}	0
c_6	8	2^{3}	0
g_1	0.0351	$2^{-5}+2^{-8}-2^{-11}$	2
g_2	0.1341	$2^{-3}+2^{-7}$	1
g_3	0.2652	$2^{-2}+2^{-7}+2^{-8}$	2

The fixed-point arithmetic model performs digital operations exactly as a VHLD design does. Thus the fixed-point arithmetic model can be directly used to judge the complexity of the $\Sigma\Delta$ modulator. Taking the Matlab fixed-point models and calculating the FOM according to Eq.1 gives data and FOM in Tab. II, clearly showing better (lower) FOM compared to the design of [4] and of Fig. 1 [9]. Expressing the current consumption of the back-end as sum of the currents needed in individual blocks we write:

$$I_{total} = I_{int} + I_{SDM} + I_{DPWM} + I_{dr} \qquad (2)$$

Where I_{int} is the current needed in the interpolation filter (see Fig. 1), I_{SDM} is the current of the $\Sigma\Delta$ modulator, I_{DPWM} is the current of the DPWM block and I_{dr} is the current of the

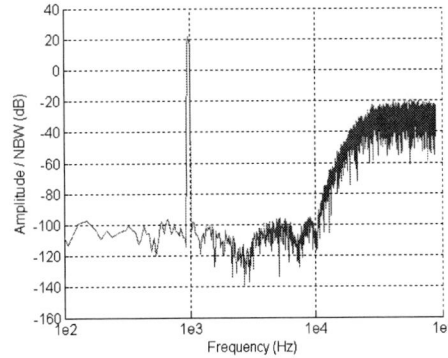

Figure 6. FFT spectrum of the $\Sigma\Delta$ modulators output signal. For the FFT Hann window was used. The FFT is 8192 points (NBW = 1.8311e-04)

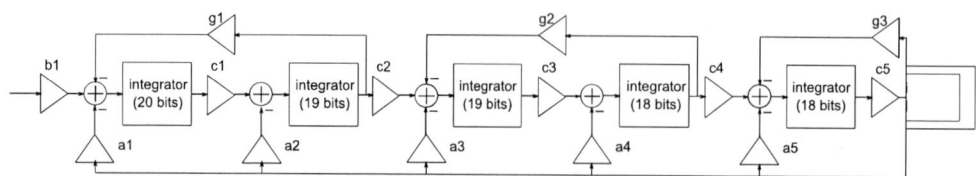

Figure 7. CRFB architecture of 6th order $\Sigma\Delta$ modulator with OSR = 8 and 5 bit quantizer.

Class D driver (power amplifier). In Section III we explained that using the proposed optimization I_{int} will be lowered by 30%, I_{dr} will be lowered by 75% and I_{DPWM} will remain the same compared to Fig. 1 [9]. Table II shows that I_{SDM} will be lowered by 60%. Thus in total there are considerable power savings achieved by the proposed optimization approach.

Table III. shows a comparison with other audio DAC designs for low-voltage low power applications. Exact comparison can not be performed as the FOM used in the reference works requires finished design. Moreover [1, 2, 3] use Class-AB power stage and require analog $\Sigma\Delta$ modulator which further complicates comparison at early design stage. Nevertheless trends of the low-voltage low power audio back-end designs can be seen in Table III. We note that one of the trends is to target SNDR = 90 dB [4, 6, 7] at the total output of the system. What most of the $\Sigma\Delta$ modulator reference designs have in common is the choice of system-level parameters of 3rd order and OSR around 64 with 3 bit quantizer [2, 3, 4, 6]. In case 1 bit quantizer is used a tradeoff is made and order of the modulator is increased from 3 to 4 to achieve the same audio quality [2, 7].

TABLE II. $\Sigma\Delta$ MODULATOR COMPARISON WITH THE DESIGN OF [4] AND [9].

	Order	Bit	OSR	H_{inf}	Adders	Pk. SQNR [dB] ideal	Pk. SQNR [dB] quantized	FOM
[4]	3	3	64	1.5	12	106	98	193
[9]	6	3	32	1.5	22	105	98	192
This work	6	5	8	5	29	100	98	77

TABLE III.   SYSTEM COMPARISON.

Design	Analog/ Digital	Power Stage	BW [kHz]	OSR	Order	Bit	SNDR [dB]
[1]	Analog	Class AB	24	128	3	3	69
[3]	Analog	Class AB	20	64	3	3	82
[2]	Analog	Class AB	20	50	4	1	73
[4]	Digital	Class D	10	64	3	3	Target is 90
[6]	Digital	Class D	20	64	3	3	90
[7]	Digital	Class D	10	64	4	1	85

We note that a lower OSR directly reduces the operating frequency of the $\Sigma\Delta$ modulator, simplifies the interpolation filter and reduces the switching frequency of the Class D power amplifier. Thus designs with lower OSR, such as proposed in

this work, clearly consume less power. If, at the same time, the audio quality is kept unchanged the design is more efficient and has lower power consumption in total.

V.   CONCLUSION

In this work we optimized the back-end path of the audio signal processing path with respect to power consumption. Lower OSR directly reduces the operating frequency of the $\Sigma\Delta$ modulator, simplifies the interpolation filter and reduces the switching frequency of the Class D power amplifier. If, at the same time, the audio quality is kept unchanged, the audio back-end is more efficient and clearly consumes less power. We trade lower OSR of the $\Sigma\Delta$ modulator for higher number of bits in its quantizer and higher maximum gain of the modulator NTF. Overall the power consumption of the entire back-end system is considerably reduced showing that trading lower OSR for higher number of bits in the quantizer and higher maximal NTF gain is an approach to be considered in low-voltage, low-power portable audio applications.

REFERENCES

[1] K. Lee, Q. Meng, T. Sugimoto, K. Hamashita, K. Takasuka, S. Takeuchi, U. Moon, G. C. Temes, "A 0.8 V, 2.6 mW, 88 dB Dual-Channel Audio Delta-Sigma D/A Converter With Headphone Driver" IEEE Journal of Solid-State Circuits, Vol. 44, No. 3, Mar. 2009.

[2] J. Roh, S. Byun, Y. Choi, H. Roh, Y. Kim, J. Kwon, "A 0.9-V 60-uW 1-Bit Fourth-Order Delta-Sigma Modulator with 83-dB Dynamic Range" IEEE Journal of Solid-State Circuits, Vol. 43, No. 2, Feb. 2008.

[3] K. Wong, K. Lei, S. U, R. P. Martins, "A 1-V 90dB Audio Stereo DAC with embedding Headphone Driver" IEEE Asia Pacific Conference Circuits snd Systems (APCCAS), Dec. 2008.

[4] P. Pracný, E. Bruun, "$\Sigma\Delta$ Modulator System-Level Considerations for Hearing-Aid Audio Class-D Output Stage Application," Proc. 2012 8th Conf. on Ph.D. Research in Microelectronics and Electronics (PRIME), pp. 103-106, Aachen, Jun. 2012.

[5] P. Pracný, P. M. Llimós, E. Bruun, "Interpolation filter design for hearing-aid audio class-D output stage application," Proc. 19th IEEE Int. Conf. on Electronics, Circuits, and Systems, Seville, Spain, Dec. 2012.

[6] T. Forzley, R. Mason, "A Scalable Class D Audio Amplifier for Low Power Applications," Proc. AES 37th International Conference, Aug. 2009.

[7] X. Yao, L. Liu, D. Li, L. Chen, Z. Wang "A 90dB DR Audio Delta-Sigma DAC with Headphone Driver for Hearing Aid" 3rd International Congress on Image and Signal Processing (CISP), 2010.

[8] R. Schreier, G. C. Temes, "Understanding Delta-Sigma Data Converters," Wiley-IEEE Press, 2005.

[9] P. Pracný, I. H. H. Jørgensen, E. Bruun, "System-Level Optimization of a DAC for Hearing-Aid Audio Class D Output Stage" 4th Doctoral Conference on Computing, Electrical and Industrial Systems (DoCEIS), Feb. 2013.

[10] R. Mehboob, S. A. Khan, R. Quamar, "FIR filter design methodology for hardware optimized implementation" IEEE Trans. Consumer Electronics, 2009, 55, (3), pp. 1669-1673.

978-1-4673-4579-8/13 $31.00 © 2013 IEEE

CMOS Current Amplifier for AFM Impedance Sensing on Chip with ZeptoFarad Resolution

Davide Bianchi, Marco Carminati, Giorgio Ferrari, Marco Sampietro

Dipartimento di Elettronica, Informazione e Bioingegneria
Politecnico di Milano
Milano, Italy
{bianchi, carminati, ferrari, sampietr}@elet.polimi.it

Abstract—**A new sensing configuration based on the direct use of the metal input pad of a dedicated CMOS amplifier as a conductive substrate electrode for nanoscale high-performance electrical measurement with a standard conductive AFM is proposed and experimentally validated. Combining the state-of-the-art performance of an integrated current preamplifier (5MHz bandwidth, 14fA/sqrt(Hz) noise at 1MHz) with pad and setup optimization, it has been possible to straightforwardly achieve 14zF total capacitive rms resolution (0.5V applied at 1MHz with 1s averaging time) that corresponds to 75pm resolution in tracking the piezo vertical movement (allowing the detection of 6nm steps of the cantilever with a SNR of ~40dB).**

Keywords—nanoscale; atomic force microscope; low-noise; impedance.

I. INTRODUCTION

The Atomic Force Microscope (AFM) has become in the last decades a consolidated tool in several nanotechnological fields, spanning from material science to biology [1]. Thanks to the widespread diffusion of this instruments (in particular of affordable entry-level models), it can be observed that, in parallel to cutting-edge research efforts devoted to pushing the limits of performance (atomic resolution [2], fast imaging rates [3], etc...), the AFM has been penetrating in several laboratories (in particular biomedical ones) as a very sophisticated but nowadays "routinary" sample characterization tool, despite the expertise still required to properly prepare samples, interpret data and avoid artifacts.

Beyond its main use for nanoscale surface topographic imaging and along with nano-mechanical probing, when mounting a conductive tip, the AFM provides researchers with a very versatile and spatially resolved tool for electrical sensing. In fact, by placing the sample on top of a conductive substrate, the conductive tip can then be used as a movable nanoelectrode that can locally probe the electrical properties of the sample area underneath the tip (either forcing or reading a voltage/current signal with respect to the substrate) for organic [4] and biological samples [5]. Thus, the majority of commercial AFMs comprise an internal module for electrical measurements, but the offered performance are generally poor in terms of bandwidth (<kHz, mainly for static I/V curves) and current resolution.

Here we present an ultra-low-noise dedicated CMOS current preamplifier whose input pad acts as the conductive substrate to be used for electrical probing, while minimizing the input stray capacitance and the capacitive coupling with the tip holder. Although more delicate than standard passive electrodes (in terms of max. current range, EMI sensitivity and sample preparation/deposition), this active substrate aims at exploring the ultimate detection limits of this technique, without requiring any modification of the commercial AFM instrument. Beyond faster temporal resolution, a wide bandwidth allows performing impedance spectroscopy and thus inferring information about the local material composition, for instance through the extraction of the local dielectric constant [5].

II. AMPLIFIER DESIGN

A. Working Principle

The particular current preamplifier topology [6] has been chosen in order to satisfy at the same time low-noise and wide bandwidth requirements, linear and bipolar dynamic and the possibility of DC current handling. The scheme is reported in Fig. 1: the current to be read (I_{IN}) is forced to flow into a device F placed into the feedback loop of an OpAmp (OA1), producing a voltage drop proportional to the current to be sensed. This same voltage is then applied across the parallel of N replicas of the same device, obtaining the input current amplified by a factor N.

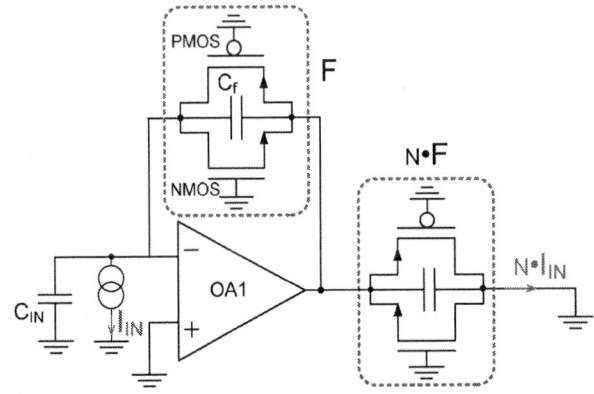

Fig. 1. Topology of the integrated current amplification stage.

Partial financial support from Fondazione Cariplo and EU under project BOND (grant no. 228685-2) is gratefully acknowledged.

Paper M4A1   *PRIME 2013, Villach, Austria*

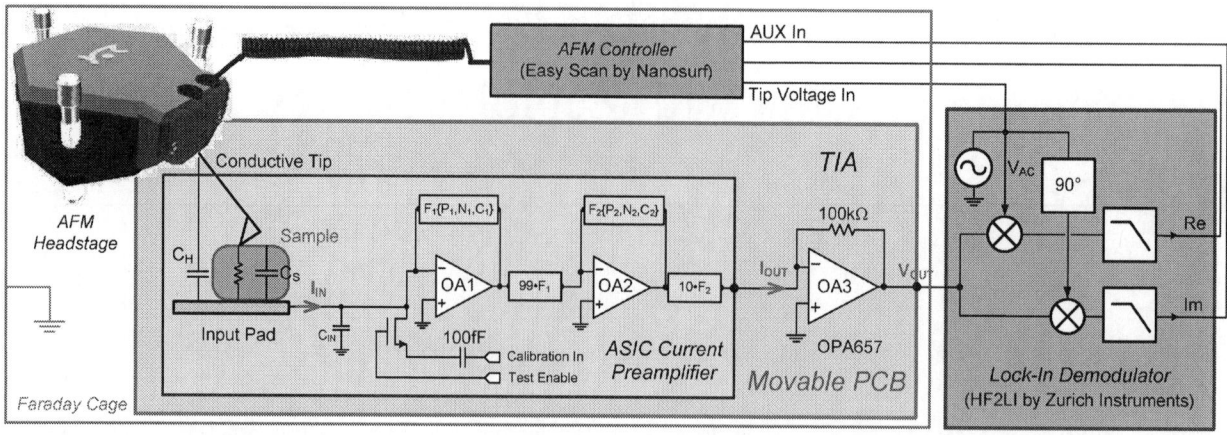

Fig. 2. Scheme of the nanoscale impedance sensing archiecture based on integrated current preamplifier whose input pad serves as a metallic sample substrate coupled with a conductive AFM.

Since the gain is only due to the geometric ratio between the feedback and the output elements, a nonlinear device can be chosen and implemented, thus allowing the other circuit requirements to be met, in particular those concerning noise. The implemented device consists of the parallel of an NMOS and a PMOS in common gate configuration: their sources are driven by the amplifier, thus allowing either a positive (through the NMOS) or a negative (through the PMOS) input current, either a DC or high frequency input current to be accepted. The transistors are always working in sub-threshold regime, thus contributing with low flicker noise (frequency corner of only few Hz) and a shot noise term that scales adaptively with the input current. Since the stability of the system must be granted independently of the MOS biasing, a capacitor has been put in parallel to the transistors, bypassing them and defining at high frequency a loop gain independent on the input DC current. The circuit bandwidth can be then calculated as $GBWP \cdot C_F/(C_F+C_{IN})$, where GBWP is the gain-bandwidth product of the OpAmp, and C_{IN} is the capacitance connected to the input node. Matching between the feedback device and the N replicas is a critical issue for ensuring a linear transfer function: all the elements must have the same nonlinear characteristic. Since the current is flowing into three different paths (PMOS for negative, NMOS for positive, capacitor for high-frequency current), transistors have been sized in order to ensure the same geometric ratio and matching of the capacitor path. Other sizing issues concerning stability (low parasitic capacitances, low channel conductivity) and noise (sub-threshold regime is preferred): very high channel length ($L \geq 7\mu m$), and low aspect ratio ($W/L \leq 3$) have been chosen for both PMOS and NMOS.

B. Implementation

The ASIC, implemented in AMS 0.35μm technology, has been designed paying particular attention to the reduction of the parasitic input capacitance, key parameter for the circuit noise performance. The input pad, 50μm x 50μm wide, has been realized using a single top metal layer: this solution not only allows to strongly reduce its parasitic capacitance, but it

also avoids to put contact vias underneath it, thus minimizing the surface roughness.

The most relevant circuit noise sources are the amplifier input transistors: their contributions are transferred to the output proportionally to the value of the total input capacitance C_{IN}. The noise optimization cannot be simply performed by increasing the input transconductance, discarding the consequent amplifier input capacitance increase, thus C_{IN} increase. The optimum solution can be found sizing the input transistor in order to have the same gate capacitance as the parasitic input capacitance due to pad, stray and feedback capacitance; in this implementation its value is close to 250fF.

The implemented ASIC (Fig. 2) is composed of two amplification stages of gain equal to 99 and 10 respectively; their MOS and capacitor sizing is different because of different dynamic and noise requirements: C_1= 100fF, $W_{P1} = 20\mu m$, $L_{P1} = 20\mu m$, $W_{N1} = 20\mu m$, $L_{N1} = 7\mu m$, $C_2 = 1pF$, $W_{P2} = 15\mu m$, $L_{P2} = 60\mu m$, $W_{N2} = 5\mu m$, $L_{N2} = 60\mu m$. As circuit calibrator, a 100fF capacitor has been connected to the input; an externally controlled switch allows to disconnect it while measuring, thus avoiding to increase the parasitic input capacitance. The ASIC layout has been studied in order to reduce the coupling between the input and the output (because of stability issues). The ASIC pads have been put as far as possible from the input pad (1.3mm in this implementation) to leave space between the bonding and the input pads for functionalization and sample deposition during experiments.

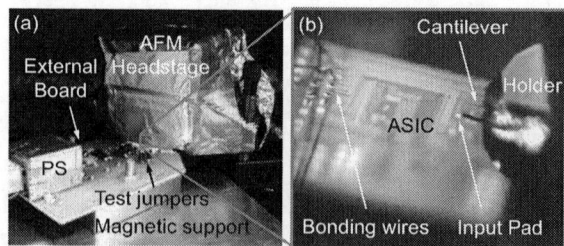

Fig. 3. Full setup image (a) and zoom (b) of the AFM tip approaching the ASIC, showing the separted input pad distant from the other I/O pads.

978-1-4673-4579-8/13 $31.00 © 2013 IEEE   62

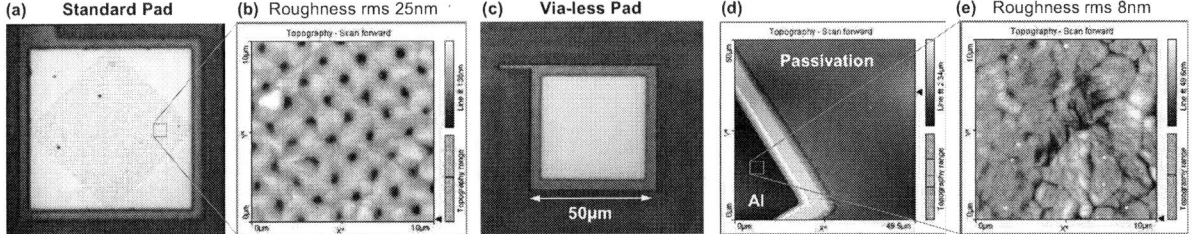

Fig. 4. Characterization of custom pad roughness: 25nm for a standard CMOS pad with vias (a-b) vs. 8nm for the dedicated via-less pad (c-e) used in the tests.

The readout system is completed with an external transimpedance amplifier (TIA) on the PCB with a feedback resistor of 100kΩ. Its noise is negligible, thanks to the ASIC current amplification of 990.

III. SYSTEM AND SETUP REALIZATION

The assembled system is shown in Fig. 3. The PCB is mounted on two magnetic supports that allow manual positioning of the assembly on the ant-vibration table in the same way of the passive magnetic sample holders. The board hosts the ASIC, the external TIA and filters. The ASIC is glued at the free end of the PCB to be placed under the AFM and it is wire-bonded directly to the copper tracks of the PCB (avoiding chip packaging to reduce parasitics). On the opposite side, a shielded RJ45 connector (PS) brings regulated power supplies (±5V and ±1.5V) to the circuit, while the output analog signal is available through a separate LEMO connector. Two jumpers activate the connection of the test capacitance and its excitation through an external signal generator. The active substrate system, the damped table and the AFM headstage are inserted in a grounded Faraday cage, fundamental for interferences shielding. The AFM headstage is shielded as well and the internal EMI produced by the switching power supply represent the major source of interferences.

IV. EXPERIMENTAL RESULTS AND DISCUSSION

A. Pad Characterization

Tapping mode imaging (CDT-NCLR diamond-coated conductive tips, force constant ~100N/m, resonance frequency ~200kHz, set-point 50%) has been used for the characterization of the pad surface. As expected and reported in Fig. 4, the removal of the vias from the pad has a beneficial effect, increasing the smoothness of the Al surface. The rms roughness decreases from 25nm (standard pad, Fig. 4a-b) to 8nm (smooth pad, Fig. 4c-e). Although encouraging and acceptable in some applications, this value is still quite high if compared with the sub-nm flatness of graphite substrates. Further improvement can be envisioned for instance by means of post-CMOS processing for the deposition of an additional thick and smoother metal layer (preferably Au).

B. Amplifier Performance

Thanks to the presence of the integrated 100fF test capacitance, it possible to test the circuit performance and calibrate its transfer function, even in the absence of the wire bonding connection of the input pad. The measured transfer function (with Agilent E5061B-3L5) is shown if Fig. 5a. The total (ASIC + external TIA) gain is 90MΩ (10% smaller than the nominal value to due component tolerances) and the -3dB bandwidth is 5MHz, corresponding to the expected bandwidth of the integrated amplifier (see simulation). The external TIA adds another singularity at 8MHz.

Fig. 5b reports the measured input-referred current noise spectrum, highlighting the excellent performance achievable with this topology. The high frequency noise raising the with frequency is fitted by a 4nV/sqrt(Hz) input equivalent voltage generator of OA1 acting on a total input capacitance ~600fF that is consistent with the value of $C_{IN} = C_1 + pad + OA1$ input capacitance. With a zero DC input signal, the shot noise of the transistor is absent and the only white noise term is due to the

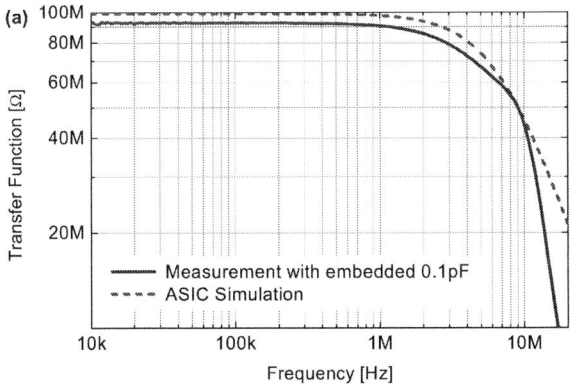

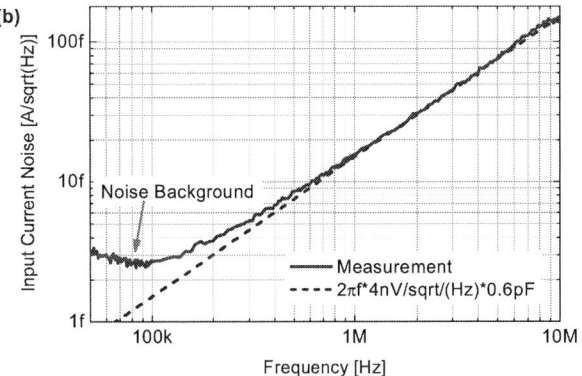

Fig. 5. Transfer function (a) and input-referred current noise spectrum (b) of the realized preamplifier offering 5MHz bandwidth and sub-fA current resolution.

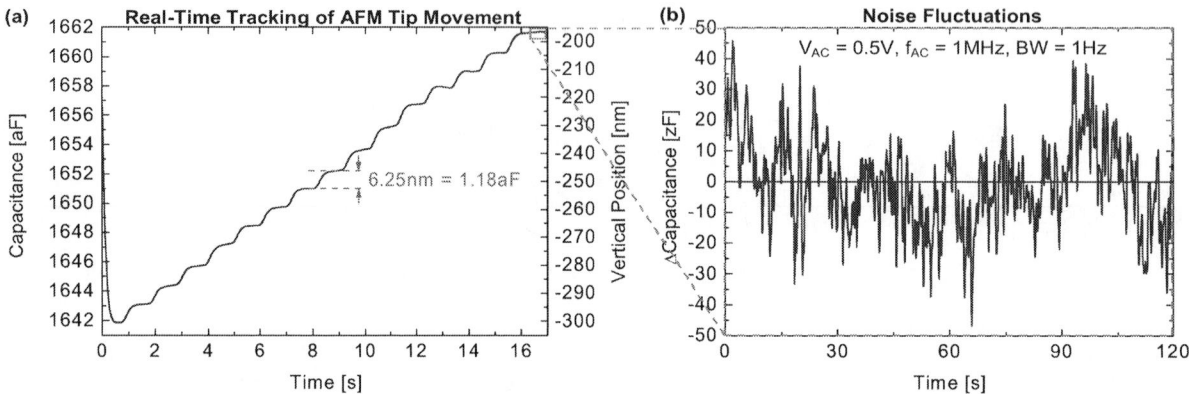

Fig. 6. Measurement of the capacitance between the AFM cantilever and the pad during vertical piezo movement in air with nanometric resolution (16 steps of 6.25nm corresponding to 1.18aF variation detected with a SNR = 84) made possible by a total noise rms level of only 14.4zF achieved applying 0.5V at 1MHz.

thermal noise of the TIA feedback resistor, that referred to the current input would give 0.4fA/sqrt(Hz). Here, the white noise ~3fA/sqrt(Hz) visible at low frequency is instead due to the noise background of the spectrum analyzer (Agilent N9020A MXA)

At 1MHz, the noise density is 14aF/sqrt(Hz) dominated by the input capacitive load, allowing fA current resolution for fast (~ms timescale) current sensing and zeptoFarad capacitance resolution (~s timescale) [7] when combined with a lock-in demodulator (here the HF2LI by Zurich Instruments).

C. Capacitance Measurements with the AFM

Fig. 6 reports the preliminary impedance measurements performed with the complete setup shown in Fig. 2. Thanks to the small area of the electrode and minimum overlap with the tip holder, the stray coupling between the holder and the input pad C_H is significantly reduced to only 1.6fF. This ultra-low baseline allows applying large V_{AC} signals (up to 2.5V) still complying with the limited input ±25nA current range. As a first test, after approaching the pad surface in contact mode (500nN set-point), the tip is moved in the vertical direction from 300nm to 200nm distance from the surface in 16 steps. As visible in Fig. 6a, the capacitive measurement allows a very clear tracking of the cantilever translation. Since the movement is small, the distance/capacitance relation is linear with a ratio of ~0.19aF/nm, that is in good agreement with the ~0.3aF/nm reported in the literature [8] with a slightly different setup (larger 1cm x 1cm graphite substrate).

In the chosen operating conditions (f_{AC} = 1MHz, V_{AC} = 0.5V, lock-in bandwidth BW = 1Hz), the rms capacitance fluctuations is 14.4zF (Fig. 6b). The theoretical capacitance noise in these conditions, according to the noise analysis reported in [7], is expected to be ~7zF; the increase is attributed to thermo-mechanical fluctuations of the setup. This excellent resolution would correspond to a vertical movement resolution of 75pm. Thus, each step of the vertical piezo movement equal to 6.25nm and corresponding to a capacitance variation of 1.18aF, is easily detected with a SNR = 84. This allows for instance to monitor independently and in real time the cantilever movement with unprecedented resolution.

V. CONCLUSIONS

Despite the limited experimental drawbacks caused by smaller (less flat, harder to functionalize or clean) and less "robust" substrate, the use of a CMOS metallic pad as an active substrate electrode is a viable approach to obtain significant performance improvement. In fact, thanks to the lower noise demonstrated with this novel setup (for instance a 140-fold improvement with respect to the 2aF noise level achieved in [8] with a custom discrete-component amplifier and higher applied voltages) better resolution in shorter measurement time is achievable, thus opening new perspectives for fast nanoscale impedance imaging and impedance spectroscopy over the DC-MHz range for non-invasive local characterization of nano-(bio)-samples.

ACKNOWLEDGMENT

The authors thank Laura Fumagalli and Gabriel Gomila for stimulating discussions and Sergio Masci for chip bonding.

REFERENCES

[1] L. W. Francis, P. D. Lewis, C. J. Wright and R. S. Conlan, "Atomic force microscopy comes of age," Biol. Cell, vol. 102, pp. 133-143, 2010.

[2] R. Garcia and E. T. Herruzo, "The emergence of multifrequency force microscopy," Nature Nanotech., vol. 7, pp. 217-226, 2012.

[3] T. Ando, "High-speed atomic force microscopy coming of age," Nanotechnol., vol. 23, pp. 062001/1-27, 2012.

[4] A. Alexeev, J. Loos, M.M. Koetse, "Nanoscale electrical characterization of semiconducting polymer blends by conductive atomic force microscopy," Ultramicroscopy, vol. 106, pp. 191-199, 2006.

[5] L. Fumagalli, G. Ferrari, M. Sampietro, G. Gomila, "Quantitative nanoscale dielectric microscopy of single-layer supported biomembranes," Nano Lett., vol. 9, pp. 1604–1608, 2009.

[6] G. Ferrari, M. Farina, F. Guagliardo, M. Carminati, M. Sampietro, "Ultra low noise CMOS current preamplifier from DC to 1 MHz," Electron. Lett., vol. 45, pp. 1278–1280, 2009.

[7] M. Carminati, G. Ferrari, F. Guagliardo, M. Sampietro, "ZeptoFarad capacitance detection with a miniaturized CMOS current front-end for nanoscale sensors," Sens. Actuators A, vol. 172, pp. 117-123, 2011.

[8] L. Fumagalli, G. Ferrari, M. Sampietro, G. Gomila, "Dielectric-constant measurement of thin insulating films at low frequency by nanoscale capacitance microscopy," Appl. Phys. Lett., vol 91, pp. 243110/1-3, 2007.

A wide input range instrumentation amplifier for impedance spectroscopy applications

F. Del Cesta, A. N. Longhitano, P. Bruschi
Dipartimento di Ingegneria dell'Informazione
University of Pisa
Pisa - Italy
francesco.delcesta@for.unipi.it

R. Simmarano
Sensichips srl
Latina - Italy

roberto.simmarano@sensichips.com

Abstract — This paper presents a fully-differential instrumentation amplifier, designed for low voltage, integrated impedance spectroscopy systems. The proposed architecture combines relative simplicity with an almost rail-to-rail input range. A synchronous demodulator is embedded into the output stage. These characteristics make it suitable for AC current sensing applications. A prototype has been designed using the UMC 0.18um MM/RF CMOS process. Simulations performed with a supply voltage of 1.5 V, showed that a maximum relative error of nearly 1 % can be achieved for stimulation frequencies up to 1 MHz and with a DC input common-mode voltage variable across a nearly 1 V interval. Power consumption is 150 uW.

Keywords—instrumentation amplifier; impedance spectroscopy; wide common mode range

I. INTRODUCTION

Instrumentation amplifiers are fundamental blocks for sensor interfacing. In many applications they are used to read DC voltages, as in the case of thermoelectric sensors [1] and Wheatstone bridges. Emerging applications require the detection and synchronous demodulation of AC voltages. An important example is represented by complex impedance measurements that are being proposed to characterize biological specimens [2] and read chemical sensors [3]. An impedance measurements system is sketched in Fig. 1. A sinusoidal generator applies a voltage stimulus to the device under test (DUT). A DC voltage is also often necessary to correctly bias the DUT, as in Electrochemical Impedance Spectroscopy (EIS). Use of two DC sources as in the figure, facilitates bipolar scans of the DC bias voltage in single power supply systems. The current is sensed across R_S and amplified by the in-amp. Further in-phase (I) and quadrature (Q) demodulation allow estimation of the current vector and hence the impedance.

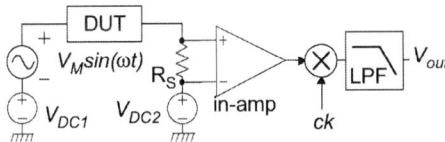

Fig. 1. Sketch of an impedance spectroscopy system.

In order to maintain maximum flexibility in the application of the DC bias, the input common mode range of the in-amp should be as wide as possible. A differential output with fixed common mode voltage is also desirable to facilitate the implementation of switch-based demodulators. The traditional three-operational amplifier configuration is not well suitable for low cost applications, since it requires precise resistor trimming to achieve acceptable common mode rejection. Difference Differential Amplifiers [4] do not tolerate large common mode mismatches between the input and feedback ports. To our knowledge, wide input range in-amps, based on current conveyors [5], are not present in the literature. Other interesting architectures, based on current feedback schemes, actually offer wide input ranges but at the cost of complexity and area consumption [6].

In this work we propose a fully differential instrumentation amplifier with an input range that can approach the power rails with only a margin of two drain-source saturation voltages. A prototype has been designed using the UMC 0.18um MM/RF CMOS process and its functionality is illustrated by means of detailed electrical simulations.

II. AMPLIFIER TOPOLOGY

A. Principle of operation

Fig. 2 shows a simplified schematic view of the amplifier. The amplifiers OTA$_{1-2}$, cascaded with the common source stages M$_1$ and M$_2$, form two-stage, high gain amplifiers closed in unity gain configuration. As a result, the input voltages V_{IP} and V_{IN} are replicated on nodes V_{SP} and V_{SN}, respectively. The current flowing through R$_1$ is than given by:

$$I_{R1} = (V_{SP} - V_{SN})/R_1 \cong (V_{IP} - V_{IN})/R_1 \qquad (1)$$

Neglecting parasitic capacitances and component mismatch, the current I_{R1} is given by

$$I_{R1} = (I_{D1} - I_{D2})/2 \qquad (2)$$

978-1-4673-4579-8/13 $31.00 © 2013 IEEE

The drain currents of M_1 and M_2 are replicated by nominally identical devices M_{1R} and M_{2R}, respectively. Thus, the current flowing into the resistor $R_2=R_{2A}+R_{2B}$ is:

$$I_{R2} = (I_{D1R} - I_{D2R})/2 \cong (I_{D1} - I_{D2})/2 = I_{R1} \qquad (3)$$

Combining (1) and (2) we get the output differential voltage:

$$V_{OP} - V_{ON} = R_2 \cdot I_{R2} \cong \frac{R_2}{R_1} \cdot (V_{IP} - V_{IN}) \qquad (4)$$

Conversely, the output common mode range is fixed to the reference voltage V_{CMO} by a conventional CMFB stage that controls current sources I_{OR}.

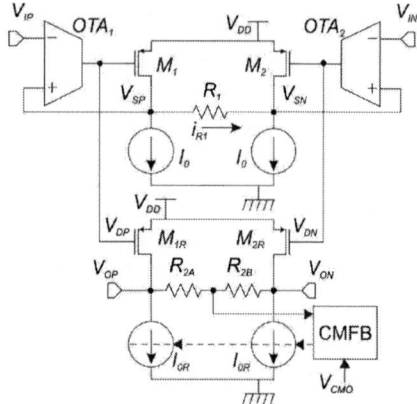

Fig. 2. Simplified schematic view of the amplifier.

This architecture is similar to a well know instrumentation amplifier topology, where the input voltages are replicated on internal nodes by means of source or emitter follower stages [7], optionally boosted with input buffers. The proposed architecture overcomes the input range limitations by substituting the source followers with common source stages(M_{1-2}). In this way, nodes V_{SP} and V_{SN} can get much closer to the supply rails, allowing an almost rail-to-rail performance. Clearly, in order to obtain this result, also OTA_1 and OTA_2 amplifiers should be designed for rail-to-rail input range. For the proposed prototype, folded cascode OTAs with *p-n* complementary input stages have been used [8]. Frequency response and stability issues are addressed in Sect. II-C.

B. Implementation of the embedded demodulator

The output demodulator has been embedded into the topology of Fig. 2, as shown in Fig. 3. Note that the actual amplifier topology uses cascode stages in order to improve current matching between the input and output (replica) stage. Demodulation of the signal is performed by chopper modulator CH_1 (*p-mos* switch matrix). Modulator CH_2 (*n-mos* matrix) is used to apply chopper modulation also to bias currents I_{OR}, reducing offset and low frequency noise contributions. Dummy switches DS, formed by two devices identical to those used in

CH_1, are introduced to make M_1 and M_2 work in the same operating conditions as M_{1R} and M_{2R}, improving precision in the replicated currents.

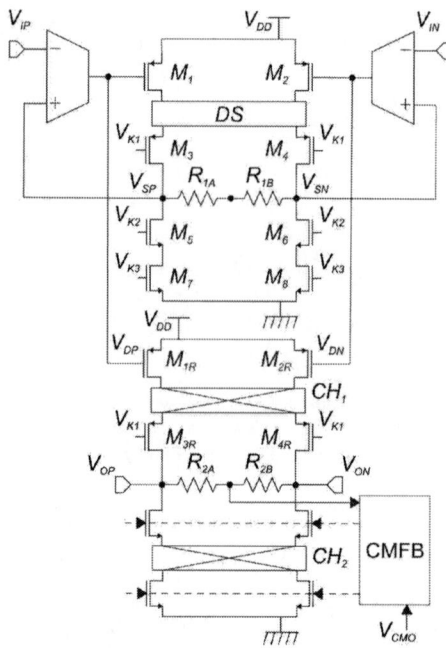

Fig. 3. Schematic view of the amplifier with chopper modulators.

C. Frequency response optimization

1) Stability of the feedback loop.

Fig. 4 shows a simplified view of the input loop. Resistor R_1 is split in two identical resistors, R_{1A} and R_{1B}. Miller compensation is used for the two stage architecture. Capacitors C_{CD} and zero nulling resistors R_{CD} compensate the differential mode loop. Unfortunately, these components are not sufficient for common-mode stability, since the gain-reduction effect of R_1 does not apply to common mode signals. In order to avoid over-compensating the differential loop gain, an additional common-mode compensation path has been introduced.

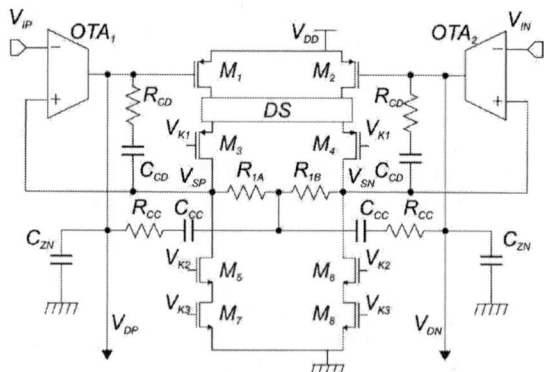

Fig. 4. Input stage of the amplifier comprehensive of the compensation paths.

2) Overall amplifier frequency response

With reference to Fig. 3, the currents flowing in M_{1R} and M_{2R} are chopped before they reach resistor R_2, so that the transfer function from the input to the output voltage is meaningless. For the amplifier performance, it is important to study the frequency response of M_{1R} and M_{2R} currents with respect to the input differential voltage. We will assume that, due to the cascode configuration and the presence of dummy switches DS, I_{D1R} and I_{D2R} are identical to I_{D1} and I_{D2}, respectively. Furthermore, considering only a differential input voltage, the half-circuit of Fig. 5(a) can be used. Capacitor C_P represents the contribution of all the parasitic capacitances connected from node V_{SP} to ground; r_O is the output resistance of the bias mirror, neglected in this analysis. Since V_{DP} variations can be assumed to be much smaller than V_{SP} ones, the circuit can be further simplified as shown in Fig. 5(b).

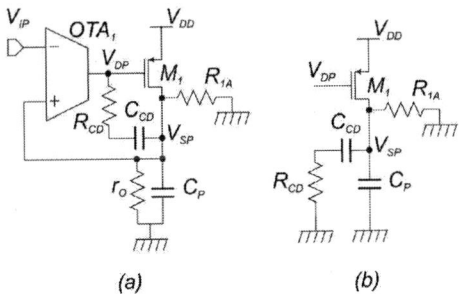

Fig. 5. Circuit for studying the frequency response of the amplifier (a) and its semplified scheme (b).

Considering, for simplicity, also that R_{CD} is small enough to be neglected with respect to C_{CD} impedance, the small signal current i_{d1} in the Laplace domain is given by:

$$i_{d1} = \frac{v_{sp}}{R_{1A} // \dfrac{1}{C_T \cdot s}} = v_{sp} \cdot \frac{1 + R_{1A} C_T \cdot s}{R_{1A}} \qquad (5)$$

where $C_T = C_P + C_{CD}$. Equation (5) shows that the frequency response includes a zero that make i_{d1} increase at high frequencies with respect to the DC value V_{SP}/R_{1A}. This effect, together with the related phase lead, sets an upper frequency limit for the amplifier. Note that this zero cannot be compensated by adding proper capacitances to the output node, since the DC component produced by demodulation, which is the signal of interest, would be unaffected.

The problem has been solved using a pole-zero compensation. Indicating with f_0 the unity gain frequency of the input feedback loop and approximating the latter with a dominant pole behavior, the following frequency response between v_{sp} and the input signal v_{ip} can be written:

$$\frac{v_{sp}}{v_{ip}} = \frac{1}{1 + j \dfrac{f}{f_0}} \qquad (6)$$

Then, combining (5) and (6):

$$\frac{i_{d1}}{v_{ip}} = \frac{1 + j \cdot 2\pi \cdot R_{1A} C_T \cdot f}{R_{1A}} \cdot \frac{1}{1 + j \dfrac{f}{f_0}} \qquad (7)$$

In order to cancel the contribution of the zero, the unity gain frequency f_0 has been decreased to make it coincide with the zero frequency. This could not be accomplished by increasing the compensation capacitor C_{CD}, since it affects also the zero. Therefore, as shown in Fig. 4, a capacitor C_{ZN} was added between M1 (and M2) gate and ground, shifting back the dominant pole of the open loop response and, in the same proportion, the first pole of the closed loop response. The correct value of this additional capacitor has been chosen by means of a simulator-assisted empirical procedure.

III. BLOCK SIZING AND SIMULATIONS

The effectiveness of the proposed approach has been proven by means of a prototype, designed using the 0.18μm CMOS MM/RF process of UMC. Circuit sizing has been carried out with bandwidth and input common mode range as principal targets. Resistors R_1 and R_2 have been set to 16.5 kΩ and 330 kΩ, respectively, so that the nominal amplifier gain is 20. The power supply voltage was set to 1.5 V, resulting in a supply current of 100 μA.

The frequency response of $(i_{d1R} - i_{d2R})$ with respect to the input differential voltage is shown in Fig. 6(a). The pole-zero cancellation approach implemented by capacitors C_{ZN} and described in previous section, allowed the frequency response to be flat up to 1 MHz with only a 4 degrees phase delay. The power spectral density (PSD) of the amplifier noise, referred to the input, is shown in Fig 6(b).

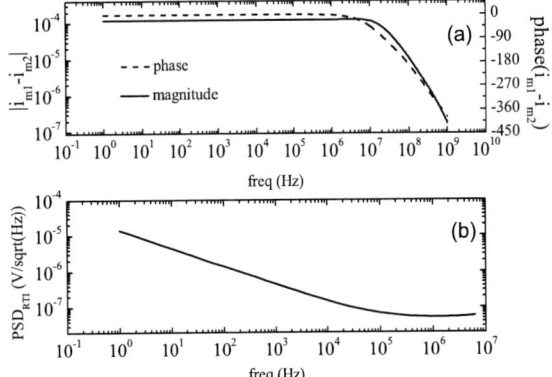

Fig. 6. Overall amplifier frequency response (a) and PSD of the amplifier noise referred to the input (b).

The functionality of the amplifier as a demodulator has been characterized by applying a sinusoidal differential input voltage superimposed to a DC common mode voltage. Chopper modulators CH_1 and CH_2 were driven with a clock signal with 50 % duty-cycle and same frequency as the stimulus. The result of a transient simulation (output voltage) is shown in Fig. 7, relative to in-phase demodulation, 100 kHz frequency and 5 mV input amplitude, 0.75 V common mode voltage. The dashed line indicates the mean value of the output waveform.

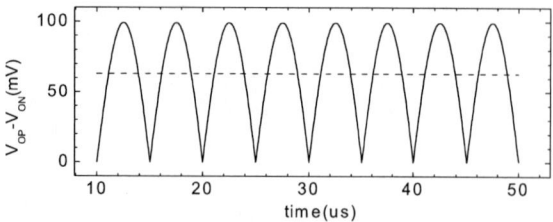

Fig. 7. Demodulated differential output voltage of the amplifier at a frequency of 100 kHz. The dashed line represents the mean value.

Fig. 8(a) shows the mean value of the in-phase component of the output differential voltage, as a function of the input frequency. Deviation from the ideal value, indicated in the figure, is less than 1% up to a frequency of about 1 MHz. Fig. 8(b) is analogous to Fig 8(a) but with a quadrature demodulation clock. The output voltage (mean value) stays below 1 % of the in-phase component up to 2 MHz.

The effect of the DC input common mode component is shown in Fig. 9(a) for the same input stimulus as in Fig. 9 and in-phase demodulation. The relative error with respect to the ideal value is within 1.1 % over the 0.2 − 1.16 V interval. Linearity with respect to the differential voltage amplitude is shown in Fig. 9(b), where the common mode voltage was maintained at $V_{dd}/2$.

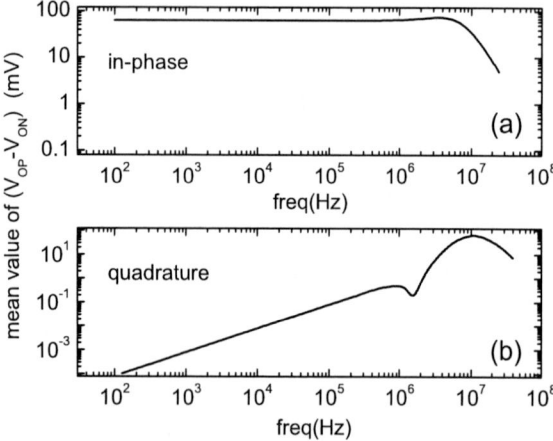

Fig. 8. Mean value of the differential output voltage of the amplifier as a function of the frequency of the input signal. The input signal has a magnitude of 5mV and a common mode of 750mV.

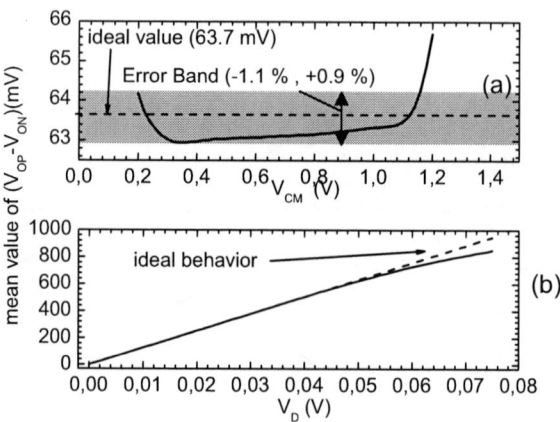

Fig. 9. Mean value of the differential output voltage as a function of the DC input common-mode voltage (a) and of the differential input voltage amplitude (b) for a 100 kHz stimulus.

IV. CONCLUSIONS

A compact structure for the implementation of instrumentation amplifiers with wide input common mode range has been described. Simulations confirmed that in-phase and quadrature detection of the input voltage with a maximum relative error of 1.1% can be achieved for frequencies up to 1 MHz over an input common mode interval of nearly 1 V with 1.5 V single power supply. The architectural simplicity and compactness of the proposed architecture make it suitable to be used in multichannel integrated impedance spectroscopy systems.

REFERENCES

[1] Butti F., Bruschi P, Dei M, Piotto M. "A compact instrumentation amplifier for MEMS thermal sensor interfacing." Analog Integrated Circuits And Signal Processing, vol. 72, p. 585-594, 2012

[2] Manickam, A. "A CMOS electrochemical impedance spectroscopy biosensor array for label-free biomolecular detection", Solid-State Circuits Conference Digest of Technical Papers (ISSCC), 2010 IEEE International

[3] Chonlatid Sontimuang, Roongnapa Suedee, Franz Dickert "Interdigitated capacitive biosensor based on molecularly imprinted polymer for rapid detection of Hev b1 latex allergen", Analytical Biochemistry 410 (2011) 224–233

[4] Michiel A. P. Pertijs, Senior Member, IEEE, and Wilko J. Kindt, "A 140 dB-CMRR Current-Feedback InstrumentationAmplifier Employing Ping-Pong Auto-Zeroing and Chopping", IEEE Journal of Solid-State Circuits, vol. 45, no. 10, oct. 2010

[5] C. Toumazou and F. Lidgey, "Novel current-mode instrumentation amplifier," Electron. Lett., vol. 25, pp. 228–230, Feb. 1989.

[6] Rong Wu, Johan H. Huijsing, and Kofi A. A. Makinwa "Current-Feedback Instrumentation Amplifier With a Gain Error Reduction Loop and 0.06 Untrimmed Gain Error," IEEE Journal of Solid State Circuits, 46(12), 2011.

[7] H. Krabbe, "A high performance monolithic instrumentation amplifier," in ISSCCDig. Tech. Papers, Feb. 1971, pp. 186-187

[8] R. J. Baker, H. W. Li, and D. E. Boyce, CMOS - Circuit Design, Layout and Simulation. Piscataway, NJ: IEEE Press, 1998, p.658

A 36μW Rail-to-Rail-Input Chopper Stabilized Amplifier using Correlated Double Sampling

A. Pipino*, M. De Blasi*[†], M. De Matteis*[†], S. D'Amico*, A. Fornasari[‡] and A. Baschirotto*

[†]Dept. of Physics, University of Milano-Bicocca, Milano, Italy
*Dept. of Innovation Engineering, University of Salento, Lecce, Italy;
Email: alessandra.pipino@gmail.com, marco.deblasi, marcello.dematteis, andrea.baschirotto@unimib.it,stefano.damico@unisalento.it
[‡]Texas Instruments

Abstract—This paper describes a chopper stabilized amplifier, obtained using an input chopper for modulation, an AC coupling for offset rejection and a correlated double sampling structure for demodulation, avoiding ripple spurs. The operational amplifier has a rail-to-rail input stage, with a supply voltage range from 1.8V to 5V. It is characterized by a multipath nested Miller compensation network with double pole-zero cancellation. The circuit achieves a simulated 130dB-dc-gain with a current consumption of 20μA, suitable for low-power application. Equivalent input offset has been evaluated by Montecarlo simulations demonstrating a maximum standard deviation of about 4μV.

Index Terms—chopper technique; modulation; CDS; Miller compensation

I. INTRODUCTION

In many instrumentation applications, such as sensors, biomedical and optoelectronics, the analog input signals to amplify are usually very small (at the microvolt level). For this reason very high gain, low offset amplifiers are a basic building block in these systems. Moreover, low power is a key feature since, reducing power, battery-life increases and portability is improved [1, 2, 3, 4].

Several analog techniques are available in literature, suitable to reduce flicker noise, and attenuate input voltage offset in instrumentations amplifiers. Among them auto-zero (AZ) or chopper techniques. The basic idea of AZ [5] is sampling unwanted signal at low frequency (the offset and 1/f noise) and subtracting them to input signal. In the sampling phase the signal path is disconnected so the amplifier cannot be used to amplify the input signal, which makes auto-zeroing difficult to implement in a continuous time-amplifier. Moreover this topology suffers from folding back of the wideband noise, increasing input referred noise. Finally storing capacitor may be large causing power increasing to drive it.

Chopping [5, 6], instead, is a modulation technique in which the signal and the offset are modulated to different frequencies, in this way they may be distinguished and the offset may be filtered out . A drawback of this technique is represented by ripple spurs at multiples of chopping-frequency causing residual offset. Alternative schemes has been realized to avoid ripples in [7, 8, 9, 10].

This design avoids ripple spurs combining chopper stabilization technique (CHS) with the correlated double sampling (CDS) method as reported in [8]. CDS replaces signal demodulation, while offset is removed through AC coupling. In particular in this work it is applied to a chopper-stabilized operational amplifier instead of the chopper basic one used in [8]. In chopper stabilization, as shown in figure 1, the main operational amplifier G_m, with an offset V_{os}, is being offset-stabilized by a stabilizing amplifier G_n with a hypothetical offset of 0V [11] . In this way an auxiliary amplifier is used to compensate for the offset of main amplifier which is never disconnected from the signal path. This topology can be seen as a multipath amplifier, where the auxiliary opamp forms the high gain and low frequency path, while the main opamp represents the high frequency path.

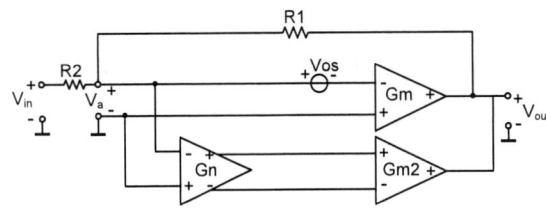

Fig. 1: Offset stabilized amplifier concept.

Stability analysis is necessary before applying offset reduction to stabilize amplifier by chopping and CDS. Moreover this work is characterized by a rail-to-rail input stage, designed in order to ensure a wide input common-mode range. Rail-to-rail input/output operation is a necessary condition to obtain acceptable signal-to-noise ratio in low voltage environments, but if on one hand rail-to-rail output is not a great obstacle on the other one more trouble rises with rail-to-rail input.

Ensuring rail-to-rail input operation with the wide variation of the supply voltage in the range (1.8V 5V) required a design effort to guarantee the safety of the devices and the robustness of operational amplifier to process spread and mismatch.

This paper is organized as follow. Section II describes the proposed CHS-CDS method. Section III shows the topology of the continuous time operational amplifier focused on the compensation network. Section IV presents the complete CHS operational amplifier, with attention to the first stage. Finally section V shows simulation results.

II. CHS-CDS TECHNIQUE

Working principle is shown in figure 2. Input chopper transposes signal to high frequency using a modulating signal $m(t)$, that is a square wave carrier with period $T = \frac{1}{f_{chop}}$. In this way input signal moves to the odd harmonics frequencies of the modulation signal; offset and 1/f noise are added and amplified with the signal. Their contribution is then removed by AC coupling, while CDS structure detects the signal.

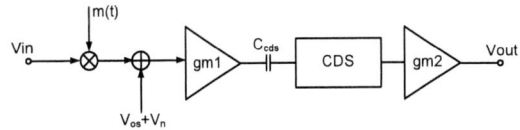

Fig. 2: Simplified CHS-CDS block diagram.

In particular, observing figure 3, during phase Φ_1 C_{cds} is charged

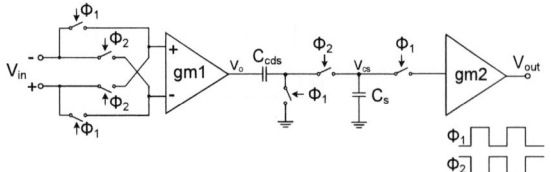

Fig. 3: Chopper and CDS scheme.

to the output voltage V_o, while C_s is discharged and disconnected; during phase Φ_2 C_{ds} is in series with C_s and voltage V_{cs} across it is the input of the second stage. Because of the chopping of input signal, expressions for V_o is in the two phases are:

$$V_o\left(\Phi_1\right) = A_1\left(-V_{in}\left(\Phi_1\right) + V_{os} + V_n\left(\Phi_1\right)\right) \quad (1)$$

$$V_o\left(\Phi_2\right) = A_1\left(V_{in}\left(\Phi_2\right) + V_{os} + V_n\left(\Phi_2\right)\right) \quad (2)$$

while during phase Φ_2 voltage V_{cs} is:

$$
\begin{aligned}
V_{cs}\left(\Phi_2\right) &= C_{cds}\frac{-V_o\left(\Phi_1\right) + V_o\left(\Phi_2\right)}{C_{cds} + C_s} \\
&= \frac{A_1 C_{cds}}{C_{cds} + C_s}\left(V_{in}\left(\Phi_1\right) + V_{in}\left(\Phi_2\right) + V_n\left(\Phi_2\right) - V_n\left(\Phi_1\right)\right)
\end{aligned}
\quad (3)
$$

From the above expressions, we can see that offset is removed, signal is low-pass filtered $\left(1 + z^{-1}\right)$, while 1/f noise passes through a high-pass function $\left(1 - z^{-1}\right)$. In this way ripple spurs are avoided because demodulation is replaced by the CDS. Since the output of stage 1 is fully differential, four CDS structures are used. They work in a ping-pong configuration: during phase Φ_1 two of the four C_{cds} are charged by output voltages, while the other two discharge the stored voltage signals in the virtual ground; viceversa during phase Φ_2.

III. OPAMP CIRCUITAL TOPOLOGY

Figure 4 shows the scheme of the continuous-time opamp only, characterized by a multipath nested Miller compensation network with double cancellation, described in this section.

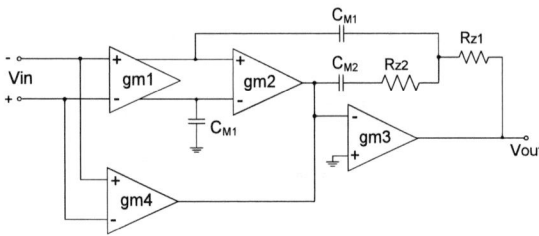

Fig. 4: Continuous time opamp.

A. Compensation network

Classical nested Miller compensation [12, 13] uses two compensation capacitors, C_{M1} and C_{M2}, that close the second and third stages. In this way one of the three poles (due to high impedance node at the output of every stage) is pushed to lower frequency, while to other two non-dominant poles are moved to higher frequency, after UGB. However the two capacitors introduce two zeros, a LHP zero and a RHP zero.

In order to ensure stability, a feedforward transconductance stage

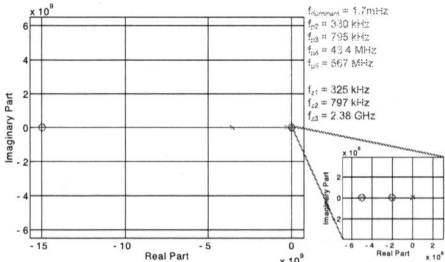

Fig. 5: Pole-zero map of continuous time opamp.

g_{m4} is added. This stage moves LHP zero to lower frequency, near the first non-dominant pole, and avoids the degradation of phase margin. This topology, called multipath nested Miller compensation, can be thought as a high gain path (g_{m1}, g_{m2} and g_{m3}) that determines DC behavior of frequency response (and so in this path the offset reduction will be applied), in parallel with a wideband path (g_{m4} and g_{m3}) that dominates at high frequency, establishing UGB and phase margin, forming the wide band main amplifier. A nulling resistor R_{z1} connected in series with the compensation capacitors C_{M1} and C_{M2} is added to move to higher frequency the RHP zero. This resistor increases impedance of the direct path through capacitors at high frequency, reducing corresponding direct current and pushing away the RHP zero. Another resistor R_{z2} is added in series with C_{M2} to obtain a pole-zero cancellation also for the second non-dominant pole. In particular R_{z2} generates a LHP zero that can be set near the other pole.

B. Circuit design

The starting point to design the compensation network is represented by the desired UGB of 350kHz and g_{m1} equal to 15μA/V, value obtained from noise considerations. Moreover g_{m3} has to be bigger than g_{m1} and g_{m2} (also 15μA/V), since load capacitor is very large (100pF); it has been set to 500μA/V. C_{M1} can be calculated from unity gain bandwidth (UGB) consideration since

$$C_{M1} = \frac{g_{m1}}{2\pi \cdot UGB} = 7pF \quad (4)$$

In order to have the same GBW for the direct path and for the feedforward path:

$$\frac{g_{m1}}{C_{M1}} = \frac{g_{m4}}{C_{M2}} \quad (5)$$

that yields:

$$g_{m4} = \frac{C_{M2}}{C_{M1}}g_{m1} \quad (6)$$

For good matching and less area occupation, it is convenient to choose $g_{m4} = g_{m1} = 15$μA/V and $C_{M1} = C_{M2} = 7$pF. Nulling resistor R_{z1}, in order to move the RHP zero to higher frequency, must be:

$$R_{z1} = \frac{1}{g_{m3}} = 2k\Omega \quad (7)$$

Finally, in order to compensate the second non-dominant pole, R_{z2} must be:

$$R_{z2} = \frac{C_L}{g_{m3}C_{M2}} = 28.5k\Omega \quad (8)$$

Figure 5 shows poles and zeros position in the Gauss plane for the designed opamp, with a zoom at low frequencies to show the dominant pole and the poles-zeros cancellation.

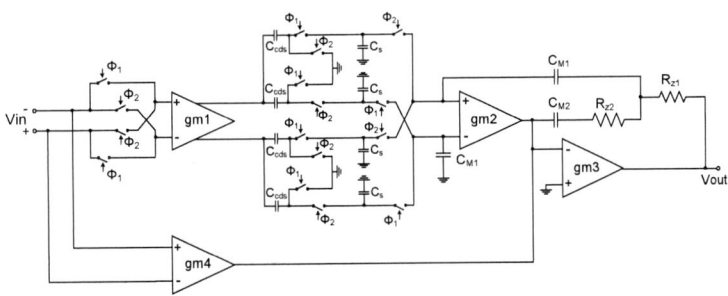

Fig. 6: Complete scheme of the chopper stabilized amplifier

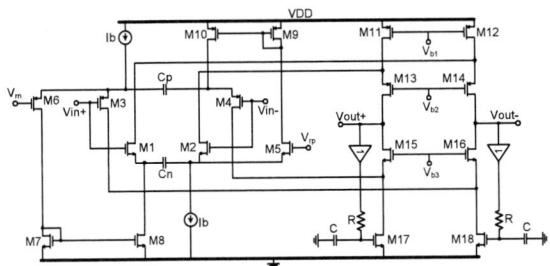

Fig. 7: First gain stage schematic.

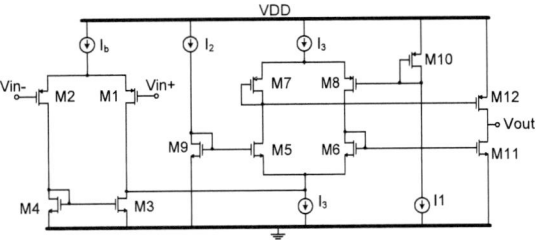

Fig. 8: Third gain stage schematic.

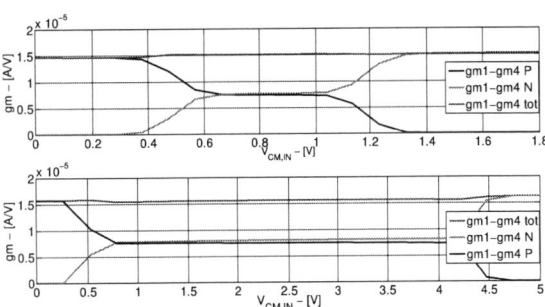

Fig. 9: Rail to rail operation of input stages for $V_{DD} = 1.8$ and $V_{DD} = 5$.

IV. OVERALL CHS OPAMP

Figure 6 shows the overall CHS amplifier. As discussed before, AC coupling removes offset, but it opens loop at DC, so it is necessary to keep input referred offset low and reduce its amplification, in order not to saturate g_{m1}. The solution is a bandpass amplifier as first gain stage [8]; it provides low gain at DC and high gain at f_{chop}. The schematic of first stage is shown in figure 7. It's a folded cascode with complementary input stage that ensures a rail to rail input range with constant gm thanks to the control of bias currents made by two current switches [14]. The high-pass transfer function is obtained adding the two capacitors C_n and C_p. At DC the two capacitors are open circuits and so there is a source follower configuration in both branches. The gain is R_{out}/R_E, where R_{out} is the output resistance of the folded cascode and R_E is the output resistance of the current sources. In this situation, gain is further attenuated by the twin-T band reject filter that opens the path of output voltages. At f_{chop} C_n and C_p are low impedances and the gain is that of a common source, $g_{m1}R_{out}$, where g_{m1} is the total transconductance of input pairs. Frequency response has a zero at $1/4\pi R_E C$ ($C = C_n = C_p$) and a pole at $g_{m1}/2\pi C$. It is important that the pole is lower than f_{chop}, in order to have gain at that frequency; but at the same time if zero and pole frequencies are close to the opamp UGB, they can change the expected phase profile and phase margin. The second stage g_{m2} is a simple differential stage with active load. It is followed by a class-AB biased rail to rail output stage as third stage [15], shown in figure 8. Stage g_{m4} is a folded cascode with rail-to-rail input stage, similar to g_{m1}.

V. SIMULATION RESULTS

The proposed chopper stabilized amplifier has been implemented using a 0.35μm technology with a supply voltage varying from 1.8V to 5V. The extended range of power supply is possible since technology provides transistor MOS operating at 5V, in addition to standard MOS operating at 3.3V. The performance over the wide available range are constant thanks to rail-to-rail input stages (g_{m1} and g_{m4}) that maintain a constant gm, as showed in figure 9 for the two extreme cases.

Figure 10 shows simulated opamp transfer function; it has a DC gain of 130dB, a bandwidth of 312kHz with a good phase margin of about 83°. Figure 11 shows a detail of a transient simulation to highlight the reduction of ripples due to CDS. The black line is the output of the amplifier just considering a chopper at the input and output of g_{m1}, while the red one is the output of the presented operational amplifier which replaces the output chopper with the CDS. A ripple reduction and the consequent reduction of residual offset is obtained. Figure 12 shows the simulated offset distribution after 50 Montecarlo runs; the obtained standard deviation is lower than 4μ. Figure 13 shows a low frequency periodic noise simulation in the case of continuous time opamp and in the case with chopper and CDS. The

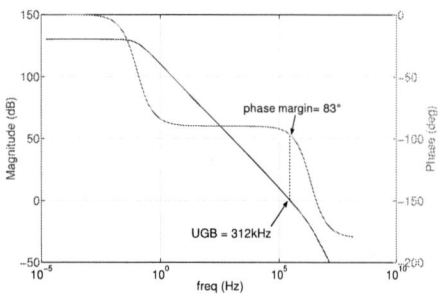

Fig. 10: Operational amplifier transfer function.

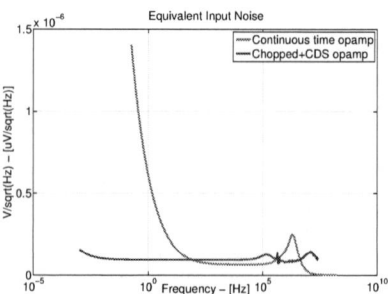

Fig. 13: Equivalent input noise.

result is a lower corner frequency thanks to the CDS structure but an increasing of the white noise that leads to final input noise of $70nV/\sqrt{Hz}$. The overall input noise of the structure depends mainly from the first gain stage noise and the kT/C noise introduced by the CDS block. A higher chopping frequency spreads the kT/C noise due to the CDS network, but increases the power consumption; therefore a trade-off for f_{chop} in terms of input noise, gain and current is 500 kHz [8].

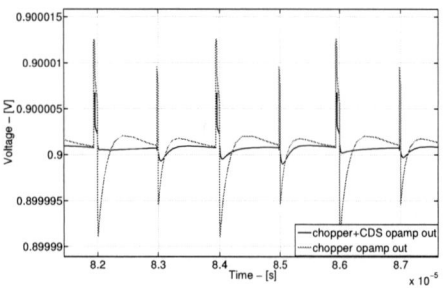

Fig. 11: Transient simulation zoom showing ripple reduction due to CDS.

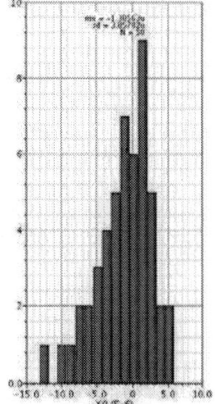

Fig. 12: Simulated offset voltage distribution.

VI. CONCLUSION

In this work a chopper stabilized amplifier is presented, obtained with a combination of the input chopping and the CDS structure, that avoids ripple spurs due to demodulation, reducing the offset voltage and the low-frequency noise spectrum. Moreover the compensation network implemented ensures a good phase margin (about 83°). The standard deviation of offset is lower than 4μV with a current consumption of 20μA. Finally the NEF [8] is:

$$NEF = v_{n,in}\sqrt{\frac{2I_{tot}}{\pi U_T 4kT}} = 12 \qquad (9)$$

REFERENCES

[1] M. De Matteis, S. D'Amico, V. Giannini, and A. Baschirotto, "A 550mV 8dBm IIP3 4th order analog base band filter for WLAN receivers," in *Solid State Circuits Conference, 2007. ESSCIRC 2007. 33rd European*, pp. 504–507, 2007.

[2] S. D'Amico, M. Matteis, and A. Baschirotto, "A 6.4 mW, 4.9 nV/$\sqrt{Hz}$, 24 dBm IIP3 VGA for a multi-standard (WLAN, UMTS, and Bluetooth) receiver," *Analog Integr. Circuits Signal Process.*, vol. 61, pp. 1–7, Oct. 2009.

[3] S. D'Amico, M. De Blasi, M. De Matteis, and A. Baschirotto, "A 255 MHz Programmable Gain Amplifier and Low-Pass Filter for Ultra Low Power Impulse-Radio UWB Receivers," *Circuits and Systems I: Regular Papers, IEEE Transactions on*, vol. 59, no. 2, pp. 337–345, 2012.

[4] P. Carniti, M. De Matteis, A. Giachero, C. Gotti, M. Maino, and G. Pessina, "CLARO-CMOS, a very low power ASIC for fast Photon counting with pixellated photodetectors," *IOPscience*, vol. 7, no. 2, pp. 1–24, 2012.

[5] C. Enz and G. Temes, "Circuit techniques for reducing the effects of op-amp imperfections: autozeroing, correlated double sampling, and chopper stabilization," *Proceedings of the IEEE*, vol. 84, no. 11, pp. 1584–1614, 1996.

[6] C. Enz, E. Vittoz, and F. Krummenacher, "A CMOS chopper amplifier," *Solid-State Circuits, IEEE Journal of*, vol. 22, no. 3, pp. 335–342, 1987.

[7] R. Burt and J. Zhang, "A Micropower Chopper-Stabilized Operational Amplifier Using a SC Notch Filter With Synchronous Integration Inside the Continuous-Time Signal Path," *Solid-State Circuits, IEEE Journal of*, vol. 41, no. 12, pp. 2729–2736, 2006.

[8] M. Belloni, E. Bonizzoni, A. Fornasari, and F. Maloberti, "A Micropower Chopper-CDS Operational Amplifier," *Solid-State Circuits, IEEE Journal of*, vol. 45, no. 12, pp. 2521–2529, 2010.

[9] Q. Fan, J. Huijsing, and K. Makinwa, "A multi-path chopper-stabilized capacitively coupled operational amplifier with 20V-input-common-mode range and 3μ V offset," in *Solid-State Circuits Conference Digest of Technical Papers (ISSCC), 2013 IEEE International*, pp. 176–177, 2013.

[10] R. Wu, K. A. A. Makinwa, and J. Huijsing, "A Chopper Current-Feedback Instrumentation Amplifier With a 1 mHz Noise Corner and an AC-Coupled Ripple Reduction Loop," *Solid-State Circuits, IEEE Journal of*, vol. 44, no. 12, pp. 3232–3243, 2009.

[11] K. A. Makinwa, *Dynamic offset compensated CMOS amplifiers*. Springer, 2009.

[12] S. P. A. D. Grasso, G. Palumbo, "Analytical comparison of frequency compensation techniques in three-stage amplifiers," *Solid-State Circuits, IEEE Journal of*, vol. 36, no. 1, pp. 53–80, 2008.

[13] K. N. Leung and P. K. T. Mok, "Analysis of multistage amplifier-frequency compensation," *Circuits and Systems I: Fundamental Theory and Applications, IEEE Transactions on*, vol. 48, no. 9, pp. 1041–1056, 2001.

[14] S. Yan, J. Hu, T. Song, and E. Sanchez-Sinencio, "Constant-gm techniques for rail-to-rail CMOS amplifier input stages: a comparative study," in *Circuits and Systems, 2005. ISCAS 2005. IEEE International Symposium on*, pp. 2571–2574 Vol 3, 2005.

[15] K.-J. de Langen and J. Huijsing, "Compact low-voltage power-efficient operational amplifier cells for VLSI," *Solid-State Circuits, IEEE Journal of*, vol. 33, no. 10, pp. 1482–1496, 1998.

Integrated Class-D Audio Amplifier Virtual Test for Output EMI Filter Performance

Roberto Mrad*[†‡], Florent Morel*, Gael Pillonnet[†], Christian Vollaire* and Angelo Nagari[‡]

*University of Lyon, ECL
Ampere, CNRS UMR5005, 36 avenue Guy de Collongue, 69134 Ecully, France
Email: firstname.lastname@ec-lyon.fr
[†]University of Lyon, CPE
INL, CNRS UMR5270, 43 bd de 11 Novembre 1918, 69616 Villeurbanne, France
Email: firstname.lastname@cpe.fr
[‡]ST Ericsson, Grenoble, France
Email: firstname.lastname@stericsson.com

Abstract—This paper proposes a model based on the segmentation technique that can be used for electromagnetic interference (EMI) filter test by simulation. It is dedicated for systems with a differential output. The present approach proposes virtual measurement of the filter performance in the final application. Thus, the source emissions and the load impedance are taken into account. The proposed approach uses impedance matrices to model the passive parts of the system. A matrix calculation permits to associate these matrices and create a compact model. The impedance matrices and the converter output voltages are used for currents and voltages computation at all the system points. Finally, the proposed method is applied and validated on a Class-D amplification system. This method shows good accuracy on the [1kHz-120MHz] frequency band.

Index Terms—Integrated Class-D amplifier, Electromagnetic compatibility, EMI modelling, Filter design, Impedance matrix.

I. Introduction

In embedded systems, battery life is a major issue. Thus, switching power management circuits are generally used thanks to their power efficiency. For loudspeaker audio applications, integrated Class-D amplifiers are a good trade off between power efficiency and audio quality. However, such switching circuits, generates serious ElectroMagnetic (EM) disturbances for the surrounding electronics. Therefore, designers use control techniques, such as spread spectrum [1]–[4], slew rate control [5], [6], pulse position [7] to reduce the power converter emissions. However, these techniques effects on the low and medium frequency range. Therefore, an ElectroMagnetic Interference (EMI) filter is inserted between the filter and the load (loudspeaker) in order to reduce the high frequency emissions in the Printed Circuit Board (PCB) track.

EMI filters are made from bulky and costly passive components. Fig. 1 shows a Class-D amplifier and an EMI filter in a smart-phone application. As can be seen, the filter can occupy an area on PCB twice the one occupied by the amplifier chip itself. Thus, the filter components should be well chosen so it will not be over-dimensioned. Moreover, High frequency EMI filter response strongly depends on components and PCB stray elements. Therefore, when studying ElectroMagnetic Compatibility (EMC) of such systems and filters, designers rely on

Fig. 1. Class-D amplifier and an EMI filter in the final application.

prototypes and measurements. Hence, a method helping the EMI filter design before production can reduce the production time and cost.

For EMI studies, simulations must have a good accuracy at high frequencies where the Passive Distribution Network (PDN) became very complex. Thus, traditional time domain simulations require a highly extended model. The main challenge for building such models is the knowledge of the circuit stray elements. However, measuring or computing those parasitic elements can be difficult, impossible or require dedicated simulations or measuring procedure. In addition, such models can take days or even weeks to make one single simulation. Therefore it cannot be repeated so many times for EMI filter optimization. Otherwise, when using already built chips, system integrators have no access on the circuit accurate models. For all these reasons a model dedicated for EMI issues is needed. Thus, a black box model in the frequency domain can be very useful.

In previous works, a method has been presented in [8]–[10]. The latter allows to model a Class-D amplification system in the frequency domain in order to compute the spectra of the amplifier output currents which are the same as the filter input currents (see Fig. 2 and Fig. 3). The method consists on decomposing the system into separate blocks and model the passive ones by impedance matrices. Hereafter, the blocks are concatenated to reconstitute the system into a compact model represented by a single impedance matrix. Finally, using the resulting matrix and the amplifier output voltages, the amplifier output current spectra are computed. There are many advantages for this method. First, this method takes into account the load impedance and source emissions to predict the filter behavior in the final application. Second, it takes into account the Common Mode (CM), the Differential Mode (DM) and the mode conversion. Third, using impedance matrices permits to easily include the high frequency effect of the circuit stray elements, therefore, to obtain a good accuracy at high frequencies. Finally, to be used in filter optimization, this method has a short simulation time (few minutes) compared to transient simulations (many days).

This paper extends the method presented above in order to obtain the currents and voltages at the EMI filter output. Thus, the filter EMI performance can be evaluated.

Section II presents the method principles as well as the required theory details. Section III presents an experimental application on a Class-D amplifier system and discuss the results. Finally, section IV concludes the paper and proposes some perspectives.

II. Modeling approach

A. Frequency model

The system in use is shown in Fig. 2. It contains a Class-D amplifier, an EMI filter and a speaker load. The considered approach lies on decomposing the system into functional blocks as shown in Fig. 3. The passive blocks are modeled by impedance matrices which relate the block voltages to the block currents as shown in (1).

$$
\begin{pmatrix} V_1 \\ \vdots \\ V_N \end{pmatrix} = \begin{pmatrix} z_{11} & \cdots & z_{1N} \\ \vdots & \ddots & \vdots \\ z_{N1} & \cdots & z_{NN} \end{pmatrix}_{(N \times N)} \cdot \begin{pmatrix} I_1 \\ \vdots \\ I_N \end{pmatrix} \quad (1)
$$

Since integrated Class-D amplifiers have a H-bridge output power stage, a 4×4 impedance matrix is needed to model the filter which correspond to 4-port block (two ports for the inputs and two others for the outputs). However, the speaker load have only two electrical input ports and an acoustic output. Thus, the speaker is modeled by a 2×2 impedance matrix which correspond to a 2-port block. The impedance matrix can be determined using an impedance analyzer [8], a vector network analyzer or by electrical simulation, such as Advanced System Design (ADS) for Agilent.

Once the impedance matrices of the filter and the load are determined, (2) is used to merge the two matrices [8].

$$
Z_R = Z_{F11} - Z_{F12}(Z_L + Z_{F22})^{-1} Z_{F21} \quad (2)
$$

with

$$
Z_F = \left(\begin{array}{c|c} Z_{F11(2\times2)} & Z_{F12(2\times2)} \\ \hline Z_{F21(2\times2)} & Z_{F22(2\times2)} \end{array} \right)_{(4\times4)} \quad \text{is the filter}
$$

impedance matrix, $Z_{L(2\times2)}$ is the load impedance matrix, Z_R is the a resulting 2×2 impedance matrix modeling all the passive blocks of the system.

Note that, if the amplifier output impedance is not negligible, it should be concatenated with the rest of the passive blocks [9]. Note as well that, the filter, load and resulting matrix are symmetric with respect to the diagonal ($Z_{ij} = Z_{ji}$; $i = 1 \ldots 4$; $j = 1 \ldots 4$; $i \neq j$) due to the reciprocity theorem.

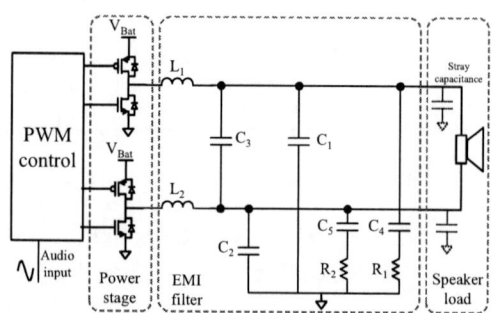

Fig. 2. Considered Class D amplifier system.

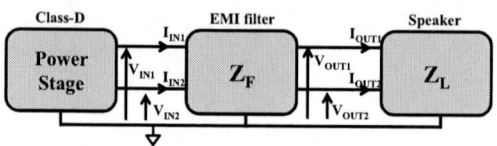

Fig. 3. Amplification system bloc diagram.

Due to the differential output, the Class-D amplifier is modeled by two AC voltage sources and an impedance matrix to model the converter output impedance as shown in Fig. 4. However, in this case, the amplifier output impedance is negligible compared to the filter and load impedance over the considered frequency range. It has been tested by measuring the converter output voltages with different loads: open circuit, speaker and speaker with EMI filter. In the three cases the output voltages remains unchanged. Determining the AC voltage sources can be done by measurement or by simulation. In case of measurements, the two voltages should be measured in the time domain at the same time to obtain a correct phase between them, then a FFT convert them to the frequency domain.

B. Currents and voltages computation

Using the amplifier model, the filter matrix, the load matrix and the resulting matrix, it is possible to compute all the spectra of the currents and voltages in the system. Thus, the filter and load can be evaluated in terms of EMI.

Equation (3) permits to compute the currents at the converter output (same as the filter input) [9].

$$
\begin{pmatrix} I_{IN1} \\ I_{IN2} \end{pmatrix} = (Z_R)^{-1} \begin{pmatrix} V_{IN1} \\ V_{IN2} \end{pmatrix} \quad (3)
$$

Where V_{IN1}, V_{IN2} are the amplifier output voltages, I_{IN1}, I_{IN2} are the amplifier output currents and Z_R is the filter and speaker resulting impedance matrix computed using (2).

The output voltages (V_{OUT1}, V_{OUT2}) and currents (I_{OUT1}, I_{OUT2}) can be computed using equation (4) and (5) respectively.

$$
\begin{pmatrix} V_{OUT1} \\ V_{OUT2} \end{pmatrix} = (I + Z_{F22} Z_L^{-1})^{-1} Z_{F21} Z_R^{-1} \begin{pmatrix} V_{IN1} \\ V_{IN2} \end{pmatrix} \quad (4)
$$

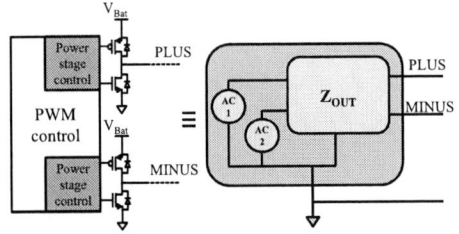

Fig. 4.   Amplifier frequency domain model.

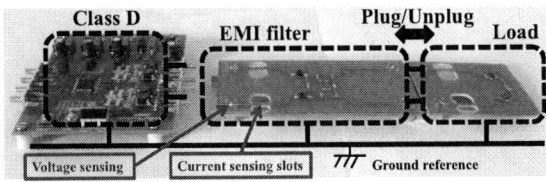

Fig. 5.   Boards used for method validation.

$$\begin{pmatrix} I_{OUT1} \\ I_{OUT2} \end{pmatrix} = Z_L^{-1} \begin{pmatrix} V_{OUT1} \\ V_{OUT2} \end{pmatrix} \qquad (5)$$

Where V_{IN1}, V_{IN2} and V_{OUT1}, V_{OUT2} are the filter input and output voltages respectively, I_{OUT1}, I_{OUT2} are the filter output currents, I is a 2×2 eye matrix, Z_F is the filter matrix presented above, Z_L is the load matrix and Z_R is the resulting matrix.

The proof of equation (4) can explained as follow.

By definition (see Fig. 3):

$$\begin{pmatrix} V_{IN1} \\ V_{IN2} \end{pmatrix} = Z_{F11} \begin{pmatrix} I_{IN1} \\ I_{IN2} \end{pmatrix} + Z_{F12} \begin{pmatrix} I_{OUT1} \\ I_{OUT2} \end{pmatrix} \qquad (6)$$

$$\begin{pmatrix} V_{OUT1} \\ V_{OUT2} \end{pmatrix} = Z_{F21} \begin{pmatrix} I_{IN1} \\ I_{IN2} \end{pmatrix} - Z_{F22} \begin{pmatrix} I_{OUT1} \\ I_{OUT2} \end{pmatrix} \qquad (7)$$

In (7), $\begin{pmatrix} I_{IN1} \\ I_{IN2} \end{pmatrix}$ and $\begin{pmatrix} I_{OUT1} \\ I_{OUT2} \end{pmatrix}$ can be replaced using equation (3) and 5. Thus, equation (4) can be computed.

III. Application

The method presented in section II is applied on amplification system for validation. The amplifier itself is a Class-D type amplifier controlled with a Pulse Width Modulation (PWM). The filter in use is a common structure for Class-D amplifiers and presented in [11]. The filter schematic is the same as the one shown in Fig. 2. The load is a dummy load which is equivalent to a loudspeaker used in cellphone applications. It contains two coils of $15\mu H$ and an 8Ω resistor. The three system parts are mounted on separate printed circuit boards, equipped with SMB connectors for an easy association/disassociation with a simple plug/unplug. Moreover, the boards are designed for an easy currents and voltages sensing on the filter and load inputs as shown in Fig. 5.

The filter and the load were simulated using ADS. Passive components were replaced by models from suppliers libraries

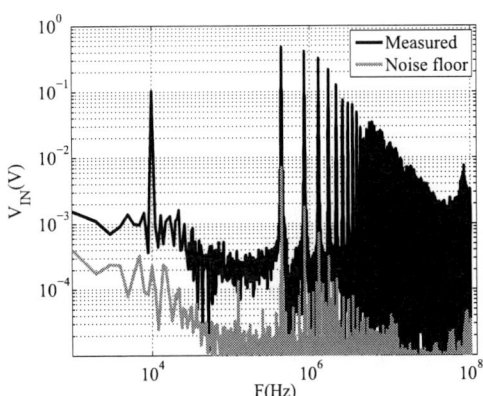

Fig. 6.   The Class-D amplifier input voltage measured.

dedicated to ADS environment. Thus, the high frequency component behavior is taken into account. PCB tracks are modeled by strip-lines and the PCB material values are included in the model. Thus, the impedance of the PCB tracks and the PCB layout are taken into account. Finally, the filter and load impedance matrices were obtained which are called Z_F and Z_L respectively. Using (2) the filter and load were concatenated into Z_R, the resulting matrix.

A test-bench was placed in an anechoic chamber to reduce the environmental noise. Then, a 12-bit quantization oscilloscope is used to measure the amplifier output voltages at the same time. Hereafter a FFT transforms the time domain signals into the frequency domain. In Fig. 6, the measurement noise floor is lower than the voltage spectrum over the entire frequency band. Thus, the voltages are valid to be used in the computation. Therefore, Z_F, Z_L, Z_R, and the measured input voltages were used in equation (3), equation (4) and equation (5) to compute the input current, the output voltages and the output current respectively. Note that, the input currents and the output current and voltage are measured to be compared with the computed ones. As the current probe used for current measurement has a bandwidth of $[1kHz - 120MHz]$, Fig. 7, Fig. 8 and Fig. 9 shows the results up to $120MHz$.

Fig. 7 shows that the computed input current have a good accordance with the measured one on the considered frequency range. The only difference between the two spectra is an extra peak at $30MHz$ in case of measurement and this is due to the current noise floor. However, in Fig. 8 the computed output voltage have a good accordance with the measured one up to $6MHz$; this is due to the the noise floor for voltage measurement. The instrumentation in use do not allow to do voltage measurements below than $15\mu V$. Finally, Fig 9 shows as well that the instrumentation in use do not allow any current measurement below $0.5\mu A$, thus the computed current is in a good accordance with the measured one up to $2MHz$. The output voltages and currents cannot be validated beyond $6MHz$ and $2MHz$ respectively, because they cannot be compared to the measurements due to the measurement noise floor.

978-1-4673-4579-8/13 $31.00 © 2013 IEEE

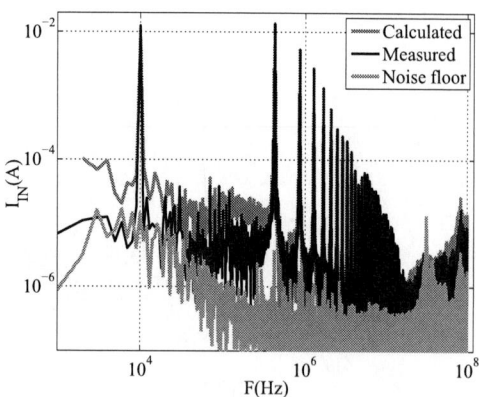

Fig. 7. Computed and measured input current.

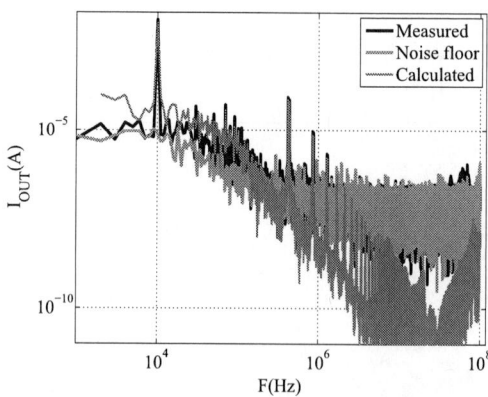

Fig. 9. computed and measured output current.

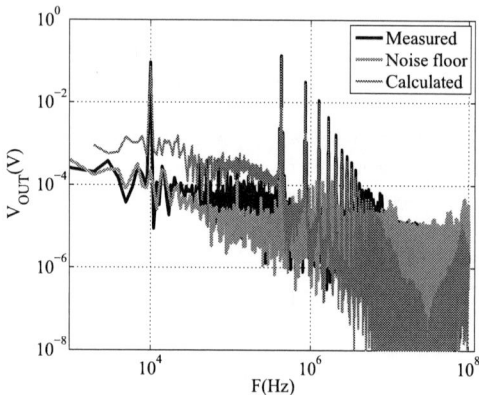

Fig. 8. Computed and measured output voltage.

IV. CONCLUSION

A modeling method in the frequency domain was presented. It is dedicated to EMI virtual measurements for integrated Class-D amplifiers which have a differential output. The system is decomposed into blocks, where the passive blocks are modeled with impedance matrices and active blocks such as Class-D amplifiers are modeled with AC voltage sources and an impedance matrix if needed. Once determined, the impedance matrices are concatenated in order to obtain a compact model for the system. The impedance matrices and the converter output voltages can be used for current and voltages computation in all the system points.

The approach is applied on a Class-D amplification system for validation. Good results are seen over the frequency ranges where the signal levels are greater than the noise floor. However, if the signal is lower than the noise floor, it means that the current is lower than $0.5\mu A$ or the voltage is lower than $15\mu V$. In these conditions, the EM emissions of the system have a very low impact on the surrounding electronics. Thus, the present method can still be helpful in EMI filter design.

The present approach allows system designers to take into account the EMI at the design stage. It is possible to make a library of blocks which can lead the EMI studies at the early development. Otherwise, this method have a short simulation time. Thus, it can be integrated in an optimization process for filter improvement.

The future work will be focused on developing and optimization process that uses this method for filter improvement.

REFERENCES

[1] A. Nagari, E. Allier, F. Amiard, V. Binet, and C. Fraisse, "An 8Ω ; 2.5W 1%-THD 104dB(A)-dynamic-range Class-D audio amplifier with an ultra-low EMI system and current sensing for speaker protection," in *Solid-State Circuits Conference Digest of Technical Papers (ISSCC), 2012 IEEE International*, feb. 2012, pp. 92 –94.

[2] M.-L. Yeh, W.-R. Liou, H.-P. Hsieh, and Y.-J. Lin, "An electromagnetic interference (EMI) reduced high-efficiency switching power amplifier," *Power Electronics, IEEE Transactions on*, vol. 25, no. 3, pp. 710 –718, mar. 2010.

[3] M. Xin, C. Zao, Z. Ze-kun, and Z. Bo, "An advanced spread spectrum architecture to improve EMI emissions in Class D amplifier," in *Communications, Circuits and Systems, 2009. ICCCAS 2009. International Conference on*, jul. 2009, pp. 661 –665.

[4] G. Dousoky, M. Shoyama, and T. Ninomiya, "On factors affecting EMI-performance of conducted-noise-mitigating digital controllers in DC-DC converters - An experimental investigation," in *Energy Conversion Congress and Exposition (ECCE), 2010 IEEE*, sept. 2010, pp. 1239 –1245.

[5] C. F. Edwards, "Efficient low EMI switching output stages and methods," no. US 7190225, mar. 2007, patent : US 7190225.

[6] T. Lim, H. Muir, S. Finney, and B. Williams, "Adaptive voltage slew control used to limit the magnitude of broadband conducted noise emissions for buck derived dc-dc converters," in *Electromagnetic Compatibility (APEMC), 2012 Asia-Pacific Symposium on*, may 2012, pp. 93 –96.

[7] T. Nakagawa and K. Osada, "Pulse position/width collaborative control for reducing EMI noise and ripple voltage in switching DC-DC converter," in *Applied Power Electronics Conference and Exposition (APEC), 2012 Twenty-Seventh Annual IEEE*, feb. 2012, pp. 1112 –1116.

[8] R. Mrad, F. Morel, G. Pillonnet, C. Vollaire, and D. Labrousse, "Differential passive circuit modelling with pentapole impedance matrices - application to an integrated audio switching amplifier for portable devices," in *EMC Europe 2011 York*, sept. 2011, pp. 304 –309.

[9] R. Mrad, F. Morel, G. Pillonnet, C. Vollaire, and A. Nagari, "Conducted EMI prediction for integrated Class D audio amplifier," in *Electronics, Circuits and Systems (ICECS), 2011 18th IEEE International Conference on*, dec. 2011, pp. 390 –393.

[10] R. Mrad, F. Morel, G. Pillonnet, C. Vollaire, P. Lombard, and A. Nagari, "N-conductor passive circuit modeling for power converter current prediction and emi aspect," *Electromagnetic Compatibility, IEEE Transactions on*, 2013 - in press.

[11] "MAXIM : MAX9700B Evaluation Kit," accessed: 08/01/2012.

Design of Mixers for a 130-GHz Transceiver in 28-nm CMOS

Dristy Parveg, Mikko Varonen, Mikko Kärkkäinen, and Kari A. I. Halonen
Aalto University School of Electrical Engineering, Department of Micro-and Nanosciences/SMARAD-2
Otakaari 5A, 02150 Espoo, Finland
dristy.parveg@aalto.fi

Abstract—A compact and 3-dB bandwidth of 118-145-GHz Gilbert-cell mixer for up-conversion and a 1-dB bandwidth of 106-143-GHz image-rejection (IR) resistive mixer for down-conversion are designed for a 130-GHz transceiver in 28-nm CMOS technology. A wide 10-GHz intermediate frequency (IF) tuning range is obtained for both mixers. The simulated results show a +1.6-dB conversion gain for the Gilbert-cell mixer with a layout size of $720{\times}633\mu m^2$ and 6.1 mW of DC power consumption. An 11-dB conversion loss and 30-dB IR ratio are simulated for the resistive mixer with a layout size of $845{\times}794\mu m^2$. The simulated 1-dB output compression point is -8 dBm for the Gilbert-cell mixer and 1-dB input compression point is +9 dBm for the resistive mixer.

Keywords—MMIC mixer; CMOS mixer; 130-GHz resistive mixer; 130-GHz Gilbert-cell mixer; 130-GHz transceiver

I. INTRODUCTION

The requirement of high-speed and wide-bandwidth wireless communication has pushed silicon technologies to higher operating frequencies. Good results have been demonstrated in nanoscale CMOS technology at millimeter (mm)-wave frequencies [1]-[3].

In this paper we present wideband 130-GHz up and down-conversion mixers designed for a transceiver suitable for both FMCW radar and millimeter-wave identification (MMID) applications. This system, which sets specific requirements for the mixers, is described in [4]. Here, we focus on detailed circuit design of a compact Gilbert-cell mixer for up conversion and balanced image-rejection (IR) resistive mixer for down conversion in 28-nm CMOS technology.

II. GILBERT-CELL MIXER

A. Circuit Design

The upconverter is designed for slow data-rate up-conversion and when operated in FMCW mode the local oscillator signal is fed through the mixer by unbalancing the mixer [4]. Furthermore, the output power of the up-conversion mixer should be high to reduce the gain requirements of the millimeter-wave power amplifier. Therefore, the active Gilbert-cell configuration, shown in Figure 1, was chosen for up-conversion.

The transistor Q1 and Q2 ($2{\times}35\mu m$) commonly known as transconductance stage are biased at saturation region, which convert IF signal voltage to current. The switching stage, transistors Q3, Q4, Q5, and Q6 ($2{\times}25\mu m$) are biased near pinch-off region to act as a switch at LO frequency.

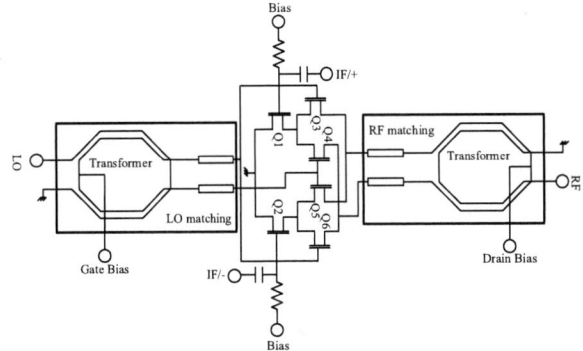

Fig. 1. Simplified schematic of the designed up-conversion Gilbert-cell mixer.

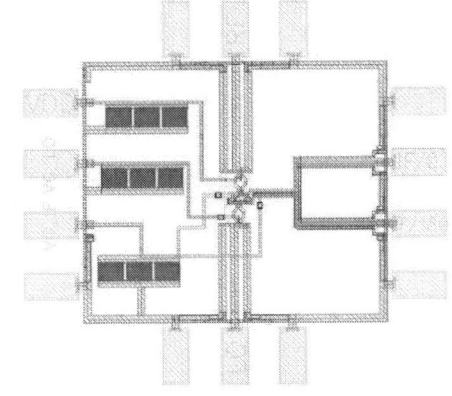

Fig. 2. Layout of the designed Gilbert-cell mixer. The chip area including the pad is $720{\times}633\mu m^2$. The active core area is $52{\times}92\mu m^2$ only.

In this design, an on-chip transformer is used at RF-port to combine the differential output signal to single-ended signal and another on-chip transformer is used at LO-port for converting the external single-ended mm-wave signals to differential signals. The transformers together with short pieces of transmission lines also provide the required impedance matching at the RF and LO ports. DC biasing is provided through the center tap of the transformers. The transformers are realized with two top metal layers from the 28-nm technology. The line width and the inner diameter of the used transformers are $4\mu m$ and $14\mu m$ respectively. A chip layout with a compact chip size of $720{\times}633\mu m^2$ in 28-nm CMOS technology is shown in Fig. 2.

978-1-4673-4579-8/13 $31.00 © 2013 IEEE

Paper M1B1

B. Simulated Results

The mm-wave transistor modeling is based on RC parasitic extractions and electromagnetic simulations of gate, drain, and source access parasitics. The passive components and transformers used for designing the Gilbert-cell mixer are simulated in ADS Momentum (2.5D EM simulator). The completed circuit is simulated in Mentor's ELDO. In all simulations, a 0-dBm LO power is used. The simulated DC power consumption is only 6.1 mW for the designed active mixer. The conversion gain, upper-side band (USB) RF power, and lower-side band (LSB) RF power as a function of the IF input power are shown in Fig. 3.

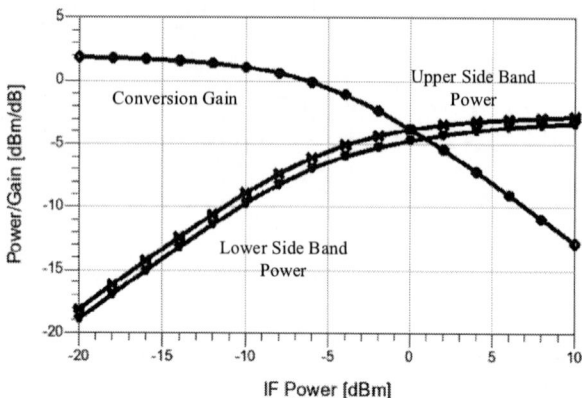

Fig. 3. Simulated up-conversion gain (circle), upper side band (USB) RF power (star), and lower side band (LSB) RF power (triangle) as a function of the IF input power. The IF and LO are fixed at 2 GHz and 130 GHz respectively.

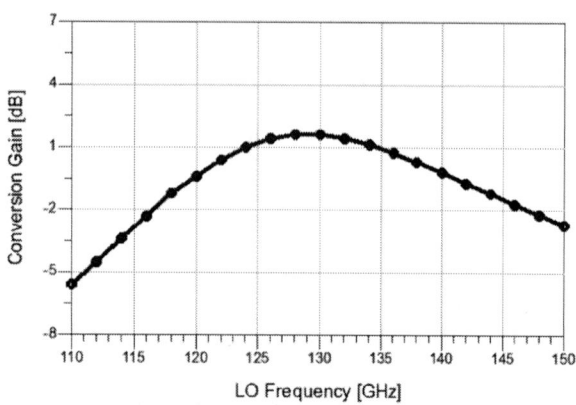

Fig. 4. Simulated up-conversion gain as a function of the LO frequency. The IF is fixed at 2 GHz with a -20-dBm power.

The conversion gain is around +1.6 dB when the IF power is -20 dBm. The simulated 1-dB output compression point is at a -8-dBm power level. A 3-dB bandwidth of 118-145-GHz is achieved as shown in Fig. 4. Simulated conversion gain over IF frequency sweep at fixed 130-GHz LO frequency is illustrated in Fig. 5. A wide 10-GHz IF frequency tuning range is achieved. For an up-conversion mixer LO-to-RF isolation is an important figure of merit. Figure 6 shows the

simulated LO-to-RF and RF-to-LO isolation which are better than 30 dB.

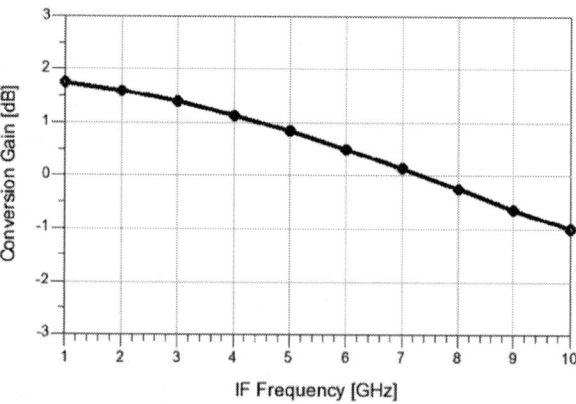

Fig. 5. Simulated up-conversion gain with a fixed LO frequency at 130 GHz as a function of IF frequency.

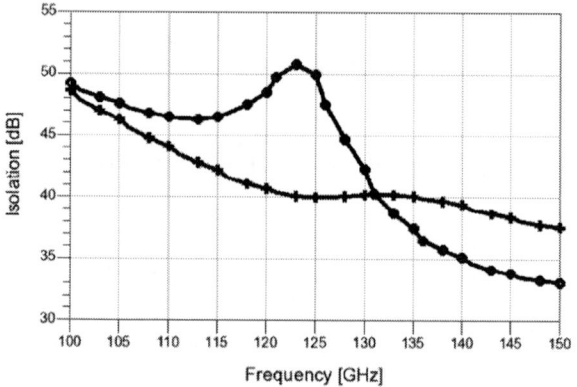

Fig. 6. Simulated LO-to-RF isolation (circle) and RF-to-LO isolation (cross) of the designed CMOS Gilbert-cell mixer.

III. RESISTIVE MIXER

A. Circuit Design

The downconverter is designed for high data rate transfer. Furthermore, the system in [4] utilizes a directional coupler which has limited isolation. This makes the design of the downconverter mixer challenging since the high transmitted output power leaks through the coupler and can compress the receiver. Therefore, we utilize singly balanced resistive unit mixers in our IQ-mixer to fulfill the linearity requirements in both MMID and FMCW modes. The design principle of the image rejection (IR) resistive mixer is similar to the one published in [5] and the simplified circuit of the balanced resistive IR mixer is shown in Fig. 7. It should be noted that the topology is suitable for both the up and down-conversion. The balanced IR resistive mixer consists of four 2×25μm transistors in 28-nm CMOS technology. The RF signal is fed through a Lange coupler to generate the required 90-degree phase shift for the two singly balanced resistive unit mixers. The Lange coupler provides accurate and wideband 90 degree phase shift with good amplitude balance between the coupled and the through port which is important for good image rejection performance. The Lange coupler is implemented in microstrip environment with finger width of 1.2μm and spacing of 0.8μm.

The LO signal is fed from a coplanar waveguide and the in-phase power is divided into two spiral transmission line baluns. The spiral transmission line balun provides 180^0 phase shift for the transistor gates of a unit mixer [6]. The direct matching technique used in [5] for the LO-port can lead to relatively high LO-power requirement. In [7] a buffer amplifier was added to relax the impedance matching of the LO-port in order to lower the LO-power. However, the buffer amplifier at 130 GHz can be inefficient and may increase the chip area. In this work, a new straight-forward approach is suggested for LO-port matching. A series line together with a short-circuited differential stub is used to transform the low-impedance of the transistor gate node to higher impedance level which relaxes the impedance matching of the LO-port. The short-circuited differential shunt stub is further used for biasing the transistor gates of the unit mixer. This method also leads to lower coupling requirement for the spiral balun, thus, making it easier to fulfill the metal density requirements in nanoscale CMOS technologies. In this design, the width of the transmission line of the balun is 2μm and the lines are separated by 9.6μm gap. For preventing the DC from bypassing to ground, finger capacitors are used at the ground connection of the spiral balun. All finger capacitors used in this work are custom designed slow-wave plate capacitors proposed in [8].

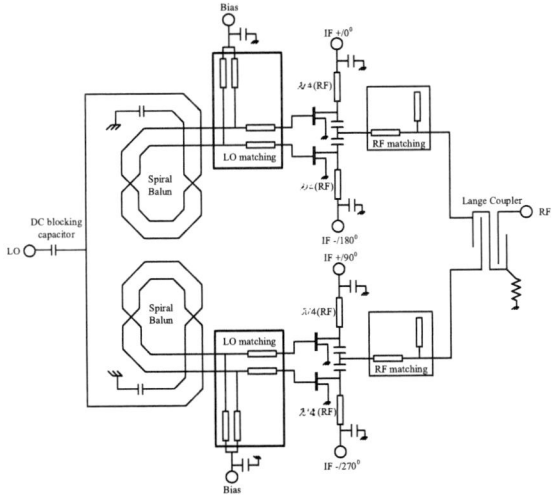

Fig. 7. Simplified schematic of the designed down-conversion balanced IR resistive mixer.

The transistor drains from each singly balance mixer are connected together with small valued finger-capacitors. The connection points act as an open circuit for the IF and as a virtual ground for the LO. The IF signals are extracted with an off-chip differential 90^0 combiner and 180^0 hybrids. Quarter-wavelength short-circuited stubs at RF frequency are used in the IF line, which isolate the millimeter-wave signals from the IF circuitry. The quarter wavelength line in the IF line is

realized in slow-wave coplanar waveguide (SW-CPW). This type of line provides substrate shielding and increased electrical length, which helps to reduce circuit area without sacrificing the performance [9]. The center conductor width of the used SW-CPW is 12μm and the signal-to-ground spacing is 9μm. The shielding is with the lower most two metal layers. The simulated SW-CPW provides the characteristic impedance of around 48Ω. The RF-port is matched to 50 ohm with a series SW-CPW line and an open-circuited shunt SW-CPW stub. A chip layout with a compact chip size of $845 \times 794 \mu m^2$ in 28-nm CMOS technology is shown in Fig. 8.

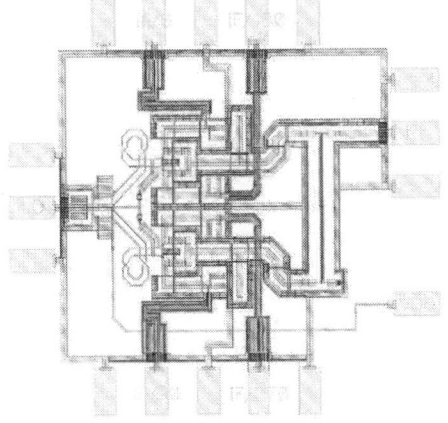

Fig. 8. Layout of the designed resistive mixer. The chip area including pads is $845 \times 794 \mu m^2$.

B. Simulated Results

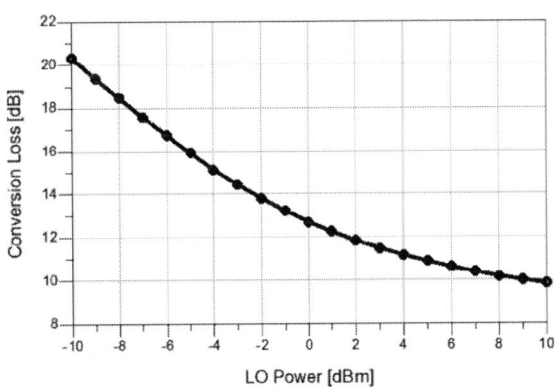

Fig. 9. Simulated down-conversion loss with a fixed LO frequency at 130 GHz as a function of LO power. IF is fixed at 5-GHz.

Same transistor modeling and passive components simulations as used for designing the Gilbert-cell mixer is also used for the resistive mixer designing.

In all simulations a +4-dBm LO-power is used, although, Fig. 9 shows that we can reduce the LO power to 0 dBm while sacrificing only 1.5-dB conversion loss, which suggest that the proposed new LO-matching approach is useful. The simulated large-signal performance is shown in Fig. 10. The conversion gain is around -11 dB in the linear region and a 30-dB IR ratio is achieved. The simulated input 1-dB compression point

Paper M1B1

(ICP) is at +9dBm. A wideband RF frequency tuning range at fixed 5-GHz IF frequency is obtained with a flatness of 1-dB for frequencies 106-143-GHz as shown in Fig. 11. Figure 11 also present the simulated LO-to-RF isolation and IR-ratio which are better than 25 dB. Simulated conversion loss over IF frequency sweep at fixed 130-GHz LO frequency is illustrated in Fig. 12. A wide 10 GHz IF frequency tuning range is achieved.

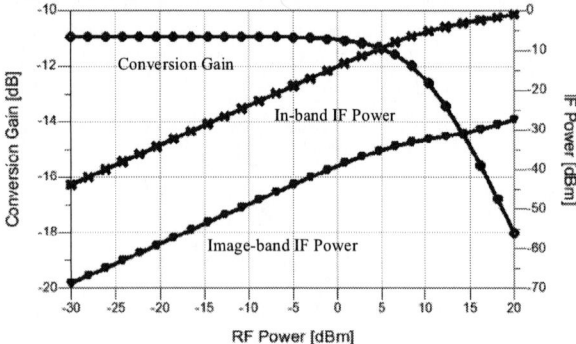

Fig. 10. Simulated down-conversion gain (circle), in-band IF output power (star), and image IF power (triangle) as a function of the RF input power. RF is fixed at 135 GHz. A +4 dBm LO power is used at 130 GHz.

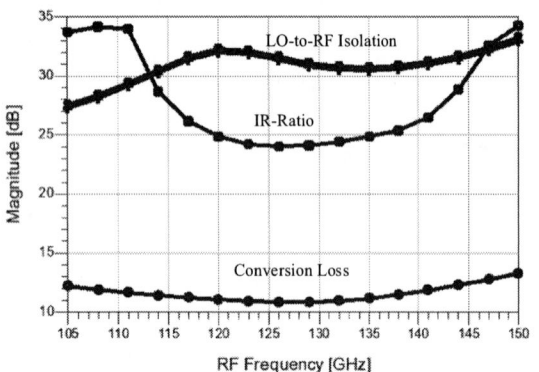

Fig. 11. Simulated LO-to-RF isolation, IR-ratio, and down-conversion loss with a fixed IF frequency at 5-GHz as a function of RF frequency.

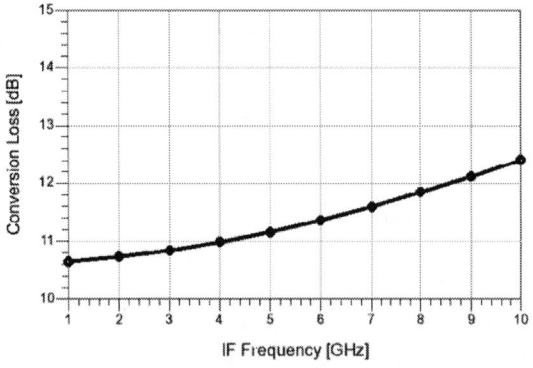

Fig. 12. Simulated down-conversion loss with a fixed LO frequency at 130 GHz as a function of IF frequency.

IV. CONCLUSION

This paper describes the design of MMIC up and down-conversion mixers for a 130-GHz transceiver. Simulated results of both mixers are presented and discussed. The presented results show promising conversion characteristics at the operating frequency and a wideband LO and IF frequency tuning range is achieved. To the best of our knowledge this is the first report describing the design of an IR resistive and Gilbert-cell mixer operating around 130 GHz and implemented in 28-nm CMOS technology. The obtained results show that the chosen design methods and the 28-nm CMOS technology are useful at this operating frequency.

ACKNOWLEDGMENT

The authors of this paper would like to thank all the project partners. This work was supported by the Finnish Funding Agency for Technology and Innovation (Tekes) under the BEAMS and SIMIDS projects, by the Academy of Finland under the FAMOS project, and by ENIAC JU project MIRANDELA. Authors are also grateful to STMicroelectronics for silicon donation.

REFERENCES

[1] Sandström, D.; Martineau, B.; Varonen, M.; Kärkkäinen, M.; Cathelin, A.; Halonen, K.A.I.; , "94GHz power-combining power amplifier with +13dBm saturated output power in 65nm CMOS," in *IEEE Radio Frequency Integrated Circuits Symposium Dig.*, Montreal, Canada, June 2011, pp.1-4.

[2] Inac, O.; Fung, A.; Rebeiz, G.M., "Double-balanced 130–180 GHz passive and balanced 145–165 GHz active mixers in 45 nm CMOS,"in *IEEE Custom Integrated Circuits Conference*, Sept. 2011, pp. 1-4.

[3] Laskin, E.; Khanpour, M.; Nicolson, S.T.; Tomkins, A.; Garcia, P.; Cathelin, A.; Belot, D.; Voinigescu, S.P., "Nanoscale CMOS Transceiver Design in the 90–170-GHz Range," *IEEE Transactions on Microwave Theory and Techniques*, vol.57, no.12, pp. 3477-3490, Dec. 2009.

[4] A. Tamminen, J. Ala-Laurinaho, D. Gomes-Martins, J. Häkli, P. Koivisto, M. Kärkkäinen, S. Mäkelä, P. Pursula, P. Rantakari, M. Sipilä, J. Säily, R. Tuovinen, M. Varonen, K.A.I. Halonen, A. Luukanen, and A.V. Räisänen, "Reflectarray for 120-GHz beam steering application: design, simulations, and measurements," *SPIE Defense, Security, and Sensing 2012, Passive and Active Millimeter-Wave Imaging XV*, Baltimore, April 26th 2012.

[5] Varonen, M.; Kärkkäinen, M.; Riska, J.; Kangaslahti, P.; Halonen, K.A.I.; , "Resistive HEMT mixers for 60-GHz broad-band telecommunication," *IEEE Transactions on Microwave Theory and Techniques*, vol.53, no.4, pp. 1322- 1330, April 2005.

[6] Maas, S.; Kintis, M.; Fong, F.; Tan, M., "A broadband planar monolithic ring mixer," in *IEEE Microwave and Millimeter-Wave Monolithic Circuits Symposium Dig.*, June 1996, pp. 51-54.

[7] Sandstrom, D.; Varonen, M.; Karkkainen, M.; Halonen, K.A.I., "A W-band 65nm CMOS transmitter front-end with 8GHz IF bandwidth and 20dB IR-ratio," in *IEEE Solid-State Circuits Conference Dig.*, pp.418,419, Feb. 2010.

[8] Sandstrom, D.; Varonen, M.; Karkkainen, M.; Halonen, K., "W-band CMOS amplifiers achieving +10dBm saturated output oower and 7.5dB NF," *IEEE Journal of Solid-State Circuits*, vol. 44, pp. 3403-3409, Dec. 2009.

[9] Cheung, T.S.D.; Long, J.R., "Shielded passive devices for silicon-based monolithic microwave and millimeter-wave integrated circuits," *IEEE Journal of Solid-State Circuits*, vol.41, no.5, pp.1183-1200, May 2006.

A 12dBm IIP3 Reconfigurable Mixer for High/Low Band IR-UWB Receivers

Mirko Pasca, Vincenzo Chironi, Stefano D'Amico
Dept. of Innovation Engineering
University of Salento, Lecce-Italy
mirkopasca@gmail.com

Marcello De Matteis, Andrea Baschirotto
Dept.of Physics
University of Milan-Bicocca. Milano-Italy

Abstract—**This paper presents a highly linear low power fully differential downconversion mixer for impulse radio ultra wideband (IR-UWB) receivers. The downconversion mixer is designed for IR-UWB IEEE 802.15.4a standard compliant receivers. It can be reconfigured according to the selected operation channel. In fact, it enables the downconversion of the #3 mandatory channel in low band (4.4928 GHz carrier frequency, 499.2 MHz channel bandwidth), or #9 mandatory channel in high band (7.9872 GHz carrier frequency, 499.2 MHz channel bandwidth), or #11 optional channel in high band (same carrier frequency of channel #9 but 1.331 GHz channel bandwidth).**
Linearity of the proposed mixer is improved utilizing derivative superposition method and source degenerations at the input stage. The proposed mixer has been designed in a 65 nm CMOS technology. Post layout simulations result in 12 dBm IIP3, 16.8 dB minimum noise figure while consuming 2.7 mW from 1.2 V supply voltage.

I. INTRODUCTION

In the last years IR-UWB transceivers have been used in many kind of applications like sensor networks and short range communications due to their low power consumption and high operating frequencies [1], [2], [3].
Most downconversion mixer from literature ([4], [5], [6]) work only in the low or in the high band of the IEEE802.15.4a standard. The proposed mixer is designed for a standard compliant IR-UWB IEEE802.15.4a receiver able to operate in #3 low band mandatory channel (4.4928 GHz carrier frequency, 499.2 MHz channel bandwidth), in #9 high band mandatory channel (7.9872 GHz carrier frequency, 499.2 MHz channel bandwidth), and in #11 high band optional channel (7.9872 GHz carrier frequency, 1331.2 MHz channel bandwidth). Reconfigurability is obtained by using a digitally programmable input stage and resistive arrays load.
Downconversion mixer is the second block in the IR-UWB IEEE 802.15.4a receiver chain [7], [8] and is considered one of the most critical blocks because of strict linearity specification [9].
In common double balanced mixers like Gilbert mixer, input referred third order intercept point (IIP3) is directly proportional to bias current of mixer. However too large currents cause large voltage drop on load resistors reducing output signal swing and increasing power consumption.
The presented mixer has been designed for a fully digital IR-UWB receiver chain. It uses Gilbert switching stage and source degenerated RF input stage, in order to achieve high linearity. To evaluate conversion gain, on-off time and linearity performances, post layout simulations have been done.

This paper is organized as follows: Section II shows mixer topology and explains circuit design choices. Section III describes linearization method. Section IV shows post layout simulation results. Conclusions are drawn in section V.

II. MIXER TOPOLOGY

Mixer schematic is illustrated in Figure 1. The mixer is designed to be embedded in a IR-UWB IEEE802.15.4a receiver. Mixer specifications are shown in Table I. They have been obtained starting from a system level analysis. A 10 m maximum distance between receiver and transmitter in a indoor residential scenario, and the mandatory 0.85 Mbps datarate have been supposed.

TABLE I.   DESIGN SPECIFICATIONS

Channel #	Channel Bandwidth [MHz]	Frequency [MHz]	Gain [dB]	INBAND IIP3 [dBm]	P-1dB [dBm]	NF DSB [dB]
3	449.2	4492.8	4±1	>11	>-1	<25
9	449.2	7987.2	2±1	>11	>-1	<20
11	1331.2	7987.2	2±1	>11	>-1	<20

The proposed mixer is a Gilbert cell with passive output load, operating in IEEE 802.15.4a #3 and #9 mandatory channels at 4492.8 MHz and 7987.2 MHz center frequency respectively (with 499.2 MHz bandwidth), and in channel #11 at 7987.2 MHz (with 1331.2 MHz bandwidth). Source degenerated RF input stage, is made by transitors M_1, M_2, M_3 and M_4, degeneration source resistors R_5, R_6, R_7, R_8 and on-off switches M_{14} and M_{15}. This input stage is composed of a CMOS inverter amplifier. This kind of amplifer is particulary suitable to obtain high gain at low supply voltage and, at the same time, to improve linearity allowing second order distorsion cancellation.
In a conventional CMOS balanced Gilbert mixer, conversion gain is equal to:

$$A_V = 2/\pi \cdot gm \cdot R_L \qquad (1)$$

Where gm is transconductance of input transistors and R_L is value of load resistors. In an inverting input stage, conversion gain is given by:

$$A_V = 2/\pi \cdot (gm_n + gm_p) \cdot R_L \qquad (2)$$

Where gm_n and gm_p are respectively NMOS and PMOS transistor transconductances.

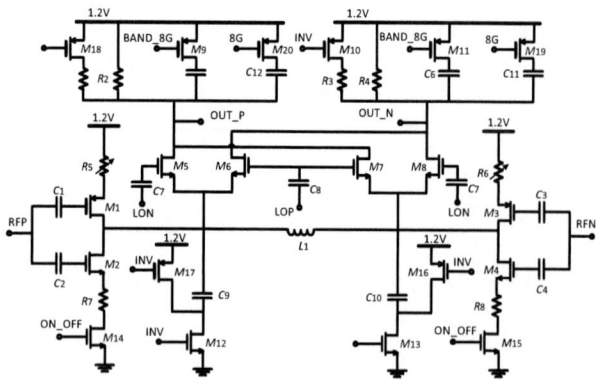

Fig. 1. Mixer Schematic.

The proposed mixer input stage uses degeneration resistors R_5, R_6, R_7 and R_8. Source degeneration resistors R_5-R_8 permit to improve linearity with a small loss in conversion gain. R_5 and R_6 are made by two series switchable resistors that allow to modify PMOS transistor degeneration tranconductance in order to work correctly at 4.4928 GHz or 7.9872 GHz.

Transistors M_2 and M_4 are biased in saturation region providing high gain to input signal.

Transistors M_1 and M_3 are biased in weak inversion region to cancel second order non linear currents improving IIP3 performance.

To limit load resistors size, we use inductor L_1, interposed between RF input stages, in order to tune out parasitic capacitance improving conversion gain.

C_9 and C_{10} capacitances (enabled by M_{12}, M_{13}, M_{16} and M_{17} switches) serve to modify resonance frequency in order to work in IR-UWB lower or upper band. M_{16} and M_{17} are necessary to avoid leaving C_9 and C_{10} floating when M_{12} and M_{13} are off. LC resonance peak frequency is equal to:

$$f_{res} = 1/2\pi\sqrt{LC} \qquad (3)$$

Where $C = C_{pM_1} + C_{pM_2}$ operating in channel #3 and $C = C_{pM_1} + C_{pM_2} + C_9$ operating in channels #9 and #11. C_{pM_1} and C_{pM_2} are M_1 and M_2 parasitic drain capacitances respectively. M_5, M_6, M_7 and M_8 transistors act as switches of Gilbert stage and are biased in weak inversion region. They have to be small to ensure small drain to drain capacitance, avoiding output signal swing reduction. M_5, M_6, M_7 and M_8 are driven by -4 dBm local oscillator (LO) power.

Output load resistors R_1, R_2 and R_3, R_4 are parallel switchable resistors, providing 2 dB conversion gain at 7.9872 GHz and 4 dB conversion gain at 4.4928 GHz with 12 dBm and 11.8 dBm IIP3. Transistors M_9, M_{11}, M_{19}, M_{20} act as switches in order to modify channel bandwidth working at 7.9872 GHz. Capacitors C_1, C_2, C_3, C_4 are used to remove DC input component. Table II shows active elements sizes.

III. LINEARITY ANALYSIS

The most important non linearity components in a down-conversion mixer, are second and third order intermodulation distortions. In zero IF architectures, switching pair non linearity and mismatches are sources of second order distortions generation [10].

TABLE II. TRANSISTOR SIZES

Transistor Name	W[μm]	L[nm]
M_1, M_3	160	120
M_2, M_4	70	240
M_{12}, M_{13}	900	60
M_{14}, M_{15}	100	60
M_{17}, M_{16}	0.8	60
M_5, M_6, M_7, M_8	40	100
M_9, M_{10}, M_{11}, M_{18}, M_{19}, M_{20}	7.5	60

Principal source of third order non linearity in active mixers is given by transconductance of RF input stage.

As demonstrated in [11], second order distortion and third order distortion can generate themselves reciprocally by sundry mechanisms. Relation by second and third order intermodulation products is given by: $V_{IIP2} = V_{IIP3}^3/4V_{os}$, where V_{os} is an offset voltage due to switching stage (or to the input stage), MOS threshold voltage mismatch, or to unbalanced currents in two sides of symmetrical mixer circuit.

Using Taylor's series expansion we can express RF section output current as:

$$i_{out} = i_{ds,n} - i_{ds,p} = \left(gmd_{M_1}(v_{in}) + \frac{gmd'_{M_1}}{2}(v_{in})^2 + \frac{gmd''_{M_1}}{6}(v_{in})^3 \right)$$
$$- \left(gmd_{M_2}(-v_{in}) + \frac{gmd'_{M_2}}{2}(-v_{in})^2 + \frac{gmd''_{M_2}}{6}(-v_{in})^3 \right) =$$
$$= \ldots + \frac{(gmd'_{M_1} - gmd'_{M_2})}{2}(v_{in})^2 - \frac{(gmd''_{M_1} + gmd''_{M_2})}{6}(v_{in})^3$$
$$(4)$$

Where gmd is the degeneration transconductance and is equal to: $gmd = 1/(1 + gm \cdot r_s)$, ($r_s$ is source degeneration resistor). gmd' and gmd'' are respectively gmd first and second order derivative. As we can see in 4, second order distorsion is related to input transistor transconductance first derivative, so, if gmd_{M_1} and gmd_{M_2} are equal or comparable, second order harmonic distorsion is attenuated.

IV. SIMULATION RESULTS

The proposed mixer is designed in CMOS 65 nm technology. Its layout, shown in Figure 2 is simmetrical and its sizes are 180 μm x 200 μm.

Most room occupation is due to inductor L_1 which is 1.94 nH with a 12.15 Q factor. Its occupation area is 46% of total occupation area. Inner radius measures 15 μm and there are 5.5 turns spaced by 2 μm.

A two tone test, with 10 MHz separation, has been performed to evaluate linearity parameters. At channel #3 center frequency, in-band tones have been set at 4368 MHz and 4358 MHz. RF signal power was set to -25 dBm. LO provided a sinusoid at 4492.8 MHz with 200 mV peak amplitude.

TABLE III. MIXER PERFORMANCES

Channel	#3	#9	#11
RF-frequency [MHz]	4498.2	7987.2	7987.2
IF-frequency [MHz]	0	0	0
Gain [dB]	4	2	2
IIP3 [dBm]	12	11.8	11.8
IIP2 [dBm]	59	47	47
P-1dB [dBm]	-2	-0.1	-0.1
Power [mW]	2.4	2.7	2.7
NF DSB	17.9	16.8	16.8

PRIME 2013, Villach, Austria   *Session M1B – Mixers and Oscillators*

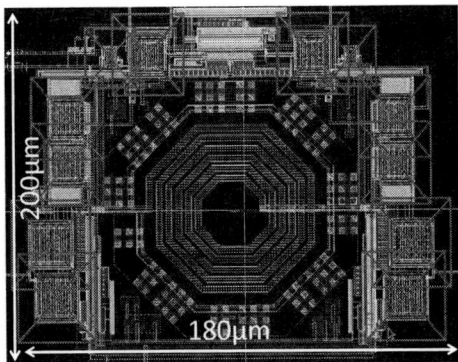

Fig. 2.   Mixer Layout (180 μm x 200 μm).

At channel #9 center frequency, in-band tones have been set at 7987.2 MHz and 7977.2 MHz. LO power has been set to -4 dBm.

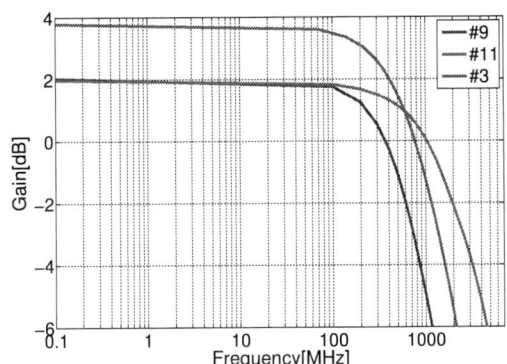

Fig. 3.   Mixer Conversion Gain.

Figure 6 shows 12 dBm IIP3 operating in channel #3 and 11.8 dBm IIP3 operating in channels #9 and #11. Mixer conversion gain is equal to 4 dB at 4.4928 GHz center frequency and 2 dB conversion gain at 7.9872 MHz center frequency. Figure 3 shows mixer output bandwidth at channels #3, #9 and #11 center frequencies. Resulting bandwidth is 700 MHz operating in channel #3, 500 MHz operating in channel #9 and 1.050 GHz operating in #11.

Figure 4 shows transient simulation results with a 4.4928 GHz RF input signal and 4.368 GHz LO signal frequency.

RF signal power has been set at -0.55 dBm and LO output power at -4 dBm. Output mixer signal is at 124.8 MHz frequency. Switching on and off transistors M_{14} and M_{15}, less than 6 ns on-off time has been obtained.

In Figure 5, noise figure performance in channel #3 is shown. It decreases as local oscillator output power increases and reaches his minimum at 2 dBm. At -4 dBm LO power (200 mV peak), double sideband noise figure (DSBNF) is equal to 17.9 dB operating in channel #3, and 16.8 dB operating in channels #9 and #11.

The 1 dB compression point is equal to -2 dBm at 4.4928 GHz and -0.1 dBm at 7.9872 GHz.

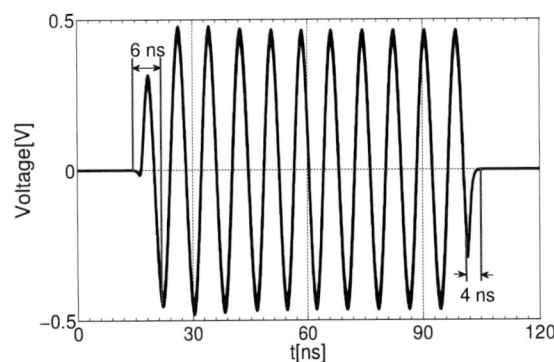

Fig. 4.   On-off performance for channel #3.

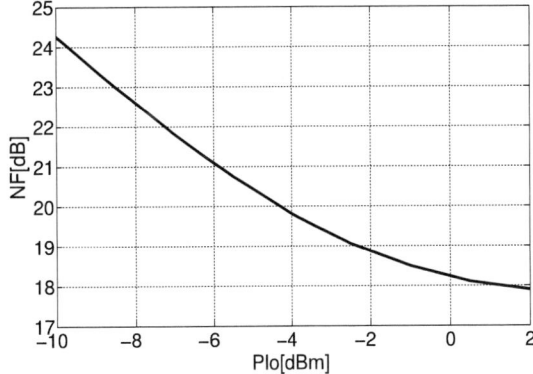

Fig. 5.   Noise Figure for channel #3.

Power consumption results in 2.4 mW working at 4.4928 GHz central frequency and 2.7 mW working at 7.9872 GHz carrier frequency. Mixer performances are illustrated in Table III. Table IV shows a comparison between the proposed mixer and downconversion mixers at state of the art. The proposed mixer is well compared in terms of linearity and power consumption.

V.   CONCLUSIONS

In this work a downconversion mixer in TSMC 65 nm CMOS technology has been presented. Presented mixer operates in channels #3, #9 and #11 of IEEE 802.15.4a standard and shows high linearity, using a derivative superposition method to perform second order distortion cancellation.

This cancellation is obtained using an inverting with degenerated source RF input stage. It permits to obtain simultaneously second order distortion cancellation and conversion gain improvement.

Switched output load and LC resonant load interposed between RF input stages allow it to work at 4.4928 GHz and 7.9872 GHz center frequencies, with 4 dB and 2 dB gain respectively.

Mixer IIP3 is equal to 11.8 dBm working at 7.9872 GHz and 12 dBm at 4.4928 GHz, high values compared with state of the art works. Post layout simulations of IIP2 results in 59 dBm IIP2 operating in channel #3 and 47 dBm operating in channels #9 and #11.

978-1-4673-4579-8/13 $31.00 © 2013 IEEE   83

TABLE IV. STATE OF THE ART COMPARISON

	RF-freq [GHz]	IF-freq [MHz]	Gain [dB]	IIP3 [dBm]	IIP2 [dBm]	P-1dB [dBm]	Power [mW]	NFdsb [dB]
[12]	3-10.6	250	9.5-12	0-4	-	>-8	8.5	8
[13]*	0.2-13	264	7.6-9.9	-10	-	-	0.88	11.7-13.9
[14]*	3.1-4.8	4-252	8.2-9.7	-3.2 - -1.2	-	-12.5 - -10.3	3.7	10.4-12.4
[15]*	2.1	0	16	9	78	-	4	-
[16]*	3-7	10.25	5.3-8.2	-3.2 - -0.3	37.3-43.4	-	2.5-5.8	6.6-10.5
This Work	4.5-8	0	4-2	12-11.8	59-47	-2 - -0.1	2.4-2.7	17.9-16.8

*measured

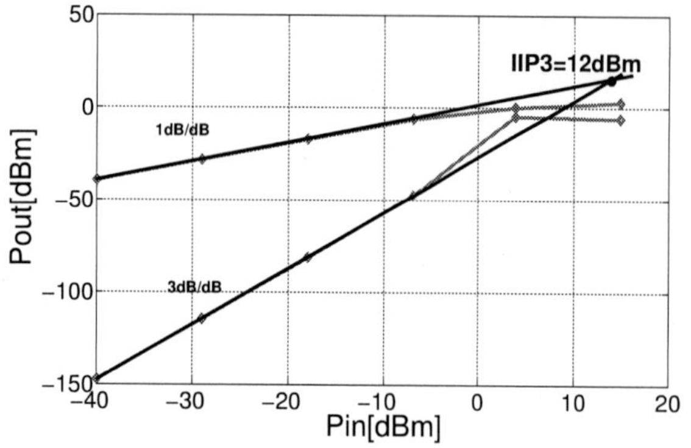

Fig. 6. IIP3 working for channel #3.

NF is equal to 16.8 dB at 7.9872 GHz and 17.9 dB at 4.4928 GHz. Power consumption is 2.4 mW working at 4.4928 GHz and 2.7 mW working at 7.9872 GHz.
The proposed mixer has been designed in 65 nm CMOS technology. Layout sizes (including bias circuit) are 180 μm x 200 μm.

ACKNOWLEDGMENT

This research has been partially supported by Regione Puglia, Italy (research project "Rete di sensori distribuita ad elevata efficienza energetica per monitoraggio industriale ed avionico operante in banda ultralarga con radio a impulsi", "RENDEZ VOUS" PO 2007 2013 Asse I Linea di Intervento 1.2 Azione 1.2.4 "Aiuto a sostegno dei Partenariati Regionali per l'Innovazione").

REFERENCES

[1] J. Fernandes and D. Wentzloff, "Recent Advances in IR-UWB Transceivers: an Overview," in Circuits and Systems (ISCAS), Proceedings of 2010 IEEE International Symposium on, pp. 3284–3287, 30 2010-June 2.

[2] X. Wang, K. Philips, C. Zhou, B. Busze, H. Pflug, A. Young, J. Romme, P. Harpe, S. Bagga, S. D'Amico, M. De Matteis, A. Baschirotto, and H. De Groot, "A High-Band IR-UWB Chipset for Real-Time Duty-Cycled Communication and Localization Systems," in Solid State Circuits Conference (A-SSCC), 2011 IEEE Asian, pp. 381–384, 2011.

[3] A. Costantini, A. Pezzotta, A. Baschirotto, M. De Matteis, S. D'Amico, F. Murtas, G. Gorini, "A CMOS 0.13um Low Power Front-end for GEM Detectors," in 19th IEEE International Conference on Electronics, Circuits and Systems (ICECS), pp. 193–196, Jan. 2012.

[4] H.-Y. Wang, K. F. Wei, J.-S. Lin, and H. R. Chuang, "A 1.2-V Low LO-Power 3-5 GHz Broadband CMOS Folded-Switching Mixer for UWB Receiver," in Radio Frequency Integrated Circuits Symposium, 2008. RFIC 2008. IEEE, pp. 621–624, 17 2008-April 17.

[5] C.-H. Wu and H.-T. Chou, "A 1.2-V High-Gain UWB Mixer Utilizing Current Mirror Topology," in Ultra-Wideband (ICUWB), 2010 IEEE International Conference on, vol. 1, pp. 1–4, Sept.

[6] M. Lei, H. Zhang, and C. Ma, "A 6-9 GHz Bi-Quadrature Folded-Switching Down-Conversion Mixer for MB-OFDM UWB Application in 0.18μm CMOS Technology," in Solid-State and Integrated-Circuit Technology, 2008. ICSICT 2008. 9th International Conference on, pp. 1496–1499, Oct.

[7] M. De Matteis, S. D'Amico, A. Costantini, A. Pezzotta, and A. Baschirotto, "A 1.25mW 3rd-order Active-Gm-RC 250MHz-Bandwidth Analog Filter Based on Power-Stability Optimization," in Electronics, Circuits and Systems (ICECS), 2012 19th IEEE International Conference on, pp. 260–263, 2012.

[8] A. Baschirotto, G. Cocciolo, M. De Matteis, A. Giachero, C. Gotti, M. Maino, G. Pessina, "A Fast and Low Noise Charge Sensitive Preamplifier in 90 nm CMOS Technology," in Journal of Instrumentation, pp. 1–8, Vol. 7, Jan. 2012.

[9] D. Fu, L. Huang, H. Du, and H. Yuan, "A 0.18nm CMOS High Linearity Flat Conversion Gain Down-Conversion Mixer for UWB Receiver," in Solid-State and Integrated-Circuit Technology, 2008. ICSICT 2008. 9th International Conference on, pp. 1492–1495, Oct.

[10] D. Manstretta, M. Brandolini, and F. Svelto, "Second-Order Intermodulation Mechanisms in CMOS Downconverters," Solid-State Circuits, IEEE Journal of, vol. 38, no. 3, pp. 394–406, Mar.

[11] A. Abidi, "General Relations Between IP2, IP3, and Offsets in Differential Circuits and the Effects of Feedback," Microwave Theory and Techniques, IEEE Transactions on, vol. 51, no. 5, pp. 1610–1612, May.

[12] M. Jouri, A. Golmakani, M. Yahyabadi, and H. Khosrowjerdi, "Design and Simulation of a Down-Conversion CMOS Mixer for UWB Applications," in Electrical Engineering/Electronics Computer Telecommunications and Information Technology (ECTI-CON), 2010 International Conference on, pp. 937–940, May.

[13] M.-G. Kim, H.-W. An, Y.-M. Kang, J.-Y. Lee, and T.-Y. Yun, "A Low-Voltage, Low-Power, and Low-Noise UWB Mixer Using Bulk-Injection and Switched Biasing Techniques," Microwave Theory and Techniques, IEEE Transactions on, vol. 60, no. 8, pp. 2486–2493, Aug.

[14] Y. Gao, F. Huang, L. Wu, and J. Cheng, "A Low-Power Reconfigurable Mixer for MB-OFDM UWB Receivers," in Microelectronics Electronics, 2009. PrimeAsia 2009. Asia Pacific Conference on Postgraduate Research in, pp. 97–100, Jan.

[15] M. Brandolini, P. Rossi, D. Sanzogni, and F. Svelto, "A CMOS Direct Down-Converter With +78dBm Minimum IIP2 for 3G Cell-Phones," in Solid-State Circuits Conference, 2005. Digest of Technical Papers. ISSCC. 2005 IEEE International, pp. 320–601 Vol. 1, Feb.

[16] K. Choi, D. H. Shin, and C. Yue, "A 1.2-V, 5.8-mW, Ultra-Wideband Folded Mixer in 0.13-μm CMOS," in Radio Frequency Integrated Circuits (RFIC) Symposium, 2007 IEEE, pp. 489–492, June.

Phase Noise Comparative Analysis of LC Oscillators in 28-nm CMOS through the Impulse Sensitivity Function

Ilias Chlis[1,2], Domenico Pepe[1], and Domenico Zito[1,2]

[1]Tyndall National Institute, "Lee Maltings", Dyke Parade, Cork, Ireland
[2]Department of Electrical and Electronic Engineering, University College Cork, Cork, Ireland
Email: {ilias.chlis, domenico.pepe, domenico.zito}@tyndall.ie

Abstract—Comparative Phase Noise (PN) analysis of Hartley, Colpitts and Cross-coupled LC oscillators at 10 GHz in 28-nm CMOS technology is reported. The results of the PN direct plots in the Cadence-SpectreRF design environment are compared with the results obtained by the Impulse Sensitivity Function (ISF). The steps for deriving accurately the ISF are reported and discussed. The results show a very good agreement in a set of conditions, and confirm that the LC cross-coupled differential oscillator exhibits superior PN performance with respect to Colpitts and Hartley, and also that Colpitts exhibits slightly superior performance with respect to Hartley.

Keywords—impulse sensitivity function, phase noise, oscillators

I. INTRODUCTION

Phase Noise (PN) of integrated oscillators is the main bottleneck regarding the number of users that could use a given bandwidth [1]. This imposes severe requirements for the local oscillators in wireless communication transceivers [2, 3] and their implementation on silicon technologies, especially at very high frequency [4]. Oscillator phase noise has been studied extensively over the last decades [5-8]. Although these studies contributed significantly to a better understanding, most of them use the linear time-invariant (LTI) oscillator models. LTI phase noise theories provide important qualitative design insights, but are limited in their quantitative prediction of power spectral density levels [9], in some cases addressed by adopting nonlinear approaches [10].

A linear time variant (LTV) model for oscillators has been introduced in [9]. This allows a quantitative understanding of oscillator phase noise. The concept of Impulse Sensitivity Function (ISF), represented as $\Gamma(x)$, is also introduced in [9]. Since the oscillator is assumed as a linear time-varying circuit, its phase sensitivity to noise perturbations can be described in terms of its (time-varying) impulse response.

The evaluation of the ISF involves a significant amount of transient simulations and data extractions, resulting in time consuming calculations, potentially prone to inaccuracy. Despite the fact that new efficient methods operating directly in the frequency domain were recently proposed [11, 12], it is worth consolidating the best practice in evaluating accurately the ISF in relation to the simulator settings.

A comparative study of the phase noise prediction in LC oscillators based on ISF, can be helpful in gaining an insight about the phase noise injections and then contributing towards

the design of low phase noise oscillators.

A comparative analysis of cross-coupled and Colpitts differential LC oscillators at 2.9 GHz in 0.35-μm CMOS technology has been carried in [13], showing the superior performance of the cross-coupled topology. In this perspective, it could be interesting to extend the comparison also to other topologies and technology nodes.

A detailed procedure for ISF computation and phase noise prediction in a linear time-varying system, in particular a source-coupled CMOS multi-vibrator frequency reference up to 2 MHz is presented in [14]. All the results however, are derived only for a single amplitude value of an injected pulse.

This paper reports a comparative study of PN for three common oscillator topologies: Hartley, Colpitts and top-biased differential LC oscillators operating at 10 GHz. The steps for accurate evaluations of the ISF are detailed and the predictions for a wide set of amplitudes of the injected current pulse are compared with the results obtained by the direct plots of SpectreRF Periodic Steady State (PSS) analysis. An insight regarding the phase noise contribution to the total phase noise from each noise source is provided. Considerations on the proper choice of pulse amplitude and duration for the evaluation of ISF are also provided.

The paper is organized as follows. Section II summarizes the key analytical expressions for PN predictions through the ISF. Section III reports the design of the oscillators at 10 GHz in 28-nm CMOS technology. Section IV highlights the key steps and settings for accurate evaluations of the ISF. Section V reports the results of the comparative analyses. Finally, in Section VI, the conclusions are drawn.

II. PHASE NOISE PREDICTION THROUGH THE ISF

The impulse response from each current noise source to the oscillator's output phase can be written as [9, 15]:

$$h_\phi(t,\tau) = \frac{\Gamma(\omega_0\tau)}{q_{max}} u(t-\tau) \tag{1}$$

where q_{max} is the charge injected into a specific oscillator node at time $t=\tau$, $u(t)$ is the unit step function and $\Gamma(\omega_0\tau)$ is a dimensionless periodic function and can be expressed as a Fourier series [9, 15]:

$$\Gamma(\omega_0\tau) = \frac{c_0}{2} + \sum_{n=1}^{\infty} c_n \cos(n\omega_0\tau + \theta_n) \tag{2}$$

This work is supported by Science Foundation Ireland (SFI).

978-1-4673-4579-8/13 $31.00 © 2013 IEEE

The contribution to the $1/f^2$ region of the phase noise spectrum for any oscillator, traditionally indicated with $\mathcal{L}$, from each given noise source with a white power spectral density, can now be expressed as [9, 15]:

$$\mathcal{L}\{\Delta\omega\} = 10\log\left[\frac{\Gamma_{rms}^2}{q_{max}^2}\frac{\left(\frac{\overline{i_n^2}}{\Delta f}\right)}{2\Delta\omega^2}\right] \qquad (3)$$

where q_{max} is the charge injected into an oscillator node by the noise source i_n and $\Delta\omega$ is the offset from the carrier angular frequency. The contribution to the $1/f^3$ region of the phase noise spectrum for any oscillator, from each given noise source with a $1/f$ spectrum can be expressed as follows [9, 15], where $\omega_{1/f}$ is the flicker noise corner of the device:

$$\mathcal{L}\{\Delta\omega\} = 10\log\left[\frac{c_0^2}{q_{max}^2}\frac{\left(\frac{\overline{i_n^2}}{\Delta f}\right)}{8\Delta\omega^2}\frac{\omega_{1/f}}{\Delta\omega}\right] \qquad (4)$$

III. CIRCUIT TOPOLOGIES

Three LC oscillators have been designed and simulated: single-ended Colpitts, single-ended Hartley and top-biased cross-coupled differential oscillators, as shown in Fig. 1. The circuits have been designed in 28-nm CMOS technology by ST-Microelectronics. The frequency of operation is 10 GHz. The sizes of the transistors and the value of the inductors and capacitors used are reported in Table I. The quality factor of the tank has been chosen equal to 10, by considering ideal capacitors and including a parasitic resistance in the ideal inductor model. In all cases the DC power consumption is 6.3 mW. The output of the oscillators is taken after a 100 nF capacitor (not shown in Fig. 1) in order to remove the DC component.

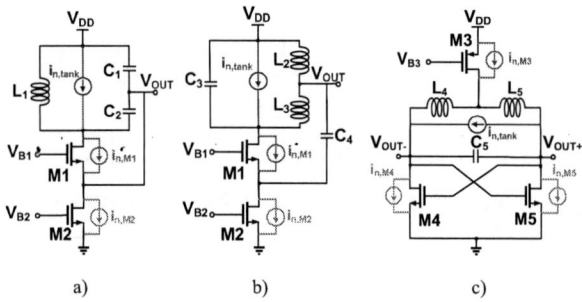

a)   b)   c)

Fig. 1. Schematic of the oscillators: a) single-ended Colpitts; b) single-ended Hartley; c) top-biased cross-coupled differential. V_{B1}, V_{B2} and V_{B3} are DC bias voltages.

IV. SIMULATION STEPS AND SETTINGS

The ISF of the oscillators has been evaluated as described in [14]. First we run a transient simulation in order to observe

TABLE I. OSCILLATOR CIRCUIT SIZING

Transistor	Width [μm]	Inductor	Value [pH]	Capacitor	Value [fF]
M1	30	L1	500	C1	970
M2	30	L2	250	C2	970
M3	30	L3	250	C3	495
M4	15	L4	500	C4	800
M5	15	L5	500	C5	229

when the amplitude reaches the steady state regime. In our case, this occurs with large margins after 5 ns. Afterwards, we apply the current impulsive sources placed in parallel to the inherent current noise sources of the LC tank and transistors, by activating one noise source at a time. The pulse duration of each current source has been chosen equal to 1 ps, with rise and fall time of 0.1 ps. The simulation has been repeated for amplitudes of injected current equal to 1, 10, 100 μA and 1, 10 mA. Each transient analysis is performed using the conservative mode and a maximum time step of 10 fs (i.e. 10^4 times smaller than the oscillation period), in order to have a good accuracy even in the case of the smallest amplitude of the injected current. The charge q_{max} injected in each case is the area under each pulse:

$$q_{max} = I_{pulse}(1.1)10^{-12} \qquad (5)$$

where I_{pulse} is the value of each source.

This is repeated for all the N noise sources connected in parallel, for the M instants of time included in a period of oscillation, where N=3 and M=40 in our case. The time instants have been chosen to be equally spaced in the oscillation period. The time shift caused by the impulse injection is then extracted.

After the oscillation has reached the steady state, the time shift Δt_i of the zero-crossing instant of the perturbed oscillation with respect to the unperturbed one, i.e. in the case in which no impulse is applied, is calculated.

These time shifts are then converted into phase shifts using the following relationship:

$$\Gamma(x) = 2\pi\frac{\Delta t_i(t)}{T} \qquad (6)$$

In order to take into account the cyclostationary nature of the noise sources, we multiply $\Gamma(x)$ with $\alpha(x)$, where $\alpha(x)$ is the absolute value of the unperturbed current flowing in the respective node where the impulses are injected, and at the same time when they are injected, normalized to its maximum value in one period.

Then we calculate the DC and root mean square (rms) components of the product $\Gamma(x)\alpha(x)$ as follows:

$$\Gamma_{DC} = \frac{\sum_{i=1}^{40}[\Gamma(x)a(x)]}{40} \qquad (7)$$

PRIME 2013, Villach, Austria

Session M1B – Mixers and Oscillators

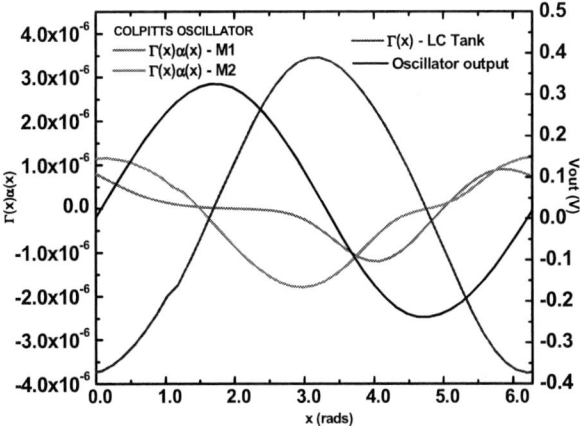

Fig. 2. $\Gamma(x)\alpha(x)$ vs. phase, for the Colpitts oscillator for a 1 µA current impulse.

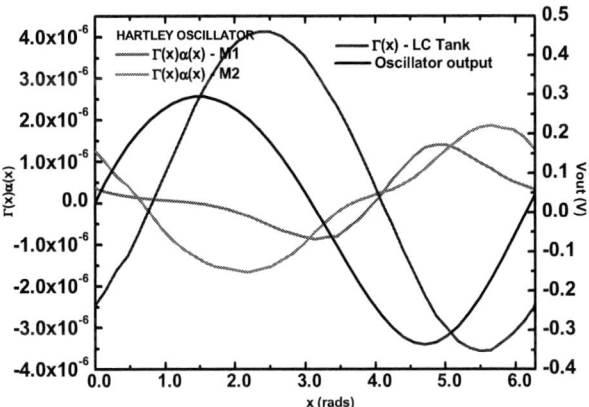

Fig. 3. $\Gamma(x)\alpha(x)$ vs. phase, for the Hartley oscillator for a 1 µA current impulse.

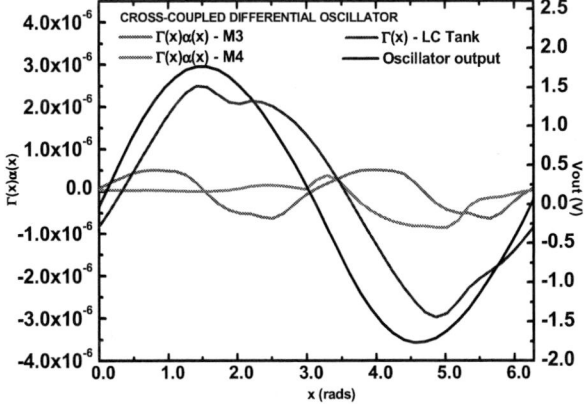

Fig. 4. $\Gamma(x)\alpha(x)$ vs. phase, for the cross-coupled differential oscillator for a 1 µA current impulse.

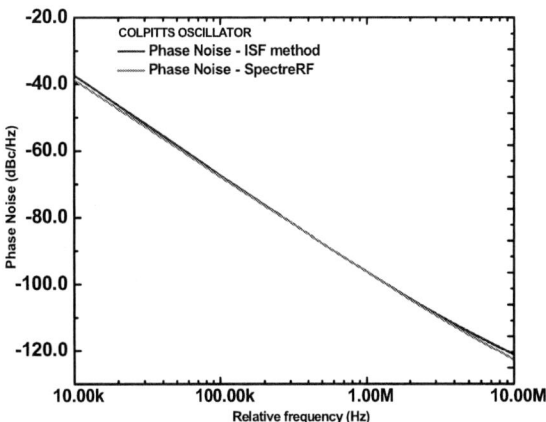

Fig. 5. PN vs. frequency offset for the Colpitts oscillator obtained through the ISF for a 1 µA current impulse and direct plot from PSS and periodic noise SpectreRF simulations.

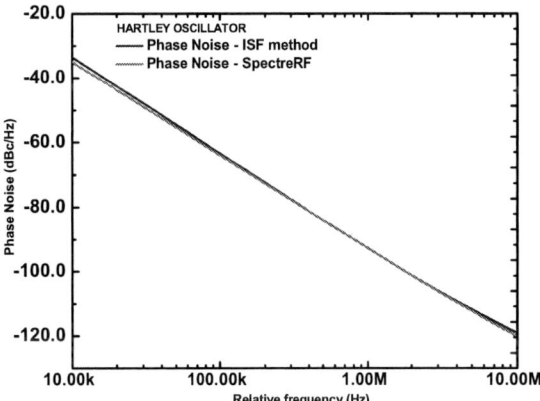

Fig. 6. PN vs. frequency offset for the Hartley oscillator obtained through the ISF for a 1 µA current impulse and direct plot from PSS and periodic noise SpectreRF simulations.

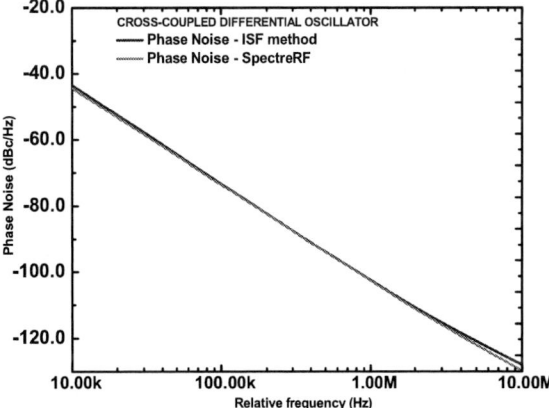

Fig. 7. PN vs. frequency offset for the cross-coupled differential oscillator obtained through the ISF for a 1 µA current impulse and direct plot from PSS and periodic noise SpectreRF simulations.

978-1-4673-4579-8/13 $31.00 © 2013 IEEE   87

$$\Gamma_{rms} = \sqrt{\frac{\sum_{i=1}^{40}\left\{\left[\Gamma(x)\,a(x)\right]^2\right\}}{40}} \qquad (8)$$

Finally, the total PN of the oscillator is computed by adding the contributions of all the noise sources acting in the circuit, according to (3) and (4). In particular, the noise of the transistors contributes to both the $1/f^3$ and $1/f^2$ regions of the oscillator's phase noise, whereas the noise of the LC tank contributes to only the $1/f^2$ region. The equation giving the total phase noise for each of the three oscillators is reported in (9), where m is the number of transistors of the oscillator.

$$\mathcal{L}\{\Delta\omega\}=10\log\left\{\sum_{i=1}^{m+1}\left[\frac{\Gamma_{rms}^{\,2}}{q_{max}^{\,2}}\frac{\left(\frac{\overline{i_n^{\,2}}}{\Delta f}\right)}{2\Delta\omega^2}\right]+\sum_{i=1}^{m}\left[\frac{\Gamma_{DC}^{\,2}}{q_{max}^{\,2}}\frac{\left(\frac{\overline{i_n^{\,2}}}{\Delta f}\right)}{8\Delta\omega^2}\frac{(\omega_{1/f})_i}{\Delta\omega}\right]\right\} \qquad (9)$$

V. RESULTS

Figures 2-4 report $\Gamma(x)a(x)$ versus the phase for the injected noise sources during one oscillation period, for an injected current pulse amplitude of 1 µA.

Figures 5-7 report the comparison between the PN obtained through the ISF and the PN obtained by direct plots from PSS and periodic noise simulations, for all the oscillators. Note that the PN predicted by the ISF method is very close to the values obtained by means of SpectreRF simulations. Table II provides the results for all the current impulse amplitude values, for a 1-MHz frequency offset from the carrier. Note that the PN of the cross-coupled oscillator is superior to the PN of the Colpitts oscillator (in agreement with [13]), which is superior to the PN exhibited by the Hartley oscillator. Moreover, note that the agreement degrades for higher pulse amplitudes, when the current-to-phase transfer function becomes nonlinear. The amplitude in which this occurs is slightly different for each oscillator, but for the injected current impulse of 1 µA, the percent difference between the phase noise predicted by the ISF method and the one given by PSS and periodic noise analysis is less than 1% at a 1-MHz frequency offset.

TABLE II. SUMMARY OF THE RESULTS

PN [dBc/Hz] @ 1 MHz frequency offset						
Oscillator	PSS	ISF				
		1 µA	10 µA	100 µA	1 mA	10 mA
Colpitts	-96.25	-96.20	-98.33	-98.49	-98.50	-98.45
Hartley	-92.75	-92.79	-95.18	-94.36	-94.85	-95.29
Cross-coupled	-102.66	-102.69	-102.84	-102.83	-102.84	-102.94

VI. CONCLUSION

Phase noise comparative analyses have been carried out for Colpitts, Hartley and cross-coupled differential LC oscillators at 10 GHz in 28-nm CMOS technology. All the steps and settings for accurate evaluations of the impulse sensitivity function (ISF) have been highlighted and discussed. The phase noise has been calculated directly from periodic steady state (PSS) simulations in Cadence-SpectreRF environment and compared with the results obtained through the ISF for a wide set of amplitudes of the injected current pulse. The predicted phase noise results are in a good agreement with the results obtained by PSS simulations performed by SpectreRF, especially for the pulse amplitude of 1 µA. Moreover, the results show that, under the adopted design conditions, the cross-coupled oscillator is superior in terms of phase noise performance to the Colpitts oscillator, confirming the results already achieved for the two oscillator topologies in 0.35-µm technology node, whereas the Colpitts oscillator is superior to the Hartley oscillator.

REFERENCES

[1] M.H.Madani, A. Abdipour, A. Mohammadi, "Analysis of performance degradation due to non-linearity and phase noise in orthogonal frequency division multiplexing systems" IET Communications, vol. 4, is. 10, 2010, pp.1226 – 1237.

[2] D. Pepe, D. Zito, "60GHz Transceivers in Nano-scale CMOS technology for wireless hd standard applications", IET Irish Signal and Systems Conference 2012, Maynooth, 28-29 June.

[3] H. Rashtian, S. Mirabbasi, "Using body biasing to control phase-noise of CMOS LC Oscillators", IET Electronics Letters, February 2012, vol.48, no. 3, pp. 168-169.

[4] S. T. Nicolson, K.H.K. Yau, P. Chevalier, A. Chantre, B. Sautreuil, K.W. Tang, S.P. Voinigescu, "Design and scaling of w-band SiGe BiCMOS vcos", IEEE JSSC, vol. 42, no.9, Sept. 2007, pp.1821-1833.

[5] D.B. Leeson, "A simple model of feedback oscillator noise spectrum", Proc. IEEE, vol. 54, pp. 329-330, Feb. 1966.

[6] J. Rutman, "Characterization of phase and frequency instabilities in precision frequency sources: fifteen years of progress", Proc. IEEE, vol. 66, pp. 1048-1174, Sept. 1978.

[7] A.A. Abidi, R.G. Meyer, "Noise in relaxation oscillators", IEEE JSSC, vol. SC-18, pp. 794-802, Dec. 1983.

[8] B. Razavi, "A study of phase noise in CMOS oscillators", IEEE JSSC, vol. 31, pp. 331-343, March 1996.

[9] A. Hajimiri, T.H. Lee, "A general theory of phase noise in electrical oscillators", IEEE JSSC, vol. 33, No. 2, pp. 179-194, Feb.1998.

[10] A. Buonomo, "Nonlinear analysis of voltage-controlled oscillators: a systematic approach", IEEE Trans. On Circuits and Systems I, vol. 55, no. 6, July 2008, pp. 1659-1670.

[11] P. Maffezzoni, "Analysis of oscillator injection locking through phase-domain impulse-response", IEEE Transactions on Circuits and Systems I: Regular Papers, vol. 55, no. 5, May 2008, pp. 1297-1305.

[12] S. Levantino, P. Maffezzoni, F. Pepe, A. Bonfanti, C. Samori, A.L. Lacaita, "Efficient calculation of the impulse sensitivity function in oscillators", IEEE Transactions on Circuits and Systems II: Express Briefs, vol. 59, no. 10, Oct. 2012, pp. 628-632.

[13] P. Andreani, X. Wang, "A study of phase noise in Colpitts and LC-tank CMOS oscillators", IEEE JSSC, vol. 40, May 2005, pp. 1107-1118.

[14] M. Paavola, M. Laiho, M. Saukoski, M. Kamarainen, K.A.I. Halonen, "Impulse sensitivity function-based phase noise study for low-power source-coupled CMOS multivibrators", Analog Integrated Circuits and Signal Processing, vol. 62, no. 1, pp. 29-41, Jan. 2010

[15] A.Hajimiri, T.H.Lee, "The design of low noise oscillators", Kluwer academic publishers, 2003.

Investigation of Two-Point Modulation to Increase the GFSK Data Rate of PLL-Based Wireless Transceivers of Wireless Sensor Nodes

Hartwig Unterassinger and Wolfgang Pribyl
Institute of Electronics
Graz University of Technology
Inffeldgasse 12, 8010 Graz, Austria
Email: {hartwig.unterassinger,wolfgang.pribyl}@tugraz.at

Martin Flatscher and Tony Gschier
Infineon Technologies Austria AG
Design Center Graz
Babenberger Straße, 8020 Graz, Austria
Email: {martin.flatscher,tony.gschier}@infineon.com

Abstract—Gaussian Frequency-shift-keying (GFSK) is a widely used modulation scheme in state-of-the-art wireless transceivers. Most often a phase-locked-loop (PLL) is utilized for carrier generation and GFSK modulation. Using this architecture the maximum applicable data rate is limited by the PLL's bandwidth. This paper investigates the possibility of using two-point modulation in order to be able to choose the GFSK data rate independently of the bandwidth. The most important aspect of two-point modulation is the mismatch of the gain of the two modulation paths. Two methods which allow close gain matching are presented in this work. Time-domain simulations using Verilog-A models of the analog and RTL models of the digital building blocks of a wireless transceiver have been carried out and confirm the capability of extending its GFSK data rate to 1 Mbps while using a PLL closed-loop bandwidth of 44.5 kHz.

I. INTRODUCTION

Nowadays GFSK modulation is one of the most popular digital modulation schemes used in wireless transceivers. In most state-of-the-art systems it is implemented by modulating the PLL's output frequency during closed-loop operation. That means the the PLL's bandwidth is a limiting factor concerning the maximum data rate which may be applied. The transceiver is intended to be used in a wireless sensor node requiring energy-efficient communication. Applying higher data rates leads to a shorter on-time of the whole system. [1] presents an energy-efficient communication protocol which requires transceivers supporting data rates of 1 Mbps. Two-point modulation has been investigated in [2], [3] as a method to enable data rates exceeding the PLL's bandwidth. The necessity of gain calibration of the two modulation paths has been established and addressed in [3], [4] and [5].

This paper presents two methods which allow using GFSK data rates exceeding the PLL bandwidth and it is organized in the following way. Section II features a short description of the transceiver. Section III explains the two-point modulation scheme and introduces two methods to calibrate the two modulation paths. Finally simulation results are shown in Section IV and the work is summarized in Section V.

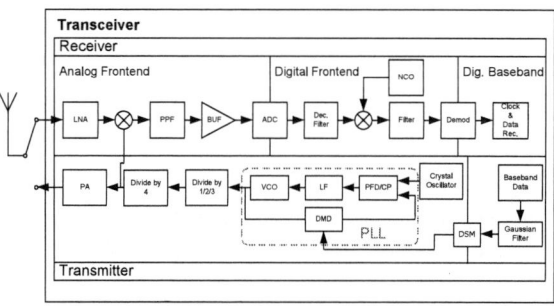

Fig. 1. Block Diagram of the Presented Transceiver

II. SYSTEM OVERVIEW

The transceiver is shown in Fig. 1. In transmit mode the PLL is used to translate the reference frequency provided by a crystal oscillator to the desired RF transmission frequency. Furthermore the PLL is used for GFSK modulation, which is implemented by changing the fractional value of the feedback divider via the $\Delta\Sigma$-modulator. The VCO's output frequency is then divided in order to generate an RF carrier in the desired frequency band, then it is amplified by the Power Amplifier (PA), and finally it is transmitted via an antenna.

In receive mode, the RF input signal is down-converted to an intermediate frequency (IF) after amplification by a low-noise amplifier (LNA). This is achieved by mixing the RF input signal with a signal provided by the PLL. Due to the fact that the mixer requires in-phase and quadrature-phase signals which are not directly provided by the PLL an additional division of the VCO output signal by a factor of four is necessary. After filtering, a $\Delta\Sigma$ analog-to-digital converter (ADC), samples the IF signal, thereby producing a digital IF signal. In the digital front end down-conversion to the baseband is done by mixing the digital IF signal with the output signal of an NCO. After further filtering the base-band signal is demodulated and processed in the digital base-band.

The transceiver is built to operate in the license free frequency bands for industrial-scientific-medical (ISM) purposes below a frequency of 1 GHz which leads to a VCO frequency range of 3.3 GHz to 4 GHz.

III. INCREASING THE GFSK DATA RATE BY USING TWO-POINT MODULATION

Fig. 2 shows a $\Sigma\Delta$ Fractional-N PLL which comprises a voltage-controlled-oscillator (VCO), a phase-frequency detector (PFD), a charge-pump (CP), a third order loop filter (LF), a dual-modulus divider (DMD) in the feedback path, and a $\Sigma\Delta$-modulator (DSM). The PLL is used to generate RF signals with a frequency which is a multiple of the reference frequency $f_{\mathrm{out}} = f_{\mathrm{ref}}(N + \frac{k}{2^b})$ where N is the integer multiplication factor, k the fractional multiplication factor, and b represents the bit width of the DSM. A GFSK modulation scheme can be implemented by using different values for k, resulting in different output frequencies.

Using this approach the maximum applicable data rate is limited by the PLL's closed-loop bandwidth. In order to increase the data rate the PLL bandwidth has to be increased but it has to be kept in mind that the bandwidth has to remain within certain limits in order to fulfill specifications like phase noise and lock time.

A. Two-Point Modulation

In principle the GFSK modulated data may be applied at three different points of the PLL, directly at the VCO, at the feedback divider, or at the reference oscillator. Modulating simultaneously at the feedback divider and directly at the VCO potentially cancels the PLL low-pass behavior and the maximum data rate is independent of the PLL loop bandwidth. The existing PLL has to be modified in order to enable modulation at the VCO. A second tuning input needs to be added to the VCO, where a second tuning voltage V_{MOD} can be applied in order to perform modulation. As shown in Fig. 2 this can be modeled as the summation of V_{LF} and V_{MOD}. Additional circuitry in the form of a digital-to-analog converter (DAC) is needed to generate V_{MOD} and a low-pass filter is required to suppress high-frequency components of the DAC. According to the gain of the VCO (K_{VCO}) the necessary V_{MOD} has to be provided by the DAC in order to ensure the necessary frequency deviation of the VCO output frequency and thus matching of the gain of the modulation paths.

If the two modulation paths match perfectly there is no perturbation of the GFSK data. In practice however there are gain (M_{gain}) and timing mismatches (M_{timing}) between the two modulation paths. In [2] and [3] PLL architectures using two-point modulation and the influence of mismatches between the two modulation paths are investigated. It has been shown that the influence of the M_{timing} is much smaller compared to M_{gain}'s significant impact.

In order to match the two gains the following considerations have to be kept in mind. If modulation is applied at the feedback divider the transfer function exhibits a low-pass behavior and is given by

$$H_{\mathrm{FB}}(s) = \frac{K_{\mathrm{PFD}}K_{\mathrm{VCO}}LF(s)N}{Ns + K_{\mathrm{PFD}}K_{\mathrm{VCO}}LF(s)} \qquad (1)$$

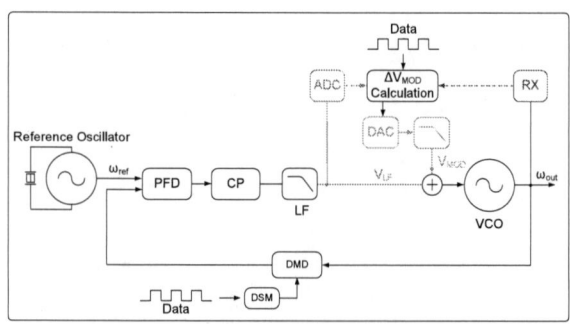

Fig. 2.   PLL with Two-Point modulation capability

and the frequency deviation can be calculated with $f_{\mathrm{dev}} = \frac{\Delta k}{2^b - 1} f_{\mathrm{ref}}$. Modulation at the VCO leads to a high-pass behavior given by

$$H_{\mathrm{VCO}}(s) = \frac{Ns}{Ns + K_{\mathrm{PFD}}K_{\mathrm{VCO}}LF(s)} \qquad (2)$$

and the frequency deviation is obtained by $f_{\mathrm{dev}} = \Delta V_{\mathrm{MOD}}K_{\mathrm{VCO}}$. Combining the two transfer functions shall lead to a flat transfer function which means that both modulation paths need to cause the exact same frequency deviation. This results in $\Delta V_{\mathrm{MOD}} = \frac{\Delta k}{2^b - 1}\frac{f_{\mathrm{ref}}}{K_{\mathrm{VCO}}}$. That means that the necessary V_{MOD} depends on the exact value of K_{VCO}. Unfortunately K_{VCO} is not known beforehand and varies significantly due to variations of the process, temperature, supply voltage and the VCO's operating frequency. Therefore determining K_{VCO} and thus adapting ΔV_{MOD} is necessary and two methods to achieve this task are presented next.

B. Matching the Modulation Paths

In order to match the gain of the two modulation paths two methods are introduced. They make use of existing building blocks of the transceiver depicted in Fig. 1. Both methods monitor the PLL's response when FSK data is applied and were developed in order to measure and calibrate the PLL bandwidth and are presented in [6]. After the start-up of the PLL is completed and it is locked to its carrier frequency a dedicated calibration phase is carried out before the actual data transmission takes place. During the calibration phase the applied data alternate between two binary FSK states with a given data rate. The Gaussian filter is turned off in order to maximize the PLL's response under closed-loop operation.

If ΔV_{MOD} is exact, the gain of both modulation paths match perfectly and V_{LF} should not be perturbed at all if data is applied. However, deviations of ΔV_{MOD} from its ideal value will be compensated by the PLL by adapting V_{LF}. On the one hand, if ΔV_{MOD} is too low V_{LF} will increase if a '1' and decrease if a '-1' is applied. V_{MOD} and V_{LF} change in the same direction. On the other hand, if ΔV_{MOD} is too high, the exact opposite behavior is caused. Thus by observing V_{LF} and iteratively changing ΔV_{MOD}, M_{gain} between the two modulation paths can be minimized.

The first method uses the $\Delta\Sigma$-ADC of the transceiver front-end which samples V_{LF} at a sampling frequency of 50 MHz. The resulting digital version of V_{LF} is subsequently processed in the digital core. This method integrates V_{LF} over one symbol period and compares the result of this integration to the expected value. The expected value is given by integration of V_{LF} when no data is applied. An integration result larger than the expected value represents a too small ΔV_{MOD} whereas a smaller result means that ΔV_{MOD} was too large. Depending on the integration result ΔV_{MOD} is increased or decreased. This procedure is illustrated in Fig. 3. In Fig. 3a ΔV_{MOD} is too low initially, thus V_{MOD} and V_{LF} change in the same direction. By increasing ΔV_{MOD}, M_{gain} is minimized. Fig. 3b shows a sequence where the initial ΔV_{MOD} is too high and thus V_{MOD} and V_{LF} change in opposite directions. By decreasing ΔV_{MOD}, M_{gain} is minimized.

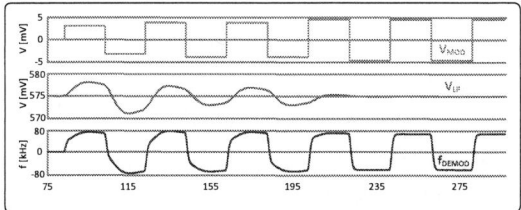

(a) Initial K_{VCO} estimation too low

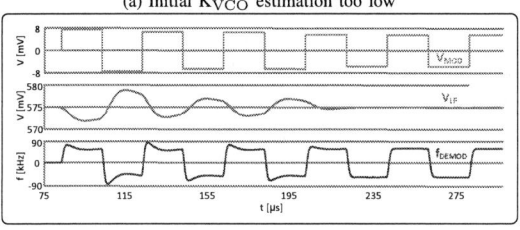

(b) Initial K_{VCO} estimation too high

Fig. 3.   Calibration of two gain paths

The second method to calibrate the modulation paths uses the receiver chain of the system but its basic operation has to be adapted. Instead of providing the LO signal the PLL is modulated and a second RF source is used to down-convert the PLL's output signal. The demodulator in the digital receiver provides a digital version of the PLL's output frequency. Ideally the shape of this demodulated signal f_{DEMOD} is a perfect rectangle and its amplitude equals f_{dev}.

Mismatches between the modulation paths lead to deviations from the ideal rectangular waveform in the form of over- and undershoots which can be detected. This approach is illustrated in Fig. 3. If ΔV_{MOD} is too low, as depicted in Fig. 3a, the initial frequency jump is too low and ΔV_{MOD} needs to be increased. In Fig. 3b the initial ΔV_{MOD} is too high and thus an overshoot occurs and ΔV_{MOD} is to be decreased. By integrating f_{DEMOD} over one symbol period and comparing the result to the ideal value, the over- and undershoots can be detected and a decision can be made whether ΔV_{MOD} is too high or too low. Again, M_{gain} between the two modulation paths can be minimized by adapting ΔV_{MOD}.

The second RF source required for this method is a still ongoing research topic and beyond the scope of this work. Currently methods using a higher harmonic of a crystal oscillator are investigated. Concerning this work the second RF source is generated by an external source.

Some publications dealing with two-point modulation cover the topic of gain calibration. Similar to the method using the ADC [4] evaluates V_{LF} with an ADC at two different voltages applied at the second analog tuning input of the VCO. [3] monitors V_{LF} using a dedicated comparator circuit. The work presented in [5] uses a PFD and a CP to monitor the phase difference between the reference and the feedback signal in response to modulation applied at the VCO.

IV. SIMULATION RESULTS

Several simulations have been carried out in order to confirm the possible GFSK data rate increase using the two-point modulation technique. In order to enable time-efficient simulations the required analog building blocks have been modeled in Verilog-A and the digital receiver is implemented in VHDL on RTL level. The PLL's analog building blocks have been modeled on different abstraction levels including purely functional models for fastest simulation time, more realistic models according to the work presented in [7] and transistor-level implementations of the actual fabricated transceiver system. For simulation each building block's abstraction level can be chosen independently. All simulation results presented in this Section have been simulated with a nominal PLL bandwidth of 44.5 kHz, a transmission frequency of 868 MHz, a GFSK data rate of 1 Mbps and an f_{dev} of 64 kHz. During calibration the Gaussian filter is bypassed whereas it is activated during normal operation.

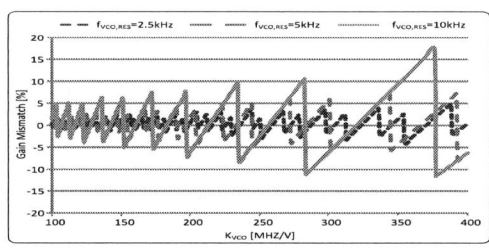

(a) Calibration with ADC

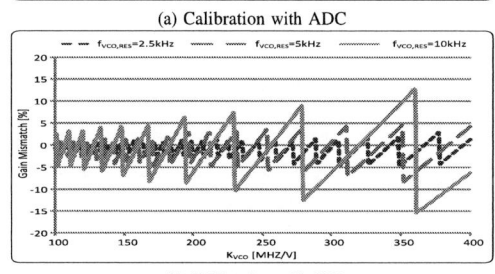

(b) Calibration with RX

Fig. 4.   Remaining Gain Error After Calibration

Co-simulation of the Verilog-A models according to [7] and RTL models of the digital receiver has been used in order to determine the remaining M_{gain} after calibration of the two modulation paths using the two methods presented in Section III-B. The results of these simulations are shown in Fig. 4 for three different VCO frequency resolutions $f_{\text{VCO,RES}} = K_{\text{VCO}}V_{\text{DAC,LSB}}$, determined by K_{VCO} and the DAC voltage caused by an LSB change. The nominal frequency resolution is based on the assumption that $K_{\text{VCO}} = 200\frac{\text{MHz}}{\text{V}}$. Therefore the frequency resolution increases for lower values of K_{VCO} and the remaining M_{gain} decreases whereas the opposite behavior can be observed for higher values of K_{VCO}. These simulations indicate that an M_{gain} of below 5 % can be achieved which matches well with the measurements of the silicon implementations of the bandwidth calibration methods presented in [6].

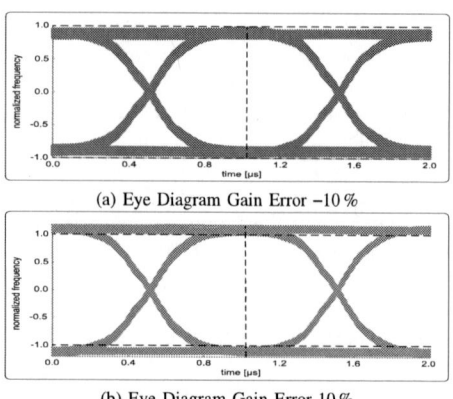

(a) Eye Diagram Gain Error −10 %

(b) Eye Diagram Gain Error 10 %

Fig. 5. TPM: Eye Diagrams

Fig. 5 shows the simulated eye diagrams of the two-point modulated PLL with an M_{gain} of −10 % (Fig. 5a) and 10 % (Fig. 5a) respectively. It represents the demodulated GFSK signal normalized to the f_{dev}. The dashed black vertical lines represent the sampling points at which a receiver would decide whether the received data represents a '1' or a '-1' and is ideally chosen to be at the maximum opening of the eye diagram. The sampled values exhibit mean values μ_1 and μ_{-1} and standard deviations σ_1 and σ_{-1}. The difference $\mu_1 - \mu_{-1}$ represents the eye opening. The quality of an eye diagram can be assessed with the quality factor Qt according to [8]

$$Qt = \frac{|\mu_1 - \mu_{-1}|}{\sigma_1 + \sigma_{-1}}. \tag{3}$$

The influence of M_{gain} and M_{timing} on the eye opening is shown in Fig. 6a. Qt in dependence of M_{gain} and M_{timing} is depicted in Fig. 6b. These results clearly show the dominating influence of M_{gain} compared to M_{timing}. If $M_{\text{gain}} < 10\%$ can be achieved an eye opening of over 85 % is the result which represents a sufficient value for successful transmission and reception. Therefore the simulation results presented in Fig. 6 confirm the capability of adapting the existing transceiver by using the presented methods in order to increase the GFSK data rate to 1 Mpbs while still operating the PLL with a loop bandwidth of 44.5 kHz.

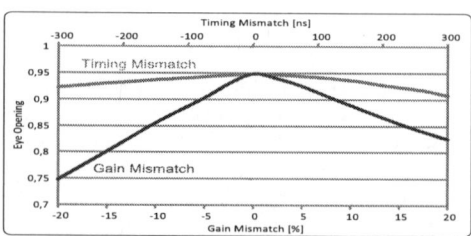

(a) Eye Opening

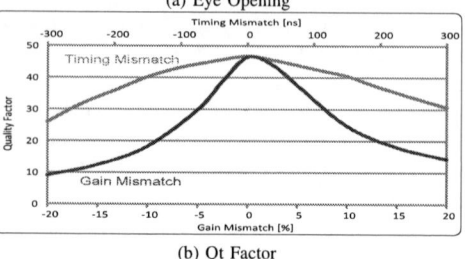

(b) Qt Factor

Fig. 6. Simulation Results of Two-Point Modulation

V. CONCLUSION

Two-point modulation has been investigated as a means of increasing the GFSK data rate beyond the bandwidth of the PLL. Two methods used to minimize the mismatches between the two modulation paths have been presented. Simulation results have been shown which confirm the capability of the presented methods of using a PLL exhibiting a nominal bandwidth of 44.5 kHz and generating GFSK modulated signals with a data rate of 1 Mbps.

ACKNOWLEDGMENT

The work presented in this paper has been partly funded by the Artemis-JU project Pollux.

REFERENCES

[1] S. Mahlknecht and M. Bock, "CSMA-MPS: a minimum preamble sampling MAC protocol for low power wireless sensor networks," in *Factory Communication Systems, 2004. Proceedings. 2004 IEEE International Workshop on*, sept. 2004, pp. 73 – 80.

[2] K.-C. Peng, C.-H. Huang, C.-J. Li, and T.-S. Horng, "High-performance frequency-hopping transmitters using two-point delta-sigma modulation," *Microwave Theory and Techniques, IEEE Transactions on*, vol. 52, no. 11, pp. 2529 – 2535, nov. 2004.

[3] S. Lee, J. Lee, H. Park, K.-Y. Lee, and S. Nam, "Self-Calibrated Two-Point $\Delta\Sigma Modulation$ Technique for RF Transmitters," *Microwave Theory and Techniques, IEEE Transactions on*, vol. 58, no. 7, pp. 1748 –1757, july 2010.

[4] S.-A. Yu and P. Kinget, "A 0.65-V 2.5-GHz Fractional-N Synthesizer With Two-Point 2-Mb/s GFSK Data Modulation," *Solid-State Circuits, IEEE Journal of*, vol. 44, no. 9, pp. 2411 –2425, sept. 2009.

[5] R. Yu, T.-T. Yeo, K.-H. Tan, S. Mou, Y. Cui, H. Wang, H.-S. Yap, E. Ting, and M. Itoh, "A 5.5mA 2.4-GHz two-point modulation Zigbee transmitter with modulation gain calibration," in *Custom Integrated Circuits Conference, 2009. CICC '09. IEEE*, sept. 2009, pp. 375 –378.

[6] H. Unterassinger, M. Flatscher, T. Gschier, J.Jongsma, and W.Pribyl, "Two On-Chip Bandwidth Calibration Methods for Phase-Locked Loops Used in Wireless Transceiver Applications," accepted and to be published at EuroCon 2013.

[7] K. Kundert, "Modeling Jitter in PLL Based Frequency Synthesizers."

[8] I. Shake, H. Takara, and S. Kawanishi, "Simple measurement of eye diagram and BER using high-speed asynchronous sampling," *Lightwave Technology, Journal of*, vol. 22, no. 5, pp. 1296 – 1302, may 2004.

Transient Analysis applying S-Parameters as Operators

Heinrich Nuszkowski, Robert Wolf, Frank Ellinger, *Senior Member, IEEE*, Gerhard Fettweis, *Fellow, IEEE*

Abstract—We propose a calculation method for transient analysis of linear time-invariant systems basing on the Mikusinski operator. Since signal-flow charts and scattering parameters (S-parameters) are used to describe the networks to analyze, the method is very vivid. The applied S-parameters form operators describing transmission and reflection of power waves from one node to an other of the network's signal-flow chart. Although the usage of operators exhibits some similarities to the Laplace transform, the effect of the operators is interpreted in time domain, and a signal transformation is not required. Thus, the presented calculation method is suitable to vividly analyze dynamic processes on electrical lines.

Index Terms—Scattering parameters, S-parameters, transient analysis, operators

I. INTRODUCTION

THE characterization of radio frequency (RF) circuits by scattering parameters (S-parameters) became state-of-the-art since the publications of Penfield [1], Youla [2] and Kurokawa [3]. In contrast to other n-port parameter sets like Z- or Y-parameters, S-parameters can even be measured very precisely for very high frequencies. Usually, S-parameters are complex numbers describing the relationship of magnitude and phase between incoming and outgoing power waves at the ports in steady state. A transient analysis using power waves is also possible [4]–[6]. In this case, the S-parameters S_{ij} are no complex numbers anymore but become impulse responses $s_{ij}(t)$ in time domain and transfer functions $S_{ij}(p)$ dependent on the complex frequency p in frequency domain. The outgoing power waves $b_i(t)$ can be derived from the incoming ones $a_j(t)$ by convolution in time domain or by multiplication of the transformed signals in frequency domain.

In this publication, we propose a method for transient analysis of dynamic processes using the S-parameters as operators introduced by Mikusinski [7], [8]. Within the book *Lineare Netzwerke* of Peter Vielhauer [9], the operators are introduced to describe and calculate dynamic processes of voltages and currents on transmission lines. Power waves and S-parameters are not considered. This extension is what we would like to present here.

Therefore, the basics of the usage of operators are presented in section II. Section III introduces the description of electrical networks by signal-flow charts and S-parameters forming operators. The utilization of this method is demonstrated by

Manuscript received March 12, 2013. This work was partly funded by the Federal Ministry of Education and Research (BMBF) in the excellence cluster CoolSilicon, project CoolBroadcastRepeater.

H. Nuszkowski and G. Fettweis are with the Vodafone Chair Mobile Communications Systems of the Technische Universität Dresden, Germany

R. Wolf and F. Ellinger are with the Chair for Circuit Design and Network Theory of the Technische Universität Dresden, Germany

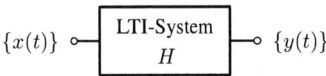

Fig. 1. Transmission model

basic examples in section IV followed by a summary in section V.

II. MATHEMATICAL BASICS

A. Transmission model

Fig. 1 shows the transmission model consisting of input function $\{x(t)\}$, the output function $\{y(t)\}$ and the transmission operator H. The curly brackets mean that the corresponding functions are causal and thereby equal to zero for negative time t, hence

$$\{x(t)\} = \begin{cases} 0 & t < 0 \\ x(t) & t \geq 0 \end{cases}. \tag{1}$$

The output function can be derived by $\{y(t)\} = H\{x(t)\}$, where H stands for the transmission operator in time domain. In other words, the transmission operator specifies which operations have to be performed on the input function in time domain to obtain the output signal. Thus, it is not required to transform the time functions into frequency domain.

The transmission operator of a linear time-invariant system

- can be derived from four basic operators: addition, subtraction, multiplication and division,
- can be formed as a quotient of two causal and for $t \geq 0$ continuous time functions, and
- can be converted into the complex transfer function by substitution of the basic operators by the corresponding transfer functions.

B. The four basic operators

The four basic operators are listed in Table I and are explained in detail below like they are introduced in [9]. It is important to mention that all quantities are normalized and thereby dimensionless.

1) Weighting function: The product operator relating two time functions $\{h(t)\}\{x(t)\}$ has to be interpreted as convolution. The notation as product is eligible since the convolution fulfills all axioms of the linear algebra for fields and rings e.g. commutativity, associativity, distributivity. An important special case reveals if a function is convoluted with the unity step function $\{1\}$ which leads to

$$\{1\}\{x(t)\} = \left\{ \int_0^t x(t-\tau)d\tau \right\} = \left\{ \int_0^t x(\tau)d\tau \right\}. \tag{2}$$

978-1-4673-4579-8/13 $31.00 © 2013 IEEE

Paper M2B2   *PRIME 2013, Villach, Austria*

TABLE I
BASIC OPERATORS

name	operator	effect
weighting function	$\{h(t)\}$	$\{h(t)\}\{x(t)\} = \{\int_0^t h(t-\tau)x(\tau)\}$
scaling operator	k	$k\{x(t)\} = \{kx(t)\}$
differentiation op.	s	$s\{x(t)\} = \{x'(t)\} + x(+0)$
shifting operator	e^{-sT}	$\mathrm{e}^{-sT}\{x(t)\} = \begin{cases} 0 & t < T \\ x(t-T) & t \geq T \end{cases}$

The output function is the integral of the input function. Thus, the unity step function acts as the integration operator. Since the integration operator is inverse to the differentiation operator, which is discussed further below, and due to the parallels to the Laplace transform, the integration operator can be noted as

$$\{1\} = s^{-1} = \frac{1}{s}. \tag{3}$$

2) Scaling operator: The scaling operator describes a proportionality between output and input function

$$\{y(t)\} = k\{x(t)\} = \{kx(t)\}. \tag{4}$$

Considering the definition $\{h(t)\} = k$ the dirac distribution $\delta(t)$ is expendable in contrast to time domain description without operators given by

$$(x * h)(t) = (x * k\delta)(t) = kx(t). \tag{5}$$

But it has to be carefully distinguished between the scaling operator k and the step function $\{k\} = k\{1\}$.

3) Differentiation operator: The differentiation operator is defined by

$$s\{x(t)\} = \{x'(t)\} + x(+0). \tag{6}$$

Meaning that applying the differentiation operator to a function $\{x(t)\}$ results in its derivative plus its right-hand limit at zero. Due to this definition, some important conclusions can be drawn. It is

$$s\{1\} = s\frac{1}{s} = 1. \tag{7}$$

This relation can be derived either by employing (6) on the unity step function $\{1\}$ or using the definition (3).

Employing (6) on the function $\mathrm{e}^{-\alpha t}$ leads to

$$s\{\mathrm{e}^{-\alpha t}\} = -\{\alpha \mathrm{e}^{-\alpha t}\} + 1 \Rightarrow \{\mathrm{e}^{-\alpha t}\} = \frac{1}{s+\alpha}. \tag{8}$$

In a more general way, it can also be derived that

$$\frac{1}{(s+\alpha)^n} = \left\{ \frac{t^{n-1}}{(n-1)!}\mathrm{e}^{-\alpha t} \right\}. \tag{9}$$

By (8), exponentially damped sinusoidal and cosinusoidal signals can be noted as operators in s by

$$\{\mathrm{e}^{-\alpha t}\cos(\omega t)\} = \frac{1}{2}\left\{\mathrm{e}^{(-\alpha+\mathrm{j}\omega t)} + \mathrm{e}^{(-\alpha-\mathrm{j}\omega t)}\right\}$$
$$= \frac{1}{2}\left(\frac{1}{s+\alpha-\mathrm{j}\omega} + \frac{1}{s+\alpha+\mathrm{j}\omega}\right) = \frac{s+\alpha}{(s+\alpha)^2+\omega^2} \tag{10}$$

and

$$\{\mathrm{e}^{-\alpha t}\sin(\omega t)\} = \frac{\omega}{(s+\alpha)^2+\omega^2}. \tag{11}$$

4) Shifting operator: The shifting operator delays a function. It also fulfills all axioms of the linear algebra. Particularly, the power laws can be applied to describe the overall delay of two cascaded delay transmission blocks,

$$\mathrm{e}^{-sT_1}\mathrm{e}^{-sT_2} = \mathrm{e}^{-s(T_1+T_2)}. \tag{12}$$

The rectangular function $\{\mathrm{rect}(t/T - 1)\}$, which is an important test function, can be easily noted by unity step functions and shifting operators

$$\{\mathrm{rect}(t/T - 1)\} = \{1\}(1 - \mathrm{e}^{-sT}). \tag{13}$$

III. NETWORK DESCRIPTION BY SIGNAL-FLOW CHARTS

The signal-flow chart is a very powerful and vivid method to describe linear networks. Especially, the combination of S-parameters and signal-flow charts for the process of power transmission e.g. in RF circuits is very helpful, since discontinuities, which lead to power reflection, and feedback loops, which may result in instability, can be found. The advantage of the proposed calculation method is, that it is relatively easy to formulate the solution approach for a transient analysis. All, what has to be done, is replacing the classical S-parameters of the signal-flow chart used for the steady state analysis by the corresponding operator relations. Thus, the reflexion, transmission and propagation processes between the notes of the signal-flow chart can easily be understood.

By the operators for resistance, inductance, and capacitance, the operators for impedances $Z(s)$ or reflexion coefficients $\Gamma(s)$ can be derived. For a passive one-port, which is fully characterized by the impedance operator $Z_l(s)$, the signal-flow chart is shown in Fig. 2a. The power waves $\{a(t)\}$ and $\{b(t)\}$ are going into and coming out of the one-port, respectively. The operator $\Gamma_l(s)$ stands for the reflexion coefficient of the load impedance $Z_l(s)$ which can be calculated by

$$\Gamma(s) = \frac{Z(s) - Z_0}{Z(s) - Z_0} \tag{14}$$

where Z_0 is the appropriate chosen reference impedance. The passive one-port is finally characterized by

$$\{b(t)\} = \Gamma(s)\{a(t)\}. \tag{15}$$

To get an active one-port the signal-flow chart has to be modified like shown in Fig. 2b. The description changes to

$$\{b(t)\} = \{b_g(t)\} + \Gamma_g(s)\{a(t)\}, \tag{16}$$

where the operator $\Gamma_g(s)$ stands for the the reflexion coefficient corresponding to the generator's internal impedance $Z_g(s)$. The wave $\{b_g(t)\}$ is the injected power wave which the generator would deliver if it is terminated by the reference impedance $Z_0(s)$, and which can be calculated by

$$\{b_g(t)\} = \{u_g(t)\}\frac{\sqrt{Z_0}}{Z_g(s) + Z_0}. \tag{17}$$

Fig. 2c illustrates the signal-flow chart of a two-port, described by the operator S-parameter set $S_{ij}(s)$, $i, j = 1, 2$

$$\begin{bmatrix} \{b_1(t)\} \\ \{b_2(t)\} \end{bmatrix} = \begin{bmatrix} S_{11}(s) & S_{12}(s) \\ S_{21}(s) & S_{22}(s) \end{bmatrix} \begin{bmatrix} \{a_1(t)\} \\ \{a_2(t)\} \end{bmatrix}. \tag{18}$$

978-1-4673-4579-8/13 $31.00 © 2013 IEEE

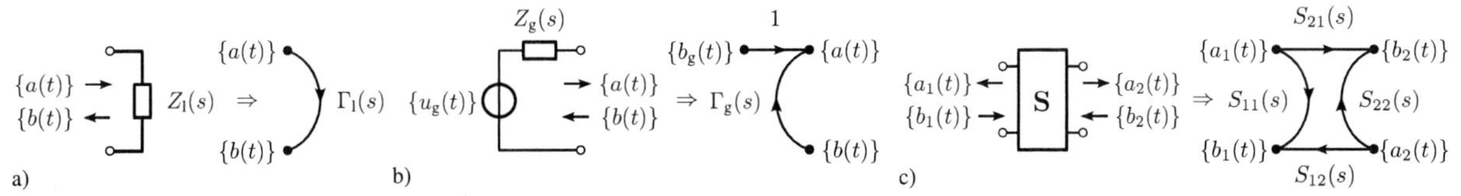

PRIME 2013, Villach, Austria

Session M2B – RF Techniques 1

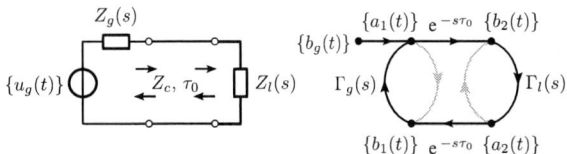

Fig. 2. circuit and signal-flow charts for a) a passive one-port, b) an active one-port, and c) a two-port

Since the proposed calculation method is particularly suitable to analyze dynamic processes on transmission lines, the S-parameter matrix of a transmission line is an important special case for a two-port. The matrix for a transmission line of the length l is

$$\mathbf{S} = \frac{1}{1 - \Gamma^2 e^{-2\gamma l}} \begin{bmatrix} \Gamma(1 - e^{-2\gamma l}) & (1 - \Gamma^2) e^{-\gamma l} \\ (1 - \Gamma^2) e^{-\gamma l} & \Gamma(1 - e^{-2\gamma l}) \end{bmatrix}, \quad (19)$$

where γ and Γ denote the propagation operator and the reflexion operator, respectively, given by

$$\gamma(s) = \sqrt{(R' + sL')(G' + sC')}, \quad \text{and} \quad (20)$$

$$\Gamma(s) = \frac{Z_c(s) - Z_0}{Z_c(s) + Z_0}, \quad (21)$$

where $Z_c(s)$ is the characteristic impedance operator

$$Z_c(s) = \sqrt{\frac{R' + sL'}{G' + sC'}}. \quad (22)$$

The parameters R', L', G', and C' are the characteristic constants of the transmission line which are also used in the Telegrapher's Equations. For $R' = 0$ and $G' = 0$, the transmission line is lossless, and the characteristic impedance Z_c becomes a scaling operator and the term (19) a shifting operator

$$e^{-\gamma(s)l} = e^{-s\sqrt{L'C'}l} = e^{-s\tau_0}. \quad (23)$$

If the reference impedance is chosen equal to the characteristic impedance $Z_0 = Z_c$ then is $\Gamma = 0$ and the S-parameter matrix can be simplified to

$$\mathbf{S} = \begin{bmatrix} 0 & e^{-s\tau_0} \\ e^{-s\tau_0} & 0 \end{bmatrix}. \quad (24)$$

For transverse electromagnetic (TEM) waves, voltages and currents at the ports can be calculated from the power waves by

$$\{u(t)\} = (\{a(t)\} + \{b(t)\}) \sqrt{Z_0}, \quad \text{and} \quad (25)$$

$$\{i(t)\} = (\{a(t)\} - \{b(t)\}) / \sqrt{Z_0}. \quad (26)$$

IV. EXAMPLES

For the next examples, the transmission line is considered lossless which reduces the effect of the transmission line to an ideal delay without any distortion. By this constraint, the handling of S-parameter operators for transient analysis can be shown on simple examples. Lossy lines can also be examined but the examples get more complicated and more computationally intensive. For more details we would like to refer on the chapter *Dynamische Vorgänge auf verlustbehafteten Leitungen* in [9]. There is shown that the operator $e^{-\gamma(s)}$,

Fig. 3. both-sided mismatched, lossless line; circuit and SFC

which can delay, damp, and distort a signal, of a lossy line can be rearranged to

$$e^{-\gamma(s)} = e^{-\alpha_0 l} e^{-s\tau_0} (1 + \{v(t)\}). \quad (27)$$

The factor $e^{-\alpha_0 l}$ forms a scaling operator and describes a linear damping, the factor $e^{-s\tau_0}$ is a shifting operator and works as the delay of the line, and $\{v(t)\}$ is a weighting function which describes the linear distortion of the lossy transmission line.

Fig. 3 depicts the circuit and the signal-flow chart for a lossless transmission line being terminated by impedances Z_g, $Z_l \neq Z_c$. The reference impedance is chosen equal to the characteristic impedance $Z_0 = Z_c$. The evaluation of the signal-flow chart e.g. for the power wave at the output reveals

$$\{a_2(t)\} = \{b_g(t)\} \Gamma_l(s) e^{-s\tau_0} / D, \quad \text{and} \quad (28)$$

$$\{b_2(t)\} = \{b_g(t)\} e^{-s\tau_0} / D \quad (29)$$

involving the signal-flow chart determinant D given by

$$D = 1 - \Gamma_g(s)\Gamma_l(s) e^{-2s\tau_0}. \quad (30)$$

Applying (17) and (25), the voltage at the output of the line can be determined to be

$$\{u_2(t)\}$$
$$= \{u_g(t)\} \frac{Z_c(1 + \Gamma_l(s)) e^{-s\tau_0}}{Z_g(s) + Z_c} \frac{1}{1 - \Gamma_g(s)\Gamma_l(s) e^{-2s\tau_0}}$$
$$= \frac{\{u_g(t)\}}{2} (1 - \Gamma_g(s))(1 + \Gamma_l(s)) e^{-s\tau_0}$$
$$\sum_{n=0}^{\infty} (\Gamma_g(s)\Gamma_l(s))^n e^{-2ns\tau_0}. \quad (31)$$

For further investigations, the voltage transmission operator $H_u(s) = \{h_u(t)\}$ is used so that the equation results in

$$\{u_2(t)\} = H_u(s)\{u_g(t)\} = \{h_u(t)\}\{u_g(t)\}. \quad (32)$$

Similarly, the current transmission operator $H_i(s) = \{h_i(t)\}$ can be defined by using (26).

A. Both-sided matching

In this case, the voltage transmission operator $H_u(s)$ can be reduced to $H_u(s) = \frac{1}{2} e^{-s\tau_0}$, since $\Gamma_g = \Gamma_l = 0$. Meaning that the output voltage is half of the generator voltage and delayed by the transmission delay τ_0.

978-1-4673-4579-8/13 $31.00 © 2013 IEEE

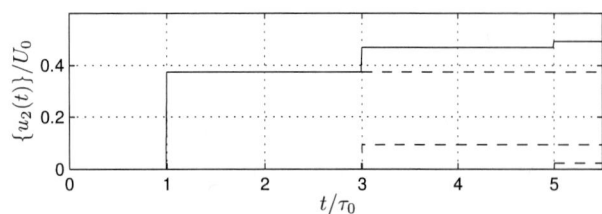

Fig. 4. First example with $Z_g = Z_l = 5Z_c$

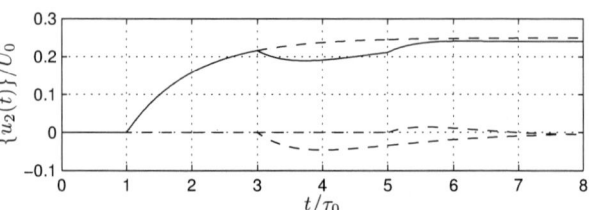

Fig. 5. Second example $Z_g = 3Z_c$ and $Y_l = 1/Z_l = 1/Z_c + sC$

B. Termination by arbitrary resistances

In this case, the reflexion coefficients are arbitrary scaling operators and (31) leads to

$$H_u(s) = \sum_{n=0}^{\infty} c_n e^{-(2n+1)s\tau_0}, \quad \text{with} \tag{33}$$

$$c_n = 0.5(1 - \Gamma_g)(1 + \Gamma_l)(\Gamma_g\Gamma_l)^n. \tag{34}$$

The input signal appears for the first time after the transmission delay τ_0 and afterwards periodically by $2\tau_0$, whereas the echoes are exponentially damped expressed by

$$\{u_2(t)\} = c_0\{u_g(t - \tau_0)\} + c_1\{u_g(t - 3\tau_0)\} + \cdots. \tag{35}$$

Fig. 4 shows the first example with $Z_g = Z_l = 5Z_c$. The generator signal is a step function $U_0\{1\}$. After an initial transient of approximately $t/\tau_0 = 5$ the normalized final value of 0.5 is almost reached.

C. Termination by arbitrary impedances

If the lossless transmission line is terminated by an arbitrary network of concentrated, linear, time invariant components the corresponding impedance operator and the corresponding reflexion coefficient is a rational function in s. The voltage transmission operator is

$$H_u(s) = \sum_{n=0}^{\infty} c_n(s) e^{-(2n+1)s\tau_0}, \tag{36}$$

where the factors $c_n(s)$

$$c_n(s) = 0.5(1 - \Gamma_g(s))(1 + \Gamma_l(s))(\Gamma_g(s)\Gamma_l(s))^n \tag{37}$$

are also rational functions in s. An interpretation is possible by partial fraction decomposition which requires that the degree of the numerator is less than the degree of the denominator. If this is not the case a polynomial long division has to be executed to ensure this criteria. Thereby, (35) results in

$$\{u_2(t)\} = c_0(s)\{u_g(t - \tau_0)\} + c_1(s)\{u_g(t - 3\tau_0)\} + \cdots. \tag{38}$$

After ascertainment of the operators $c_n(s)$, their effect on the input signal can systematically be analyzed and visualized. For the second example the generator's internal impedance is $Z_g = 3Z_c$, and the load admittance is $Y_l = 1/Z_l = 1/Z_c + sC$. For a stimulus by a step function, (37) and (38) lead to

$$\frac{\{u_2(t)\}}{U_0} = \frac{1}{4}\{1 - e^{-\alpha t}\}e^{-s\tau_0} - \frac{1}{8}\{\alpha t e^{-\alpha t}\}e^{-3s\tau_0} -$$

$$+ \frac{1}{32}\{(2 - \alpha t)\alpha t e^{-\alpha t}\}e^{-5s\tau_0}\cdots, \quad \text{with} \tag{39}$$

$$\alpha = \frac{1}{\tau_0} = \frac{2}{CZ_c}. \tag{40}$$

It can be seen that after a initial transition of approximately $t/\tau_0 = 6$ the output voltage is almost equal to its final value $\{u_2(t)\}/U_0 = 0.25$.

V. CONCLUSION

The transmission operator $H(s)$ introduced by Mikusinski is an analogon to the complex transfer function in the frequency domain of the Laplace transform. The different approaches complement each other and allow a vivid interpretation of transmission processes in time and in frequency domain. The pole–zero plot resulting from the Laplace transform is a valuable and indispensable tool for network analysis and synthesis. For considerations in time domain, the usage of operators exhibits some advantages [9] which are for instance:

- The operator emerges directly from the described system.
- A detour via a frequency domain is not required. Signals and operations remain in time domain.
- Convergence investigations and resulting constrains can be omitted.
- The problems revealing from the dirac distribution are solved very easily and exactly.

Since the focus is on the transient analysis in time domain, the presented method applying signal-flow charts and S-parameter operators is very convenient. The method proved its applicability and its vividness for teaching dynamic processes on transmission lines within the course *Lineare Netzwerke* at the Technische Universität Dresden.

REFERENCES

[1] P. Penfield, Jr., *Noise in negative resistance amplifiers*, IRE Trans. on Circuit Theory, vol. CT-7, June 1960, pp 166-170.
[2] D. C. Youla, *On scattering matrices normalized to comlex port numbers*, Proc. IRE, vol. 49, July 1961, p. 1221.
[3] K. Kurokawa, *Power waves and the scattering matrix*, IEEE Trans. on Microwave Theory and Tech., March 1965, pp 194-202
[4] J. E. Schutt-Aine, R. Mittra, *Scattering parameter transient analysis of transmission lines loaded with nonlinear teminations*, IEEE Trans. on Microwave Theory and Tech., vol. 36, No. 3, March 1988, pp 529-536.
[5] D. Winklestein, M. B. Steer, R. Pomerleau, *Simulation of arbitrary transmission line networks with nonlinear terminations* IEEE Trans. on Circuits and Systems, vol. 38., No. 4, April 1991, pp 418-422.
[6] T. Dhaene, L. Martens, D. De Zutter, *Transient simulation of arbitrary nonuniform interconnection structures characterized by scattering parameers*, IEEE Trans. on Circuits and Systems, vol. 39, No. 11, Nov. 1992, pp 928-937.
[7] J. Mikusinski, *Operatorenrechnung*, VEB Deutscher Verlag der Wissenschaften, Berlin 1957.
[8] *The Operational Calculus*, Pergamon Press, Oxford 1983
[9] P. Vielhauer, *Lineare Netzwerke*, VEB Verlag Technik, Berlin 1982.

PRIME 2013, Villach, Austria

Session M2B – RF Techniques 1

Design of a W-Band 2-bit Differential CMOS Phase Shifter

Ali Vahdati, Mikko Varonen, Mikko Kärkkäinen, Kari. A. I. Halonen

Department of Micro- and Nanosciences, SMARAD-2
Aalto University School of Electrical Engineering
Espoo, FINLAND

Abstract—**This paper presents the design of a W-Band I-Q phase shifter and the corresponding simulation results in 28-nm CMOS technology. The design of passive components like a 90° hybrid and transformers needed to realize differential I-Q operation is shown. The phase of the RF signal can be varied from 0° to 270° in steps of 90°. The simulated input and output matching are better than -10 dB, and the maximum phase error is roughly 10° with a maximum output imbalance of 0.5 dB at 90 GHz using a 1-V supply voltage. The total power consumption is 41 mW and the die area is 0.46 mm².**

Keywords—Phased Arrays; CMOS Active Switches; W-Band Phase Shifters; Millimeter Waves

I. INTRODUCTION

Today active phased-array antenna systems are becoming more and more important for satellite communications and radar applications because they cover a higher channel capacity and enable to improve the SNR in a transceiver. Commercial applications such as terrestrial wireless links and automotive radars can be realized by the use of phased-array elements. In addition, the process of realizing the commercial products as low-cost and small-size is highly considerable [1]. Today progress in CMOS technology enables to realize the millimetre-wave radio front-end systems with higher level of integration [2] and cheaper commercial products when aimed for mass production. However, CMOS technology for millimetre-wave applications faces some challenges like high silicon substrate loss and high loss of the lines. Moreover at millimetre-waves, other challenges like line-of-sight (LOS) and non line-of-sight (NLOS) applications, and high path-loss in transceiver have to be considered. A beam-steerable phased array is utilized as a potential solution to overcome to some of these challenges.

Phase shifter (PS) is an essential component in a phased array system [3]. Electronic phase shifters are used in electronic beam-steering systems and in many cases they are capable of producing any arbitrary phase shift across the entire 360°, for instance, in phased-array antennas. A major classification of phase shifters is to divide them into active and passive.

At millimeter-wave frequencies, phase shifters can be realized by different techniques. Some of the common phase-shifting architectures are reflection-type phase shifters [4], switched-line phase shifters [5], loaded-line phase shifters [6] and vector sum based phase shifters [7].

Each type is chosen based on performance specifications such as area, phase and amplitude imbalance, insertion loss and power. In this paper, a 2-bit controllable differential I-Q phase shifter has been designed for W-band based on 28-nm CMOS technology.

II. PHASE SHIFTER DESIGN

Fig. 1 depicts the block diagram of the 2-bit controllable phase shifter, designed to operate at W-band. The phase shifter is designed for differential operation; however for measurement as a stand-alone block, transformers are used both at the input and output to convert the single-ended RF signal to differential signal and vice versa. There is also a fully differential passive I-Q generator. The differential I and Q signals feed the active switches of each branch separately. Each active switch provides 0° and 180° phase shift when the control bits are changed. Hence, the phase shifter of Fig.1 is capable to generate 0°, 90°, 180°, and 270° phase shift at the output. The schematic of the active switch based on a Gilbert-cell is shown in Fig. 2 [8]-[9].

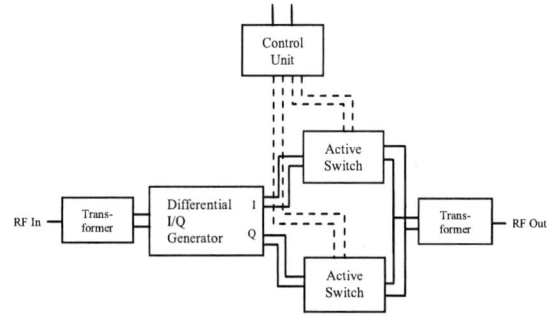

Fig. 1: Block diagram of a 2-bit digitally controlled differential I-Q phase shifter.

As Fig. 2 illustrates, the duty of the active switch is to generate 0° and 180° phase shift of the input signal. The differential input signal generated by I-Q branch feeds to nodes +in, −in and the differential output signal are obtained from nodes +out, −out. 0° and 180° phase shift are generated when the control bits (bit 1, bit 2) are changed.

978-1-4673-4579-8/13 $31.00 © 2013 IEEE

97

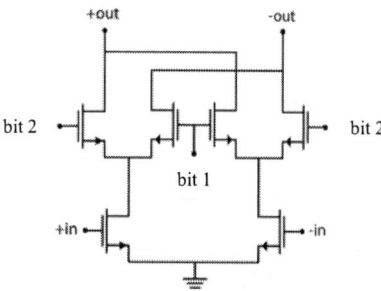

Fig. 2: Active switch with 2 control bits.

As represented in Fig. 3, the I-Q generator is a differential branch-line coupler with slow-wave shields underneath the lines. The slow-wave transmission lines have an electrical length of quarter-wavelength ($\lambda/4$) and are loaded by capacitors. As explained in [10], using slow-wave transmission line results lower phase velocity and shorter wavelengths. Hence, the design can be realized in smaller size of the chip-area with less attenuation for the signal lines. In addition, capacitors are used to increase the slow-wave effect and decrease the impedance level of the lines [11]. The top metal layer is used for the signal lines and the capacitors are formed as slabs where lower metal layers are also used to increase the capacitance. This approach enables the design of the branch-line coupler in a very compact chip-area. Area saving compared to a conventional branch-line coupler is around 75%.

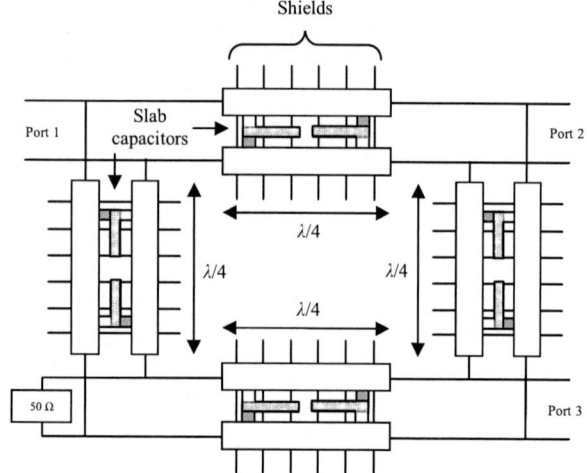

Fig. 3: Simplified schematic of the differential slow-wave shielded branch-line coupler utilizing capacitor loading.

Fig. 4 depicts the differential simulation results of the designed branch-line coupler across the entire W-band. At the operating frequency (90 GHz), there is only 2° phase error and the input matching (S_{11}) is –17 dB. The amplitude imbalance between port 2 (S_{21}) and port 3 (S_{31}) is negligible at 90 GHz.

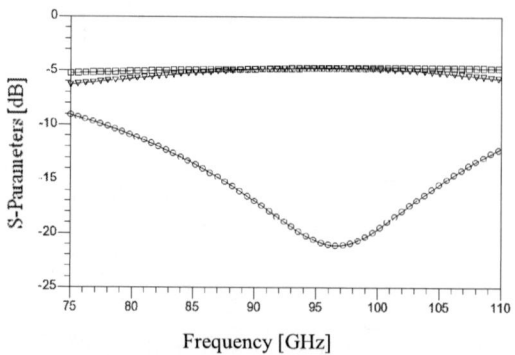

Fig. 4: Simulated S-parameters of the differential branch-line coupler; circle-line: S_{11}, square-line: S_{21}, and triangle-line: S_{31}.

Fig. 5 shows the layout of the phase shifter designed to operate at 90 GHz.

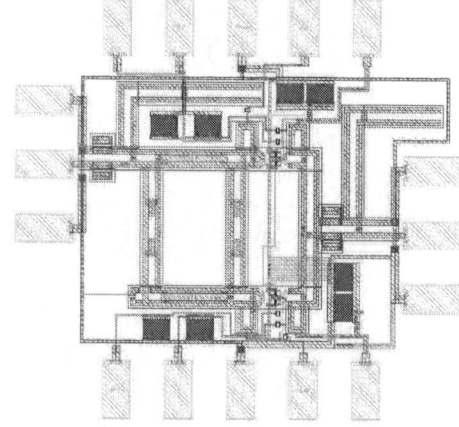

Fig. 5: Layout of the designed differential phase shifter in 28-nm CMOS.

III. SIMULATION RESULTS OF THE PHASE SHIFTER

Fig. 6 illustrates the simulated output phase shift across the entire of W-band when the control bits are changed. For both active switches, the meaningful bit combination set is 10-10, 10-01, 01-01 and 01-10. At 90 GHz which is the target frequency of this design and simulation, the maximum phase error is around 10°.

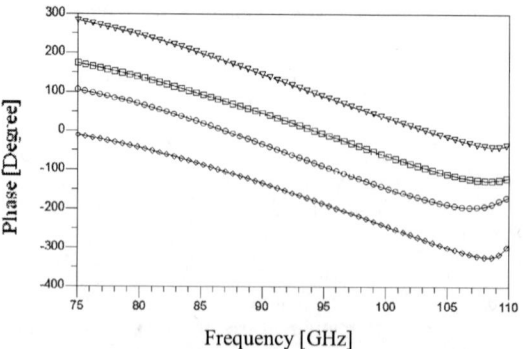

Fig. 6: Simulation results by changing control bits; diamond-line: 01-10, circle-line: 10-10, square-line: 10-01, triangle-line: 01-01.

Fig. 7 displays the expected simulation results of the return loss at the input of the phase shifter when the combination set of the control bits is selected. For the W-band, the S_{11} is below –10 dB from 85 GHz to 96 GHz. It can also be seen that for the combination set of different control bits, the input return loss curves coincide on top of each other with the maximum 0.5 dB difference for the entire W-band.

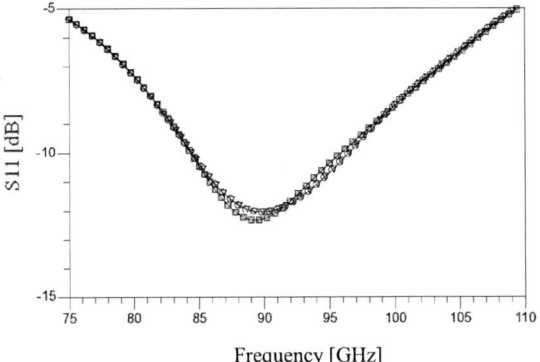

Fig. 7: Simulation results of input matching; diamond-line: 01-10, circle-line: 10-10, square-line: 10-01, triangle-line: 01-01.

Fig. 8 depicts the simulation results of the return loss at the output of the phase shifter. For the W-band, the S_{22} is below –10 dB between 86 GHz and 95.5 GHz. For the entire W-band, the output return loss curves coincide nicely on top of each other when different control bits are selected.

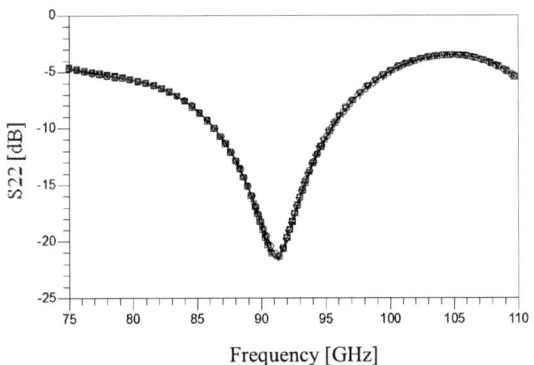

Fig. 8: Simulation results of the output matching when bits are changed; diamond-line: 01-10, circle-line: 10-10, square-line: 10-01, triangle-line: 01-01.

Fig. 9 shows the simulated signal losses of the phase shifter across the entire W-band. At 90 GHz, the maximum output imbalance between phases is 0.5 dB although it becomes larger when the frequencies of both ends are considered. The maximum acceptable output imbalance between phases is an important factor which limits the bandwidth of the phase shifter operation.

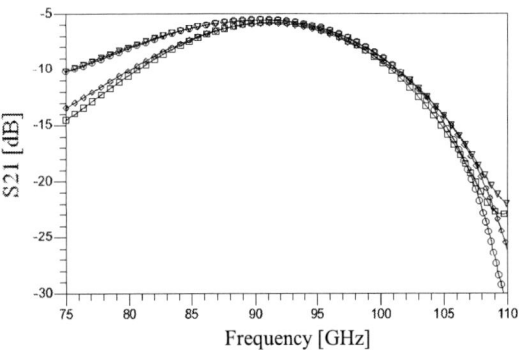

Fig. 9: Simulation results of output imbalance; diamond-line: 01-10, circle-line: 10-10, square-line: 10-01, triangle-line: 01-01.

IV. CONCLUSION

A W-band 2-bit active phase shifter in 28-nm CMOS technology has been designed and simulated in this paper. The simulation results show promising results for millimeter-wave beam-steering applications. A slow-wave structure has been used in this design to reduce the substrate loss. The simulation results show maximum 10° phase deviation, 5.8 dB signal loss, and 0.5 dB output imbalance at 90 GHz. Although the phase shifter has been designed for wideband operation, the input and output matching together with the maximum acceptable output imbalance limit the bandwidth to around 10-GHz from 85–95 GHz. The input and output matching are satisfied properly within 10-GHz bandwidth and are –12.3 dB and –18.6 dB, respectively at operating frequency.

ACKNOWLEDGMENT

The authors of this paper would like to thank all the project partners. This work was supported by the Finnish Funding Agency for Technology and Innovation (Tekes) under the BEAMS and SIMIDS projects, by the Academy of Finland under the FAMOS project, and by the ENIAC JU project MIRANDELA. The authors also thank STMicroelectronics for silicon donation.

REFERENCES

[1] R. A. York and T. Itoh, "Injection- and phase-locking techniques forbeam control", *IEEE Trans. Microw. Theory Tech.*, vol. 46, no. 11, pp. 1920–1929, Nov. 1998.

[2] M. Varonen, M. Kaltiokallio, V. Saari, O. Viitala, M. Kärkkäinen, S. Lindfors, J. Ryynänen, K. A. I. Halonen, "A 60-GHz CMOS receiver with an on-chip ADC", *IEEE. Radio Frequency Integrated Circuits Symposium, RFIC*, pp. 445-448, June 2009.

[3] S. K. Koul and B. Bhat, *Microwave and millimeter-wave phase shifters*, vol. 1, Artech House, Norwood, MA, 1991.

[4] K. Miyaguchi, "An Ultra-Broad-Band Reflection-Type Phase-Shifter MMIC With Series and Parallel *LC* Circuits", *IEEE Trans. Microwave Theory Tech.*, vol. 49, no. 12, pp. 2446-2452, Dec. 2001.

[5] D. W. Kang, "Ku-Band MMIC Phase Shifter Using a Parallel Resonator with 0.18-µm CMOS Technology", *IEEE Trans. Microwave Theory Tech.*, vol. 54, no. 1, pp. 294-300, Jan. 2006.

[6] Y. Yu, "A 60 GHz Digitally Controlled Phase Shifter in CMOS", *ESSCIRC.*, pp. 250-253, Sep. 2008.

[7] K-J. Koh, "0.13-μm CMOS Phase Shifters for X-, K$_u$-, and K-Band Phased Arrays", *IEEE. J. Solid-State Circuits*, vol. 42, no. 11, pp. 2535-2546, Nov. 2007.

[8] I. Sarkas, M. Khanpour, A. Tomkins, P. Chevalier, P. Garcia, and S. P. Voinigescu, "W-band 65-nm CMOS and SiGe BiCMOS Transmitter and Receiver with Lumped I-Q Phase Shifters", *IEEE. Radio Frequency Integrated Circuits Symposium, RFIC*, pp. 441-444, 2009.

[9] M. Elkhouly, S. Glisic, F. Ellinger, J. C. Schyett, "A 60 GHz Four Channel Beamforming Transmitter in 0.25 μm SiGe BiCMOS Technology", *Proceeding of the 6th European Microwave Integrated Circuits Conference*, Oct. 2011.

[10] N. Deferm, P. Reynaert, "A 120GHz 10Gb/s phase-modulating transmitter in 65nm LP CMOS", *Solid-State Circuits Conference Digest of Technical Papers (ISSCC), IEEE International*, pp. 290-292, Feb. 2011.

[11] I. Haroun, J. Wight, C. Plett, A. Fathy, Da-Chiang Chang, "Experimental Analysis of a 60 GHz Compact EC-CPW Branch-Line Coupler for mm-Wave CMOS Radios", *IEEE Microwave and Wireless Components Letters*, vol. 20, no. 4, pp. 211-213, April 2010.

Impact of PA class on reconstruction filters sizing for a WCDMA base station LINC transmitter

Nico Prou , Fabian Rivière
Crismat / LaMIPS
Caen, France
prou.20707725@etu.unicaen.fr

Corinne Berland
Université Paris-Est, ESYCOM, ESIEE Paris,
Noisy Le Grand, France
c.berland@esiee.fr

Abstract—The LInear amplification with Nonlinear Component transmitter is a high power efficient architecture for radio communication systems. Nevertheless, this architecture is sensitive to mismatches between the two paths and its efficiency performances are dependent on the power amplification stage topology. Additively, careful attention has to be paid to baseband performances in terms of sampling rate, Digital to Analog Converters (DACs) resolution, reconstruction filter bandwidth... In this paper, we present the specification of the reconstruction filters for a single carrier WCDMA 3GPP base station transmitter and this, according to the Power Amplifier (PA) class (switched mode (SMPA) or AB class). We address the impact of the class of the PAs on the overall performances and on the reconstruction filters specifications. Admissible time delays and gain mismatches between the two paths will also be quantified.

Keywords-component; LINC; base station; WCDMA; bandwidth; reconstruction filter; mismatches; power amplifier; switch mode; saturated; AB class; PAPR.

I.   INTRODUCTION

The LINC architecture, LInear amplification with Nonlinear Component, is a good candidate for the realization of power efficient transmitter for radio communication systems. The architecture principle relies on the separation of any modulated signal into two constant envelope phase-modulated signals. Hence, these two signals can be efficiently power amplified before being recombined to recover initial modulated signal, see Fig.1.

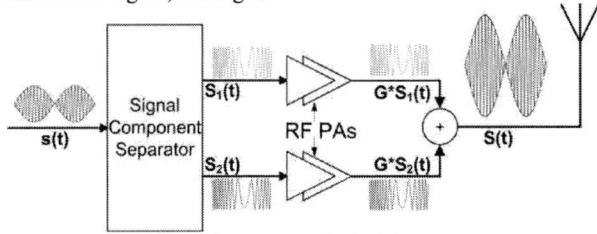

Figure 1.   LINC principle

This architecture is considered as a promising solution for high power efficiency transmitter and, as a result, its feasibility is under the focus of many research teams. Indeed, the performances of the LINC transmitter are impacted by the impairments of its analog baseband and RF components. The architecture is sensitive to gain, phase and delay mismatches between the two paths, I/Q modulator misalignment [1, 2]. In

order to compensate for these issues, calibration algorithms are presented in [4-6]. Moreover, the transmitter performances depend also on the choice of the output combiner (isolated or nonisolated), and the class of the PAs. Chireix combiner with efficient PAs presents the best efficiency at the expense of linearity degradation that can be recovered by predistorsion [7]. The solution based on the Wilkinson combiner is more linear but less efficient: multi level LINC solutions are then proposed [8]. Few papers are dealing with the sizing of the baseband part of the transmitter, constituted by the DACs and reconstruction filters [10].

In this paper, the principle of the LINC architecture applied to a 3G base station transmitter [9] is recalled in part II. The part III analyzes the impact of the effect of the bandwidth reduction of signals at the DACs output and the impact of the choice of the reconstruction filter. The performances are given in terms of Output RF Spectrum (ORFS) according to Spectrum Emission Mask (SEM), ACLR (Adjacent Channel Leakage Ratio), EVM (Error Vector Magnitude) and spurious emission. In the fourth part of this paper, the impact of the PA type (switched mode or saturated) is modeled and specification of the transmitter is given for both studied case. The conclusion is then drawn in the fifth part.

II.   LINC PRINCIPLE AND STUDY METHOD

A. LINC transmitter principle

Let us consider a narrow band representation of a modulated signal s(t) with a(t) and ϕ(t) respectively the envelope and the phase of the modulated signal.

$$s(t) = a(t) \cdot \cos[\omega_0 t + \phi(t)] \quad (Eq.1)$$

with

$$0 \le a(t) \le A_m \quad (Eq.2)$$

This signal can be split into two constant amplitude signals

$$s(t) = S_1(t) + S_2(t)$$
$$S_1(t) = A_m/2 \cdot \sin[\omega_0 t + \phi(t) - \theta(t)] \quad (Eq.\ 3 - Eq.\ 4 - Eq.\ 5)$$
$$S_2(t) = A_m/2 \cdot \sin[\omega_0 t + \phi(t) + \theta(t)]$$

with

$$\theta(t) = \arcsin\left[\frac{a(t)}{A_m}\right] \quad (Eq.6)$$

These two signals are digitally generated with a Signal Component Separator (SCS) as demonstrated in [11], before being transferred to the analog domain through DACs. Previous equations are represented by vectors, in Fig.2.

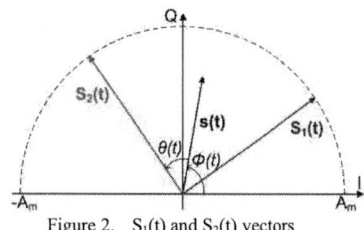

Figure 2. $S_1(t)$ and $S_2(t)$ vectors

B. Simulation environnement

In our simulation environment, ADS (Advanced Design Simulator [14]), the 3GPP signal source generator is sampled at an output rate of 64 samples per chip. It represents a sampling frequency of 3.84 Mcps x 64 = 245.76 MHz. This sampling frequency is commonly used by the most recent generation of commercial clock cleaner circuits [12].

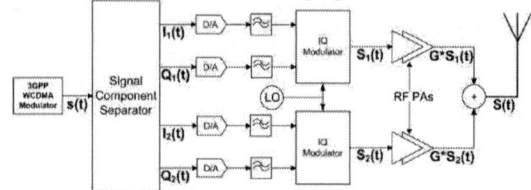

Figure 3. LINC transmitter

C. Specifications

In order to fulfill the requirements of the transmitter, a specification margin to standard values has to be taken to budget degradation contribution caused by each transmit chain blocks. As an initial specification, a 20-dB margin is introduced for EVM and a 10dB-margin for emission mask, spurious and ACLR (Table 1). For the sizing of the overall chain, those margins have to be modulated depending on the difficulty to achieve the requested specifications.

Criteria	3GPP Specification	Specification with margin
EVM	-15.14 dB	-35.14 dB
ACLR @ 5 MHz	45 dBc	55 dBc
ACLR @ 10 MHz	50 dBc	60 dBc
Spurious emission	-90dBm / Hz	-100dBm / Hz
ORFS @ 4 MHz	-73dBm / Hz	-83dBm / Hz

Table 1. Standard specifications with and without margin

III. RECONSTRUCTION FILTERS

A. Reconstruction filters selection

After digital to analog conversion, reconstruction filters are needed to clean-up signal spectrum from replicas at multiple of sampling frequency. Moreover the SCS output signals bandwidths are enlarged compared to SCS input signal. We want the filter to have a maximally-flat behavior in the band pass, and efficient attenuation of the replicas. Moreover it must be paid careful attention on the group delay as it introduces phase distortion.

In the context of base station transmitter, LC Ladder filters are generally employed for linearity constraints. For cost and process spread reasons, the number of passive elements has to

be limited. Moreover the Self Resonance Frequency (SRF) of component mustn't disturb the filter template. As a consequence, the lowest cut-off frequency as well as the lowest order has to be considered as additive criteria for the final choice.

Filter selection study covers Butterworth, Chebyshev type I, Chebyshev type II and Bessel filters for several orders and different ripples (for Chebyshev filters), see Fig.4. Specification metrics to meet are EVM, Spectrum Emission Mask, Spurious emission and ACLR at 5 MHz and 10 MHz from the carrier.

It is important to keep in mind that the filtering reintroduces an envelope variation in $S_1(t)$ and $S_2(t)$. The associated metric is the Peak to Average Power Ratio (PAPR).

On next parts, the cut-off frequency of the filter will be normalized to the 5 MHz 3GPP channel bandwidth to indicate the bandwidth expansion defined as:

$$BW = \text{bandwidth expansion} \times 2.5\,\text{MHz (Eq. 7)}$$

B. PAPR of filtered signals

The filtering of both $S_1(t)$ and $S_2(t)$ signals generates an envelope variation and thus an unwanted PAPR increase on these supposed constant envelope signals. Fig.4 shows the variation of the PAPR according to reconstruction filters bandwidth.

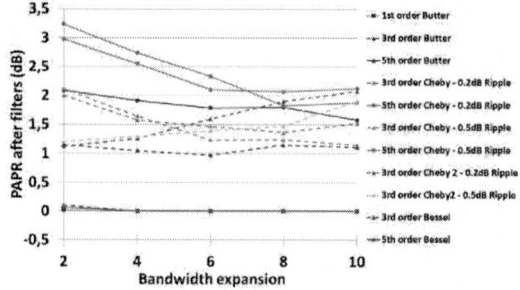

Figure 4. PAPR vs. bandwidth expansion

This figure points out it's better to select a maximally flat filter (best case is Bessel) to strongly limits PAPR increase.

It quantifies the performances of filters but it cannot be the only. Group delay variation within the pass band (phase deviation) has also to be considered. Group delay impact is observed by analyzing the recombined spectrum.

It's known that type II Chebyshev filters, as Bessel ones, have a quite constant group delay characteristic [13] while type I Chebyshev has a poor one.

IV. IMPACT OF PA CLASS

A. Switched mode PA

For the set of simulation, a theoretical switch mode PA is modeled by suppressing the envelope variation. Fig.5 presents the used model, at the output of the IQ modulators.

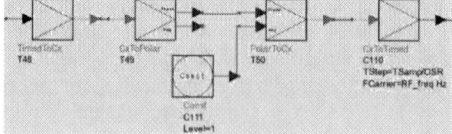

Figure 5. SMPA Model

Study results show that ORFS at a 4 MHz offset and 10 MHz ACLR are the most difficult specification points to achieve. These two parameters are respectively presented Fig.6 and Fig.7 according to bandwidth expansion (Eq.7).

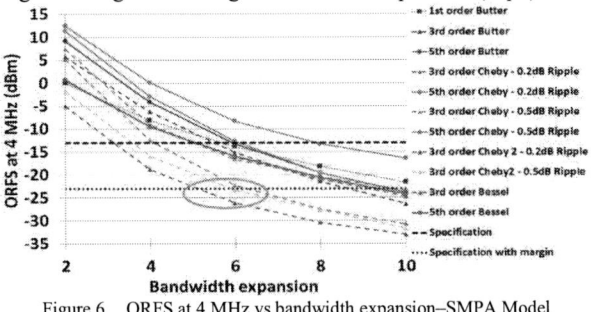

Figure 6.   ORFS at 4 MHz vs bandwidth expansion–SMPA Model

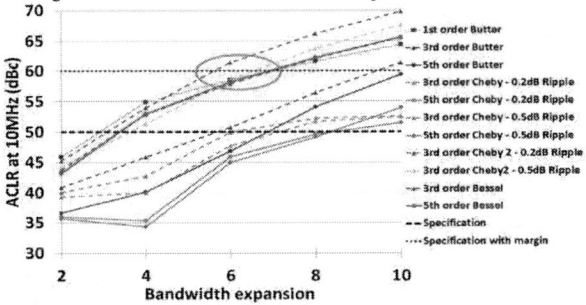

Figure 7.   ACLR at 10 MHz vs bandwidth expansion–SMPA Model

Filters that give the best trade-off in terms of ACLR, ORFS and minimum required bandwidth are 3rd order type II Chebyshev, Bessel and Butterworth. On the other hand, type I Chebyshev filter is less efficient because it recreated AM modulation that cannot be transmitted by the SMPA.

As a conclusion, 3rd order type II Chebyshev with 0.2 dB ripple and 15 MHz cut-off frequency offers the best trade-off in terms of bandwidth and filter complexity.

B. AB Class Satured PA

It is interesting to check how evolve the previous filtering requirements by using a 1-dB compressed PA (multi-level LINC solution [8]) instead of a SMPA. Indeed the unexpected envelope variation of $S_1(t)$ and $S_2(t)$ can be partially transmitted. The power at PA driver output is:

$$P_{OUT\_DRIVER} = P_{IN\_PA} = P_{OUT\_PA} - Gain\_Ampli + 1dB \qquad (Eq. 8)$$

The classical AM / AM function has been used for the AB Class PA model and is illustrated Fig.8.

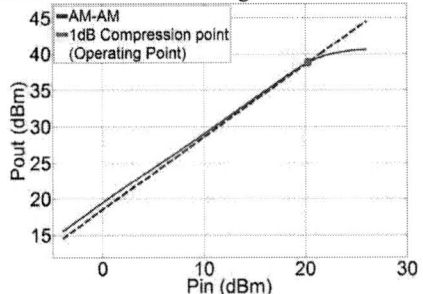

Figure 8.   AB Class PA AM/AM function

In the same way as previously studied, ORFS at a 4 MHz offset and 10 MHz ACLR are respectively presented Fig.9 and Fig.10.

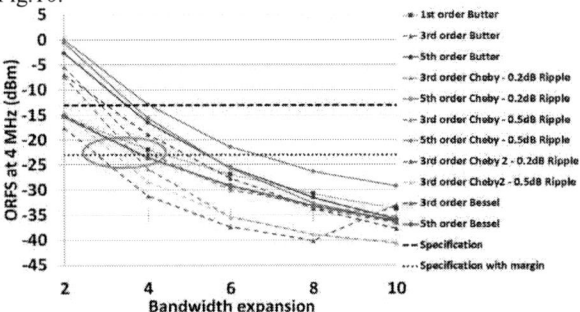

Figure 9.   ORFS at 4 MHz vs bandwidth expansion–Saturated AB Class PA

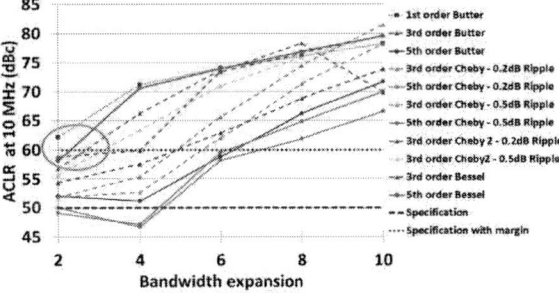

Figure 10. ACLR at 10 MHz vs bandwidth expansion–Saturated AB Class PA

3rd order type II Chebyshev with 0.2 dB ripple is still the best filter to provide the trade-off between complexity and bandwidth.

By reducing the cut-off frequency down to 10 MHz instead of 15 MHz, the same performances are achieved.

By keeping the same 15 MHz cut-off frequency then ACLR performance is 12 dB improved and ORFS at 4 MHz from the carrier is 10 dB better.

Thus AB Class PA working at compression point usage relaxes filtering cut-off frequency constraint or leads to ACLR and ORFS performance improvement.

C. AB Class Backed-off PA

For sake of completion, the test case using an AB Class PA working in its linear operation area, so in backed-off condition, is explored by selecting the best filter selected in SMPA study case. In this case, Back-off is equal to PAPR of the input signals (after filtering).

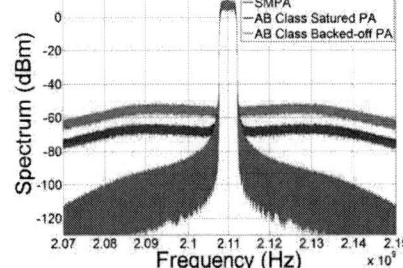

Figure 11. ORFS for different PA models with a 3rd order type II Chebyshev filter – 15 MHz BW – 0.2 dB ripple

As expected, Fig.11 shows that the ORFS is nearly not impacted using backed-off PA because the PA doesn't clip the recreated amplitude modulation due to reconstruction filters. Reconstruction filter template will depend on attenuation needed for achieving spurious emissions specification.

Of course, SMPA and saturated PA are definitely the practical cases for power efficiency.

V. MISMATCHES

One of the limiting points of this architecture is the sensitivity to mismatches between S_1 and S_2 paths. It is a well-known statement but not quantified in the case of a WCDMA base station transmitter.

Admissible gain mismatches and time delays are evaluated by looking at ORFS at 4 and 5 MHz (most critical points):

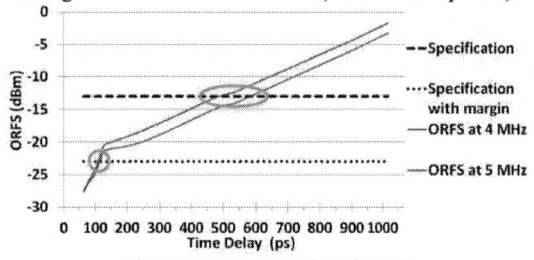

Figure 12. ORFS vs. Delay mismatch

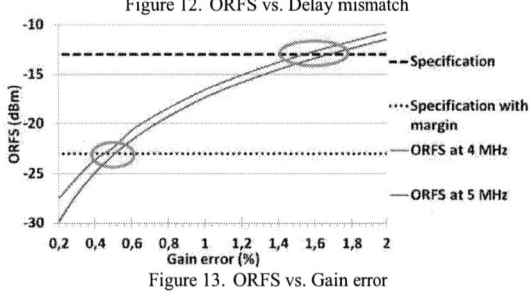

Figure 13. ORFS vs. Gain error

We can see some limits of this architecture, especially concerning delay mismatches, that is equivalent to a phase error. To reach specification with 10 dB margin, minimal time mismatch is close to 100ps while for gain error, the minimal value is 0.45%. These figures have to be considered during the design of the transmitter and further calibration system requirements.

VI. CONCLUSION AND DISCUSSION

The impact of the PA class on the baseband filtering sizing within LINC architecture context for a single carrier WCDMA base station is studied in this paper.

Clearly, depending on the PA class, filter sizing requirements change. In the use case of SMPA, those requirements are very stringent. This is because amplitude modulation - recreated by reconstruction filters - is clipped. On the other hand, with PA at compression point or in linear region, amplitude variation is respectively partially or completely transmitted. It involves less degradation of the spectrum after recombination.

As a consequence lower cut-off frequency is requested (from 15 MHz down to 10 MHz) when using a saturated AB

class PA compared to SMPA usage. In other words, for a given filter bandwidth, the performances are really improved with an AB Class PA working at 1-dB compression point (12 dB better for ACLR @ 10 MHz and 10 dB better for ORFS @ 4 MHz), in comparison with a SMPA use case.

With a SMPA, the most suitable reconstruction filter is the 3rd order type II Chebyshev with 0.2 dB ripple with a 15 MHz cut-off frequency. This filter matches because its frequency response is flat in the pass-band and its group delay variation is low enough. The recreated amplitude variation (≈ 25 % that corresponds to 1 dB PAPR) fully fulfills requirements. Type II Chebyshev filter is really area consuming (LC networks). Bessel one can be an alternative if replicas attenuation requirements are less stringent.

This paper also quantifies admissible gain and delay mismatches between the two LINC paths: 1.5% or 480ps to achieve 3GPP specifications.

REFERENCES

[1] Choffrut, A.; Van Veen, B.D.; Booske, J.H.; , "Effects of modulator misalignment on LINC transmission with traveling wave tube amplifiers," Vacuum Electronics Conference, 2002. IVEC 2002. Third IEEE International , vol., no., pp. 344- 345, 2002

[2] Sundstrom, L.; , "Spectral sensitivity of LINC transmitters to quadrature modulator misalignments," Vehicular Technology, IEEE Transactions on , vol.49, no.4, pp.1474-1487, Jul 2000

[3] Laskar, J.; Lim, K.; Hur, J.; Kim, K.W.; Lee, O.; Lee, C.-H.; , "Emerging multi-level architectures and unbalanced mismatch calibration technique for high-efficient and high-linear LINC systems," Circuits and Systems (ISCAS), Proceedings of 2010 IEEE International Symposium on , vol., no., pp.821-824, May 30 2010-June 2 2010

[4] Garcia, P.; de Mingo, J.; Valdovinos, A.; Ortega, A.; , "Adaptive digital correction of gain and phase imbalances in LINC transmitters," Vehicular Technology Conference, 2004. VTC 2004-Spring. 2004 IEEE 59th , vol.3, no., pp. 1237- 1241 Vol.3, 17-19 May 2004

[5] Helaoui, M.; Boumaiza, S.; Ghannouchi, F.M.; , "On the outphasing power amplifier nonlinearity analysis and correction using digital predistortion technique," Radio and Wireless Symposium, 2008 IEEE , vol., no., pp.751-754, 22-24 Jan. 2008

[6] Berland, C.; Bercher, J.; Venard, O.; , "Gain and delay mismatches cancellation in LINC and polar transmitters," Circuits and Systems (ISCAS), Proceedings of 2010 IEEE International Symposium on , vol., no., pp.1017-1020, May 30 2010-June 2 2010

[7] Van der Heijden, M. P.; Acar, M.; Vromans, J. S.; Calvillo-Cortes, D. A.; , "A 19W high-efficiency wide-band CMOS-GaN class-E Chireix RF outphasing power amplifier," Microwave Symposium Digest (MTT), 2011 IEEE MTT-S International , vol., no., pp.1, 5-10 June 2011

[8] Helaoui, M.; Ghannouchi, F.M.; , "Linearization of Power Amplifiers Using the Reverse MM-LINC Technique," Circuits and Systems II: Express Briefs, IEEE Transactions on , vol.57, no.1, pp.6-10, Jan. 2010

[9] 3GPP TS 25.104", version 11.0.0, 2011-12

[10] Sundstrom, L.; , "Effects of reconstruction filters and sampling rate for a digital signal component separator on LINC transmitter performance" Electronics Letters , vol.31, no.14, pp.1124-1125, 6 Jul 1995

[11] Gerhard, W.; Knoechel, R.H.; , "LINC digital component separator for single and multicarrier W-CDMA signals," Microwave Theory and Techniques, IEEE Transactions on , vol.53, no.1, pp. 274- 282, Jan. 2005

[12] NATIONAL SEMICONDUCTOR: LMK04800 Family – Low-Noise Clock Jitter Cleaner with Dual Loop PLLs, Mar. 2011

[13] Rolf Schaumann; Mac E. Van Valkenburg, "Design of analog filters", 2010

[14] Advanced Design System – Agilent; http://www.home.agilent.com/en/pc-1297113/advanced-design-system-ads

Scalable Hybrid CORDIC-LUT Architectures for CG-FFT Processors

Andrea Congiu, Andrea Picciau, Massimo Barbaro
DIEE - Dept. of Electrical and Electronic Engineering
University of Cagliari
Cagliari, Italy
Email: andrea.congiu, barbaro@diee.unica.it

Emanuele Bodano
Infineon Technologies
Villach, Austria
Email: Emanuele.Bodano@infineon.com

Abstract—In this work we introduce Processing Element (PE) scalability in twiddle factor generators for FFT processors. First the twiddle factor indexing scheme for Constant Geometry FFT is analyzed and a CORDIC-based novel algorithm is deduced. It uses single-step rotations and does not need any CORDIC gain correction. Then, two architectures implementing the algorithm are presented with the goal of scalability. The first (*shared core*) is characterized by both low register count and variable throughput, while the second (*pipelined*) achieves the maximum throughput during the whole computation. Our hybrid models use both one ROM and multiplier-based CORDIC modules. The designs are then evaluated in terms of register usage and output error, showing scalability of register bits as a function of the number of PEs if compared to other architectures. Architectures were coded in VHDL and synthesized on a Xilinx Virtex-5 330T FPGA.

I. INTRODUCTION

Digital signal processing (DSP) has emerged as a common solution in a large domain of applications, ranging from radar to biomedical, to telecommunications and control systems. Being most of signal processing approaches based on frequency domain analysis, a wide range of custom FFT processor architectures can be found in literature. Pipelined structures have both high throughput and computational efficiency, but when area constraints are stronger than timing, non-pipelined counterparts are usually preferred.

Scalable architectures in terms of number of processing elements (PEs), are flexible solutions. Astola and Akopian [1] introduced a family of hardware-oriented algorithms resulting in scalable constant geometry (CG) structures. In [2] a not-in-place architecture targeted for ASIC implementation is proposed. This solution uses data shuffling registers, thus it does not require a memory for intermediate results.

Classic approaches to FFT designs require a large amount of memory for storing precomputed twiddle factors. The CORDIC iterative algorithm presented in [3] allows computation of twiddle factors at runtime. To this purpose many architectures introduce this algorithm inside the PEs by substituting the complex multipliers with iterative phase rotations. With this approach, in [4] the number of iterations has been reduced by using optimized sequences and corresponding scale factors both stored in a LUT. In [5] a multi-bank RAM structure to reduce memory logic is presented. On the other hand, some systems replace the twiddle factor storage ROM with

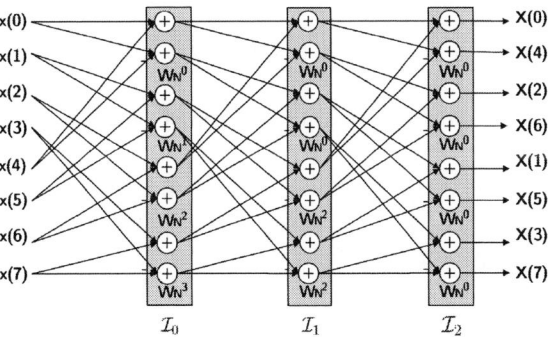

Fig. 1. Dataflow diagram of an 8-point Radix-2 DIF CG-FFT

a CORDIC-based generation system. Nonetheless, CORDIC hardware implementations can be very expensive in terms of area usage [6], [7], consequently hardly suitable for a scalable FFT approach.

In this paper we introduce PE scalability on two novel hybrid CORDIC-LUT systems designed for generating twiddle factors in FFTs. In Section II mathematical considerations are drawn on the addressing scheme for twiddle factors in Constant Geometry FFT (CG-FFT), in order to reduce the dimension of the considered set of twiddle factors. In Section III two PE-scalable CORDIC architectures are presented. In Section IV results are discussed. Finally, conclusions are drawn in section V.

II. MATHEMATICAL CONSIDERATIONS

A. Properties of the CG-FFT

A classic distinction among FFT algorithms is done considering both radix and decimation type [8] while CG-FFT algorithms are classified by choosing PE implementation. Our designs are based on Radix-2 Decimation-In-Frequency CG-FFT as in [2].

A main feature of CG-FFT is the sequence of twiddle factors in the dataflow in Fig. 1. In general, the twiddle factor of index i is defined as

$$W_N^i = e^{-j\frac{2\pi}{N}i}, \tag{1}$$

where N is the number of input points to the FFT. i ranges from 0 to $\frac{N}{2} - 1$. In Radix-2 implementations, the FFT is computed in $\log_2 N$ stages, indexed considering s ranging from 0 to $\log_2 N - 1$. For each stage $\frac{N}{2}$ twiddle factors must be considered according to the sequence of indexes $\mathcal{I}_s$:

$$\mathcal{I}_s = \{\underbrace{0,\ldots,0}_{2^s}, \underbrace{1 \cdot 2^s,\ldots,1 \cdot 2^s}_{2^s}, \underbrace{2 \cdot 2^s,\ldots,2 \cdot 2^s}_{2^s}, \ldots\}, \quad (2)$$

that can be summarized in $\mathcal{I}_s = \{i^{(0)}, \ldots, i^{(N/2-1)}\}$. Let us introduce the scalability factor $P \in \{x \in \mathbb{N} \mid x = 2^y, y \geq 0\}$ equal to the number of parallel PEs, enumerated using index p ranging from 0 to $P-1$. The twiddle factors generation system must output P values at the same instant, one for each PE. Every stage s is thus split in $C = \frac{N}{2P}$ cycles, indexed as $c = 0, \ldots, C-1$, and similarly sequence $\mathcal{I}_s$ is divided into C subsequences $\mathcal{J}_{sc}$ of indexes:

$$\mathcal{I}_s = \{\mathcal{J}_{s0}, \ldots, \mathcal{J}_{sc}, \ldots, \mathcal{J}_{s(C-1)}\}, \quad (3)$$

where $\mathcal{J}_{sc} = \{i_{sc0}, \ldots, i_{scp}, \ldots, i_{sc(P-1)}\}$. The twiddle factor needed on step c of stage s by the p-th PE has index

$$i_{scp} = \left\lfloor \frac{p + cP}{2^s} \right\rfloor 2^s. \quad (4)$$

The number of considered indexes can be reduced from $\frac{N}{2}$ to $\log_2 N - 2$, observing that:

A1) for every i, W_N^i is a complex number with unitary module falling either in the third or fourth quadrant of the Gauss plane. With basic trigonometry all $\frac{N}{2}$ different twiddle factor values necessary during computation can be calculated using indexes in the set $M = \{0, \ldots, \frac{N}{8}\}$.

A2) Consecutive elements $i^{(k)}$ in $\mathcal{I}_s$ are either the same index or differ by 2^s.

A3) The first elements i_{sc0} of each subsequence $\mathcal{J}_{sc}$ (which represent inputs to the first PE) can be the same or differ by a power of 2.

A4) Multiplication by W_N^i is an angular rotation of $-j\frac{2\pi}{N}i$.

The smallest set of indexes that we consider in our models is the subset $M' \subset M$

$$M' = \{m = 0 \vee m = 2^x, x = 0, 1, \ldots, \log_2 N - 3\}. \quad (5)$$

B. Scalable Rotational Algorithms

We present here a scalable rotational algorithm for computing the twiddle factor sequence $\{W_N^{(k)}\}$. The procedure execution requires a set $W(M')$ of precomputed twiddle factors with indexes in M'.

Input: $\mathcal{I}_s, N, C, P, W(M')$
    $k \leftarrow 0$
    for $c = 0$ to $C-1$ do
3:      for $p = 0$ to $P-1$ do
        $q_{scp} \leftarrow i_{scp}$ reflected in $\{0, \ldots, \frac{N}{8}\}$
        if $c = 0$ and $p = 0$ then
6:        $l \leftarrow 0$
        else if $p = 0$ then
          $l \leftarrow q_{s(c-1)(P-1)}$ or $l \leftarrow q_{s(c-1)0}$
9:        else
          $l \leftarrow q_{sc(p-1)}$
        end if
12:      if $q_{scp} \in M'$ then
        $W_N^{(k)} \leftarrow W(q_{scp})$
        else if $q_{scp} = l$ then
15:      $W_N^{(k)} \leftarrow W_N^{(k-1)}$
        else {CORDIC iteration}
          $\alpha \leftarrow \pm(q_{scp} - l), |\alpha| \in M'$
18:        $W_N^{(k)} \leftarrow W_N^{(k-1)} \cdot W(\alpha)$
        end if
        $W_N^{(k)} \leftarrow W_N^{(k)}$ reflected in $(-\pi, 0)$
21:      $k \leftarrow k + 1$
      end for
    end for
Output: $\{W_N^{(k)}\}$.

In most approaches, CORDIC algorithms need precomputed scale factors and converge towards results in a number of steps depending on the resolution bits B [3], [4]. Taking advantage of statement A4, each element of the twiddle factor output sequence is obtained in only one iteration, without the need of any gain correction. Also, observation A1 has been used in line 4, in order reduce the number of indexes to $\frac{N}{8} + 1$. Moreover, exploiting considerations given in A2, the presented algorithm (lines 12 and 14) does not require any arithmetic operation both when reflected index q_{scp} belongs to M' and when q_{scp} is equal to the index l of the previous step.

We distinguish between two versions of the same algorithm: if we decide to exploit the observation given in A2 then $l \leftarrow q_{s(c-1)(P-1)}$, otherwise considering A3, $l \leftarrow q_{s(c-1)0}$.

III. PE-SCALABLE CORDIC APPROACHES

The architectures presented in this section are twiddle factor generation modules implementing the two versions of the algorithm in section II-B respectively. Both these modules receive P parallel unsigned signals corresponding to subsequences $\mathcal{J}_{sc}$ and output P twiddle factors belonging to sequence $\{W_N^{(k)}\}$ in signed fixed-point format.

We refer to *shared core* (Fig. 2) when $l \leftarrow q_{s(c-1)(P-1)}$. This means that the computation on subsequence $\mathcal{J}_{s(c-1)}$ must be completed before the start of calculations on $\mathcal{J}_{sc}$.

When $l \leftarrow q_{s(c-1)0}$ the first element of subsequence $\mathcal{J}_{sc}$ can be processed before the end of calculations on subsequence $\mathcal{J}_{s(c-1)}$, using only the data from the computation of the first element ($p = 0$) of $\mathcal{J}_{s(c-1)}$. This leads to another architecture, referred to as *pipelined* (Fig. 3), it parallelizes the for loop on line 2 of the algorithm by using different pipeline stages.

Algorithm II-B can require to perform a rotation by using twiddle factors resulting from CORDIC iterations themselves. This can be an issue in terms of error propagation. In order to compensate this effect, we perform internal operations by using $B' \geq B$ bits and then rounding to B-bit words, where B is the resolution of the output twiddle factors.

A. Hybrid CORDIC-LUT architecture shared-core

Module *mapper* realizes the first algorithmic step (line 4) in order to output the mapping of the incoming requested array of P indexes to the set M, then representing the portion $(-\frac{\pi}{4}, \ldots, 0)$ of the complex plane.

Then *reference builder* scans these outputs comparing every value with the preceding ones, determining if this array needs computing of a new twiddle factor or not. If computation is

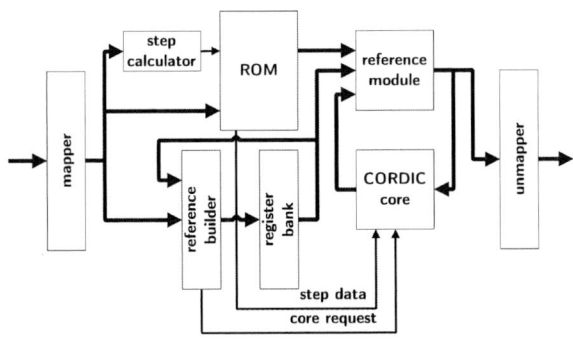

Fig. 2. The *shared core* scalable CORDIC architecture

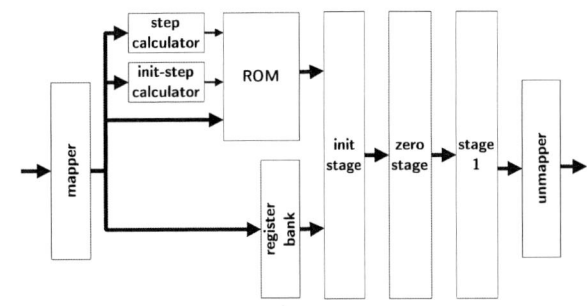

Fig. 3. The *pipelined* scalable CORDIC architecture. $P = 2$.

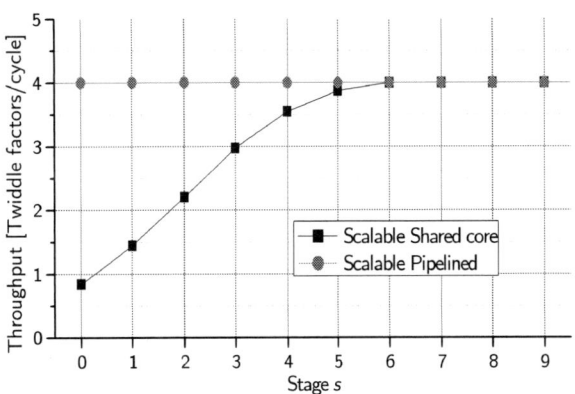

Fig. 4. Pipelined architecture speedup. $N = 1024$, $P = 4$.

not necessary data can be immediately retrieved because either the twiddle factor has already been computed by CORDIC iteration or it belongs to $W(M')$. In these cases, the *reference builder* outputs to *register bank* the information for selecting data used by *reference module*. Request of CORDIC core usage is done by driving signal *core request*.

The output array of *mapper* is also analized by block *step calculator* with the purpose of determining α (line 17), that is the ROM address where *step data* is stored, needed to compute the CORDIC iteration.

The $P+1$ port ROM stores the set $W(M')$: it outputs a nonzero value only when address belongs to M'.

Combinatorial block *reference module* selects data from the ROM or from CORDIC results and manages the case of consecutive same-valued indexes (line 14) by reading reference informations in *register bank*.

The CORDIC core reads the *core request* signal. Then, it serves the pending requests using a complex multiplier, composed of 3 real multipliers. Initialization data is taken from the output of *reference module* and the product is performed using *step data* according to line 18 of the algorithm.

When computations are executed, twiddle factors feed block *unmapper* reflecting the obtained results to the third and fourth quadrant of the complex plane. This block implements line 20 of the forementioned procedure.

B. Hybrid CORDIC-LUT architecture pipelined

Modules *mapper*, *step calculator* and *register bank* are the same used in architecture *shared core*. Compared to the former, the extra block *init-step calculator* behaves similarly to *step calculator* but it considers only the first elements between adjacent $\mathcal{J}_{sc}$ request arrays. That is, it computes α using $l = q_{s(c-1)0}$ by taking advantage of A3. Thus, a $P+2$ port ROM is needed for this solution.

The pipeline is composed of $P+2$ stages. *Init stage* performs if needed the inversion of the imaginary part of α used in CORDIC iterations, depending on the direction of rotation. *Zero stage* serves the requests for $p = 0$, by either reading the ROM or performing a CORDIC iteration if needed. In the latter case, step data conforms to the address computed by

init-step calculator. The behaviour of the following stages is similar, but each stage matches a different value of p, from 1 to $P - 1$. For these stages iteration step data accords to the output of *step calculator*. Similarly to architecture *shared core*, the last pipeline stage is the *unmapper* block.

IV. RESULTS

We implemented a Radix-2 Decimation-In-Frequency FFT processor as in [2], inserting our models of hybrid PE-scalable twiddle factors generators. The entire design has been coded by using VHDL RTL.

The *pipelined* architecture can generate P twiddle factors per clock cycle, while the *shared core* architecture has a variable throughput until it reaches the maximum value (P). This is because the immediate retrieval of data becomes more effective as s increases. The trend is shown in Fig. 4.

The total number of bits needed for computational data storage is $2(2B' + B)P$ for *shared core* and $2(B'(P + 1) + B)P$ for *pipelined*. In classic CORDIC approaches or even in the particularly efficient solution [7] at least $2B^2P$ bits need to be stored. Fig. 5 shows the accuracy of twiddle factors output for our architectures and for the *cordicsincos* MATLAB classic CORDIC built-in function. Every curve is obtained by varying B' with $B = 16$ and $s = 0$. The classic CORDIC curve is obtained by fixing the number of iterations to B. The

TABLE I
FPGA SYNTHESIS RESULTS. $N = 1024$, $P = 4$, $B = 16$.

	Classic	Shared core	Pipelined
B'	16	20	18
Slice registers	2691	732	1387
Slice LUTs	3100	2822	2104
Multipliers	4	3	12
Maximum Frequency [MHz]	477	92	114

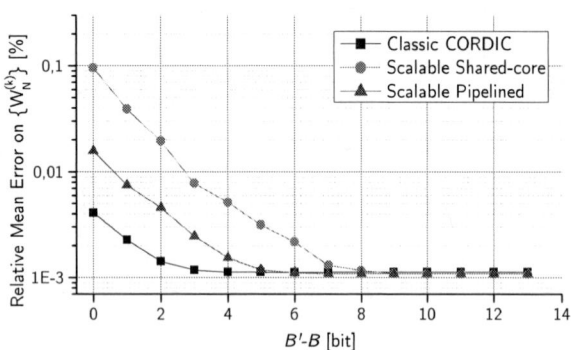

Fig. 5. Relative Mean Error on $\{W_N^{(k)}\}$. $B = 16$, $s = 0$.

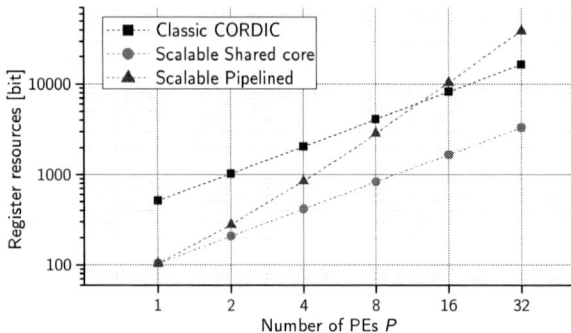

Fig. 6. Register bits as a function of the number of PEs. $B = 16$, $B' = 18$.

error mean $|W_{N,\text{obtained}}^{(k)} - W_{N,\text{exact}}^{(k)}|$ is evaluated with respect to precomputed exact values. The classic CORDIC approach performs successive micro-rotations [6], [7]. In our approaches we have only one CORDIC iteration, but data coming from previous steps are involved. Anyway the introduced error proves to be easily compensated.

Figure 6 evaluates scalability by showing computational data storage in registers ($B = 16$, $B' = 18$) as a function of P. The plot is obtained with the previous theoretical considerations. As P increases, *shared core* always uses less registers while *pipelined* saves registers until the number of PEs reaches B.

The number of real multipliers is 3 for *shared core* and $3P$ for *pipelined*. The usage of multipliers to compute the single iteration can lead to a tradeoff in terms of maximum frequency and scalability. In table I, we compare synthesis results of our architectures to a classic CORDIC approach freely accessible at OpenCores.com [9]. The used FPGA is a Xilinx Virtex-5 XC5VLX330T. We chose different values of B' in order to take into account accuracy considerations. We can conclude that our models achieve about the same mean error but saving resources. On the counterpart both have a lower frequency, as expected from architectures intended for applications in which resource saving is a stronger constraint than timing.

V. CONCLUSIONS

We introduced a novel PE-scalability oriented algorithm by analizing the properties of the CG-FFT. The procedure for twiddle factor generation uses single-step rotations, therefore, unlike other CORDIC approaches [4], our solution does not need any gain correction. The algorithm was implemented in two VHDL models. The hybrid CORDIC-LUT designs use a $log_2 N$-sized ROM to store the smallest set of twiddle factors. Both architectures show a better PE-scalability in terms of register bits if compared with pipelined CORDIC approaches [7], [9]. The mean relative error on output twiddle factors was used as a parameter to assess the accuracy depending on bit resolution. The *shared core* design has the advantage of using less resources as a tradeoff with accuracy, and its throughput increases during the processing of the FFT until it reaches the maximum. The *pipelined* design achieves the maximum throughput of P twiddle factors per cycle during the entire processing, but using more multipliers than the former. Both solutions have a maximum frequency of about 100 MHz on Virtex-5. Our architectures can be preferred when constraints on resource saving are stronger than the one on clock frequency.

REFERENCES

[1] J. Astola and D. Akopian, "Architecture-oriented regular algorithms for discrete sine and cosine transforms," *Signal Processing, IEEE Transactions on*, vol. 47, no. 4, pp. 1109–1124, Apr. 1999.

[2] A. Suleiman, A. Hussein, K. Bataineh, and D. Akopian, "Scalable fft architecture vs. multiple pipeline fft architectures; hardware implementation and cost," in *Systems, Man and Cybernetics, 2009. SMC 2009. IEEE International Conference on*, Oct., pp. 3792–3796.

[3] J. E. Volder, "The cordic trigonometric computing technique," *Electronic Computers, IRE Transactions on*, vol. EC-8, no. 3, pp. 330–334, Sept. 1959.

[4] C.-Y. Yu, S.-G. Chen, and J.-C. Chih, "Efficient cordic designs for multi-mode ofdm fft," in *Acoustics, Speech and Signal Processing, 2006. ICASSP 2006 Proceedings. 2006 IEEE International Conference on*, vol. 3, May, pp. III–III.

[5] X. Xiao, E. Oruklu, and J. Saniie, "Reduced memory architecture for cordic-based fft," in *Circuits and Systems (ISCAS), Proceedings of 2010 IEEE International Symposium on*, 30 2010-June 2, pp. 2690–2693.

[6] R. Bhakthavatchalu, N. Abdul Kareem, and J. Arya, "Comparison of reconfigurable fft processor implementation using cordic and multipliers," in *Recent Advances in Intelligent Computational Systems (RAICS), 2011 IEEE*, Sept., pp. 343–347.

[7] M. Garrido and J. Grajal, "Efficient memoryless cordic for fft computation," in *Acoustics, Speech and Signal Processing, 2007. ICASSP 2007. IEEE International Conference on*, vol. 2, April, pp. II–113–II–116.

[8] E. Chu and A. George, *Inside the FFT Black Box. Serial and Parallel Fast Fourier Transform Algorithms.* CRC Press, 2000.

[9] Opencores.org. [Online]. Available: http://opencores.org

978-1-4673-4579-8/13 $31.00 © 2013 IEEE

Off-line BIST in Watt Hour Meters

Rok Ribnikar
ISKRAEMECO d.d.
Savska loka 4, 4000 Kranj, Slovenia
rok.ribnikar@iskraemeco.si

Drago Strle
University of Ljubljana, Faculty of Electrical Engineering
Tržaška 25, 1000 Ljubljana, Slovenia
drago.strle@fe.uni-lj.si

Abstract—In this work the concept, circuits and algorithms for an efficient off-line Built-In Self-Test (BIST) in watt hour meters are presented. On-chip signal generator is needed and additional digital hardware is necessary to evaluate the result. For a good signal to noise ratio (SNR) as well as good frequency resolution, a large number of samples would be needed to accurately calculate the FFT of the result. This approach is time-consuming and inappropriate for BIST, since it requires too much silicon area. We tried to avoid FFT calculation and replace it with small digital hardware and hardware algorithm running on the chip. The most important errors, which may happen in measurement path during the operation of the watt hour meter, can be efficiently detected. The parameters obtained by the suggested BIST hardware/algorithm were compared to the results obtained by the FFT for verification. The results confirm that the proposed concept of BIST could be used for an efficient on chip and off-line built-in self-test.

Keywords—off-line BIST, testing watt hour meter, sigma-delta modulator, sigma-delta oscillator

I. INTRODUCTION

The off-line testing is required in many mixed-signal-circuits. Usually, this is performed during start-up, or from time to time, to prove reliability of the system. There are many different solutions how off-line testing can be performed in the laboratory or at the end of the manufacturing process. The testing of measurement channels of watt hour meter, which is located in the house and measures power consumption, is a challenging task.

Watt hour meters are composed of two or more measurement channels [1] dependent on the number of phases and other functions. At least one channel is used for current measurements and another for voltage measurements. The goal of this work is to find a way to test one-phase watt hour meter's measurement channel and get as much information as possible with the smallest possible amount of additional silicon area.

The measurement channel consists of a sensor, analog to digital converter (ADC) and digital signal processing (DSP) [1]. The ADC is usually composed of a sigma-delta modulator (SDM) due to high accuracy performances and appropriate decimation filter [2]. In this paper a SIMULINK model of the 2nd order SDM, including main non-idealities, and the DSP, including all main digital filters, is used to verify the BIST method. The other blocks of the watt hour meter such as

microprocessor, display, communication protocols and others will be excluded from our considerations.

The paper is organized as follows. In section II the block diagram of the off-line BIST integrated in watt hour meter is presented, where we describe the concept of testing with built-in stimulus. In sub section II the main analogue and digital parts of the measurement channel in watt hour meter that need to be measured are presented. The algorithms and the system level simulation results are presented in sections III and IV The last section concludes the article.

II. WATT HOUR METER WITH OFF-LINE BIST

For a reliable off-line BIST, it is necessary to generate precision sine-wave signal with known frequency and amplitude that requires as small silicon area as possible for the implementation. The best way to analyze the response would be to implement on chip calculation of Discrete Fourier Transform (DFT); unfortunately such calculation would require a very big silicon area and is therefore not applicable. Fig. 1 presents a block diagram of the measurement part of a watt hour meter. The voltage and current measurement channels are built from standard components: anti-aliasing filter (AAF), ADC built from modulator and cascaded CIC filter, compensation filter (Comp filter) to correct the phase and amplitude response, high pass (HP) filter to eliminate offset of the measurement paths including the sensor and root mean square function (RMS) for calculation of RMS values of voltage and current and apparent power [3],[4]. The active and reactive power or energy is calculated in the block power and energy calculation.

The purpose of the off–line BIST methodology is to measure the main parameters that define the accuracy of the measurement path. During the life time of a watt hour meter, different components can fail, for example: signal-to-noise ratio may decrease, wrong gain of measurement path can cause wrong measurement of power consumption etc. With an off-line test the electricity distributor can perform simple test on demand during appropriate time, if watt hour meter supports a communication possibility, to see if the metering device's accuracy fulfills the requirements. The off-line BIST is initiated by disconnecting measured signals (voltage and current) from the input of the measurement channels and connecting appropriate internal sine-wave signal to the channel inputs. The sine-wave signal at the input of the measurement channels is the reference signal with minimum noise and

---

Operation was partially financed by the European Union, European Social Fund.

978-1-4673-4579-8/13 $31.00 © 2013 IEEE

harmonics. The sine-wave with predefined parameters can be generated by digital sigma-delta oscillator, which generates single tone signal with desired frequency and amplitude. The switches are added to deactivate measurements and activate off-line BIST. To evaluate the correct operation of the measurement channels, existing RMS calculation component could be used. This component is already built into the measurement channel and calculates mean and standard deviation of consecutive RMS values. From this information it is possible to determine, if the measurement channels are operating correctly.

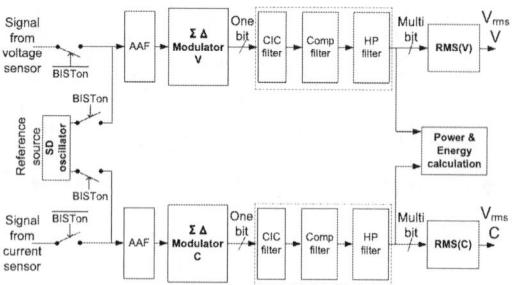

Fig. 1. Measurement part in watt hour meter with integrated built in stimulus

There are many possibilities for generating a reference signal on chip. One possibility is to generate a square wave signal with an adjustable amplitude and frequency using DAC and chopper. The problem of such signal is in higher harmonics. Using multi-tone reference, the signal requires more complicated evaluation of measured results [5]. However, the problem remains if we use a triangular reference signal [6], which can also be built from digital counter and a suitable DAC; it also contains unwanted higher harmonics, in addition to the fundamental spectral component. The most suitable solution for an off-line BIST in watt hour meter, would be to use an adjustable single-tone sine wave reference signal using digital sigma-delta (SD) oscillator adopted by [7], which is built of two digital integrators connected in the loop and two multi-bit multiplications that provide appropriate gain in the loop, as suggested on Fig. 2a. The bit-stream of the SD oscillator can be brought to the input of a SD modulator through 1-bit DAC and analog low pass filter or through reference voltage of modulator and small modification of the input switched capacitor (SC) stage [9]. The use of 1- bit DAC is desirable due to linear characteristic and simple structure.

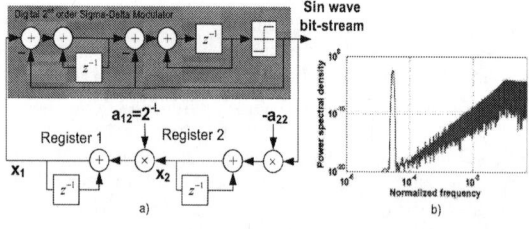

Fig. 2. a) Sigma-Delta oscillator with digital 2nd order sigma delta oscillator and b) power spectral density of the output of oscillator

The characteristic equation of sigma-delta oscillator is given in (1).

$$z^2 + (a_{12} \cdot a_{21} - 2)z + 1 = 0 \qquad (1)$$

One solution of the quadratic equation (1) where the poles are located in the right-half part of the unit circle is:

$$z_{1,2} = e^{\pm j\cos\left(1 - \frac{a_{12}a_{21}}{2}\right)}; \ 0 < a_{12}a_{21} \leq 2 \qquad (2)$$

From equation (2) we can derivate the oscillation frequency (f_{in}), where the product of a_{12} and a_{21} must be between 0 and 2 ($0 < a_{12}a_{21} \leq 2$) to satisfy oscillation conditions.

To simplify the multiplication, a_{12} can be selected in such a way that it corresponds to one of the integers corresponding to 2^{-L}. The multiplication of digital signal from the Register 2 with a constant a_{12} is thus simplified. The multiplication of digital bit-stream with constant a_{21} is simple if the digital SD modulator has a one bit quantizer; in this case the constant a_{21} is multiplied with ± 1 and defines the oscillation frequency (f_{in}). It can be calculated using (3), where f_{os} is oversampling frequency and f_{in} is oscillation frequency.

$$a_{21} = \frac{2}{a_{12}} \cdot \left[1 - \cos(2\pi \cdot \frac{f_{in}}{f_{os}}) \right] \qquad (3)$$

To keep a_{12} as suggested above, only a discrete set of frequencies is possible; for example for $a_{12}=2^{-6}$, $a_{21}=1.5791367e-6$ and $f_{os}=4$MHz the frequency of oscillation is $f_{in}=100$Hz. To set the amplitude of a sine-wave oscillation, the initial values of registers 1 and 2 must be determined by $x_1(0)$ and $x_2(0)$.

$$Ampl = \frac{(1 - a_{12}a_{21})x_1(0) + a_{12}x_2(0)}{\sin(\omega_{in}T)} \qquad (4)$$

If the initial value of Register 1 is set to zero ($x_1(0)=0$) than the initial value of register 2 for the required amplitude is calculated by (5).

$$x_2(0) = \frac{Ampl \cdot \sin\left(2\pi \cdot \frac{f_{in}}{f_{os}}\right)}{a_{12}} \qquad (5)$$

The parameter *Ampl* represents the desired amplitude of the sine-wave signal. The frequency spectrum of sin-wave bit-stream is presented in Fig. 2b. The output from SD oscillator contains sine-wave signal and shaped quantization noise that is removed by the CIC filter, which attenuates high frequency components [2].

III. Signal Procesing

Usually, for testing and characterization of the sigma-delta modulator the FFT algorithm is used. In this way we can calculate signal power, noise power and the power of harmonic components in the band of interest. The main problem of the FFT algorithm is the calculation time and the hardware/software needed to perform the calculation in real time on the chip. One possibility for speeding up the testing procedures and for saving additional hardware, is to calculate RMS value using (6).

$$V_{rms} = \sqrt{\frac{1}{n} \cdot \sum_{i=1}^{n} x^2(iT)} \qquad (6)$$

This equation calculates RMS value of the signal with a known signal frequency, which must be coherent with the sampling frequency. The signal frequency is equal to the input frequency (f_{in}) from the output of the SD oscillator. The period (n) of the signal in (6) is defined as the ratio of sampling period of the decimation filter T_{samp} divided by the input period T_{in} ($n = T_{in} / T_{samp}$), where $x(iT)$ is the discrete time signal from the filters. Execution of the RMS function returns $N \cdot V_{rms}$ values from which we are able to calculate the mean value ($\overline{V_{rms}}$ $\overline{V_{rms}}$) (7) and standard deviation (σ) (8).

$$\overline{V_{rms}} = \frac{1}{N} \cdot \sum_{i=1}^{N} V_{rms}(i) \qquad (7)$$

$$\sigma = \sqrt{\frac{1}{N-1} \cdot \sum_{i=1}^{N} \left(V_{rms}(i) - \overline{V_{rms}} \right)^2} \qquad (8)$$

The RMS value can be computed in the time domain and in the frequency domain (9) and compared using Parseval's theorem.

$$V_{rms} = \sqrt{\frac{1}{n} \cdot \sum_{i=1}^{n} x^2(iT)} = \sqrt{\frac{1}{n} \cdot \sum_{i=1}^{n} |X(f)|^2} \qquad (9)$$

The accuracy of the calculations depends on the number of samples taken from the process. For example, to get high resolution frequency spectrum from the filtered signal, we have to compute the FFT of at least 2^{12} samples. In order to get the same accuracy from the calculated RMS value, the number of samples has to be of the same order. In our case, just to compute a single RMS value we need 64 samples ($n = T_{sig}/T_{samp}$). On the basis of all samples (2^{12}) of the filtered signal we can get a vector of RMS values. The mean value ($\overline{V_{rms}}$) of the RMS vector and V_{rms} calculated from the FFT spectrum are the same. The mean of the RMS value expresses the sum of the signal and noise power. To evaluate the noise power we only need to compute a standard deviation of the RMS vector. Relation between the noise level, which we can estimate from the FFT, and the standard deviation, is defined in (10).

$$V_{nd} = \frac{\sigma}{\sqrt{BW/n}}; \quad BW = f_{p1} - f_{p2} = 3584 Hz \qquad (10)$$

In this equation the standard deviation is indicated by the parameter σ, $BW = fp1 - fp2$, which is defined by pole frequencies of the corresponding filters (Fig. 3) and n, which is the ratio of input and sampling period ($n = T_{in} / T_{samp}$).

IV. VERIFICATION

To verify the BIST effectiveness, a SIMULINK model of the 2nd order sigma-delta modulator was built. The mathematical model includes all major non-ideal effects [8]

like open-loop gain, gain bandwidth, slew-rate, saturation nonlinearity, offset voltage and input noise density of the amplifiers etc. The bit-stream from SD modulator is filtered with DSP containing the 3rd order CIC filter, followed by compensation FIR filter, which narrows the signal pass band as presented in Fig. 3 (CIC + Comp filter) and adjusts the phase response. With a high pass filter we attenuate additional DC component [4]. The frequency response of the complete channel is presented in Fig. 3, using oversampling frequency 4MHz; the resulting pass band width is 3.584 kHz.

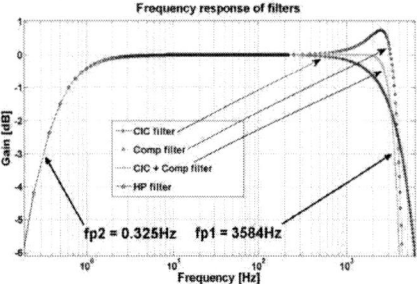

Fig. 3. Frequency response of the complete measurement path

To prove the method and the algorithm, the circuit was modeled using Matlab/Simulink environment that includes most important non-idealities of the modulator and also bit-true model of the on-chip DSP calculations. To quantify the quality of our BIST methodology, the results of the simulations performed with on-chip RMS calculations are compared to the results obtained by FFT calculations. The noise level can be compared in a similar way. To verify the influence of important circuit parameters to the accuracy of the measurement channel, the RMS and σ values were calculated and compared to the results obtained with the FFT. Many different simulations were performed for many different parameter values. The resolution of the Sigma-Delta ADC depends mostly on the quality of the amplifier in the first integrator [8]. The main parameters that influence the behavior are open-loop gain, slew-rate, unity gain bandwidth, saturation region and input noise level. With extended simulations we quantified the influence of important parameters to the RMS results as suggested in section III.

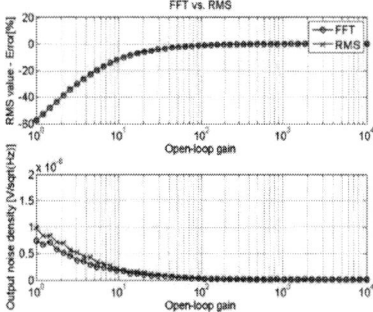

Fig. 4. FFT compared to RMS vs. open-loop gain

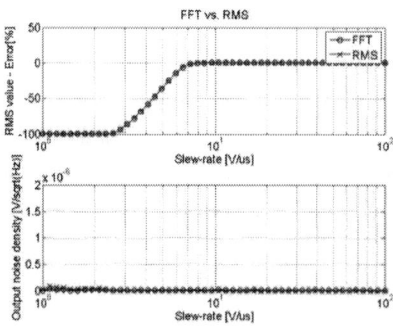

Fig. 5. FFT compared to RMS vs. slew – rate

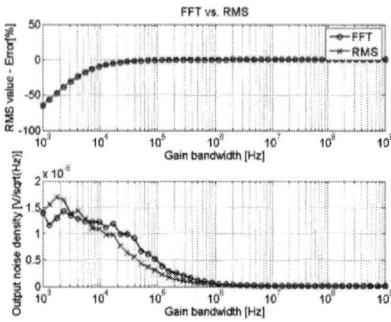

Fig. 6. FFT compared to RMS vs. unity gain bandwidth

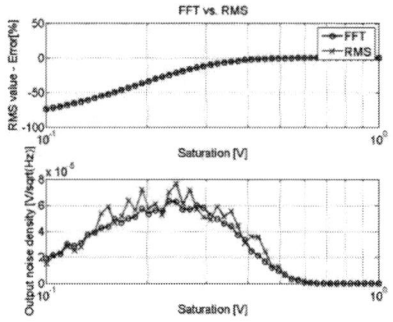

Fig. 7. FFT compared to RMS vs. saturation region of the output of amplifier

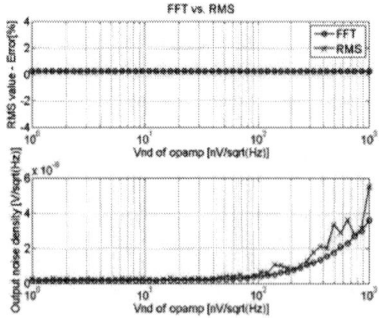

Fig. 8. FFT compared to RMS vs. input noise voltage

Fig. 4 to 8 show the different parameters calculated from the FFT and the mean RMS value calculated with (6) and (7). The difference of both values is very small. Therefore, the proposed algorithm and BIST methodology could be used as a method for discriminating a good chip from a bad one without using expensive testing equipment or a lot of additional silicon area. The standard deviation of the RMS results show an estimate of the noise level due to noise of the amplifier and/or indicate whether or not the saturation region in the integrators is reached.

V. CONCLUSION

In this article a simple concept is presented for an off-line, on-chip testing of watt hour meter measurement channels. The chip needs small amount of additional digital and analog hardware to implement an efficient built-in self-test of high precision measurement channels. The use of expensive on-chip FFT calculations is avoided. Additional hardware can be easily integrated and would not require a lot of additional silicon area of the ASIC. From simulations carried out on the system level we can confirm our expectation that the results from FFT algorithm and RMS function give almost the same results, so this methodology could be used for on chip BIST methodology. In the future we still have to prove all these results with measurements.

ACKNOWLEDGEMENT

This work was partially financed by the European Union, European Social Fund.

REFERENCES

[1] Uroš Bizjak, Drago Strle, "IEC 0.5 electronic watt-hour meter implemented with first-order sigma-delta converters", AEU – International Journal of Electronics and Communications, Volume 59, Issue 8, pp. 447-453, Dec 2005.

[2] Richard Schreier, Gabor C. Themes, "Understanding Delta-Sigma Data Converters", Wiley Interscience, New Jersey, 2005.

[3] Prabhakar S. Naidu, "Modern Spectrum Analysis of Time Series", CRC Press, Inc., 1996.

[4] Alojzij Kunčič, Magistrsko delo, "Izboljšave obstoječega digitalnega vezja za merjenje električne energije in njegova izvedba v vezju FPGA", Univerza v Ljubljani, Fakulteta za Elektrotehniko, Ljubljana, 2009.

[5] RIBNIKAR, Rok, BIZJAK, Uroš, STRLE, Drago. Efficient built-in self-test of a high-precision electronic watt-hour meter. *Inf. MIDEM*, 2011.

[6] Hongzhi Li, "A BIST (Built-In Self-Test) strategy for mixed-signal integrated circuits," Der Technischen Fakultat der Universitat Erlangen-Nurnberg, Erlangen, 2004.

[7] Lu, A.K.; Roberts, G.W.; Johns, D.A., "A high-quality analog oscillator using oversampling D/A conversion techniques," Circuits and Systems II: Analog and Digital Signal Processing, IEEE Transactions on , vol.41, no.7, pp.437,444, Jul 1994.

[8] Rok Ribnikar, Drago Strle, "Modeling of $\Sigma\Delta$ Modulator Non-idealities", Proc. MIDEM, International Conference of Microelectronics, Device and Materials, vol. 42 pp. 319-324, Strunjan, 2006.

[9] HONG, Hao-Chiao; LIANG, Sheng-Chuan ; SONG, Hong-Chin, "A Built-in-Self-Test Sigma-Delta ADC Prototype", In: *J. Electronic Testing*, 25 (2009), Nr. 2-3, S. 145-156.

A MEMS BPSK Demodulator

Micromechanical Mixing and Filtering using MetalMUMPs

Jeremy Scerri, Ivan Grech, Edward Gatt, Owen Casha

Microelectronics & Nanoelectronics Department,
Faculty of ICT, University of Malta,
Msida, Malta.
jsce0003@um.edu.mt

Abstract—In this paper two MEMS structures that are intended to demodulate a BPSK low data rate signal are compared. They are designed in conformity with the MetalMUMPs process. Target applications for such devices would be low-rate, wireless personal area networks (LR-WPANs) such as those described in the IEEE 802.15.4 standard onto which the ZigBee standard is built. The presented MEMS structures mix the Radio Frequency (RF) and Local Oscillator (LO) signals electrostatically and also filter this mixed signal prior to electrostatic sensing. These MEMS structures can also be used as the building blocks to demodulate higher order PSK schemes. The first structures' mode of vibration is torsional in nature and the use of torsional vibrations for BPSK demodulation is innovative. The first structure has an undamped resonant frequency of 2 MHz. The second structure is an extension of the first and is also supported on tethers exhibiting torsional vibrations. This structure resonates at 680 kHz.

Keywords— MEMS, phase detector; mixer; torsional osscilator; PSK demodulator;

I. INTRODUCTION

Electrostatically driven torsional structures have been used extensively since one of their first successful appearance [1] which involved driving a mirror in the 1980's. In [1], the plate undergoing torsional vibrations was used both as the mirror and as the electrostatic actuator.

As shown in [2], the switching speed of such micro-mirrors has come a long way since then. The first structure considered in this paper has dimensions within the same order of magnitude of [2]. The major differences are the manufacturing process, the application, and in [2], the mirror undergoing angular movement has mechanical stops.

Modelling of electrostatic torsional actuation is described in [3]. One of the results given in [3] is the maximum angular rotation before pull-in occurs. As described in Section III, the plate dimensions and voltages were selected such that pull-in does not occur.

A considerable amount of work to achieve signal mixing has also been done using CMOS commercial fabrication [4 - 7]. These are called CMOS-MEMS devices. These solutions have the advantage that they can be embedded within CMOS circuitry. The process involves standard CMOS fabrication followed by an anisotropic etch. Subsequently, a

final release step involving a combination of DRIE and isotropic silicon etch is used to release the structure. In [5], it was claimed that mixing for frequencies in the range of 10 MHz to 3.2 GHz was successfully demonstrated. The electrostatic gap used was 1.3 µm. In all these CMOS-MEMS structures, the mixing is performed electrostatically and the output is then mechanically filtered with the cantilever structure.

In [8], mixing is also achieved electrostatically, but here, a clamped-clamped beam is used. This beam is mechanically coupled with a similar clamped-clamped beam to achieve filtering. In this case, a polysilicon surface-micromachining process was used with electrostatic gaps ranging from 325 Å to 1000 Å. Successful mixing at 200 MHz was reported.

The mixing and filtering structure presented here employs a relatively large capacitive gap at 1.45 µm. Simulation results demonstrate the feasibility of such a structure to demodulate BPSK as described in the IEEE 802.15.4 standard.

To adhere to this standard the carrier frequency of 868MHz must be filtered out while the data signal (300kchips/s) needs to be extracted successfully. This requires the mechanical structure to resonate, at least, at three times the data rate and to settle in less than 1.7 µs.

II. PROPOSED STRUCTURES

A. Constraints Imposed by the MetalMUMPs process

The smallest achievable gap between conductors in the MetalMUMPs fabrication process is 1.45 µm. This is the gap between the nickel layer, hereon referred to as metal, and the polysilicon layer. The polysilicon is encapsulated within top and bottom nitride layers.

The smallest gaps achievable in the horizontal directions, between metals and/or polysilicon, are larger at 5 µm. Hence, a structure that vibrated in the vertical gap between the metal and the polysilicon was designed (Figure 1). The electrostatic gap g is 1.45 µm, but only 1.1 µm of vertical movement is allowed as there is 0.35 µm of nitride above the polysilicon.

B. Structure 1 – Torsional Oscillations of a Plate

Figure 2 shows how actuation and sensing were achieved in Structure 1 (S1). On the actuation side, the biased RF signal at the metal interacts with the biased LO signal to produce a force on the plate. In turn, this force drives the plate into torsional oscillations. On the sensing side, a current is generated at the polysilicon output electrode by interacting with the metal above it which is DC biased. The output signal is also filtered by the low-pass vibration characteristics of the plate structure. The overall dimensions of the nitride plate is 60 μm x 35 μm with the axis of rotation being parallel to the longer side. The area of the polysilicon electrodes is 250 μm² and the width of the tether is 15 μm.

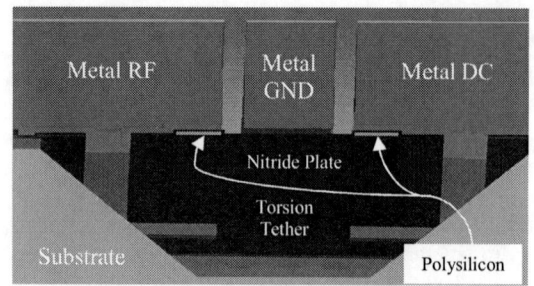

Fig. 1. Section through S1; view from bottom showing only one tether.

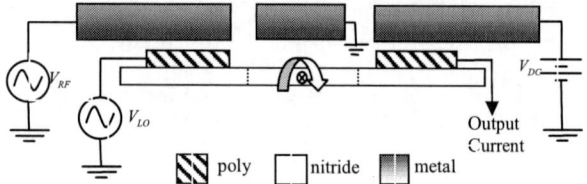

Fig. 2. Schematic diagram of the torsional BPSK demodulator depicting the bias and excitation scheme required for mixing, filtering and sensing.

C. Structure 2 – Oscillations of a Free-Free beam

Here the only difference from S1 is that the electrostatic force generated between the RF metal and LO polysilicon is driving a free-free beam into oscillations. As will be shown in Section IV, the four tethers are positioned at nodes of vibration. Sensing is done as described in S1, but this time the vibration characteristics of the FF-beam is filtering the input force.

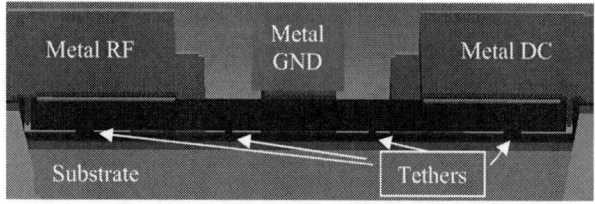

Fig. 3. Section through S; view from the bottom. Z-scale is exaggerated such that the polysilicon is visible.

The dimensions of the nitride beam are 445 μm x 60 μm. As shown in Figure 3, the tethers are positioned at 45 μm and 165 μm from each edge. The polysilicon electrode area is 5500 μm².

III. MODELLING AND ANALYSIS

In the first part of this section, some theoretical background on electrostatic mixing is presented. Then, this theory is

applied to the mixing of the LO signal and the BPSK modulated RF signal, and the frequency components of the resulting input force are discussed. In the second part, the mechanical filtering is used to shape the input spectrum and in the last part, current sensing at the output electrode is shown to provide the required data signal.

A. Frequency content of the Input Force

In general, the LO and RF are not necessarily in phase hence $\alpha \neq 0$ in (1) shows that this is noncoherent detection. The LO signal is DC biased by V_{dcLO} volts and ω rad/s is the carrier frequency. For BPSK, the phase in the RF signal has two distinct values and the modulated BPSK signal with DC bias can be formulated as in (2).

$$V_{LO}(t) = V_{LO}\cos(\omega t + \alpha) + V_{dcLO} \qquad (1)$$

$$V_{RF}(t) = V_{RF}\left(2d(t) - 1\right)\cos(\omega t) + V_{dcRF} \qquad (2)$$

where,

$d(t)$ ($\in [0, 1]$) represents the binary data signal for integer durations of nT. For the lowest data rate specified in the 802.15.4 standard which has a chip rate of 300 kchips/s, the pulse duration $T = 1/300000s$ and $\omega = 2\pi(868)\ Mrad/s$. If the vertical displacement of the electrodes is denoted by $z(t)$ μm, ε is the permittivity in air and A is the electrode area in μm, then the force at the input side would be described by (3).

$$F_{in}(t) = -\frac{\varepsilon A}{2(g + z(t))^2}\left[V_{LO}(t) - V_{RF}(t)\right]^2 \qquad (3)$$

If we further simplify by assuming that $z(t) << g$, i.e. no spring softening effects occur, and letting $V_{LO} = V_{dcLO} = V_{RF} = V_{dcRF} = V$ and substituting (1) and (2) in (3), we get (4):

$$F_{in}(t) = E\left\{\left(\frac{D^2 - B^2}{2}\right)\cos(2\omega t) + BD\sin(2\omega t) + \left(\frac{D^2 + B^2}{2}\right)\right\} \qquad (4)$$

where, $D = 2d(t) - 1 - \cos\alpha, B = \sin\alpha\ \&\ E = -\dfrac{\varepsilon A V^2}{2g^2}$

Equation 4 has three terms; the first two have bandwidths centered on 2ω while the last term is centered on 0 Hz. All these components can be seen in Figure 4. We are interested in the baseband part of the input force as the other components are very high and will be mechanically filtered out. Table 1 further breaks down the last term. In (4), this last term can be expressed as (5):

$$E[2d^2(t) - 2(1 - \cos\alpha)d(t) + 1 + \cos\alpha] \qquad (5)$$

The bandwidth of the data signal $d(t)$ is inversely proportional to the pulse duration T. While theoretically, the bandwidth of the data signal is infinite, we aim to capture information up to its 3rd harmonic.

Table 1: Breakdown of force components around 0 Hz as in equation 5.

Centre Freq.	Term	Bandwidth
0 Hz	$E[2d^2(t)+1+\cos\alpha]$	Bandwidth of $d(t)$
0 Hz	$-2E[1-\cos\alpha]d(t)$	Bandwidth of $d(t)$

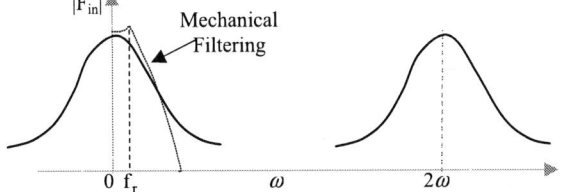

Fig. 4. The spectrum of the electrostatic force generated, and the required spectrum for decent reconstruction.

B. Mechanical Filtering

Considering S1, for small oscillations the rectangular tether's torsional spring constant can be calculated using (6) [9].

$$k_t = k_1 G(2a)^3(2b) \qquad (6)$$

where, $b \times a$ are the tether section of 15 x 0.8 μm, I the moment of inertia and G the modulus of rigidity of the plate. The parameter k_1 is a function of the ratio b/a and is tabled in [9]. The dimensions of the plate were selected such that the plate's natural frequency is around $f_n = \frac{1}{2\pi}\sqrt{k_t/I} = 2$ MHz. As required, for the data rate under consideration, this is higher than the 3rd harmonic of 900 MHz. A simplified 2nd order linear model which includes damping is described in (7).

$$I\ddot{\theta} + c\dot{\theta} + k_t\theta = F_{in}(t) \qquad (7)$$

When damping is taken into consideration, the resonant frequency f_r would be lower (see results section). Having selected f_n at least two times greater than the 3rd harmonic of $d(t)$ would still provide a substantial safety margin.

From [3], pull-in would occur when $\theta_p \approx 0.4404g/L$, with L being half the width of the plate i.e. 17.5 μm. Hence, for S1, $\theta_P = 0.027^r$, meaning that approximately a maximum vertical displacement of 470 nm is permissible.

S2 is a larger structure which results in a smaller resonant frequency for the first mode of vibration. On the other hand, there are two clear advantages of S2 over S1. The first benefit is the area of the electrostatic actuation/sensing plates which produces larger forces and movements with smaller voltages. The second benefit is that there would be less parasitics between RF and LO electrodes and the polysilicon output sensing electrode.

C. Current Sensing at the Polysilicon electrode.

At the output side, the polysilicon electrode is moving within the electrostatic field generated between itself and the metal above which is DC biased by V_{DC}. This induces a current at the polysilicon output electrode which can be eventually converted into voltage via a suitable amplifier.

The distances between the two high frequency inputs i.e. the LO and RF inputs and the output electrode are small, especially in S1. Parasitic coupling at high frequency is bound to occur.

Hence, together with the required current generated due to oscillations, a current proportional to the LO/RF frequency is also generated at the output electrode.

One solution to cancel these parasitic couplings is to use a differential topology as shown in Figure 5, in which case, two output currents from two mechanically identical structures are fed into a differential amplifier. The only difference in the structures would be that for one structure, S1$_a$, the Metal-DC would be biased by $+V_{DC}$, while for the second, S1$_b$, the biasing would be $-V_{DC}$.

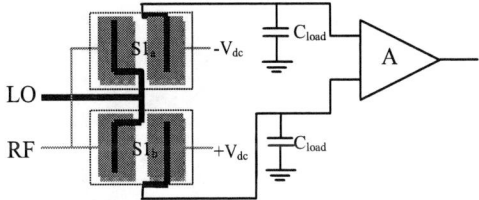

Fig. 5. Differential Setup for S1.

Although the analysis presented here is greatly simplified, it still provides the fundamentals required such that one can wisely choose the electrical and mechanical parameters. As will be demonstrated in the next section, the setup shown in Figure 5 using S1 is successful in demodulating low data rate BPSK signals.

IV. SIMULATIONS AND RESULTS

This section presents the results of the simulations using Coventor® for both S1 and S2. Table 2 summarizes the mechanical results for both structures.

Table 2: Mode types and frequency, Q factor and damped resonant frequency.

Structure	Mode Freq. (MHz)	Mode Type	Q factor / Damped Resonant Freq. (MHz)
S1	1.57	Not Torsional	unneeded
	2.01	**Torsional**	60 / 1.54
	4.05	Not Torsional	unneeded
	4.96	Not Torsional	unneeded
	7.15	Not Torsional	unneeded
S2	0.57	Not in Phase	unneeded
	0.61	Not in Phase	unneeded
	0.66	**In Phase**	6 / 0.54
	0.68	**In Phase**	6 / 0.57
	0.98	Not in Phase	unneeded

It is important to note that in S1, the required mode should be a torsional mode. In S2, the driving plate (LO) which consists of two rectangular sections connected with a polysilicon track, must have both sections in phase as the RF metal would be driving them together. This requirement together with the tethers at the nodes provide sufficient constraints such that other modes of vibration are kept attenuated.

Hence, for S2, the mode type 'In Phase' refers to the fact that both sections of the driving plate are in phase. Figure 6 shows the modes of vibration under investigation for both structures. Figure 7 shows the harmonic response of S1. The maximum displacement of 25 nm in the structure (<470 nm for pull-in) occurs at the edge of the plate and at the damped natural frequency of 1.54 MHz.

Due to the strong parasitic coupling between the output electrode and the RF/LO signals, the other two current graphs are almost linear in nature. These are the output currents from $S1_a$ and $S1_b$ respectively, as shown in Figure 5. This linearity breaks down at resonance due to the 'large' displacements. If these currents are subtracted (and taking into account the phase information which is not shown) the differential output current required (Figure 8) can be converted into voltage across a load capacitance.

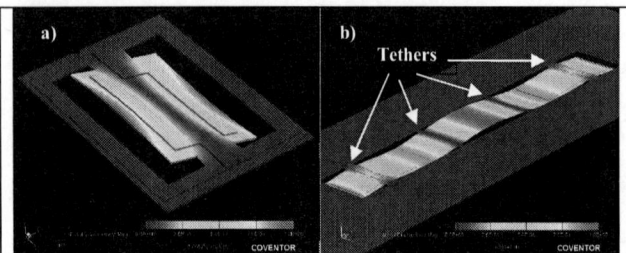

Fig. 6: a) Mode of interest of S1. (b) Mode of interest of S2

For S1, this voltage would contain enough harmonics from $d(t)$ for successful reconstruction / demodulation. Even though the DC content is highly attenuated due to the substantial Q factor, this does not present a problem as the 32 bit chips used in the 802.15.4 modulation scheme do not have any DC content.

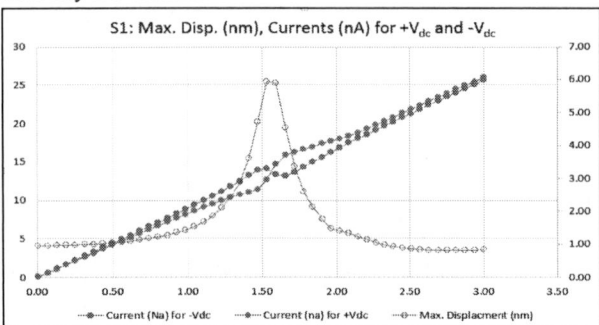

Fig. 7: Current at Output Electrode for both positive and negative DC biasing and maximum displacement v.s frequency for S1

For S2, a similar response is obtained, the differences being:

i. the damped resonant frequency is not high enough to capture the 3rd harmonic;

ii. voltages could be lowered to less than 5 v as the electrode area is much larger;

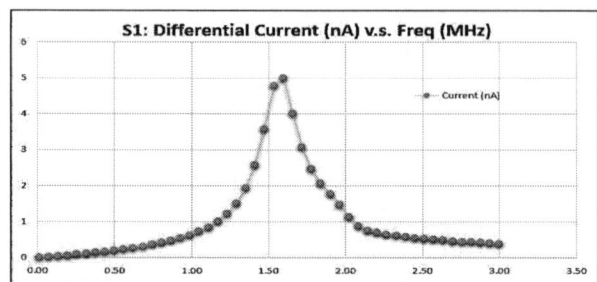

Fig. 8: Differential Current at the Output Electrode for S1.

iii. parasitics are still strong and hence a differential setup is still required but the current at resonance is greater which gives a pronounced double peak at the two modes which are excited.

V. CONCLUSIONS

Through the MetalMUMPs fabrication process, we have demonstrated that a torsional vibratory structure can be built to successfully demodulate a low data rate BPSK signal with a carrier frequency of 868 MHz and a chip rate of 300 kchips/s.

The structure of choice S1 has a damped resonant frequency of 1.54 MHz (>3rd harmonic of data), a Q-factor of 60, DC bias voltages of 20 v and AC signal amplitudes of 20 v. The current generated at resonance is 5 nA and is operated far from the pull-in region.

Parasitics were eliminated by using a differential setup for common mode rejection. Further work with this device can provide solutions for higher order PSK demodulators.

REFERENCES

[1] K.E. Peterson, "Silicon torsional scanning mirror," IBM J. Res. Dev. 24 (5), 1980, pp. 631–637.

[2] Nielson Gregory N. et al., "High-Speed MEMS Micromirror Switching," Lasers and Electro-Optics, CLEO 2007, 2007, pp.1-2.

[3] Nielson Gregory N. and Barbastathis George, "Dynamic pull-in of parallel-plate and torsional electrostatic MEMS actuators," Microelectromechanical Systems, Journal of, 2006, Volume: 15, Issue: 4, pp 811- 821.

[4] Lopez, J.L. et al., "Mixing in a 220MHz CMOS-MEMS," Circuits and Systems, 2007. ISCAS 2007. IEEE International Symposium on , vol., no., pp.2630,2633.

[5] Fang Chen et al., "CMOS-MEMS resonant RF mixer-filters," Micro Electro Mechanical Systems, 2005. MEMS 2005. 18th IEEE International Conference on , vol., no., pp.24,27.

[6] J. Stillman, "CMOS MEMS Resonant Mixer-Filters", M.S. thesis, Dept. Electron. Eng., Carnegie Mellon University, Pennsylvania, 2003.

[7] U. Arslan, "CMOS-MEMS Downconversion Mixer-Filters", M.S. thesis, Dept. Electron. Eng., Carnegie Mellon University, Pennsylvania, 2005.

[8] Ark-Chew Wong, Nguyen, C.T.C, "Micromechanical mixer-filters ("mixlers")," Microelectromechanical Systems, Journal of , vol.13, no.1, pp.100,112, 2004.

[9] Louis L. Bucciarelli Jr., "Engineering Mechanics of Solids: A first course in Engineering", Chapter 6 [Online]. Available: http://web.mit.edu/emech/dontindex-build/,2013.

A MEMS-based oscillator with synchronous amplitude control

Jakub Gronicz[*], Nikolai Chekurov[†], Lasse Aaltonen[‡] and Kari Halonen[*]

[*]SMARAD2-Department of Micro and Nanosciences, Aalto University
School of Science and Technology, Espoo, Finland
[†] KTH Royal Institute of Technolgy, Stockholm, Sweden
[‡] Murata Electronics Oy, Vantaa, Finland

Abstract—This paper describes the design and simulation of a MEMS-based oscillator with a silicon tuning fork as frequency selective element. The interface electronics include a synchronous amplitude control circuit that allows for precise control of oscillation amplitude. The nominal oscillation frequency is 1.8 MHz. The structure has been implemented using a 0.35 μm High Voltage CMOS process and operates with a nominal supply of 3.3V.

I. INTRODUCTION

Oscillators give the heart-beat to most modern electronic circuits. Providing a stable and accurate clock is essential e.g. in communication systems where poor frequency sources may lead to transmission errors. Generating high quality clock signals with frequencies in the MHz range requires either LC tanks or quartz-based circuits. Both solutions are impractical from integration standpoint due to their size when low frequencies are of interest. LC tanks require quality inductors that would consume large chip area when integrated with the rest of the interface electronics. On the other hand, quartz resonators pose another challenge as their manufacturing process is not compatible with that of CMOS integrated circuits. In order to achieve a low frequency, high quality integrated frequency source, other alternatives need to be investigated. One of them are oscillators using MEMS resonators as frequency selective elements. They can exhibit high Q-factors, comparable with quartz elements while maintaining small form factors, even for low frequencies. The reason behind it is the relation between mechanical resonance of a MEMS resonator and electrical signal that is read by the interface electronics [1], [2]. An additional benefit of micromechanical resonators is the possibility to adjust the resonance frequency by varying the DC bias voltage [1], [2]. This gives an opportunity to create a voltage controlled oscillator that generates high quality signal, maintains small form factor and can be tuned using voltage levels generated on-chip.

MEMS-based VCOs typically incorporate micromechanical varactors, that are used to tune the oscillation frequency. Their main drawback however, is the need for inductors. In applications operating at RF frequencies, this is a reasonable solution as the size of integrated passives is acceptable. For low frequency oscillators however, integration of such a circuit would require passive elements of considerable size making the process unfeasible.

The purpose of this work is to develop a high-Q MEMS-based oscillator that could be used as a VCO tuned with voltages below 20 V and operate at low-MHz frequencies. The choice of operating voltage is dictated by the requirements for biasing and tuning of the MEMS resonator [3]. Such approach does not require bulky passive components and could be a step towards a fully integrated high quality clock source. In the following sections the design of such oscillator together with simulation results showing the system operation are presented.

II. CIRCUIT OVERVIEW

The oscillator is built using a silicon single-ended tuning fork as frequency setting component and amplifier loop with synchronous amplitude control as shown in Fig. 1. The drive and read electrodes are biased with a constant DC voltage V_{bias}.

A. MEMS resonator

The resonator structure has been optimized to achieve maximum motional current while maintaining a relatively low tuning voltage. The resonator topology has been chosen to be a Clamped-Free (C-F) beam structure. The resonance frequency of a Clamped-Free resonator can be tuned over a wide range using DC bias voltage Assuming the same component size, C-F resonators offer lower mechanical spring constant than e.g. Clamped-Clamped (C-C) or Bulk Acoustic Wave (BAW) [3] and thus are more tunable than the other structures. The resonance frequency can be described as

$$f = f_0 \left(1 - \sqrt{\frac{k_e}{k_m}} \right), \qquad (1)$$

where f_0 is the mechanical resonance frequency and k_m, k_e are mechanical and electrical spring constants, respectively. The relation between k_e and DC bias voltage V_{bias} is shown in eq. (2):

$$k_e = \frac{C_0 V_{bias}^2}{d^2}, \qquad (2)$$

C_0 being the electrode capacitance and d – the resonator gap size.

Assuming a certain operating frequency, gap size (determined by process limitations) and DC bias voltage, the component size has to be minimized in order to achieve highest motional current. The chosen topology for the resonator allows to have the tuning bias decoupled from the signal path [3].

978-1-4673-4579-8/13 $31.00 © 2013 IEEE

Paper M4B2

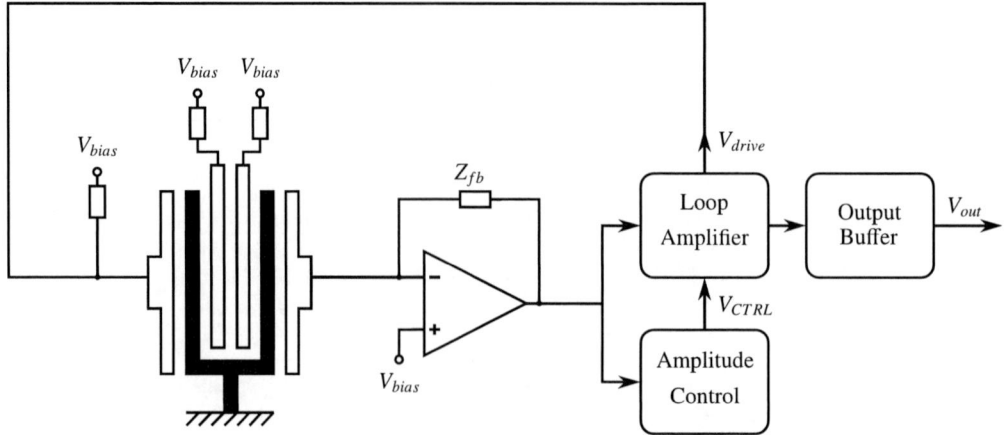

Fig. 1.   MEMS oscillator block diagram

TABLE I.   RESONATOR PARAMETER SUMMARY

parameter	Value	parameter	Value
bias voltage	16.6 V	strucutre height	2.5 μm
fork arm length	20 μm	fork arm width	800 nm
electrode gap	800 nm	Q	> 1500

B. Resonator modeling

For the purpose of ASIC development a model shown in Figure 2 has been prepared in MEMS+ by Coventor. It allows to co-simulate the mechanical model of the silicon resonator in the environment used to develop the integrated electronics. It has been prepared using the dimensions and material properties as used for the fabrication of the resonator. This model however, does not include secondary effects such as the coupling between drive and read electrodes, or the effect of the substrate, bonding pads, and interconnect cross-coupling on the interconnect parasitics. To improve this model and take into account the above mentioned phenomena, an RC network was used to include the effects of the parasitics, as shown in Fig. 3. A set of fabricated components has been measured. Further, the RC values have been adjusted to match the simulated frequency response of the resonator model to the measured response of the fabricated devices.

C. Readout Electronics

The readout electronics has been implemented in AMS 0.35 μm High Voltage CMOS technology. The HV option was required due to the biasing scheme of the read side of the resonator. The transimpedance amplifier has been implemented using isolated transistors that tolerate voltages up to 20 V DC with respect to the circuit substrate.

The motional current induced by the vibratory motion of the resonator mass with respect to fixed electrodes is converted to voltage, buffered and amplified further with a variable-gain amplifier (VGA) block. The transimpedance amplifier (TIA) has dual function, it converts the motional current to voltage and it is used to bias the read side of the resonator. The virtual ground of the TIA is set to the bias potential V_{bias} with respect

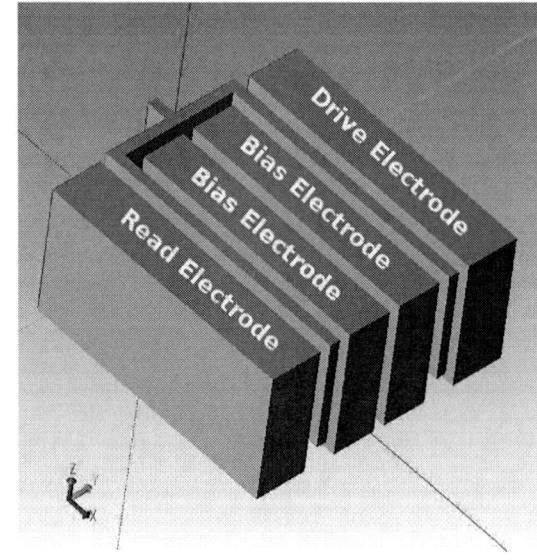

Fig. 2.   Resonator model in Coventor MEMS+

to ground. Such a readout and biasing scheme reduces the impact of the parasitic capacitance C_p (Fig. 3) which would otherwise create a voltage divider with a DC block capacitor and drastically reduce the available signal.

The motional current amplitude strongly depends on the bias voltage according to eq. (3) [2].

$$i_m \approx Q \frac{V_{bias}^2 v_{ac} \omega C_0^2}{kd^2}, \qquad (3)$$

where Q is the resonator quality factor, V_{bias} is the bias voltage, v_{ac} the drive signal amplitude, C_0 the nominal capacitance, k the spring constant and d the nominal gap between moving mass and electrode. It is therefore desirable to use as high DC bias voltage as possible. The limiting factors are the pull-

978-1-4673-4579-8/13 $31.00 © 2013 IEEE

in voltage of the MEMS and the voltage tolerance of the IC process for the ASIC.

For the element used in this design the nominal bias of 16.6 V has been chosen. A higher DC bias could be possible as the resonator used has pull-in voltage V_P over 30 V, however, the goal was to use voltage levels that could be generated on-chip with reasonable efficiency [4]. This was also the reason to use a HV CMOS process with devices that tolerate voltages up to 20 V.

The VGA path comprises three high pass filter stages implemented current starved inverters. The gain of the path is controlled by limiting the current to the two last stages. The first stage operates with constant gain. The output of the first high pass stage is fed to the amplitude controller therefore the gain is maintained constant to avoid potential stability issues. At power-on the loop gain is highest to reduce oscillator start-up time.

III. AMPLITUDE DETECTOR AND GAIN CONTROL

The amplitude detector has been designed as a synchronous switched-capacitor peak detector. It utilizes the clock derived from the main oscillator loop. The input signal is sampled at peak amplitude, buffered and passed to a proportional+integral stage. It is then compared to a preset control voltage level. The difference between those two values is amplified and used to adjust the gain of the VGA increasing or reducing the oscillation amplitude.

The main benefit of having an amplitude detector and a gain control circuit is the high accuracy with which the oscillator amplitude can be controlled. This is especially important if the MEMS is sensitive to deviations in the drive

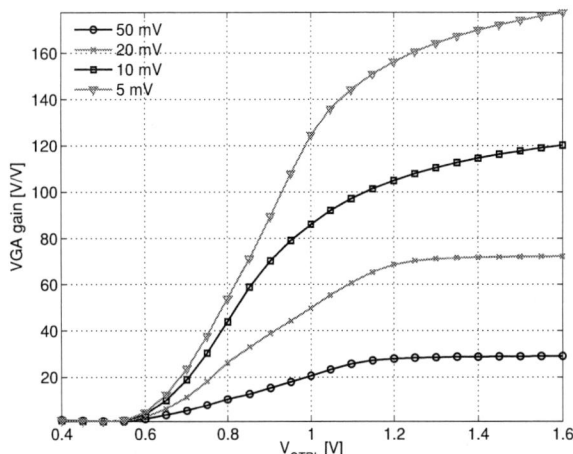

Fig. 5. Loop amplifier gain vs. control voltage for various input signal amplitudes. High gain at low amplitudes helps to reduce the start-up time of the oscillator.

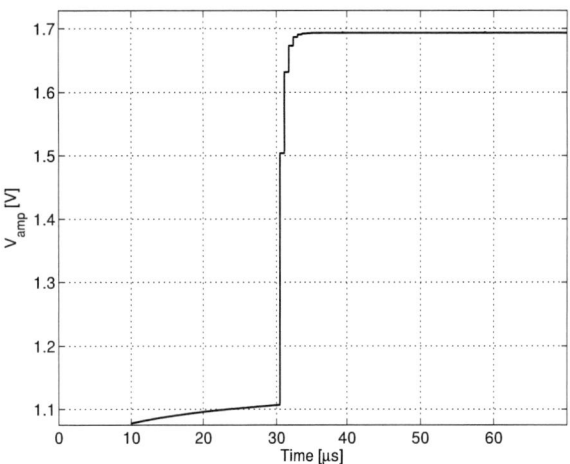

Fig. 7. Simulated example amplitude detector output. The circuit has been enabled $30\,\mu s$ from the start of the simulation. Input voltage amplitude is 50 mV and the offset voltage is 1.65 V.

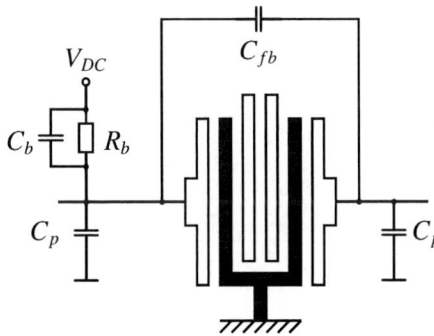

Fig. 3. Bonding pad and interconnect parasitics modeled, R_b is the biasing resistor, C_P, C_b and C_{fb} are parasitic components due to bonding pads and interconnects.

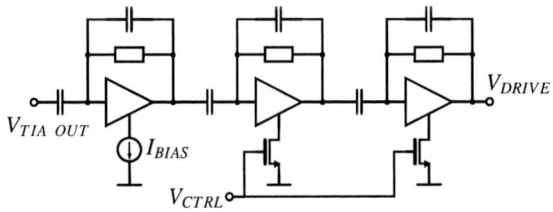

Fig. 4. Loop amplifier diagram

voltage, as is the case with a Clamped-Free structure due to the minimized mechanical spring constant. The vibration amplitude at resonance is related to the drive voltage amplitude as per equation (4) [2]:

$$x_R = Q\frac{\eta v_{ac}}{k}, \tag{4}$$

where η is the electromechanical coupling factor, $\eta \approx V_{bias}\frac{C_0}{d}$

The relation in eq. (4) shows that the drive voltage amplitude needs to be controlled in order to keep the mechanical vibration amplitude small and keep the device operating in linear regime. If the drive signal level is not precisely maintained, the capacitive nonlinearity may cause unstable output signal frequency [1].

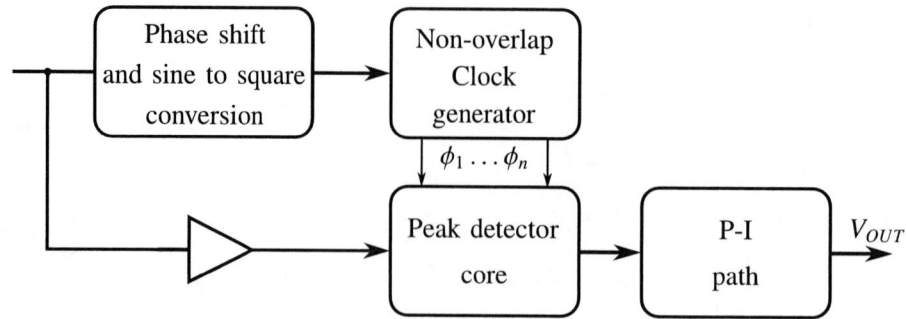

Fig. 6.   Amplitude detector block diagram

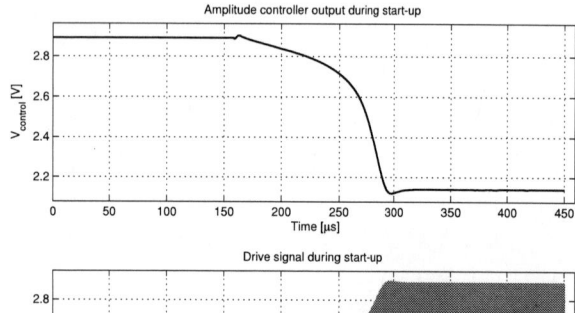

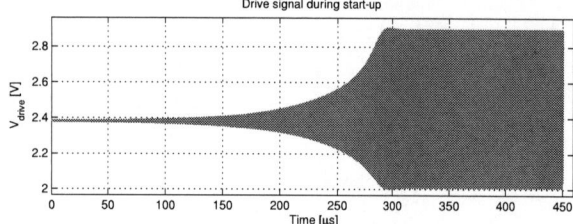

Fig. 8.   Simulated closed-loop operation during start-up

TABLE II.   SYSTEM PARAMETER SUMMARY

parameter	Value
IC supply voltage	3.3 V
f_{osc}	1.8 MHz
resonator bias voltage	16.6 V
active device area	1.2mm x 0.6mm

IV.   SUMMARY

The work presented in this paper tackles the design of a MEMS-base oscillator with 1.8 MHz nominal operating frequency. The implemented interface can be used to precisely control the drive signal, avoiding nonlinear operation of the MEMS resonator. The performance needs to be verified by measurements, however, simulated results show that the proposed circuit could be used to develop a fully integrated frequency source with a silicon tuning fork resonator.

ACKNOWLEDGMENTS

The project has been funded by the Academy of Finland (project no. 134506).

REFERENCES

[1]   V. Kaajakari, *Practical MEMS*. Las Vegas, Nev.: Small Gear Publishing, 2009.

[2]   I. Tittonen and M. Koskenvuori, *Electrostatic and RF Properties of MEMS Structures in: Handbook of Silicon Based MEMS Materials and Technologies*. Elsevier, 2010, ch. 12, pp. 221–237.

[3]   N. Chekurov, L. Aaltonen, J. Gronicz, M. Kosunen, and I. Tittonen, "Design and fabrication of a tuning fork shaped voltage controlled resonator with additional tuning electrodes for low-voltage applications," in *Proceedings of Eurosensors XXIV*, 2010.

[4]   L. Aaltonen, M. Saukoski, and K. Halonen, "On-chip digitally tunable high voltage generator for electrostatic control of micromechanical devices," in *IEEE Custom Integrated Circuits Conference, 2006. CICC '06*, 2006, pp. 583–586.

Fully electrical test procedure for inertial MEMS characterization at wafer-level.

A. Sisto*, O. Schwarzelbach**, L. Fanucci*

*Università di Pisa, Italy

**Fraunhofer Institute for Silicon Technology ISIT, Fraunhoferstrasse 1, 25524 Itzehoe, Germany

Abstract—The fast growth of MEMS technologies for the production of inertial sensors in the last decade makes the characterization at wafer-level very important. In this paper is presented a test setup for measuring electrical and mechanical parameters of capacitive MEMS inertial sensors. The test setup is used in the production for automotive and consumer applications. It is fully electrical (i.e. none of the stimuli to the sensors is mechanical). The core of the test setup is a test algorithm. The design of the test algorithm was aimed at a fast, reliable and repeatable wafer-sort test. With the test setup described in this paper, it is possible to measure electrical and mechanical parameters of inertial sensors with up-to-6 dimensions.

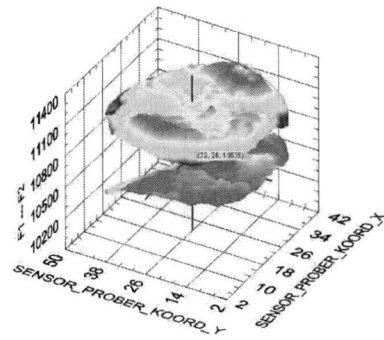

Fig. 1: Resonance frequencies of a gyroscope mapped along a wafer.

INTRODUCTION

MEMS technologies are a continuously growing field. One of the most important applications of MEMS technologies are the inertial sensors. With a "micromachining" MEMS process it is possible to produce accelerometers and gyroscopes with good performances and sub-millimeter dimensions. Many MEMS inertial sensors are based on a capacitive read-out. The movement of an inertial mass causes a capacitance variation, which is read by an electronic front-end and then computed, to get the movement information.

The MEMS-device performances are largely dependent on the stability of the production process. It is necessary to check process parameters which affect micromachining. Sensor performances strongly depend on the variation of geometrical dimensions (e.g. line width, thickness variations). Sensor features are frequently dependent on the position of the die on the wafer.

For a MEMS-sensor producer it is very important to get informations about mechanical and electrical parameters at wafer-level. Since process parameters affect the functioning of inertial sensors, it is necessary to select good-working sensors, and to map some key parameters along the wafer plane (Fig. 1). Furthermore the package cost is a big part of the whole sensor cost([1]): a wafer-level test can prevent packaging of bad sensors.

In the case of MEMS inertial sensors the number of devices per wafer is in the order of some thousands. So the test time-per-DUT is a key parameter: a good test algorithm has to be fast and precise. The test speed depends on the set of parameters to be measured, and on the kind of stimuli the sensors are excited with.

There are two stimuli strategies: one is to mechanically move the sensor and then measure the electrical output, and the other is to supply electrical signals to the sensor.

The first strategy has good features, e.g. the more intuitive approach to a movable micro-structure, the easy measurement of sensor sensitivity and the availability of ready-to-use test systems on the market, which have only to be programmed. But this solution have an high cost, especially when the test has to be automatic, and must stimulate all possible sensor's degrees of freedom. Many case studies have been presented in literature, e.g. [2].

A mechanical test approach makes the measurement of electrical parameters very difficult, e.g. the capacitance values in the case of capacitive MEMS.

A capacitive MEMS stucture has the duality property: a mobile-plate capacitance can be both a sensor and an actuator. The second strategy refers to the actuator mode.

A fully electrical test has better performances about speed, tests at wafer-level, and cost effectiveness. An electrical signal supply is almost instantly in operating phase. On the other hand the mechanical stimuli to a sensor cannot be applied very quickly, because the generation systems has engines which have some latency. Bt this approach has some problems when the sensitivity has to be measured. It is an important parameter for the sensor characterization, and it has been demonstrate that it is possible to calculate the sensitivity parameter with a fully electrical test setup, too [3].

In this paper a test setup is presented; its objective was to provide feedback to manufacturing process and design engineers: (i) to identify the process steps, which more affect MEMS functionality; (ii) to evaluate design choices by measurements, since MEMS simulation results are not fully reliable.

I. PROPOSED TEST SETUP

The test setup presented was targeted for wafer-sort tests in production. It was developed for inertial measurement units (IMU), which detect movements for 6 degrees of freedom (DOF). The test setup can automatically measure electro-mechanical parameters of MEMS inertial sensors for automotive.

The test setup consists of a wafer probe, a test board, a switch matrix, an FPGA card and an elaboration unit, i.e. a PC, on which the test algorithm runs (Fig).

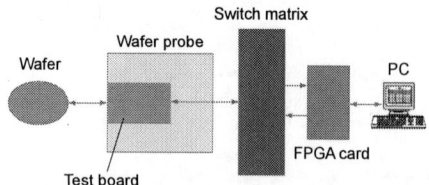

Fig. 2: Block diagram of the test setup

The wafer probe used in production test is an *UF200A* from *Prober Solutions*.

The test board interfaces with the DUT-sensor. It features two channels, one for the gyroscope output signal and one for accelerometers. Each channel has a read-out block for capacitive sensors and an amplification stage.

The FPGA board is the core of the test algorithm: it is configured to generate electrical stimuli for the sensor input, to acquire output signal from the sensor and for the post-processing of acquired signals. A National Instruments PXI-7854R board was used.

The FPGA generates the driving and the modulation voltages as well as it performs a real-time post-processing.

The test procedure is a LabView program which controls the FPGA card and the switch matrix. A complete test procedure measures for each sensor all the parameters listed below for each DOF. The algorithm is automatic and was integrated in the wafer probe system.

The results of the test procedure are exported in a test log.

The cost effectiveness of this test setup is largely lower than the cost of a MEMS-test ATE (automatic test equipment) which supplies mechanical stimuli.

With this test setup the following parameters can be measured:

- **Capacitance absolute values and mismatch** of the differential capacitance transducer structure.
- **Resonance frequency** of the MEMS structures.
- **Q-factors**, which are correlated to the residual pressure in the vacuum cavities of gyroscopes.
- **quadrature error (QUAD bias)**, which is a mechanical coupling between different DOF of a gyroscope. The QUAD bias is caused by process imperfections of the micro-mechanical structure [4], [5].

II. CAPACITIVE READ-OUT BASICS

A basic introduction about capacitive read-out has to be introduced to better understand the test procedure.

The electrical equivalent of a single-DOF capacitive transducer is shown in figure 3. It is basically a capacitance pair; coupled capacitances have differential variations. The electronic front-end of the processing system is aimed at the detection of those variations.

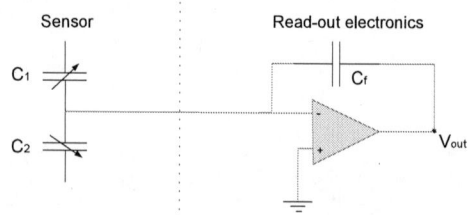

Fig. 3: Sensor's electrical equivalent and charge amplifier.

The hardware solution for this configuration is a charge amplifier; its output signal is:

$$V_{out} \propto \frac{\Delta C}{C_f} sin(\omega_m t) \qquad (1)$$

where ΔC is the differential capacitance variation and ω_m is the angular frequency of the modulation signal. A standard high-frequency modulation is used to shift in the frequency domain the "movement" signal around the modulation frequency.

In test procedure the sensors inputs are usually a driving signal, which causes a movement of the inertial mass, and a modulation carrier.

III. THE TEST PROCEDURE

The test procedure is a collection of single test steps. In the next pages the single-test blocks are presented.

A. Capacitance values and mismatch

The differential capacitances of an inertial MEMS can be in the range of hundreds of femtofarad. The measurement of small capacitances is strongly affected by test board parasitic effects. For that reason a calibration step was needed in the measurement.

The configuration to measure C_1 is shown in Fig. 4. The source signal is a sine wave at a modulation frequency (so that the inertial mass doesn't move)[1].

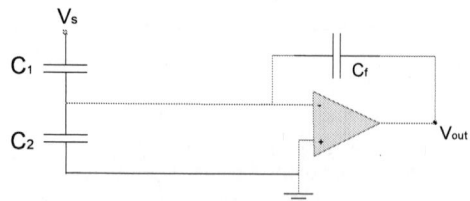

Fig. 4: Configuration to measure C_1.

To measure C_2 one has only to switch the source signal to the right pin, and connect the other to ground.

A value for the input capacitance can be evaluated from the output voltage amplitude.

$$C_{in} = \frac{C_f \cdot V_{out}}{V_S} \qquad (2)$$

---

[1]The low-pass filtering behaviour of the mass-spring system ensures that an high-frequency electric signal is strongly attenuated.

PRIME 2013, Villach, Austria

Session M4B – MEMS

The capacitance values are mainly useful for checking the process quality. Designers can evaluate the nominal capacitance values by geometry. An accurate and early measure of actual values provides important feedback of the silicon manufacturing process.

The mismatch between coupled capacitances is an unwanted occurence. Coupled capacitors are nominally identical. If they are not, the output signal is affected by offset. For actuator structures, it also causes the inertial mass to be dynamically unbalanced. This is a problem for high-precision sensors.

The procedure to measure the capacitance values and the mismatch between two coupled capacititances of a transducer has a preliminary stage, which returns the values of stray capacitances. The procedure starts with a measurement of capacitance in an open-configuration: if the wafer probe doesn't contact the sensor's pads, stray capacitances are measured. This "open" test was repeated for each pin pair. Those values were stored and used for the whole test procedure, by subtracting the stored value to the measured capacitance of the corresponding pin[2].

The measured values of capacitances are affected by an error, given by the C_f tolerance and by the gain of the signal processing chain. This error can be corrected with a scale factor. But this error doesn't affect the mismatch value, since it is the result of a ratio:

$$MM = 2 \cdot \frac{C_2 - C_1}{C_2 + C_1} \qquad (3)$$

and its effect is the same for the numerator and the denominator.

Measurements show capacitance values in the range of $100fF$, and maximum mismatches of around 7%.

B. Resonance frequency

The accelerometers and the gyroscopes have by design very different resonance frequencies. An accelerometer resonance frequency is in the range of $1kHz$, whereas for gyro the range is over $10kHz$. For that reason two different procedures were implemented.

For the accelerometers, a point-by-point diagram is built, for frequencies between $100Hz$ and $3kHz$, by stimulating a capacitances pair with a carrier ($156kHz$) and a sinusoidal wave at a driving frequency. An offset voltage was needed to bias the mechanical structure. The input signals are:

$$v_{S1/2} = V_m sin(\omega_m) \pm V_d sin(\omega_d) + V_{offset} \qquad (4)$$

The driving signal inversion is needed to make the inertial mass move. The output signal of the sensor is then demodulated by the FPGA card, which returns a wave at the driving frequency. The driving signal frequencies are swept with a programmable step size.

The algorithm returns the resonance frequency as the maximum value of the amplitudes. It also provides the

[2]The stray capacitance of a pin is intended to be the parasitic capacitance between that pin and the one which contacts the output pad of the sensor.

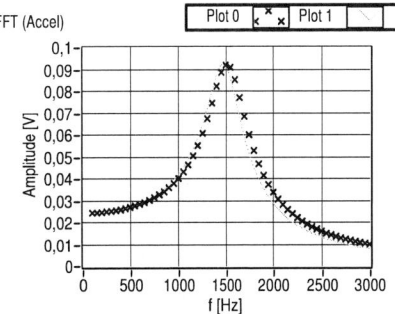

Fig. 5: Frequency response of an accelerometer. Plot 1 is an ideal curve.

possibility to repeat the sweep, but in a narrower range, to obtain a more accurate value for the resonance frequency. But for the measurements a resolution of $50Hz$ is usually enough.

For gyroscopes the resonance frequency is obtained from the frequency response of the system to a modulated chirp signal. The procedure repeats three times a chirp driving. The range of the chirp was increasingly narrow, so that the peak became higher. For this test the accuracy is about $1Hz$.

Acquired data are FFT-transformed, then the maximum amplitude value is sought in a range between $(f_m + f_{chirp,min})$ and $(f_m + f_{chirp,max})$. This pair of values (amplitude, frequency) is stored and the chirp generation is repeated in a narrower range around the frequency of maximum amplitude.

The FFT plot shows resonance frequency peaks around the carrier (Fig. 6).

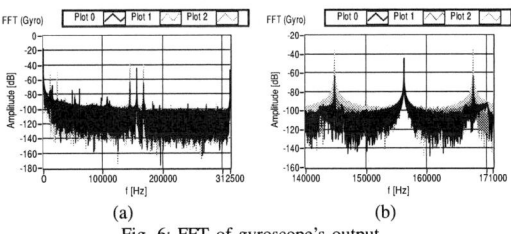

(a) (b)

Fig. 6: FFT of gyroscope's output.

For a N-DOF gyroscope, N+1 resonance frequencies have to be measured: one for each DOF, and one for the driving-mode structure (which is needed to generate the Coriolis acceleration along the sensing axes). The measure accuracy of the driving-mode resonance frequency is really important: the gyro must be driven exactly at the resonance frequency, to ensure the maximum output ampltude.

Resonance frequency test are very repeatable. 100 measurements on the same DUT show always the same result for the accelerometer resonance frequency, which has a large resolution of $50Hz$, and values in the range of $11kHz$ for gyro movable structures, with variations of $1Hz$, which is the resolution.

C. Q-factor

In a complete test procedure the Q-factor measurement always follows a resonance frequency test. The procedures to measure the Q-factor of accelerometers and gyroscopes are different: the Q-factor of an accelerometer structure is in the range of some units (or less than one, for z-axis accelerometers);

for gyroscopes it is more than 10000. It is because a gyroscopes are placed in a vacuum cavity, whereas accelerometers work in a damped environment ([6]).

The accelerometer Q-factor test calculates the Q-factor from the data acquired during the resonance frequency test: by definition the Q-factor can be expressed as ratio between the amplitude response at resonance and the amplitude in the flat zone (DC):

$$Q = \frac{A_{res}}{A_{DC}} \quad (5)$$

The value of Q-factor is used to plot the ideal curve in Fig. 5.

The Q-factor of gyroscopes is calculated from the ring-down curve of the movable structure: it is connected to a modulated signal at the resonance frequency; the stimuli is stopped and the data acquisition started. The on-board demodulation returns the ring-down curve (Fig. 7).

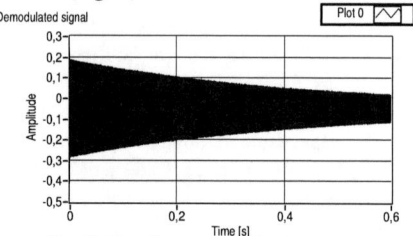

Fig. 7: Ring-down curve of a gyroscope.

The Q-factor was computed from the expression of the ring-down curve for under-damped systems, where δ is the damping ratio.

$$c_{rd} \propto e^{-\delta \omega_0 t} sin(\omega_0 t) \quad (6)$$

$$Q = \frac{1}{2\delta} \quad (7)$$

Measurements show that this approach is very precise, as the standard deviation for 100 repeated tests was about 0.2% of the mean value.

D. QUAD bias

The QUAD bias test requires signal generation with multiple carrier frequencies. The driving axis of a gyroscope is stimulated with a standard modulated signal (carrier at f_{m1} and driving signal). Then another high-frequency signal at f_{m2} is connected to a target sensing axes.

The cross-axes coupling causes the movable structure to move along the sensing axis. This movement is detected by observing side-band peaks around f_{m2}. Fig. 8 shows the FFT of the output signal.

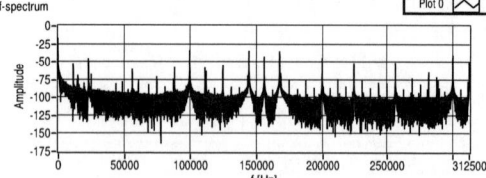

Fig. 8: FFT with multi-DOF modulation.

The QUAD bias is correlated to the ratio between the amplitudes of side-band peaks. A good expression is

$$Q_b = k_c \cdot \frac{A_s}{A_d} \quad (8)$$

where A_s is the amplitude of the side-band peak for the sensing axis (whose movements were modulated at f_{m2}), A_d is the amplitude of the side-band peak for the driving axis, and k_c is the "correlation factor", which is expressed in $\frac{\circ}{s}$. The correlation factor is a value in the range of $40000\frac{\circ}{s}$. A good estimation for this factor is given by a mapping of QUAD bias measured by module-tests. It has to be uploaded for each lot of wafers.

Measurements show precise results, with a coefficient of variation (defined as ratio between standard deviation and mean value) in the range of 0.02. The error increases when the QUAD bias is smaller, because the detection of side-band peak is more affected by noise.

IV. CONCLUSIONS

This paper presents a test setup that was targeted to wafer-sort tests. It allows a characterization of MEMS inertial sensors, up to 6-DOF sensors (3D gyroscopes and 3D accelerometers on the same chip). The mechanical behaviour of the movable structures was characterized by using only electrical stimuli.

The test procedure consisted of different blocks which configure the signal generation and the data acquisition and compute reliable values for the target parameters. It returns electrical parameters, i.e. capacitance values and coupled-capacitances mismatch, and mechanical parameters, i.e. resonance frequencies, Q-factors and QUAD bias.

Measurements showed a good precision level and fast execution time, that make the test setup suitable for production test. The execution time depends on the number of DOFs; it is in the range of 4-10 seconds for each sensor.

The architecture of this system was designed to allow the parallelization of the test procedure; it is possible to upgrade the system in order to test up-to-16 devices simultaneously. This feature largely increases test time effectiveness.

REFERENCES

[1] C. Song, "Commercial vision of silicon based inertial sensors," in *Solid State Sensors and Actuators, 1997. TRANSDUCERS'97 Chicago., 1997 International Conference on*, vol. 2. IEEE, 1997, pp. 839–842.

[2] L. M. Ciganda Brasca, P. Bernardi, M. Sonza Reorda, D. Barbieri, L. Bonaria, R. Losco, L. Marcigot, and M. Straiotto, "A parallel tester architecture for accelerometer and gyroscope mems calibration and test," *Journal of Electronic Testing*, vol. 27, no. 3, pp. 389–402, 2011.

[3] N. Dumas, F. Azaïs, F. Mailly, A. Richardson, and P. Nouet, "A novel method for test and calibration of capacitive accelerometers with a fully electrical setup," in *Design and Diagnostics of Electronic Circuits and Systems, 2008. DDECS 2008. 11th IEEE Workshop on*. IEEE, 2008, pp. 1–6.

[4] B. Yeh and Y.-C. Liang, "Modelling and compensation of quadrature error for silicon mems microgyroscope," in *Power Electronics and Drive Systems, 2001. Proceedings., 2001 4th IEEE International Conference on*, vol. 2, Oct., pp. 871–876 vol.2.

[5] V. Kempe, *Inertial MEMS: Principles and Practice*. Cambridge University Press, 2011.

[6] P. Merz, K. Reimer, M. Weiss, O. Schwarzelbach, C. Schroder, A. Giambastiani, A. Rocchi, and M. Heller, "Combined mems inertial sensors for imu applications," in *Micro Electro Mechanical Systems (MEMS), 2010 IEEE 23rd International Conference on*, jan. 2010, pp. 488 –491.

978-1-4673-4579-8/13 $31.00 © 2013 IEEE

Capacitive out of plane large stroke MEMS structure

Christoph Glacer*, Rainer Laur
Institute for Electromagnetic Theory
and Microelectronics (ITEM),
University of Bremen,
Bremen, Germany
*glacer.external@infineon.com

Alfons Dehé
Infineon Technologies AG
Munich, Germany

David Tumpold
Institute of Mechanics and Mechatronics,
University of Technology,
Vienna, Austria

Abstract—This paper introduces a way to increase the air-gap of a capacitive MEMS sensor or actuator. Usually, a thicker sacrificial layer composed of silicon dioxide is used to increase the distance between two parallel plates. In our case, both poly-silicon planes are being produced with a small gap filled with silicon dioxide. Corrugation grooves lined with highly tensile silicon nitride cause a stress induced displacement of the stator after the release etch was done. We will present first results, gained from FEM-simulations and fabricated wafers. An insight into the optimization of the deflection profile and the mechanical stability of the buckling electrode is discussed.

I. INTRODUCTION

Microelectromechanical systems (MEMS) are strongly upcoming in the last decade. Most common applications are pressure sensors, accelerometers, gyroscopes, BAW filters/duplexers, RF switches and silicon microphones [1].

A frequently used processing variant to fabricate MEMS actuators and sensors is the surface micromechanical sacrificial layer process. In this case two layers are defined, enclosing a sacrificial layer. Through perforation holes in one or both of layers the sacrificial layer will be removed and an air gap remains. The possible gap between those two plates is thereby defined by the thickness of the sacrificial layer, e.g. made from silicon dioxide. Increasing the thickness of those layers affords a longer process time and can cause tension-conditioned cracks.

In this paper we present a MEMS structure consisting of one or two perforated stator plates and a moveable membrane. A sacrificial silicon dioxide layer of $2\,\mu m$ thickness is defining the initial air gap between those plates. Besides that, the poly-silicon stator plates are corrugated and partially coated with silicon nitride which causes the plates to raise after the release etch and enables a much higher air gap.

Such a structure could be used e.g. for a silicon micro loudspeaker. This system affords a large membrane stroke and a big enclosed air volume between the plates. Problems of large electric field strength needed for the actuation can partially be overcome by the dome shaped stator profile.

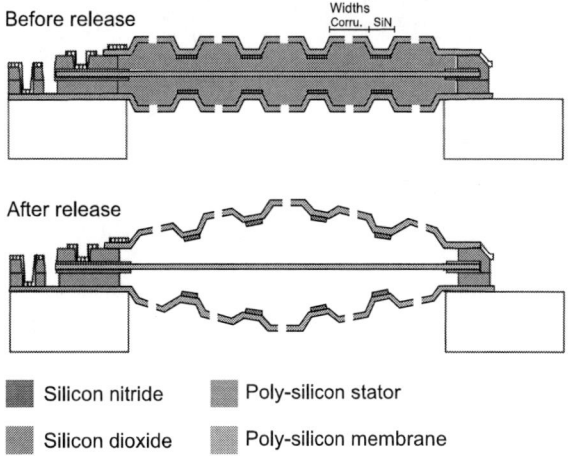

Fig. 1. Schematic of the buckling stator system, as introduced in this work before and after the release etch. A wet chemical etch is used to release the corrugated stators and the membrane through perforation holes in the stators. The tension of the silicon nitride causes the stators to buckle in a certain direction, so that an out of plane displacement of several micrometers can be reached.

II. SYSTEM REALIZATION

A. Principle of stress induced self-raising

The principle of self-raising or buckling of silicon layers due to an induced bending moment is applicable to circumferentially clamped membranes and beams that are clamped on both ends. In our case, the system consists of two fixed and highly perforated and acoustically transparent electrodes called the stators as well as the movable membrane, which is sandwiched between both stators. The diameters of the systems discussed in this work differ from $500\,\mu m$ to $3.5\,mm$. A micrograph of this transducer system, taken from the topside is shown in figure 2.

A doped poly-silicon with a tensile stress of approximate $40\,MPa$ is used as the raw material for the stators. Silicon dioxide (oxide) rings underneath the stator are generating corrugation grooves with a height of $600\,nm$ in the stator. On the membrane facing sides silicon nitride rings with a tensile

Fig. 2. Micrograph of a partially fabricated system, containing only the lower stator on a DRIE hole. The diameter of the system is $958\,\mu m$. A strong deflection of the inner area can be seen.

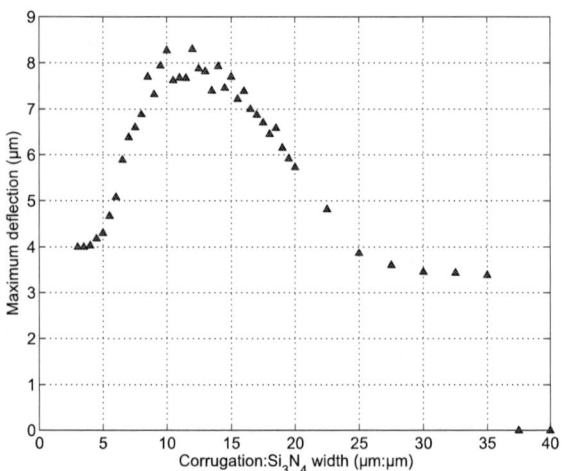

Fig. 3. Measurement of maximum deflection of test structures using clamped-clamped beams with increasing nitride and corrugation width (1:1) to extract the ideal layout. An optimum for this structures can be seen between 10 and $12\,\mu m$.

stress of $1\,GPa$ are being deposited. Using the walls of the corrugation as a hinge, those highly tensile nitride rings are causing a change in length of the stator after the release etch has happened. This results in a buckling of the stator with an out of plane displacement of up to $30\,\mu m$ on a $1\,mm$ diameter system.

A scheme of the buckling stator system before and after the release etch can be seen in figure 1.

B. Influence of process and layout parameters

To optimize the deflection and the stability of the stators, various parameters can be tuned in fabrication process and layout. For this purpose test structures composed of various double-sided clamped beams (cc-beams) have been created. Those structures afford less space than circular membrane-stator systems and allow a fast analysis with a white-light interferometer.

A mayor influence by layout occurs when the width of the silicon nitride and corrugation rings/patches is changed. A half-automated, script based layout generation unveils e.g. the ideal width of nitride and corrugation patches to be between 10 and $12\,\mu m$, when the ratio is kept at 1:1. For this purpose, 43 cc-beams with a length of $470\,\mu m$ each and an increasing width of nitride/corrugation have been compared. The result can be seen in figure 3.

The buckling in z-direction is proportional to the radius of the stator. Attributable to this, the enclosed air volume after the release is strongly influenced by the diameter of the structure.

A rectangular system consisting of beams offers more degrees of freedom and allows a stronger buckling. The enclosed air volume gets strongly increased but the mechanical stability suffers at the same time.

III. RESULTS

A. Simulations

FEM simulations were done to investigate the mayor influence factors, coming from the fabrication process. Variations in silicon nitride and poly-silicon thickness, a different corrugation groove height and a differing pre-stress of the poly-silicon layer are possible. Here an effective pre-stress of the poly-silicon layer was assumed, due to the perforation holes weakening the layer.

Figure 4 shows that biggest influence to the buckling arises from the height of the corrugation grooves. An increase from 600 to $900\,nm$ already leads to a doubled max. deflection. Altering the pre-stress of the poly-silicon, induced by a doping with phosphor as well as the poly-silicon thickness only resulting in a slight change.

B. Optical measurements of the defection profile

Figure 5 shows a 3-dimensional deflection profile of a system with $280\,nm$ thick and $10\,\mu m$ wide nitride on $10\,\mu m$ wide corrugation grooves. This system reaches with a diameter of $958\,\mu m$ a top z-deflection of $19.5\,\mu m$. The buckling begins after one third of the stators radius, while the outer regions are relatively flat.

In fig. 6 a rectangular system is shown. The maximum deflection is comparable to a circular system, but the clamping on only two sides, leads to a different profile. No flat regions appear and every beam of the membrane is showing the same buckling profile.

Several systems with different diameters, nitride/corrugation widths and perforation densities have been produced. Table I shows the difference in buckling. In this table d: is the diameter of the stator (μm), SiN: width of silicon nitride (μm), Corr.: width of corrugation (μm), ρ_{hole}: perforation density, z_B:

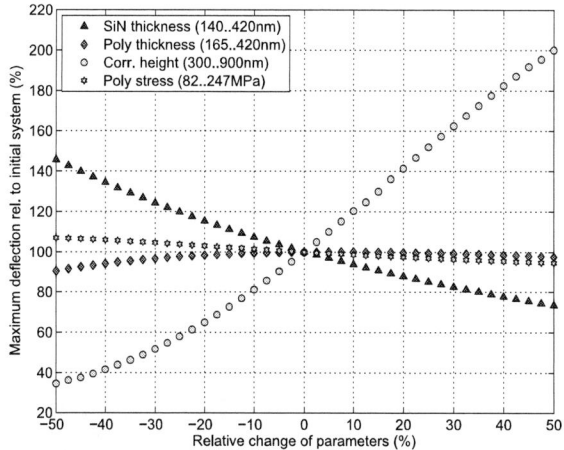

Fig. 4. FEM simulation results indicating the maximum change in deflection relative to the initial system with $330\,nm$ poly-silicon thickness, $280\,nm$ silicon nitride thickness, $165\,MPa$ effective pre-stress of the poly-silicon and $600\,nm$ high corrugation grooves.

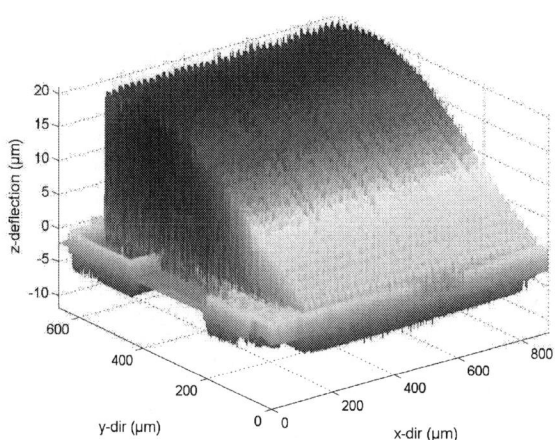

Fig. 6. Deflection profile of a rectangular system, measured with a white-light interferometer. The enclosed air volume of this design is much bigger than of the circular system with nearly the same area.

TABLE I
MEASURED MAXIMUM DISPLACEMENT OF PRODUCED CIRCULAR
SYSTEMS WITH $280nm$ NITRIDE THICKNESS. ALL DATA IN μm UNLESS
OTHERWISE STATED.

d	SiN	Corr.	ρ_{hole}	z_B	z_B/d
502	10	10	40.4%	9.5	1.89%
502	5	5	36.2%	5.5	1.1%
958	10	10	39.2%	18.5	1.93%
958	20	20	39.1%	3	0.31%
958	20	10	39.1%	12	1.25%
958	10	20	39.1%	12	1.25%
958	20	20	16.3%	0	-
958	20	20	55.2%	9	0.94%
2158	10	10	40.6%	43	1.99%
2158	20	20	40.6%	11	0.51%
2158	10	40	40.6%	17	0.79%
3358	10	10	41.1%	68	2.03%
3358	20	20	41.1%	24	0.71%
3358	10	20	41.1%	49	1.46%
3358	10	40	41.1%	21	0.63%

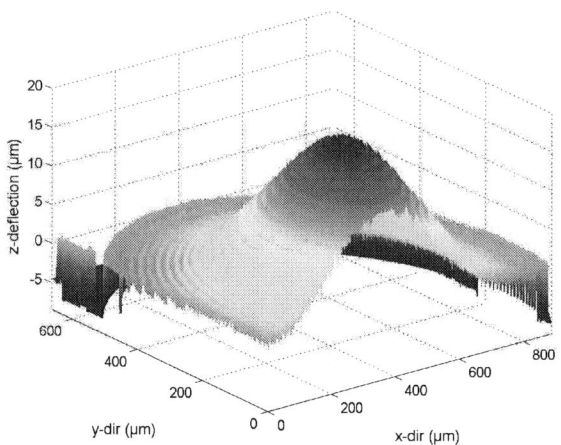

Fig. 5. Deflection of the upper stator of a circular system, measured with a white-light interferometer. This particular system shows a maximum deflection in z-direction of around $20\,\mu m$.

maximum deflection in z-direction (μm) and z_B/d: the ratio of buckling, respective to the diameter. A maximum deflection of $68\,\mu m$ is reached by the biggest systems.

C. Electrical measurements

Electrical measurements of the capacitance with an LCR-meter can provide information about the initial air-gap, the behavior when the plates are in contact, as well as the mechanical compliance.

The acting Coulomb force whose value is proportional to the square of the value of the input voltage, causes a quadratic increase of the capacitance C. When the acting electrostatic force equals the mechanic restoring force, the so called pull-in

event happens. This causes the movable plate, to snap quickly against the fixed electrode [2]. It can be detected by a sharp rise in capacitance at a certain bias voltage.

Because of the special geometry of the systems produced in this work, regular mathematical descriptions as used for parallel-plate capacitors are no longer valid here. Estimations of the pull-in voltage(s) were made using FEM-simulations. Figure 7 shows a step-wise pull in, starting at the flat region of the circular system. It was observed, that single or multiple grooves pulling in at certain voltages. The softer the plates of the system are, the earlier the first pull-in happens. Also the maximum capacitance is increased, because of the greater contact area of the plates. The silicon nitride is acting also as an electrical insulator in this case.

From the geometry of the plates fully in contact, a maximum capacitance of the system can be calculated. Assuming dielectric constants of $\varepsilon_{r,SiO_2} = 3.9$ for oxide and $\varepsilon_{r,Si_3N_4} = 7.5$ for the nitride, a capacitances of around $181\,pF$ respectively $94\,pF$ can be calculated for the 140 or the $280\,nm$ thick nitride.

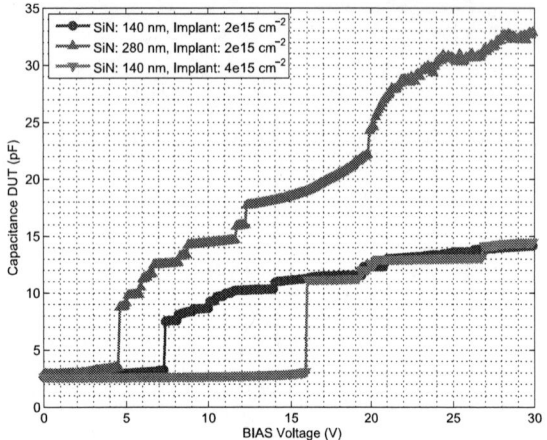

Fig. 7. Measured capacitance over bias voltage, indicating pull-in voltages. A step-wise pull-in of single/multiple rings can be seen.

IV. DISCUSSION

Because of the all-around clamping, the circular systems starting with a flat region, proceeding to a conical shape. With the used parameters, the intended and more stable dome-shaped deflection profile was not achieved yet. From this an earlier pull-in and a virtually softer stator arises.

After the flat region of the stator is in contact with the membrane, a higher voltage is required to pull-in the remaining part of the membrane. With the existing systems, it was not possible to bring the inner area into contact.

The ratio of deflection to diameter reaches from $1.89\,\%$ to $2.03\,\%$ for systems with 502 up to $3358\,\mu m$. Clamping effects are carrying weight the smaller the structure gets. Nevertheless, to keep the stator stable, it will be necessary to find an optimum between deflection/enclosed volume and mechanical stability.

Beside the nitride to corrugation ratio, the perforation of the stators, which is crucial for the release etch, was varied. A higher perforation density causes a softer poly-silicon in the stator. This enables a larger maximum deflection and also a higher acoustic transparency, needed e.g. for microphone or loudspeaker applications.

To monitor the mechanical behavior, systems only consisting of a stator were made. This showed, that the membrane, which is also under compressive stress causes a decrease of the total buckling. A possible explanation could be that the membrane is contracting the clamping of the stators and

is hence influencing the deflection profile. The maximum deflection is reduced for system including the membrane from around 30 to $18.5\,\mu m$.

The least influence on the profile can be seen on rectangular systems. The more effective use of the floor area and no flat area causing a much larger enclosed air volume. The pull-in voltage of this system could not be determined yet, so that no prediction of the stability of those beams can be made.

FEM simulations, as well as observations on fabricated samples showed that a big improvement of the maximum buckling and deflection profile can be reached by increasing the corrugation groove height.

V. CONCLUSION

Measurements and optical observations showed that it is possible to create a large air-gap between two plates, without the use of a thick sacrificial layer by this principle.

With the intended dome-shaped stator, large membrane strokes will be possible. The maximum deflection of the movable membrane will be in the central region were also the deflection of the stator is largest.

Yet, this structure offers an interesting way of increasing the maximum air-gap and the enclosed volume of capacitive MEMS without the use of a thicker sacrificial layer. Nevertheless, more efforts have to be made to increase the mechanical stability of the stator and optimize the profile of the buckling.

ACKNOWLEDGMENT

The authors would like to thank Dr. A. Kenda from the Carinthian Tech Research AG for providing white light interferometer and Doppler laser vibrometer measurements.

REFERENCES

[1] Yole Developpement. (2010) MEMS & Sensors for Smartphones: -2010 edition-. Lyon, France.
[2] V. Kaajakari, *Practical MEMS*, 2nd ed. Las Vegas, US: Small Gear Publishing, 2009.

PRIME 2013, Villach, Austria

Session M1C – FPGA

Graph Coverage: an FPGA-targeted Implementation

Alessandro Cinti, Antonello Rizzi

Department of Information Engineering, Electronics and Telecommunications,
University of Rome "La Sapienza"
Via Eudossiana 18, 00184 Rome, Italy
alessandro.cinti@uniroma1.it, antonello.rizzi@uniroma1.it

Abstract—**Classification systems specifically designed to deal with fully labeled graphs are gaining importance in many application fields. The main computational bottleneck in such systems is the dissimilarity measure between pairs of graphs. In this paper we propose to accelerate in hardware such computations, relying on the Graph Coverage as the core inexact graph matching procedure, targeting the design to FPGA as an inexpensive way to design specific co-processing devices. A comparison in terms of computational time between the proposed system and a software implementation on a standard workstation shows encouraging results.**

Keywords—graph coverage; tensor product; FPGA; parallel computing

I. INTRODUCTION

Pattern recognition is extensively approached when facing real world applications that deal with highly structured patterns. Since graphs are powerful data structures instruments, commonly used to model both relational and semantic information, each pattern can be conveniently represented with a labeled graph. Applications employing labeled graphs to represent patterns include, for instance, computer vision, biometric identification problems, chemical compounds behavior prediction, mechanical structures diagnostic systems, smart grids monitoring and control. During the years, our research team has proposed several pattern recognition systems able to deal with structured domains [1], [2], [3] and [4].

The generalization capability of any classification system depending on a particular data driven modeling problem is strictly dependent on the definition and calculation of the inductive logic inference, based on the choice of a dissimilarity measure between input patterns. Such a measure is therefore the most important procedure in any inductive modeling system. In fact, such a dissimilarity measure should take into account both the topological information and the information stored in node and edge labels.

An effective classification system should be designed to map objects belonging to the same class into (a few) compact clusters or, equivalently, in the same decision region. This fundamental characteristic can be obtained adapting the underling dissimilarity measure by defining the graph matching procedure on the basis of the particular modeling problem at hand. Efficient solutions are based on soft computing techniques, where algorithms are usually characterized by a remarkable computational complexity. Thus, since the

nowadays pattern recognition applications deal with more and more complex patterns often organized in huge databases, it is worth to dedicate a special effort for finding efficient ways to calculate dissimilarity measures. Recently, we are considering the opportunity of accelerating in hardware dissimilarity measures between fully labeled graphs, usually characterized by a high computational cost, developing low cost custom devices, to be used as co-processors of a standard workstation. Especially when dealing with real time recognition systems, delegating to a slave device high demanding computations needed to classify the graph at hand, can be convenient in terms of classification accuracy. In fact, this approach allows a considerable amount of computational resources to be available for more complex preprocessing procedures on the next incoming pattern and/or to implement advanced feature extraction modules for the node and edge labels computing. FPGA (Field Programmable Gate Array) is a user-programmable integrated circuit that, thanks to the inherent parallelism of its logic resources, reveals high flexibility for the parallel implementation of given algorithms and allows considerable computational throughputs even at low MHz clock rates. In particular, we describe an FPGA targeted implementation of a particular dissimilarity measures between graphs, based on tensor product computation. The tensor product hardware acceleration problem has been faced in [5], [6].

II. GRAPH COVERAGE AS DISSIMILARITY MEASURE

A. Labeled graphs

We define a labeled graph as $G = G(V, E, \mu, v)$, where

- V is the set of vertices with cardinality N_V,

- E is the set of edges with cardinality N_E,

- μ is the vertex labeling function that associates a label (l_V-dimensional vector) to a vertex,

- v is the edge labeling function that associates a label (l_E-dimensional vector) to an edge.

The adjacency matrix A_G associated to a graph G is a square matrix of order N_V, whose elements are

$$a_{i,j} = \begin{cases} 1, & \forall\, i,j \mid \exists\, e_{i,j} \\ 0, & otherwise \end{cases} \quad (1)$$

978-1-4673-4579-8/13 $31.00 © 2013 IEEE

129

where $e_{i,j}$ represents the edge between the nodes v_i and v_j, with $i, j = 0, \ldots, N_V - 1$.

B. Tensor product

Labeled graphs are powerful data structures for representing a set of object together with their own relationships, but they lack of a strong and well established mathematical framework for some specific operations. In particular defining how much two given graphs are similar (or dissimilar) is not a trivial task [1]. In [4] we have defined a dissimilarity measure, i.e. the Graph Coverage, well suited to be implemented in special parallel computing devices, such as GP-GPUs and FPGAs. Given two graphs G_1 and G_2, this measure is based on the notion of tensor product between them and is defined as:

$$
\begin{aligned}
G_\otimes &= G_1 \otimes G_2 = G(V_\otimes, E_\otimes) \\
V_\otimes &= \left\{ \left(v_1^p, v_2^r\right) : v_1^p \in V_1, v_2^r \in V_2 \right\} \\
E_\otimes &= \left\{ \left(\left(v_1^p, v_2^r\right), \left(v_1^q, v_2^s\right)\right) : \left(v_1^p, v_1^q\right) \in E_1 \wedge \left(v_2^r, v_2^s\right) \in E_2 \right\}
\end{aligned} \quad (2)
$$

In the pattern recognition context, it is mandatory to define a dissimilarity measure between two given graphs in terms of topological information and similarities between the labels of both nodes and edges. For this purpose, the basic tensor product formulation has been appropriately redefined as follows. The adjacency matrix $A_{G_\otimes}$ associated to the tensor product of two labeled graphs is a square matrix of order $N_{V_\otimes} = N_{V_1} \cdot N_{V_2}$ with $N_{E_\otimes} = N_{E_1} \cdot N_{E_2}$ non-zero elements

$$
a_{\otimes|i,j} = \begin{cases} k_{1,2}^{(p,q),(r,s)}, & \left(\left(v_1^p, v_2^r\right), \left(v_1^q, v_2^s\right)\right) \in E_\otimes \\ 0, & otherwise \end{cases}, \quad (3)
$$

where $k_{1,2}^{(p,q),(r,s)}$ is a positive definite kernel function [4].

C. Graph coverage

Given two input graphs G_1 and G_2, it is possible to calculate three different tensor product graphs, namely, $G_\otimes^{(1,1)}$, $G_\otimes^{(2,2)}$ and $G_\otimes^{(1,2)}$, where the first two represents the best matching tensor product graphs. If G_1 and G_2 are exactly equal, their similarity must be the maximum achievable and, conversely, the minimum if they are completely different. The graph coverage C_G [4] is used to measure the overlap degree between the two graphs G_1 and G_2 (considering both topological information and labels content) by comparing the weight of the tensor product graph $G_\otimes^{(1,2)}$ to the maximum achievable between $G_\otimes^{(1,1)}$ and $G_\otimes^{(2,2)}$:

$$
C_G(G_1, G_2) = \frac{W_G\left(G_\otimes^{(1,2)}\right)}{\max\left(W_G\left(G_\otimes^{(1,1)}\right), W_G\left(G_\otimes^{(2,2)}\right)\right)}, \quad (4)
$$

where W_G represents the weight of a graph and is defined as the summation of the elements of the adjacency matrix associated to the graph itself. For example, the weight of $G_\otimes^{(1,2)}$ is

$$
W_G^{(1,2)} = W_G\left(G_\otimes^{(1,2)}\right) = \sum_{i=0}^{N_{V_1}-1} \sum_{j=0}^{N_{V_2}-1} a_{\otimes|i,j}. \quad (5)
$$

The main contribution to the overall computational complexity of the graph coverage is dominated by the computation of the tensor product between the two input graphs. If n and m are, respectively, the order of G_1 and G_2, then the computational complexity of the tensor product between G_1 and G_2 is of order $O(n^2 \cdot m^2)$. Therefore, the computational complexity of the graph coverage as a whole is $O(n^4 + m^4 + n^2 \cdot m^2)$, which is asymptotically dominated by the maximum of the order of the two input graphs.

III. COPROCESSING UNIT: PROPOSED IMPLEMENTATION

The implementation of a system for calculating in parallel the graph coverages between one labeled graph G_0 and N-1 sample labeled graphs G_i. ($i = 1, \ldots, N$-1) is proposed in this Section. Thanks to the positive definiteness of the kernel functions [4], for each pair of labeled graphs G_l and G_m ($l, m = 0, \ldots, N$-1), one element of the matrix $A_{G_\otimes}$ can be defined as the product of three different valid kernel functions specialized for node (k_V) and edge (k_E) labels:

$$
k_{l,m}^{(p,q),(r,s)} = k_V\left(v_l^p, v_m^r\right) \cdot k_E\left(e_l^{p,q}, e_m^{r,s}\right) \cdot k_V\left(v_l^q, v_m^s\right). \quad (6)
$$

Gaussian Radial Basis kernels have been chosen for both k_V and k_E

$$
k(x,y) = \exp\left(-\left(\frac{d(x,y)}{\sqrt{2}\sigma}\right)^2\right), \quad d(x,y) = \sqrt{\sum_{i=0}^{N-1}(x_i - y_i)^2}, \quad (7)
$$

where σ represents the standard deviation of the Gaussian distribution and d represents the Euclidean distance as dissimilarity measure over the specific set of labels. By substituting (7) in (6) and properly rearranging, it can be easily obtained the following result

$$
\begin{aligned}
k_{l,m}^{(p,q),(r,s)} = \exp\Bigg(&- \sum_{i=0}^{l_V-1}\left(\frac{\mu\left(v_{l,i}^p\right) - \mu\left(v_{m,i}^r\right)}{\sqrt{2}\sigma_V}\right)^2 \\
&- \sum_{i=0}^{l_E-1}\left(\frac{\nu\left(e_{l,i}^{p,q}\right) - \nu\left(e_{m,i}^{r,s}\right)}{\sqrt{2}\sigma_E}\right)^2 - \sum_{i=0}^{l_V-1}\left(\frac{\mu\left(v_{l,i}^q\right) - \mu\left(v_{m,i}^s\right)}{\sqrt{2}\sigma_V}\right)^2\Bigg)
\end{aligned} \quad (8)
$$

The labels needed to calculate $k_{l,m}^{(p,q),(r,s)}$ can be collected into two input vectors having size $l_r = 2 \cdot l_V + l_E$.

In Fig. 1 the architecture of the implemented system is presented. An input/output interface (I/F) is the only point that a central processing unit (PC) uses to access the proposed co-processing unit (CU), both for its configuration and for the data insertion (the graph to be measured) and for the data extraction (the calculated graph coverage values). The PC communicates with the CU by packing the information in a custom frame. A finite state machine FSM behaves as a bridge between PC and CU. Fig. 2 shows the main states that the CU can assume. The CU is in Idle state until a new frame is not received. The FSM moves to the Analyzer state where the incoming frame is analyzed. The only two actions that can be executed are the configuration (graph prototypes loading) and the classification (loading a graph to be compared with the prototypes and dissimilarities calculation). A faulty frame

PRIME 2013, Villach, Austria

Session M1C – FPGA

takes the state machine returns to the Idle state. In the Conf state the system is configured. The RAMs of each sample graph (in Fig.1 from DPRAM 1 to DPRAM N-1) are fulfilled with the proper data, constant parameters (such as the standard deviations σ) are stored in internal registers and the $W_G^{(i,i)}$ values (i = 1, ..., N-1) are calculated and stored. Then the system returns to the Idle state. When the graph to be processed is sent to the CU, the system passes from the Analyzer state to the Load state and both the DPRAM 0 and the proper constant registers are loaded. The system is so ready to go in the Calc state where all the $W_G^{(0,i)}$ values (i = 0, ..., N-1) are calculated. The CMP block is a "greater than" comparator that selects the denominator in (4) for all the N-1 graph coverage values that are calculated by the final divider in Fig. 1. The system moves to the Result state and the computed values are passed to the FSM block that packs them into the required frame format to be sent back to the PC.

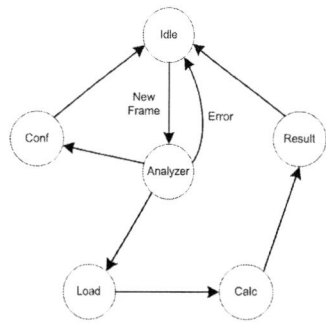

Fig. 2. Simplified version of the FSM state diagram.

The system returns to the Idle state waiting for either a new configuration or a new graph to be measured.

The DPRAM blocks are dual-port RAM to allow the calculation of the $W_G^{(i,i)}$ values (i = 0, ..., N-1) and for each graph the DPRAMs have to be three separate RAMs: the adjacency matrix RAM (A), the node labels RAM (V) and the edge labels RAMs (E). Even though the values that are used in the calculations are the ones stored in the V-RAM and in the E-RAM, the presence of the A-RAM is justified to act as an indirect addressing memory (CAM-like approach). It introduces a formal redundancy that prevents the hardware implementation of high-cost search algorithms.

When aiming to a real hardware implementation, the latency of each operation block must be taken into account and pipelining is a way to increase the operative frequency f_S. When considering an accumulator, the finite latency L_{ADD} of the adder block allows to combine in the loop only data that are temporally distant L_{ADD} clock cycles. Once l_τ data enters the accumulation stage, L_{ADD} partial sums are ready after $L_{ADD} \cdot \lfloor l_\tau / L_{ADD} \rfloor$ clock cycles and they loop without combining into a single value. In order to avoid this limitation in the accumulator block, it has been provided with a dedicated simple finite state machine ADD FSM that properly delays (by using the tapped delay-line DLY) the fed-back data in order to combine them in $L_{ADD} \cdot (\lceil \log_2 L_{ADD} \rceil + 2) - 1$ clock cycles. The ADD FSM is activated by the FSM and then manages the accumulator stage by itself. Fig. 3 clarifies how the accumulator stage works showing the evolution of the data at the inputs and the output ports of the adder.

The value of $k_{l,m}^{(p,q),(r,s)}$ is calculated in

$$
\begin{aligned}
L_k &= L_{SUB} + L_{DIV} + L_{SQR} + \\
&+ L_{ADD} \cdot \left(\lfloor l_\tau / L_{ADD} \rfloor + \lceil \log_2 L_{ADD} \rceil + 2 \right) - 1 \\
&+ L_{EXP} = L_{SUB} + L_{DIV} + L_{SQR} + l \cdot L_{ADD} - 1 + L_{EXP} \cong \\
&\cong L_{SUB} + L_{DIV} + L_{SQR} + l \cdot L_{ADD} + L_{EXP}
\end{aligned}
\tag{9}
$$

clock cycles. With the second accumulator stage the W_G value is obtained and since it doesn't work at full-rate f_S, the implementation with the dedicate ADD FSM is not required if $l_\tau \geq L_{ADD}$ (always true) because the throughput of the incoming data is f_S / l_τ.

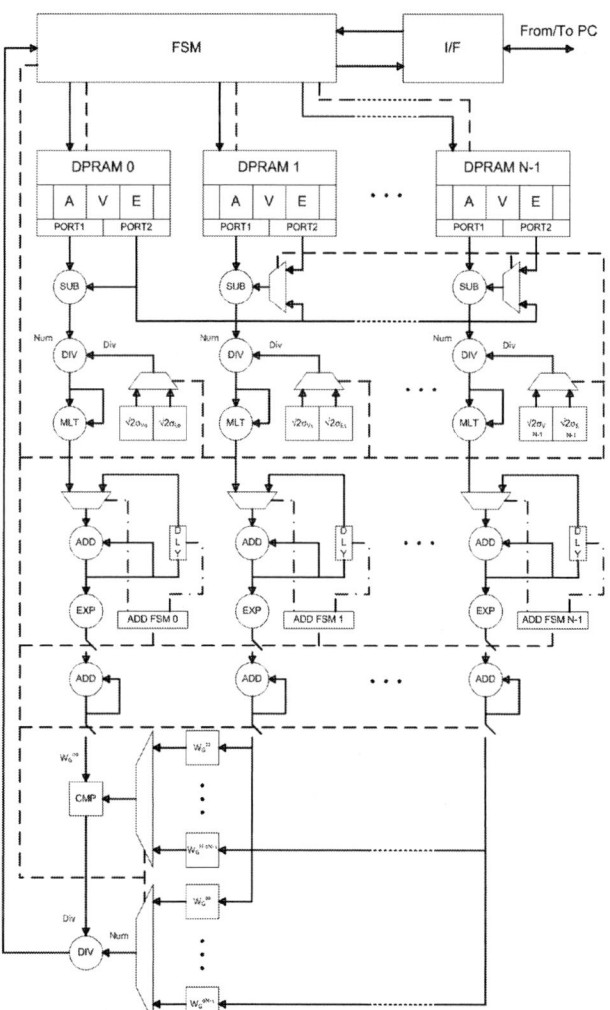

Fig. 1. Conceptual scheme of the proposed implementation showing both the data path (solid lines) and the control path (dashed lines).

978-1-4673-4579-8/13 $31.00 © 2013 IEEE   131

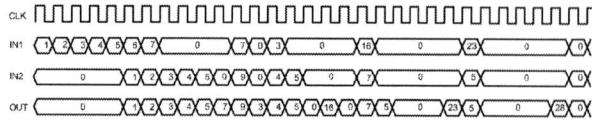

Fig. 3. Example of accumulator stage with $l_\tau = 7$ and $L_{ADD} = 5$. Input IN1 comes from the multiplexer 2:1 of Fig. 1.

The time required for calculating W_G depends on the number of the non-zero elements of the tensor product adjacency matrix

$$T_{W_G} = \left(L_k + l_\tau \cdot \left(N_{E_\otimes} - 1\right)\right) \cdot T_S \cong \left(L_k + l_\tau \cdot N_{E_\otimes}\right) \cdot T_S =$$
$$= \left(L_{SUB} + L_{DIV} + L_{SQR} + l \cdot L_{ADD} + L_{EXP} + l_\tau \cdot N_{E_\otimes}\right) \cdot T_S \quad (10)$$

where $T_S = 1 / f_S$ is the period of system clock and data rate. It can be noticed that (10) is dominated by the contribution of $l_\tau \cdot N_{E_\otimes}$ when dealing with large graphs characterized by a great number of features in both the node and edge labels. Finally, the time needed to obtain the graph coverage can be calculated as follows:

$$T_{C_G} = \max_i \left(T_{W_G}^{G_\otimes^{(0,i)}} \right) + L_{DIV} \cdot T_S . \quad (11)$$

IV. CASE STUDY

The presented case study consists in comparing the time required for calculating the graph coverage between one input graph G and $N - 1 = 15$ sample graphs, where the chosen data representation for the labels is the floating point single precision (IEEE 754). The architecture described in the previous section has been implemented on an Altera Cyclone IV E FPGA device. The architecture has been dimensioned to handle labeled graphs with a maximum of $N_V = 16$ nodes and a maximum of $N_E = 240$ non-oriented edges (all the possible combinations). For each node, a maximum of $l_V = 16$ real valued features is allowed, whereas for each edge the maximum number of features is $l_E = 4$. Each labeled graph is stored into the FPGA internal RAMs with size(A) + size(V) + size(E) $\approx$ 45kbit of memory occupation. In Table I are reported the values of latencies, number of LUTs and number of registers used for implementing each operation block. Being $l_\tau = 36$ and $N_{E_\otimes} = 240^2 = 57600$ with an operative frequency of $f_S = 100$ MHz, substituting the values of TABLE I in (10) and (11) the time that the system need to calculate the

15 graph coverages is 20.73 ms.

The same calculations have been performed running a C++ graph coverage code on a PC with the following characterics: Intel® Core™ i7-2670QM CPU @ 2.20GHz × 4 processors, 8 GByte of DDR3 1,66 GHz, Ubuntu 12.04 operating system. Averaging on 20000 random runs, the same result has been obtained in 50.31 ms. Using a USB 2.0 peripheral at full rate speed (12 Mbps) to interface the PC to the CU, the time overhead for transferring one graph to be processed into the CU is 3.7 ms. The proposed systems is able to calculate 15 graph coverage values every 25ms.

V. CONCLUSION

In this paper we propose a hardware implementation of a specific inexact graph matching procedure, namely the Graph Coverage, to be used as the core inductive inference engine of pattern recognition systems able to deal with fully labeled graphs as input patterns. Performance comparison with a software implementation suggests that the proposed architecture is better in terms of computing times, achieving a speedup of more than 100%. It is worth to underline that such a result has been obtained on a low cost line FPGA, clocked at a safe 100MHz frequency. Moreover, the availability of a higher level line FPGA, with additional hardware resources and higher operative frequencies, could easily allow to increase the number of features in edge and node labels and/or the number of reference graphs, greatly enhancing computing performances. Obviously our implementation should be considered as a first proposal, subject to further optimization.

Moreover, we are currently considering to design FPGA-targeted systems implementing different graph matching procedures, such as the TWEC [7], belonging to the edit distance dissimilarity measures family.

REFERENCES

[1] L. Livi, A. Rizzi. The Graph Matching Problem. Pattern Analysis and Applications, Springer. DOI: 10.1007/s10044-012-0284-8.

[2] L. Livi, G. Del Vescovo, A. Rizzi. Combining Graph Seriation and Substructures Mining for Graph Recognition. In Advances in Intelligent and Soft Computing, Springer. DOI: 10.1007/978-3-642-36530-0_7.

[3] L. Livi, A. Rizzi. Parallel Algorithms for Tensor Product-based Inexact Graph Matching. In Proceeding of the 2012 IEEE IJCNN, pages 2276-2283, June 2012, Brisbane, Australia. ISBN: 978-1-4673-1489-3. DOI: 10.1109/IJCNN.2012.6252681.

[4] L. Livi, G. Del Vescovo, A. Rizzi. Inexact Graph Matching Through Graph Coverage. In Proceeding of the First International Conference on Pattern Recognition Applications and Methods, volume 1, pages 269-272, February 2012, Vilamoura, Algarve, Portugal. ISBN: 978-989-8425-98-0. DOI: 10.5220/0003732802690272.

[5] A. Elnaggar, H.M. Alnuweiri, M.R. Ito "Mapping tensor products onto VLSI networks with reduced I/O" GLSV '94, Proceedings., Fourth Great Lakes Symposium on Design Automation of High Performance VLSI Systems, 04 March 1994, D.O.I.: 10.1109/GLSV.1994.289978

[6] A. Elnaggar, H.M. Alnuweiri, M.R. Ito "Highly parallel VLSI architectures for linear convolution" ISCAS '95, IEEE International Symposium on Circuits and Systems, pp. 1424 - 1427, vol.2, D.O.I: 10.1109/ISCAS.1995

[7] A. Rizzi and G. Del Vescovo. "Automatic Image Classification by a Granular Computing Approach". In Proceedings of the 2006 16th IEEE Signal Processing Society Workshop on Machine Learning for Signal Processing, pages 33–38, September 2006, doi: 10.1109/MLSP.2006.275517.

TABLE I.   CHARATERISTICS OF THE USED OPERATION BLOCKS

Operation block	Characteristics		
	Latency	LUTs	Registers
Subtraction (SUB)	8	168	467
Division (DIV)	6	194 + 16 dsp_9bit + 74 mux 2:1	339 + 1 RAM 9K
Squaring (SQR)	5	111 + 7 dsp_9bit	209
Summation (ADD)	8	168	467
Exponentiation (EXP)	17	2118 + 31 dsp_9bit + 124 mux 2:1	873

Design-Space Exploration of an eFPGA Soft-Core based on Multi-Stages Switching Networks

Matteo Cuppini, Eleonora Franchi Scarselli
ARCES – University of Bologna
Viale Pepoli 3/2, 40123 Bologna - Italy
e-mail: {mcuppini, efranchi}@arces.unibo.it

Claudio Mucci
STMicroelectronics
Agrate Brianza, Italy
e-mail: claudio.mucci@st.com

Abstract—**Embedded FPGAs are becoming appealing IPs to enhance modern SoCs, since technology scaling is enabling reconfigurability at lower area impact. This notwithstanding, to become effective eFPGAs should be highly adaptable to support application-specific optimization, in terms of DSP blocks, technology options and floorplan requirements. For that, in this paper, we analyse a soft-core eFPGA template based on Multi-Stage Switching Network which couples high flexibility with a modular design approach based on the regular replication of few simple switch modules for the programmable routing. Implementation on 65nm technology showed the existence of a significantly wide design space which allows to quickly optimize the device for area, speed and/or leakage power. Results show that depending on architectural and technology options adopted, performance can vary in terms of area (~50%), speed (+/-30%) and leakage (~90%) with respect to a reference design.**

Keywords—eFPGA, SoC, Multi-Stage Switching Networks

I. INTRODUCTION

One of the reasons of System on Chips (SoCs) success is their capability to couple performance and miniaturization with time-to-market. Implementation approaches based on Standard-Cell design allows designers to select the most appropriate technology flavors for a given application in order to find the best trade-off among speed, area and power. The availability of a rich portfolio of libraries allow the designer to explore quite easily different implementation scenarios, including libraries with different transistors thresholds (Vt), to balance speed and leakage, or libraries for high-density design, to balance area and speed. This design style ensures time-to-market and risk-reduction inheriting its robustness by the pre-verification and re-use of cells and IPs. Usually, a SoC features a processor-based environment enhanced with a set of accelerators and I/O peripherals. Flexibility and upgradability are ensured adopting Application Specific Standard Processors (ASSPs) and Digital Signal Processors (DSPs) whereas performance requirements allow designers to plan a software-programmable solution. So far, SoC market improves flexibility through the increase of processor-based computation, be it on the processor core (e.g. adopting multicore processors) or be it on the acceleration domains (e.g. by parallel architectures based on GPU-like structures).

With the increase of manufacturing and NREs costs of SoCs, FPGAs are becoming more appealing, although their cost model appears suitable only for low-volume markets. The flexibility advantage of FPGAs allows implementing a whole system on a programmable hardware, enabling changes on-the-field. On the other hand, full hardware programmability introduces significant area & performance penalties which could limit their adoption. To overcome this, many FPGAs – especially the high-end solutions – are integrating many hard-macros (multipliers, high-performance processors, high-speed interfaces...). It appears that while SoCs are becoming more flexible thanks to programmability, FPGAs are becoming more effective thanks to hardening. Lot of works has been done in the past to bridge this gap, opening the way for reconfigurable computing. Many factors limited the success of those solutions on the marketplace. In fact, reconfigurable devices have been often proposed as monolithic macros to optimize area and performance by full-custom design, resulting in devices poorly matching specific integration needs (e.g. size, shape, pins location, power, application-specific blocks), too general-purpose to replace hardware accelerators and too area demanding to replace ASSPs/DSPs.

This notwithstanding, technology scaling is pressing on reconfigurability, enabling SoCs with increased complexity and thus with the possibility to embed programmable devices at lower area impact on the overall SoC. To make this sustainable, coupling programmability with application-specific optimization, embedded FPGAs based on soft-core (i.e. synthesizable) appear a very promising approach. In this paper, we will show how a soft-core eFPGA template can be optimized for different area-speed-leakage trade-off points through a standard-cell based design flow. Key point of this work is the adoption of a multi-stage switching network as foundation of the programmable interconnects, potentially avoiding the need for custom circuit design or, in alternative, limiting it to the design of few simple switch blocks.

II. BACKGROUND

FPGA structures are usually associated to the traditional 2D island-style architecture proposed many time ago by [1]. This kind of programmable routing was based on channels with bi-directional routing segments interconnected through pass-transistors or tri-state buffers [2]. Many research activities have been proposed to overcome the issues associated to the bit-level programmability, as summarized in the following.

978-1-4673-4579-8/13 $31.00 © 2013 IEEE

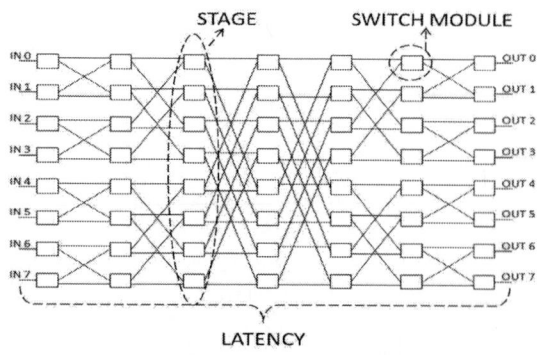

Figure 1 Example of 8x8 MSSN back-to-back butterfly

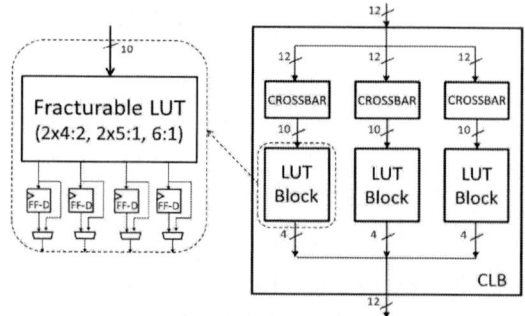

Figure 2 Simplified CLB structure

Bi-directional segmented routing requiring pass-transistor logic has been replaced by uni-directional segmented routing and multiplexers [3]. Congestion issues in the middle of the device were addressed either increasing channels size or adopting non-uniform channels with the drawback of a significant area increase [4]. Alternatively, hierarchical interconnects (e.g. based on H-meshes) were proposed, featuring local crossbars (sized through the Rent's rule) to locally connect the clusters at the various level of the hierarchy [5].

On the Configurable Logic Blocks (CLB) side, to increase the computational density, pure LUT-based approaches were enhanced adding carry- and combinational-chains to exploit the locality to speed-up some typical circuits. Recently, special-purpose CLBs have been proposed with hard-macros implementing multipliers, application-specific DSP blocks, SRAM banks, processors and so on [6], opening the way for heterogeneous architectures.

Focusing now on the devices suitable for embedding on SoCs, the critical point is usually represented by the area overhead introduced by programmable routing, which traditionally represents ~80-90% of the area [7]. To address this issue, hierarchical structures have been adopted in the past in eFPGAs. AboundLogic (former M2000) proposed a hierarchical interconnect with local crossbars based on Clos Networks, ensuring a local connectivity with non-blocking properties [8]. Other examples of hierarchical networks have been proposed by Leopard Logic [9] and, more recently, by UCLA [10] which adopt Multi-Stage Switching Networks (MSSN) based on Benes/Butterfly topology.

This kind of topology, also known as logarithmic interconnect, has been widely used on different application fields like high performance computing and communication networks. The architecture of a MSSN is based on the utilization of small basic switch modules organized in multiple levels or stages, as shown in Figure 1. Each switch module is an MxN crossbar (in Figure 1, $M=N=2$) and every stage features a fixed number of connections with a pattern typical of the level. The number of levels to be crossed to connect inputs to outputs represents the latency of the network. In addition, depending on their topology, MSSNs can be classified as non-blocking, if any connection can be realized, or blocking, if there are paths that cannot be satisfied [11].

For eFPGA application, where network configuration is static, the adoption of a MSSN shows two straightforward advantages: the first one is that blocking properties are mostly defined and predictable at topology level, thus simplifying the routability analysis; the second and most important advantage is the modularity of the approach, based on the regular replication of one or few switch modules, that can be eventually optimized at circuit level (as coarse standard cells) without compromising the soft-core approach. On the other hand, the latency of the network "as-is" not allow to exploit the locality of the paths, leading designers towards the research of some architectural improvements at the basic MSSN exploiting, for example, its symmetry properties through folding and U-turns connections [8].

III. EFPGA ARCHITECTURE TEMPLATE WITH MSSN

The architecture template we take into account for our exploration is based on a set of CLBs connected through a butterfly-based MSSN, focusing our analysis on the interconnect that usually represents the critical point for both implementation and figures of merits.

Configuration bit-cells and logic have been made synthesizable and in particular all the bit-cells have been implemented with simple latches (without reset) to minimize the area, while the configuration scan-chains, separated for CLBs and MSSN, have been implemented with flip-flops. CLB architecture has been kept simple, excluding from this analysis any kind of application-specific accelerators, memories or dedicated chains that usually have a great impact on performance without a significant impact on layout. This notwithstanding, in order to work with a CLB featuring a reasonable complexity, also in terms of area and bit-cells count, we used the CLB model adopted in the VTR chain [12], which support fracturable-LUTs [13] and inputs equivalence through local crossbars. After a rough analysis with VTR, we selected a CLB featuring 12 inputs, 12 outputs and including 3 independent LUT6:1 fracturable down to 6 independent LUT4:2. Inputs are thus partially shared and distributed to the LUTs by 3 12x10 crossbars, as sketched out in Figure 2.

As anticipated, we adopted a MSSN based on butterfly topology, thus featuring 2x2 switches organized in a double "back-to-back" network, as shown in Figure 1. This architecture is characterized by logarithmic latency ($O(log_2K)$, with K=number of I/Os) and thanks to its structure is proven to

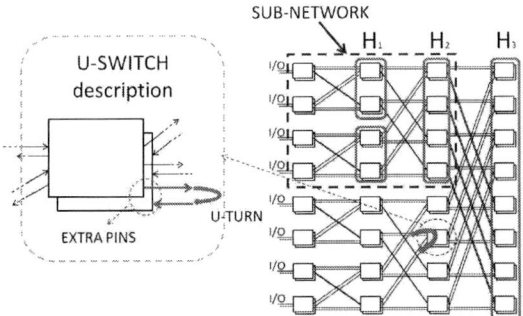

Figure 3 Folded MSSN supporting *U-turn* connections

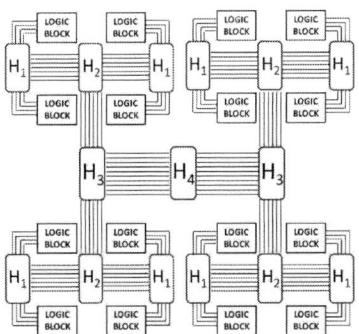

Figure 4 Hierarchy-aware 2D-layout of butterfly MSSN

be rearrangeably non-blocking for multicast connections. For a statically configured device, such as an eFPGA, this property implies that all the connection patterns can be made without any congestion issues. Since butterfly topology features power-of-two I/Os and since each CLB requires 12+12 I/Os, this structure leaves empty 4 pins per CLB that can be used for primary eFPGA I/Os. As an example, for 16 CLBs, 192+192 I/O pins are required for CLBs connections and the remaining 64+64 for the primary eFPGA I/Os.

Two topologies have been analyzed in our exploration, a simple back-to-back butterfly and an enhanced version exploiting locality thanks to hierarchy-aware connectivity. As depicted in Figure 3, a folded back-to-back butterfly shows an intrinsic hierarchical structure in which each group H_i defines a non-blocking sub-network if a set of dedicated connections allows us to bypass upper levels of the hierarchy (e.g. in Figure 3, by the *U-turn* bypass). Figure 4 shows a possible hierarchical 2D layout of a butterfly folded network. This enhancement imposes extra pins and logic for each switch module, but it allows faster paths on near CLBs without compromising the routability. Such architecture is fully manageable through a standard implementation flow, since fully-synthesizable, under the following guidelines:

- Bit-cells arrays are implemented with latches and configured by 2 scan-chains (one for data word and one for rows write-enable). All the write enables should be managed as generated clocks, while the configuration sub-system is mostly asynchronous with respect to the functional clock driving CLB's flip-flops.
- Interconnect programmability and CLBs with potentially combinational outputs cause a combinational feedback, resulting in a loop. This condition is peculiar for the uncommitted device since all the bit-cells can be either 0 or 1. Such condition is managed during implementation "breaking" the loops by forcing all the CLBs to be configured as sequential (e.g. through *set_case_analysis* or *set_disable_timing* constraints).

Following this approach, a programmable device based on MSSN, be it bypassed or not, is made fully synthesizable. In a "brute-force" approach, timing constraints are referred to the longest paths (i.e. full latency through MSSN), while a multi-scenario analysis can be used to budget accurately all the levels of the hierarchy.

IV. EXPERIMENTAL RESULTS

The aim of this work is to explore the design space of an eFPGA soft-core template based on MSSN, analyzing the different area-speed-leakage trade-off points achievable. We take into account a test-case featuring 16 CLBs connected by a 256+256 I/Os MSSN. Implementation analyses have been performed using STM CMOS 65nm LP technology, starting from a Verilog RTL description. Synthesis trials have been performed with Synopsys Design Compiler Graphical [14] using standard cells libraries featuring transistors with 3 different Vt (High, Standard and Low Voltage Threshold - HVT, SVT, LVT). This approach allows us to work with real floorplan and physical coarse placement to quickly estimate parasitics on wires and potential congestion issues, thus improving the correlation with respect to place-and-route phase. In order to further verify the feasibility of the implementation, we realized complete place-and-route flows for some relevant test-cases, analyzing the results at sign-off level. Back-annotated simulations have been carried out to verify the constraints (e.g. loops breaks, false paths on asynchronous domains...) with some test-patterns.

Reference implementation is the SVT-only design summarized in Table 1, where frequency is the maximum achievable in synthesis. A verification place-and-route has been done at 200MHz, without reporting violations (timing and DRC) and with comparable area and power. Hence, our analysis covered implementation with different mix of cells:

- HVT-only, to minimize the leakage;
- HVT+SVT, a near-to-SVT option to reduce leakage without significant performance penalties;
- HVT+SVT+LVT, for top-speed applications.

For each test-case, we implemented up to 4 different syntheses, varying the target frequency up to their technology limit. In addition, each trial has area and leakage values

Table 1 SVT post-synthesis summary
16 CLBs with butterfly MSSN without bypass

Area	~0.3 mm^2
Frequency	248.7 MHz
Leakage	328 μW
% Sequential cells area	33.0 %
% Interconnect area	56.9 %
% Seq. cells on interconnect area	32.3 %

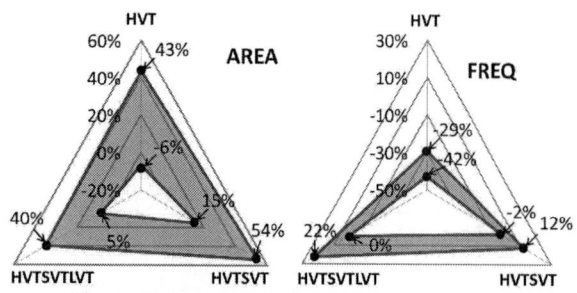

Figure 5 Area and Frequency ranges wrt Vt

minimized by constraints. As an educated guess, I/Os have been constrained with a delay of ~1/3 of the target clock period. Figure 5 shows the achieved design space in terms of area and frequency ranges. For example, HVT-only designs show an area ranging from -6.6% to +43% compared to SVT reference, while the corresponding frequency range decreases from -42.8% to -29.2%. Table 2 shows the leakage for the various implementations, reporting the correspondent *Vt*-mix. Compared to the SVT design, leakage power decreases by one order of magnitude with the HVT-only low-speed solution or increases up to ~400% leveraging high-speed cells utilization.

We have also evaluated a bypass-enhanced eFPGA, as in Figure 3, achieving 61% area and 72% leakage increases, without significant frequency penalties. To evaluate the bypass benefit on effective performance, we have implemented a test-case with 256 CLBs and 4096-points MSSN (for CLBs and I/Os). Bypass-enhancement needs to distinguish between implementation frequency and effective working frequency. The first one is related to the full-latency MSSN we use for implementation. The second one takes into account bypasses, which can provide faster paths to near CLBs, and thus depends on placement and design topology (i.e. maximum number of combinational CLBs). Although the brute-force approach we followed for physical synthesis showed some limitation especially for run-time, results achieved are interesting. Table 3 shows the effective working frequency considering the utilization of bypasses at a given hierarchy level and the number of cascaded combinational CLBs. Frequency is normalized with respect to the implementation frequency that is the working frequency without using bypasses (in this case, 125MHz, with 10% LVT cells area). Timings have been extracted configuring the MSSN through *case analysis* constraints. Table 3 shows that bypass exploitation allows performance easily improved by ~50-100% with respect to un-bypassed MSSN.

V. CONCLUSIONS

In this paper we presented a soft-core eFPGA template based on MSSN featuring butterfly topology. We proved its

Table 2 Leakage power design space summary

VT	Max Leak	% cells			Min Leak	% cells		
		H	S	L		H	S	L
H	-87.1%	100	/	/	-91.9%	100	/	/
H,S	57.1%	17.3	82.7	/	-1.5%	30.4	69.6	/
H,S,L	396.8%	17.3	26.5	56.2	46.8%	35.2	53.1	11.8

Table 3 Normalized frequencies with bypass exploitation

256 CLBs with folded MSSN featuring bypass

Bypass Level	# of cascaded combinational CLBs			
	1	4	5	10
1	3.21	0.80	0.64	0.32
3	2.27	0.57	0.45	0.23
5	2.02	0.50	0.40	0.20
8	1.53	0.38	0.31	0.15
no bypass	1	0.25	0.20	0.10

flexibility in terms of area-speed-power trade-off points implementing the device on STM CMOS 65nm technology and showing the existence of a significant design space that can be effectively explored. Results show that area can vary up to ~50% with respect to a simple SVT design, as well as the speed can range +/-30% leveraging on LVT cells. On the other hand, when leakage is an issue, this approach allows saving up to 90% with full HVT design. The same design space is available improving the network with bypass structures, which can improve the effective frequency of ~50%. When the device growth in size, some issues in terms of run-time don't allow to implement the device using a brute-force approach, requiring improvements on the implementation methodology.

Acknowledgment

Authors would like to thanks L. Cali', S. Pucillo, V. Nardone, R. Canegallo and P. Rolandi for their support. This work has been partially funded by STMicroelectronics.

References

[1] J. Rose, A. El Gamal, A. Sangiovanni-Vincentelli, "Architecture of field-programmable gate arrays," in *Proceedings of the IEEE*, 1993.

[2] G. Lemieux and D. Lewis, "Circuit design of routing switches," *Proc. ACM Int'l Symposium on FPGA*, Feb. 2002.

[3] G. Lemieux, E. Lee, M. Tom, A. Yu, "Directional and single-driver wires in FPGA interconnect," *IEEE Int'l Conf. on FPT*, Dec. 2004.

[4] V. Betz, J. Rose, A. Marquardt, "Architecture and CAD for deep-submicron FPGAs", Kluwer Academy Publishers, Feb. 1999.

[5] W. Tsu et al., "HSRA: High-speed, hierarchical synchronous reconfigurable array," *Proc. ACM Int'l Symposium on FPGA*, Feb. 1999.

[6] I. Kuon, R. Tessier, J. Rose, "FPGA architecture: survey and challenges", Foundations and Trends in EDA, Feb. 2008.

[7] A. DeHon, "Balancing interconnect and computation in a reconfigurable computing array" *Proc. of the Int'l Symposium on FPGA*, 1999.

[8] F. Reblewski, O. Lepape, "Reconfigurable integrated circuit with a scalable architecture", U.S. Patent n. 6594810, Filed Oct. 4, 2001.

[9] D. Wong, "Interconnection network for a field programmable gate array", U.S. Patent n. 6693456, Filed Aug. 3, 2001.

[10] C. Wang et al., "A 1.1 GOPS/mW FPGA chip with hierarchical interconnect fabric", *IEEE Int'l Symposium on VLSI Circuits*, 2011.

[11] W. Dally, B. Towles, "Principles and practice of interconnection networks", Morgan Kaufmann Ed., Jan. 2004.

[12] J. Rose et al. "The VTR project: architecture and CAD for FPGAs from verilog to routing," in *Proc. ACM Int'l Symposium on FPGA*, Feb. 2012.

[13] M. Hutton et al. "Improving FPGA Performance and Area Using an Adaptive Logic Module", *Proc. Int'l Conference on FPL*, May 2006.

[14] Synopsys. *Design CompilerTool* in *Graphical mode*. http://www.synopsys.com/ .

Direct Digital Frequency Synthesizers implemented on high end FPGA devices

Mariangela Genovese, Ettore Napoli

Dept. of Electrical Engineering and Information Technology
University of Napoli Federico II, Napoli, Italy
Email: mariangela.genovese@unina.it; ettore.napoli@unina.it

Abstract—Direct Digital Frequency Synthesizer (DDFS) circuits are routinely implemented in many electronic systems. Advanced DDFS design techniques have been proposed and optimized for ASIC (Application Specific Integrated Circuits) implementations. Nowadays, FPGA devices are frequently chosen as target for digital circuits. This paper presents the FPGA implementation of state of the art DDFS architectures and compares their performance providing hints on optimal design as a function of the chosen performance parameter.

Keywords—*Direct Digital Frequency Synthesizer, Field programmable gate arrays, Application specific integrated circuits.*

I. INTRODUCTION

Modern communication systems, mixers, modulators, and measurement instruments require DDFS circuits able to generate single-phase or quadrature sinusoids with excellent frequency resolution, good spectral purity, very fast frequency switching and phase continuity on switching, [1].

Several DDFS architectures have been proposed to date, [2]. The majority of these architectures has been implemented using state of the art ASIC technologies, [3]-[26].

Nowadays, FPGA devices are always more frequently chosen as target technology for digital circuits since they provide a fast time to market, are reprogrammable, and are available in a wide range of performance and cost. Moreover, FPGA devices are implemented in state of the art ASIC technologies that are not available to most designers.

The scientific literature focuses on the performance of the DDFS when implemented using standard cell ASIC technologies. The study of the literature is therefore useless for the selection of the best performing DDFS architecture when the FPGA is the design target since not only the absolute performance can vary but the performance trend can be different.

This paper presents a detailed analysis of the performances provided by the state of the art DDFS architectures when implemented on FPGA devices.

II. DDFS ARCHITECTURES

The basic architecture of the DDFS circuits is shown in Fig. 1. The 'Phase Accumulator' unit accumulates the Frequency Control Word (FCW) signal to produce a digital sweep with a slope imposed by the FCW value. The 'Phase Accumulator' is allowed to freely overflow and wrap around and each overflow corresponds to a sinusoidal signal period. The output of the 'Phase Accumulator' feeds the address line

of a $2^N \times S$ location ROM ('Phase to Sine Mapper') storing the amplitude values of the output signal. The frequency of the output sinusoidal waveform is:

$$f_{out} = \text{FCW}\frac{f_{Clk}}{2^N} = \text{FCW}f_{res} \qquad (1)$$

where N is the FCW word length, f_{Clk} is the frequency of the input clock signal *Clk*, and $f_{res}=f_{Clk}/2^N$ is the frequency resolution of the DDFS. Note that f_{res} improves with N but the ROM size for the 'Phase to Sine Mapper' also grows with N.

Several ROM compression methods have been proposed during the years, [2] in order to reduce the ROM size while keeping a good output spectral purity, [27]. The parameter that characterizes DDFS output spectral purity is the spurious-free dynamic range (SFDR) defined as the ratio (in dBc) of the amplitude of the fundamental frequency to the amplitude of the highest undesired frequency component.

The truncation of the 'Phase to Sine Mapper' input signal to P bits with P<N reduces the ROM size to $2^P \times S$ bits while spurious noise is introduced in the DDFS output, so that the P value is usually chosen according to the required SFDR, [27].

Quadrant compression technique allows a further shrinking of the ROM by storing the sine amplitude values for angles in $[0, \pi/2)$ and exploiting the trigonometric functions to generate the sine values for the full range $[0, 2\pi)$ of the input phase. Quadrature DDFS (that generate both sine and cosine), exploit the eighth-wave symmetry of sine and cosine to store the waveforms for angles belonging to $[0, \pi/4)$.

Even after performing phase truncation and quadrant compression, the ROM size is usually large. Several alternative approaches for the implementation of the 'Phase to Sine Mapper' have been proposed, [3]-[26].

III. PERFORMANCE PARAMETERS AND DESIGN PROCEDURE FOR THE FPGA IMPLEMENTATIONS

In this paper several DDFS architectures are implemented and compared by fixing the target SFDR value (considered values are 60, 80, 100, and 120dBc) and evaluating maximum working frequency, dynamic power dissipation, and logic resource occupation.

Selected devices are the StratixIV-GX230KF40C2 (Altera), and the Virtex5-lx50 (Xilinx) FPGA.

Synthesis and Place and Route have been conducted by using ISE (Xilinx) and Quartus (Altera) while Modelsim PE

Paper M1C3

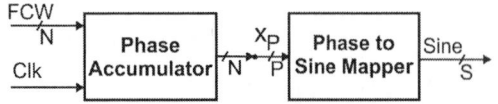

Fig. 1. Simplified architecture of a single phase DDFS circuit.

has been employed to simulate the circuits and to generate '.vcd' files used for the power analysis. The power analysis has been conducted by using XPower Analyzer for Xilinx and PowerPlay Power Analyzer for Altera.

DDFS circuits have been implemented with a latency of four clock cycles and a structure similar to the one of Fig. 2.

The SFDR has been obtained simulating a whole numerical period of the output waveforms choosing the FCW that maximizes the phase error. The spectrum of the output waveform has then been analyzed obtaining the actual SFDR value. Note that the resulting SFDR is usually higher than the imposed target due to the limited flexibility of most design techniques.

IV. ROM BASED DDFS

The straightforward implementation of a quadrature ROM-based DDFS is obtained by storing in a ROM the amplitude values of an entire period of sine and cosine signals.

ROM-based DDFS (RD) circuits have been designed for 60, 80, 100, and 120 dBc SFDR values. The word lengths P and S for input and output of the 'Phase to Sine Mapper' have been chosen referring to DDFS theory, [8],[28], and imposing a minimum numerical period of 64 samples. The chosen P and S values are reported in Table I.

As visible in Table I the required memory exponentially increases with the target SFDR. Actually, due to the large ROM requirements both the Virtex5 (XC5 in the following) and StratixIV FPGA (EP4 in the following) can only fit 60dBc and 80dBc designs. An RD DDFS with SFDR of 100 dBc could be implemented on selected EP4 by using the majority of the memory resources of the FPGA.

RD implementations use the blocks of RAM (BRAM) embedded on the FPGA devices. Table I also reports the BRAM usage, the frequency and the power dissipation of the implemented RD circuits on XC5 and EP4 FPGA.

The BRAM blocks for XC5 have size and maximum working frequency of 36Kb and 550MHz, respectively. EP4 FPGA has 14283Kb of embedded memory organized in two type of memory blocks, M9K and M144K, operating at up 600 MHz and with size 9Kb and 144Kb, respectively. The

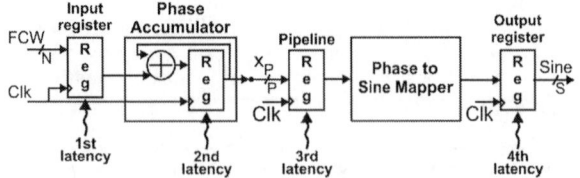

Fig. 2. Distribution of the registers responsible of the clock latency in the implemented DDFS circuits.

frequency of the RD DDFS is very high and close to the BRAM frequency for low SFDR values, Table I.

The memory requirement exponentially increases with the target SFDR. The result is a steep increase of power dissipation as a function of the SFDR. As an example, for the XC5 implementation, the power dissipation at 60dBc and 80dBc is 56.2μW/MHz and 424.2μW/MHz, respectively.

A. Quadrant compression

The main advantage of quadrant compression techniques is the reduction of memory usage of about one-eighth. The technique also allows a reduction of the power dissipation. The reduction of memory requirements allows the implementation of ROM-based DDFS circuits with SFDR higher than 80dBc on both XC5 and EP4 FPGA. As an example, Table I reports the implementation of a quadrant compression DDFS (RD(QC)) with SFDR of 100dBc.

The disadvantage of quadrant compression techniques is the use of further combinatorial logic in addition to the ROM that reduces the maximum working frequency of the circuit. As an example, as show Table I, when an 80dBc SFDR is considered, a XC5 implementation of a RD(QC) DDFS with respect to the RD circuit reduces the power dissipation and the frequency by 67% and 35%, respectively.

V. STATE OF THE ART DDFS TOPOLOGIES

A. ASIC implementations

Various DDFS implementation techniques that improve the performances of the RD DDFS when implemented in standard cell ASIC technologies have been proposed in the literature. Among these, top notch performances are provided by [8], [16], and [19] in which the implementation results refer to a 0.25μm ASIC technology. The performances taken from the above cited papers are reported in Table II.

The optimized Multipartite Table (MTM) architectures of [8] are implemented by using 3 Table of Offset (MTM1, MTM2) and 4 Table of Offset (MTM3, MTM4). MTM1 and MTM3 are designed for 80dBc SFDR while MTM2 and MTM4 are designed to achieve an SFDR equal to 100dBc.

PL1, PL2, PQ1, and PQ2 architectures implement the Piecewise-Linear (PL) and Piecewise-Quadratic (PQ) polynomial approximation techniques of [16]. PL architectures have been implemented for SFDR values of 60dBc and 80dBc while PQ circuits have been implemented for 100dBc, and 120dBc SFDR.

The architecture indicated with DS is Dual Slope DDFS of [19] designed for an 80dBc SFDR.

B. FPGA implementations

In the proposed paper, all the architectures reported in Tab. II have been implemented on FPGA devices. Selected FPGA devices for the implementations are XC5 and EP4.The occupation of logic resources is reported in Table III. The frequency and power are reported in Fig. 3.

It is worth highlighting that the DS architecture is the only one taken from the literature that implements the pipeline

978-1-4673-4579-8/13 $31.00 © 2013 IEEE   138

TABLE I. PERFORMANCE AND CHARACTERISTICS FOR ROM BASED DDFS. NO ROM REDUCTION TECHNIQUES ARE IMPLEMENTED.

SFDR (dBc)	circuit	P (# bit)	S (# bit)	memory (kb)	BRAM (#)	Frequency (MHz)	Dynamic Power (μW/MHz)
60	RD	11	8	32	1 (XC5); 4 M9K (EP4)	550.1 (XC5); 600.2 (EP4)	56.2 (XC5); 66.8 (EP4)
80	RD	14	11	352	11 (XC5); 3 M144K (EP4)	505.3 (XC5); 496.0 (EP4)	424.2 (XC5); 153.6 (EP4)
80	RD(QC)	14	11	352	2 (XC5); 6 M9K (EP4)	326.2 (XC5); 228.3 (EP4)	138.2 (XC5); 97.0 (EP4)
100	RD(QC)	17	14	3584	14 (XC5) ; 4 M144K (EP4)	214.0 (XC5); 238.1 (EP4)	568.0 (XC5); 222.2 (EP4)

TABLE II. PERFORMANCES OF THE ASIC IMPLEMENTATIONS (CMOS 0.25μm) OF THE STATE OF THE ART DDFS TAKEN FROM THE LITERATURE

Reference	Architecture	Accumulator/Phase/Output (# bit)	SFDR (dBc)	Pipeline levels	Maximum Frequency (MHz)	Dynamic Power (μW/MHz)	Silicon Area ($10^3 \mu m^2$)
[8]	MTM1	24 / 14 / 11	80.5	0	251	39.1	22.2
Multipartite	MTM2	24 / 18 / 16	100.6	0	200	60.6	36.7
tables	MTM3	24 / 14 / 11	80.8	0	250	43.2	23.8
	MTM4	24 / 18 / 16	100.7	0	201	61.7	36.1
[16]	PL1	24 / 11 / 9	60.0	0	224	39.3	16.8
Piecewise	PL2	24 / 14 / 12	80.2	0	166	57.2	23.6
polynomial	PQ1	24 / 21 / 19	121.4	0	101	137.9	64.3
	PQ2	24 / 18 / 16	104.5	0	118	98.4	40.5
[19] Dual slope	DS	24 / 14 / 12	83.2	6	600	127.0	90.0

registers in order to increase the frequency, Table II. In order to perform a fair comparison, the DS architecture has been modified and implemented eliminating 3 clock cycles of latency to obtain a structure similar to Fig. 2.

The first result that can be devised from the FPGA implementations is that, on both XC5 and EP4 devices, the trend of circuit performances is similar to what is known for the ASIC implementations of Table II even if some noticeable variations can be devised.

Among the DDFS circuits with 80 dBc SFDR the MTM circuits are the less power hungry circuits while, oppositely to what observed for the ASIC implementation, when implemented on XC5, the PL2 circuit provides higher frequency than the MTM circuits. When compared against the MTM1, MTM3, and PL2 architectures, the working frequency of the DS architecture is higher for both XC5 and EP4 implementations but the power dissipation is higher, too.

With reference to the logic resource occupation, it is still valid that the MTM architectures require less resources while the DS architecture use more logic and, on XC5, an additional BRAM block.

It can be concluded that, at 80dBc, the DS implementation, on both XC5 and EP4 FPGA is the preferred choice when the frequency is the main target. The MTM architectures provide the best performance in terms of power and logic resource occupation. The behavior of the PL2 architecture is dependent on the chosen FPGA device but is never the best choice.

When a 100dBc SFDR value is required, the MTM architectures represent the best choice since they provide higher frequency, and lower power dissipation and area utilization than PQ2. This is valid for both XC5 and EP4 implementations.

The PQ1 is the only circuit designed for 120dBc SFDR and, as expected, is the slowest and the least energy efficient circuit.

The PL1 is the only circuit designed for 60dBc SFDR. In agreement with the lower SFDR value the PL1 circuit is

the fastest among the considered DDFS, however the energy dissipation of the circuit is similar to the MTM1 at 100dBC.

C. Comparing the state of the art implementations with RD circuits

The second result presented in the paper is obtained comparing the RD implementations of Table I with the implementations of the advanced architectures whose performance are in Fig. 3.

Differently from what reported in the scientific literature oriented to ASIC implementations, the ROM-based DDFS (RD DDFS) provide, for a given SFDR value, the highest working frequency. The result is explained from the tendency of RD DDFS to exploit the optimized BRAM blocks available in modern FPGA devices that also work at the highest frequency available for the FPGA devices. The advantage of the high working frequency is contrasted by the high values of power dissipation and by the fact that high SFDR implementations are not possible due to the exponential increase of the number of occupied BRAM blocks.

VI. CONCLUSION

The paper has presented various DDFS architectures implemented on XC5 and EP4 devices. Considered architectures are conventional ROM based (RD) and advanced, high performance architectures taken from the scientific literature (MTM, PL, PQ, and DS).

On XC5, at 60dBc the best performing architecture is the RD. At 80dBc the RD architecture has still the highest frequency while minimum power dissipation is obtained by using the MTM1 circuit. At 100dBc the MTM2 is the best architecture in terms of both power dissipation and frequency.

On EP4 the RD architectures provides the highest frequency at 60 and 80dBc while the minimum power dissipation is given by the PL1 and MTM1 architectures, respectively. When the SFDR of 100dBc is required the architectures MTM and PQ2 present the best performances in power and frequency, respectively.

978-1-4673-4579-8/13 $31.00 © 2013 IEEE

TABLE III.   RESOURCE UTILIZATION FOR THE IMPLEMENTED DDFS.

SFDR	Architecture	XC5 (Virtex5-lx50)		EP4 (StratixIV-GX230)	
		Slice	DSP	ALM	DSP
60dBc	PL1	38	0	79	0
80dBc	MTM1	38	0	82	0
	MTM3	52	0	83	0
	PL2	47	0	132	0
	DS	58	0	123	0
100dBc	MTM2	53	0	154	0
	MTM4	67	0	146	0
	PQ2	45	2	306	0
120dBc	PQ1	57	5	399	1

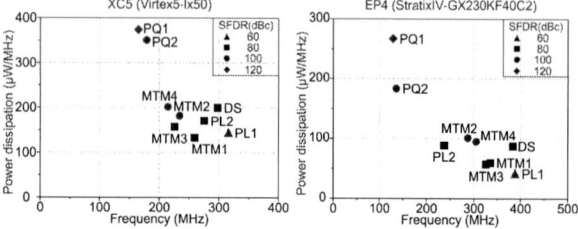

Fig. 3.   Performances of the DDFS architectures of [8], [16], and [19] when implemented on Virtex5 (lx50) FPGA and StratixIV (GX230KF40C2) FPGA.

REFERENCES

[1]   J. Tierney, C.M. Rader, and B. Gold, "A digital frequency synthesizer," in IEEE Trans. Audio Electroacoustics, vol.AU-19, no.1, 1971, pp.48-57.

[2]   J.M.P. Langlois, and D. Al-Khalili, "Phase to sinusoid amplitude conversion techniques for direct digital frequency synthesis," Inst. Proc. Elect. Eng. Circuits Devices Syst., vol.151, no. 6, pp. 519-528, 2004.

[3]   B. H. Hutchinson : "Contemporary frequency synthesis techniques," in Gorski-Pcpicl, J. (Ed.): "Frequency synthesis: techniques and applications" (IEEE Press, 1975), pp. 25-45.

[4]   D.A. Sunderland, R.A. Strauch, S.S. Wharfield, H.T. Peterson, and C.R. Cole: "CMOS/SOS frequency synthesizer LSI circuit for spread spectrum communications," IEEE J. Solid-State Circuits, 1984, vol.19, pp.497-505

[5]   H. T. Nicholas, and H. Samueli, "A 150-MHz direct digital frequency synthesizer in 1.25-micron CMOS with 90dBc spurious-free dynamic range," IEEE J. Solid-State Circuits, vol.26, no.12, pp.1959-1969, 1991.

[6]   F. De Dinechin, and A. Tisserand, "Multipartite table methods," IEEE Trans. Comput., vol.54, no. 3, pp. 319-330, Mar. 2005.

[7]   A.G.M. Strollo, D. De Caro, N. Petra, "A 630 MHz, 76 mW Direct Digital Frequency Synthesizer Using Enhanced ROM Compression Technique," Solid-State Circuits, IEEE Journal of , vol.42, no.2, pp.350,360, Feb. 2007 doi: 10.1109/JSSC.2006.889382

[8]   D. De Caro, N. Petra, and A.G.M. Strollo: "Reducing Lookup-Table Size in Direct Digital Frequency Synthesizers Using Optimized Multipartite Table Method," IEEE Trans. on Circuits and Systems I, vol.55, no. 7, pp. 2116-2127, 2008, doi: 10.1109/TCSI.2008.918008.

[9]   D. De Caro, E. Napoli, and A.G.M. Strollo, "Direct digital frequency synthesizers with polynomial hyperfolding technique," IEEE Trans. Circuits Syst. II, Exp. Briefs, vol.51, no. 7, pp. 337-344, Jul. 2004, doi: 10.1109/TCSII.2004.829553.

[10]   Y. H. Chen, and Y. A. Chau, "A direct digital frequency synthesizer based on a new form of polynomial approximations," IEEE Trans. Consum. Electron., vol.56, no. 2, pp. 436-440, May 2010.

[11]   D. De Caro, E. Napoli, and A.G.M. Strollo, "ROM-less direct digital frequency synthesizers exploiting polynomial approximation," Int. Conf. on Electronics Circuits and Systems, 2002, vol.2, pp. 481-484, doi: 10.1109/ICECS.2002.1046202.

[12]   A.G.M. Strollo. E. Napoli, and D. De Caro, "Direct Digital Frequency Synthesizers using First-Order Polynomial Chebyshev Approximation," in Proc. Eur. Solid State Circuits Conf. (ESSCIRC), 2002, pp. 527-530.

[13]   A.G.M. Strollo, E. Napoli, and D. De Caro, "Direct digital frequency synthesizers using high-order polynomial approximation," in Proc. IEEE Int. Solid-State Circuits Conf., (ISSCC'02), vol.1, Feb. 2002, pp.134-135, doi: 10.1109/ISSCC.2002.992972.

[14]   J. M. P. Langlois and D. Al-Khalili, "Novel approach to the design of direct digital frequency synthesizers based on linear interpolation," IEEE Trans. Circuits Syst. II, Analog Digit. Signal Process., vol.50, no. 9, pp. 567-578, Sep. 2003.

[15]   A. Ashrafi, R. Adhami, and A. Milenkovic, "A direct digital frequency synthesizer based on the Quasi-linear interpolation method," IEEE Trans. Circuits Syst. I, vol.57, no.4, pp.863-872, 2010.

[16]   D. De Caro, and A.G.M. Strollo, High-performance direct digital frequency synthesizers using piecewise-polynomial approximation," IEEE Trans. Circuits Syst. I, Reg. Papers, vol.52, pp. 324-337, Feb. 2005, doi: 10.1109/TCSI.2004.841592.

[17]   A. Ashrafi, and R. Adhami, "Theoretical upperbound of the spurious free dynamic range in direct digital frequency synthesizers realized by polynomial interpolation methods, IEEE Trans. Circuits Syst. I, Reg. Papers, vol.54, no. 10, pp. 2252-2261, Oct. 2007.

[18]   D. De Caro, N. Petra, and A.G.M. Strollo, "Direct Digital Frequency Synthesizer Using Nonuniform Piecewise-Linear Approximation," IEEE Trans. on Circuits and Systems I, Oct. 2011, vol.58, no. 10, pp. 2409-2419, doi: 10.1109/TCSI.2011.2123730.

[19]   D. De Caro, and A.G.M. Strollo, "High performance direct digital frequency synthesizers in 0.25 μm CMOS using dual-slope approximation," IEEE J. Solid-State Circuits, vol.40, no. 11, pp. 2220-2227, Nov. 2005, doi: 10.1109/JSSC.2005.857371.

[20]   J.M.P. Langlois, and D. Al Khalili, "Low power direct digital frequency synthesizer in 0.18 μm CMOS," in Proc. Custom Integr. Circuits Conf., Sep. 2003, pp. 2124.

[21]   A.G.M. Strollo, D. De Caro, E. Napoli, and N. Petra, "Direct digital frequency synthesis with dual slope approach", in Proc. Eur. Solid State Circuits Conf. (ESSCIRC 2003), Sep. 2003, pp. 397-400, doi: 10.1109/ESSCIRC.2003.1257156.

[22]   A.G.M. Strollo, D. De Caro, E. Napoli, and N. Petra, "High-speed direct digital frequency synthesizers in 0.25-μm CMOS," Proc. of Custom Integrated Circuits Conference, 2004, pp.163-166, doi: 10.1109/CICC.2004.1358764.

[23]   Y. Song, and B. Kim, "Quadrature direct digital frequency synthesizers using interpolation-based angle rotation," IEEE Trans. Very Large Scale Integr. (VLSI) Syst., vol.12, no. 7, pp. 701-710, Jul. 2004.

[24]   A. Torosyan, D. Fu, and A. N. Willson Jr., "A 300-MHz quadrature direct digital synthesizer/mixer in 0.25- μm CMOS, IEEE J. Solid-State Circuits, vol.38, no. 6, pp. 875-887, Jun. 2003.

[25]   D. De Caro, N. Petra, and A.G.M. Strollo, "A 380 MHz direct digital synthesizer/mixer with hybrid CORDIC architecture in 0.25 μm CMOS," IEEE J. Solid-State Circuits, vol.42, no. 1, pp. 151-160, Jan. 2007, doi: 10.1109/JSSC.2006.886527.

[26]   J.M.P. Langlois, and D. Al-Khalili, "Hardware optimized direct digital frequency synthesizer architecture with 60 dBc spectral purity," IEEE International Symposium on Circuits and Systems, 2002, vol.5, pp. V361-V364, 2002.

[27]   B. G. Goldberg, Direct Digital Frequency Synthesis Demystified. Eagle Rock, VA: LLH Technol. Publ., 1999

[28]   D. De Caro, "Architetture innovative per la sintesi digitale diretta di frequenza", Chapter 2, 2002.

[29]   N. Petra, D. De Caro, V. Garofalo, E. Napoli, A. G. M. Strollo, "Truncated Binary Multipliers With Variable Correction and Minimum Mean Square Error," IEEE Transactions On Circuits And Systems-I: Regular Papers, vol.57, no.6, June 2010, pp.1312-1325, doi: 10.1109/TCSI.2009.2033536.

[30]   N. Petra, D. De Caro, V. Garofalo, E. Napoli, A. G. M. Strollo, "Design of Fixed-Width Multipliers With Linear Compensation Function," IEEE Transactions on Circuits and Systems I: Regular Papers, vol.58, no.5, May 2011, pp.947-960, doi: 10.1109/TCSI.2010.2090572.

[31]   V. Garofalo, N. Petra, E. Napoli, "Analytical Calculation of the Maximum Error for a Family of Truncated Multipliers Providing Minimum Mean Square Error," IEEE Transactions on Computers, vol.60, no.9, Sept. 2011, pp.1366-1371, doi: 10.1109/TC.2010.236.

978-1-4673-4579-8/13 $31.00 © 2013 IEEE

Gap in pagination due to withheld paper.

Pages 141-144

Random Interleaved Pipeline Countermeasure Against Power Analysis Attacks

Renato Menicocci
Fondazione Ugo Bordoni
Roma, Italy
rmenicocci@fub.it

Alessandro Trifiletti and Francesco Trotta
Dipartimento di Ingegneria dell'Informazione, elettronica e Telecomunicazioni (DIET)
Sapienza Università di Roma, Italy
trafiletti@die.uniroma1.it, trottaf@die.uniroma1.it

Abstract—**An RTL countermeasure intended to protect the AddRoundKey and SubByte steps of the AES algorithm against DPA or CPA attacks has been proposed and tested on an AES encoding coprocessor implemented on FPGA. Experimental results based on first order CPA attacks confirmed the effectiveness of the proposed countermeasure, especially in protecting the SBOX output, showing that even with the acquisition of 300000 power curves, the absolute value of correlation function is embedded in the measured noise floor and there are no peaks able to reveal the encryption key.**

Index Terms—**Side Channel Attack, DPA, CPA, RTL countermeasure, AES, FPGA.**

I. INTRODUCTION

Differential Power Analysis Attacks [1] are a serious threat to security of smart-card and other cryptographic processors, because they allow to violate the secrets hidden in such devices, by exploiting their power consumption characteristics rather than mathematical properties of cryptographic algorithms. Power analysis attacks are non-invasive and can be performed with simple and low-cost equipment. To counteract them, several types of countermeasures have been invented. The goal of a countermeasure against power analysis is to make the power consumption of a cryptographic device independent from the input data and the intermediate values of the cryptographic algorithm. Some countermeasures use special logical styles, such as, for example, SABL [2] that makes the power consumption of the single logic gates almost uncorrelated with processed data. Other approaches use particular architectural arrangements, for example some countermeasures are based on the so called "masking" [3] that through operations involving random-generated data, make the power consumption of the entire encryption processor uncorrelated with the processed data. Also algorithmic countermeasures are available, i.e. special algorithms designed to conceal the secrets contained in the cryptographic devices [4]. All these methods are expensive in term of chip area and,

in the case of logical styles, are not implementable in programmable logics. An alternative is the implementation of countermeasures at register transfer level. Among this type of countermeasure a couple of examples is represented by the random data insertion and the random interleaved pipeline [5] [6]. Also in these cases there is a random value that is internally generated, but it is not processed by the logics that implement the algorithm's operations, but only inserted in the cryptographic coprocessor's pipeline, in the aim to add noise.

II. ABOUT DPA AND CPA

Differential Power Analysis (DPA) and Correlation Power Analysis (CPA) are currently regarded as a major danger to the security of cryptographic devices as Smart Cards and the development of countermeasures against them is the subject of a great number of research works. Common electronic devices leak information on processed data through their power consumption, i.e. through the current absorbed by the ground terminal or the power supply terminal. Most digital circuits, including cryptographic devices use CMOS technology. A CMOS gate absorbs current only in the presence of a switching output (neglecting the sub-threshold current) and the absorption peak is greater the greater is the load capacity. As shown in Figure 1, the inverter, the simplest CMOS cell and thus the best example, absorbs power when the output has a transition to a new logic state: if the transition is from 0 to 1, the output is loaded at the same voltage of the power supply bar, if the transition is 1-0, the capacity is discharged by dissipating the energy previously stored during the charging, while in the other two situations, 0-0 and 1-1, there is no consumption of energy because none of the two transistors becomes conductive. This provides useful information for the purposes of power analysis, because from each absorption peak can be deduced that a transition occurs in the output's logic state, while in the case of absence of peaks it can be concluded that the output has not undergone any transition. Regarding the mathematical processing involved in DPA, the

978-1-4673-4579-8/13 $31.00 © 2013 IEEE

reader should refer to the work of Kocher et al [1]. CPA is based on the correlation between the power consumption traces and the data processed by the device, using the correlation between the consumption W of the device and its model of energy consumption F. The most widely used model is given in linear form as Hamming weight H (or the Hamming distance) of the intermediate data. The correlation factor between W and F is proportional to the correlation factor between W and H. Let be, moreover, the k-th key Kk, the value Hk of H assumed for the k-th key, the correlation factor of the CPA is given by the following formula [7]:

$$\rho_{WH,k}(B) = \frac{E(WH_k) - E(W)E(H_k)}{\sigma_W \sigma_{H_k}} \qquad (1)$$

Where $E(W)$, $E(H\kappa)$, $E(WH\kappa)$ are respectively the expected values of W, $H\kappa$, and $WH\kappa$, while σW and σHk are the variances of W and Hk. The choice of the number of points, in both CPA and DPA, depends on various factors, such as clock frequency, instrument resolution, acquisition time (a larger number of samples involves a greater expenditure of time).

III. AES Algorithm

Nowadays, Advanced Encryption Standard (AES) is the most popular cryptographic algorithm. It is a block cipher, ie the data to be coded are processed in blocks of fixed size. The simplest version uses 128-bit data blocks with 128-bit key. More powerful versions use data sizes of 192 or 256 bits, with keys of the same size. The 128-bit version it is based on 11 round transformations, R0, R1, . . . , R10, each transforming a 128-bit input into a 128-bit output under the control of a 128-bit round key (192 and 256-bit versions use a greater number of rounds). The input at R0 is the plaintext, and the output from round R10 is the cyphertext. Each round from R1 to R10 takes as input the output from the previous round and uses a so called round key. Out of the 11 round keys 10, RK1,...RK10 are generated from the external key K by a specific algorithm (Key Expansion), while RK0 is the external key. The first round, R0 performs a mere XOR operation of the 16 bytes of P with the corresponding 16 bytes of RK0 (the step that performs the XOR operation is called *AddRoundKey*). The rounds from R1 to R9, instead, operate as follows: each of the 16 input bytes is mapped into a new value by an invertible function called SBOX (*SubBytes*), this is the only non-linear function in the AES algorithm and it ensures the encoding effectiveness. Then the new 16 bytes are permuted according

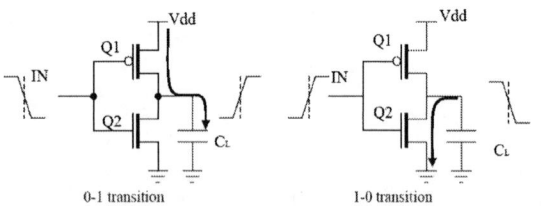

Figure 1. Output state transition and power consumption of the CMOS inverter.

to a fixed rule called *ShiftRows*, subsequently, the 16 bytes are divided into four columns of four bytes, and for each byte of any column, a new value is computed by a specific linear combination, over GF(2^8), of the four bytes in the column *(MixColumns)*. These 16 bytes are finally XORed with the corresponding 16 bytes of the specific round key (*AddRoundKey*). R10 differs in that the MixColumns step is omitted. The decryption algorithm works in reverse, and it is based on the inverse version of the operations SubByte (based on the inverse SBOX) ShiftRows and MixColumn. For further information on AES, see the article by J. Daemen. and V. Rijmen [8].

IV. Experimental Setup

The encryption coprocessor to be tested has been synthesized in an Altera Cyclone FPGA, along with additional functional blocks managing the input-output operations. The FPGA power supply is separated from the board main power supply, so it is possible to measure the power absorption of the FPGA without noise due to the other on-board components. This replicates the situation of an attack on a smart-card which is a mere chip, not soldered to any other component. For putting into practice the experiment were needed the FPGA board, a personal computer running the LabView development environment, a digital oscilloscope, an inductive current probe, through which passes the wire of the positive supply of the FPGA, and a regulated linear power supply. In the experiments that will be described below, each trace contains 500 samples. For each experiment 300000 traces were collected and a first order CPA analysis based on the Hamming weight was performed. The targets of the attacks were the output of the XOR between the input plaintext and the key in the first round (R0) of the AES algorithm and the output of the second round (R1) SBOX. Experimental results will be described in the following paragraphs.

V. Proposal for a DPA/CPA RTL Countermeasure

Here a simple RTL countermeasure against DPA/CPA working both on AddRoundKey and SubByte steps is proposed. This countermeasure is easy to apply in a pipelined architecture. An idea for limiting the dependence of the current absorbed by the AddRoundKey step from processed data may consist in the duplication of the combinatory logic (namely the XOR gates) in such a way that the output of each bit of the logic is the negated of the output of the other one. Then also the fan-out of the two logics will be duplicated. This design is inspired by that of dual rail logics, even if, in this case, the implementation is at register transfer level instead of gate level. A preliminary way for implement such concept may be that shown in figure 2. In this implementation both AddRoundKey and SubByte blocks are duplicated. A more feasible implementation, avoiding the duplication of entire functional blocks is shown in figure 3. This implementation has a further benefit due to random data insertion in the pipeline, this is inspired by the countermeasure described by Bucci et al. in [5]. In addition to the duplicated XOR gates there are two couples of cascaded registers, one of them has as input the negated random data, while the other has the same

PRIME 2013, Villach, Austria　　　　　　　　　　　　　　　　*Session M2C – Digital Techniques 1*

data but asserted. In this way each register of the pair contains the negated content of the corresponding register of the other pair. The registers are intended to add load capacity, in so that the two branches, direct and negated, have the same fan-out. This is much more practical than duplicating the *SubByte* stage. The cascaded registers make that the real data and random ones are processed simultaneously, making it difficult to reconstruct, through the absorption of a single register, the intermediate data. At downstream of the registers there is a multiplexer that balances the fan-out of each of the two output registers of the two pairs. In this way also the SubByte stage is countermeasured. Applying the trick of the cascaded registers to the entire pipeline allows to obtain the interleaved pipeline countermeasure described in [5] of which this one represents an enhanced version. The (pseudo) random data obtained with a simple linear feedback shift register are inserted through a multiplexer in each of the pipeline stages. In this way the countermeasure acts not only on the Subbyte stage but also on the successive stages which are also vulnerable to an attack toward the SBOX since they only perform a mix up of the bytes coming out of the SBOX, via the ShiftRows and MixColumn operations, but without changing them. The only drawback of the cascaded registers, besides of occupied area increase, is the need of a doubled clock frequency.

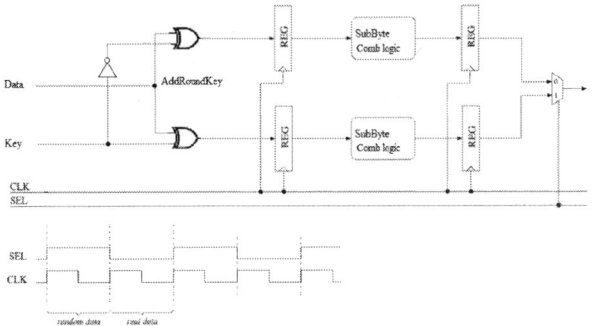

Figure 2. A preliminary concept of the proposed countermeasure, in such a way that are duplicated both the AddRoundKey and SubByte blocks.

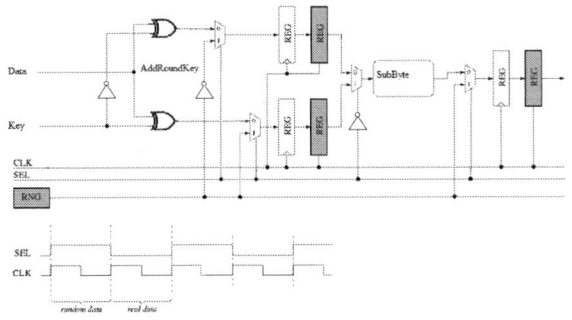

Figure 3. Proposed AddRoundKey countermeasure, as described in the text, the AddRoundKey step is merely a XOR operation between plaintext data and the internal key, here it is duplicated and each XOR gate is followed by a multiplexer and the cascaded registers. Other pairs of cascaded registers replace the single accumulator registers of the non countermeasured version of the pipeline.

VI. EXPERIMENTAL RESULTS

For the experiments here described a pipelined 128-bit AES encoder was used [9]. The pipeline has 4 stages, each performing one of the four AES elementary operations, namely AddRoundKey (XOR operation), SubByte (SBOX), ShiftRows and MixColumns. The same pipeline executes all the 11 rounds iteratively, each time taking as input the output of the previous round. Two CPA attacks were conducted, one on the original version of the AES encoder and another on a countermeasured version of the same encoder, implementing the countermeasure visible in figure 3. Both experiments required the acquisition of 300000 power traces. In figures 4 and 5 the correlation analysis for non countermeasured and countermeasured version is shown. In Figure 6, furthermore, a comparison between the area occupation of the two versions is shown, note how the addition of the countermeasure results in an increase of area occupation but this can be considered acceptable. With the same synthesis settings and the same peripheral functional blocks, the countermeasured core required 2692 so called "logic elements" (LE), compared with 1524 LE required by the original version. The most noticeable improvement involved the SBOX, implemented by the SubByte stage. In the former case the curve corresponding to the correct key stands out clearly on the others, while in the latter case this curve is indistinguishable from the noise floor. As regards the effectiveness in protecting the AddRoundKey operation, the effect is less evident, since even without countermeasure such an attack is rather difficult on this particular design. The greatest difficulty in attacking this step, compared to SBOX, derives from the linearity of the XOR operation, but in some designs this operation is attackable. In this case, even without countermeasure, the curve corresponding to the correct key is still immersed in the noise, however, in the case of the countermeasure, the absolute value of the correlation is more than halved.

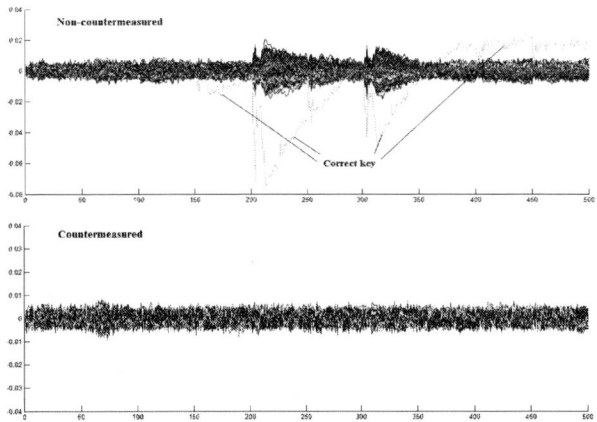

Figure 4. Correlation on the SubByte output (SBOX), above, non countermeasured version of the AES encoder, bottom, countermeasured version, according to the schematic in figure 3. The time window includes the first two rounds of the AES algorithm. Note the dramatic improvement in power analysis resistance.

978-1-4673-4579-8/13 $31.00 © 2013 IEEE

Paper M2C2   *PRIME 2013, Villach, Austria*

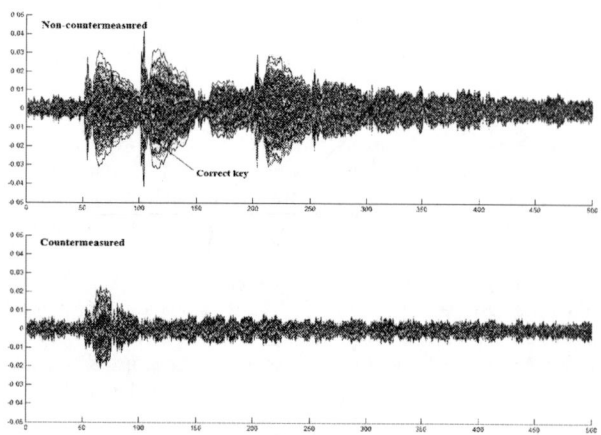

Figure 5. Correlation on AddRoundKey. This encoder is not easy to attack In the addRoundKey step even in the non-countermeasured version (above), where the correct-key curve (in green) is hardly distinguishable, but the countermeasured version has a lower correlation absolute value. In ordinate has been used the same scale for easy comparison.

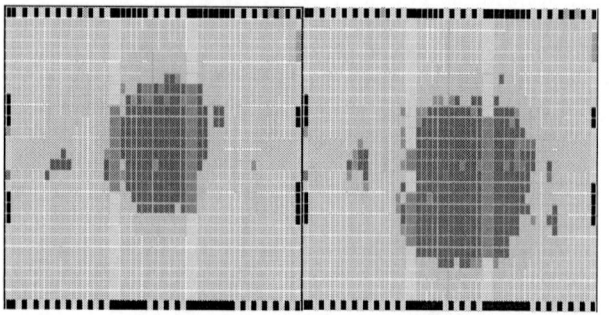

Figure 6. Layout of the FPGA synthesis, left, the original version of the AES encoder (1524 logic elements), right, the countermeasured version (2692 logic elements). The synthesis was performed with the same settings and all the peripheral circuitry, mainly control and I/O, are the same in both cases. Note that the countermeasure caused an acceptable increase in occupied area.

VII.   CONCLUSION

The SubByte operation in the second round of the AES encoder considered in this article, is rather attackable with correlation power analysis. The simple countermeasure here described can be an effective obstacle to such attacks making the coding key extremely hard to find, and its implementation does not imply an excessive waste of chip area and does not require the use of particular logic families, to such an extent that it can also be realized in FPGA, as in the case discussed here. As regards the AddRoundKey operation in round 0, the results are very encouraging, even if, given the difficulty in attacking this operation in this particular AES encoder, even in the absence of countermeasures, this experiment is not conclusive about the effectiveness of this countermeasure in protecting also this particular operation from power analysis attacks.

REFERENCES

[1]   P. Kocher, J. Jaffe, and B. Jun, *"Differential Power Analysis"*, Proc. Advances in Cryptology (CRYPTO '99), Lecture Notes in Computer Science, vol. 1666, Springer-Verlag, 1999.

[2]   K. Tiri and I. Verbauwhede, "Charge recycling sense amplifier based logic: securing low power security ICs against DPA [differential power analysis]". ESSCIRC 2004. Proceeding of the30th European Solid-State Circuits Conference, 2004.

[3]   S. Chari, C. S. Jutla, J. R. Rao, and P. Rohatgi. *"Towards Sound Approaches to Counteract Power-Analysis Attacks"* In Michael J. Wiener, editor, Advances in Cryptology - CRYPTO '99, 19th Annual International Cryptology Conference, Santa Barbara, California, USA, August 15-19, 1999, Proceedings, volume 1666 of Lecture Notes in Computer Science, pages 398-412. Springer, 1999.

[4]   S. Mangard, E. Oswald, T. Popp, *" power Analysis Attacks, Revealing the Secrets of Smart card"* Springer, 2007, ISBN: 978-0-387-30857-9.

[5]   M. Bucci, M. Guglielmo, R. Luzzi and A. Trifiletti,*" A Power Consumption Randomization Countermeasure for DPA-Resistant Cryptographic Processors"*, Integrated Circuit and System Design. Power and Timing Modeling, Optimization and Simulation, Lecture Notes in Computer Science, 2004, Volume 3254/2004, 481-490, DOI: 10.1007/978-3-540-30205-6_50.

[6]   M. Bucci, R. Luzzi, F. Menichelli, R. Menicocci, M. Olivieri, and A. Trifiletti, *"Testing power-anolysis attack susceptibility in Register-Transfer Level designs"*, IET Information Security (ex IEEE Proceedings Information Security), vol. 1 (3) pp.128-133, 2007.

[7]   Thanh-Ha Le , C. Canovas, J. Clédière, *"An overview of side channel analysis attacks"*, in Proceedings of the 2008 ACM symposium on Information, computer and communications security (ASIACCS '08).

[8]   J. Daemen. and V. Rijmen, *"AES Proposal: Rijndael"*, National Institute of Standards and Technology (NIST), July 2001.

[9]   The VHDL code of this cryptographic core was downloaded from the site *inmcm-hdl.googlecode.com*, Copyright 2010 Michael Calvin McCoy.

Fast Constant Time Memory Allocator
for Inter Task Communication
in Ultra Low Energy Embedded Systems

Gregor Rebel* (*Author*), Francisco J. Estevez*,
Ingo Schulz**, Peter Glösekötter*

* Dept. of Electrical Engineering and Computer
Science, University of Applied Sciences Munster,
Germany
** Dept. of Computer Science Chair IV, TU
Dortmund University, Germany

Abstract—Modern microcontrollers provide enough processing power to benefit from the advantages of multitasking schedulers or operating systems even in the area of small, battery based or energy self-sustaining devices. Many of these devices communicate with other devices via different interfaces. For a multitasking operating system, communication means to collect individual bytes in memory blocks and to transport these blocks between tasks. This paper describes how to use a combination of memory pools and memory headers to provide a fast, constant time memory allocator with low internal fragmentation. The proposed memory allocator is fast enough and has so few internal fragmentation, that it is applicable even in ultra low energy embedded systems with few kilobytes of ram. It can provide memory blocks of equal size at high frequency.

Keywords—*Dynamic, Memory, Allocator, Software, Embedded, Low Power*

I. Introduction

A. Ultra Low Energy Embedded Realtime Systems (ULERS)

Today's embedded systems range from smart, self-sustaining sensors with own data pre-processing up to line powered, number crunching multi-core architectures. This paper focuses mainly on software optimization for embedded systems at the low energy end of this spectrum. These devices typically offer one processing core and a single level memory interface. Fore these devices, the amount of available RAM typically ranges from 128 bytes (E.g. MSP430G2x31) to 66 KB (E.g. MSP430F5xx). Most dynamic memory allocators are optimized to manage much larger amounts of ram. This is why many software developers, in this field, implement their own memory management instead of using an external dynamic memory allocator.

B. Software Techniques of Data Transport

Whenever a microcontroller communicates, data has to be transported between a low-level interface driver and the main application. In multitasking operating systems, data also has to be transported between different tasks or threads. Two main types of software data transport are known: Transport by value or transport by reference. Transport by value means to copy all bytes of transported data between each adjacent layers of the software stack. This transport mechanism slows down with each additional software layer. This disadvantage can be overcome if data is transported by reference. Once the low-level driver has received a data packet, only a packet reference is passed to higher layers.

C. Conventional Dynamic Memory Allocators (CDMA)

In the context of this paper, a conventional dynamic memory allocator is a software scheme that allows to dynamically manage a fixed amount of memory at runtime. An alloc() operation provides reference to a memory block of requested size. The inverse operation free() returns the memory block for later reallocation to the memory allocator.

Many algorithms and implementations have been demonstrated for CDMAs that are optimized for different hardware architectures. Some examples of CDMAs especially for embedded systems are described in [1], [2] and [3].

Operating systems for desktop computers are mostly optimized for maximum data throughput. For these systems, the average case runtime of alloc()- and free()-operations should be minimized. This optimization typically increases memory usage or worst case runtime. On the contrast, operating systems for realtime systems often have to fulfill hard time constrains. These systems require low worst case runtimes for alloc() and free(). When it comes to ULERS, both worst-case and average-case runtime plus memory overhead should be as small as possible. In practice, the requirements of such systems often forbid the use of a CDMA.

According to [1], most of the CDMAs described in literature cannot be used in embedded systems. Either the runtime of alloc() or free() are unbounded or their memory overhead is not acceptable. The *Smart CDMA* described in [1] claims to fulfill the requirements of small embedded systems. In fact every CDMA requires its own data structures in memory. The occupied memory cannot be used by the application. This overhead is called internal fragmentation. The initialization and use of these structures increases runtime.

Paper M2C3   PRIME 2013, Villach, Austria

The biggest problem of all CDMAs is that their block management algorithms are heuristics. For every heuristic, a scenario can be constructed in which the optimization fails . In the case of CDMAs this means an increase of external fragmentation (non allocatable bytes of memory due to small holes between allocated blocks) and increased runtime for alloc() and free(). Some algorithms state O(1) runtime requirement for their operations. For small embedded systems, not only the asymptotic but also the effective runtime is important. In practice, even function calls can have a measurable impact on overall performance.

According to [1], the main aspects of a CDMA are response time, fragmentation, cache pollution, mutual exclusion and synchronization. The aspects handled in this paper are response time and fragmentation. The microcontrollers in the intended area of applications typically provide very simple architectures when compared to desktop computers. Multi level caches and multi core devices are not typical for ULERS.

D. Main Aspects of the proposed Algorithm

The algorithm proposed in this paper is not a complete CDMA in the sense of the cited sources. It is an extension that can base on any CDMA. The lowest management overhead can be achieved by using a minimal DMA that can only allocate memory. The main idea is to use memory pools of equal sized memory blocks and memory headers to provide a general, flexible, fast, robust and realtime memory allocator with very little internal and without external fragmentation. Internal fragmentation is very little when compared to CDMAs. Allocate() and free() operations can take place in small, constant time. The algorithms have been designed for use in multi- and single-tasking environments. References to allocated memory blocks can be freely passed around in an application. The Algorithm is optimized for small sized memory blocks of a few 100 bytes but can be used for larger block sizes too.

E. Simple Dynamic Memory Allocator (SimpleDMA)

The implementation of proposed memory allocator can be based on any CDMA. The most simple DMA (SimpleDMA) can only allocate memory. This requires

1. A static array of individual bytes Heap[HEAP_SIZE]

2. A byte pointer NextFree at start points to Heap[0]

3. A function void* alloc(unsigned int Size) That checks if enough space remains in Heap[] for required amount of bytes. If enough space is available, the function increases NextFree by Size and returns its old previous value. Otherwise, NULL is returned.

The further description is based on SimpleDMA. SimpleDMA can be easily replaced by any other CDMA implementation.

II. DESIGN OF A FAST CONSTANT TIME MEMORY ALLOCATOR BASED ON MEMORY POOLS AND HEADERS

A. Main Components of FCTMA

1. Operation **allocBuffer()** returns pointer P to allocated memory. P can be directly casted for use as any structure. Prototype: struct memory_s* allocBuffer(Bytes)

2. Operation **freeBuffer()** pushes buffer at front of list of unused memory blocks in corresponding memory pool. Prototype: void freeBuffer(struct memory_s* Buffer)

3. Operation **createPool()** allocates and intializes a new pool_s structure. If MaxBlocks is given, all required blocks can be allocated immediately to ensure fast constant response times for all further memory operations. Prototype: struct pool_s* createPool(unsigned int BlockSize, unsigned char MaxBlocks)

4. A data structure **struct memory_s** at address P-1 of each memory block stores at least two entries.
   1. struct memory_s* Next
      Memory blocks can form a single linked list.
   2. struct pool_s* Pool
      Points to the memory pool that manages this block

5. A data structure **struct pool_s** manages all memory blocks of same size by storing four entries.
   1. mutex_t PoolLock
      A mutual exclusion lock protects concurrent accesses to the pool.
   2. semaphore_t BlocksAvailable
      A semaphore initialized by createPool() with the value of MaxBlocks. AllocBuffer() takes 1 from this semaphore before obtaining the mutex PoolLock. freeBuffer() gives 1 to it after releasing PoolLock.
   3. unsigned int BlockSize
      All memory blocks of same pool have same size.
   4. struct memory_s* FirstFree
      Points to start of a single linked list of unused memory blocks in this pool.

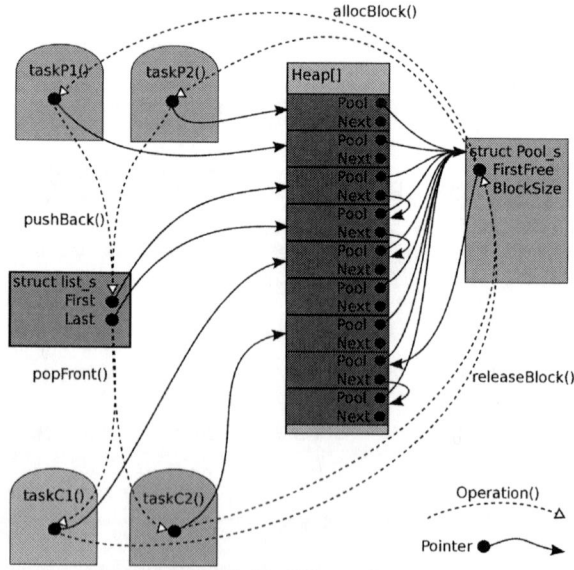

Fig. 1: Dataflow and Memory Structures

978-1-4673-4579-8/13 $31.00 © 2013 IEEE

PRIME 2013, Villach, Austria

Session M2C – Digital Techniques 1

III. Performance Comparison

A. Performance Metric

The response times of a CDMA are further defined as the minimum (Tmin), maximum (Tmax) and average (Tavg) runtimes of alloc() and free() operations. The amount of internal fragmentation is measured as ratio of allocated user bytes and total bytes. User bytes means the amount of bytes requested by user application and total bytes being the total amount of bytes allocated from heap plus static allocations like global and local static variables in RAM required by the individual implementation.

The described algorithm has been implemented and tested in an artificial test application. The multitasking scheduler used was *FreeRTOS* [4]. This scheduler is available as open source and has been already described by other papers like [5].

Fig. 1: Dataflow and Memory Structures shows the data flow operations between m Producer tasks taskP1(), ... taskPm() and n Consumer tasks taskC1(), ... taskCn(). It also shows the pointers that build up the complex data structure. At start, the memory allocator under test is initialized. The list is initialized too. Then 10 producer and 10 consumer tasks are spawned. Each producer task allocates twenty memory blocks of 127 bytes (the maximum size of an IEEE 802.15.4 data packet). The task issues pushBack() operations on the list to send all memory blocks to consumer tasks. After push back and if the total amount of allocated memory has reach 16 KB, the producer task sleeps for some random milliseconds. Each consumer task issues a pop_front() operation on the list to retrieve up to ten memory blocks. The retrieved memory blocks are given back to the pool by free() operations. The consumer task then waits for a pseudo random time.

A simple pseudo random number generator is used to calculate waiting times. Fig. 2 shows its implementation.

```
static unsigned long sr_Seed = 1;
sr_Seed = sr_Seed * 1103515245 + 12345;
WaitingTime = ((sr_Seed>>16)) % MaxTime;
```
Fig. 2: Pseudo Random Number Generator

The random waiting time simulates a random processing time to produce and consume each memory block. Fine tuning MaxTime value of producers and consumers lets converge the total amount of allocated buffers around 100 blocks after some minutes.

Comparing different implementations needs exact measure of runtime for each alloc() and free() operation. This is achieved by inserting extra probe-commands at begin and end of these functions to set and clear an individual GPIO pin. Pin ALLOC is set at begin of alloc() and cleared at its end. Pin FREE is controlled by free(). The extra probe-commands add constant runtime of <1µs to each operation. The minimum, average and maximum runtime can then be measured by use of a digital oscilloscope.

Multitasking and interrupt requests are disabled during alloc() and free() operations for more precise measures. Task synchronization, task switching and interrupt service routines add extra uncertainty to the runtime of any implemented algorithm and would blur runtime measures.

B. Candidates in Comparison

1) Algorithm FCTMA

The algorithm described above has been implemented in a self developed software toolchain "The ToolChain" [6] for STM32 based microcontrollers. The ToolChain makes use of the *FreeRTOS* [4] multitasking scheduler as a SimpleDMA. Task synchronization is provided by special, 32-bit optimized mutexes, semaphores, queues and lists. FCTMA is used for memory pools in The ToolChain beginning at revision 1.0.52.

2) FreeRTOS Heap4

The *FreeRTOS* multitasking scheduler provides different implementations of DMAs. The most advanced DMA implementation is called *Heap4*. It consolidates adjacent memory blocks as they are freed to reduce memory fragmentation. The benchmark was run against *FreeRTOS* revision 7.3.0. The extra probe-commands have been inserted directly into pvPortMalloc() and vPortFree() inside the vTaskSuspendAll() block.

3) Algorithm SDMA

The *Smart Dynamic Memory Allocator* described in [2] has been implemented by use of The ToolChain [6]. At initialization, a chunk of 32 KB is allocated using *FreeRTOS Heap4*. All further memory allocations are then served by the SDMA implementation from this memory chunk. The amount of free list classes has been reduced to 15 to avoid wasting memory. 15 classes are enough to represent memory allocation of up to 32 KB. The lookup tables LTB1[] and LTB2[] have been implemented as constant char arrays which places them into flash memory and saves an additional 512 bytes of ram. The BlkPredMask has been initialized as 255 to set the first eight classes as short lived. This allows to treat all blocks allocated by the benchmark as short lived and avoids block split and merge operations. The pseudocode implementation of second-level-index calculation was not used as it seemed to be incomplete. Instead an implementation based on bit-shifting was used according to the textual description.

C. Test Setup

The software implementation was tested on a CortexM3 based STM32F103ZET6 microcontroller manufactured by ST Microelectronics Corp. The chip provides a 32-bit architecture with a uniform 4 GB address space. This uniform address space allows to create C-pointers to every memory region or register. The microcontroller is equipped with 64 KB of RAM and 512 KB of FLASH memory. It is clocked at 72 MHz. When compared to older 8- and 16-bit microcontrollers, the CortexM3 provides more computing power, a linear address space and more ram. This allows to run more tasks to stress test the implementation than on smaller architectures. Despite of its computing power, the chosen microcontroller is still a better representation of a ULERS than a desktop computer.

Runtime analysis is done by attaching a Tektronix MSO4104 Digital Oscilloscope to Pins ALLOC and FREE. Its signal analyzer allows to measure minimum, maximum and average high-time of signals on both pins. Measures have been taken after 1 minute of continuos simulation. All sources have been compiled by gcc (GNU for ARM Tools) v4.7.1 without any optimizations and with enabled debugging support.

978-1-4673-4579-8/13 $31.00 © 2013 IEEE

Paper M2C3

D. Performance results

TABLE I. DURATION OF ALLOC() OPERATION FOR DIFFERENT MEMORY ALLOCATOR IMPLEMENTATIONS

Algorithm	T_{avg}	T_{min}	T_{max}	Std Deviation	1000 * Std Deviation / $T_{average}$
FCTMA	2196ns	2174ns	2211ns	14ns	6.38 ‰
FreeRTOS Heap4	3300ns	2706ns	7111ns	1070ns	324.24 ‰
SDMA	171600ns	171600	171600	13.5ns	0.08 ‰

TABLE II. DURATION OF FREE() OPERATION FOR DIFFERENT MEMORY ALLOCATOR IMPLEMENTATIONS

Algorithm	T_{avg}	T_{min}	T_{max}	Std Deviation	1000 * Std Deviation / $T_{average}$
FCTMA	957ns	955ns	958ns	666.3ps	0.7 ‰
FreeRTOS Heap4	2847ns	1943ns	6874ns	1095ns	384.62 ‰
SDMA	165800ns	165800ns	166100ns	26ns	0.1 ‰

TABLE III. INTERNAL FRAGMENTATION FOR DIFFERENT MEMORY ALLOCATOR IMPLEMENTATIONS

Algorithm	User Bytes allocated	Total Bytes allocated	Internal Fragmentation (%)
FCTMA	20558 (21198)	21878	6.03 % (3.11 %)
FreeRTOS Heap4	26114	26939	3.06 %
SDMA	13152	15936	17.47 %

Table 1 shows average, minimum and maximum runtime measures of different memory allocator implementations. It also shows the standard deviation as absolute and per mill value. The FCTMA implementation is in average 50% faster than *FreeRTOS Heap4*. Still *Heap4* provides a very fast alloc() operation. But the standard deviation of *Heap4* is more than 50 times bigger than FCTMA. This effect is caused by the block splitting algorithm of *Heap4*. The SDMA algorithm is 78 times slower than FCTMA. This is due to large computational overhead in calculation of first- and second level index values for each operation. The very low standard deviation of SDMA is caused by the basic implementation done for this comparison. Block splitting and merging have been disabled for this test.

Table 2 basically shows a similar picture as Table 1. The FCTMA free() operation is nearly 3 times faster than *Heap4* because of its minimal management overhead. Its standard deviation is practically zero. The *FreeRTOS Heap4* implementation merges adjacent blocks which explains its high standard deviation similar to it alloc() performance. The SDMA implementation takes nearly as long for free() as for

alloc() because of its disabled block merging. SDMA is now 173 times slower than FCTMA.

Table 3 compares memory usage and internal fragmentation of the three implementations. FCTMA shows 6.03 % of internal fragmentation. This value results from 32 bytes for each memory pool and 8 bytes for each memory header. If the application makes use of the next-pointer in each memory block, then we can add 4 bytes for each memory header to User Bytes and achieve an effective internal fragmentation of 3.11% (numbers in brackets). This benchmark clearly shows, that the SDMA algorithm is optimized to manage larger amount of memory. [2] was benchmarked on a Pentium D 3.4Ghz with 1 GB RAM with up to 4798 MB of allocated memory. But we must also be fair to say that internal fragmentation of SDMA would benefit from block merge like *Heap4* does.

IV. CONCLUSION & OUTLOOK

This paper has compared three different implementations of memory allocators in an artificial, inter task communication benchmark on an up to date microcontroller. The introduced FCTMA algorithm has proven that it can provide dynamic memory of constant size in near constant time and faster than any CDMA implementation. The internal fragmentation of FCTMA is equal or comparable to its competitors. The algorithm makes software development of embedded systems easier and safer by replacing self made memory managers. FCTMA can be used on top of any CDMA and is easy to implement. It is ideally suited to provide memory blocks of equal size in high frequency.

Future research may concentrate on extending FCTMA with support for interrupt service routines. In contrast to a task, an interrupt service routine cannot wait for other tasks to unlock a mutex to gain access to a memory pool.

ACKNOWLEDGEMENT

I wish to thank Oliver Wesch for proof reading this paper.

REFERENCES

[1] M. Masmano, I. Ripoll, A. Crespo, and J. Real. TLSF: a New Dynamic Memory Allocator for Real-Time Systems, 2004

[2] Ramakrishna M, Jisung Kim, Woohyong Lee and Youngki Chung. Smart Dynamic Memory Allocator for Embedded Systems, 2008

[3] *David A. Barret, Benjamin Zom. Using Lifetime Predictors to improve Memory Allocation Performance, 1993*

[4] R. Barry. FreeRTOS. *http://www.freertos.org/*, 2013

[5] *Jan Tobias Muhlberg, Leo Freitas.Verifying FreeRTOS: from requirements to binary code. Proceedings of the 11th International Workshop on Automated Verification of Critical Systems (AvoCS 2011), 2011*

[6] *Gregor Rebel. The ToolChain – An open source software toolchain for CortexM3 microcontrollers. http://thetoolchain.com, 2013*

978-1-4673-4579-8/13 $31.00 © 2013 IEEE

Novel Field Programmable Embryonic Cell for Adder and Multiplier

Gayatri Malhotra, Joachim Becker, Maurits Ortmanns
Institute of Microelectronics
University of Ulm
D-89081 Ulm, Germany
Email: gayatri.malhotra@uni-ulm.de

Abstract—A novel approach for the implementation of a field programmable embryonic cell is proposed. This is inspired from the concept of cloning used by biological cells to grow a bigger organism. An exemplary full adder and multiplier circuit evolution is simulated using Verilog and verified. An automated generation of configuration data for the embryonic cells from a behavioral description of the overall organism can be done using a genetic algorithm approach in the future.

Index Terms - Embryonics, Field Programmable Gate Array, Adder, Multiplier

I. INTRODUCTION

Field Programmable Gate Arrays (FPGA) contain logic blocks connected through reconfigurable interconnects to implement different circuit configurations. There is an array structure of logic blocks within the FPGA. Each logic block comprises of few logical cells made of look up tables (LUT), adders and flip-flops. These components can be programmed for the desired circuit functionality by setting of configuration bits for each individual part.

The emerging field of embryonic electronics is inspired from biological cells and therefore it is centered on the creation and configuration of cells. The cellular differentiation and division are the two main processes in the organism development [1].

The self healing property of these bio-inspired circuits are depicted in many ways [2][3]. Like the genome of an organism contains its hereditary information, the electronic cell can be consisting of the complete circuit data in its configuration bits. The electronic genome can be made of numerous genes each consisting of different LUT data.

This paper proposes a novel embryonic cell and shows the viability by an exemplary implementation of a 1-bit full adder and its cloning to a bigger full adder (e.g.10 bit) as well as the implementation of a single multiplier embryonic cell cloned to a bigger multiplier (e.g. 4x4).

II. EMBRYONIC CELL FABRIC ARCHITECTURE

The deterministic array structure of present FPGAs needs a fixed amount of configuration bits. This project aims to significantly reduce the number of configuration bits by exploitation of redundancy in regular cell structures. The first cell configuration is done externally and the next cells configuration is done during runtime by cloning the configuration bits to the

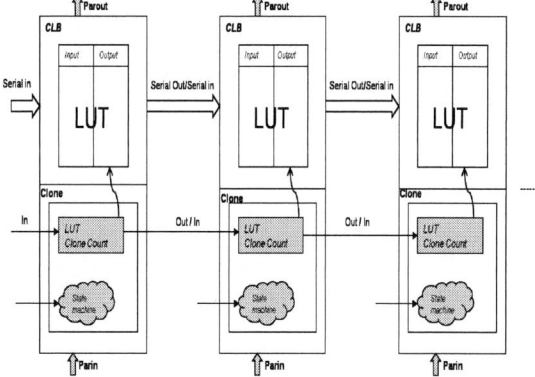

Fig. 1. The Embryonic Cell Fabric of three regular cells.

next cells. This can lead to evolution of a regular structure since the configuration bits are same for all the cells.

The embryonic cell fabric is a set of cloned cells from a single cell. The different cells are created by the cloning process defined by a state machine implemented within the cell.

The embryonic cell is made of a CLB and a cloning circuit which takes care of copying the genome from one to other cell. The fabric with three cells for a regular structure like an adder and multiplier is shown in Fig. 1.

The genome data stored in the first cell will be copied to the next cells till the clone count becomes zero. This data configures the cells for the function to be done.

This novel genome consists of configuration bits for the LUT and cloning bits for copying information. It can comprise data for one or more number of genes of used LUT and corresponding cloning data. The LUT data is used to create the functionality of a Configurable Logic Block (CLB).

III. DIFFERENT LUT IMPLEMENTATION AND CLONING PROCESS

For a regular structure like adder and multiplier, the parallel and serial I/O for each cell is considered as two. The possible LUTs are verified for this structure like feedthrough, swap, adder and multiplier. It can be further implemented for

ip1	ip0	is1	is0	op1	op0	os1	os0
0	0	0	0	0	0	0	0
0	0	0	1	0	0	0	1
0	0	1	0	0	0	1	0
0	0	1	1	0	0	1	1
0	1	0	0	0	1	0	0
0	1	0	1	0	1	0	1
0	1	1	0	0	1	1	0
0	1	1	1	0	1	1	1
1	0	0	0	1	0	0	0
1	0	0	1	1	0	0	1
1	0	1	0	1	0	1	0
1	0	1	1	1	0	1	1
1	1	0	0	1	1	0	0
1	1	0	1	1	1	0	1
1	1	1	0	1	1	1	0
1	1	1	1	1	1	1	1

Fig. 2. Feedthrough LUT.

ip1	ip0	is1	is0	op1	op0	os1	os0
0	0	0	0	0	0	0	0
0	0	0	1	0	1	0	0
0	0	1	0	0	0	0	0
0	0	1	1	0	1	0	0
0	1	0	0	0	1	0	0
0	1	0	1	0	0	0	1
0	1	1	0	0	1	0	0
0	1	1	1	0	0	0	1
1	0	0	0	0	1	0	0
1	0	0	1	0	0	0	1
1	0	1	0	0	1	0	0
1	0	1	1	0	0	0	1
1	1	0	0	0	0	0	1
1	1	0	1	0	1	0	1
1	1	1	0	0	0	0	1
1	1	1	1	0	1	0	1

Fig. 3. Full Adder LUT.

other structures like a divider. The feedthrough is to forward parallel/serial input to output. Feedthrough LUT is shown in Fig. 2. Configuration data for the LUT is assembled in four arrays of 16-bits of data. It is called as op1, op0, os1 and os0 for two parallel and two serial outputs. Swap LUT is for swapping input data to the output.

Cloning is initiated based on the clone data in the genome and the configuration flag. Clone data includes LUT data and the number of times cloning is required. The first cell is configured initially and cloning of cells follows for the number of times asked. After the cells are configured, the serial and parallel inputs are applied and the output is verified for circuit functionality. This way of configuration has many advantages over regular FPGA configuration. Since cloning multiplies the information internally, only the data for one cell has to be transmitted, instead of a complete set of individual configuration bits. Even if the data for one cell contains multiple LUT data for cell differentiation and additional cloning count data, there is still a very high potential for savings.

IV. ADDER IMPLEMENTATION

An exemplary embryonic fabric is designed for 16 cells. From a single 1-bit full adder cell, ten cells are cloned for a 10-bit full adder. Cloning data contains ten as the number of times cloning required. Parallel inputs ip0 and ip1 are mapped to two input bits to be added and serial input is0 is for previous carry. Sum is mapped to parallel output op0 and carry is to serial output os0. The LUT for a full adder is shown in Fig. 3. The simulation results of the adder are verified for different input data.

In this fabric the total configuration data comprise of 64 bits of LUT data and 4 bits of clone data, all together 68 bits. Thus 68 bits are all needed to implement maximum 16 bit adder in this fabric. Though this is not the limit of the fabric for bigger circuit implementation. It has clear advantage as it needs one time 68 bits than ten times of 68 bits for ten bit adder for FPGA implementation.

V. MULTIPLE LUTs AND DIFFERENTIATION

To further verify this concept, a genome with multiple genes is implemented. It is called as operative genome and assigned with four genes. Each gene represents LUT data

and its count for cloning. Each cell is assigned a gene to be configured for. This selection can be cell position dependent. Direction or position based selection is thought for next to be implemented. There are four genes like adder, feedthrough, swap and constant output, each cell can be simulated for. The simulation is shown in Fig. 4. One gene for adder is cloned for ten times and another gene for feedthrough is cloned for seven times.

Initially the LUT data for adder gene is cloned for some cells, afterwards the same cells are configured with the feedthrough gene. This shows that the same hardware can be reconfigured for different functions.

VI. 2D ARRAY AND MULTIPLIER GENOME

The configuration of an embryonic cell can be extended to a 2-D array structure. Cloning can be done in two directions simultaneously or one after another. Here the digital multiplier circuit of dimension 4x4 for example is considered for implementation. It needs 16 cells with the same LUT data so it can be mapped to the cellular fabric. A(3:0) and B(3:0) are the two 4-bit input data to be multiplied, and the product is saved in P(7:0). The basic block for the multiplier is shown in Fig. 5 and the multiplier LUT is shown in Fig. 7.

The 2-D cell array structure for a 4x4 multiplier is shown in Fig. 6. The cloning has to be performed for 16 cells, starting from first row and right most vertical column fully then to next adjacent column and so on.

In Fig. 6, S denotes sumin/sumout and C denotes carryin/carryout of the cell. A and B are mapped to each single bit parallel input of the cell. The sumin/sumout and carryin/carryout are mapped to serial inputs/outputs of the cell. As cloning is required in two directions horizontal or vertical, the array needs to arranged without diagonal direction. The new orthogonal structure is shown in Fig. 8. The cloning will start from rightmost column to leftmost and in each column from top row to bottom one.

The multiplier genome in this example consists of a LUT and cloning data for 16 cells. The serial input and output data movement direction for each cell is hard coded, thus limits the scope to a small multiplier. The cloning process and a simulation result for product of A =11(1011), B= 13(1101) is shown in Fig. 9. The product is 143(10001111).

978-1-4673-4579-8/13 $31.00 © 2013 IEEE

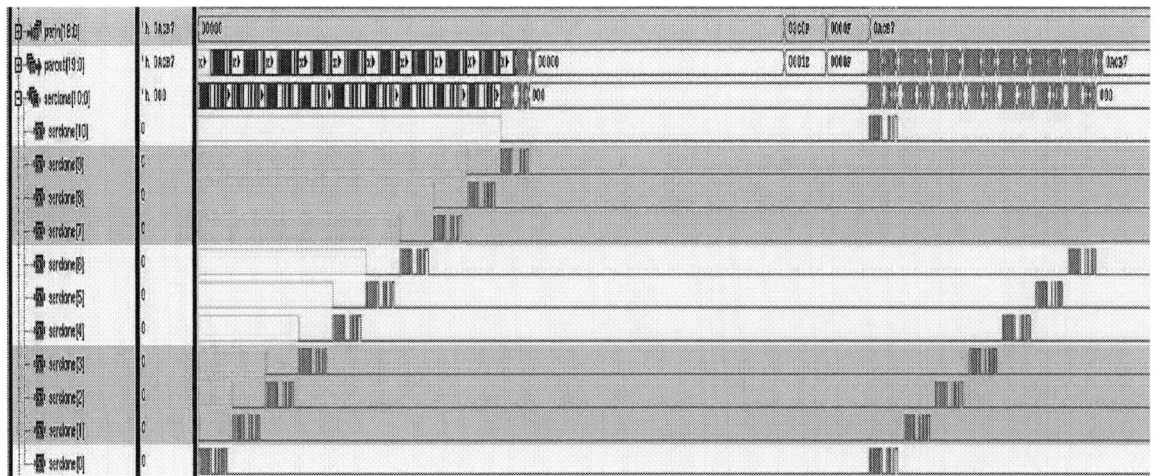

Fig. 4. Cloning with different gene data.

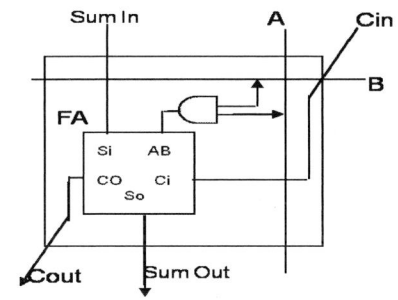

Fig. 5. Multiplier basic block

A	B	Sum In	Cin	AB	Sum Out	Cout
0	0	0	0	0	0	0
0	0	0	1	0	1	0
0	0	1	0	0	1	0
0	0	1	1	0	0	1
0	1	0	0	0	0	0
0	1	0	1	0	1	0
0	1	1	0	0	1	0
0	1	1	1	0	0	1
1	0	0	0	0	0	0
1	0	0	1	0	1	0
1	0	1	0	0	1	0
1	0	1	1	0	0	1
1	1	0	0	1	1	0
1	1	0	1	1	0	1
1	1	1	0	1	0	1
1	1	1	1	1	1	1

Fig. 7. Multiplier LUT

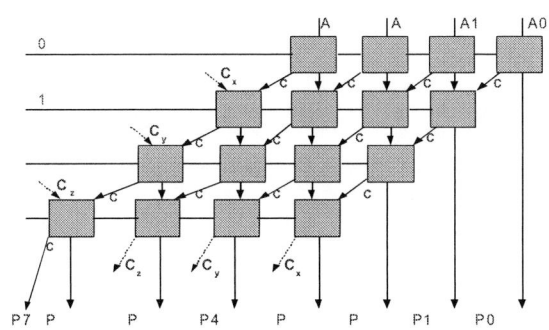

Fig. 6. 2D Array of 4x4 multiplier.

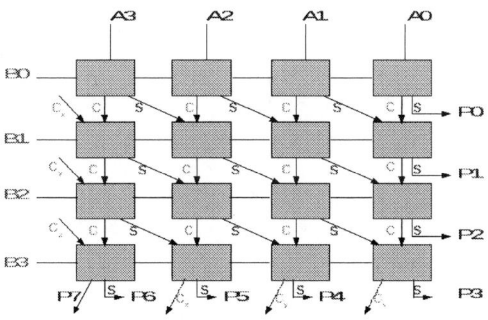

Fig. 8. Orthogonal Array for multiplier.

VII. CONCLUSION

An embryonic cell is designed and simulated for a 10 bit adder and a 4x4 multiplier as an example. The concept can further be modified for synchronous components like flip-flop, included in the cell. There is a good example for possible growth of embryonic cell in [4].

The connections of serial inputs and outputs of the 4x4 multiplier cell is done manually in this design, while it needs to be automated for a bigger multiplier implementation. It is also required to have automatic LUT data generation for complex circuit structures. These automation requirement needs for example a genetic algorithm to be integrated with the design.

The genetic algorithm [5] can be used for random generation of genome data resulting the desired LUT. There are some

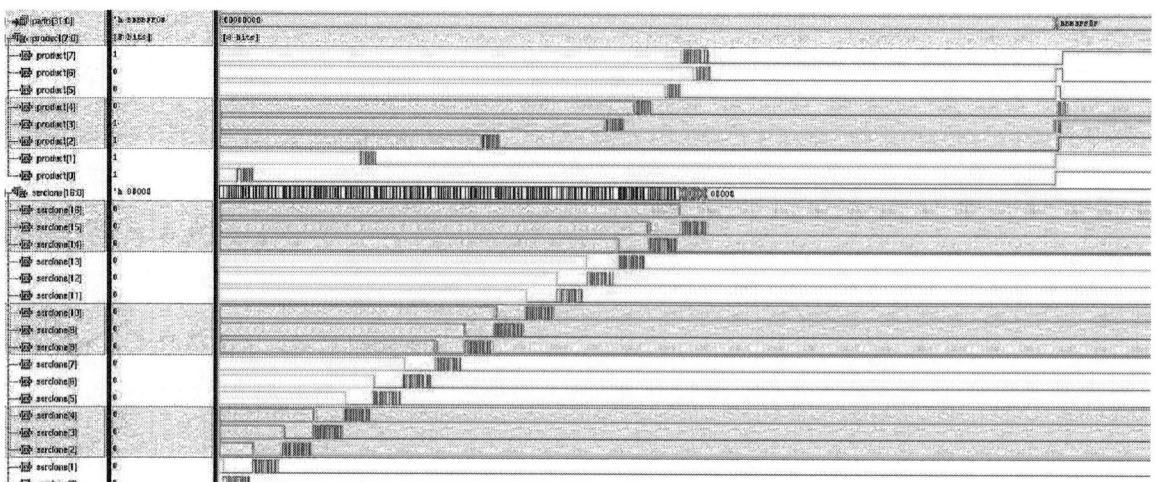

Fig. 9. Cloning process and product of multiplier.

genetic algorithms designed especially for hardware implementation [6].

The advantage of the concept of reconfiguration during runtime is that the configuration bits required are much reduced. As the same data is copied/cloned to the next cell depending on the clone number, it has the flexibility to extend the circuit size with only change in clone data bits. This process is now verified for regular structures only, to be modified for irregular ones. The self-repair for a cell can be achieved if the process can program the next available cell with same gene, though the routing has to be redefined. This will lead to improved fault tolerance.

The hardware development board planned to be used is Xilinx Zynq board. It is an embedded platform with processing system and logic fabric together in the same chip. This platform can have fast fitness evaluation as the genetic algorithm is running in the same chip.

ACKNOWLWDGEMENT

The Author is on study leave from ISRO Satellite Center, Bangalore, India to pursue her PhD at Ulm University.

REFERENCES

[1] D. Mange, A. Stauffer, L. Peparolo, G. Tempesti, A Macroscopic View of Self-Replication, Proceedings of the IEEE, Vol. 92, No. 12, Dec 2004.

[2] A. Stauffer, J. Rossier, Self-Testable and Self-Repairable Bio-Inspired Configurable Circuits, NASA/ESA Conference on Adaptive Hardware and Systems, 2009.

[3] A. Stauffer, D. Mange, and G. Tempesti, Embryonic Machines that Grow, Self-Replicate and Self-Repair, Proceedings of the 2005 NASA/DoD Conference of Evolution Hardware.

[4] M. Reibel Boesen, J. Madsen and D. Keymeulen, Autonomous Distributed Self-Organizing and Self-Healing Hardware Architecture- the eDNA Concept, IEEE Aerospace Conference, 2011.

[5] E. Benkhelifa, A. Pipe, G. Dragffy, M. Nibouche, Towards Evolving Fault Tolerant Biologically Inspired Hardware Using Evolutionary Algorithms, IEEE Congress on Evolutionary Computation, CEC 2007.

[6] Z. Zhu, D. Mulvaney, and V. Chouliaras, A Novel Genetic Algorithm Designed for Hardware Implementation, International Journal of Computational Intelligence 3;4, 2007.

Design and Applications of Magnetic Tunnel Junction Based Logic Circuits

Hiwa Mahmoudi, *Student Member, IEEE*, Thomas Windbacher, Viktor Sverdlov,
and Siegfried Selberherr, *Fellow, IEEE*

Institute for Microelectronics, Technische Universität Wien, Gußhausstraße 27–29/E360,
A-1040 Wien, Austria, Phone: +43-1-58801-36049, Fax: +43-1-58801-36099
E-mail: {mahmoudi|windbacher|sverdlov|selberherr}@iue.tuwien.ac.at

Abstract—By offering zero standby power, non-volatile logic is a promising solution to overcome the leakage current issue which has become an important obstacle, when CMOS technology is shrunk. Magnetic tunnel junction (MTJ)-based logic has a great potential, because of unlimited endurance, CMOS compatibility, and fast switching speed. Recently, several non-volatile MTJ-based circuits have been presented which inherently realize logic-in-memory circuit concepts by using MTJ devices as both memory and the main computing elements. In this work we present a reliability simulation method for designing MTJ-based logic gates integrated with CMOS. As an application example, we study the reliability of a magnetic full adder in two different designs based on the implication and the reprogrammable MTJ logic gates.

Index Terms—Logic-in-memory, material implication (IMP), magnetic tunnel junction (MTJ), non-volatile logic, spin transfer torque (STT)

I. INTRODUCTION

High standby power due to leakage currents has become an important obstacle for scaling CMOS logic circuits (Fig. 1) at sub-100nm technologies [1]. A possible solution to overcome this problem is introducing non-volatility into the logic circuits [2]. The Spin-transfer torque (STT) [3] switching magnetic tunnel junction (STT-MTJ) is one of the most promising non-volatile storage technologies, which combines the advantages of CMOS compatibility, high speed, high density, unlimited endurance, and scalability [4].

As shown in Fig. 2, distributing non-volatile memory elements over the CMOS logic circuit plane (so-called logic-in-memory architecture [5]) can provide extremely low standby power consumption and instant start-up by holding the information in the MTJs and eliminating the need for refreshing pulses which are critical for CMOS-based memory elements [6], [7], [8]. Furthermore, by using the MTJ technology the effective area and interconnections delay (the data traffic on a main data bus between separated logic and memory modules as shown in Fig. 1) can be reduced due to easy three-dimensional integration of the MTJs on top of the CMOS layers (Fig. 2). However, in hybrid CMOS/MTJ circuits the MTJs are used only as ancillary devices which store the computation results [9]. Therefore, sensing amplifiers [10] are required for reading the data at each logic stage and providing

the next stage with an appropriate voltage or current signal as input. This limitation increases the device count, delay, and power consumption. Furthermore, the generalization to large-scale logic systems is problematic.

Recently, it has been demonstrated that STT-MTJs can be directly connected to perform logic operations [11], [12], [13], intrinsically enabling logic-in-memory architectures with no need for extra hardware (also known as "stateful" logic [14]). Here, we show how the MTJ logic gates can be generalized to large-scale logic systems based on one-transistor/one-MTJ (1T/1MTJ) cells (Fig. 3b) which are utilized as the basic memory cells in the STT magnetoresistive random-access memory (STT-MRAM) structure [15].

II. INTRINSIC MTJ-BASED LOGIC-IN-MEMORY

The MTJ device includes a fixed and a free ferromagnetic layer separated by an oxide barrier as shown in Fig. 3a. The magnetization of one layer is pinned (reference layer), while the magnetization of the other layer (free layer) can be switched freely using an external magnetic field or the STT effect. The STT switching technique improves significantly its scalability and the STT-MTJ exhibits pure electrical switching [4]. The MTJ resistance depends on the relative orientation of the magnetization directions of the ferromagnetic layers.

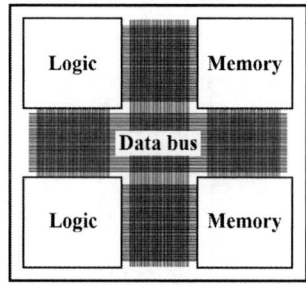

Fig. 1. Separated logic and memory units in a two-dimensional structure of CMOS logic. The interconnections delay dominates chip performance. The continuously powered memory units cause large static power consumption due to the leakage currents.

The (anti)parallel alignment results in a low (high) resistance state across the barrier R_P (R_AP), which is mapped to logic '0' ('1'). The resistance modulation is described by the tunnel magnetoresistance (TMR) ratio, defined as $(R_\text{AP} - R_\text{P})/R_\text{P}$.

The non-volatile magnetic logic-in-memory concept (Fig. 2) has been recently tested as a potential candidate for logic circuit design. For example, a magnetic full adder has been demonstrated in [6] for non-volatile arithmetic applications. However, the logic operation is still performed by CMOS logic elements, which requires 26 transistors for logic, 8 for MTJ writing, and 4 MTJs. Therefore, there is no benefit regarding the transistor count compared to the CMOS full adder. Furthermore, a key limitation of this magnetic full adder is the necessity of different kind of inputs and outputs for which some inputs or outputs are voltage signals, whereas the others are the resistance state of the MTJ elements. This mismatch causes the need for extra hardware and increases complexity.

Recently, MTJ-based logic gates have been demonstrated which use MTJs as main devices for logical computations and intrinsically enable logic-in-memory architectures with no need for extra hardware. In fact, in a logic mode the MTJs are used as the basic elements for computations and in a memory mode they are used for non-volatile storage. This enables extending non-volatile electronics from memory to logical computing applications and eliminates the need for sensing amplifiers and intermediate circuitry as compared to the hybrid CMOS/MTJ non-volatile logic circuits.

In [11] and [12], STT-MTJ-based reprogrammable logic gates (Fig. 4a and Fig. 4b) are demonstrated to realize the conventional Boolean logic operations including AND, OR, NAND, and NOR. By applying the voltage V_A, the resistance states of the input MTJs modulate the critical current required for the STT-switching of the output MTJ. In [13], we proposed a STT-MTJ-based implication logic gate (Fig. 4c) to realize a fundamental Boolean logic operation called material implication (IMP). Depending on the initial logic (resistance) states of the source and the target MTJs a conditional switching behavior on the target MTJ is provided, when the current I_IMP is applied to the IMP gate. The final logic state of the

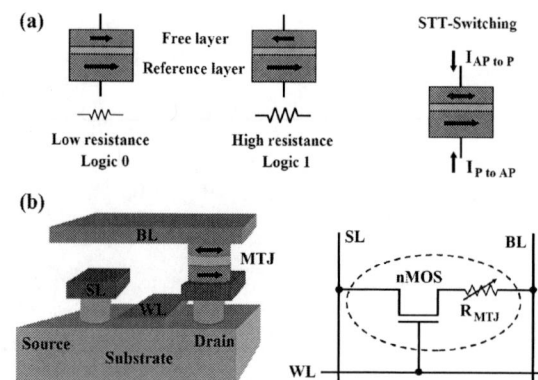

Fig. 3. (a) MTJ basic structure. (b) 1T/1MTJ structure and the equivalent circuit diagrams.

target MTJ represents the logical output of the IMP operation. By replacing the MTJ devices with 1T/1MTJ cells (Fig. 4d), the IMP logic gates can be extend to large-scale non-volatile magnetic circuits. Since the 1T/1MTJ cell is the basic element for STT-MRAMs, this concept provides magnetic logic circuits which also can be used for random-access memory applications (Fig. 5). In the next section we present a method for reliability analysis of the MTJ-based logic circuits and we consider in particular the reliability of a magnetic full adder.

III. MODELING AND RESULTS

The realization of MTJ-based intrinsic logic-in-memory circuits relies on a conditional STT-switching behavior of the output (target) MTJ for each logic operation. It has been shown that a reliable MTJ-based logic behavior requires a high enough TMR ratio [13]. A typical 1T/1MTJ memory cell of the STT-MRAM structure consists of an access transistor and a MTJ as its storage element (Fig. 3b). The access transistor acts as on/off switch to control the current flowing through the MTJ. When a selecting voltage signal (V_s) is applied to a word line (WL), the current $I_\text{AP-P}$ ($I_\text{P-AP}$) applied to the bit line (BL) switches the MTJ (corresponding to the selected access transistor by the WL) from AP to P (P to AP) state.

For the reliability analysis of the intrinsic logic-in-memory circuits, it can be shown that the reliability of a specific logic function (f) is determined by

$$R(f) = 1 - P_\text{err}(f) = \prod_{i=1}^{n_\text{f}} [1 - P_\text{err}(i)], \qquad (1)$$

where $P_\text{err}(i)$ is the average error probability of the ith MTJ-based logical step and n_f is equal to the total number of the MTJ-based operations (conditional switching evens) required for performing f using either implication or reprogrammable gates. For example, a NAND-based design of a full adder using the reprogrammable gate includes 9 NAND operations ($n_\text{f} = 9$). Our design for an IMP-based full adder includes 18 IMP operations ($n_\text{f} = 18$) [16].

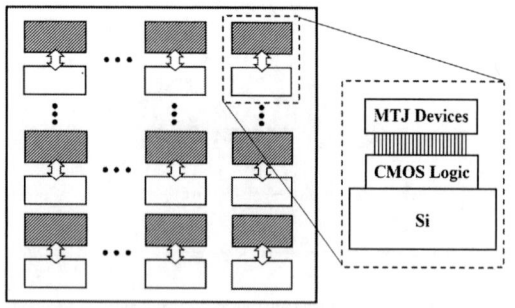

Fig. 2. Logic-in-memory architecture and the three-dimensional structure of the magnetic logic circuits.

For a basic logic operation using MTJ gates, P_{err} is proportional to the sum of the switching probabilities for undesired switching events (P_u) and the term $(1 - P_d)$ for the desired switching events as [13]

$$P_{err} = \frac{1}{k+h}\left\{\sum_{i=1}^{k} P_u(i) + \sum_{j=1}^{h} 1 - P_d(j)\right\}, \qquad (2)$$

where k and h are the number of the possible undesired and desired switching events. According to the theoretical model [17] and the measurements [15], the switching probability (P) of an MTJ in the thermally-activated switching regime (switching time $t > 10$ ns) is given by

$$P = 1 - \exp\left\{-\frac{t}{\tau_0}\exp\left[-\Delta_0\left(1 - \frac{I}{I_{C0}}\right)\right]\right\}, \qquad (3)$$

where $\Delta_0 = E/k_B T$ is the MTJ thermal stability factor, I is the current flowing through the MTJ, t is the current pulse duration, $\tau_0 \sim 1$ns, and I_{C0} is the critical switching current extrapolated to τ_0 [18].

In order to calculate the current passing through the MTJs in the 1T/1MTJ-based gate (Fig. 4d), we have

$$\begin{aligned}
V_{GS}^i &= V_{WL}, \\
V_{GS}^{j'} &= V_{WL}, \\
V_{DS}^i &= V_{BL} - V_M^i = V_{BL} - R_M^i I_M^i, \\
V_{DS}^{j'} &= V_{BL} - R_G I_M^{j'} - V_M^{j'} = V_{BL} - (R_G + R_M^{j'})I_M^{j'},
\end{aligned} \qquad (4)$$

when the current I_{IMP} is applied and the target and the source cells are selected by a voltage V_{WL} applied to the ith and j'^{th}

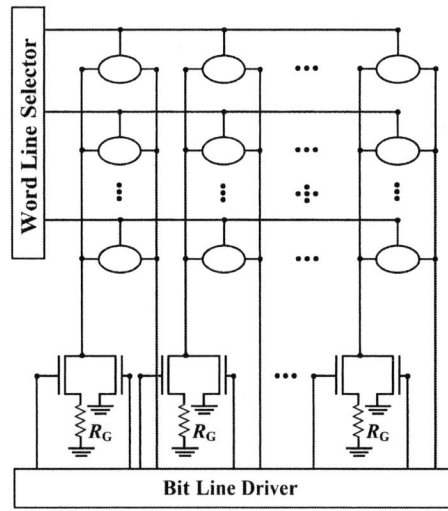

Fig. 5. The magnetic logic circuit based on the STT-MRAM architecture [13] with 1T/1MTJ cells (Fig. 3b) to realize intrinsic logic-in-memory for large-scale logic applications.

word lines (WLi and WLj'), respectively. Here, V_{GS} (V_{DS}) is the voltage difference between the gate (drain) and the source of the access transistor, R_M is the MTJ resistance, and I_M is the current through the MTJ device.

The $R - V$ characteristics of the MTJ device is determined by the voltage-dependent effective TMR model [19], if the MTJ is in the antiparallel state, as

$$R_M = R_{AP} = (1 + TMR_{eff})R_P = (1 + \frac{TMR_0}{1 + \frac{V_M^2}{V_h^2}})R_P, \quad (5)$$

where TMR_0 (TMR_{eff}) is the TMR ratio under zero (non-zero) bias voltage (V_M), and V_h is the bias voltage equivalent to $TMR_{eff} = TMR_0/2$. The current passing through the access transistor (I_M) satisfies [20]

$$I_{DS} = I_M = \mu_n C_{ox}\frac{W}{L}\left[(V_{GS} - V_{TH})V_{DS} - V_{DS}^2\right] \quad (6)$$

when the transistor is working in the triode (ohmic) region ($V_{GS} > V_{TH}$ and $V_{DS} < V_{GS} - V_{TH}$). Here, μ_n is the mobility of electrons, C_{ox} is the oxide thickness, and W (L) is the channel width (length). For $V_{DS} > V_{GS} - V_{TH}$ (saturation region), the channel exhibits pinch-off near drain and the current can be approximated by the maximum value of (6) as [20]

$$I_{DS} = I_M = \frac{1}{2}\mu_n C_{ox}\frac{W}{L}(V_{GS} - V_{TH})^2. \quad (7)$$

From a circuit point of view, for given MTJ device characteristics the value of the circuit parameters (V_A, I_{IMP}, and R_G in Fig. 4) can be optimized to minimize P_{err}. An example of such an optimization for the MTJ-based gate is presented in our previous work [13]. Based on (1)-(7), the

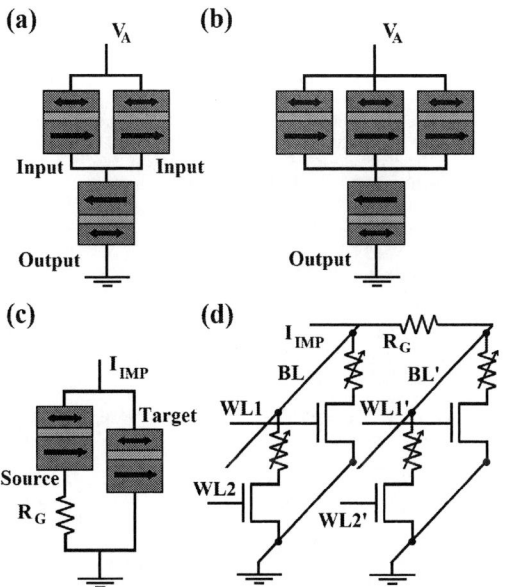

Fig. 4. Reprogrammable two-input (a) and three-input (b) MTJ-based logic gates [11], [12]. (c) Current-controlled MTJ-based implication logic gate [13]. (d) 1T/1MTJ-based IMP logic circuit.

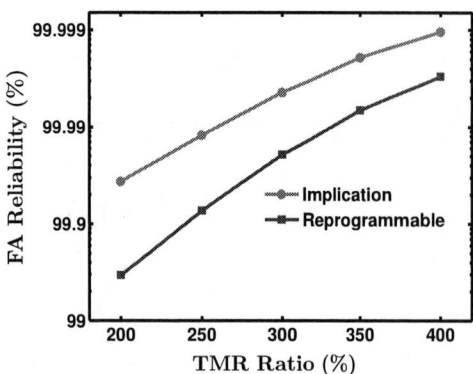

Fig. 6.   Reliability of the MTJ-based magnetic full adder as a function of the TMR ratio plotted for optimized circuit parameters and $\Delta_0 = 60$. The record room temperature TMR of 604% [21] promises highly reliable MTJ-based logic applications.

reliability of a magnetic full adder as a function of the TMR ratio is plotted in Fig. 6 for IMP-based and reprogrammable-based implementations with optimized circuit parameters. It demonstrates that the reliability increases with the TMR for both circuits. Due to a more reliable conditional switching behavior exhibited by the IMP gate, it provides a magnetic full adder with smaller error probability compared to the reprogrammable gates.

The 1T/1MTJ-based implementation of the intrinsic logic-in-memory gates can extend MRAM electronics from memory to logical computing applications for complex logic functions. An IMP-based full adder design involves 27 subsequent FALSE and IMP operations [16] on 6 1T/1MTJ cells which already exist within the STT-MRAM structure (Fig. 5). As compared to the non-volatile full adder in [6], the need for CMOS-based logic is eliminated and superior logic density can be achieved.

IV. CONCLUSION

A method for reliability analysis of the MTJ-based logic circuits is presented. As an example, the reliability of a magnetic full adder using IMP and reprogrammable gates is considered. It is shown that the reliability improves with the TMR for both designs and the IMP-based implementation exhibits a more reliable logic behavior. These structures can be used to build more complex circuits which implement more complex logic functions. Since the 1T/1MTJ cell can be utilized as both memory and the main computing element, non-volatile intrinsic logic-in-memory can be realized with no need for intermediate circuitry. Circuits which exhibit low power consumption, high logic density, and high speed operation simultaneously can thus be designed.

ACKNOWLEDGMENT

The work is supported by the European Research Council through the grant #247056 MOSILSPIN.

REFERENCES

[1] N. S. Kim, T. Austin, D. Baauw, T. Mudge, K. Flautner, J. S. Hu, M. J. Irwin, M. Kandemir, and V. Narayanan, "Leakage Current: Moore's Law Meets the Static Power," *Computer*, vol. 36, no. 12, pp. 68–75, 2003.

[2] S. Matsunaga, J. Hayakawa, S. Ikeda, K. Miura, T. Endoh, H. Ohno, and T. Hanyu, "MTJ-Based Nonvolatile Logic-in-Memory Circuit, Future Prospects and Issues," *Proc. Des. Autom. Test Eur. Conf. (DATE)*, pp. 433–435, 2009.

[3] L. Berger, "Emission of Spin Waves by a Magnetic Multilayer Traversed by a Current," *Phys. Rev. B, Condens. Matter*, vol. 54, pp. 9353–9358, 1996.

[4] C. Chappert, A. Fert, and F. N. V. Dau, "The Emergence of Spin Electronics in Data Storage," *Nat. Mater.*, vol. 6, pp. 813–823, 2007.

[5] W. H. Kautz, "Cellular Logic-in-Memory Arrays," *IEEE Trans. Comput.*, vol. C-18, no. 8, pp. 719–727, 1969.

[6] S. Matsunaga, J. Hayakawa, S. Ikeda, K. Miura, H. Hasegawa, T. Endoh, H. Ohno, and T. Hanyu, "Fabrication of a Nonvolatile Full Adder based on Logic-in-Memory Architecture using Magnetic Tunnel Junctions," *Appl. Phys. Express*, vol. 1, p. 091301, 2008.

[7] W. Zhao, E. Belhaire, C. Chappert, F. Jacquet, and P. Mazoyer, "New Non-Volatile Logic Based on Spin-MTJ," *Phys. Status Solidi (a)*, vol. 205, pp. 1373–1377, 2008.

[8] Y. Gang, W. Zhao, J. O. Klein, C. Chappert, and P. Mazoyer, "A High-Reliability, Low-Power Magnetic Full Adder," *IEEE Trans. Magn.*, vol. 47, no. 11, pp. 4611–4616, 2011.

[9] W. Zhao, L. Torres, Y. Guillemenet, L. V. Cargnini, Y. Lakys, J.-O. Klein, D. Ravelosona, G. Sassatelli, and C. Chappert, "Design of mram based logic circuits and its applications," in *ACM Great Lakes Symposium on VLSI*, 2011, pp. 431–436.

[10] W. Zhao, C. Chappert, V. Javerliac, and J.-P. Nozie, "High Speed, High Stability and Low Power Sensing Amplifier for MTJ/CMOS Hybrid Logic Circuits," *IEEE Trans. Magn.*, vol. 45, pp. 3784–3787, 2009.

[11] A. Lyle, J. Harms, S. Patil, X. Yao, D. Lilja, and J. P. Wang, "Direct Communication Between Magnetic Tunnel Junctions for Nonvolatile Logic Fan-out Architecture," *Appl. Phys. Lett.*, vol. 97, p. 152504, 2010.

[12] A. Lyle, S. Patil, J. Harms, B. Glass, X. Yao, D. Lilja, and J. P. Wang, "Magnetic Tunnel Junction Logic Architecture for Realization of Simultaneous Computation and Communication," *IEEE Trans. Magn.*, vol. 47, pp. 2970–2973, 2011.

[13] H. Mahmoudi, T. Windbacher, V. Sverdlov, and S. Selberherr, "Implication Logic Gates using Spin-Transfer-Torque-Operated Magnetic Tunnel Junctions for Intrinsic Logic-in-Memory," *Solid-State Electron.*, vol. 84, pp. 191–197, 2013.

[14] J. Borghetti, G. S. Snider, P. J. Kuekes, J. J. Yang, D. R. Stewart, and R. S. Williams, "Memristive Switches Enable Stateful Logic Operations via Material Implication," *Nature*, vol. 464, no. 7290, pp. 873–876, 2010.

[15] M. Hosomi, H. Yamagishi, T. Yamamoto, K. Bessho, Y. Higo, K. Yamane, H. Yamada, M. Shoji, H. Hachinoa, C. Fukumoto, H. Nagao, and H. Kano, "A Novel Nonvolatile Memory with Spin Torque Transfer Magnetization Switching: Spin-RAM," *IEDM Tech. Dig.*, pp. 459–462, 2005.

[16] H. Mahmoudi, V. Sverdlov, and S. Selberherr, "A Robust and Efficient MTJ-based Spintronic IMP Gate for New Logic Circuits and Large-Scale Integration," *Proceedings of the 17th International Conference on Simulation of Semiconductor Processes and Devices (SISPAD)*, pp. 225–228, 2012.

[17] Y. Higo, K. Yamane, K. Ohba, H. Narisawa, K. Bessho, M. Hosomi, and H. Kano, "Thermal Activation Effect on Spin Transfer Switching in Magnetic Tunnel Junctions," *Appl. Phys. Lett.*, vol. 87, p. 082502, 2005.

[18] J. D. Harms, F. Ebrahimi, X. F. Yao, and J. P. Wang, "SPICE Macromodel of Spin-Torque-Transfer-Operated Magnetic Tunnel Junctions," *IEEE Trans. Electron Devices*, vol. 57, no. 6, pp. 1425–1430, 2010.

[19] Y. Zhang, W. Zhao, Y. Lakys, J. O. Klein, J. V. Kim, D. Ravelosona, and C. Chappert, "Compact Modeling of Perpendicular-Anisotropy CoFeB/MgO Magnetic Tunnel Junctions," *IEEE Trans. Electron Devices*, vol. 59, pp. 819–826, 2012.

[20] B. Razavi, *Fundamentals of Microelectronics*. Wiley, 2006.

[21] S. Ikeda, J. Hayakawa, Y. Ashizawa, Y. M. Lee, K. Miura, H. Hasegawa, M. Tsunoda, F. Matsukura, and H. Ohno, "Tunnel Magnetoresistance of 604% at 300 K by Suppression of Ta Diffusion in CoFeB/MgO/CoFeB Pseudo-Spin-Valves Annealed at High Temperature," *Appl. Phys. Lett.*, vol. 93, p. 082508, 2008.

PRIME 2013, Villach, Austria
Session M3C – Digital Circuits

Optimal Deployment of Shadowed Registers in Systems with Serial Clock Distribution

Mario Trifković, Dušan Raič, Drago Strle
Faculty of electrical engineering, CO NAMASTE
University of Ljubljana
Ljubljana, Slovenia
mario.trifkovic@fe.uni-lj.si

Abstract – **Digital sections in several high-precision mixed systems operate at a relatively low frequency, allowing the trading of speed for the mitigation of switching noise and power distribution problems. While the serial clock distribution potentially solves most of these problems, the standard library cells and synthesis tools do not provide much, if any support for the serial clock tree implementation. In our work we propose the necessary design flow modification and seek the optimal clock signal allocation for the deployment of shadowed registers subject to timing constraints and minimal area overhead.**

Keywords—Switching noise; Serial Clock Distribution; Clock tree; Timing model

I. INTRODUCTION

Synchronous digital circuits implemented by standard synthesis tools are known to produce large supply current spikes due to simultaneous switching of registers, activated by the global clock. Large I*R and L*di/dt voltage drops on the power grid accelerate electro-migration and force the designer to oversize power lines and use multiple power pins, which increase manufacturing costs. Voltage fluctuation on supply lines is of particular importance in high-precision mixed systems where the need for low noise comes together with large data words, typically in the range of 24...48 bits. On the other hand, the processing speed is usually low, and is dictated by the physical limitations of analog circuits and sensors. Many precision systems, like inertial MEMS and data logging systems for seismic and medical applications have low output data rates and range from DC to about 150 Hz since the measurement results are highly filtered with the decimation process. As a consequence, the ratio of the peak supply current over the average current value may be very large, indicating the feasibility of trading the speed for the solution of power and switching noise related problems.

Synthesis tools put most of their effort into optimizing speed, power, and/or area. When it comes to switching noise they do not provide much support, and designers have to apply additional measures. Several approaches on the basis of clock skew scheduling have been proposed [1, 2]. Still, due to timing limitations the efficiency is relatively low, typically in the range of 10-30%. Our work is based on serial clock distribution which allows much better results, provided that enough timing slack is given. We address the serially clocked system in a fully synchronous environment, starting from the RTL description to

assure compatibility with standard synthesis tools, cell libraries and file formats. The required timing slack may be provided either by the long clock period or by processing in multi-cycle data paths. The general timing solution of data paths with negative hold time is assured by the application of shadowed registers [3]. We contribute the concept for an optimal division between standard and shadowed registers that is supplemented by the synthesis of the serial clock tree. The number of shadowed registers is minimized by appropriate allocation of the delayed clock signals in order to reduce the power and silicon area.

Timing and modeling issues related to the serial clock tree implementation with standard tools are presented in Section 2. Section 3 describes the method for the optimal deployment of shadowed registers and the construction of a serial clock tree. In Section 4 we illustrate the standard design flow with necessary modifications to implement the proposed methodology. Finally the conclusion and directions for future work are given in Section 5.

II. TIMING AND MODELING ISSUES

The concept of serial clocking was presented in [1] for the case of cascaded registers and clock buffers with the delay Δt. While the registers in regular synchronous systems switch almost simultaneously, causing a large current spike in a small time interval, cascaded registers (Fig. 1) distribute the switching events in time intervals so that the accumulated peak supply current is much lower; a reduction of ~80% has been reported in [1]. For proper circuit function condition (1) has to be fulfilled, where D_{Pmin} and D_{Pmax} are data delay paths between register cells, S_T is the setup time and Δt is the delay of the delay cell. It is worth mentioning that this condition is always met in cascaded digital structures with reverse serial clocking. If we consider serial clock distribution in a general case, it becomes clear that all data paths do not satisfy condition (1). The most critical paths have the high multiplier n between the register pair $(n\Delta t \gg \langle D_{Pmin}^{0n}, D_{Pmax}^{0n}\rangle - S_T^n)$, meaning that the

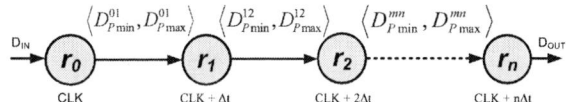

Fig. 1. Serial clock distribution

978-1-4673-4579-8/13 $31.00 © 2013 IEEE 161

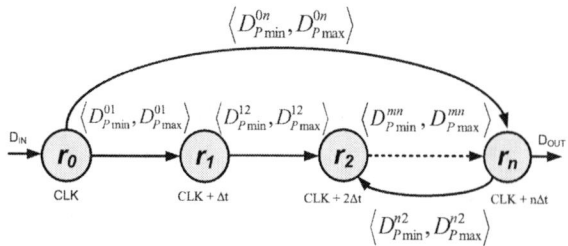

Fig. 2. Serial clock distribution on a general circuit

next register r_n in the register pair $r_0 \rightarrow r_n$ takes the value which in synchronous circuit would be taken in the next clock cycle.

$$\sum_{i=m}^{n} \Delta t^i < \left\langle D_{Pmin}^{mn}, D_{Pmax}^{mn} \right\rangle - S_T^n < T \qquad (1)$$

The application of a shadowed register gives the possibility of preforming a serial clock distribution in a general case. The shadow register has a triple-latch structure to strobe input data at the positive clock edge and a set output at the negative clock edge [3]. The data processing time is reduced to half of the clock period (2), while the other half remains to be used for clock distribution.

$$\sum_{i=m}^{n} \Delta t^i < \left\langle D_{Pmin}^{mn}, D_{Pmax}^{mn} \right\rangle - S_T^n < \frac{T}{2} \qquad (2)$$

For proper operation of a general digital circuit all critical registers have to be replaced by shadowed registers. An example of such a replacement (r_0) is shown in Fig. 3. The path delay of the critical register pair is extended for a half clock cycle period, which in this case satisfies condition (1).

In terms of power consumption it is not economical to place delay buffers in front of every register cell. Considering that all buffers in the serial clock tree drive equal loads and that the loads are small we can replace the delay buffers by register-internal clock buffers [4]. The standard, minimum clock skew tree is therefore eliminated, enabling a large power saving in combination with the peak supply current reduction. The resulting clock tree structure is presented in Fig. 4.

Standard synthesis tools do not accept register models with clock data path output. In order to implement the serial clock tree presented in Fig. 4 we have to split the register model into separate latch and clock buffer cells. The synthesis of the complete clock tree then requires only a few additional buffers outside of the register cells.

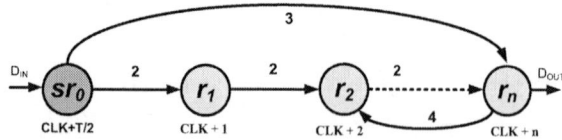

Fig. 3. Example of using a shadowed register

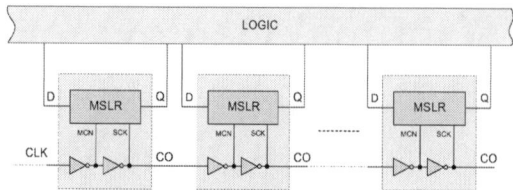

Fig. 4. The clock skew scheduling method

Register models with access to the delayed clock output signal must be prepared in the standard Liberty format [5, 6]. In the case of 'normal' registers the characterization tool recognizes the logical structure from the SPICE netlist and creates the gate file with the simulation plan for timing arcs. The non-standard structure from Fig. 4 is not recognized automatically and must be written by hand. It is necessary to specify complementary clock inputs *MCN* and *SCK*, using the Liberty model with the *clocked_on_also* statement [3] and eliminate redundant timing arcs from the automatically generated model. The adapted register model then allows the synthesis and static timing checking with standard tools, as in any other case of synchronous design.

III. SERIAL CLOCK DISTRIBUTION AND OPTIMAL DEPLOYMENT OF SHADOWED REGISTERS

As we have shown in Section 2, the shadowed register is used in critical register pairs to meet the timing constraints. In the general case there are many connections among registers scattered across the entire circuit so that a large number of shadowed registers would be needed for serial clocking. Since shadowed registers consume more power and area than standard registers the optimization tends to replace as many shadowed registers with standard registers as possible, without compromising the timing constraints. This can be done with an appropriate distribution of clock delay times among registers. The idea is to provide clock delays in proportion to data delays so that register pairs with large data delay would be given a more relaxed clock delay.

Circuit timing relations are described by the directed graph and the adjacency matrix [7]. We start with the initial adjacency matrix and timing details about slack time, path delay, setup time and hold time. The application of a serial clock defines two new matrices. The first matrix D_S reflects the distribution of clock buffers in increasing order, according to the matrix size. This matrix is principally needed to find the necessary number of parallel branches so that the total branch delay does not exceed the clock cycle time. We start to build D_S from two auxiliary matrices D_{Slr} and D_{Srl} representing the two possible clock data path orientations in increasing and decreasing index order, respectively.

$$D_{Srl} = \sum_{i=0}^{n} \sum_{j=0}^{n} n - j, \quad D_{Slr} = \sum_{i=0}^{n} \sum_{j=0}^{n} j - 1 \qquad (3)$$

The matrix D_S is then expressed as the sum of D_{Slr} and D_{Srl} and is masked by the upper and lower triangular mask matrices A_u and A_l. The second matrix D_C represents the initial clock delay distribution.

$$D_S = A_l \cdot D_{Srl} + A_u \cdot D_{Slr} \tag{4}$$

Therefore, it can be expressed as the absolute value of the difference between D_{Slr} and its transpose matrix D_{Slr} (5).

$$D_C = \left| D_{Slr} - D_{Slr}{}^T \right| \tag{5}$$

Depending on the clock delay line direction, the adjacency matrix can be interpreted in two different ways. If the clock signal path is directed in increasing indexing order then the upper elements of the adjacency matrix A_{dd} relate to the data path, while the lower elements relate to feedback connections (Fig. 5a). The opposite orientation A_{df} yields the feedback paths in the upper triangular part and the data in the lower one (Fig. 5b). We take both implementations in consideration to find a possible local optimum.

The delay matrix between register pairs without clock distribution is given as D_P:

$$D_P = T - t_{SLACK} - t_{SETUP} - t_{Sunc} \tag{6}$$

where T is the clock period, t_{SLACK} is the slack time, t_{SETUP} is the setup time and t_{Sunc} is the setup time uncertainty. In the next step we calculate the data and feedback path delay matrices, including the clock distribution. We assume that all clock delay cells have the same delay value. The data path delay matrix D_{dp} represents the delay of the register cell and the combinational logic, where t_{BUFF} is the delay time of the delay cell:

$$D_{dp} = D_P - (D_C \cdot t_{BUFF}) - t_{HOLD} - t_{Hunc} \tag{7}$$

The feedback matrix D_{fp} describes the available slack time. It is defined as the difference between the slack time and the clock distribution delay.

$$D_{fp} = t_{SLACK} - D_C \cdot t_{BUFF} \tag{8}$$

If all elements in matrices D_{dp} and D_{fp} are positive, then the matrix D_C represents an acceptable solution for clock distribution. However, this is generally not the case and it is necessary to use shadowed registers in critical register pairs as shown in Section 2.

Circuit complexity is reflected in the adjacency matrix edges where the need for shadowed registers is most probable.

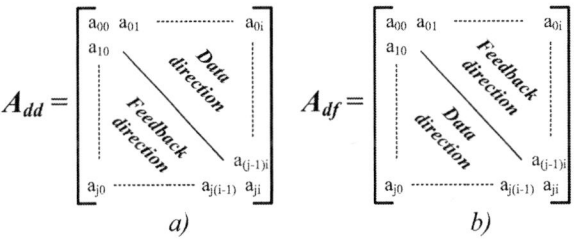

Fig. 5. Directed adjacency matrix interpretation

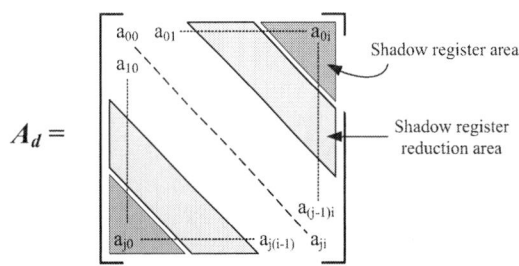

Fig. 6. Possible position of shadowed registers

This probability is illustrated by dark areas in Fig. 6. Register pairs in the dark area are therefore less likely to be optimized than those in the gray area.

Global optimum would be found by considering all possible reordering vectors. However, the number of all reordering vectors is too large to be implemented in reasonable computing time. To find an acceptable local optimum we first reduce the matrix bandwidth on the basis of well-known methods, such as the reverse Cuthill–McKee reordering algorithm [8]. This procedure may be improved by taking into consideration the difference between the upper and lower triangles. More effort should be put into the optimization of the data path triangle than into the feedback triangle. Additional refinement is possible by fine rearrangement of elements close to the diagonal.

The optimization concept is illustrated by a simple circuit shown in Fig. 7. After considering both orientations we find A_{dd} to be better than A_{df}. Serial clock distribution is in the direction of the data path $D_{dp} \rightarrow D_{dp} \cdot A_u$ (A_u is upper mask matrix). Negative values in the upper part of matrix D_{dp} indicate the need for shadowed registers. The reordering of the D_P matrix yields the reduction of shadowed registers so that problematic connections are moved from the gray area to the unmarked area (Fig. 6).

D_P elements (1, 3) and (2, 4) in our example represent problematic connections. As the register r_2 does not have any connection towards data the elements in the third row are equal to zero. This makes the swapping of the first and the third row possible. To preserve matrix symmetry the related columns are swapped as well. Similarly the second negative element is eliminated by swapping the second and third row and column, respectively. After insertion of the serial clock distribution in the reordered D_P matrix we get positive elements in $D_{dp} \cdot A_u$ (Fig. 8). The final circuit arrangement is presented in Fig. 9. Long data connections from Fig. 7 now represent feedback connections, while short feedbacks turn into data connections.

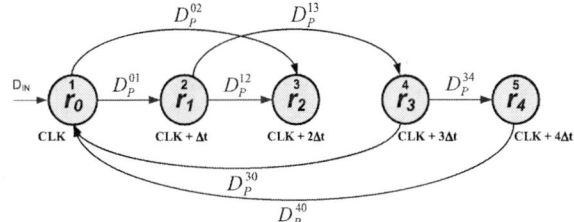

Fig. 7. Example circuit with initial clock distribution

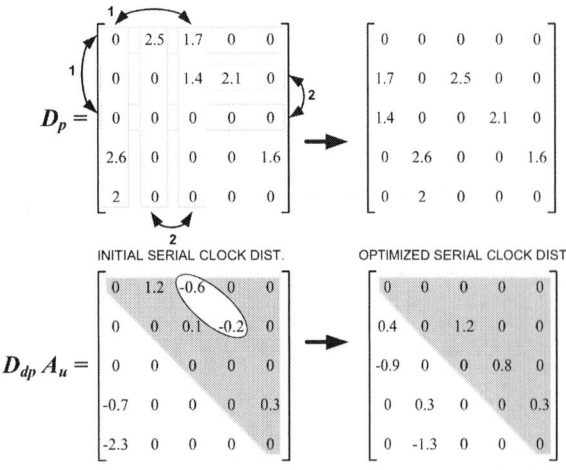

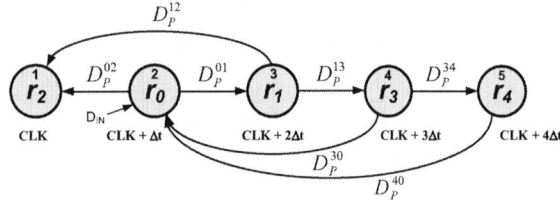

Fig. 8. Switching columns and rows of the adjacency matrix

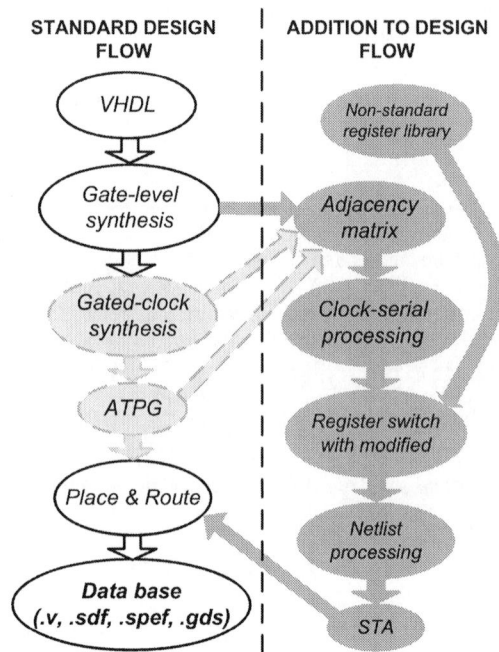

Fig. 9. An example circuit with optimized clock distribution

In this simple example the reordering vector of adjacency matrix gives the optimal solution.

IV. DESIGN FLOW

The proposed design flow is similar to the standard flow with the exception of additional steps at certain points. After the initial gate-level synthesis we extract the directed adjacency matrix from the tool database, using specific Tcl commands. The matrix is saved for separate mathematical processing and upgraded with the serial clock data. The matrix generator may be upgraded to take into consideration the specific requirements of the gated-clock synthesis and automatic test pattern generator (ATPG). In the next step the register swapping is applied to the netlist, where new connections are added for clock distribution based on the clock distribution matrix. For correct circuit timing, the STA (Static Timing Analysis) has to be done. In next steps the placement and routing are implemented, followed by database export using the standard synthesis tool procedures as shown in Fig. 10.

V. CONCLUSION

The concept of serial clock synthesis by optimal deployment of standard and shadow registers has been presented. As standard tools do not support the serial clock distribution and tend to work in the opposite direction (minimal skew), interference with non-standard steps is required.

Fig. 10. Design flow modification

However, tools have a wide set of commands which can be used to manipulate the design like it has been done in our case. The presented concept is applicable to the flat netlist structure. It has been tested on a few small-sized examples with satisfactory results. Future work will continue to upgrade the method to consider multi-cycle paths and ATPG-specific requirements, as indicated in Fig. 10 by dashed lines.

ACKNOWLEDGMENTS

The authors acknowledge Center of Excellence NAMASTE for supporting this research.

REFERENCES

[1] A. Vittal, H. Ha, F. Brewer, and M. Marek-Sadovska, "Clock skew optimization for ground bounce control," in Proc. IEEE/ACM Int. Conf. Computer-Aided Design., Nov. 1996, pp. 395-399.

[2] Benini L., Vuillod P., Bogliolo A. and De Micheli G., 'Clock-skew optimization for peak current reduction', Journal of VLSI Signal Processing, Vol. 16, 1997, pp.117-130

[3] Raič D., Shadowed register cell for skew-resistant timing, Electronics Letters, 8th July 2010 Vol. 46, No. 14

[4] Raič D., Patent application PCT/SI2008/000027, May 2008

[5] Cadence, "Encounter Library Characterizer User Guide", Product Version 9.1.2, July 2010

[6] E. Brunvand, Digital VLSI Chip Design With Cadence and Synopsis CAD Tools, Pearson 2010.

[7] I. S. Kourtev, E. G. Friedman, "Timing Optimization Through Clock Skew Scheduling". ISBN 0-7923-7796-6, Kluwer Academic Publishers, Norwell, Massachusetts USA, 2000.

[8] E. Cuthill, J. McKee, "Reducing the bandwidth of sparce symmetric matrices," in Proc. ACM, 24th national conference (1969), pp. 157-172.

Monte Carlo Based Post-Silicon Verification Considering Automotive Application Variances

Manuel Harrant
Thomas Nirmaier
Jérôme Kirscher
Infineon Technologies AG
Neubiberg, Germany
Manuel.Harrant@infineon.com

Christoph Grimm
Technische Universität Kaiserslautern
Kaiserslautern, Germany

Georg Pelz
Infineon Technologies AG
Neubiberg, Germany

Abstract—**In this paper we present a fully automated approach to consider device-to-device variances of automotive power applications during post-silicon verification. Due to the high complexity of target applications for automotive smart power microelectronics, it is not sufficient to affirm compliance to their specification. Car manufacturers therefore push for more extensive application robustness beyond classical methods. To cope with this requirement a FPGA platform is used to evaluate physical equations of automotive power application components in real-time together with a dynamic power amplifier to interface the digital FPGA outputs to the analog world. The functionality and the advantage of this approach is evaluated based on several Monte Carlo experiments by using an Advanced Front Lighting system as an example.**

Keywords: Automotive Smart Power IC, Automotive Application Emulation, DC Motor Modelling

I. INTRODUCTION

The complexity of automotive power micro-electronic devices is mainly driven by a continuously rising customer demand for energy efficiency and safety. Verifying whether the automotive power device is working under all possible operating conditions cannot be covered by traditional post-silicon verification methods, which are mainly oriented to check their compliance to specification with either directed or statistical verification methodologies.

Nowadays, car manufacturers more and more push for assessing robustness inside their target application beside traditional compliance to specification. In the past, automotive power microelectronics were verified completely independent from their target application and its robustness was rarely checked during application test or sometimes called system verification. System verification mainly relies on known issues from former device failures, specific test cases or requirements coming from the car manufacturer and the supplier itselves. Hence, there is room for deeper investigations inside this area to achieve more coverage within the application space.

For that reason we want to explore the device's verification space by emulating the dynamic behavior of several external components from the target automotive application, including correlations between these components. Statistical Monte

This research project is supported by the German Government, Federal Ministry of Education and Research under the grant number 01M3195.

Carlo experiments should give feedback regarding the behavior of the micro-electronic device inside its target application.

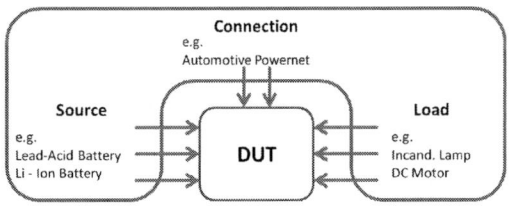

Fig. 1.   Application-driven post-silicon verification

II. STATE OF THE ART

Automotive smart power products have to operate reliable and safe in a wide range of target applications and, moreover, the target application is often not completely known during post-silicon verification of the micro-electronic device. This leads to a broad field for research topics that is rarely covered nowadays. Verification with respect to target application is mostly done during pre-silicon verification but there are also some investigations found to be used for lab measurements.

Most common methods in this field of study are related to hardware-in-the-loop concepts [2], [3], which are closely related to the presented approach. Hardware-in-the-loop concepts are primarily used for testing complete electronic control units within their environment (e.g. bus communication to other control units, correct control of external components by the device under test, etc.) but more or less unusable for an application-driven post-silicon verification approach of smart power devices due to missing power levels.

There are other approaches, going in a similar direction as the presented one, considering single parts of the target automotive applications. Some solutions are found regarding battery emulation, especially lithium-ion batteries [4], or for emulation of specific automotive loads [5].

But these solutions do not allow to go deeper into the broad portfolio of automotive application that may influence the correct behavior of automotive microelectronics as far as they only consider single functional blocks.

978-1-4673-4579-8/13 $31.00 © 2013 IEEE

III. Application Emulation Concept

The basic hardware concept, as shown in Fig. 2, used for this kind of automotive application emulation, was already presented in [12], [13] and used for emulation of either incandescent lamps or Lithium-Ion battery cells. Considering as many automotive application components as possible, we did some further development in this hardware concept as well as the software architecture. We established a connection to a host PC for automated Monte Carlo experiments with access to the full application space.

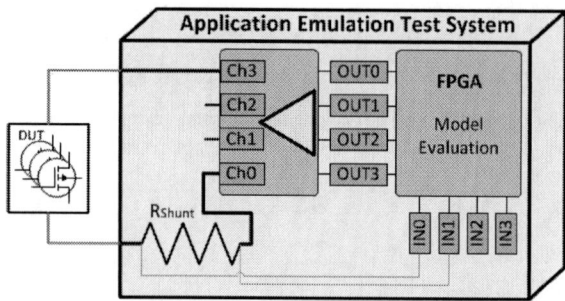

Fig. 2.   Hardware setup for application emulation

The amplifier's inputs can be configured dynamically working as a voltage source or as a current sink inside the FPGA. In the presented configuration, CH3 is configured as voltage source according to [13] to generate the DUT's supply voltage and CH0 is configured as current sink according to [12] to sink the calculated load current.

This concept allows to consider the following relation between power source and power load connected to the micro-electronic device:

$$v_{Bat}(t) = f(i_{Load}, BatteryModel) \qquad (1)$$

where v_{Bat} is the time-variant battery voltage dependent on the discharge current and the model parameters from the battery itself and the power net architecture.

$$i_{Load}(t) = f(v_{Bat}(t), LoadModel) \qquad (2)$$

where i_{Load} is the time-variant load current dependent on the battery voltage and the load model.

According to the presented concept it is possible to support single-channel devices such as smart power switches or linear voltage regulators as well as half bridges or airbag driver ICs consisting of an high-side/low-side switch combination.

IV. Implementation Example

As an automotive application example, as presented in Fig. 3, a battery (1) combined to a part of the automotive power net architecture (2) and a DC motor (e.g. used within Advanced Front Lighting application) as an automotive power load (3) have been chosen. In our test case the micro-electronic device was an integrated motor driver.

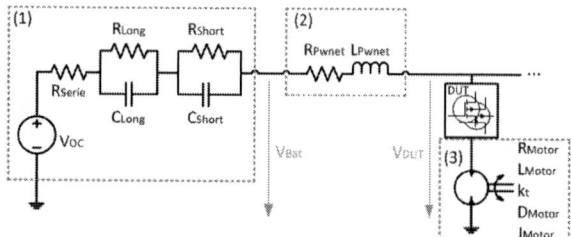

Fig. 3.   Automotive application example

(1) Battery

The battery model consists of the electrical equivalent circuitry containing a serie resistor for the ohmic voltage drop and two time constants for dynamic behavior. The battery voltage can be calculated with:

$$v_{Bat}(t) = v_{oc} - v_{R_{Serie}} - v_{\tau_{Long}} - v_{\tau_{Short}} \qquad (3)$$

where v_{oc} is a non-linear voltage source, $v_{R_{Serie}}$ is the ohmic voltage drop and $v_{\tau_{Long}}$ as well as $v_{\tau_{Short}}$ describe the dynamic behavior of the battery.

(2) Automotive Power net

A basic implementation of the automotive power net is the serial connection of wire resistance R_{Pwnet} and wire inductance L_{Pwnet}. Based on Kirchhoff's voltage law, the output voltage $v_{DUT}(t)$ can be derived from the time-variant battery voltage $v_{Bat}(t)$ as follows:

$$v_{DUT}(t) = v_{Bat}(t) - R_{Pwnet} \cdot i(t) - L_{Pwnet} \cdot \frac{di}{dt} \qquad (4)$$

There are also more complex models available considering a parasitic capacitance to the ground net. Since this capacitance is pretty low, it was ignored for this first implementation.

(3) DC Motor

The model of the DC motor is based on an electro-mechanical model, considering the armature resistance, armature inductance and the back electro-motive force (Back-EMF) for the electrical part and the generated torque and the power losses for the mechanical part. This behavioral model was already presented in [10] and the related parameters were extracted in [11] to be used for simulations in VHDL-AMS.

$$v_{DUT}(t) = k_t \cdot \omega + R_{Motor} \cdot i(t) + L_{Motor} \cdot \frac{di}{dt} \qquad (5)$$

$$T = -k_t \cdot i(t) + D_{Motor} \cdot \omega + J_{Motor} \cdot \frac{d\omega}{dt} \qquad (6)$$

where $v_{DUT}(t)$ is the time-variant voltage at the DUT output, $i(t)$ is the motor current, T is the generated torque and ω is the angular velocity. R_{Motor}, L_{Motor}, D_{Motor}, J_{Motor} and k_t are motor parameters.

The differential equations were transformed, as described in [12], to be used for real-time evaluation inside the FPGA target.

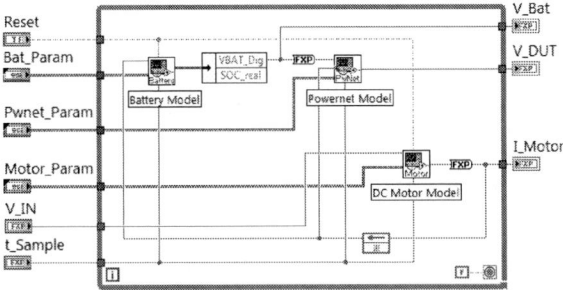

Fig. 4. FPGA block circuit in LabView toolbox

The transformed equations from (3), (4), (5) and (6) are implemented inside the three subsystems shown in Fig. 4. On the left side all model parameters for battery, power net and motor are delivered within cluster variables to the block circuit. V_{DUT} and I_{Motor} are converted to analog and control the amplifier's inputs.

The presented approach allows to access and adjust all application-relevant parameters, listed in table I, during post-silicon verification of the Automotive Smart Power micro-electronic device.

TABLE I
VARIABLE PARAMETERS OF APPLICATION EXAMPLE

(1) Battery		
Parameter	**Description**	**Performance Range**
C_{BAT}	Battery Capacity	50 Ah ... 90 Ah
SOC	Battery State of Charge	0 ... 1
(2) Power net		
Parameter	**Description**	**Performance Range**
R_{Pwnet}	Power net Resistance	20 mΩ ...120 mΩ
L_{Pwnet}	Power net Inductance	2 μH ...6 μH
(3) DC Motor		
Parameter	**Description**	**Nominal Value**
R_{Motor}	Armature Resistance	35 Ω
L_{Motor}	Armature Inductance	34.1 mH
k_t	Torque Constant	9.95 $\frac{mV}{rad/s}$
D_{Motor}	Damping Constant	$1.56 \cdot 10^{-7} \frac{Nm}{rad/s}$
J_{Motor}	Moment of Inertia	$1.62 \cdot 10^{-7} kg \cdot m^2$

The dynamic behavior of the battery is dependent on the discharge current and the state of charge. That is why only the battery capacity and the initial state of charge was identified as real input parameter for the battery model.

Experimental results were made for automotive Advanced Front Lighting applications. The presented values in table I of the identified motor parameters are nominal values, characterized in [11], while values for the power net are general standard values.

V. EXPERIMENTAL RESULTS

The great advantage of using this application-driven characterization approach is to access and to adjust all identified component parameters within the emulated automotive application. However, it is necessary to think about reasonable distributions of all parameters listed in table I.

The specific motor for Advanced Front Lighting will have a nominal set of motor parameters including a spread from their nominal values coming from manufacturing tolerances. Usually low performance and low cost motors are used for this application and the spread can be significant for the behavior of the motor driver. For the set of experiments carried out here Gaussian distributions were chosen for the motor parameters.

The same goes for the battery, which suffers from temperature dependencies and aging effects. This is why battery parameters also deviate from their nominal values.

The dynamic behavior of the automotive power net is mainly influenced by length and thickness of the wire, implementation inside the vehicle, etc. These parameters have a wide range which fits to an uniform distribution to be as close to reality as possible.

First results, as they are shown in Fig. 5, were made with normal distributed motor parameters and standard deviation $\sigma = \pm 5$ % from its nominal value. The current waveform was monitored to evaluate its dependency to these identified application parameters.

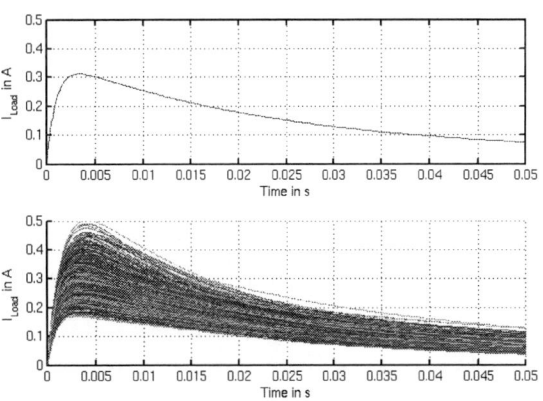

Fig. 5. Load current for nominal motor parameter (top) and for gaussian distribution of motor parameters (bottom)

It is obvious that even the small spread of motor parameters has a significant influence on the output of interest. The current waveform differs from the nominal waveform with a deviation of about 50 %.

For verification tasks, and especially for an application-oriented approach, of such an integrated motor driver the behavior during switching off inductive loads is of great importance. Indeed the electro-motive force voltage and the induced current are in such a direction that they tend to oppose the change according to Lenz's law. Therefore, when a high-side switch attempts to stop sensing current, the inductor will discharge in some finite time and a reverse polarity voltage pulse across it will occur. This means that the output voltage of the high-side driver will fall below ground and eventually reaches the avalanche break-down of the micro-electronic device.

That is why second results were obtained for the maximal energy dissipated at the inverse diode caused by the modelled DC motor considering a Gaussian distribution for motor parameter spread.

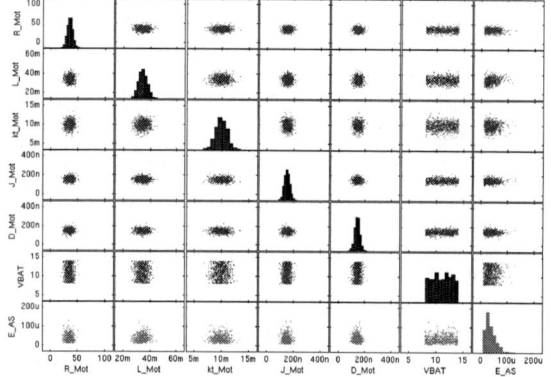

Fig. 6. Matrix Plot of 1000 Monte Carlo experiments for maximum energy dissipation (E_{AS}) related to different distributions of application-related input parameters

Matrix plots were chosen to visualize dependencies between on the one side the output of interest, and input parameters and on the other side between input parameters themselves. Figure 6 shows the matrix plot for the maximal energy dissipation E_{AS} dependent on all application-relevant input factors like armature resistance (R_{Mot}) and inductance (L_{Mot}), torque constant (kt_{Mot}), damping constant (D_{Mot}) and moment of inertia (J_{Mot}).

For a better overview, the influences coming from the battery and the power net architecture were compressed into one single parameter called V_{BAT}. According to the low current consumption of the motor only the initial state of charge of the battery will have a visible influence on the system behavior.

The remaining energy in the armature inductance was calculated according to:

$$E_{AS} = \frac{1}{2} \cdot L_{Motor} \cdot i(t)^2 \tag{7}$$

Due to the fact that all application-related parameters, except the compressed battery voltage, were Gaussian distributed, the standard deviation σ was only about 5 % and the energy dissipation E_{AS} at the inverse diode is pretty small, it is not possible to see significant dependencies between either input parameters nor the output.

The mean value of the energy dissipation lies at approximately $40\mu J$ but the variance in application-relevant parameters leads to a maximum value of more than $100\ \mu J$, which is 2.5 times higher than its mean value, for special combinations.

VI. CONCLUSION AND OUTLOOK

In this paper we present an fully automated approach to consider the variance of target application components during post-silicon verification of automotive smart power devices. The concept has been proven on the basis of a real automotive application and several Monte Carlo experiments were made during first tests. The approach can help to design automotive smart power products more in the direction of their target applications (e.g. to size clamping diode).

Future investigations to further improve this concept can be the implementation of additional components to extend the actual example. This can be an alternator which is working in parallel to the automotive battery to stabilize the vehicle electrical system voltage or additional mechanical parts like gear boxes connected to the motor. Other directions can be to simulate switching behavior of other electronic content inside the vehicle or the influence of transients, such as short circuits in other electronic parts, that influences the device under test.

REFERENCES

[1] T. Nirmaier, V. Meyer zu Bexten, M. Tristl, M. Harrant, M. Kunze, M. Rafaila, J. Lau, G. Pelz
Measuring and Improving the Robustness of Automotive Smart Power Microelectronics, Design, Automation and Test in Europe (DATE) 2012
[2] C. Dufour, J. Belanger, V. Lapointe
FPGA-based Ultra-Low Latency HIL Fault Testing of a Permanent Magnet Motor Drive using RT-LAB-XSG, POWERCON 2008, pp. 1-7
[3] M.T. Shih, H.C. Chang
A DSP FPGA Based Hardware-in-the-Loop Testing Platform International Conference on Control, Automation, Robotics and Visions (ICARCV) 2010, pp. 1980-1985
[4] A. Thanheiser, T.P. Kohler, H.-G. Herzog
Battery emulation considering thermal behavior, Vehicle Power and Propulsion Conference (VPPC) 2011, pp. 1-5
[5] O.A. Mohammed, N.Y. Abed, S.C. Ganu
Real-Time Simulations of Electrical Machine Drives with Hardware-in-the-Loop, IEEE Power Engineering Society General Meeting 2007
[6] M. Chen, G.A. Rincon-Mora
Accurate Electrical Battery Model Capable of Predicting Runtime and IŬV Performance, IEEE Transactions on Energy Conversion 2006, pp. 504-511
[7] L. Gao, S. Liu, R.A. Dougal
Dynamic Lithium-Ion Battery Model for System Simulation IEEE Transactions on Components and Packaging Technologies 2002, pp. 495-505
[8] O. Erdinc, B. Vural, M. Uzunoglu
A dynamic lithium-ion battery model considering the effects of temperature and capacity fading Conference on Clean Electrical Power 2009, pp. 383-386
[9] O. Tremblay, L.-A. Dessaint, A.-I. Dekkiche
A Generic Battery Model for the Dynamic Simulation of Hybrid Electric Vehicles, Vehicle Power and Propulsion Conference (VPPC) 2007, pp. 284-289
[10] A. Pirker-Fruehauf, K. Schonherr, A. Laroche, G. Pelz
Worst-Case Modeling and Simulation of an Automotive Throttle in VHDL-AMS Behavioral Modeling and Simulation Workshop (BMAS) 2007, pp. 142-152
[11] J. Kirscher, M. Lenz, D. Metzner, G. Pelz
Verification of an Automotive Headlight Leveling Circuit and Application Using Smart Component Property Extraction, Behavioral Modeling and Simulation Workshop (BMAS) 2008, pp. 1-6
[12] M. Harrant, T. Nirmaier, G. Pelz, D. Fona, C. Grimm
Configurable Load Emulation using FPGA and Power Amplifiers for Automotive Power ICs, Forum on Specification and Design Languages (FDL) 2012, pp. 84-89
[13] M. Harrant, T. Nirmaier, G. Schwarzberger, C. Grimm, G. Pelz
Battery Cell Stack Emulation for Battery Management IC Characterization, Workshop Test und Zuverlässigkeit von Schaltungen und Systemen (TUZ) 2013, pp. 29-32

tLIFTING: an Open-Source Multi-Level Fault Simulator for Ionizing Effects

Feng Lu, Giorgio Di Natale, Marie-Lise Flottes, Bruno Rouzeyre

LIRMM (Université Montpellier II /CNRS UMR 5506)
Montpellier, France
{lu, dinatale, flottes, rouzeyre}@lirmm.fr

Abstract—This paper presents a multi-level simulator tLIFTING for fault simulation in digital circuits. Multi-level simulation is used for precision of the information of the fine grain transistor level and the conciseness of the coarse grain logic level. This multi-level process allows handling natural and maliciously induced physical phenomenon leading to circuit misbehavior, while dealing with large circuits.

Keywords – Multi-level Fault Simulation; Ionizing Effects

I. INTRODUCTION

Fault Simulation has been widely developed for analysis of the operation of a circuit in the presence of faults and thus for evaluation of test sequences, construction of fault dictionaries, and for its important role in test generation.

Analysis of operating conditions in presence of faults is particularly important for reliability evaluation. Fault simulation indeed permits to verify correct operation of error-detection/correction mechanisms implemented with the intention of coping with phenomenon that cause the system to fail. These mechanisms were originally implemented in high-reliability systems to cope with natural and non-predictable phenomenon. More recently, error detection/correction mechanisms are also developed for secure applications because related circuits may be the targets of numerous attacks, and namely the fault attacks where faults are intentionally injected in the structures for retrieving sensible information [1].

The complexity of fault simulators depends on the accuracy of the models (defects and circuit). Abstraction of physical perturbations into higher-level fault models is the common practice to improve the performances of fault simulators and other conception/test tools. For instance, the Stuck-At fault model together with circuit modeling at logic level allows dealing with very large circuits. Besides, test patterns for production testing of digital devices are generated from these models. However, the stuck-at fault model does not cover numerous low-level defects (e.g. defects that lead a transistor to be always on or off whatever the gate value). The stuck-at model is not appropriate either for modeling misbehaviors related to bulk ionization after particle strikes or laser illumination. Ionization phenomenon behaves as current sources in transistors' bulk. This transient phenomenon may result in a voltage transient at the output of the incriminate gate, the so call Single-Event-Transient (SET) model, or a bit flip in a memory element, the Single-Event-Upset (SEU). Both models have been extended to multi-site situations: Multiple-Event-Transient and Multiple-Event-Upset. However, even simple MET and MBU cannot fully model ionizing phenomena since lacking of information on: effects of very long transients; neighboring locations of affected gates; the number of failing gates/memory elements; their physical location on a layout; the duration of simultaneous transient faults. Therefore, the relation of the fault effect on multiple gates cannot be easily described with classic fault models. A straightforward solution would be the use of fully electrical- or physical-level simulators. However, the execution time required for this fine-grained simulation would not be acceptable for today's large circuits.

Multi-level simulation intends to provide low-level information handling for simulation of transistor-level phenomenon and higher-level information handling for simulation of logic-level events. The low level is used for fault injection at given spots, while logic-level is used for computation of the rest of the circuit for run time optimization.

In this paper we present tLIFTING (timing LIRMM Fault Simulator), a multi-level fault simulator able to perform delay-annotated fault simulations. It integrates transistor-level simulations for part of the circuit affected by user-defined electrical-level fault models, and transparently automates the link between the different abstraction levels.

The paper is organized as follows: Section 2 provides background and state-of-the-art. Section 3 describes the architecture of the fault simulator, whereas Section 4 describes the multi-level approach. Section 5 presents some experimental simulation results. Eventually, Section 6 concludes the paper and draws future perspectives of this work.

II. STATE-OF-THE-ART

Fault simulations at different levels (e.g., transistor level, gate level, RTL level ...) are represented in numerous approaches. Among them, the multi-level which includes low-level and delay-annotated fault simulation is a short-board. Now due to the increase in the transistor density and the clock frequency, the ionizing effect induced transient fault simulation, such as laser induced fault simulation, is becoming a hot topic for assessing the effectiveness of fault-tolerant mechanisms at design time. We focus here on simulators that handle transient faulty phenomena, and thus require delay-annotated computations, and multi-level descriptions for precision of low-level information and conciseness of high level description.

The work in [2] presents a method for simulating SET in VDSM (very deep submicron) ICs. At first, a double-exponential current pulse is generated and transformed to voltage pulse with the stricken node parameters by analog simulator (SPICE). Then, this voltage pulse is digitized with the logic threshold. Finally, the digital transient pulse is injected by modern simulators. Since the authors' focus was on SET fault simulation, the MET fault and multi-electrical fault simulation are not mentioned. In addition, through the experiment, we found that the accuracy is reduced if the analog fault pulse is digitized just at the stricken node.

Paper [3] presents a gate-level simulation environment for alpha-particle-induced transient faults. It includes two

978-1-4673-4579-8/13 $31.00 © 2013 IEEE

simulation engines for both annotated-delay and zero-delay simulations. Author defined a set of customized "latching windows" corresponding to the different widths of transient pulse for each flip-flop of the standard cell. The timing logic-level simulation is operated with the arrival time of the transient pulse and the "latching windows". It has been shown that an improvement in simulation time is achieved by using the fault-driven algorithm as opposed to the standard event-driven algorithm to perform fault injection. Once the transient faults have been latched, the timing simulator is no longer needed and further speedup is achieved by using a zero-delay parallel fault simulator. Unfortunately, as a gate-level fault simulator, the fault model is defined as a logic pulse with different widths for the analogic phenomenon of α-particle injection. Moreover, for their experiments, charges are injected only at the output nodes of gates. All the constraints of simulation will not hold in reality, it represents a reasonable compromise between modeling effort and usefulness of fault simulation results.

A multi-level fault simulator operating at switch- and gate-levels is described in [4]. It supports the inertial delay model which is the minimum pulse width for an input pulse to cause a gate output switch. A transistor is modeled as a switch in series with a resistor. In this approach, transistor conductance and node capacitance are modeled by discrete strengths in order to respect both switch-level fault-models and timing information. But by modeling as a switch for transistor, switch-level simulation operates logic-level values (1, 0 and X). Thus, it is not sufficient for the electrical ionizing-induced fault which is modeled as an analog pulse.

A multi-level fault-simulator for bridging faults is presented in [5]. It supports the gate-, switch- and electrical-level fault simulation. For the fault simulation, a region is defined by the actual fault sites plus some digital levels forward in the circuit, and in which low-level simulation is applied, while the rest of the circuit that is not directly affected by a fault is simulated at gate-level. For the low level simulation, switch-level simulation is the inaccuracy incurred whenever there is a conflict in CMOS circuits which have a path from power to ground. Thus, it uses electrical-level simulation whenever there is a path from power to ground, and uses switch-level analysis otherwise, within the low level region. Unfortunately, the implementation is insensitive to timing, and the fault model considered was quite limited for the simulation of different faults injection. However, the concept of low level region is instructive to multi-level simulation.

III. LOGIC SIMULATOR ARCHITECTURE

tLIFTING is an open-source fault simulator tool which allows both 0-delay and delay-annotated simulations for digital circuit described in Verilog. Figure 1 sketches its architecture.

The fault simulator reads the netlist of the circuit described in verilog (.v), and the input test sequence described in a proprietary format (.ts).

Moreover, the simulator can read the Delay-Annotated file, which provides information related to the delays of each gate in the Standard Delay Format (.sdf), and the fault list, which explicitly define the faults to take into account.

In order to handle any technological library, the simulator integrates a converter that, starting from the information of each cell (I/O number, pin names, truth table, and input capacitance), generates a corresponding

tLifting-compatible description (i.e., in C++).

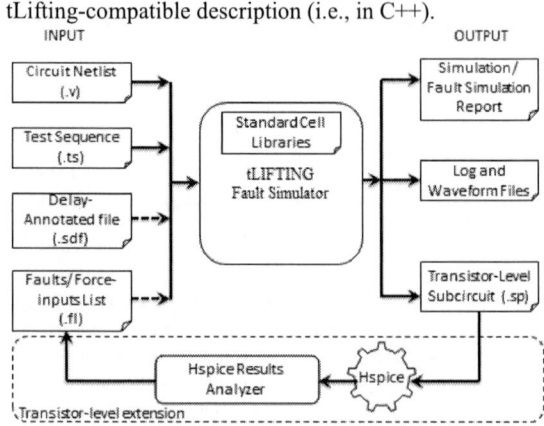

Figure 1. tLIFTING architecture

A. delay logic simulation

The basic idea of the simulator is that each circuit gate is modeled as a C++ class. The root of the object model hierarchy is the abstract class *generic_gate*. It includes the information of a common digital gate such as its name, number and current values of inputs and outputs, and the list of gates connected to each output. Moreover, a method is implemented to set the value of an input (*set_input*) while a virtual method (*calculate_output*) is used to define the logic function of the gate. Each standard cell is written as a class that inherits from *generic_gate* and that implements this method in order to specify the functionality of the gate.

When the simulator reads the netlist of the circuit, *generic_gates* are instantiated and linked one to the others by building the proper *fan_out* array. This array stores, for each *generic_gate*'s output port, all the input ports of the gates connected to it.

The main method of *generic_gate* is *set_input*. It resorts to the *calculate_output* method to determine the new output value of the gate. When the new value is different from the old one, it propagates the new value to each gate connected at the output by invoking the *set_input* method of those gates. This recursive method allows propagating any signal variation up to primary outputs of the circuit.

B. Timing Simulation

tLIFTING implements an event-driven simulator engine to allow delay-annotated simulations. It is based on a priority queue to store simulation events. Each event is characterized by the time and the type. The queue is sorted according to the event time, and a new element inserted in the queue will be placed accordingly. While the priority queue is not empty, the simulator pops the first event and executes the corresponding action according to type of event.

The simulator manages 4 types of event: Primary Input (PI) event, Propagation event, Fault Injection (FI) event and Fault Release (FR) event. PI events are generated when reading the input stimuli file. This event means that a primary input value is presenting. Propagation events are created whenever the output of a gate changes its value. The corresponding event time is set to the current time plus the propagation delay of the gate. FI and FR events are used to handle faulty behaviors. FI event set the value of a specific port to 0 or 1 during the simulation, no matter the actual value of the signal. On the contrary, FR event removes the previously applied FI fault and restores the proper value of the specific port.

The delay information, which is given by design tools, can be more or less precise based on the design level. For instance, before placement and routing there is not the delay of wires while the delay between the instant the gate input switches and the instant the gate output switches is included.

Before starting the simulation, all events related to faults are created according to what defined in the fault list. These events are then inserted in the priority queue.

IV. MULTI-LEVEL SIMULATION METHOD

tLIFTING enables multi-level fault simulation to analyze the effect of transient faults generated by ionizing effects, which can be described only at electrical level. It automatically performs the simulation of the whole circuit at gate level before the fault appearance, then the fault simulation at electrical level of only the gates involved in the fault, to move again to the gate level to finish the simulation. It improves the simulation run time compared to full transistor-level fault simulation.

A. Electrical Fault Model

A transient fault is not the result of a circuit defect but results from an external phenomenon such as a particle strike or a laser pulse. When an energetic particle or a beam of laser passes through a micro-electronic device, it will have an ionizing effect that triggers the formation of a dense track of electron-hole pairs. The separation and diffusion of these electron-hole pairs can produce a current transient which is subsequently observed on the signal node as a transient fault pulse. Normally, the transient fault is not considered permanently damaging to the transistor's or circuits' functionality.

The Figure 2 presents a simple electrical model of fault induced by laser. Recently, many approaches about electrical model of Photoelectrical Laser Stimulation (PLS) are published. According to paper [6], PN junction in CMOS is susceptible to a photoelectrical effect if exposed to a laser beam, and this effect can be modeled as a current source

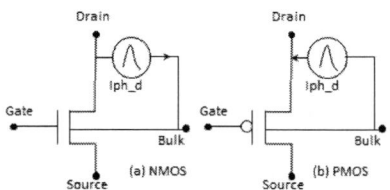

Figure 2. The simple electrical model for the fault induced by laser injection

Obviously its electrical model is based on a photocurrent source that corresponds to the photoelectrical effect in drain junction. For calculating amplitudes of this photocurrent source, it is necessary to give the value of laser's power. In this paper, we take this simple example just to explain non-permanent fault injection of our simulator.

B. Transistor-level Fault List

Transistor-level fault list is defined by user for specifying the fault parameters and its location. With this fault list, the fault descriptions are modeled and injected into the transistor-level circuit element description for the transistor-level fault simulation. With the example shown in

Figure 2, a fault induced by laser can be injected by the statement with the key word "*ionize*":

```
ionize cmos_name  current_value
```

The type of cmos can be automatically identified as nmos or pmos to insert a double exponential current source (Figure 3) with the current value of *current_value* between the drain and the bulk.

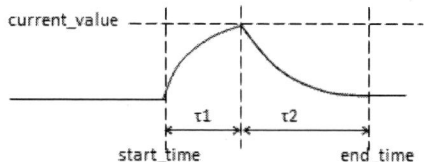

time_constant_ratio = τ1/τ2

Figure 3. The schematic diagram of exponential source

The parameters of start and end time of the fault pulse and the ratio of rise/fall time constant are defined by additional key word "*time*" for each faulty cmos in the fault list:

```
time start_time  end_time  time_constant_ratio
```

The following is an example in which a laser spot illumines two inverters (Figure 4.a) and its complete fault list (Figure 4.b).

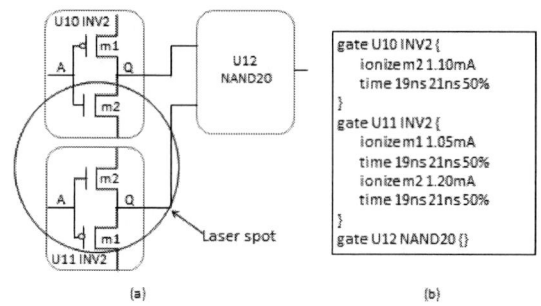

Figure 4. An example of laser induced fault, (a) fault location with laser spot, (b) complete fault list

These faults which are modeled by inserting different current sources are injected in the nmos named "*m2*" of the inverters "*U10, U11*" and in the pmos "*m1*" of the inverters "*U11*" from 19 to 21 ns.

In the fault list, user can also define fault-free gate in order to create the faulty transistor-level sub-circuit with a certain fault-free part, such as the nand gate "*U12*" in the example above.

C. Cross-level Simulation Process

The justification of the multi-level simulation resides in the fact that the fault model cannot be abstracted at a logic level but requires a dedicated simulation at electrical level. On the other hand, the whole circuit does not necessitate a complete simulation at electrical level since most part of the circuit is not affected and behaves as in the fault-free scenario. This multi-level fault simulation flow thus intends to improve simulation run time compared to full transistor-level fault simulation.

Figure 5.a shows the topological partitioning of the circuit for the multi-level simulation. The gates directly affected by the perturbation are shown in the sub-circuit $\mathbb{G}$, which has n input bits and m output bits. We define $\mathbb{C}$ as the set of gates in the input and output cones of $\mathbb{G}$. Finally, $\mathbb{S}$ is the whole system. Figure 5.b shows the timing partitioning, which will be used for the overall simulation process. The simulation starts at T_0. The perturbation affecting $\mathbb{G}$ lasts from T_{Pon} to T_{Poff}. To accurately take into account the effects of the perturbation on the whole circuit, we decided to consider the clock limits around the perturbation for the fine-grained simulation (instants T_{Start} and T_{Stop}).

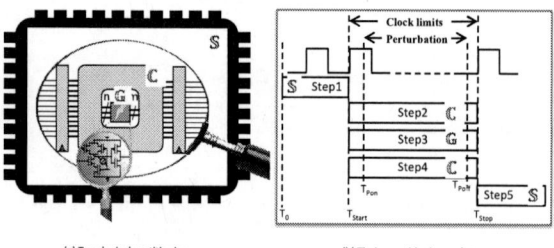

(a) Topological partitioning   (b) Timing partitioning and steps

Figure 5. Multi-level fault simulation

The proposed multi-level fault simulation involves 5 steps during which the different parts of the system are simulated at different time periods, abstraction levels and accuracy as follows:

1. The whole system $\mathbb{S}$ is simulated from T_0 to T_{Start} in order to compute the state of $\mathbb{C}$ just before fault injection.
2. The sub-circuit $\mathbb{C}$ is simulated from T_{Start} to T_{Stop} by taking into account the delay of each gate and nets. This simulation can be more or less accurate based on the precision of the delays. The goal of this step is to extract the waveforms for the input and output bits of $\mathbb{G}$ during the perturbation.
3. $\mathbb{G}$ is modified to include the electrical model of the perturbation. It is then simulated at electrical level by using the input waveforms obtained in step 2. The output analog waveform is translated to logical levels and compared with the nominal logic values obtained in step 2 to create a list of timed logic faults.
4. $\mathbb{C}$ is fault simulated from T_{Start} to T_{Stop} using the faults defined in step 3 and taking into account the delay of nets and gates. The goal is to compute the state of $\mathbb{C}$. If the state is equal to the one obtained in Step 2, the perturbation has not effect on the circuit and the simulation is stopped.
5. If, on the contrary, the states are different, the whole system $\mathbb{S}$ is simulated starting from the faulty state up to the end of the simulation at logic level.

V. EXPERIMENTAL RESULTS

We compare the execution time of the multi-level simulation with respect to pure Hspice simulation. Since performing steps 1 and 5 with Hspice simulation would be both very long and without any interest, we conducted our experiments on combinational circuits for which steps 1 and 5 are not needed.

We have set up a fault simulation experiment for each circuit taken from the combinational *ISCAS85* benchmarks, which we have considered as the $\mathbb{C}$ sub-circuit. We injected a current pulse lasting 2ns within a PMOS transistor of an inverter in the circuit, as the one shown in Figure 2. For each benchmark we properly selected the inverter and the energy of the pulse such that it modifies the final state of the circuit.

Table 1. Experimental results of *ISCAS85* Circuits

Circuit information				Simulation [s]		Fault simulation [s]	
Circuit	Size	In	Out	Vectors	0-delay	Delay	Multi-Level	Hspice
c17	5	5	2	6	0.000	0.004	0.028	0.16
c432	112	36	7	77	0.008	0.048	0.116	55.78
c499	133	43	32	77	0.004	0.056	0.132	89.05
c1355	162	41	32	72	0.000	0.056	0.132	86.48
c1908	169	35	25	80	0.012	0.104	0.228	129.12
c880	204	62	26	77	0.008	0.192	0.404	119.22
c2670	292	159	64	136	0.024	1.016	2.052	393.78
c3540	476	52	22	171	0.040	0.948	1.916	676.7
c5315	608	180	123	113	0.044	2.092	4.204	705.05
c7552	705	208	107	167	0.056	4.704	9.428	1356.56
c6288	1286	34	32	63	11.621	4.144	8.308	1248.75

Table 1 summarizes the experimental results. For each benchmark, we reported its size in gates, the number of inputs and outputs of the circuit, and the number of simulated vectors. We then show the execution time of the pure simulation of the circuit by using the 0-delay and the delay-annotated engines (we used the SDF information obtained after circuit synthesis). Finally, the last two columns compare the execution times of the fault simulation between the multi-level approach and the full Hspice simulation. The increase in speed ranges from 10x for the smallest circuit, up to 150x for the biggest ISCAS85 benchmark. This magnification shall grow even more for bigger circuits.

CONCLUSIONS

In this paper we presented tLIFTING and its multi-level extension for the simulation of faults that need to be described at electrical level. The proposed approach is based on a combination of electrical-level simulation for the sub-circuit affected by the fault and logic level simulation of the rest of the circuit to speed up the whole process. We implemented the proposed method in our open-source fault simulator tLIFTING. This tool will be extensively exploited in the process of evaluating countermeasures against fault attacks, as well as in the reliability evaluation of deep sub-micron devices.

REFERENCES

[1] Marc Joye, Michael Tunstall, "Fault Analysis in Cryptography," 2012, Springer, ISBN: 978-3-642-29655-0
[2] D. Alexandrescu, L. Anghel, M. Nicolaidis "Simulating Single Event Transients in VDSM ICs for Ground Level Radiation", Journal of Electronic Testing and Apllications, JETA, 2004, Vol. 20, pp : 413-421
[3] C. Hungse, E.M. Rudnick, J.H. Patel, R.K. Iyer, G.S. Choi, "a gate-level simulation environment for alpha-particle-induced transient faults," IEEE Transactions on Computers, Vol. 45, No. 11, Nov. 1996, pp. 1248-1256
[4] Meyer W, Camposano R, "Active timing multilevel fault-simulation with switch-level accuracy," IEEE Trans CAD Integr Circ Syst, Vol. 14, 1995, pp. 1241–1256.
[5] G. S. Greenstein, J. H. Patel, "E-PROOFS: A CMOS bridging fault simulator," in Proc. Int. Conf. Computer-Aided Design, 1992, pp.268-271.
[6] A. Sarafianos, R. Llido, J.M. Dutertre, O. Gagliano, V. Serradeil, M. Lisart, V. Goubier, A. Tria, V. Pouget, D. Lewis, "Building the electrical model of the Photoelectric Laser Stimulation of a PMOS transistor in 90 nm technology," Microelectronics Reliability, Volume 52, Issues 9–10, September–October 2012, pp. 2035–2038.

CIRSIUM: A Circuit Simulator in MATLAB® with Object Oriented Design

A. Gokcen Mahmutoglu
Koc University
Sariyer, Istanbul, Turkey
Email: amahmutoglu@ku.edu.tr

Alper Demir
Koc University
Sariyer, Istanbul, Turkey
Email: aldemir@ku.edu.tr

Abstract—We present CIRSIUM, a CIRcuit SImulator Using MATLAB® and its object oriented programming language. CIRSIUM has been developed as a flexible and modular framework in order to enable the rapid development of new device models and prototyping of new circuit analysis paradigms and algorithms. The modular core code includes the PSP MOSFET compact model and uses SUNDIALS for the solution of dynamical circuit equations. New device models can be added by supplying their stamps in vector/matrix form, and existing models can be translated from the Verilog-A hardware description language using ADMS that automatically computes the necessary Jacobians.

CIRSIUM has been initially developed for the analysis of random telegraph signal (RTS) noise in nano-scale integrated circuits. For this purpose, the PSP MOSFET compact model has been expanded with a detailed, non-stationary RTS noise model, that includes a physical description of gate oxide defects and the calculation of their voltage and position dependent capture/emission time constants. Future applications of CIRSIUM will include non-electronic systems, such as noise modeling and analysis for biological neurons, neuronal networks and the nervous system.

I. INTRODUCTION

Numerical simulation is an indispensable tool in the product development cycles and design processes in various engineering disciplines. This is particularly true in integrated circuit (IC) design, since the development costs due to multiple design iterations amount to the larger part of the end product's cost for low volume chip fabrication. This aspect led to an interest and intense work in circuit simulators, proportional to the rate of rapid growth in this field in early the 70's. The public domain simulation tool SPICE and its derivatives have been used and enhanced extensively since then. These enhancements were directed to extend the capabilities of SPICE and also to deal with its shortcomings regarding simulation speed and the complexity of the original code.

One shortcoming of the original SPICE code is the lack of a clean interface between device models and the numerical simulation algorithms: The device model code and the numerical simulation code are unnecessarily intertwined with each other, rendering the implementation of new device models and new numerical simulation algorithms very cumbersome. Attempts have been made in order to deal with this issue, e.g., the Verilog-A hardware description language (HDL) [1]. Our work follows the spirit of recent work on ModSpec [2]. Here we introduce an effort towards a flexible and modular circuit simulator infrastructure developed in the widely used computing environment of MATLAB® and designed to sub-

stantially reduce the complexity of expanding the simulator core with new device models and simulation algorithms. An additional tool we have developed based on ADMS [3] allows the translation of existing device models from Verilog-A HDL and automatically generates the code for the computation of Jacobians. CIRSIUM uses the IDA solver from the SUNDIALS suite [4] that is based on multi-step methods with time-step control for the efficient and robust numerical integration of differential-algebraic equations (DAE).

In the second part of this paper, we present an application that illustrates the capabilities of CIRSIUM: We expand the translated code of the PSP MOSFET model [5] to include a physical description of gate oxide defects and a non-stationary model that enables the simulation of noise phenomena employing a uniformization [6] based algorithm previously discussed in [7]. The capture and emission of electrons from the channel by the traps in the oxide cause a change in the flatband voltage of the transistor and thus affect the dynamical behaviour of the circuit. This effect is known as random telegraph signal (RTS) noise because of its discrete nature. Conversely, the position dependent voltage in the channel changes the rate for the capture and emission events, establishing a two-way coupling between the node voltages of the circuit and the occupation states of the traps. As an example, we demonstrate the effects of these dynamics on the periodic waveforms generated by a ring oscillator circuit. Furthermore, based on this foundation we are currently developing non-Monte Carlo methods and novel algorithms for noise analysis in nano-scale ICs. We plan to apply these techniques to non-electronic systems as well. Prospective uses of CIRSIUM include noise modeling and analysis for biological neurons, neuronal networks and the nervous system.

II. SIMULATOR ARCHITECTURE AND PRINCIPLES

Electrical circuits are best represented in an hierarchical manner. At the bottom of the hierarchy, there are the devices without internal nodes, such as resistors and capacitors. The next level, subcircuits, are composed of these devices and possibly other subcircuits. At the top most level, the circuit serves as a controller for all of its components. This hierarchical structure suggests a partitioning of the simulator core into individual units, each with isolated functionalities. CIRSIUM uses MATLAB®'s object oriented programing language to realize this partitioning in terms of software classes which is depicted in a class diagram in Figure 1. On the other hand, these units

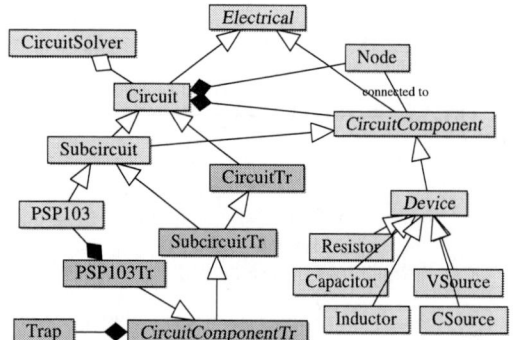

Fig. 1: Class diagram of the CIRSIUM core.

bear some similarities to each other. For instance, during the generation of KCL/KVL equations, subcircuits and devices are handled almost in the same manner by the circuit and hence, a subcircuit can be construed as a circuit component with internal nodes. These relationships between different parts of the simulator translate into inheritance relations among classes which are shown by arrows (from sub to super-class) in Figure 1. In the same diagram, lines with filled and hollow diamond heads represent composition and aggregation relationships respectively, e.g., a circuit consists of circuit components and nodes.

This object oriented framework constitutes the basis for a modular structure that makes adding new functionality to CIRSIUM quite easy. As an example, Listing 1 shows the definition of a non-linear resistor class with a quadratic I-V characteristic, $I = V^2$.

```
classdef resistor < device
    properties (Constant)
        numTerms = 2; numCurrentVars = 0;
    end

    methods
        function obj = resistor(varargin)
            obj = obj@device(varargin{:});
        end

        function s = I(thisResistor, x, t)
            i = (x(1) - x(2))^2;
            s = [ i; -i];
        end

        function s = dI(thisResistor, x, t)
            G1 = 2*x(1); G2 = 2*x(2);
            s = [ G1, -G2; -G1,  G2];
        end
    end
    % charge and source functions are not shown.
end
```

Listing 1: Definition of a class for a non-linear resistor.

The functions I(x,t) and dI(x,t) define the dependence of the device current on the voltages of its two terminals and the Jacobian of the current wrt. these voltages. The definition of a device in its most general form would also include the charge storage function Q(x,t), its Jacobian dQ(x,t) and the source function J(x,t). These functions collectively constitute the *stamp* of the device and are used by the circuit class in the solution of the dynamical KCL/KVL

equations

$$\frac{\mathrm{d}}{\mathrm{d}t}\mathbf{q}(\mathbf{x}, t) + \mathbf{i}(\mathbf{x}, t) = \mathbf{j}(\mathbf{x}, t) \qquad (1)$$

Here, the vector $\mathbf{q}$ can also include the magnetic flux (instead of charge) for inductive devices. The variable vector $\mathbf{x}$ will have entries for branch currents (instead of node voltages) in this case. The Jacobian matrices

$$\mathbf{dQ} = \frac{\partial \mathbf{q}}{\partial \mathbf{x}} \quad , \quad \mathbf{dI} = \frac{\partial \mathbf{i}}{\partial \mathbf{x}} \qquad (2)$$

are usually sparse and are stored in MATLAB®'s sparse data structures which enables the use of fast linear system solution algorithms in the implicit numerical integration methods for equation (1). These methods discretize the time variable t in an adaptive manner and find the vectors $\mathbf{x}_k$ at time points t_k that satisfy the discretized version of the circuit equations. The discretization of the time variable implies that we must use a discrete approximation for the time derivatives as well. This can be done using the charge Jacobian, i.e., $\dot{\mathbf{q}} = \mathbf{dQ}\,\dot{\mathbf{x}}$, and then discretizing the variable vector $\mathbf{x}$. This is equivalent to linearizing the non-linear capacitors in the circuit, which introduces an additional source of numerical error in the solution that has to do with charge conservation [8]. In order to remedy this issue, we expand the DAE system (1) to include a vector of auxiliary variables, $\hat{\mathbf{q}}$, for charges

$$\frac{\mathrm{d}}{\mathrm{d}t}\hat{\mathbf{q}} + \mathbf{i}(\mathbf{x}, t) = \mathbf{j}(\mathbf{x}, t) \qquad (3)$$

$$\hat{\mathbf{q}} = \mathbf{q}(\mathbf{x}, t) \qquad (4)$$

This new DAE system is then solved for the new variable vector $\hat{\mathbf{x}} = [\mathbf{x}\ \hat{\mathbf{q}}]^T$. CIRSIUM also includes an option to find a solution for a reduced set of variables. In circuits without inductive components, the variable for the current through an independent voltage source together with the voltage of one of its nodes can be eliminated. This is especially useful in cases where abrupt changes in the circuit cause large deviations in these currents.

III. AN APPLICATION: RTS NOISE SIMULATION

We now describe an application which uses CIRSIUM in order to simulate the effects of RTS noise in ICs. We first provide a short overview on the emergence of RTS noise through defects in the gate oxide of a transistor and how these defects can be modeled within the PSP compact model. We then present an example simulation of the transient behavior of a ring oscillator built with transistors suffering from RTS noise.

A. Oxide Traps

Oxide defects and the RTS noise caused by them is a major concern in integrated circuits fabricated in modern nano-scale process technologies. These defects act as traps and capture charge carriers from the channel of the transistor resulting in a change in the currents through the transistor. The capture and the subsequent emission of the charge carriers occur randomly governed by the statistics of a non-stationary Poisson process. The classical Shockley-Read-Hall (SRH) theory [9] estimates

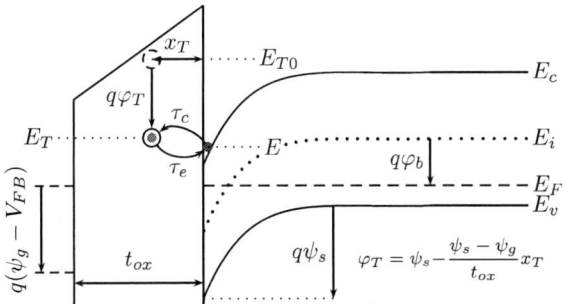

Fig. 2: Capture and Emission process of an electron by a trap located close to the conduction band edge.

the capture and emission time constants, τ_c and τ_e, of this random process as

$$\tau_c = \frac{1}{n\bar{v}\bar{\sigma}} \quad , \quad \tau_e = \tau_c \, e^{-(E_T - E_F)/kT} \quad (5)$$

Here, n is the carrier concentration at the trap location, $\bar{v}$ is the average velocity of charge carriers, $\bar{\sigma}$ is a constant called the capture cross section of the trap, E_T is the trap energy level and E_F is the Fermi level. The capture/emission process of an electron and the associated balance among the quantities given above is shown on a band diagram in Figure 2.

B. Embedding Traps into the PSP Compact Model

PSP is a state of the art compact MOSFET model actively developed by the *Arizona State University* and NPX *Semiconductors* [10]. The physics of the PSP model is based on the surface potential calculations at the channel of the transistor. The intermediate variables, such as the position dependent sheet charge, can be used to build an extension to PSP and calculate the trap capture/emission rates using SRH statistics. The time constants of each trap will depend –among other parameters of the trap– on its position along the channel and into the oxide as well as on the voltages across the transistor. Moreover, the carrier capture and emission processes governed by these time constants will cause a change in the behavior of the whole transistor. We model this effect in a simple, straightforward manner.

According to equation (5), the time constants τ_c and τ_e can be calculated based on the electron concentration at the location of the trap. For this purpose, we use the location dependent normalized sheet charge information, q_i, from the PSP model. This quantity is calculated using the auxiliary variable q_{im} with [5]

$$q_i(y) = q_{im} - \alpha(\psi_s(y) - \psi_m) \quad (6)$$

Here, y is the location along the channel, α is a linearization coefficient and $\psi_s(y)$ and ψ_m are the surface potential at y and the potential mid-point respectively. The position dependent surface potential is calculated using a method called symmetric linearization (SLM) and is given by

$$\psi_s(y) = \psi_m + H\left[1 - \sqrt{\frac{2\Delta\psi}{HL}(y - y_m)}\right] \quad (7)$$

The definitions of the symbols used in equation (7) can be found in [5]. These variables are all supplied by the PSP code with the following names: Normalized drain and source potentials are called x_d and x_s and they have to be multiplied with the thermal voltage ϕ_T in order to obtain the surface potentials. The mid-point potential, ψ_m, is the arithmetic mean of the two. $\Delta\psi$ is called dps_dc and H is simply called H_dc. L is the channel length and the position of the surface potential mid-point, y_m, is given by

$$y_m = \begin{cases} \frac{L}{2}\left(1 + \frac{\Delta\psi}{4H}\right) & \text{for} \quad V_G > 0 \\ \frac{L}{2} & \text{otherwise} \end{cases} \quad (8)$$

The remaining variables in equation (6) are named qim_dc and alpha_dc in the calculations for DC-characteristics. For the AC-characteristics of the model, the _dc suffixes have to be switched with _ac.

With the equations above, we can calculate the sheet charge at the trap location per unit area via $Q_i(y) = q_i(y)C_{ox}$, where C_{ox} is the oxide capacitance per unit area. The knowledge of Q_i enables us to determine the carrier concentration. For this purpose, we define an effective channel thickness, d_{eff}, assuming an exponential decay of the carrier numbers as a function of the distance from the oxide interface, $n(t) = n_0 \exp(-\phi(t)/\phi_T)$, and drawing the line for the end of the channel at the distance where the relationship $E_\perp d_{eff} = 3/2\phi_T$ is satisfied [11]. Here, $E_\perp$ is the electric field component perpendicular to the oxide interface and it is found by introducing a proportionality constant η which is given by the normalized vertical electric field parameter of PSP, Eeffm: $E_\perp = \eta|\psi_s - \psi_g|/t_{ox}$. Thus, for the effective channel thickness and consequently for the volume concentration of carriers we have

$$d_{eff} = \frac{3\phi_T}{2\eta|\psi_s - \psi_g|}t_{ox} \ , \quad n(y) = \frac{Q_i(y)}{d_{eff}} \quad (9)$$

Finally, for the mean capture time we can write

$$\tau_c = \frac{1}{n(y_T)\,\bar{v}\,\bar{\sigma}(x_T)} \quad (10)$$

This quantity depends on the bias voltage and the trap position along the channel through $n(y_T)$ and on the position into the oxide through $\bar{\sigma}(x_T) = \bar{\sigma}_0 \exp(-x_T/\lambda)$ with a characteristic attenuation length λ [12]. For the calculation of the mean emission time through equation (5), we need the trap energy level under the bias conditions of the transistor. In this case, the trap energy is simply modified by the potential at the trap location as shown in Figure 2:

$$E_T = E_{T0} - q\left[\psi_s - (\psi_s - \psi_g)\frac{x_T}{t_{ox}}\right] \quad (11)$$

Finally, the effect of a captured charge carrier in the channel of the MOSFET is modeled by a change in the flatband voltage, ΔV_{FB}, with the simple relationship [13]

$$\Delta V_{FB} = \frac{q}{C_{ox} W_{eff} L_{eff}} \quad (12)$$

This effect is easily integrated into PSP by changing the instance parameter VFB_i.

978-1-4673-4579-8/13 $31.00 © 2013 IEEE

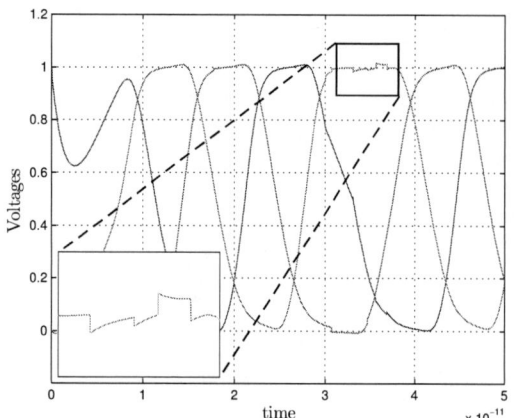

Fig. 3: Output of a transient simulation sample for a 3-stage ring oscillator with oxide traps. The inset shows a time span of high trap activity.

C. Simulation Example - Ring Oscillator with Oxide Traps

The colored (darker shaded) boxes in Figure 1 stand for classes specifically created for RTS noise simulation. The abstract class `circuitComponentTr` models a container that can hold a device, such as PSP, together with objects representing the oxide traps. It also defines the necessary functions for RTS noise simulation, e.g., time constant calculations. The `PSP103Tr` class implements these functions as described in the previous section. A minimal piece of code for the simulation of a ring oscillator built with noisy transistors is given in Listing 2. Note that this script also illustrates the use of subcircuits in CIRSIUM. Finally, Figure 3 shows the output of a sample run of this code and the effects of the traps on the waveforms which were artificially amplified for demonstration purposes.

```
nmos = psp103Tr('n', 35e-9, 100e-9);
pmos = psp103Tr('p', 35e-9, 180e-9);

inv = subcircuitTr('inverter');
inv.addComponent(nmos, 'out', 'in', 'vss', 'vss');
inv.addComponent(pmos, 'out', 'in', 'vdd', 'vdd');
inv.setExternalNodes('in', 'out', 'vdd', 'vss');

ckt = circuitTr('ring oscillator');
ckt.addComponent(inv.copy, 'in', 'out1', 'vdd', 'gnd');
ckt.addComponent(inv.copy, 'out1', 'out2', 'vdd', 'gnd');
ckt.addComponent(inv.copy, 'out2', 'in', 'vdd', 'gnd');
ckt.addComponent(voltageSourceDC(1), 'vdd', 'gnd');

ckt.setGroundNode('gnd');
ckt.seal;

solver = circuitSolver(ckt, 'y0', [1;0;0],...
                            't0', 0,...
                            'tend', 5e-11);
solver.solve();
```

Listing 2: Minimal code for the simulation of a 3-stage ring oscillator.

IV. CONCLUSION

We have presented CIRSIUM, a new circuit simulator developed in the object oriented programming language of MATLAB®. The modular and flexible structure of CIRSIUM enables the rapid development of new device models as well as the translation of existing ones from Verilog-A HDL. Other features of CIRSIUM include subcircuit capabilities and integration with SUNDIALS. CIRSIUM has been developed as a part of an on-going project on non-Monte Carlo methods and novel algorithms for noise modeling in electronic and neuronal systems. In future work, it will be employed for simulations of realistic circuits and biological neuronal networks. The source code of CIRSIUM will be released under GPL and we believe that it will be useful in a wide variety of scenarios in research as well as in teaching.

ACKNOWLEDGMENT

This work was supported by the Scientific and Technological Research Council of Turkey (TÜBİTAK) under project 111E188. The authors would also like to thank Osman Dülek for suggesting the name CIRSIUM and Jaijeet Roychowdhury for valuable discussions and collaboration.

REFERENCES

[1] D. FitzPatrick and I. Miller, *Analog Behavioral Modeling with the Verilog-A Language.* Springer, 1998.

[2] D. Amsallem and J. Roychowdhury, "ModSpec: an open, flexible specification framework for multi-domain device modelling," in *2011 IEEE/ACM International Conference on Computer-Aided Design (IC-CAD),* Nov. 2011, pp. 367 –374.

[3] L. Lemaitre, C. McAndrew, and S. Hamm, "ADMS-automatic device model synthesizer," in *Custom Integrated Circuits Conference, 2002. Proceedings of the IEEE 2002,* 2002, pp. 27 – 30.

[4] A. C. Hindmarsh, P. N. Brown, K. E. Grant, S. L. Lee, R. Serban, D. E. Shumaker, and C. S. Woodward, "SUNDIALS: suite of nonlinear and differential/algebraic equation solvers," *ACM Trans. Math. Softw.,* vol. 31, no. 3, p. 363–396, Sep. 2005.

[5] G. Gildenblat, X. Li, W. Wu, H. Wang, A. Jha, R. van Langevelde, G. Smit, A. Scholten, and D. Klaassen, "PSP: an advanced surface-potential-based MOSFET model for circuit simulation," *IEEE Transactions on Electron Devices,* vol. 53, no. 9, pp. 1979 –1993, Sep. 2006.

[6] A. P. A. van Moorsel and W. H. Sanders, "Adaptive uniformization," *Communications in Statistics. Stochastic Models,* vol. 10, no. 3, pp. 619–647, 1994.

[7] K. Aadithya, A. Demir, S. Venugopalan, and J. Roychowdhury, "Accurate prediction of random telegraph noise effects in SRAMs and DRAMs," *IEEE Transactions on Computer-Aided Design of Integrated Circuits and Systems,* vol. 32, no. 1, pp. 73 –86, Jan. 2013.

[8] K. S. Kundert, *The Designer's Guide to Spice and Spectre.* Norwell, MA, USA: Kluwer Academic Publishers, 1995.

[9] M. J. Kirton and M. J. Uren, "Noise in solid-state microstructures: A new perspective on individual defects, interface states and low-frequency (1/f) noise," *Advances in Physics,* vol. 38, no. 4, pp. 367–468, 1989.

[10] G. Gildenblat, X. Li, H. Wang, W. Wu, R. Van Langevelde, A. J. Scholten, G. D. J. Smit, and D. B. M. Klaassen, "Introduction to PSP MOSFET model," in *Proc. the MSM 2005 Int. Conf., Nanotech 2005,* 2005.

[11] S. Schwarz and S. Russek, "Semi-empirical equations for electron velocity in silicon: Part II–MOS inversion layer," *IEEE Transactions on Electron Devices,* vol. 30, no. 12, pp. 1634 – 1639, Dec. 1983.

[12] Y. Manéglia, F. Rahmoune, and D. Bauza, "On the Si–SiO2 interface trap time constant distribution in metal-oxide-semiconductor transistors," *Journal of Applied Physics,* vol. 97, no. 1, pp. 014 502–014 502–8, Dec. 2004.

[13] S. Martin, G. Li, E. Worley, and J. White, "The gate bias and geometry dependence of random telegraph signal amplitudes [MOSFET]," *IEEE Electron Device Letters,* vol. 18, no. 9, pp. 444 –446, Sep. 1997.

Circuit Simulation Using State Space Equations

Kai Chi Alex Lam, Mark Zwolinski
School of Electronics and Computer Science
University of Southampton
Southampton, UK SO17 1BJ
{kcal1g10,mz}@soton.ac.uk

Abstract—**This paper proposes a method to recast the classic circuit simulation algorithm to maximize parallelism by formulating the circuit equations as state variable equations and using an integration method such as Runge Kutta as the foundation to break the time dependencies. A brief introduction to the SPICE algorithm and previous attempts to exploit parallelism is given. The implementation of the new algorithm for both linear and non-linear circuits is presented and results are compared with the traditional method. The opportunities for extracting parallelism are discussed.**

I. Introduction

Circuit simulation remains a vital tool for the design of integrated circuits. As transistors decrease in size with each generation of CMOS technology, their variability is increasing, so there is a need to characterize digital systems at the circuit level. Conventional algorithms, such as those used in SPICE, have many features that are inherently sequential and that have proved very difficult to parallelize. With the speed of conventional processors limited to less than 4GHz, it is essential that parallel algorithms are found.

In recent years, a number of attempts to accelerate circuit simulation algorithms in new computing architectures such as GPGPUs [1], [2], [3], multi-core CPUs or clusters [4], [5], [6] have been published, but all of them are based on the traditional methods as described in [7], [8]. Although the device evaluation phase can be naturally executed in parallel, matrix solution is more difficult and there are barriers between these two phases. According to Amdahl's law [9], these barriers are the bottleneck to further speedup of the algorithm.

A state space approach is proposed to solve the circuit simulation problem. The traditional method of circuit simulation is compared with the new method proposed in this paper using MATLAB which makes the foundation of parallel simulation. This paper first gives a an introduction of SPICE algorithm in section II. A review of parallelized circuit simulation literature is given in section III. Section IV describes the hypothesis of this research and how it is applied to linear circuits. Section V describes solving non-linear circuits using the new method suggested. Results and conclusions are presented in section VI and section VII, respectively.

II. The SPICE Algorithm

By converting an electronic circuit into a mathematical form and solving it using numerical methods, we can better estimate the behaviour of a circuit before manufacture. There are several types of analysis that can be performed in SPICE, including DC analysis, AC analysis and Transient analysis. They are performed by applying the following techniques as are shown in Fig 1:

- Formulate the circuit equations by Modified Nodal Analysis (MNA)
- Apply numerical integration methods to evaluate the time-dependent behaviour of the circuit
- Linearize the nonlinear circuit model using Newton-Raphson.
- LU factorization is used to solve the resulting system of linearized equations repeatedly.

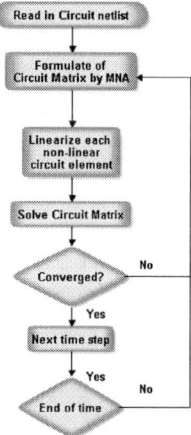

Fig. 1. Flow of SPICE

The computational effort in SPICE is focused in two phases: device evaluation and matrix solution [7].

III. Attempts to Parallelize Circuit Simulation

With the advances in high-speed computer networks, inexpensive distributed multiprocessor systems such as computer clusters have been built. In such systems, MPI (Message Passing Interface) is the standard mechanism for communication between processors. The research of [6] is the example of parallel circuit simulation using MPI. Another approach is multithreading [4] on shared memory multi-core processors using OpenMP. MPI and OpenMP enable a developer to design code that can be used in multi-core processors in small scale

978-1-4673-4579-8/13 $31.00 © 2013 IEEE

machines or large scale clusters. GPUs have been extensively used in recent years for scientific computation. Many such problems require a huge number of floating point operations and can be highly parallelized. With the release of CUDA and OpenCL platforms, developers are able to map existing serial algorithms to GPUs with a relatively low learning curve. There are three papers [1], [2], [3] that describe GPU targeting for time domain circuit simulation. Two publications [1], [2] used CUDA from NVIDIA and the third [3] used Brook+ from AMD. All of them focus on device evaluation. The speedup is generally about 10-50 times faster than the traditional sequential method on a CPU alone for various circuit topologies, but GPUs have a fixed architecture and thus the circuit size will be constrained. Overall, parallelizing circuit simulation is difficult because of the nature of the algorithm, which is separated into device modeling and matrix solving. Each phase needs to wait for the other before further processing. Therefore, the algorithm must be solved by a march-in-time approach. Therefore, we need to formulate the problem in a different way to allow exploitation in massively parallel machines.

IV. RESEARCH HYPOTHESIS

Conventional methods use predictor-corrector methods to solve differential equations. This means that simulation time must progress in a strictly monotonic order, even when further information (such as input waveforms) is known. Consequently, computations must adhere to this flow of time. If a different numerical integration method, such as Runge-Kutta or (more credibly) Burlisch-Stoer were used, computations could be performed as soon as sufficient information was available. Hence, the strict march-in-time limitation could be overcome, allowing parallelism to be exploited.

To do this, however, a state variable (SV) formulation method is needed, which is not compatible with conventional MNA formulation. It is possible to transform MNA to SV, but this has not been widely researched. Two pieces of work[10], [11] discussed this approach, but circuits with nonlinear components were not included. Here, the idea from those papers will be extended by converting a circuit matrix formulated by MNA, which includes transistor components, to an SV circuit matrix. Then, we solve the matrix by the Runge-Kutta method and compare its performance with conventional methods.

A. Convert from MNA to State Variable Form

State Variable analysis (SV) is a concise way to describe the behaviour of circuits, but SV formulation is nontrivial compared with MNA, and this has limited the use of SV in circuit simulation. A method of transforming MNA to SV [10] addressed this problem. SV is formed by eliminating excess voltage and current variables in MNA. Hence, only capacitor voltages and inductor currents are preserved. Here, an example of MNA formulation and SV formulation shows the idea before and after conversion by the method described in [10]. For a simple series RLC circuit, MNA gives $Ax = B$:

$$A \begin{pmatrix} v_L \\ v_R \\ v_C \\ i_1 \\ i_L \end{pmatrix} = B \tag{1}$$

where A is a matrix of conductances and B is a vector of stimuli. Both are formed by inspection. v_L, v_R, v_C are the voltages across elements, i_1, i_L are values of currents of other circuit elements and inductors. In SV form, the circuit equation is written in the general form: $\frac{dx}{dt} = Cx + Du$. Variable u is input which is v_1, therefore:

$$\frac{d}{dt} \begin{pmatrix} i_L \\ v_C \end{pmatrix} = C \begin{pmatrix} i_L \\ v_C \end{pmatrix} + Dv_1 \tag{2}$$

To transform equation (1) to equation (2), rewrite equation (1) as:

$$\begin{pmatrix} A_d + A_{s1} & A_{s2} \end{pmatrix} \begin{pmatrix} X_1 \\ X_2 \end{pmatrix} = \begin{pmatrix} B_1 \\ B_2 \end{pmatrix} \tag{3}$$

in which A_d is the submatrix of conductances of dynamic elements. A_{s1} and A_{s2} are the submatrices of conductances of static elements. The minimum number of state variables is r_o, which is the number of rows of submatrices A_d, A_{s1}, X_1 and B_1. X_1 is the vector of state variables. R is the number of rows of the whole matrix, therefore, $R - r_o$ is the number of excess variables represented by X_2. The matrix is reordered by converting A_{s2} to $(A_{21}I)$ and I is the unity matrix with size $(R - r_o) * (R - r_o)$ to the form:

$$\begin{pmatrix} A_d & A_{12} \\ A_{21} & I \end{pmatrix} \begin{pmatrix} X_1 \\ X_2 \end{pmatrix} = \begin{pmatrix} B_1 \\ B_2 \end{pmatrix} \tag{4}$$

The transformation is summarized in Algorithm 1. Excess variables need to be eliminated as well using (5).

$$\begin{aligned} A_{sv} &= A_d - A_{12} \times A_{21} \\ B_{sv} &= B_1 - A_{12} \times B_2 \end{aligned} \tag{5}$$

The equation without excess variables becomes

$$A_{sv} \begin{pmatrix} i_L \\ v_C \end{pmatrix} = B_{sv} \tag{6}$$

Rearranging equation (6), equation (2) is obtained and solved by the Runge Kutta or similar method.

B. Using the SV method to solve a linear circuit

This whole algorithm is demonstrated using a simple RC circuit in Fig.2. In MNA, after time discretization ($t^{n+1} - t^n = dt = h$), the left-hand side and the B matrix are:

$$Ax = \begin{pmatrix} \frac{1}{R} & -\frac{1}{R} & 1 \\ -\frac{1}{R} & \frac{1}{R} + \frac{C}{h} & 0 \\ 1 & 0 & 0 \end{pmatrix} \times \begin{pmatrix} v_1^{n+1} \\ v_c^{n+1} \\ i_{v_1}^{n+1} \end{pmatrix}, B = \begin{pmatrix} 0 \\ \frac{C}{h}v_c^n \\ v_{in}^n \end{pmatrix} \tag{7}$$

where n is the number of time steps. By the flow in Algorithm 1 and the steps in equations (5) and (6), the matrices from equation (7) converted to SV become

$$A_{sv} = \left(\frac{1}{R} + \frac{C}{h} \right), B_{sv} = \left(\frac{C}{h}v_c^n + \frac{v_{in}^n}{R} \right) \tag{8}$$

Create A and B matrix;
Count number of dynamic elements;
for $j = 1; j \leq no.dynamic\ elements; j++$ **do**
    Move the rows with dynamic elements to the top rows of A and B;
    Move the columns in matrix A with dynamic elements to the left-most columns;
end
while A_{22} *is not the end of the matrix and not a diagonal matrix* **do**
    For the first cell with value '1' in each col, pivot the whole row of the '1' to make the diagonal filled with '1';
end
while A_{22} *is not the end of the matrix and not a identity matrix* **do**
    Divide non-zero value in the diagonal by itself to obtain '1'. For non-zero values outside the diagonal, subtract them by the multiple of other row to make it 'zero';
end

Algorithm 1: MNA to SV conversion

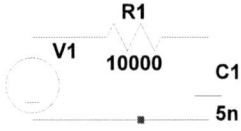

Fig. 2. Simple RC circuit

The MNA to SV conversion is as described above, where the set of equation is still written in the form $Ax = B$. In order to solve it using the Runge Kutta method, the equation should be rearranged in normal form as $\frac{dv}{dt} = f(v,t)$:

$$\left(\frac{1}{R} + \frac{C}{h}\right)\left(v_c^{n+1}\right) = C \times \frac{v_c^n}{h} + \frac{v_{in}^n}{R} \quad (9)$$

This can be written as

$$\left(\frac{1}{R} + \frac{C}{h}\right)\left(v_c^n + dv_c\right) = C \times \frac{v_c^n}{h} + \frac{v_{in}^n}{R} \quad (10)$$

As $h = t^{n+1} - t^n = dt$, rearranging gives:

$$\frac{dv_c}{h} = \frac{dv_c}{dt} = \left(\frac{v_{in}^n - v_c^{n+1}}{RC}\right) \quad (11)$$

V. SOLVING NON-LINEAR CIRCUITS

The full adder circuit[12] in Fig. 3 is considered in this section to show how the algorithm can be applied to a circuit with non-linear devices. The reduced matrix obtained by the flow from algorithm 1 must be rearranged to the set of explicit ordinary differential equations in order to use the RK4 integration method. The derivation of the A_{sv} matrix is lengthy; for instance, the entry $A_{sv}(1,1)$ with the sum of conductances is

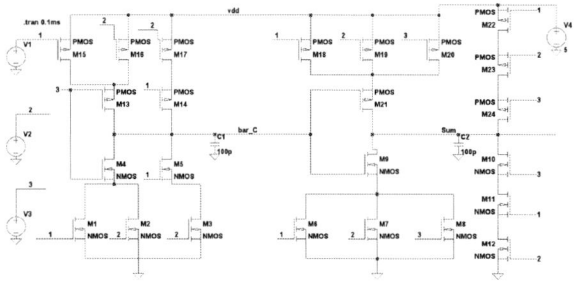

Fig. 3. Fulladder

$(G_{dsp1}+G_{dsn4}+G_{dsp2}+G_{dsn5}-(G_{dsp2}*(G_{dsp2}+G_{gsp2})\ldots$
For the sake of simplicity, let $A_{sv}(1,1)$ be written as G_1, the sum of conductances, $G_2 = A_{sv}(1,2)$ and so on. For B_{sv}, let i_1 be the sum of currents relating to the first state variable, etc. Hence, $A_{sv}X_1 = B_{sv}$

$$\begin{pmatrix} G_1 + \frac{C_1}{h} & G_2 \\ G_3 & G_4 + \frac{C_2}{h} \end{pmatrix} \begin{pmatrix} v_1^{n+1} \\ v_2^{n+1} \end{pmatrix} = \begin{pmatrix} C_1 \frac{v_1^n}{h} + i_1 \\ C_2 \frac{v_2^n}{h} + i_2 \end{pmatrix} \quad (12)$$

The matrix is transformed to state variable form $\frac{dv_1}{dt} = f(v_1,t), \frac{dv_2}{dt} = f(v_2,t)$. The first equation is written as

$$\left(G_1 + \frac{C_1}{h}\right)v_1^{n+1} + G_2 v_2^{n+1} = C_1 \times \frac{v_1^n}{h} + i_1 \quad (13)$$

Rearranging gives

$$G_1 v_1^{n+1} + \left(\frac{C_1}{h}\right)\left(v_1^n + dv_1\right) + G_2 v2^{n+1} = C_1 \times \frac{v_1^n}{h} + i_1 \quad (14)$$

Similar with equation 11, the final form of the ODE for v1 is

$$\frac{dv_1}{h} = \frac{dv_1}{dt} = \frac{i_1 - G_1 v_1^{n+1} - G_2 v_2^{n+1}}{C_1} \quad (15)$$

The second equation can be written in the same way to become

$$\frac{dv_2}{h} = \frac{dv_2}{dt} = \frac{i_2 - G_3 v_1^{n+1} - G_4 v_2^{n+1}}{C_2} \quad (16)$$

For the circuit equation with m capacitors, the matrix can be written as $\frac{dv_1}{h} = f(v_1,t), \frac{dv_2}{h} = f(v_2,t)\ldots\frac{dv_m}{dt} = f(v_m,t)$. The final form of the set of ODE equations representing the non-linear circuit appears as:

$$\begin{aligned} \frac{dv_1}{dt} &= \frac{i_1 - G_1 v_1^{n+1} - G_2 v_2^{n+1} \cdots - G_m v_m^{n+1}}{C_1} \\ \frac{dv_2}{dt} &= \frac{i_2 - G_{m+1} v_1^{n+1} - G_{m+2} v_2^{n+1} \cdots - G_{m+m} v_m^{n+1}}{C_2} \\ &\vdots \\ \frac{dv_m}{dt} &= \frac{i_n - G_{(m-1)\times m+1} v_1^{n+1} \cdots - G_{(m-1)\times m+m} v_m^{n+1}}{C_2} \end{aligned} \quad (17)$$

Where m is the number of nodes/states to be calculated and n is the number of time iterations. These equations can be solved by the Runge-Kutta method combined with Newton-Raphson iterations.

VI. RESULTS

The algorithm is feasible for circuit with dynamic and linear components. For integrated circuit simulation, non-linear elements need to be included in the circuit simulation. The waveform generated from the new algorithm by MATLAB is compared with the output by MATLAB implementation of both conventional method and the state space approach in Fig 4. The biggest difference is between the LTspice output, which is in a broken line and the MATLAB implementation by the original method appears as a dotted line. The dotted line lag varies from 0-6.5 microseconds, the main difference between them is that a fixed time step is used in the MATLAB code whereas LTspice uses variable time steps. However, by using the same simple MOS model and a fixed time step strategy, using the state space approach with Runge-Kutta method in MATLAB, which is in a solid line, is closer to the result calculated by LTspice, with 0-4 microsecond lag only. The elapsed time of state variable approach is 0.78 seconds and the execution time of the original algorithm is 0.56 on the desktop computer with Intel Xeon W3520 CPU in 2.67GHz and 12GB DRAM. The operating system is Windows 7 Enterprise SP 1. The SV approach takes about 40% longer than the conventional one since it takes more Newton iterations to converge to the solution. It is due to the stability of explicit RK4 being poor. However there are better numerical integration methods such as Burlisch-Stoer that give better stability to enable the algorithm converges faster. We can further test it with different numerical integration methods for a better performance.

VII. CONCLUSION

The architecture of the circuit simulator is presented as well as the theories underneath it. The recent approaches in parallelizing circuit simulation algorithm are reviewed. We have found that there are limitations for doing massive parallelism by putting the traditional algorithm into parallel architectures in which only the device evaluation phase can be parallelized. The matrix solver phase still needs to be solved in serial order. Therefore, the new approach converts MNA equation to the SV form and solves the equation in normal form by the Runge-Kutta method instead of the Euler method. The MNA to SV algorithm is developed in MATLAB and applied to both the simple RC circuit and the 24 transistor mirror full adder. The Runge-Kutta method is successfully implemented on the SV form equations of mirror adder. Although its performance is not as good as the original approach and the problem is still solved in a one-dimension time marching problem, the successful formulation of state space equations allows us to model the problem in a 2-D way, enabling us to explore the hypothesis that we can solve the problem without any dependency of time by partitioning the circuit into sub-circuits and considering each as states to solve it independently as far as enough information supplied. An implementation using C++ is the next step of the research and is being developed.

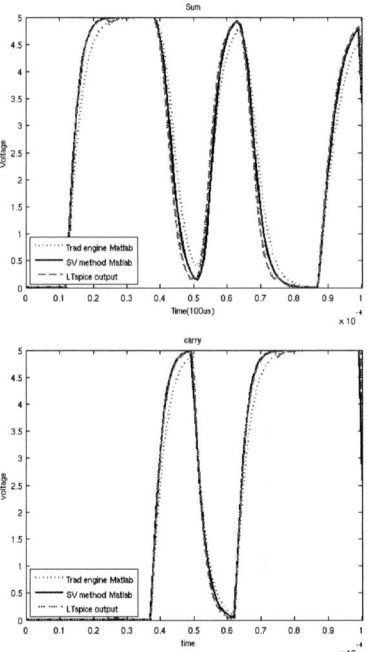

Fig. 4. Comparsion of results between LTspice (broken line),MATLAB using SV method (solid line) and MATLAB using traditional method (dotted line).

REFERENCES

[1] K. Gulati, J. Croix, S. Khatri, and R. Shastry, "Fast circuit simulation on graphics processing units," in *Design Automation Conference, 2009. ASP-DAC 2009. Asia and South Pacific*, jan. 2009, pp. 403 –408.

[2] R. Poore, "Gpu-accelerated time-domain circuit simulation," in *Custom Integrated Circuits Conference, 2009. CICC '09. IEEE*, sept. 2009, pp. 629 –632.

[3] A. M. Bayoumi and Y. Y. Hanafy, "Massive parallelization of spice device model evaluation on gpu-based simd architectures," in *Proceedings of the 1st international forum on Next-generation multicore/manycore technologies*, ser. IFMT '08, 2008, pp. 12:1–12:5.

[4] R. Perng, T. Weng, and K. Li, "On performance enhancement of circuit simulation using multithreaded techniques," in *Computational Science and Engineering, 2009. CSE'09. International Conference on*, vol. 1. IEEE, 2009, pp. 158–165.

[5] H. Peng and C. Cheng, "Parallel transistor level full-chip circuit simulation," in *Design, Automation & Test in Europe Conference & Exhibition, 2009. DATE'09*. IEEE, 2009, pp. 304–307.

[6] B. Andjelković, V. Litovski, and V. Zerbe, "Grid-enabled parallel simulation based on parallel equation formulation," *ETRI journal*, vol. 32, no. 4, 2010.

[7] L. W. Nagel, "Spice2: A computer program to simulate semiconductor circuits," Ph.D. dissertation, EECS Department, University of California, Berkeley, 1975.

[8] V. Litovski and M. Zwolinski, *VLSI circuit simulation and optimization.* Springer, 1996.

[9] M. Hill and M. Marty, "Amdahl's law in the multicore era," *Computer*, vol. 41, no. 7, pp. 33–38, 2008.

[10] Y. Kang and J. Lacy, "Conversion of mna equations to state variable form for nonlinear dynamical circuits," *Electronics Letters*, vol. 28, no. 13, pp. 1240–1241, 1992.

[11] Y. Kang, "Systematic method for obtaining state-space representation of nonlinear dynamic circuits using mna," *Electronics Letters*, vol. 28, no. 21, pp. 2028–2030, 1992.

[12] N. Weste and K. Eshraghian, "Principles of cmos vlsi design: A system perspective. 1993."

Analog Performance of PD-SOI MOSFETs at High Temperatures Using Reverse Body Bias

A. Schmidt, H. Kappert, R. Kokozinski

Fraunhofer Institute for Microelectronic Circuits and Systems (IMS), Finkenstraße 61, 47057 Duisburg, Germany
alexander.schmidt@ims.fraunhofer.de

Abstract—**The analog performance, i.e. intrinsic gain and bandwidth, of SOI (Silicon-on-Insulator) MOSFETs in a wide temperature range up to 400°C has so far been strongly affected by device leakage currents. Thereby the moderate inversion region as a preferred point of operation has been unusable as leakage currents dominate drain currents at high temperatures. In this paper we present a reverse body biasing (RBB) approach to improve the transistor's analog performance up to 400°C. Thereby operation in the lower moderate inversion region of the SOI transistor device is feasible. The method allows beneficial FD (fully depleted) device characteristics in a 1.0 µm PD (partially depleted) SOI CMOS process. NHGATE and PHGATE devices with an H-shaped gate have been investigated. Results report a significant improvement of the g_m/I_d factor and the intrinsic gain A_i in the moderate inversion region by applying RBB.**

I. INTRODUCTION

Analog SOI (Silicon-on-Insulator) circuit design for a wide operating temperature range up to 400°C is significantly affected by the transistor characteristics and their dependencies on temperature. Within the temperature range of the target application, e.g. measurement technologies for use in harsh environments, analog circuits have to meet requirements like sufficient accuracy and operating speed. Also specific circuit design techniques like the switched capacitor (SC) technique are preferred. High leakage currents within these circuits lead to reduced accuracy and also cause malfunction at high temperatures. Leakage currents in SOI transistors have to be considered as a major source of error in analog circuit design up to 400°C. Since technology improvements are only capable of reducing leakage currents by a limited amount, advanced design techniques are required to eliminate the resulting effects. This implies the reduction of leakage currents within the possibilities of circuit design in the first place and secondly the compensation of remaining leakages by compensation structures brought into the circuit. Improving intrinsic gain and bandwidth of the transistor devices at high temperatures is also to be realized by using adapted design techniques. Solving these issues will allow analog circuit design for a wide temperate range up to 400°C. The DC and RF behavior of SOI transistors has been studied up to 300°C [1-5]. Body biasing has been found useful in influencing the threshold voltage and the breakdown characteristics of SOI CMOS devices [6]. In

Figure 1. SOI NHGATE. The body contact (F) is realized on both sides of the device channel.

this paper we demonstrate that reverse body biasing (RBB) is an elementary design technique facing analog circuit design issues at high temperatures. The paper is organized as follows: Section II presents the SOI process and transistor devices used for this investigation. Section III describes the reduction of device channel leakage currents and in section IV the improvement of the analog performance is presented.

II. SOI TECHNOLOGY & DEVICES

Figure 1 shows a photograph of an NHGATE device fabricated in the Fraunhofer IMS 1.0 µm PD-SOI CMOS process. The film thickness t_{SI} is 150 nm with a gate oxide t_{OX} of 40 nm. The buried oxide thickness t_{BOX} is 400 nm. The channel surface doping concentrations N_A and N_D are $3 \cdot 10^{16}$ cm^{-3} and $1.8 \cdot 10^{16}$ cm^{-3} for the NHGATE and the PHGATE, respectively. The minimum channel length for analog devices at a supply voltage of 5V is 1.6 µm. The film (body) of the NHGATEs is connected via a P$^+$ body-contact. The film of PHGATEs is connected with an N$^+$ body-contact.

III. REDUCTION OF LEAKAGE CURRENTS

An effective way to reduce device leakage currents is to influence both, the depletion depth x_d and the threshold voltage V_{th} [10][11]. The depletion depth x_d of an NHGATE device depends on temperature and doping concentration N_A and is given in (1) [7][8].

$$ x_d = \sqrt{\frac{4\varepsilon_{SI}\Phi_F}{qN_A}} = \sqrt{\frac{4\varepsilon_{SI}\frac{kT}{q}\ln\left(\frac{N_A}{n_i}\right)}{qN_A}} \qquad (1) $$

In (1), Φ_F is the Fermi potential, ε_{SI} the dielectric constant of silicon, q the elementary charge, k the Boltzmann constant, n_i the intrinsic carrier concentration and T the temperature. The transistor turns PD as soon as x_d is smaller than the film thickness t_{SI}. In that case, depletion can also occur on the back-interface of the channel dependent on the back-gate voltage [7]. It should be noted that the transition from FD to PD also depends on the doping concentration of the film. It can be calculated from (1) that for a doping concentration of $4 \cdot 10^{16}$ cm^{-3} the device turns PD at approximately 110°C. The threshold voltage V_{th} of an N-channel SOI device including the body effect is given in (2) in which Φ_{MS} is the metal-semiconductor work function difference, C_{OXF} the front-gate oxide capacitance and V_b the body-potential [7]. Interface traps have been neglected in this case. The threshold voltage itself depends on the depletion depth x_d in case the device is PD.

$$V_{th}\big|_{Q_{ox}=0} = \Phi_{MS} + 2\Phi_F + \frac{q \cdot N_A}{C_{OXF}} \underbrace{\sqrt{\frac{2\varepsilon_{si}(2\Phi_F - V_b)}{q \cdot N_A}}}_{x_d(V_b)} \qquad (2)$$

Taking advantage of the body effect, the threshold voltage can be controlled by varying the body voltage V_b. It can be seen from (2), that V_b influences the square-root term and thereby the depletion depth x_d. Applying a negative body bias increases x_d and the device remains fully depleted in the considered temperature range. In FD state, the space charge within the channel is constant and determined by t_{SI}. The temperature dependence of the threshold voltage is then determined by the metal-semiconductor work function difference and the Fermi potential. The threshold voltage has been extracted for $V_{DS} = 0.1$ V and a back-gate voltage of $V_{BG} = 0$ V using the tangent method at the point of maximum slope in the linear I_d/V_{GS} input characteristic curve and is shown in figure 2. The threshold voltage is now investigated by applying RBB. For a body bias of $V_b = 0$V the device turns PD at approximately 150°C and the variation of x_d dominates the temperature dependency of the threshold voltage above that temperature. It should be noted here, that for a temperature of 400°C the theshold voltage is approximately 150 mV and thus too low to prevent significant device channel leakage for proper analog circuit design. By applying a negative body bias of $V_b = -1$V this transistion is moved to higher temperatures. Therefore a lower temperature dependence in good compliance with (2) has been verified by our results. At 400°C the threshold voltage still remains at 800 mV and exhibits a linear decrease with temperature as expected.

IV. Improved Analog Device Performance

The intrinsic gain A_i and intrinsic bandwidth GBW_i of a single transistor is given in (3) [1][9].

In (3), g_m is the transconductance, g_{ds} the conductance of the transistor, I_d the drain current, V_A the Early voltage and C_L the load capacitance.

$$A_i = \frac{-g_m}{g_{ds}} \approx -\left(\frac{g_m}{I_d}\right) \cdot V_A \quad , \quad GBW_i = \left(\frac{g_m}{I_d}\right) \cdot \frac{I_d}{2\pi C_L} \qquad (3)$$

The moderate inversion region of the transistor represents a good compromise between intrinsic gain and bandwidth and is

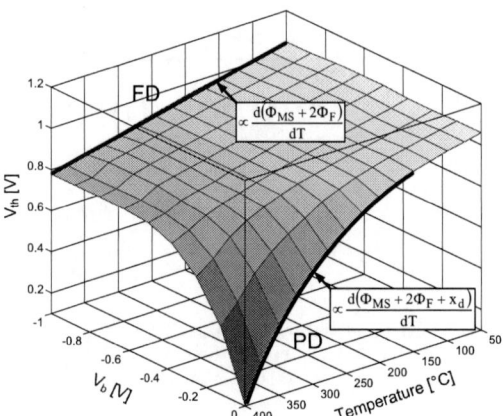

Figure 2. Measured threshold voltage V_{th} of an NHGATE transistor and its dependency on temperature and body potential V_b.

therefore preferred for analog circuit design. A standard design approach for high temperature electronics would select a high drain current in order to keep a safety margin to leakage currents. In order to operate the transistors in moderate inversion a large device width is then required. Due to the proportional increase of pn-junction leakage currents with device width, moderate inversion cannot be used at high temperatures [11].

A. Transconductance efficiency g_m/I_d

The transconductance efficiency factor g_m/I_d as a figure of merit is most suitable to determine the analog performance of a transistor device [9][1]. g_m/I_d is separately defined in weak inversion (WI) and strong inversion (SI) and is given in (4) [1].

$$\frac{g_m}{I_d}\bigg|_{(WI)} = \frac{1}{nV_t} \quad , \quad \frac{g_m}{I_d}\bigg|_{(SI)} = \sqrt{\frac{2\mu_0 C_{OXF} W/L}{n I_d}} \qquad (4)$$

In (4), V_t is the thermal voltage, μ_0 the zero-field carrier mobility and n the body effect coefficient. Since n is significantly smaller when applying a negative body bias, the g_m/I_d factor can be directly influenced by reverse body biasing [10]. Thereby two effects are responsible for an increase of g_m/I_d at high temperatures. First of all the maximum value of g_m/I_d in FD state as noted in (4) decreases only by the influence of V_t while n remains constant with increasing temperature. As a second effect, the leakage current in the device is decreased as discussed before. Decreasing the overall leakage current therefore results in a higher g_m/I_d value. The transconductance efficiency can be extracted from the input curve of the transistor by performing the calculation in (5).

$$\frac{g_m}{I_d} = \frac{\partial I_d}{\partial V_{GS}} \cdot \frac{1}{I_d} = \frac{\partial \ln(I_d)}{\partial V_{GS}} \qquad (5)$$

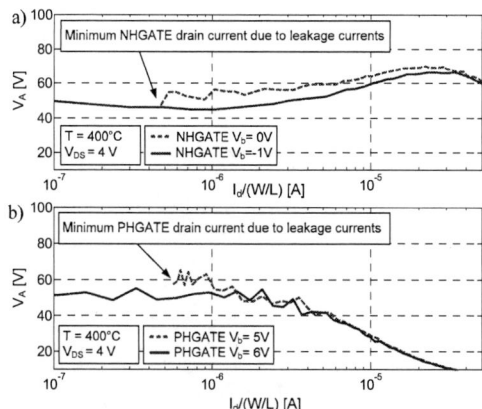

Figure 3. Early voltage over $I_d/(W/L)$ for $V_{DS}=4V$ of a) an NHGATE and b) a PHGATE at 400°C.

The g_m/I_d factor has been investigated for NHGATE and PHGATE devices and is presented in figure 4. It should be mentioned here, that the PHGATE exhibits a back-gate-source voltage of 5V and therefore is the more critical device. In case of the devices without RBB, leakage currents tend to increase intensively with temperature resulting in a very low g_m/I_d value at high temperatures. By applying RBB to the NHGATE and PHGATE devices, the normalized operating current can be reduced to 100 nA. Thereby the g_m/I_d value can be increased significantly at low and high temperatures. Choosing a drain current of 100 nA with a sufficient safety margin to the leakage current level, a g_m/I_d value of 6.7/V can be reached at 400°C. The intrinsic gain of the PHGATE at 400°C also increases from 0.55/V to 3.5/V by applying RBB at a reduced operating current. Summarizing these results the g_m/I_d factor of the NHGATE device can be increased significantly in the moderate inversion region by applying a negative body bias.

B. Early Voltage V_A

The Early voltage V_A can be extracted from the output characteristics of the transistor using the calculation in (6).

$$V_A = \frac{I_d}{g_{ds}} = \left(\frac{\partial I_d}{\partial V_{DS}} \cdot \frac{1}{I_d} \right)^{-1} = \left(\frac{\partial \ln(I_d)}{\partial V_{DS}} \right)^{-1} \quad (6)$$

The Early voltage has been investigated as a function of the normalized drain current for NHGATE and PHGATE devices for $V_{DS}=4V$ at 400°C. The results are shown in figure 3. It can be seen that V_A of both devices is slightly lower with applied RBB. It should be noted here that the Early voltage cannot be extracted at very low drain currents from devices without RBB. The reason is that channel leakage currents at 400°C limit the minimum drain current to approximately 500 nA. On the other hand, an Early voltage of 50V at a normalized drain current of 100 nA can be measured when RBB is applied. Comparing these results it is found that RBB is a suitable solution to achieve an adequate Early voltage in the lower moderate inversion region at high temperatures.

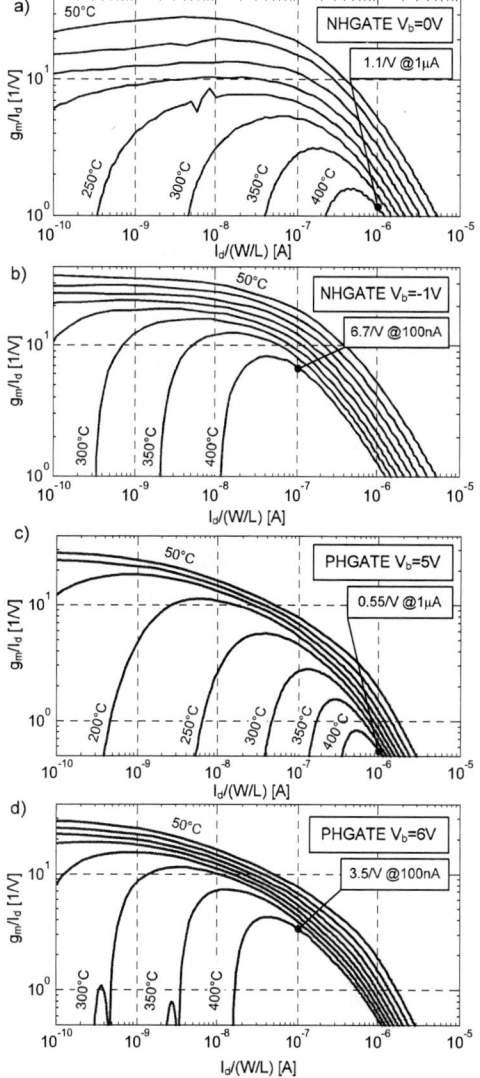

Figure 4. Investigated g_m/I_d over $I_d/(W/L)$ for $V_{DS}=0.1$ V. a) NHGATE without RBB, b) NHGATE with RBB, c) PHGATE without RBB and d) PHGATE with RBB.

C. Intrinsic gain A_i

Intrinsic gain is one of the most important parameters in analog circuit design, especially in applications where high DC gain is required. As g_m/I_d decreases with increasing temperature, also the intrinsic gain as a product of g_m/I_d and V_A decreases. The intrinsic gain of NHGATE and PHGATE devices has been investigated for $V_{DS}=3V$ and is shown in figure 5 a), b), c) and d) for the NHGATE without RBB, the NHGATE with RBB, the PHGATE without RBB and the PHGATE with RBB, respectively. In order to keep a safety margin to leakage currents, an operating point of 5 µA is usually desired without RBB. The intrinsic gain at 400°C is then 41 dB and 39 dB for the NHAGTE and PHGATE, respectively. By applying RBB, an operating point of 100 nA can be chosen for both transistors. Thereby also the intrinsic gain of the NHGATE at low temperatures increases about 18 dB and 14.5 dB for the PHGATE. At 400°C the intrinsic gain of the NHGATE is improved by 14 dB and 8 dB for the PHGATE. Considering an application where two gain stages are cascaded, e.g. a two-stage op-amp, an improvement of the overall DC gain of about 20 dB at 400°C seems feasible.

V. CONCLUSION

The experimental results presented in this work show that RBB is a proper circuit design technique allowing analog circuit design with significantly reduced leakage currents and improved device performance in a wide temperature range up to 400°C. By applying RBB to the HGATE transistor device, superior FD device characteristics are available in a PD SOI process. It effectively decreases the temperature coefficient of the threshold voltage at high temperatures and thereby reduces device leakage currents. Furthermore RBB allows the circuit designer to operate the devices in the lower moderate inversion region even at high temperatures up to 400°C. Thereby intrinsic gain and bandwidth in this region are improved by an increased g_m/I_d factor. Although further investigations are necessary to apply the presented method to analog circuits, it represents a promising approach to achieve a much better high temperature operation capability compared to standard operation. The presented method not only improves the performance of PD-SOI devices up to 300°C but also push the limit of analog circuit design in SOI technology close to 400°C.

REFERENCES

[1] J.P. Eggermont, et al., "SOI CMOS Operational Amplifiers for Applications Up To 300°C", Transactions of the Second International High Temperature Electronics Conference, HiTEC, Vol. II, pp. 21-26, 1994.

[2] M.N. Ericson, et al., "1/f Noise and DC Characterization of Partially Depleted SOI N- and P-MOSFETs from 20°C-250°C", IEEE Aerospace Conference, 2005.

[3] Guo-Wei Huang, et al., "Impact of body bias on the high frequency performance of partially depleted SOI MOSFETs", IEEE Asia-Pacific Microwave Conference, APMC, 2008.

[4] El Kaamouchi, M., et al., "Body-Biasing Control on Zero-Temperature-Coefficient in Partially Depleted SOI MOSFET", IEEE SiRF, 2008.

[5] M. Emam, "High Temperature DC and RF Behavior of Partially Depleted SOI versus Deep n-Well Protected Bulk MOSFETs", IEEE SiRF, 2009.

[6] S. Maeda, "Substrate-bias effect and source-drain breakdown characteristics in body-tied short-channel SOI MOSFET's", IEEE Transactions on Electron Devices, Vol. 46, No. 1, 1999.

[7] J.P. Colinge, Silicon-On-Insulator Technology: Materials to VLSI, 3rd ed., Kluwer–Academic, New York, 2004.

[8] S. M. Sze, K. K. Ng, Physics of Semiconductor Devices, 3rd ed., 2007.

[9] D. M. Binkley, Tradeoffs and Optimization in Analog CMOS Design, John Wiley & Sons, Chichester, UK, 2007.

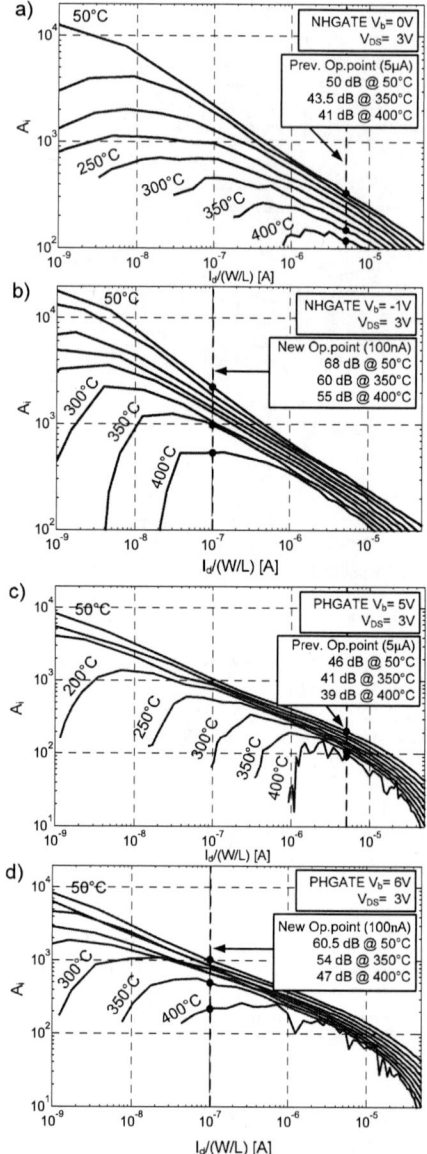

Figure 5. Intrinsic gain over $I_d/(W/L)$ for V_{DS}=3V. a) NHGATE without RBB, b) NHGATE with RBB, c) PHGATE without RBB and d) PHGATE with RBB.

[10] A. Schmidt, et al., "Precision Analog Circuit Design in SOI CMOS for a Wide Temperature Range up to 350°C", Conference on Ph.D. Research in Microelectronics and Electronics, PRIME, 2012.

[11] A. Schmidt et al., "PD-SOI MOSFET Performance Optimization for High Temperatures up to 400°C Using Reverse Body Biasing", 13. GMM/ITG Technical Meeting (ANALOG), 2013.

Technique for reducing on-resistance of high-voltage drivers based on stacked standard CMOS

Sara Pashmineh[1], Hongcheng Xu[2], Dirk Killat[1]

[1] Microelectronics Department, Brandenburg University of Technology Cottbus, Germany
[2] Institut of Microelectronics, Ulm University, Germany

Abstract—This paper presents a new technique for reducing on-resistance of high-voltage drivers, which are based on N-stacked standard CMOS. A theory to calculate gate voltages of HV-driver transistors to drive the maximum drain current for minimum on-resistance is introduced. According to the calculated gate voltages, a circuit design methodology for generating them is described. This concept is technology independent and compatible with scaled CMOS devices. The theory and circuit design are proved by simulating a 2-stack CMOS driver in 65-nm technology, demonstrating significantly improved rise and fall times of the driver, if compared to previous work.

I. INTRODUCTION

As demand on more high-voltage applications in integrated circuits, high-voltage drivers have become much more important. Various methods to design HV-circuits such as in [1] have been developed, which are technology dependent. In contrast, HV-drivers based on stacked standard CMOS transistors have better benefits because of technology independence and full integration with digital circuitries to provide system-on-chip solutions. Various techniques have been described in the literature such as in [5]-[10], which have low switching speed because of the high on-resistance of the drivers.

The aim of this paper is to research and design a HV-driver based on stacked standard CMOS with minimum on-resistance, which is technology independent. This requires a theory to calculate the gate voltages of transistors to drive the maximum drain current. The paper is organized as follows: section II describes the structure of a HV-driver based on stacked standard CMOS transistors. Section III presents a theory to calculate gate voltages of transistors to drive the maximum drain current for minimum on-resistance. In section IV, with respect to the calculated gate voltages a circuit design methodology is described to generate these voltages. The extension of the principle on N-stack drivers is explained. In Section V, the theory is proved by simulating a 2-stack CMOS driver in 65-nm technology. The simulation results demonstrate significantly improved rise and fall times of the driver, which indicate reduced on-resistances. Finally, conclusions are given.

II. SYSTEM DESCRIPTION

The schematic of a high-voltage driver is given in Fig. 1, where V_n stands for nominal operating voltage of a standard MOSFET transistor. The driver contains stacked PMOS transistors in the pull-up and NMOS transistors in the pull-down path, which charge and discharge the output by the given input signals V_{in} and V_{pin}, which is level-shifted from V_{in}. The voltage between the terminals of a transistor has to

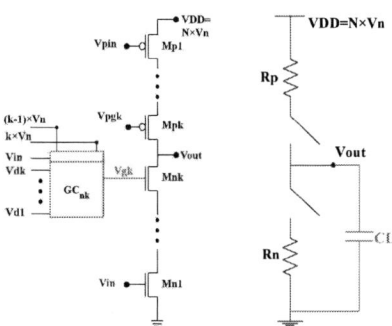

Fig. 1. N-stacked CMOS HV-driver with a gate-control circuit GC_{nk}

be within the technology limits. Due to this, the number of stacked transistors (N) depends on the supply voltage, which is N times greater than the nominal voltage. Hence the output load can be charged and discharged between $N \times V_n$ and ground. The charging and discharging can be faster when the on-resistance of pull-up and pull-down respectively is minimal. To achieve this goal, it needs external circuits to generate appropriate gate voltages of the driver transistors such as GC_{nk} in Fig. 1. This requires a theory to calculate the gate voltages of transistors to drive the maximum drain current, which indicate a minimum on-resistance.

III. GATE- CONTROL CIRCUITS

A. Mathematical Calculated Gate Voltages

In this work, the gate voltages of N transistors of a high-voltage driver at each output voltage have been calculated using a software program in two cases: (1) for a maximum drain current at an input signal of 2.5 V, which switches the driver on, and (2) at an input signal of 0 V, which switches the driver off.

In the first case, the driver output is discharged from the supply voltage ($V_{DD}=N \times V_n$) to ground. The required gate voltages were calculated for a maximum drain current. The results exhibit that the source node voltage of each NMOS transistor is proportional to the driver output voltage and the gate node has an offset equal to the nominal operating voltage (2.5 V) to the source voltage. This gate voltage can be described as a function of the source and the output voltage. Fig. 2(a) shows the characteristics of the source (V_{s2}) and gate (V_{g2}) voltages of the second transistor of 2-stacked NMOS devices vs. the driver output voltage. V_{out} is discharged from 5 V to 0 V. As can be seen, the relationship of the source voltage V_{s2} is half of the output. The gate voltage V_{g2} has an offset value of 2.5 V relative to V_{s2}:

978-1-4673-4579-8/13 $31.00 © 2013 IEEE

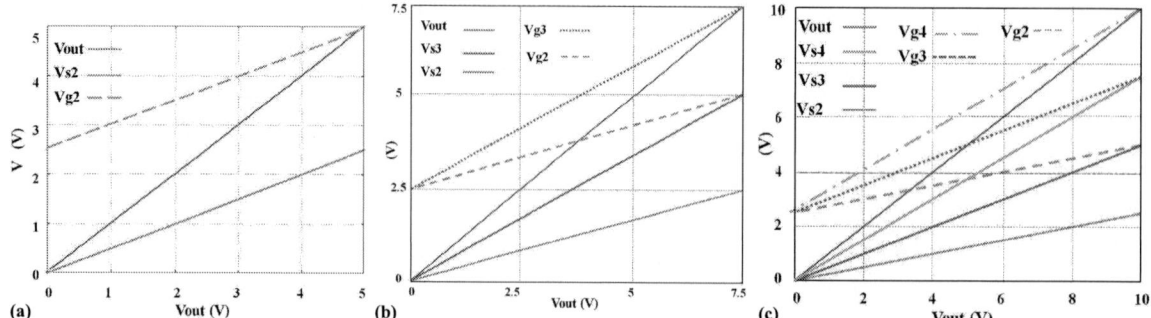

Fig. 2. Gate and source voltage charachteristics of a (a) 2- (b) 3- (c) 4-NMOS driver for a maximum drain current

$$V_{g2} = V_{s2} + 2.5 \quad , \quad V_{s2} = \frac{V_{out}}{2} \Rightarrow V_{g2} = \frac{V_{out}}{2} + 2.5\,V \qquad (1)$$

Fig. 2(b) shows the source and gate voltages of 3-stacked NMOS devices vs. the driver output with a maximum load voltage of 7.5 V. The relations between node voltages can be expressed with the following conditions:

$$V_{s2} = \frac{V_{out}}{3}, \quad V_{g2} = \frac{V_{out}}{3} + 2.5\,V \qquad (2a)$$

$$V_{s3} = \frac{2 \times V_{out}}{3}, \quad V_{g3} = \frac{2 \times V_{out}}{3} + 2.5\,V \qquad (2b)$$

According to the same procedure and the results of a 4-stacked NMOS driver with a maximum load voltage of 10 V in Fig. 2(c), the node voltages of the k-th N-stacked NMOS-driver transistor for a maximum drain current can be described as:

$$V_{sk} = \frac{(k-1) \times V_{out}}{N} \Rightarrow V_{dk} = \frac{k \times V_{out}}{N}, \quad V_{gk} = \frac{(k-1) \times V_{out}}{N} + 2.5\,V \quad (3)$$

Fig. 3 shows all possible calculated results of a gate voltage for a maximum drain current. Above a certain point (A) at a fixed output voltage (V_A), the gate has several values, which build a parallelogram-shaped region such as area **B**. The point A is the boundary of triode and saturation regions. The region **B** helps to simplify circuit design to generate gate voltages, because the required voltage does not need to track exactly the line of (3) in the saturation region.

With an input signal of 0 V, which turns off the transistor *Mn1*, the driver output switches from 0 V to the supply voltage V_{DD}. The maximum difference between drain and source node voltages at each transistor can be the nominal voltage (2.5 V). To stay in this condition, each transistor (k) has to be turned off according to the output by charging to $(k–1) \times 2.5\,V$, as depicted in Fig. 4 for a 4-stacked NMOS-driver. The second NMOS transistor turns off at an output voltage of 2.5 V, the third one by 5 V and the fourth by 7.5 V. In the off-state of each transistor, the gate-source voltage has to be equal to or less than the threshold voltage.

With an input signal of 0 V, the driver output is charged from 0 V to the supply voltage $N \times 2.5\,V$. The source and gate voltages of N stacked PMOS-transistors in the pull-up branch of a driver have been calculated for a minimum on-resistance, which drives the maximum drain current. They are

formulated as following, where k_p stands for k-th PMOS-transistor of driver:

$$V_{psk} = \frac{(k_p - 1) \times V_{out}}{N} + (N + 1 - k_p) \times 2.5\,V \qquad (4)$$

$$V_{pgk} = V_{psk} - 2.5\,V \Rightarrow V_{gk} = \frac{(k_p - 1) \times V_{out}}{N} + (N - k_p) \times 2.5\,V$$

B. Design of Gate-Control Circuit

In this section, a circuit design methodology to control the gates of driver transistors according to the calculated gate voltages is described.

Fig 5(a) shows the circuit GC_{n2} to generate V_{g2} of a 2-NMOS driver. Three PMOS transistors are connected in

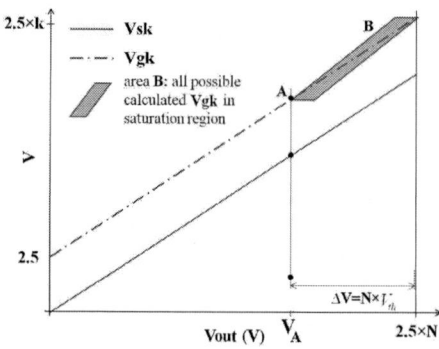

Fig. 3. V_{sk} and all possible calculated V_{gk}, the line (3) and area B (V_{in}=2.5 V)

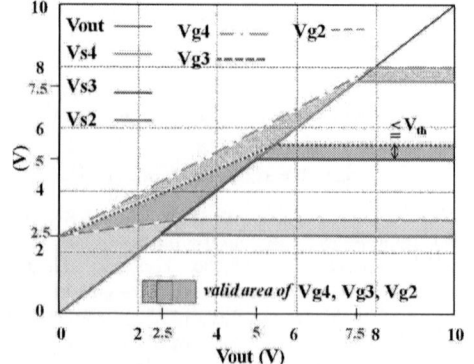

Fig. 4. V_{gk} and V_{sk} of a 4-stacked NMOS driver in off-condition (V_{DD}=7.5 V)

978-1-4673-4579-8/13 $31.00 © 2013 IEEE

series: gate-drain-connected *mp3*, *mp1* and *mp2*, whose gate nodes are controlled by the driver nodes voltages V_{d1} and V_{d2} respectively. When the driver is in on-state, the voltage V_{D2} of 5 V supplies the current I_{mp3}. The gate voltage and dimensions of *mp1* keep the transistor *mp2* in saturation region. The voltage of the node **n1** is the required V_{g2}, which follows the rule (3):

$$\left.\begin{array}{l} I_{mp3} = \dfrac{\beta_p}{2} \times (V_{D2} - V_{g2} - V_{thp})^2 \\[2mm] I_{mp2} = \dfrac{\beta_p}{2} \times (V_{g2} - V_{d2} - V_{thp})^2 \end{array}\right\} {\scriptstyle (3): V_{d2}=\frac{2\times V_{out}}{N}} \Rightarrow V_{g2} = \dfrac{V_{out}}{N} + \dfrac{V_{D2}}{2} \quad (5)$$

The current I_{mp3} starts to flow when the gate-source voltage of *mp3* exceeds its threshold voltage, which limits the desired voltage of 5 V at the node **n1** to $5\,V - V_{th}$. This problem can be solved by connecting a PMOS transistor in parallel to *mp3* such as P1 in Fig. 7, whose gate is controlled by a bias voltage (V_{b1}). On the other side, the parallelogram-shaped region above point **A** in Fig. 3 helps to simplify circuit design, because the required voltage V_{g2} does not need to track exactly the line of (3) in the saturation region. When the driver is in off-state, V_{D2} is switched from 5 V to 2.5 V. According to the gate voltages of *mp1* (V_{d1}) and *mp2* (V_{d2}), the voltage of node **n1** (V_{g2}) is changed to approx. $2.5\,V + V_{th}$ (Fig. 8).

In Fig 5(b), the circuit GC_{n3} is given, which generate the gate voltage V_{g3} of the third driver transistor. Two PMOS transistors are added in series to the described circuit GC_{n2} in Fig 5(a). One of them is gate-drain-connected and the gate of another one *mp3* is controlled by V_{d3}. The supply voltage V_{D3} switched between 5 and 7.5 V. The gate voltages and dimensions of *mp4* and *mp5* keep the transistor *mp6* in saturation area. According to the drain current equations of *mp6*, *mp7* and *mp8* in saturation region, the voltages of the nodes **n2** and **n3** can be obtained as:

$$\left.\begin{array}{l} n3: \quad V_3 = \dfrac{V_{D3}}{2} + \dfrac{V_2}{2} \\[2mm] n2: \quad V_2 = V_{g3} = \dfrac{V_{d3}}{2} + \dfrac{V_3}{2} \end{array}\right\} {\scriptstyle (3): V_{d3}=\frac{3\times V_{out}}{N}} \Rightarrow V_{g3} = \dfrac{2\times V_{out}}{N} + \dfrac{V_{D3}}{3} \quad (6)$$

With the supply voltage V_{D3} of 7.5 V, the voltage of the node **n2** (V_2) is the required gate voltage V_{g3} as in (3).

Fig. 6 and Fig. **7** present a schematic of a 4-stacked NMOS high-voltage driver with 3 gate-control circuits GC_{n2}, GC_{n3} and GC_{n4} to generate the gate voltages V_{g2}, V_{g3} and V_{g4} respectively. The supply voltage of the driver is 10 V. The length of each transistor is 280 nm and the widths are given in table 1.

The gate-control circuit GC_{n4} to generate V_{g4}, comprises of seven in series connected PMOS transistors: *mp9-mp15* (Fig. 7). The supply voltage V_{D4} is switched between 7.5 and 10 V. With the same procedure the gate-control circuit of a k-th NMOS-driver transistor consists of $k-1$ gate-drain-connected PMOS and k PMOS transistors as the same devices, whose gates are controlled by the drain nodes of the stacked driver transistors respectively Fig. 5(c). The connection of both groups of PMOS transistors generates the

desired gate voltage V_{gk}. The supply voltage V_{Dk} switches between $k \times 2.5\,V$ and $(k-1) \times 2.5\,V$. The current can only flow when the gate-source voltage of each transistor exceeds its threshold voltage, which limits the desired voltage. This problem can be solved in the same way as described by connecting of a PMOS transistor in parallel to the gate-drain-connected PMOS transistors such as transistors P1-P6, which are controlled by bias voltages V_{b1}-V_{b6} (Fig. 7).

The design circuits to generate of gate voltages in pull-up driver are the complement form of the above described circuits using NMOS transistors instead of PMOS (*mpk*).

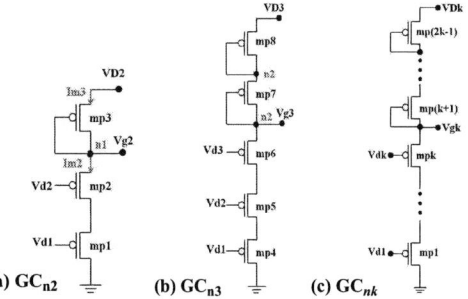

(a) GC_{n2} **(b)** GC_{n3} **(c)** GC_{nk}

Fig. 5. Circuit design GC_{nk} to generate gate voltages a) V_{g2} b) V_{g3} c) V_{gk}

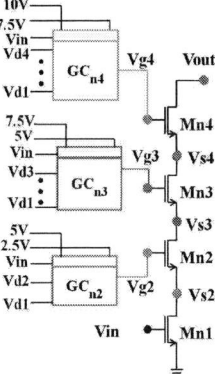

Fig. 6. Schematic of a 4-NMOS HV-driver with gate-circuits GC_{n2}, GC_{n3} and GC_{n4} to generate gate voltages V_{g2}, V_{g3} and V_{gk}

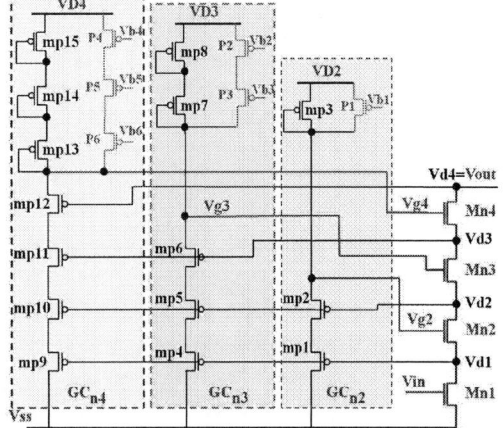

Fig. 7. 3-NMOS HV-driver with gate-circuits GC_{n2}, GC_{n3} and GC_{n4}

978-1-4673-4579-8/13 $31.00 © 2013 IEEE   187

IV. SIMULATION RESULT

The introduced principle is applied on a 2-stack CMOS high-voltage driver in 65-nm with a supply voltage of 5 V and nominal voltage of the 2.5 V. Fig. 8 depicts the simulation results of this work. In the case of an input signal of 0 V, the gate voltage V_{g2} switches the transistor $Mn2$ off by reducing the gate-source voltage below V_{th}. The output voltage V_{out} was charged from the low level (approx. 2 mV) to 5 V and the source V_{s2} to 2.5 V, which is half of the output voltage. At an input signal of 2.5 V the driver is switched on. Thereby the output is discharged from 5 V and V_{s2} from 2.5 V to the low voltage level. The voltage V_{g2} follows the rule according (3). The voltage between each transistor's terminals was kept within the technology limited range without exceeding the nominal voltage with 5% tolerance.

Fig. 9 shows the output and pull-down drain current of this work (A) in comparison to previous work (B)[5][8], in which the gate voltages of the 2. NMOS and PMOS transistors have been set at a constant value (2.5 V). Table 2 gives a comparison between the output voltage of this work (A) and the previous work (B). The rise and fall times of the output voltage of this paper are respectively improved approximately 26% and 24% with a load capacitance of 150 pF if compare to B[5][8]. During discharging and charging of the output, the initial pull-down and pull-up on-resistances of this work are respectively 30% and 39% less than B[5]. Due to this, the driver switches faster.

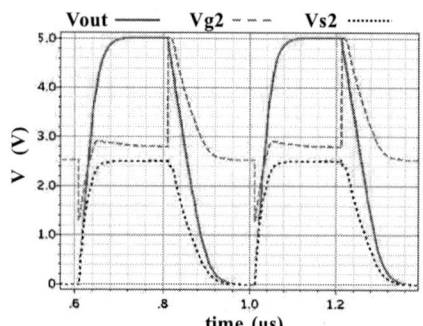

Fig. 8. Simulation results of this work (V_{DD}=5 V and V_n=2.5 V)

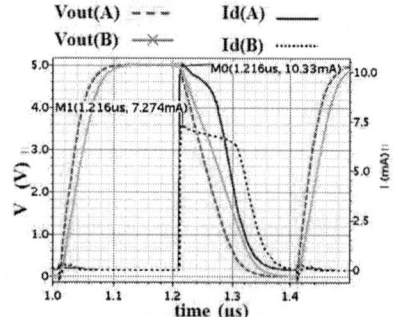

Fig. 9. Simulation results of this work (A) in comparison with B[5][8]

TABLE 1. Width of transistors in Fig. 7 [μm]

P1,m_{p2}, m_{p3}	P2, P3, m_{p6}- m_{p8}	P4, P5, P6, m_{p12}- m_{p15}	M_{nk}
20	18	16	1
m_{p1}	m_{p4}, m_{p5}	m_{p9}-m_{p11}	M_{pk}
32	40	52	3

TABLE 2. Comparison results between the Model A and Model B [5][8]

V_{DD} [V]	Model	t_{LH} [ns]	Δt_{LH} [ns]	t_{HL} [ns]	Δt_{HL} [ns]	Rn_{on} Ω	Rp_{on} Ω
5	B [5,8]	57	15	99	24	687	385
	this work	42		75		484	234
	IMPROV	26.3%		24%		30%	39%

V. CONCLUSIONS

In this paper, a theory to calculate gate voltages of N-stacked CMOS high voltage driver to drive a maximum drain current was introduced, which indicate a minimum on-resistance. According to the calculated gate voltages, a circuit design methodology to generate them was presented. The introduced technique is technology independent. The theory is applied on a 2-stack CMOS driver in 65-nm with a nominal voltage of 2.5 V and a supply voltage of 5 V. The Simulation results demonstrated that the gate and source voltages follow the theoretical optimum characteristics. The rise and fall times of the output are a considerable improvement, which indicate a minimum on-resistance driver. The principle can be applied to N-stack driver transistors with a supply voltage of $N \times V_n$.

ACKNOWLEDGMENT

This work has been funded by the German Science Foundation (DFG).

REFERENCES

[1] S. Bandyopadhyay, Y. Ramadass, and A. Chandrakasan, "20 μA to 100 mA DC-DC converter with 2.8 to 4.2 V battery supply for portable applications in 45 nm CMOS", ISSCC 2011, pp. 386-388, 2011.

[2] S. Monga, V. Kumar, "A 73μW 400Mbps stress tolerant 1.8V-3.6V driver in 40nm CMOS", ESSCIRC 2011, pp. 187 – 190, 2011.

[3] C. Menolfi, T. Toifl, M. Rueegg, M. Braendli, P. Buchmann, M. Kossel, T. Morf, "A 14Gb/s high-swing thin-oxide device SST TX in 45nm CMOS SOI", ISSCC 2011, pp. 156 – 15, 2011.

[4] Y.Choi, I. Park, H. Lim, M. Kim, C. Yoon, N. Kim, K. Yoo, L. Hutter, "A versatile 30V analog CMOS process in a 0.18μm technology for power management application", 23rd ISPSD 2011, pp. 219 – 222, 2011

[5] B. Serneels, E. Geukens, B. De Muer, T. Piessens, "A 1.5W 10V-output Class-D amplifier using a boosted supply from a single 3.3V input in standard 1.8V/3.3V 0.18μm CMOS", ISSCC 2012, pp. 94-96, 2012

[6] Y. Ahn, J. Roh, "5-V Buck Converter Using 3.3-V Standard CMOS Process With Adaptive Power Transistor Driver Increasing Efficiency and Maximum Load Capacity", ITPE, vol. 27, NO. 1, pp. 463-471, 2012

[7] S.R. Bradburn, H.L. Hess, "An integrated high-voltage buck converter realized with a low-voltage cmos process", 53rd Int. MWSCAS, pp. 1021-1024, 2010

[8] B. Serneels, M. Steyaert, "Design of High Voltage xDSL Line Drivers in Standard CMOS", Springer, 2008

[9] P. Swaroop, A. Vasani, M. Ghovanloo, "A High-Voltage Output Driver for Implantable Biomedical Stimulators and I/O Applications", 49th Int. MWSCAS'2006, pp. 566-569

[10] E. Mentze, H. Hess, K. Buck, T. Windley, "A Scalable High-Voltage Output Driver for Low-Voltage CMOS Technologies" IEEE Trans., VLSI Systems, vol.14, NO.12, pp.1347-1353, 2006.

Diode Detector with Voltage Gain

Robert Wolf, and Frank Ellinger, *Senior Member, IEEE*

Abstract—We propose a detector with the properties of a diode detector and with additional voltage gain within the same circuit. Thereto, the principle of the diode detector is implemented by an active control loop. By calculations, it is proven that the behavior is similar to the conventional diode detector. This is validated by measurements of a detector circuit which was implemented as part of an integrated circuit for an envelope following system. Additionally, we show by calculations that it is possible to modify the circuit such that it reveals better properties.

Keywords—*Envelope Detector, Diode Detector, Envelope Following.*

I. INTRODUCTION

ENVELOPE detectors are required for systems like incoherent receivers [1], [2], automatic gain control [3], and envelope following systems [4]. Especially for envelope following systems, the linearity and the response time of the envelope detector is important. A simple diode detector is very advantageous for this application. Unfortunately, the conventional diode detector has no voltage gain. Hence, an additional base-band amplifier is required [5]. This increases the circuit complexity, the energy consumption, and the response time.

We show that it is possible to combine the detector, the reference stage and the amplifier to one circuit. Therefore, the fundamental diode detector principle is implemented by an active control loop. By calculations, we derive that the transfer characteristic of the proposed architecture is the same as of the conventional one. Furthermore, we prove that it is possible to leave out the capacitor in this circuit leading to better properties. The analysis of this modification is fundamental, which allows a wide variety of implementations. By measurement results, the functionality of the circuit with the combination of the sub-blocks is proven.

II. DIODE DETECTOR

A. Conventional Diode Detector

A schematic of a typical diode detector is shown in Fig. 1a. In this schematic, the diode is substituted by a bipolar transistor which has the benefit of additional current gain so that the circuit has less influence on the source. The transfer characteristic stays the same.

The transfer characteristic of such a diode detector can be calculated by using the exponential characteristic of the bipolar transistor given by

$$I_C = I_S \exp\left(\frac{U_{BE}}{U_T}\right). \tag{1}$$

Manuscript received March 25, 2013. This work was partly funded by the Federal Ministry of Education and Research (BMBF) in the excellence cluster CoolSilicon, project CoolBroadcastRepeater. The authors are with the Chair for Circuit Design and Network Theory, Technische Universität Dresden, 01062 Dresden, Germany (e-mail: robert.wolf@tu-dresden.de).

For a sufficiently low-ohmic source and by neglecting the base current, the input voltage is

$$U_e = \hat{U}_e \cos\varphi + U_0 \tag{2}$$

for steady-state. The average emitter current $\overline{I_E}$ must be equal to the bias current I_0 which can be expressed by

$$I_0 \overset{!}{=} \overline{I_E} = \overline{I_C} = \frac{1}{\pi} \int_0^\pi I_S \exp\left(\frac{U_e - U_a}{U_T}\right) d\varphi \tag{3}$$

where U_a is the output voltage. If the capacitor is sufficiently large the output voltage U_a can be assumed to be constant for one period so that the integral can be rearranged to

$$U_a = U_0 - U_T \ln\left(\frac{I_0}{I_S}\right) + U_T \ln \mathbb{I}_0\left(\frac{\hat{U}_e}{U_T}\right) \tag{4}$$

where $\mathbb{I}_0(x)$ is the modified bessel function first kind zero order. It can be seen that the bias point just results in additive terms and the transfer characteristic is inherent.

In some publications, it is distinguished between quadratic and linear operation region of the diode detector. Considering (4), it is clear that this can just be an approximation. The transfer characteristic can be well approximated by

$$U_a = U_0 - U_T \ln\left(\frac{I_0}{I_S}\right) + \frac{U_T}{k_1}\left(\sqrt{1 + \left(\frac{\hat{U}_e}{U_T/k_2}\right)^2} - 1\right)$$

with $k_1 = 0.54$ and $k_2 = 0.52$ for $U_T = 25\,\text{mV}$, (5)

which exhibits these two operation regions for small and for large excitation.

In order to achieve voltage gain, an additional gain stage and a reference stage are required which can be implemented e.g. like shown in Fig. 1b. This increases the circuit complexity and the energy consumption. We show that it is possible to combine the detector and the amplification.

B. Combined Diode Detector

The basic principle of the diode detector is that a constant voltage at the emitter is set such that the average collector current is constant and equal to the bias current. At the basic diode detector, this happens inherently. We propose to implement this behavior by an active control loop. Thereby, new properties can be achieved. In our case the detector, the reference, the control loop and thereby the voltage gain is combined to a robust and energy efficient circuit.

The schematic of this circuit is shown in Fig. 2. The circuit has a differential input connected to two transistors operating in combination with the capacitor C_1 as a differential diode detector. The transistor, which is required as a reference in

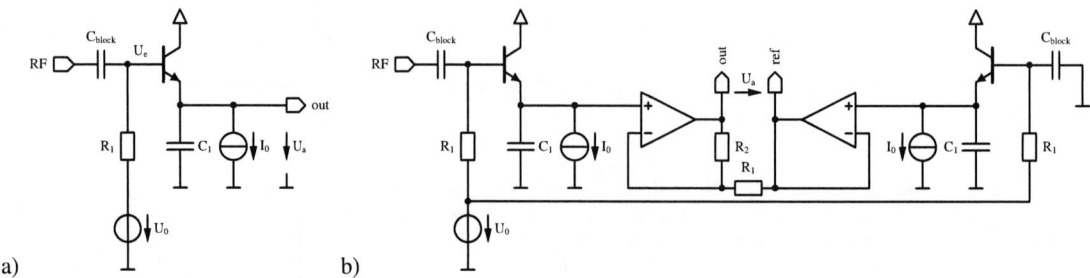

Fig. 1.   Schematic a) of a conventional diode detector and b) of a diode detector with additional gain stage with reference stage

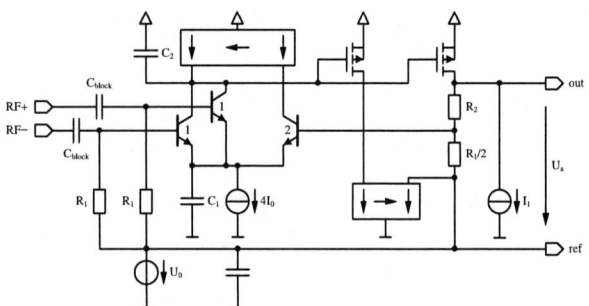

Fig. 2.   Schematic of the diode detector with voltage gain

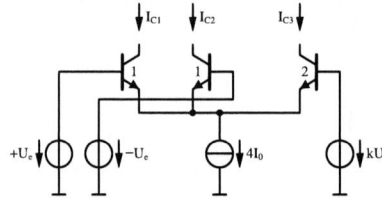

Fig. 3.   Circuit used for the calculation of the transfer characteristic without capacitor

the conventional architecture in Fig. 1b, is additionally used as one part of a differential pair for the control loop. The detector transistors form the second part of this differential pair. A current mirror and a common-source stage complete the error amplifier. By a resistive feedback network the voltage gain is set and the control loop is closed. Additional elements in the feedback network like a diode can be used to further linearize the transfer characteristic. The capacitor C_2 ensures the stability of the loop and allows the input transistors to operate as a diode detector since it provides an RF short. The second common-source stage at the output of the error amplifier in combination with the second current mirror lowers the load on the bias voltage source U_0.

All calculations from the basic diode detector still apply. Especially, the pseudo-differential architecture does not have any influence on the calculations. Thus, the transfer characteristic is still the same besides the implemented voltage gain.

C. Diode Detector without Capacitor

During the analysis of the proposed architecture, the question arises if the capacitor C_1 is still required since it slows down the response time. The result is that for a single-ended implementation of the proposed architecture the capacitor is necessary. If a differential circuit is used, the capacitor can be left out. Hence, the transfer characteristic changes. But it can still be calculated. Therefore, the circuit which is shown in Fig. 3 is used.

By the equations

$$4I_0 = I_{C1} + I_{C2} + I_{C3}, \tag{6}$$

$$0 = +\frac{U_e}{U_T} - \ln\frac{I_{C1}}{I_S} + \ln\frac{I_{C3}}{2I_S} - \frac{kU_a}{U_T}, \tag{7}$$

$$0 = -\frac{U_e}{U_T} - \ln\frac{I_{C2}}{I_S} + \ln\frac{I_{C3}}{2I_S} - \frac{kU_a}{U_T}, \tag{8}$$

$$U_e = \hat{U}_e \cos\varphi \tag{9}$$

and equivalent to (3), it can be derived that

$$2I_0 \overset{!}{=} \frac{1}{\pi}\int_0^\pi I_{C3}\mathrm{d}\varphi$$
$$= \frac{8I_0}{\pi}\int_0^\pi \frac{1}{\exp\left(\frac{U_e-kU_a}{U_T}\right) + \exp\left(\frac{-U_e-kU_a}{U_T}\right) + 2}\mathrm{d}\varphi. \tag{10}$$

This equation is independent from the bias point and it can be solved numerically. Fig. 4 shows the comparison between this solution and the conventional diode detector for $k = 1$ and Fig. 5 shows the slope $v = \mathrm{d}U_a/\mathrm{d}\hat{U}_e$ of those functions. It can be seen that for small excitations the behavior is similar. But the slope for large excitation differs. It converges to 1 and to $1/\sqrt{2}$ for the conventional diode detector and for the detector without capacitor, respectively. Furthermore, the detector without capacitor reaches linear operation already for smaller excitations than the conventional one. 90% of the final slope are reached at 86 mV and 130 mV for the two detector types. Additionally, the slope of the detector without capacitor is flatter.

Many different implementation of such a detector without capacitor are imaginable. Fig. 6 and Fig. 7 illustrate the

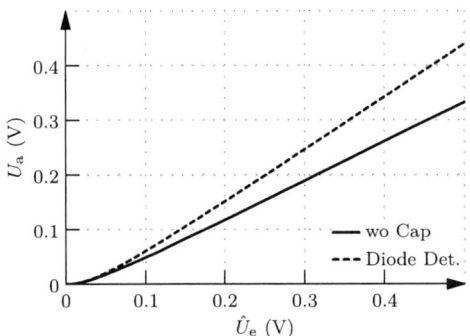

Fig. 4. Transfer characteristic of the detector without capacitor and the conventional diode detector

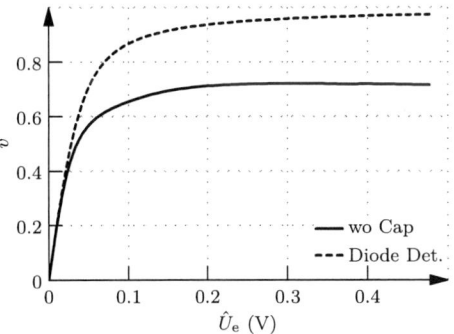

Fig. 5. Slope of the transfer characteristic for both detectors

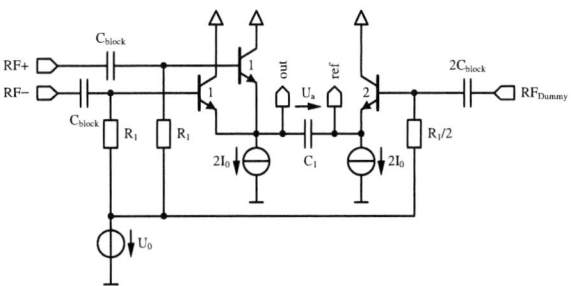

Fig. 7. A possible implementation of a true differential diode detector based on the theory of the detector without capacitor

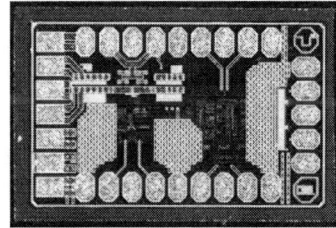

Fig. 8. Photograph of the chip which uses the circuit shown in Fig. 2

simplified schematic of two implemented circuits. Both are true differential and are hence much less sensitive to supply noise.

In the circuit which is shown in Fig. 6, the differential input is connected to two transconductance stages giving additional gain and allowing the signal adding afterwards. The added signal implements the voltage kU_a in a differential way. Thereafter, the signal passes the detector which is followed by a error amplifier. The capacitor in this circuit ensures the stability of the loop. This circuit responses very fast since the only capacitor is in the forward path of the loop. Even tough the circuit is relatively complex, neighboring circuit blocks can be less complex like the variable gain amplifier (VGA) which in this case needs less gain due to the transconductance stages at the beginning of the detector.

Although the circuit shown in Fig. 7 appears to be very similar to the conventional diode detector it bases on the theory of the detector without capacitor and it exhibits its properties. In this case the control loop is formed by a capacitor only which implements the voltage kU_a in the mesh which is used for (7) and (8). Thus, the derivation gives the same result.

III. MEASUREMENT RESULTS

Some detectors based on the approaches proposed in the last section were implemented in an envelope following system for linear power amplifiers. The detector generates the reference value for the modulation of the supply. A variable gain amplifier (VGA) is placed in front of the detector to adjust the system parameters. Test pads allow the measurement of the transfer characteristic of the combination of VGA and detector.

At the moment only the detector shown in Fig. 2 was detailedly characterized. This circuit was implemented in IHP's 250 nm BiCMOS technology. It takes 70 μA out of a 2.5 V supply which is just 175 μW. The voltage gain, which is implemented by the control loop, is 2.5. A chip photograph is shown in Fig. 8. The pads for the RF input are located at the lower edge on the right side. The test pads lie on the opposite side. The detector circuit is situated in the middle.

A Rhode&Schwarz ZVA67 was used to generate the amplitude sweep and to simultaneously measure the DC output voltage of the detector. The VGA was set to unity gain for comparison. Fig. 9 shows the theoretical result which is obtained by scaling the transfer characteristic by the voltage gain of 2.5. Furthermore, the simulation result without and with VGA, and the measurement result are plotted. The simulation differs slightly from the theoretical result. Possible reasons are the intrinsic emitter resistance and the limited size of the capacitor C_1. Adding the VGA to the simulation introduces its compression effects which can be seen in the plot. Comparing this to the measurement result, it reveals that the compression happens slightly earlier than expected. Nevertheless, the measurement results fit well to the theoretical and to the simulation result.

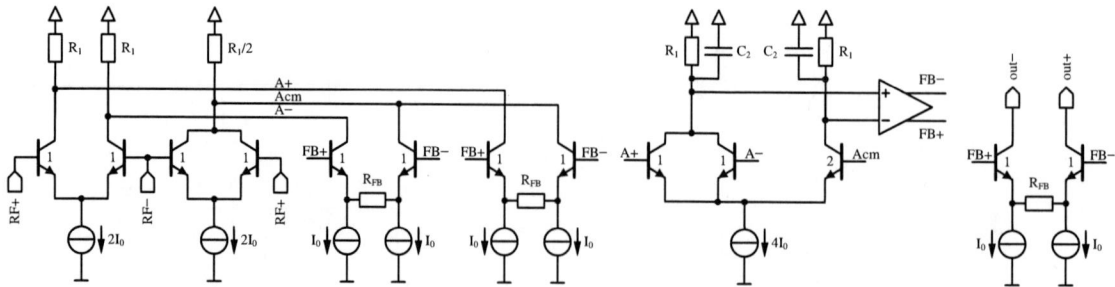

Fig. 6. A possible implementation of a diode detector without capacitor

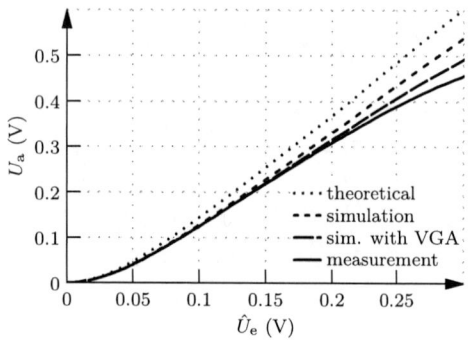

Fig. 9. Measured transfer characteristic of the circuit shown in Fig. 2

IV. CONCLUSION

We showed that it is possible to achieve the same transfer characteristic as a diode detector by an active control loop. Thereby, also additional properties like voltage gain or linearization can be implemented. This is proven by measurement results of a diode detector with a voltage gain of 2.5. Furthermore, it is derived that it is possible to leave out the capacitor of the diode detector which leads to a different and more linear transfer characteristic.

REFERENCES

[1] Fraunhofer IIS, *μRX1080*, 2010, datasheet.

[2] B. Van Liempd, M. Vidojkovic, M. Lont, C. Zhou, P. Harpe, D. Milosevic, and G. Dolmans, "A 3μW fully-differential RF envelope detector for ultra-low power receivers," in *Circuits and Systems (ISCAS), 2012 IEEE International Symposium on*, 2012, pp. 1496–1499.

[3] Analog Devices, *AD8330*, 2012, datasheet.

[4] J. Staudinger, B. Gilsdorf, D. Newman, G. Norris, G. Sadowniczak, R. Sherman, T. Quach, and V. Wang, "800 MHz power amplifier using envelope following technique," in *Radio and Wireless Conference, 1999. RAWCON 99. 1999 IEEE*, 1999, pp. 301–304.

[5] E. Nilsson and C. Svensson, "Ultra low power wake-up radio using envelope detector and transmission line voltage transformer," *Emerging and Selected Topics in Circuits and Systems, IEEE Journal on*, vol. 3, no. 1, pp. 5–12, 2013.

PRIME 2013, Villach, Austria

Session T2A – ADCs and DACs

An Analog 1:16 Demultiplexer for Time-Interleaved A/D-Converters with a Sampling Rate of up to 64 GS/s

Felix Lang, Janina Gerigk, Damir Ferenci, Markus Grözing, Manfred Berroth
Institute of Electrical and Optical Communications Engineering,
University of Stuttgart
Pfaffenwaldring 47, 70569 Stuttgart

Abstract—An analog current-based 1:16-demultiplexer with integrated sample-and-hold is presented. It is designed in a 28 nm CMOS technology and is the basis for a 16-fold time-interleaved ADC. It offers sampling rates up to 64 GS/s, while consuming only 0.9 W of power and 2.6 mm^2 of chip area.

Index Terms— Analog integrated circuits, Demultiplexing, Signal sampling, Analog-digital conversion

I. INTRODUCTION

The increasing need for higher data rates and the limits reached by simple on-off keying systems make it inevitable to use new modulation formats. State of the art 100G systems regularly require at least four 25 GS/s ADCs [1] or faster with at least about 4 bit effective resolution. The next generation systems will continue to increase the requirements on the converters at both the transmitter and the receiver side [2]. This means higher sampling rates as well as a higher effective resolution are required. Furthermore the growing number of converters per system makes it attractive to use a one chip solution in CMOS.

This paper presents a 1:16-demultiplexer structure in 28 nm low power CMOS. It may be used with 16 time interleaved ADCs at the outputs. Thus the sampling rate per channel can be significantly reduced to 4 GS/s per sub-ADC compared to 64 GS/s for the whole ADC.

In the first section of this paper the demultiplexer itself is introduced. Two different possible structures are presented. The first one is a pure tree-like 1:16 structure out of 1:2 substructures. In the second version the first two 1:2 substructures are replaced with a 1:4 structure which allows to omit the fastest clock and therefore to reduce the complexity in the clocking network. In the second section the clocking and reset circuits are described, which mainly dominate the performance for that high sampling rates. Finally simulation results show the resolution which may be achieved by the demultiplexer together with ADCs at the backend.

II. DESIGN OF THE DEMULTIPLEXER

The concept of the analog demultiplexer (DEMUX) is based on the integration of an input current by means of capacitors, which are connected to the input for a defined time slot. Fig. 1 shows a single ended block diagram of the proposed differential circuit. For simplicity the second identical path is not illustrated.

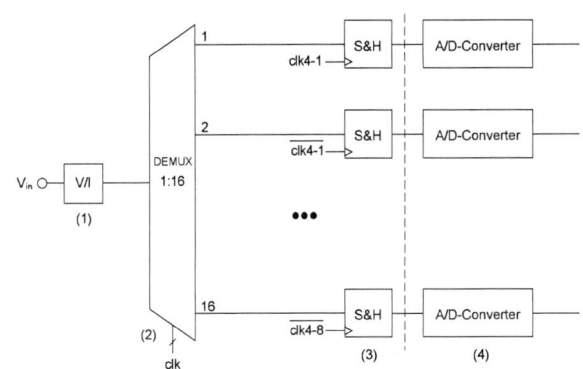

Figure 1.   Block diagramm of the 1:16 demultiplexer

The current is generated by a highly linear transconductance (1). By simple NMOS-transfer transistors in the DEMUX (2) itself the current is switched stepwise to the 16 outputs. Hence only a very small part of the connected circuits, which may here be considered as load capacitances, have to be reloaded in one sampling time interval. When the signal is completely sampled on the capacitor it is resampled by a Sample-and-Hold (S&H) circuit (3). This holds the signals constant for the following ADCs, which are by help of this decoupled from changes induced by the demultiplexing or the reset process.

Fig. 2 shows the current transmission process for one sub-channel of the differential demultiplexer. First the input voltage V_{in} is converted to a current. This is summed up in a time slot T_P on the respective load capacitance C_{Store}. Afterwards the resulting output voltage is sampled by a S&H. This allows resetting the capacitor during the hold-mode with the switch S_{Reset} [3][4].

978-1-4673-4579-8/13 $31.00 © 2013 IEEE   193

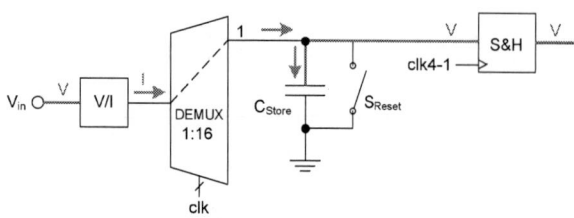

Figure 2. One current signal path of the demultiplexer

Fig. 3 shows two possible switching structures for the demultiplexer. The NMOS transfer transistors are shown as simple switches controlled by the corresponding clock signals. Switching scheme (a) requires four clock frequencies for the four stages. For 64 GS/s operation 32 GHz, 16 GHz, 8 GHz and 4 GHz clock signals are required (clk0-clk3). Those clocks have to be synchronized and shifted to the correct position to prevent currents flowing from one capacitance back to any other node.

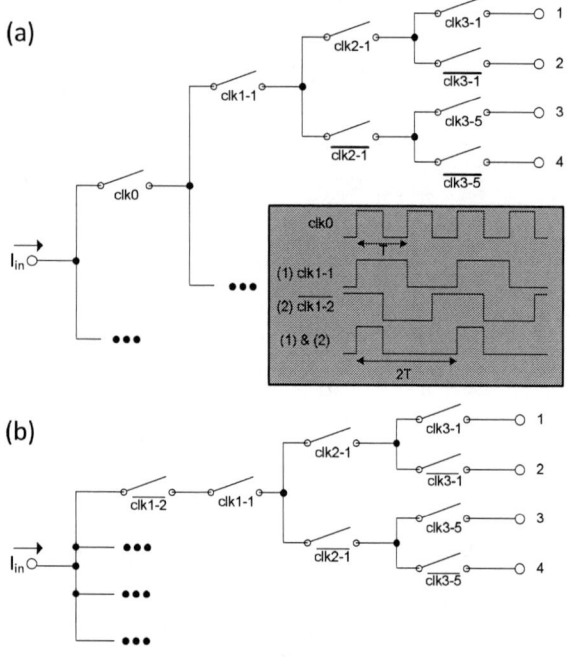

Figure 3.   Switching diagram for (a) a 1:2 demux tree structure and   (b) a 1:4 demux followed by two stages of 1:2 demux

In addition to this the structure in Fig. 3 (b) has been implemented. It has a 1:4 structure at the input. By means of this the 32 GHz clock can be omitted. It is replaced by two 16 GHz clocks overlapping with 90° phase difference. The cascaded switches operate like a logic AND. The second structure has a lower bandwidth if the MOSFET dimensions in the two structures are similar. This is because more transistors are connected to the input node in parallel. This increases the capacitive load. The on-resistance of the conductive path from the left to the right will be the same, since both structures are set up with four serial transistors.

For a good performance of the demultiplexer a highly linear current has to be supplied to the differential input. Fig. 4 shows the developed transconductance amplifier. It is linearized with linearization resistors R and the Miller effect is reduced by a cascoded NMOS-pair on top of the input-transistors. To further increase the linearity the current source of the amplifier has a value of 3.5 mA, while only $I_{out,max} = 0.75$ mA is used to load the capacitors at the end of the demultiplexer chain. The difference is subtracted by two PMOS current sources. This keeps the transistors of Fig. 4 in the same operation regions for the used output current range.

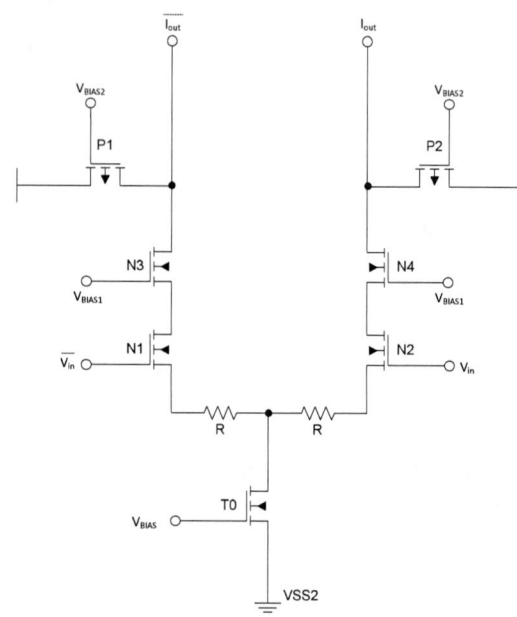

Figure 4.   Linearized differential transconductance amplifier

The amplifier requires a second lower supply voltage VSS2 of about -2.5 V in addition to the regularly used -1.4 V to ensure a fast and linear operation. Simulation results show THD-values below -36.6 dB up to 20 GHz. Expressed in effective resolution, which could be achieved by a succeeding ADC, this results in a value of about 5.8 bit up to 20 GHz.

III. CLOCK AND CONTROL CIRCUIT

The clocking network is the most critical part of the circuit. At clock rates of 32 GHz, indirectly also generated by the AND-concatenation in the second switching scheme, there is only a time slot of 16 ps per channel. The amplitude of the output signal will vary if the clock signals of the 4-stage demultiplexer do not match exactly. The impact of large clock jitter is even worse, the channels output signals will differ randomly to each other and therefore the ADC-outputs of the interleaved structure will diverge [5]. Large jitter effects can be avoided by a good choice of the dimensions of the jitter relevant parts. The timing of the switching signals is done manually by insertion of enough buffers to evolve equal delay lines for every clock rate and some fine-tuning by means of switchable capacitors.

Fig. 5 shows the clocking network for the 1:16 demultiplexer in Fig. 3 (a). The 32 GHz clock clk0 is divided stepwise using three serial clock dividers (CD). The switching transistors of the demux-stage two to four require differential clock signals equally shifted in phase. This is done by doubling the number of clock dividers stepwise from the left to the right. They use pairwise the available phase-shifted signals of a predecessor. The clock dividers cause a delay between the output- and input-clock of around 20 ps. The difference is roughly compensated by insertion of amplifiers as stated before, where every amplifier has a delay of about 3 ps. The clock signals have to be synchronized to the rising edge as illustrated in Fig. 5. If the signals would be synchronized to the falling edge or to anywhere in the center, there would be conductive paths in the demultiplexer which are not supposed to be there. This would yield to false voltages at the outputs of the demultiplexer.

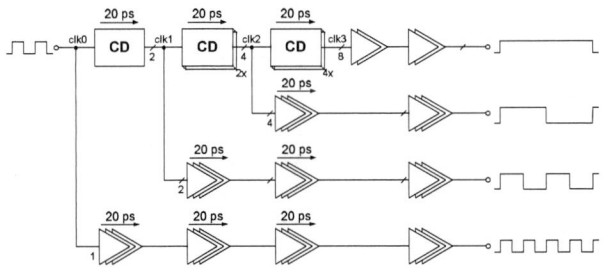

Figure 5.   Clocking network for the 1:16 demultiplexer (here: Fig. 3 (a))

The 16 single ended reset signals for the S&H at the back end of the demultiplexer are generated by a simple AND-concatenation of the slowest clock at the demux with a less delayed one (Fig. 6).

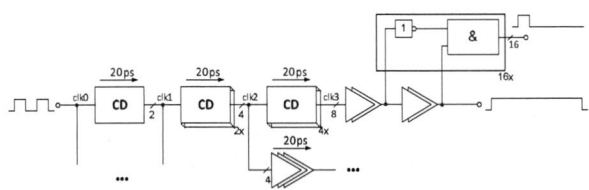

Figure 6.   Generation of the control and reset signals

To reach that high clock rates the whole circuit is designed with differential current mode logic in combination with inductive peaking [6]. The clock dividers, the very fast clock paths and the final buffers of the slow paths are assisted by that to improve the rise and fall times and in doing so the "on"-time of the transfer-transistors. In the rather slow parts of the clock tree the inductors have been omitted to reduce the circuit size of the whole demux. Additionally there are switchable capacitors in the first amplifier chain buffers to fine-tune the delay of the clock buffers.

The three output signals of the clock dividers and the input clock of Fig. 5 are shown at the top of Fig. 7. The signals are asynchronous, do not reach the desired values and have a rather large rise and fall time. By means of the synchronization in the clock tree a behavior like at the bottom of Fig. 7 is achieved. The voltage levels of the clock signals clk1 to clk3 are furthermore shifted down to improve the operation regions of the serial transfer transistors.

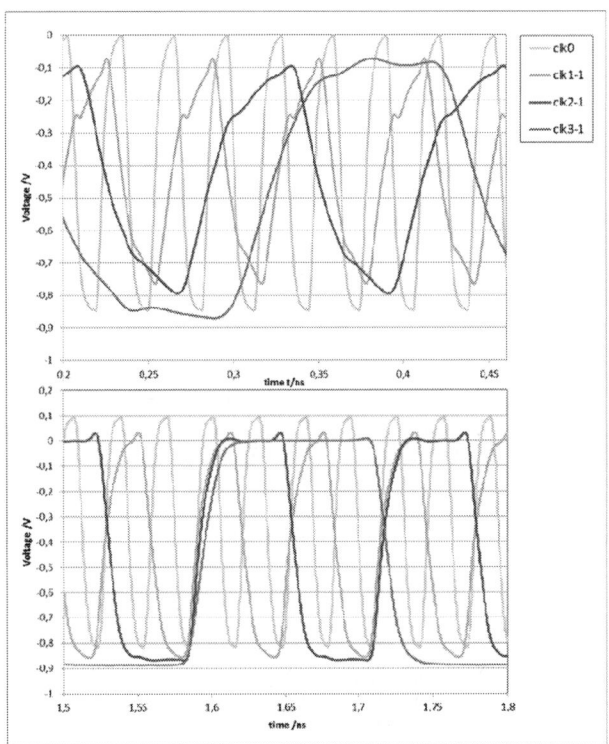

Figure 7.   Unsynchronized outputs of the clock dividers (top) and synchronized clock outputs at the transfer transistors of the demultiplexer (bottom)

IV.   SIMULATION RESULTS

In order to verify the functionality and to characterize the circuit, the demultiplexer circuit of this paper is tested together with an ideal multiplexer structure in Cadence. The dynamic behavior can be derived by application of a sinusoidal signal to the input [7]. The multiplexed time domain signal at the output can then be brought to the frequency domain to calculate the signal-to-noise and distortion ratio (SNDR).

Fig. 8 shows the simulation results for both structures at a sampling rate of 64 GS/s up to $f_s/4$. Version 1, the raw 1:2 demux structure, seems to have a little advantage for some frequencies compared to version 2. For the decision which of the two structures should be preferred, several other indicators should be considered, e.g. power consumption, input bandwidth, complexity of the clocking network and the size of the layout. Due to the expected larger bandwidth, which is a key value for a 64 GS/s demux, and similar results, the further analysis is only given for version 1.

A possible floor plan concept for the clocking network is shown in Fig. 9. Since the circuit size is dominated by the huge number of peaking inductors, the overall layout will be linked to the placement of the inductors. The area of the final peaking inductors is estimated to about 50x50 μm².

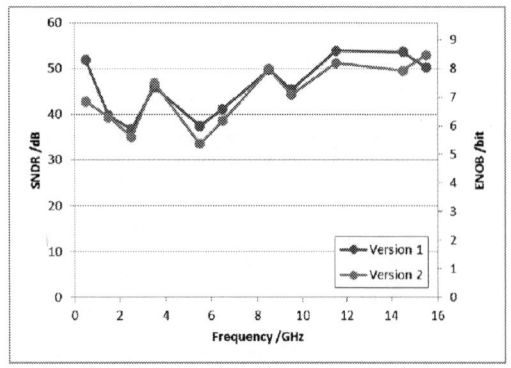

Figure 8. Comparison of the dynamic performance of the two demultiplexer versions at 64 GS/s (Version 1: Raw 4-stage 1:2-demux-structure; version 2: 1:4 demux followed by two 1:2 stages)

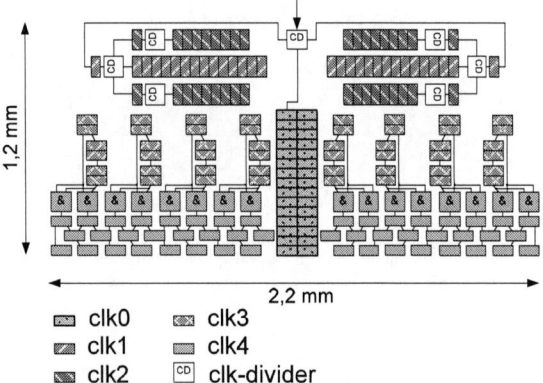

clk0   clk3
clk1   clk4
clk2   CD clk-divider

Figure 9.   Floorplan for the demultiplexer (version 1)

The final simulation results are given in Fig. 10. For a transient noise simulation neglecting the noise of the clock tree and all other parts except the demultiplexer transistors, the SNDR is above 30 dB up to the Nyquist frequency. With noise activated for all parts of the circuit including the clock tree the SNDR is decreased to values above 22 dB up to $f_S/4$ and 15 dB up to $f_S/2$. A jitter analysis according to [5] for the clock-tree results in a total jitter $\sigma_{tot} \approx 101$ fs. This yields to a theoretical loss according to Fig. 10 (3). An additional loss of about 8 dB is originated in the proceeding sample-and-holds and their timing (see Fig. 2), the transconductance amplifier and the reset switches (Fig. 10 (4)). The results may be optimized by increasing the size of the clock-dividers, which are the main source for the occurring jitter. Again a trade-off between power consumption and accuracy of the overall converter has to be found. The table finally shows the characteristics of the simulated design. After adaption of the storage capacitors the design has also been tested for 40 GS/s with similar transfer characteristics as for 64 GS/s as result.

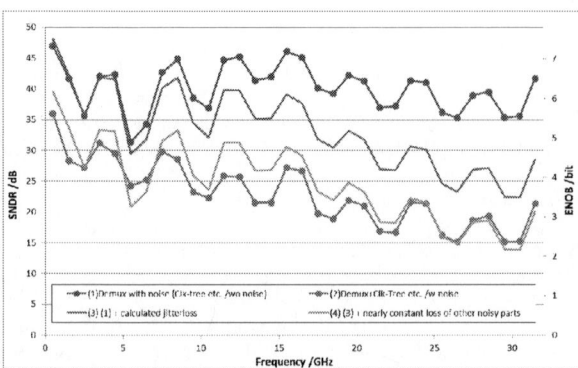

Figure 10. Simulated signal-to-noise and distortion ratio (SNDR) and effective number of bits (ENOB) up to the nyquist frequency at 64 GS/s—with simulated noise for all parts except the clock tree and with noise for all parts with the clock tree

TABLE I.   DEMULTIPLEXER CHARACTERISTICS

Technology	28 nm LP CMOS
Vdd-Vss / Vdd-Vss2 (V)	1.4 / 2.5
Sampling rate (GS/s)	40/64
SNDR (dB@Hz)	35@DC > 22@DC-16G > 15@DC-32G
Max. Input Range $V_{pp,diff}$ (V)	400 mV
Power Chip (W)	0.9
Core Size (mm^2)	~2.6

V.   CONCLUSION

An analog 1:16 demultiplexer for operation at 64 GS/s is shown in 28 nm LP CMOS. The circuit achieves SNDR values above 30 dB up to the Nyquist frequency if the clock noise effects are neglected. If considered, the SNDR will decrease to 22 dB up to $f_S/4$ and 15 dB up to $f_S/2$ for a circuit with a power consumption of 0.9 W. Higher resolutions can be achieved by increasing the clock tree size. This is accompanied with a trade-off between power and accuracy.

REFERENCES

[1] F. Lang, T. Alpert, D. Ferenci, M. Grözing , M. Berroth, "A 6 Bit 25 GS/s Flash Interpolating ADC in 90 nm CMOS Technology", PRIME 2011, Madonna di Campiglio, Italy, July 2011.

[2] Fujitsu Semiconductor Europe GMBH, ">50Gs/s CMOS ADC, DAC and DSP", http://indico.cern.ch/getFile.py/access?resId=0&materialId=slides&confId=121657.

[3] Sami Karvonen, Thomas A. D. Riley, "A CMOS Quadrature Charge-Domain Sampling Circuit With 66-dB SFDR Up to 100 MHz", in IEEE Transactions on Circuits and Systems I, vol. 53, no. 2, pp. 292 – 304, Feb. 2005.

[4] L. Richard Carley and Tamal Mukherjee, "High-speed Low-Power Integrating CMOS Sample-and-Hold Amplifier Architecture", IEEE Custom Integrated Circuits Conference (CICC) 1995, May 1995.

[5] R. J. van de Plassche, "CMOS Integrated Analog-to-Digital and Digital-to-Analog Converters", 2nd Edition, Springer-Verlag, 2003.

[6] J. Kim, J.-K. Kim, B.-J. Lee, D.-K. Jeong, "Design Optimization of On-Chip Inductive Peaking Structures for 0.13- µm CMOS 40-Gb/s Transmitter Circuits", in IEEE Transactions on Circuits and Systems I, vol. 56, no. 12, pp. 2544 – 2555, Dec. 2009.

[7] IEEE Instrumentation & Measurement Society, "IEEE Standard for Terminology and Test Methods for Analog-to-Digital Converters", IEEE Std 1241™-2010, January 2011.

A continuous time switched capacitor DAC with offset and flicker noise cancellation

A. N. Longhitano, F. del Cesta, P. Bruschi
Dipartimento di Ingegneria dell'Informazione
University of Pisa
Pisa - Italy
aurelio.longhitano@for.unipi.it

R. Simmarano
Sensichips srl
Latina - Italy

roberto.simmarano@sensichips.com

Abstract—A switched capacitor DAC, capable of producing a continuous time output signal, is presented. The multiphase conversion cycle allows cancellation of the input offset and low frequency noise of both op-amps used in the circuit. Electrical simulations, performed on a 12 bit prototype, designed by means of UMC 0.18 um MM/RF CMOS process, are shown. The main estimated performances are: 2.4 us conversion time, 0.42 mW power consumption at 1.8 V power supply, 0.49 mV rms output noise over a 400 kHz bandwidth.

Keywords—DA converters, switched capacitor, offset cancellation

I. Introduction

Data converters play a key role in mixed signal circuits, allowing the analog world to communicate with the digital world and vice-versa. These operations are necessary for sensor and actuator interfacing; incorporation of DACs (Digital to Analog Converters) and ADCs (Analog to Digital Converters) into the present and future generation of System on a Chip (SoC) is becoming a very common requirement. High speed and high resolution DACs are widely used for video, audio and communication systems [1]. DACs are also required to generate precise stimuli for sensors (e.g. in electrochemical impedance spectroscopy [2] and voltammetry) and biomedical mobile equipments.

The output signal of DACs connected to off-chip subsystems should generally be time continuous, that is time intervals where the signal is not valid are not allowed. Resistive DACs are still a popular choice, since they naturally meet this requirement. However, they always need an output buffer that, in CMOS implementations, represents a significant source of offset errors and low frequency noise. Use of auto-zero (AZ) or correlated double sampling (CDS) techniques to mitigate these non-idealities is generally not compatible with a time continuous signal, unless complicated architecture are adopted [3]. Switched capacitor DACs are generally designed to be insensitive to offset and low frequency noise from the involved amplifiers by means of CDS-like techniques. They also are marked by low power consumption, but do not provide a continuous time output signal [4], unless they are buffered with a track and hold (T/H) block, generally marked by poor offset and noise performances. Offset compensated T/H stages are available, but precise operation occurs only during the hold

phase, while the offset still appears in the track phase [5].

In this work we propose an alternative architecture for capacitive DACs, capable of producing an offset free output signal in all phases of the conversion cycle. The circuit uses an active feedback chain to control an op-amp based integrator, which can be designed to directly deliver the required current to the load. The results of electrical simulations performed on a 12 bit prototype, designed with the 0.18 μm CMOS process of UMC, are presented, together with a precise estimation of the output noise for different design choices.

II. System Architecture

A. Principle of operation

The proposed architecture is shown in Fig. 1, where OP1 and OP2 are operational amplifiers, while V_{refP} and V_{refN} are two reference voltages. V_{refN} sets the lower end of the DAC output range. Voltage V_{CM} does not affect the conversion result but is required to properly bias the inputs of the op-amps in a single power supply configuration. A conversion cycle is made up by three phases, shown in Fig. 2, where the connections of the various capacitors and the conventions used for the sign of the relevant voltages are indicated.

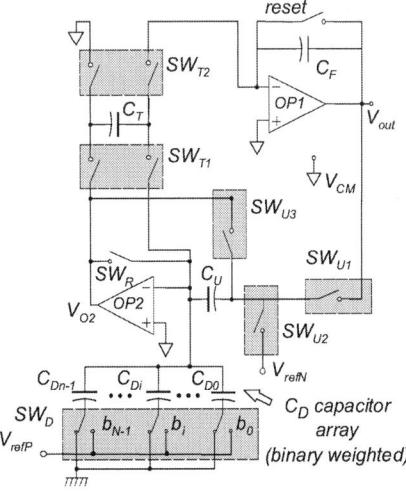

Fig. 1. Block diagram of the proposed DAC architecture.

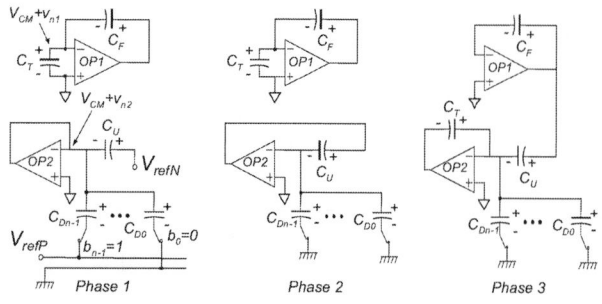

Fig. 2. Main phases of the DAC conversion cycle.

The reset switch, included in Fig.1 is optionally activated at power-up and is left open during normal operation. Phases will be indicated with superscripts, so that, for example, $V_{CT}^{(1)}$ is the voltage across C_T in phase 1. In the following discussion OP1 and OP2 noise and offset input voltages will be modelled with a single voltage source connected, as customary, in series with the input terminals. These sources will be indicated as v_{n1} and v_{n2} for OP1 and OP2, respectively. We will suppose that, in all phases, the gain of the local feedback loops is much higher than one in the whole frequency band of interest. In these conditions, we can approximate the input voltage of each amplifier with the corresponding noise/offset voltage source.

In phase 1, capacitors of the array C_D, whose corresponding bits are set to 1, are pre-charged to V_{refP}. The parallel of these capacitors will be indicated with $C_{DH}=DC_{D0}$, where:

$$D = \sum_{i=0}^{N-1} 2^i b_i \qquad (1)$$

while the remaining capacitors of the array (connected to gnd) will be indicated with C_{DL}. We can write the following expressions:

$$\begin{cases} V_{CU}^{(1)} = V_{refN} - v_{n2}^{(1)} - V_{CM} \\ V_{CDH}^{(1)} = -V_{refP} + v_{n2}^{(1)} + V_{CM} \\ V_{CDL}^{(2)} = +v_{n2}^{(1)} + V_{CM} \end{cases} \qquad (2)$$

When the system gets into phase 2, capacitor C_U is connected across OP2 and all capacitors of the C_D array are switched to ground. Pre-charged capacitors (C_{DH}) inject a charge proportional to V_{refP} into C_U, so that OP2 synthesizes the target voltage (V_{DAC}):

$$V_{DAC} = V_{refN} + \frac{C_{DH}}{C_U} V_{refP} = V_{refN} + \sum_{i=0}^{N-1} 2^i b_i \frac{C_{D0}}{C_U} V_{refP} \qquad (3)$$

Considering also the contributions of the noise/offset voltages we easily find the output voltage of OP2 (V_{O2}):

$$V_{O2}^{(2)} = V_{DAC} + v_{n2}^{(2)} - v_{n2}^{(1)} + \frac{C_{DG}}{C_U}\left(v_{n2}^{(2)} - v_{n2}^{(1)}\right) \qquad (4)$$

where $C_{DG}=C_{DH}+C_{DL}$. When the system enters phase 3, capacitor C_T is connected across OP2. In this operation C_T samples the input noise voltage of OP1. At the same time C_U is

connected to V_{out} by a terminal that, in previous phase, was at the target potential V_{DAC}. In this way C_U experiences a voltage variation that is proportional to the error between V_{out} and the target voltage. A charge proportional to this error voltage is injected into C_T. Considering also the noise contributions:

$$V_{CT}^{(3)} = v_{n1}^{(2)} - \frac{C_U}{C_T}\left(V_{DAC} - V_{out}^{(3)}\right) - \frac{C_U + C_{DG}}{C_T}\left(v_{n2}^{(3)} - v_{n2}^{(1)}\right) \qquad (5)$$

At the end of phase 3, the system begins a new cycle and goes back to phase 1. In order to distinguish this new phase 1 from the previous one, we will indicate it with phase 1N. Considering Fig. 2, we note that C_T is connected back to OP1 so that it transfers its charge into capacitor C_F, applying a correction to the output voltage V_{out} proportional to the error voltage. With trivial algebraic passages, we get the following final expression:

$$\begin{aligned} V_{out}^{(1N)} = V_{out}^{(1)} &+ \frac{C_U}{C_T}\left(V_{DAC} - V_{out}^{(1)}\right) + \\ &+ \left(1 + \frac{C_T}{C_F}\right)\left(v_{n1}^{(1N)} - v_{n1}^{(1)}\right) - \frac{C_U}{C_F}\left(v_{n1}^{(3)} - v_{n1}^{(1)}\right) - \\ &- \frac{C_T C_U}{C_F^2}\left(v_{n1}^{(2)} - v_{n1}^{(1)}\right) + \frac{C_U + C_{DG}}{C_T}\left(v_{n2}^{(3)} - v_{n2}^{(1)}\right) \end{aligned} \qquad (6)$$

It can be observed that all noise contributions consist of differences between noise samples taken at different instants. This results in cancellations of all constant components of v_{n1} and v_{n2} (i.e. offset) and rejection of correlated components, such as flicker noise contributions. A more precise noise analysis will be illustrated in next section. From (6), neglecting all noise contributions, we get:

$$V_{out}(k) = a V_{out}(k-1) + (1-a) V_{DAC}(k-1) \qquad (7)$$

where $a=1-C_U/C_F$. Voltages $V_{out}(k)$ and $V_{DAC}(k-1)$ are discrete signals produced by a sequence of conversion cycles, indexed by k. The output voltage is updated at the beginning of each conversion cycle (phase 1) and held constant across the remaining part of the cycle itself. The digital inputs that set V_{DAC} should be stable during phase 1. It can be easily shown that V_{out} asymptotically tend to V_{DAC} for a constant value of the latter, provided the system is stable ($|a|<1$). The response ΔV_{out} to a step ΔV_{DAC} of the target voltage is given by:

$$\Delta V_{out}(k) = \Delta V_{DAC}\left[1 + a^k\right]u(k-1) \qquad (8)$$

where $u(k)$ is the discrete unity step function.. The time required by V_{out} to reach the final value with a relative error ε_r is given by:

$$k(\varepsilon_r) = \frac{\ln(\varepsilon_r)}{\ln(a)} + 1 \qquad (9)$$

An interesting solution can be obtained with $a=0$, corresponding to $C_U=C_F$, for which the DAC output is updated to the final value with only a single cycle delay.

B. Noise analysis

It is possible to write the equivalent of (7) to for the noise components:

$$V_{out-n}(k) = aV_{out-n}(k-1) + V_{N1}(k) + V_{N2}(k) \quad (10)$$

where $V_{out-n}(k)$ is the output noise sequence, while $V_{N1}(k)$ and $V_{N2}(k)$ are the cumulative result of sampling of v_{n1} and v_{n2}, respectively, at different instants of the conversion cycles, as indicated in (6). Indicating with S_{N1} and S_{N2} the discrete power spectral density (D-PSD) of V_{N1} and V_{N2}, it is possible to express the D-PSD of the output sequence:

$$S_{Nout}(f) = |H(f)|^2 (S_{N1}(f) + S_{N2}(f)) \quad (11)$$

where f is the frequency and

$$H(f) = \frac{1}{1 + a \cdot e^{-j2\pi f/T}} \quad (12)$$

is a transfer function that can be easily derived from (10), while T is the total duration of a conversion cycle. The spectra S_{N1} and S_{N2}, can be calculated from v_{n1} and v_{n2} input noise PSDs by generalization of the methods in [3].

III. PROTOTYPE DESIGN AND SIMULATION

In order to demonstrate the effectiveness of the proposed approach, a prototype has been designed using the UMC 0.18 μm CMOS process. Charge injection from the switches has been taken under control using p-n complementary pass-gates with dummy switches [3] and as small as possible device areas. The output voltage range of the DAC is from V_{refN}=200 mV to $V_{refP}+V_{refN}$=1.2V. OP1 is a low voltage, p–input, class AB operational amplifier designed according to the topology described in [6] and sized for a maximum output current of 20 mA, making the circuit compatible with low load resistances. OP2 is a conventional p-input, folded cascode OTA [7], designed for maximum output range. A single power supply, V_{dd}=1.8 V was used; V_{CM} was set to $V_{DD}/2$. A resolution of 12 bits was chosen. The capacitor array was implemented using the subranging configuration of Fig.3, to reduce excessive maximum-to-minimum required capacitance ratios. MIM capacitors have been used for all capacitors in the design. Value of capacitors are: C_T=7pF, C_U=3.5pF, C_{D0}=54.7fF ($2^6 C_{D0} = C_U$=3.5fF). Capacitor C_F was varied to investigate the effect on the transient response predicted by (8). The larger value of C_T was necessary to prevent large V_{DAC} steps from temporarily saturating OP2.

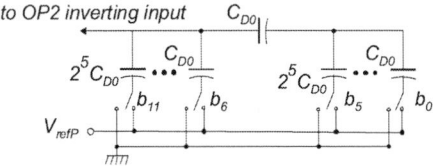

Fig. 3. Subranging implementation of the DAC capacitor matrix.

Timing for the switches is provided by a finite state machine implemented with VHDL code. Three additional phases have been added after phase 2. These phases do not alter the conversion cycle illustrated in Fig.2, but determine a particular switch timing schedule between phase 2 and 3. In this way we

eliminated small glitches arising if C_U is connected to V_{out} when the short transient produced by placing C_T across OP2 is not finished. For simplicity, all phases have same duration (one clock cycle). Since no important charge transfer is involved, the additional phases can be made much shorter than the three main ones, reducing the conversion cycle. The clock frequency was set to 2.5 MHz, resulting in a total conversion time T=2.4 μs (six phases).

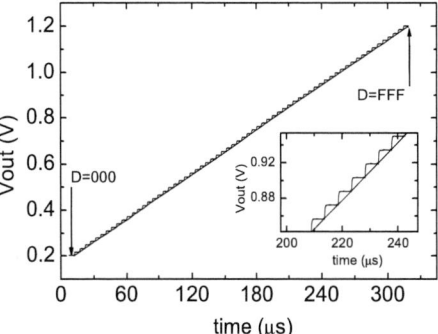

Fig. 4. Transient response obtained by progressively scanning the full digital data range by a 040 (hexadecimal) increment. A straight line, obtained by least square fitting of the values at the end of each interval is also shown.

Fig. 4 shows transient simulations performed by scanning the digital input by a 040 (hex) step. The last value (FFF), otherwise not covered by the scan, has been forced at the end of the simulation. Each value is maintained for two conversion cycles (4.8 μs). This time was sufficient to settle to the final value, since the choice a=0 (C_U/C_F=1) was adopted (see previous section). The values sampled at the end of each time step were fitted by a straight line, shown in the figure. Maximum deviation from the straight line was less than 1 LSB (244 μV) for V_{out}>0.22 V and increased to 2.5 LSB for V_{out} approaching the lower output range (0.2V) This is due to limitations of the OP2 output range.

Fig. 5 shows the transient responses to three digital input values, indicated in the figure. The response is obtained with the nominal circuit (dashed line) and with 50 mV voltage sources placed in series with OP1 and OP2 inputs to simulate the presence of offset (solid lines). Monte Carlo simulations were not possible since the corresponding models were not available for the devices used in this design. The reset signal is active at the beginning of the simulation (grey area), connecting OP1 in unity gain configuration. In this zone the two curves differs by the OP1 offset voltage. In normal operation (reset off), an effective offset cancellation is visible. The inset shows that such a large offset (50 mV) modifies V_{out} by less than 62μV. Various combinations of the simulated OP1 and OP2 offset sources were studied, with less than 1 LSB effects on the output voltage. Simulations have been repeated with a 0.7 kΩ load resistor connected between V_{out} and ground. Also in this case the effect was less than 1 LSB, proving that an additional buffer is not required. Simulations performed in all corner combinations affected the output value by less than 0.5 LSB.

Paper T2A2

Fig. 6 shows transient responses to 1 MSB increase, applied at 41 μs (vertical dashed lines), for different values of parameter $a = 1 - C_F/C_U$.

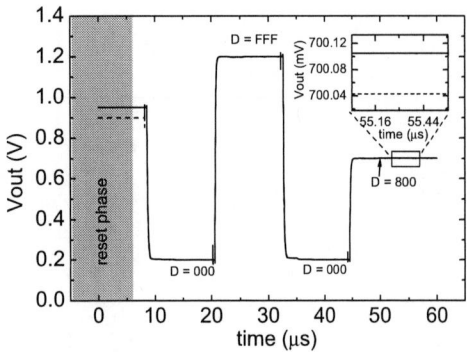

Fig. 5. Transient responses of the ideal circuit (dashed line) to three different digital inputs, compared to the response obtained by simulating the presence of offset voltage by means of 50 mV voltage sources in series with the OP input terminals (solid line).

In the case $a=0$, V_{out} reaches the asymptotic value with a single cycle delay. For $a=0.5$, V_{out} reaches the asymptotic value in a monotonic exponential fashion, while for $a=-0.5$ an oscillating exponential decay is visible. These behaviors are perfectly represented by (8).

Finally, an estimation of the output *rms* noise has been performed on the basis of the method described in Sect. II-B, implemented with a numerical program. The *rms* noise has been obtained by integration of the estimated output PSD.

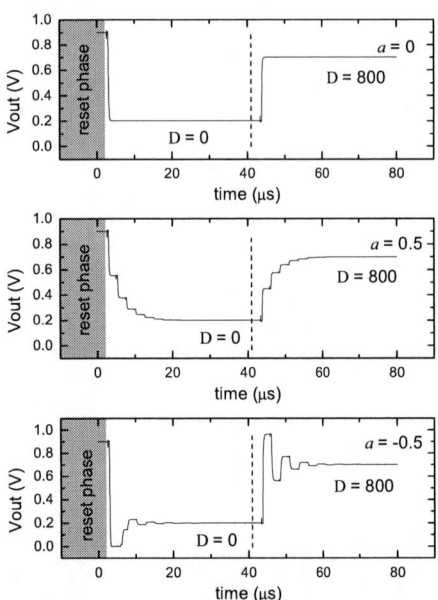

Fig. 6. Transient response for three diferrent values of a ($a=-0.5$, $a=0.5$, $a=0$).

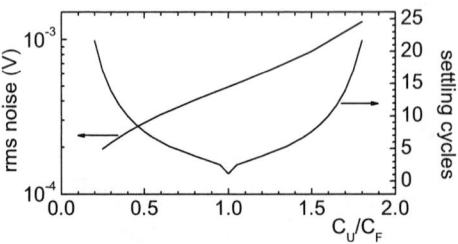

Fig. 7. Output rms noise and settling time (number of cycles) as a function of design parameter $C_U/C_F = 1 - a$.

The results are shown in Fig. 7 as a function of the design parameter C_U/C_F (C_F was varied). The *rms* noise for $C_U/C_F=1$, corresponding to the best option in terms of settling time, is 490 μV. A significant noise reduction can be obtained choosing lower C_U/C_F values, at the expenses of settling time. Values greater than one are detrimental in terms of both noise and settling time. The supply current of the circuit (with no output load) is nearly 270 μA, of which 210 μA are required by OP1, due to the complicated topology of the latter, required for low-voltage class AB operation.

IV. CONCLUSIONS

An original architecture for implementing DA converters, capable of producing a continuous time output free from errors due to offset and low frequency noise contribution, has been proposed and validated by means of electrical simulations. The extreme simplicity of the architecture makes it suitable for integration into mixed-signal SoCs for precise sensor stimulation and actuator driving. The flexibility of the architecture is enhanced by the possibility to choose between different noise/settling-time trade-offs by simply tuning a capacitor ratio (C_U/C_F).

REFERENCES

[1] F. Maloberti, "Data converters", Springer, Dordrecht (The Netherlands) 2007, pp. 79-80.

[2] X. Liu, D. Rairigh and A. Mason "A fully integrated multi-channel impedance extraction circuit for biosensor arrays", proc. of ISCAS 2010, Paris, May 30 2010-June 2 20010, pp. 3140- 3143.

[3] C.C. Enz and G.C. Temes, "Circuit techniques for reducing the effects of op-amp imperfections: autozeroing, correlated double sampling, and chopper stabilization", proc. of the IEEE, vol. 84, pp. 1584-1614, 1996

[4] M. J. Bell, "An LCD column driver using a switch capacitor DAC", IEEE J. Solid-State Circuits, vol. 40, no. 12, December 2005, pp. 2756-2765.

[5] H. Yoshizawa and G.C. Temes, "Switched-capacitor track-and-hold amplifiers with low sensitivity to op-amp imperfections", IEEE Trans. Circ. Systems, vol.54, pp. 193-199, January 2007.

[6] K. de Langen and J. H. Huijsing, "Compact low-voltage power-efficient operational amplifier cells for VLSI", IEEE J. Solid-State Circuits, vol. 33, no. 10, pp. 1482–1496, Oct.1998.

[7] R. J. Baker, "CMOS − Circuit design, layout and simulation". Piscataway, NJ: IEEE Press, 3rd edition, 2010, p. 803.

978-1-4673-4579-8/13 $31.00 © 2013 IEEE

A 10-b 100-MSPS low power Pipeline ADC for high energy physics experiments

A. Donno[1,2], S. D'Amico[1,2], M. De Matteis[3], A. Baschirotto[3]

[1]Department of Innovation Engineering, University of Salento, Italy

[2] INFN-Lecce,

[3]Department of Physics, University of Milano Bicocca, Italy

Abstract — **A 10b 100-MS/s five-stage pipeline analog-to-digital converter (ADC) is implemented in 65nm digital CMOS process with power-reduction techniques for high energy physics experiments. It achieves 56.9dB signal-to-noise and distorsion ratio (SNDR), 58.5dB signal-to-noise ratio (SNR), 9.2 effective number of bits (ENOB) for a full-scale input sine at nyquist frequency. The ADC power consumption is 12.7mW from a 1.2V supply. It occupies 0.8mm² die area.**

Index Terms—**Analog to digital converter (ADC), Pipeline.**

I. INTRODUCTION

The Pierre Auger Observatory [1] , operated by an international collaboration, is conceived with the aim of measuring the spectrum, arrival direction and composition of ultra-high energy cosmic rays (UHECR) with unprecedented statistics and with low systematic uncertainties. The Southern Observatory, completed in May 2008, is located near the town of Malargue in Argentina. Both Observatory provide a full sky coverage when combined. It consists of an array of water Čerenkov detectors covering 3000 km² with about 1600 detectors spaced 1.5 km apart in a triangular grid. The large surface detection area allows the collection of a large amount of statistics data for analysis in a reasonable amount of time. For its structure and location, the observatory need to be powered by a lot of low power on site battery, that need low power circuitry to ensure a long life durability. Čerenkov light produced by charge particles of the showers is detected by three photomultipliers (PMTs). Every signal produced by the three PMTs is amplified and converted in the digital domain by a 100MSPS 10b ADC according to the data acquisition scheme shown in Fig. 1.

Pipeline architecture is the most used architecture for the speed/resolution set of specifications [2][3][4].

A standard pipeline ADC architecture is reported in Fig. 2 represents the effective stage resolution, *r* represents the stage redundancy for the redundant-signed-digit coding (RSD), [2]. Each stage could have different resolution and the number of the stages depends on *B* and *r*. It is made by an input sample-&-hold (S&H), a sub-ADC and an MDAC, [2].

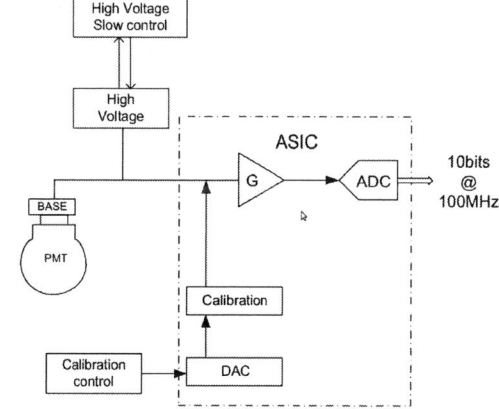

Fig.1. Data acquisition scheme.

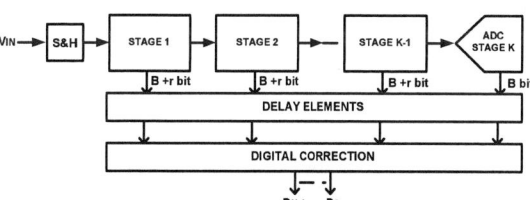

Fig. 2. Standard Pipeline ADC architecture.

In order to reduce power consumption and improve performances different low-power techniques have been used. One of them is the opamp-sharing [3], [4]. Each pipeline stage operates with a two-phase non-overlapping clock, *phi1* and *phi2*. Two successive stages use the same operational amplifier in two different time slots during the overall sampling period. Therefore the opamp is only needed during the residue generation phase. In a conventional pipeline ADC the most power hungry blocks are opamps, thus significant power and area can be saved with the opamp-sharing technique. Another technique used for low-power architecture consists in removing the input S&H stage shown in Fig. 2, [5], [6]. Since it is the first processing block, its noise & accuracy & linearity requirements are very stringent, and then it is one of the most power hungry blocks in the ADC. Using this technique aperture errors can occur. These are due to resistance-capacitance delay mismatch between the input networks for the first MDAC and sub-ADC. In this case a careful layout is needed. Another technique taken into account is the digital background calibration, [7], [8]. Digital correction techniques are suitable for compensating analog limitations in the digital domain. Moreover, these techniques can relax the analog requirements and correct many errors such as sub-ADC errors, capacitance mismatch, opamp non-idealities. For this work a different design approach has been adopted. A detailed analysis of the RDS coding has been performed. Furthermore, a total opamp sharing has been used in order to save power consumption. About the state-of-the-art, the same ADC performances have been presented in [12] but high performance analog MOS transistors are used.

A 65nm digital CMOS technology has been used in the simulation results. The paper is organized as follow: in Section II the proposed architecture is explained; in Section III the circuit implementation of the main blocks is shown; in Section IV the simulation results are reported and in the Section V the conclusions are reported.

II. PROPOSED ARCHITECTURE

Usually, the 1st pipeline stage is designed with the higher effective number of bit B. This choice improves the linearity of the overall ADC because the 1st stage linearity requirements are higher than those for the next stages. Typically 1st stage uses a 2.5b/stage architecture in order to meet power, speed and linearity requirements. Concerning the analysis performed, two different architectures have been considered: a 1.5b/stage and a 2.5b/stage. The aim of this analysis is to define the optimal configuration in terms of resolution among the pipeline stages allowing power minimization. Some simple relations can be carried out in order to estimate the amount of current consumption of each stage. In the following lines the procedure is explained. For each stage a minimum capacitance value according to the signal-to-noise ratio requirement is calculated.

$$C_{stage} = k \cdot T \cdot \left(\frac{10^{\frac{(stage\_precision+1)\cdot 6}{20}}}{\frac{FS}{2 \cdot \sqrt{2}}} \right)^2 \quad (1)$$

FS is the ADC full-scale. The stage capacitance minimum value is fixed by the technology. Then the feedback factor and the loading capacitance of each opamp stage are evaluated as follows:

$$\beta = \frac{C_{stage}}{2^{Bi} \cdot C_{stage} + C_{parasitic}} \ , \quad C_{load,i} = C_{stage,i} + 2^{B_{i+1}} \cdot C_{stage,i+1} \quad (2)$$

After that, the opamp UGB and the transconductance of each stage is calculated. A single-pole opamp has been supposed.

$$UGB = \frac{f_s \cdot (stage\_precision+1) \cdot \log 2}{\pi \cdot \beta} \quad (3)$$

$$g_m^2 = 2 \cdot \pi \cdot UGB \cdot C_{load} \quad (4)$$

In the end a normalized current consumption for a transistor in saturation region is obtained as follows:

$$I_{norm} = \left(\frac{W}{L} \right) \cdot \mu \cdot C_{ox} \propto g_m^2 \quad (5)$$

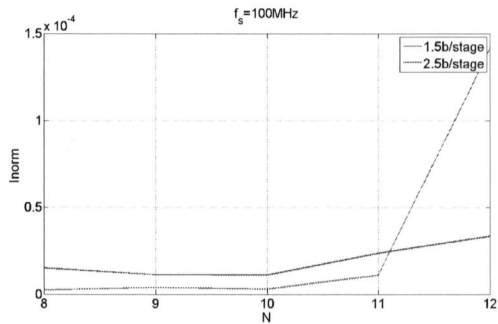

Fig. 3. *Inorm* vs. N (*fs*=100MHz).

In Fig. 3 the trend of the normalized current consumption vs. the ADC resolution is shown. The analysis results lead to a design of 1.5b/stage up to 11 bit ADC resolution. This result is strongly affected by some model simplifications. First of all, all the parameters calculated in the analysis have to be considered as minimum value (*UGB, gm*). Moreover, in scaled technology and low supply voltage (i.e. 1.2V) the opamp DC gain requirement is satisfied at least with a two-stage topology. Then, more current consumption is needed. Another non-ideality is the opamp settling behavior modified by the input switch resistance. An *UGB* higher than (3) is often needed. All these practical design considerations introduce a shift in choosing the ADC resolution stage. In Fig. 4 the overall architecture designed is shown. Fig. 4 shows a different effective bit resolution for the first two stages. The capacitance scaling between successive stages allows using the same opamp for different topologies. The choice of 1.5bit/stage allows to design an opamp with less critical bandwidth requirement than a 2.5bit/stage architecture. Concerning linearity performances, a careful design of the opamp is needed. Moreover, the stage architecture is less critical than 2.5b/stage when the input S&H is removed. The 3rd and the 4th stage have a 2.5b/stage architecture. With respect to the conventional 1.5b/stage implementation the overall ADC architecture has less number of opamps and then less power consumption is achieved.

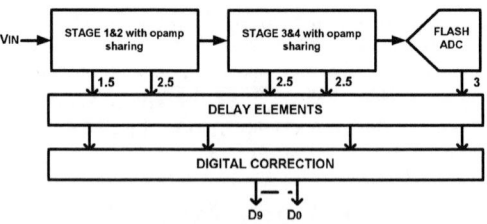

Fig. 4. Proposed Pipeline ADC architecture.

III. CIRCUIT IMPLEMENTATION

After the overall ADC architecture has been chosen a behavioral model has been implemented in Matlab/Simulink environment. The following non-idealities have been taken into account in the behavioral model: finite opamp DC gain/bandwidth/slew-rate/offset, opamp input parasitic capacitance, comparator offset, capacitance mismatch, voltage reference errors [1]. The achieved results have established the lower limits of the ADC blocks performances.

A. Input Stage

The 1.5b/stage represents the first block in the pipeline ADC designed. As previously reported, this choice allows to design an opamp with a bandwidth performance less critical than a 2.5b/stage. This is principally due to the higher feedback factor β. Moreover, in our design the first block also acts as input S&H. In Fig. 5 the single-ended stage architecture is reported. Actually, the structure is fully-differential. A detailed mathematical analysis of the architecture is reported in [5]. The aperture error tolerated for this architecture is about *Vref/4*. As reported in [6], for a 2.5b/stage the time constant mismatch tolerated is about 20%, the same relations for a 1.5b/stage leads to a maximum mismatch of about 40%. This result relaxes the layout constraints. In Fig. 5 the timing diagram is also reported. It can be obtained by clock generator circuit shown in [6]. The different duty cycle among phases allows circuit to work properly. A common flip-around residue generator architecture has been designed for all the MDACs, [11]. Concerning the sub-ADC block a charge distribution input network has been adopted [10]. The threshold voltage of the comparator has been adjusted linearly with the capacitance ratio. Only three MDAC references have been designed. This choice allows saving power.

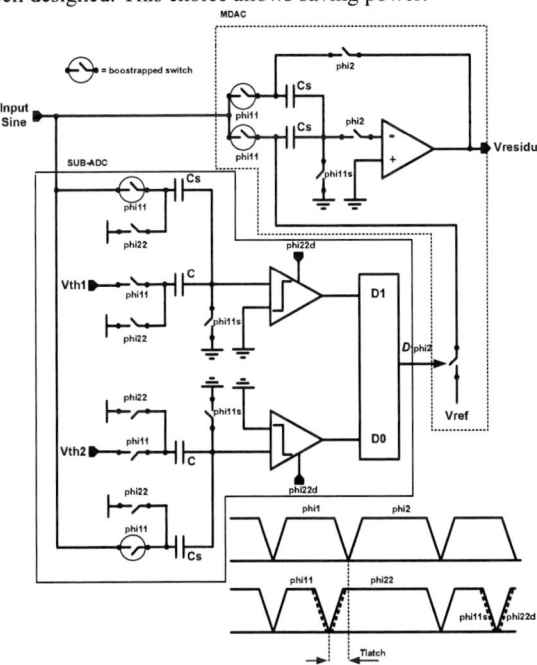

Fig. 5. Pipeline Input stage architecture.

B. Opamp

The minimum operational amplifier performances have been extracted by (3) and verified in the behavioral model. It means that in worst cases those values have to be respected. A two stage topology has been designed. The required DC-gain specification is not satisfied with a single stage in low voltage,

deep-scaled technologies. For higher gain a folded-cascode stage has been chosen. At the output a common inverting stage has been designed. A Miller-compensation between the two stages has been chosen. The opamp schematic is shown in Fig. 6. The nominal supply voltage is 1.2V and the transistor threshold voltage is about 500mV. For this reason different input and output common-mode voltages have been taken into account. A lower common-mode voltage allows a better polarization of the source-coupled input stage. An output common mode voltage of about Vsupply/2 has been chosen. Therefore, the output swing has been maximized. A continuous-time common-mode feedback circuit has been designed.

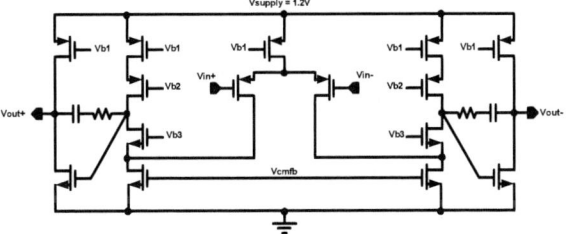

Fig. 6. Opamp schematic.

In the following table the detailed opamp performances are reported. They are referred to the nominal conditions.

TABLE I - OPAMP PERFORMANCES

	Opamp stage 1/2	Opamp stage 3/4
DCGAIN	76dB	73dB
UGB	1.5GHz (Cl=900fF) 1.7GHz (Cl=600fF)	1.2GHz (Cl=450fF) 1.9GHz (Cl=400fF)
PM	25°	15°
Rc	520Ω	1.2kΩ
Cc	380fF	200fF
Current Consumption		
Input Stage	600μA	300μA
Cascode Load	150μA	75μA
Output Stage	800μA	40 μA
Total	2.5mA	1.25mA

C. Comparator

A latch-type voltage-sense amplifier topology has been chosen for the design. The comparator is described in [9]. Due to the 1.5b/stage and 2.5b/stage architecture the input offset is not critical. In the first case a total offset value of *Vref/4* can be tolerated before errors occur. Concerning the second stage architecture a total offset value of *Vref/8* can be tolerated.

D. Reference and Phase Generator

Two different resistive-string voltage references have been designed. The first one provides reference voltages for the first two stages. The second one provides reference voltages for the stages three and four and the full-flash ADC. In this way the resistive strings can have different resolution performances and then current consumption can be saved. The first one consumes about 2mA, the second one consumes about 1mA. As previously shown in Fig. 5 a properly phase generator is needed. The topology taken into account is reported in [6].

IV. SIMULATION RESULTS

In this section the ADC transistor-level simulation results are shown. In Fig. 7 the output PSD (*Power Spectrum Density*) of the ADC for a full-scale input sine is shown. The input frequency is about 49MHz. It achieves an ENOB value of about 9.2b in nominal conditions. In Fig. 8 post-layout simulation of the output is shown. The ADC working frequency is 100MHz. Reported results are referred to nominal conditions. For these simulations a full-scale input sine has been taken into account.

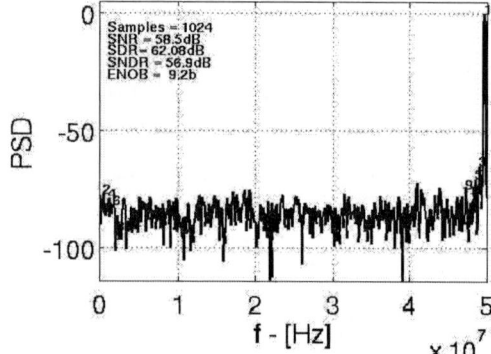

Fig. 7. Schematic ADC output PSD with input sine of 49 MHz.

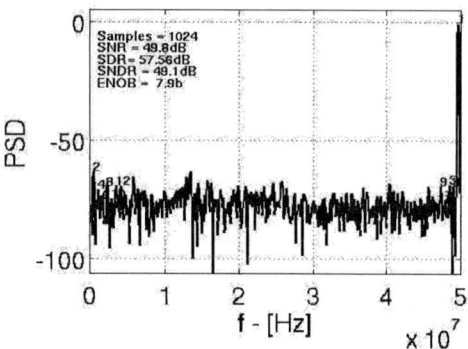

Fig. 8. Post-layout ADC output PSD with input sine of 49 MHz.

In the following table the ADC performances summary are reported. In Fig. 9 the ADC layout is shown, the occupied die area is about 0.8mm².

V. CONCLUSIONS

A 10b 100MSPS pipeline ADC with an optimized bit-stage resolution has been presented. The input sampling is performed directly on the first stage without the explicit S&H saving power consumption. The overall architecture is made by two main stages with opamp-sharing technique and a 3b full-flash ADC. The first stage has a 1.5b resolution architecture, the remaining stages have a 2.5b resolution architecture. The ADC achieves low power consumption while maintaining reasonable performances. It achieves 62dB SFDR, 58.5dB SNR, 9.2 of ENOB. The ADC power consumption is about 12.7mW from a 1.2V supply. The

technology used is 65nm digital CMOS. It occupies 0.8mm² die area.

TABLE II - ADC PERFORMANCES SUMMARY

Resolution	10b
Process	tsmc 65nm Standard CMOS
Input Range	1Vpp
Power Supply	1.2V
SNDR	56.9dB
SDR	62dB
ENOB	9.2b
ADC Power	12.7mW @ 1.2V Supply
Die Area	0.8mm²

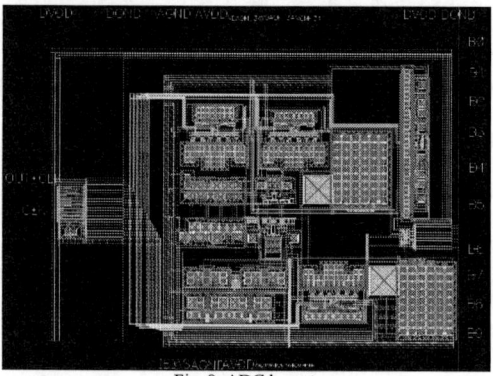

Fig. 9. ADC layout.

REFERENCES

[1] Auger Collaboration, J. Abraham et al., "Properties and performance of the prototype instrument for the Pierre Auger Observatory," *Nucl. Instrum. Meth.* A523 (2004) pp.50-95.

[2] F. Maloberti, *Data Converters*, Springer 2007.

[3] K. Nagaraj, H. S. Fetterman, J. Anidjar, S. H. Lewis, and R.G. Renninger, "A 250-mW. 8-b, 52-Msample/s parallel-pipelined A/D converter with reduced number of amplifiers," *IEEE J. Solid-State Circuits*, vol. 32, no. 3, pp.312-320., Mar. 1997.

[4] B.-M.Min, P. Kim, F. W. Bowman, D.M.Boisvert, and A.J. Aude, "A 69-mW 10-bit 80-MSample/s pipelined CMOS ADC," *IEEE J. Solid-State Circuits*, vol. 38, no.12, pp.2031-2039, Dec 2003.

[5] I. Mehr and L. Singer, "A 55-mW 10-bit 40-MSamples/s Nyquist-rate CMOS ADC," *IEEE J. Solid-State Circuits*, vol.35, no.3, pp.318-323, Mar. 2000.

[6] D.-Y. Chang, "Design Techniques for a Pipelined ADC Without Using a Front-End Sample-and-Hold Amplifier," *IEEE Trans. Circuits Syst. I*, vol. 51, no. 11, pp. 2123-2132, Nov. 2004.

[7] K.-W. Husueh, Yu-K. Chou, Yu-H. Tu, Yi-Fu Chen, Ya-L. Yang, and H.S. Li, " A 1V 11b 200MS/s Pipelined ADC with Digital Background Calibration in 65nm CMOS," in *IEEE Int. Solid-State Circuits Conf. (ISSCC) Dig. Tech. Papers*, 2008, pp. 546-634.

[8] B. J. Farahani, A. Meruva, "Low Power High Performance Digitally Assisted Pipelined ADC" in *IEEE Computer Society Annual Symposium on VLSI*, 2008, pp. 111-116.

[9] B. Wicht, T. Nirschl, and D Schmitt-Landsiedel, " Yield and Speed Optimization of a Latch-Type Voltage Sense Amplifier," *IEEE J. Solid-State Circuits*, vol. 39, no. 7, pp.1148-1158, July 2004.

[10] L. Sumanen, M. Waltari, K. Halonen, "CMOS Dynamic Comparators for Pipeline A/D Converters," in *Proceedings of the IEEE Int. Symposium on Circuits and Systems (ISCAS'02)*, May 2002, pp. V-157-160.

[11] L. Sumanen, "Pipeline analog-to-digital converters for wide-band wirelessCommunications," Helsinky University of Technology Electronic Circuit Design Laboratory, Report 35, Espoo 2002.

[12] M. Boulemnakher, E. Andre, J. Roux, F. Paillardet, " A 1.2V 4.5mW 10b 100MS/S Pipelined ADC in a 65nm CMOS," in *IEEE Int. Solid-State Circuits Conf. (ISSCC) Dig. Tech. Papers*, Feb. 2008, pp. 250-611.

An improved ultra-low-power wireless sensor-station supplied by a photovoltaic harvester

A. Lazzarini Barnabei*, Enrico Dallago*, Piero Malcovati[†], Alessandro Liberale*

*Department of Electrical, Computer and Biomedical Engineering, Power Electronics Laboratory
[†]Department of Electrical, Computer and Biomedical Engineering, Integrated MicroSystems Laboratory
University of Pavia, Pavia, Italy. Contact e-mail: alessandro.lazzarinibarna01@ateneopv.it

Abstract—In this work we describe a system optimized to use the few microwatts generated by a 4 mm^2 photovoltaic energy harvester to acquire, process and wirelessly transmit information about the environment temperature and light irradiance. The energy needed for the described power-expensive operations is obtained by charging a buffer capacitor while the system is in a low-power-consumption state. A voltage level detector senses the charging status of the capacitor and wakes-up the sensing station every time enough energy is stored. Thus, the environmental parameters are monitorized and transmitted with an asynchronous and intermittent strategy. The optimization work related to this paper has persued the target of minimize the unerasable continuous power consumption of the voltage detector during the charging of the capacitor. We reached the goal of reducing it from the 50 μW of the previous version of the system to the current 5 μW, with a significant usability improvement. Simulations and experimental evidences demonstrate that our station can work totally battery-free when just illuminated with 100 W/m^2 of light irradiance, transmitting data with a 802.15.4 compliant protocol and potentially occupies a volume of 1 cm^3.

I. INTRODUCTION

The success of the energy harvesting idea is related to the always more relevant presence of portable and low-power consumption electronic devices in the all-days life [1] [2] [3].

The main limit related to the energy harvesting is the low generation capabilities of the sources [4], insufficient to directly supply complex electronic functions, such as the wireless transmission of information.

The natural and most common approach to use energy harvesting sources is related to the energy accumulation strategy, summarized in Fig. 1. The harvester converts a physical phenomenon into an electric power source. A capacitor is charged in order to store a packet of energy. A voltage detector senses the capacitor voltage and enables the power flux to the load when a threshold is reached. The load, the sensor station in our case, is supplied for a short time window with the energy packet and performs power-expensive operations, discharging the capacitor and restarting the cycle.

Assuming that a generic harvester supply can ensure a continuos average power delivery (P_H), depending on the external conditions and on its conversion efficiency, the stored energy is given by:

$$E_{STORED} = (P_H - P_{VD} - P_{LC})t_{STORAGE}, \quad (1)$$

where P_{VD} is the average power consumption of the voltage detector, P_{LC} is the average power loss by the capacitor (due

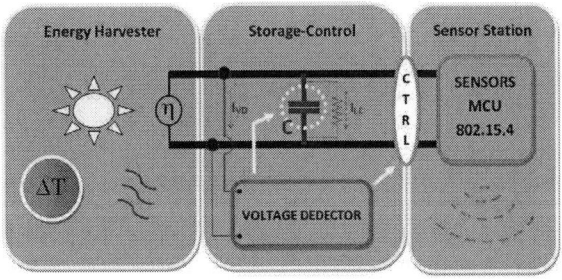

Fig. 1. Scheme

to its parassitics) and $t_{STORAGE}$ is the time needed to charge the capacitor.

The total efficiency of the system depends on the harvester yield, on the power consumption of the storage-control stage (voltage detector power consumption and capacitor leakages) and on the sensor station energy consumption (considering a single acquisition-processing-transmission operation) and defines the frequency of communication.

The minimum input power to act the system depends only on the power consumption of the storage-control stage, that is the only load permanently connected to the source.

In this work we consolidated the system realized in our previous researches [5] [6], improving its total efficiency. In particular, we focused our effort in the reduction of the minimum input power parameter, acting on the voltage detector circuit and experimenting new low-leakage capacitors.

II. PHOTOVOLTAIC INTEGRATED ENERGY HARVESTER

Both in the previous versions of the sensing station and in the presented one, the same photovoltaic energy harvesting system was used (Fig. 2).

It consists in an integrated photovoltaic power source and represents the result of a two-years research activity. It is realized in a 0.35-μm BCD SOI technology and occupies an area of 4 mm^2. The chip includes 35 photovoltaic cells and an electronic circuit, composed by a bandgap, a temperature sensor and a LDO (Low Drop Out) regulator. The PV cells are organized in a series/parallel configuration to provide two different voltage references with two different current delivery levels. The peculiarity of this integrated harvester

978-1-4673-4579-8/13 $31.00 © 2013 IEEE

Paper T3A1   *PRIME 2013, Villach, Austria*

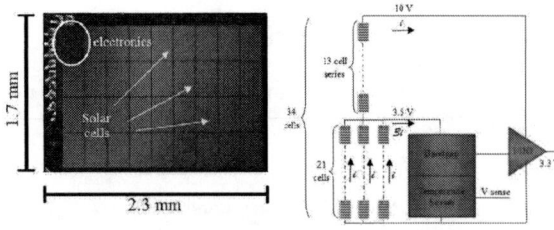

Fig. 2. Photograph and scheme of the photovoltaic energy harvesting chip

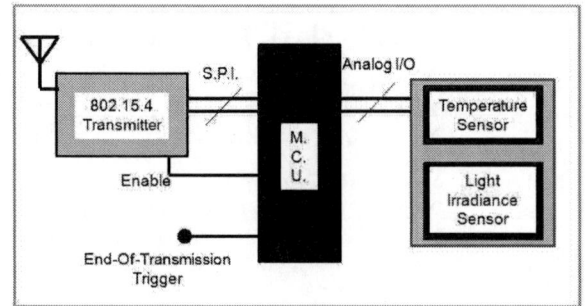

Fig. 3. Sensors station scheme

is that it produces electric power at high voltage level. This possibility is ensured by the SOI technology and makes the generator ready to supply devices as MCUs, Wireless Modules or Sensors, without voltage step-up operations. The included electronic circuits represent a multiutility ultra-low-power platform designed to support the circuitry of the system. While the first version of this system used all the integrated circuitry, the one we are analyzing uses only the harvester module and the temperature sensor.

III. SENSOR STATION

The sensors station includes two sensors, a microcontroller (MCU) and a wireless transmitting module, 802.15.4 compliant (Fig. 3). The working principle is simple: when supplied, the MCU initializes itself, enables the sensor power supply, reads their output with its internal ADC and formats the data, enables the transmission module and performs the transmission of information. This working algorithm was borrowed from the previous versions of the station, and is basically the same except for the supply management strategy. In fact, this version takes a particular care of the power consumption fluxes in order to optimize them: the start of operations corresponds to the supply of the MCU; the sensors and the wireless module are supplied later, only for the time they spend for their own functions. Another power saving strategy, already implemented in the previous versions, is the end-of-transmission triggering that forces the load switch-off internally and limits its working time window.

As in the previous works, we used the Microchips® PIC18LF13K22 as MCU and the Microchips® MRF24J40MA as wireless transmitting module.

The temperature sensor is embedded in the chip (as already mentioned) and is realized with a resistive sensing shunt, measuring a temperature-proportional current mirrored from the bandgap circuit. It presents a good linearity and achieves a measured sensitivity of 3.8 $mV/°C$ over a 60 °C dynamic range.

The light irradiance sensor exploits a debug indipendent cell, present on the chip. Its sensing element is the cell short-circuit current, that is linearly linked to the variation of incident light irradiation and is simple to measure. Fig. 4 shows both the relationship between the short-circuit current and the light irradiance and the sensor acquisition circuit.

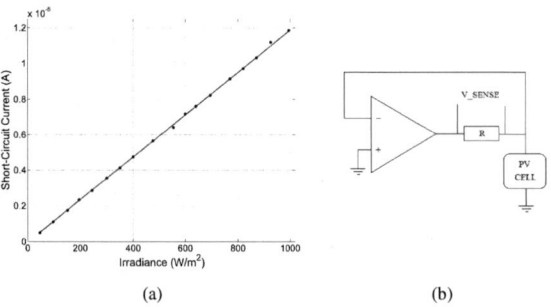

Fig. 4. Short-circuit current of a PV cell (I_{sc}) as a function of the light irradiance (a) and light sensor circuit (b).

IV. STORAGE-CONTROL CIRCUIT

The storage-control circuit shown in Fig. 1 represents the most critical part of the entire system. Its design results critical both considering it by the point of view of the power consumption and of the stability. To analyze and better explain the new circuit proposal in this paper we will follow a comparative approach, making a parallel between the previous and the new circuit topologies.

A. Storing-control circuit: old version

Fig. 5 shows our first version of the circuit, a traditional topology for energy accumulation and for voltage level detection, based on the hysteresis comparator and equipped with a low-leakage 470 μF capacitor.

The circuit is built around the energy harvesting chip and uses the two voltage generators corresponding to the cells network output, the bandgap circuit (as voltage reference) and the LDO (as power regulated output). The design of this circuit must face the fact that all the voltage levels (the harvester output, as well as the bandgap and the LDO ones) reach their steady-state values with a raise transient linked to the capacitor charge time. This problem produces instability in the comparator and requires a particular circuit topology.

In fact, the comparator needs a supply and a voltage reference to properly work and reacts to the slow variations of these voltages with a slow change of its output state. Unfortunately

978-1-4673-4579-8/13 $31.00 © 2013 IEEE   206

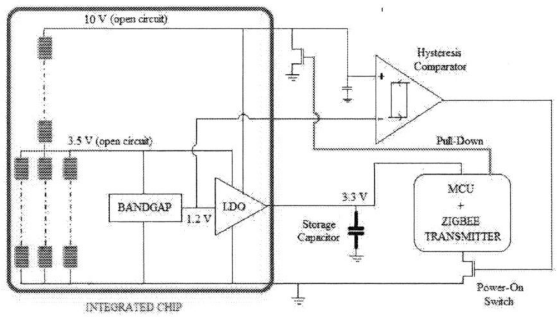

Fig. 5. Storing-control circuit: old version.

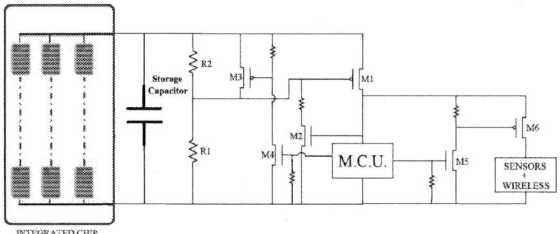

Fig. 6. Storing-control circuit: new version.

this slow state switching reverberates on the switch-on mosfet that enables the load. To solve this issue that prevents the system to work, we exploited the fact that the integrated cells are series-connected and that if a capacitor is connected to the first reference voltage node (the 3.5-V one), the second reference point (the 10-V one) cannot reach its nominal value until this capacitor is charged. When this happens, the current starts flowing throught the upper arm of the cell string and suddenly stand-up the higher reference voltage to its value. This fast voltage step is sensed by the comparator that quickly changes its output state, enabling the load.

B. Storing-control circuit: new version

Fig. 6 shows the new circuit used to store and control the energy packets. A first comparison to the previous circuit shows the complexity reduction of this topology (consider that the electric diagram of Fig. 5 is strongly semplified). The new circuit is equipped with an ultra-low leakage aluminium electrolytic capacitor (Nichicon® UKL0J102KPD) as energy storage and uses 4 MOSFETs and 5 resistors to detect the voltage level. Furtermore, this topology is voltage reference-free [7] and totally indipendent of the harvester chip, thus suitable for other different power generators and harvesters. Its working principle is similar to a latch or an astabile circuit resetted by the end-of-trasmission trigger from the MCU.

The node between resistors R1 and R2 is connected to the gate of M1. During the capacitor charging, it raises with a smaller slope with respect to the capacitor voltage, because of the voltage divider effect. The voltage drop across R2 is proportional to the supply voltage and is equal to the V_{SG} of M1: when it reaches the threshold voltage of M1, M1 turns on supplying the load, represented by the MCU. Transistor M2 is introduced to help M1 to turn on faster and to operate the load with fast transients. When the MCU is active, it performs the operations described in the Section III (enabling sensors and the wireless module through M5). After that it triggers transistor M3, turning off M1 by zeroing its V_{SG}. Transistor M4 is used only to decouple the MCU port from the voltage divider central-node to prevent unwanted threshold perturbations. When the capacitor is discharged by the load,

the potential difference across R2 falls below the threshold voltage of M1 and the cycle begins again.

Naturally, this topology must be coupled with a microcontroller to work, but we also realized a different version of the circuit in which the restart trigger is clocked with two RC low-pass networks.

The circuit shown in Fig. 6 refers actually to the entire system, comprehensive of the generator and of the sensors station. It shows a good stability and results uneffected by the harvester power generation fluctuating behaviour.

We designed and realized it by using discrete commercial components but it is fully integrable, potentially ultra-low-cost and with a reduced area occupation. The voltage threshold of the detector is fully controllable through the resistive voltage divider starting from voltages as low as 0.5 V. For the described application we choose a threshold voltage of 2.8 V, in agreement with the minimum voltage supply of the sensor station components.

V. MEASUREMENT RESULTS

The optimization persued during this work lead to an efficiency and a performance improvement of the system. The measurement campaign that we performed and that we present in this paper shows the differences between the new circuits and the old ones, without focusing on the absolute parameters of the system, already explained in previous works [5] [6].

Fig. 7 shows a comparison between the buffer capacitor voltage transients for the cases illustrated in Fig. 5 and Fig. 6. The capacitor is charged until its voltage reaches a fixed threshold; when this happens the load is supplied and quickly discharges the capacitor. Once the energy consumption of the sensor station for a single acquisition-processing-transmission operation is fixed, the frequency with which the process repeats itself depends on the instantaneous harvested power and on the instantaneous power consumption of the storage-control stage in agreement with (1). While the load energy consumption per cycle determines the discharge voltage step shown in Fig. 7, the other two parameters influence the charge time of the capacitor. The two mesurements compared in Fig. 7 were realized with the same harvester chip at the same light irradiation (850 W/m²). Therefore we can assume that the instantaneous harvested power in the two cases is equivalent. The main differences between the two plots concerns the

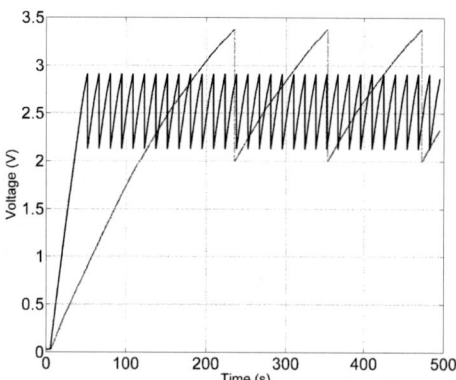

Fig. 7. Transient of the capacitor voltage drop in both the case of the old and the new system versions.

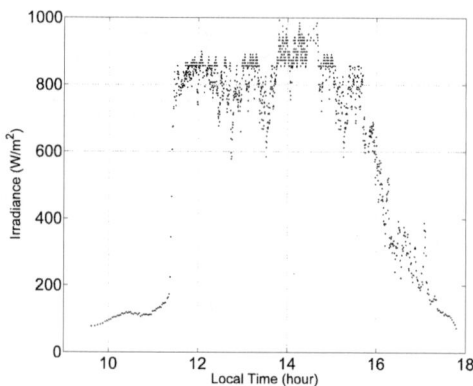

Fig. 8. Light irradiance on March 22, 2013 in Pavia, Italy.

charge slope of the capacitors and their discharge voltage step amplitude. This two differences correspond to the two optimization targets discussed in Section IV. In fact, the capacitor is charged faster with the new circuit because of the reduction of the losses in the storage-control term of (1), as a conseguence of the impoved performance of the voltage detector circuit. Furthermore, in this work, we experimented new commercial low-leakage capacitors and we found an optimum by reducing also the second leakage term of (1). On the other hand, the particular care that we reserved to the power consumption fluxes in the sensor station leads to the reduction of the voltage drop in the discharge phase of the system, thanks to the reduced energy consumption of the load. The instantaneous power consumption of the voltage detector circuit as well as the instantaneous power loss by the capacitor increase proportionally to the supply voltage and reach their maximum at the threshold voltage. The measured maximum value of the voltage detector power consumption is 1.5 μW while the maximum power loss by the capacitor is 3 μW (measurements made with the Keithley® 2000 digital multimeter at 2.8 V). The total instantaneous power consumption of the new version of the storage-control stage is lower than 5 μW while the previous one was about 50 μW. The calculated energy consumption during a single acquisition-processing-transmission operation is about 0.8 mJ while in the previous version was about 2 mJ.

Fig. 7 shows a light irradiance measurement made by the sensor station in Pavia. The measured values of light irradiance start from about 100 W/m^2 (the minimum value by which the system can work) and quickly reach higher values when the sun directly irradiate the harvester (the fast transient is due to the building shadowing).

VI. CONCLUSION

We have designed, realized and tested a wireless sensor station built for energy harvesting applications. The main innovative task of the project is related to the system ability of manage extremely low electric-power amounts, to perform complex and energy expensive operations. To reach this goal we used a simple functional paradigm that can be replicated to other harvesting applications and that can be also persued on generic energy-saving electronic strategies. The fact that the system works with few microwatt power means also that it can manage large amount of power with improved efficiency, making it suitable for a large specter of solutions. We do not consider the work as concluded, but we are exploring new ideas to enhance the device performance, as the conception of a brand-new photovoltaic chip customized on the needs of the current system.

ACKNOWLEDGMENT

Thanks to Daniele Gianluigi Finarelli and Marco Grassi for their important contributions. The BCD SOI technology has been provided by the R&D Department of STMicroelectronics, Cornaredo (Milano), Italy.

REFERENCES

[1] Z.G. Wan, Y.K. Tan, C. Yuen, "Review on Energy Harvesting and Energy Management for Sustainable Wireless Sensor Networks", in *Proc. of Communication Technology IEEE ICCT 2011*

[2] I.Y.W. Chung and Y.C. Liang, "A low-cost photovoltaic energy harvesting circuit for portable devices ", in *Proc. of Power Electronics and Drive Systems (PEDS) 2011* , Singapore, 5-8 December 2011.

[3] E. Mackensen, M. Lai, T.M. Wendt, "Bluetooth Low Energy (BLE) based wireless sensors ", in *Proc. of IEEE Sensors 2012*, Taipei, Taiwan, 28-31 October 2012.

[4] E. Dallago, A. Danioni, M. Marchesi, G. Venchi: An Autonomous Power Supply System Supporting Low Power Wireless Sensors, in *IEEE Transaction on Power Electronics, Vol. 27, no. 10, October 2012, pp. 4272-4280*.

[5] A. Lazzarini Barnabei, M. Grassi, D. Pinna, E. Dallago, P. Malcovati and G. Ricotti, "Integrated Self-Supplied System for Environmental Temperature Sensing", in *Proc. of IEEE Sensors 2011*, Limerick, Ireland, 28-31 October 2011.

[6] A. Lazzarini Barnabei, E. Dallago, G. L. Finarelli, "A Sensing Systems Platform Designed For Wireless Sensor Network (WSN) Applications", in *Proc. of IEEE PRIME 2012*, Aachen, Germany, 12-15 June 2012

[7] I. Syranidis, F. Xia, A. Yakovlev, "A reference-free voltage sensing method based on transient mode switching", in *Proc. of IEEE PRIME 2012*, Aachen, Germany, 12-15 June 2012

Wireless Energy and Data Transfer for Neural Recording and Stimulation Applications

Gurkan Yilmaz and Catherine Dehollain

RFIC Research Group
Ecole Polytechnique Federale de Lausanne
Lausanne, Switzerland
gurkan.yilmaz@epfl.ch, catherine.dehollain@epfl.ch

Abstract—**This study presents design and implementation of a wireless energy and transfer system. The system is targeted to work in accordance with neural recording and stimulation purposes, more specifically for *in-vivo* epilepsy monitoring. A brief overview of the system is given and supported with detailed explanations of functional blocks. An in-vitro experimental setup has been designed and used for performance characterization while mimicking tissue absorption. 35% power transfer efficiency has been achieved to transfer 10 mW to the implant unit from 10 mm. Moreover, 50 kbit/s and 500 kbit/s downlink data communication has been realized with 19% and 33% power transfer efficiencies, respectively.**

Keywords—wireless energy transfer; inductive link; resonance; downlink communication.

I. INTRODUCTION

Intracranial neural recording and stimulation have gained a significant importance for diagnosis and therapy of neurological disorders such as epilepsy, Parkinson's disease, and essential tremor [1], [2]. Current practice in recording neural signals is to implant flat surface electrodes onto the cortex whereas neurons are stimulated by means of needle-shaped electrodes which are implanted deeper into the brain. Both recording and stimulation require data communication between these implanted electrodes and an external instrument. Recorded neural signal is transferred to the external unit (uplink communication) for further investigation leading to diagnosis. A more specific example is to localize the epileptic focus before resectioning (removing) the part of the brain causing the epileptic seizures. On the other hand, in case of stimulation, transmission of stimulation information to the implant (downlink communication) becomes more critical. Currently these links are established via transcutaneous wires. At this point, it is worth to note that recording systems also require a downlink communication; however, at a lower data rate compared to stimulation systems.

However, using wires for communication has certain drawbacks: risk of cerebrospinal fluid (CSF) leakage or even worse infection of CSF which may have fatal consequences, reduced patient mobility and comfort, and hence obstacles for long-term implantation. These issues can be overcome by employing wireless data communication which eliminates the

transcutaneous wires. Furthermore, using active electrodes for recording improves the quality of acquired neural information since it improves signal-to-noise ratio (SNR). On top of that, miniaturizing the electrodes thanks to the improvements in microtechnology enhances the spatial resolution. All these additional features eventually result in a power demand in the implant. At first glance, implantable batteries seem to solve this problem; however, the size limitations and capacity requirements do not meet in a single product for specific applications. Even it is concluded to implant a battery, targeting long-term implants requires the battery to be charged periodically. Although there has been a great improvement in the ambient energy harvesters in the last decade, they are still not sufficient for powering neural implants requiring several milliwatts [3]. Therefore, wireless energy transfer stands out as a generic solution covering various applications.

In this study, we combine wireless energy and data transfer systems in the frame of an epilepsy monitoring system. Overall system aims to acquire neural signals via micro electrode arrays and to transmit the signals to the external base station using the wireless data link which is established on the wireless energy transfer link. This study mainly focuses on the design and implementation of the wireless energy transfer and downlink communication parts of the entire system. As an outlook, higher data rate downlink data communication is also realized aiming for stimulation purposes in the future. Note that the ultimate goal is to combine a recording system with a stimulation system in such a way that epileptic seizures are intervened as soon as they begin.

II. SYSTEM OVERVIEW

Presented system has two main functions: wireless energy transfer to the implant and downlink data communication. Energy transfer is performed via magnetic coupling by means of a 4-coil resonant inductive link. Magnetic coupling, briefly, constitutes the operation principle of the transformers. Compared to the other solutions of wireless energy transfer such as electromagnetic radiation and ultrasound, magnetic coupling is found to be more suitable for our operation frequency of 8.6 MHz, as well as the operation distance of 10 mm, which is dictated by the average thickness of the human scalp. The operation frequency has been optimized to minimize the tissue absorption [4] and to allow high data rate downlink

This project has been funded by Swiss National Science Foundation (SNSF) via "Epilepsy in-vivo" project.

978-1-4673-4579-8/13 $31.00 © 2013 IEEE

communication while preserving the power transfer efficiency as much as possible. Note that downlink data communication is performed on the energy transfer link by changing the operation frequency, which in return degrades the power transfer efficiency.

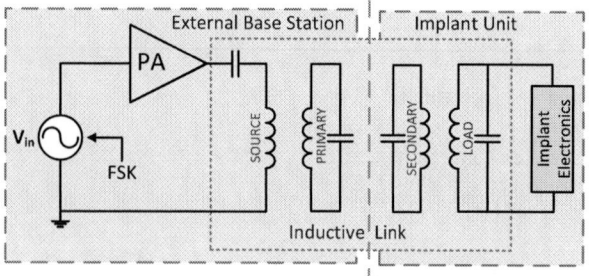

Figure 1: Representation of the wireless energy and data transfer system

Entire system is composed of an external base station, 4-coil resonant inductive link, and the implant electronics as illustrated in Figure 1. External base station is to be located outside the scalp skin and is composed of a signal generator and a power amplifier. The power amplifier drives the inductive link via source coil and the load coil drives the implant electronics which are illustrated in detail in Figure 2. The combination of a rectifier and a regulator creates a reliable power supply for micro electrode array and all other electronic circuits in the implant. According to the estimated power consumption of active micro electrode array and all other electronics, 10 mW power delivery at the output of the regulator is targeted. Modulated information in the external base station is demodulated by means of an ASK demodulator. Downlink data rate is estimated to be lower than 50 kbit/s in neural recording applications in which only operational commands are sent to the implant. However, according to the typical settings of deep brain stimulation (DBS) given in [2], almost 500 kbit/s data rate is required for 10-bit resolution stimulation amplitude and 90 μs pulse width in trains of 165 Hz stimulation frequency when 4 stimulation electrodes are employed. Moreover, a clock recovery circuit is designed to provide a clock signal to the ADCs employed for recording purposes.

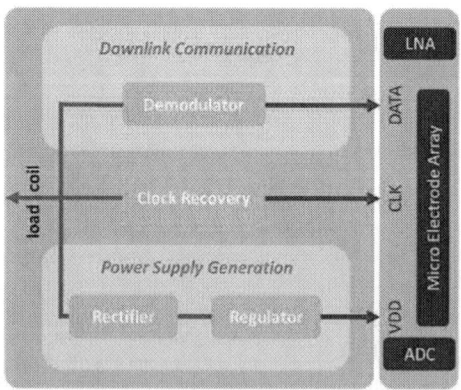

Figure 2: Implant electronics responsible from wireless energy and data transfer, as well as clock recovery.

III. WIRELESS ENERGY TRANSFER

AC power is supplied by the signal generator and transferred to the source coil of the inductive link via a power amplifier and to the primary coil of the inductive link via magnetic coupling. This creates a time varying magnetic field which induces an AC voltage on the terminals of the secondary and the load coils. Then, received AC power by the load coil is converted to a reliable DC supply via a half-wave active rectifier and a low drop-out voltage regulator. Design and characterization of the 4-coil resonant inductive link has already been discussed in [5], [6] as well as a conceptual study of the rectifier. Figure 3 depicts the schematic of the implemented half-wave active rectifier. The rectifier is composed of a pass transistor with dynamic bulk biasing, a comparator, and a multiplexer. The PMOS pass transistor works as a switch and is controlled by the comparator and the multiplexer according to the input and output voltage levels. It is turned on when the input is higher than the output, and turned off otherwise. Therefore, the reservoir capacitance at the output is charged when the switch is on, and reverse leakage to the input is minimized when the switch is off. The comparator decides to pull-up or pull-down the gate voltage of the PMOS pass transistor according to the difference between input and output voltages and the multiplexer changes the bias voltage accordingly.

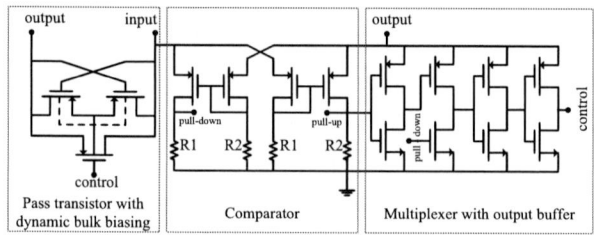

Figure 3: Schematic of the active half-wave rectifier

The output of the rectifier is composed of a DC voltage with a ripple at the operation frequency. In order to eliminate this ripple and provide a DC voltage independent of the input voltage, a regulator has to be employed. As explained in [5], the power efficiency of the blocks on the energy transfer link is the most critical design goal; therefore, a low drop-out voltage regulator is designed to provide a reliable DC supply. Figure 4 presents the schematic of the low drop-out voltage regulator biased with cascoded bootstrapped current source. Reference current is mirrored to bias the operational transconductance amplifier (OTA). OTA compares the reference voltage with the sampled voltage of the output and drives the PMOS transistor to sustain the output voltage constant at 1.8 V.

Figure 5 presents the power supply rejection ratio of the implemented regulator. Measurement is performed with the help of Texas Instruments THS3120EVM board for an input DC voltage of 2 V and an output load of 330 Ω and 100 nF. We can estimate the power efficiency of the regulator at this point by dividing the output voltage to the input voltage since the power consumption in the OTA and reference circuits are quite negligible with respect to the output power. This estimation yields 90% power efficiency for the regulator.

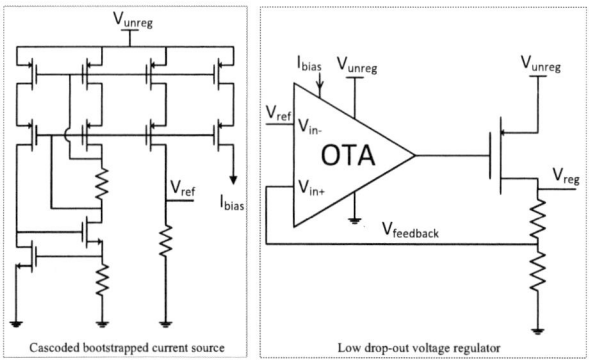

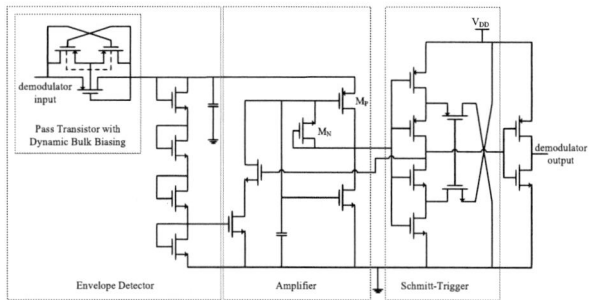

Figure 4: Schematic of the low drop-out voltage regulator and cascoded bootstrapped current source

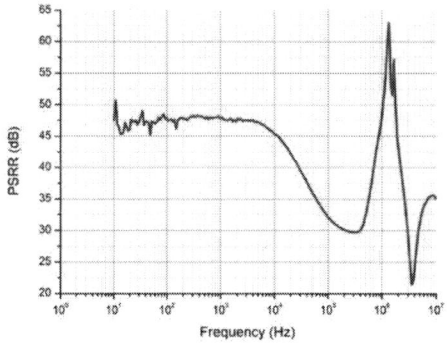

Figure 5: Power supply rejection ratio of the regulator for $V_{in} = 2\ V_{DC}$ while loaded with 330 Ω and 100 nF

IV. WIRELESS DOWNLINK COMMUNICATION

Downlink communication is built on the power transfer link. Note that power transmission is normally realized at resonance frequency to maximize power transfer efficiency, therefore changing the operation frequency results in a different response in the implant side, i.e. another logic state. Here, modulation is performed via switching the operation frequency; however, demodulation is performed according to the variations in the amplitude of the voltage induced on the load coil. In other words, modulation scheme is frequency-shift keying (FSK) and demodulation scheme is based on amplitude-shift keying (ASK).

Modulation has been realized by means of a FSK compatible signal generator in the external base station (Figure 1). An ASK demodulator is employed in the implant electronics to resolve the modulation. Accordingly, an envelope detection based ASK demodulator has been designed and implemented by modifying [7]. Figure 6 presents the schematic of the self-referenced ASK demodulator composed of a pass transistor with dynamic bulk biasing as a part of envelope detector, an amplifier, and a Schmitt-trigger. Since an inverting Schmitt-trigger is employed in the circuit, an additional inverter circuit is added to the end to be consistent with the logic state of the demodulator input. In the amplifier,

M_P is self-biased thanks to the diode-connected M_N transistor. When the envelope of the input waveform is considered as "high", Schmitt-trigger output gives "low", so gate of M_P is biased through the M_N; however when the input is "low", Schmitt-trigger output goes "high" and reduces the gate bias of M_P more via two series connected NMOS transistors to ensure that it is still conducting.

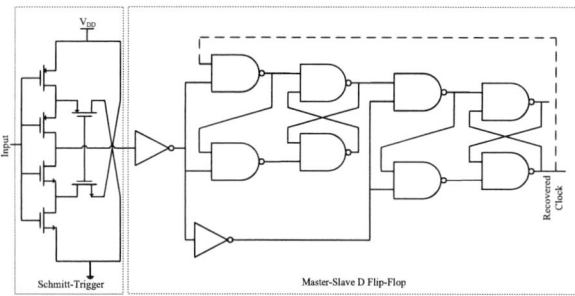

Figure 6: Schematic of the ASK Demodulator

In addition to the demodulator, a clock recovery circuit has been designed and implemented for the implant electronics. It is worth to note that acquired neural signals will be sampled and converted to digital signals. Therefore, either a clock generation in the chip or a clock recovery is required. Here, we select the latter one for the sake of simplicity. Figure 7 presents the schematic of the clock recovery circuit which is composed of a Schmitt-trigger and a master-slave configuration D flip-flop (DFF). A square wave is generated at the half frequency of operation frequency by Schmitt-trigger and employing DFF ensures 50% duty cycle of the generated square wave.

Figure 7: Schematic of the clock recovery circuit composed of a Schmitt-trigger and a master-slave configuration D flip-flop

V. MEASUREMENT RESULTS AND CONCLUSIONS

The designed circuits are manufactured using UMC 0.18 μm 1P6M MM/RF process technology. A micrograph of the implant electronics which occupy 300x350 μm² is presented in Figure 8. Figure 9 depicts the experimental setup built to characterize wireless energy and data transfer system *in-vitro*. In order to mimic the tissue absorption effects, mock cerebrospinal fluid is utilized.

First, wireless energy transfer is realized and 35% power transfer efficiency is achieved at 10 mm distance for 10 mW regulator output power. As mentioned earlier, operation frequency is set to 8.6 MHz.

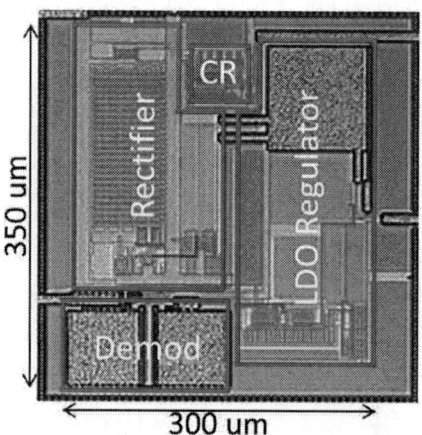

Figure 8: Micrograph of the implant electronics composed of a rectifier, a LDO regulator, a demodulator (Demod), and a clock recovery circuit (CR)

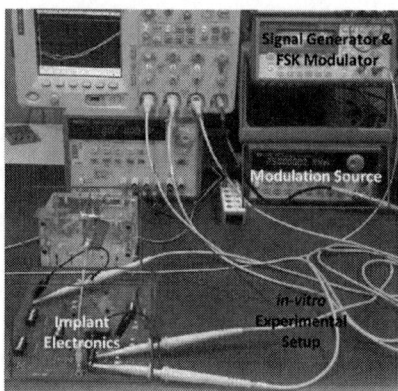

Figure 9: Measurement setup for wireless energy and data transfer system

Next, downlink data communication is performed on the power link. Hop frequency for FSK modulation has been selected as 8.8 MHz in order to preserve power transfer efficiency as much as possible. Note that as the hopping frequency gets away from the resonance frequency, overall power transfer efficiency deteriorates. Considering neural recording applications, first, we have transmitted 50 kbit/s to the implant unit with 19% power transfer efficiency. Since in a long stream 1s and 0s are equiprobable, 25 kHz square wave at 50% duty cycle is generated and fed to the signal generator to modulate it in FSK mode. Afterwards, 500 kbit/s data transmission is also realized for further stimulation applications. We achieve 33% power transfer efficiency for 500 kbit/s downlink data transfer. At first glance, it seems odd to achieve a higher efficiency while transferring higher data rate. However, this is mainly due to the lower data rate limit of the demodulator. In order to perform correct demodulation, voltage supplied by the signal generator has to be increased by 1/3, which almost doubles the input power, in other words, halves the power transfer efficiency for constant output power

dictated by the regulator. Figure 10 presents the waveforms during 50 kbit/s and 500 kbit/s downlink data communication.

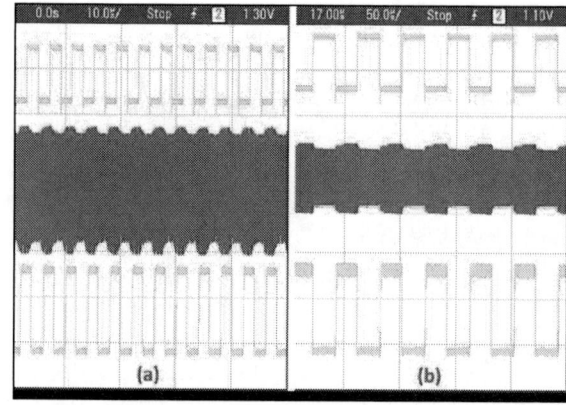

Figure 10: Downlink data communication at (a) 500 kbit/s and (b) 50 kbit/s (waveforms from top to bottom; turquoise: modulator input (5V/div), purple: demodulator input ((a)2V/div (b)5V/div), and green: demodulator output (1V/div), respectively)

ACKNOWLEDGMENT

The authors thank Oguz Atasoy, Enver Gurhan Kilinc, and Kerem Kapucu for their contributions.

REFERENCES

[1] G. Deuschl *et. al.*, "A Randomized Trial of Deep-Brain Stimulation for Parkinson's Disease," *New England Journal of Medicine*, vol. 355, no. 9, pp. 896–908, 2006.

[2] W. H. Theodore and R. S. Fisher, "Brain stimulation for epilepsy," *The Lancet Neurology*, vol. 3, no. 2, pp. 111–118, Feb. 2004.

[3] A. P. Chandrakasan, N. Verma, and D. C. Daly, "Ultralow-Power Electronics for Biomedical Applications," *Annual Review of Biomedical Engineering*, vol. 10, no. 1, pp. 247–274, 2008.

[4] P. Vaillancourt, A. Djemouai, J. F. Harvey, and M. Sawan, "EM radiation behavior upon biological tissues in a radio-frequency power transfer link for a cortical visual implant," in *EMBS, 1997. Proceedings of the 19th Annual International Conference of the IEEE*, 1997, vol. 6, pp. 2499 –2502 vol.6.

[5] G. Yilmaz and C. Dehollain, "An efficient wireless power link for implanted biomedical devices via resonant inductive coupling," in *Radio and Wireless Symposium (RWS), 2012 IEEE*, 2012, pp. 235 –238.

[6] G. Yilmaz and C. Dehollain, "A wireless power link for neural recording systems," *Ph.D. Research in Microelectronics and Electronics (PRIME), 2012 8th Conference on*, pp. 1 –4, Jun. 2012.

[7] C.-C. Wang, C.-L. Chen, R.-C. Kuo, and D. Shmilovitz, "Self-sampled all-MOS ASK demodulator for lower ISM band applications," *Circuits and Systems II: Express Briefs, IEEE Transactions on*, vol. 57, no. 4, pp. 265 –269, Apr. 2010.

Design of a Passive UHF RFID Tag for Capacitive Sensor Applications

Kerem Kapucu, Jose Luis Merino Panadés, Catherine Dehollain

Institute of Electrical Engineering, RFIC Research Group
Ecole Polytechnique Fédérale de Lausanne (EPFL)
CH-1015, Lausanne, Switzerland
e-mail: kerem.kapucu@epfl.ch

Abstract— In this work, the design of a passive UHF RFID tag for sensor applications is described. The system level design issues are discussed taking into account the possible operation scenarios. Among the blocks in the tag, the emphasis is on the design of the rectifier and the capacitive sensor interface. The three-stage, differential-input rectifier is designed to generate the supply voltage with input levels as low as -13 dBm. The power conversion efficiency of the rectifier is 70%, driving a 30 kΩ load at -13 dBm source power. The fully-digital capacitive sensor interface designed to operate with a humidity sensor, can work with supply voltages as low as 0.8 V with good linearity and a power consumption of 12 μW. The tag is capable of getting linear sensor readout under input power levels as low as -9 dBm. The circuits are implemented in 0.18 μm UMC CMOS process.

Index terms—Radio frequency identification (RFID), passive RFID, wireless humidity sensor, food quality monitoring, UHF

I. INTRODUCTION

In recent years there has been a growing interest in radio frequency identification (RFID) systems both in industry and academia. The application of RFID is very popular in many areas such as purchasing and distribution logistics, service industries, etc. [1]. RFID technology replaces the traditional barcodes by providing additional features such as adding sensing capabilities to tags. The addition of sensory capabilities to RFID tags can extend the applications of RFID to environmental monitoring and healthcare applications.

The FlexSmell European Marie-Curie Project (www.flexsmell.eu) is among the efforts to use RFID technology in food quality monitoring. The project focuses on very low cost, ultra-low-power chemical sensing RFID tags to be used for food freshness and quality traceability control. This requires the addition of temperature, humidity and gas sensors on RFID tags.

In this work, the aim is to integrate a capacitive humidity sensor on a passive UHF RFID tag which is to be used in food quality monitoring applications. The emphasis is on the system level design as well as the design of the rectifier and capacitive sensor interface. Section II will give an overview on the system level design issues. The design of the rectifier is discussed in Section III. In section IV, the capacitive sensor interface is described. Section V discusses the effects of backscattering on the performance of the rectifier and the sensor interface.

II. SYSTEM DESIGN

In order to select the frequency of the RFID tag, a detailed description of the operation scenario is necessary. For food tracking, a possible operation scenario is as follows: an operator carrying a handheld RFID reader walks through the shelves in a depot. Each box of goods is packed with an RFID tag with sensors, which can detect the spoiled goods to be removed from the inventory. Such an operation scenario requires an interrogation range of 1-1.5 m as well as the capability of reading multiple tags at once. Moreover, the system should be as immune as possible to tag-reader alignment. Finally, the tags should be remotely powered for cost minimization.

The high-frequency (HF) RFID systems that operate at e.g. 13.56 MHz, lack the long interrogation range, since it is hard to transmit power to distances on the order of meters with reasonable sized antennas. For commercial HF RFID systems the operating range is generally limited to a few centimeters [1]. Moreover, the alignment of the tag and the reader is important for good inductive coupling between the tag and the reader coils. Therefore, an ultra-high-frequency (UHF) RFID system is the obvious choice for this project considering the possible operation scenarios. A tag antenna with a broad radiation pattern can generate immunity to tag-reader disorientation. Furthermore, the long range of UHF RFID systems allows for accessing multiple tags at once.

Considering the maximum allowed RF power levels, the 865.6-867.6 MHz ISM band is chosen for the system since it offers the highest allowed reader RF power (2 W e.r.p), which will result in the highest operating range. The design of the system is optimized for 866 MHz.

This work is funded by the FP7 Marie-Curie ITN FlexSmell Project (www.flexsmell.eu)

978-1-4673-4579-8/13 $31.00 © 2013 IEEE

Paper T3A3   *PRIME 2013, Villach, Austria*

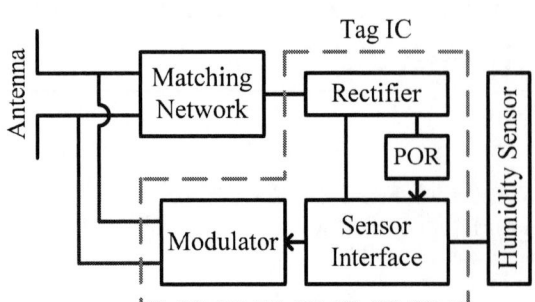

Figure 1. The block diagram of the RFID tag with humidity sensor

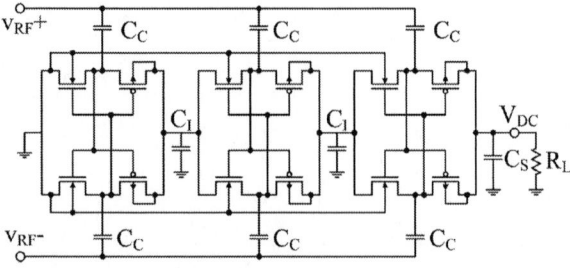

Figure 2. The circuit schematic of the three-stage differential-input rectifier with active threshold cancellation

The power available at the tag antenna terminals can be calculated using the well-known Friis' transmission equation [2]. For example, for 2 W e.r.p source power at 866 MHz, and 0 dBi tag antenna gain, the available power at the tag antenna terminals is 4 dBm and 0.5 dBm at 1 m and 1.5 m, respectively for perfectly matched tag antenna.

Fig. 1 shows the block diagram of the RFID tag. The tag antenna is an inductively-fed meandered antenna whose impedance is designed to be the complex conjugate of the input impedance of the tag circuit [3]. However, a matching circuitry is added to compensate for the impedance variations between the designed and the manufactured values. The supply voltage is generated by a rectifier. A low-dropout voltage regulator (LDO) is not included in the tag because the capacitive sensor interface is immune to the changes in supply voltage value as discussed in Section IV. When the output of the rectifier settles, the power-on reset (POR) block generates a reset signal for the sensor interface. There is no receiver block in the tag, meaning that the tag will operate without any addressing as soon as enough voltage is generated from the incoming electromagnetic wave.

The tag IC is designed to operate with a capacitive humidity sensor, which has an offset value of 5 pF and a range of 1.5 pF. The value of the capacitance increases from 5 pF to 6.5 pF as the relative humidity increases from 0% to 100%.

III. RECTIFIER

The RF-to-DC converter of the tag consists of the antenna, the matching network, and the rectifier. In this work, the emphasis is on the design and implementation of the rectifier.

The power conversion efficiency (PCE) of a rectifier is defined as the percentage ratio of the rectifier output power to the rectifier input power [4]. The PCE of a rectifier depends on several parameters such as the circuit topology, the RF input power and frequency, the output load, and the threshold voltages of the diodes (or the diode connected transistors [5]). The threshold voltage of the diodes should be minimized in order to have a high PCE. For this purpose, Schottky diodes are commonly utilized in rectifiers [6].

In this work, the utilization of exotic technology nodes such as silicon on sapphire, and expensive process options such as Schottky diodes are avoided since the aim is to achieve a low-cost system. The rectifier is implemented using

the MOSFET transistors in UMC 0.18μm technology. However, MOSFETs have higher threshold voltages than Schottky diodes, which results in lower PCEs. Therefore, the active threshold voltage compensation technique proposed in [5] is used, which performs better than the previous static threshold compensation techniques.

The circuit schematic of the three stage rectifier is given in Fig. 2. The rectifier design is optimized at 866 MHz for a maximum load (R_L) of 5 kΩ. The gate widths of the NMOS transistors are designed to be 8 μm, whereas the gate lengths are designed to be 240 nm. Keeping a width ratio of 5 between NMOS and PMOS transistors, the gate widths of the PMOS transistors are designed to be 40 μm, whereas the gate lengths are designed to be 240 nm. Interstage smoothing capacitors (C_I) are 100 fF each, whereas the coupling capacitors (C_C) are 1.13 pF. The storage capacitor (C_S) at the output is chosen to be 200 pF. The rectifier needs only 4.35 us to achieve the 95% of the final voltage with a -9dBm RF input.

For simulations, perfect impedance matching between the antenna and the rectifier at -5 dBm source power with 10 kΩ load is assumed and the RF input power is swept keeping the source impedance constant. In other words, the antenna impedance is not changed at each power level to have perfect matching at every step of the input power sweep. This is more realistic since an adaptive impedance matching scheme will not be utilized in the final RFID tag.

The variation in PCE with respect to RF input power levels between -25 dBm and 0 dBm for three different load conditions are shown in Fig. 3(a). As the load resistance increases, the highest PCE is obtained at lower input power levels. The maximum PCE of 70% is achieved for 30 kΩ load at -13 dBm RF input power, where the output of the rectifier is 0.8 V. This corresponds to a distance of 4.5 m from the reader for 2 W e.r.p source power and a perfectly matched 0 dBi tag antenna. Fig. 3(b) shows the rectifier output voltage with respect to RF input power for four different load conditions. An RF input power of -6 dBm is necessary to generate 1.8 V at the rectifier output for 30 kΩ load, which corresponds to a distance of 2.8 m from the reader.

The change in the performance of the circuit due to process variations is characterized by means of corner simulations. The power conversion efficiency at 0 dBm input power changes by less than 2% and the 1.8 V output voltage changes by less than 3 mV as a result of simulating the circuit

978-1-4673-4579-8/13 $31.00 © 2013 IEEE   214

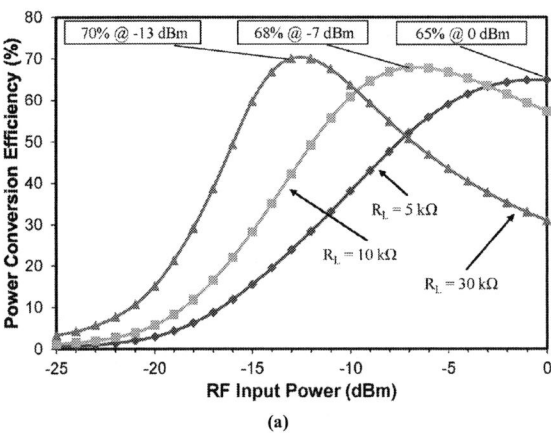

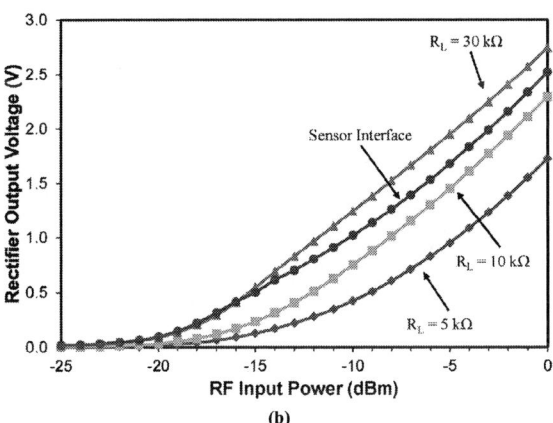

Figure 3. (a) Power conversion efficiency and (b) output voltage of the rectifier as a function of RF input power for different loads

at different technology corners with a 50 Ω power source and fixed matching. The drop in rectifier efficiency in slow-slow corner is less than 10%, for input power levels greater than -5 dBm, whereas the fast-fast corner always results in higher efficiencies.

IV. CAPACITIVE SENSOR INTERFACE

The traditional approach for a capacitive sensor interface consists of an analog signal conditioning block such as an instrumentation amplifier and an analog-to-digital converter such as in [7]. In this work, a fully-digital capacitive sensor interface similar to the one proposed in [8] is used. Fig. 4 shows the block diagram of the capacitive sensor interface. The sensor interface is a basic PLL structure consisting of two oscillators and a phase detector. Both oscillators are asymmetrically loaded, five-stage ring oscillators. The sensor oscillator has the capacitive sensor as the load of one of its stages, whereas the digitally controlled oscillator has a switchable capacitive load. The digitally switched capacitive load consists of two capacitors, denoted as C_0 and C_m. The capacitor C_0, which refers to the offset value of the sensor, is always connected to the ring oscillator. On the other hand, C_m,

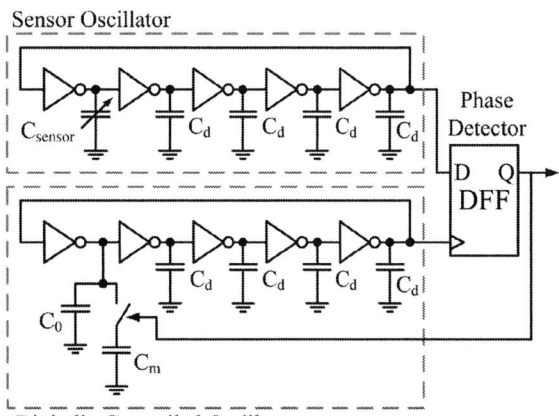

Figure 4. The block diagram of the PLL based capacitive sensor interface

which refers to the change in the sensor capacitance, is connected to and disconnected from the ring oscillator by means of a switch.

The switch is controlled in a feedback manner by the output of the phase detector, which is a pulse train that corresponds to the phase difference between the output signals of the two ring oscillators. It is clear that this structure is a simple PLL due to the feedback loop, and the output of the phase detector is a pulse train whose average value is proportional to the variation in the sensor value.

The capacitor C_0 is selected as 5 pF, which is equal to the offset value of the sensor capacitance and the digitally switched capacitor C_m is selected to be higher than the range of the sensor capacitance. The selection of the value of C_m is discussed in Section V. In the physical implementation, the digitally switched load capacitor is implemented as a programmable capacitor, whose value can be trimmed after the chip is manufactured. The dummy load capacitors (C_d) at each stage are chosen to be 1 pF.

The phase detector is simply a D Flip-Flop, with the output of the sensor oscillator connected to the data input and the output of the digitally controlled oscillator is connected to the clock input. The output of the phase detector is used as the feedback control signal that switches on and off, the digitally switched load capacitor. Although not shown in the figure, the output of the phase detector is buffered in the physical implementation to obtain a signal with sharper transitions. The control signal is actually a pulse train with a duty cycle that is proportional to the change in the sensor capacitance value [8]. In other words, the average value of the control signal is proportional to the change in the sensor capacitance. In the RFID system, this signal will be modulated and sent to the reader by backscattering.

This architecture has an inherent immunity to supply voltage variations due to the fact that the measurement is done in a differential manner, i.e., the phase difference of two oscillations that are supplied from the same voltage source is measured. This gives an advantage over the absolute

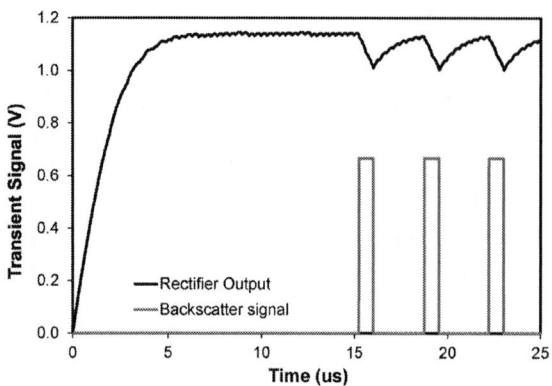

Figure 5. The output of the rectifier and the output of the sensor interface, i.e., the backscatter signal.

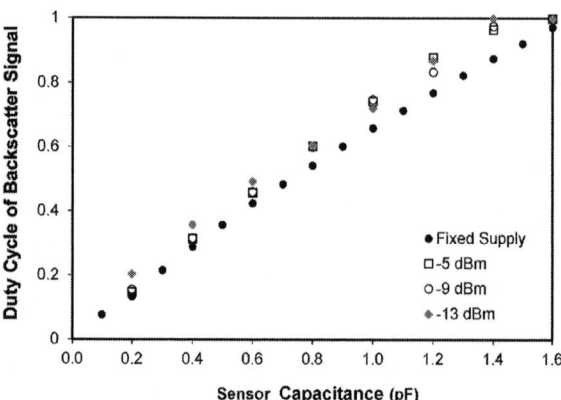

Figure 6. The response of the sensor interface with respect to the sensor value for different RF input power levels. The response is linear up to half the sensor range.

frequency measurement methods in terms of power supply rejection.

V. BACKSCATTERING EFFECTS

The data is sent from the tag to the reader by backscattering, i.e., the antenna inputs are shorted to ground when a digital high level is being sent. This results in a voltage drop at the rectifier output as seen in Fig. 5. As discussed in Section IV, the effect of the supply voltage drop on the capacitive sensor interface circuitry is minimal due to its differential nature, because the sensor oscillator and digital controlled oscillator are affected the same. The rectifier output voltage with respect to input power when the sensor interface is operating is given in Fig. 3b. The rectifier generates a stable supply voltage with input power levels as low as -13 dBm, which corresponds to a distance of 7.1 m from the reader.

The response of the sensor interface is linear with respect to the sensor value, up to half the sensor range for different RF input values and supply voltages as shown in Fig 6. However, the rectifier cannot recover its initial voltage value with a backscattering signal with more than 50% duty-cycle. This reduces the performance of the sensor interface.

The solution to maintain the linearity for all the operating range is to choose the C_m value to be 3.2 pF, twice the maximum value of C_s, to keep the duty cycle of the output lower than 50%. By this way, the rectifier output can be kept stable. Of course, there is a trade-off between linearity and sensitivity. With this solution, the sensitivity is reduced, but a highly linear response is obtained for a wide range of RF input power levels.

Using two different input power levels (-5 & -9 dBm) a sensitivity of 0.379 duty cycle per pF is obtained with a R^2 linearity factor of 0.9956. Those results allow determining the value of capacitance by analyzing the average of the

backscatter signal with average and maximum deviances of, respectively, 1.7% and 5.1% of the measured capacitance.

VI. CONCLUSION

The system level design of a passive UHF RFID tag to be used in sensor applications has been described. The design and implementation of the rectifier and the PLL-based capacitive sensor interface is detailed. It is shown by simulations that the rectifier can reach a maximum PCE of 70% for 30 kΩ load at -13 dBm RF input power. The fully-digital sensor interface can operate with 0.8 V supply voltage consuming only 12 μW. The tag can send linear sensor readout by backscattering at -9 dBm RF input power, which corresponds to a 4.5 m tag-reader distance for 2 W e.r.p source power and 0 dBi tag antenna.

REFERENCES

[1] K. Finkenzeller, *RFID Handbook: Radio-Frequency Identification Fundamentals and Applications*. John Wiley, 1999.

[2] D. M. Pozar, *Microwave Engineering*. Wiley, 2005.

[3] G. Marrocco, "The art of UHF RFID antenna design: impedance-matching and size-reduction techniques," *IEEE Antennas and Propagation Magazine*, vol. 50, no. 1, pp. 66–79, Feb. 2008.

[4] J.-P. Curty, N. Joehl, F. Krummenacher, C. Dehollain, and M. J. Declercq, "A model for μ-power rectifier analysis and design," *IEEE Transactions on Circuits and Systems I: Regular Papers*, vol. 52, no. 12, pp. 2771–2779, Dec. 2005.

[5] K. Kotani, A. Sasaki, and T. Ito, "High-Efficiency Differential-Drive CMOS Rectifier for UHF RFIDs," *IEEE Journal of Solid-State Circuits*, vol. 44, no. 11, pp. 3011–3018, Nov. 2009.

[6] U. Karthaus and M. Fischer, "Fully integrated passive UHF RFID transponder IC with 16.7-μW minimum RF input power," *IEEE Journal of Solid-State Circuits*, vol. 38, no. 10, pp. 1602–1608, Oct. 2003.

[7] N. Gay and W.-J. Fischer, "Ultra-low-power RFID-based sensor mote," in *2010 IEEE Sensors*, 2010, pp. 1293–1298.

[8] H. Danneels, K. Coddens, and G. Gielen, "A fully-digital, 0.3V, 270 nW capacitive sensor interface without external references," in *ESSCIRC (ESSCIRC), 2011 Proceedings of the*, 2011, pp. 287–290.

Full-duplex Communication and Remote Powering Implementation of an Electronic Knee Implant

Oğuz Atasoy, Catherine Dehollain

RF–IC Group, Ecole Polytechnique Fédérale de Lausanne (EPFL), CH-1015, Lausanne, Switzerland
Email: oguz.atasoy@epfl.ch

Abstract—**Full-duplex communication is realized by using ASK modulation for downlink (the channel from controller to the implant) and by using FSK modulation for uplink (from implant to the controller). The ASK modulation is done at the remote powering frequency and by using the inductive link. The class-E type power amplifier is fed by the modulated data. For the uplink communication a stand alone transmitter is designed with a crystal oscillator. The implant node is modeled by using a microcontroller which is sending data and a dummy resistor to mimic the power consumption of the sensors. This microcontroller is also powered up by remote powering.**

I. INTRODUCTION

By the emergence of the applications in biomedical domain, it becomes more critical to handle the data coming from the controller or from the sensors. Depending on the application whether it is a stimulation or a sensing application, communication plays a critical role. To be able to handle the data, high speed communicating circuit blocks are necessary. In the project SimoS, which is a sensing project, there are different kinds of sensors [1]. The data coming from these different sensors are collected by using a microcontroller which is chosen as one of the controller from MSP430 family, due to their low power characteristics.

The inductive link for SimoS application is shown in figure 1. It consists of two mutually coupled coils and a power amplifier. The power amplifiers for the inductive link are usually chosen as class E type power amplifiers due to their high power efficiencies and their ability of standing against huge induced voltages of the coils. This typical link is concluded with a rectifier and a voltage regulator at the secondary side. These are necessary blocks to create useful supply voltage for the application. In the SimoS project this voltage source is going to supply the voltage of the sensors, the analog front of those, and the microcontroller. In addition to these, a higher voltage is necessary for supplying the power of magnetometers.

In this paper, the class-E type power amplifier (PA), the inductive link, the rectifier, the demodulator, the FSK transmitter are explained. Except the PA and the link, every other design was fabricated by using CMOS 0.18μm technology. In addition to these, to close the loop two commercial regulators, and one MSP430 family microcontroller are used.

II. POWER AMPLIFIER WITH MODULATOR

The PA is chosen as a class-E type amplifier which is usually used for inductive links. Since the inductive link is

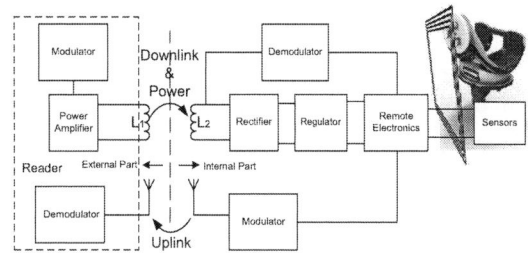

Fig. 1.   Inductive link for smart knee prosthesis

modeled as a big inductor and a resistor, the canceling series capacitor in the class-E type PA makes this kind of PA comfortable to be used in inductive links. The series capacitor resonates with the primary inductance of the link and allows to induce high voltages at this inductance by protecting the switch from high voltages.

As it is shown in figure 2 the class-E type PA consists of a switch (transistor M_1), capacitors, and inductors. The switch is a transistor which is controlled at the desired inductive link frequency. The series capacitance and inductance determines the resonance frequency. By turning the switch on and off at this frequency, and optimizing the values of the choke inductance and the parallel capacitance values, the efficiency of these amplifiers can reach to 100% theoretically, and 75% practically with parasitics [2].

In addition to remote powering, the downlink modulation is realized by changing the timing of the control voltage of the switch. This is done by applying the control voltage at two different frequencies. Since the PA is optimized to work at one desired frequency, any change in the switching frequency results slight inefficiency. Hence this change is seen as amplitude modulation at the coils. The resulted modulation index (MI) in this case is controlled by the hopping frequency difference. The MI is defined for voltage as:

$$MI = \frac{V_H - V_L}{V_H + V_L} 100\%$$

Although there are other methods to do ASK modulation, this solution is preferred for practical reasons and due to its higher efficiency compared to others. For example it is also possible to use a Darlington configuration to make the modulation through the supply of the PA. However in this case even when there is no modulation the high current should

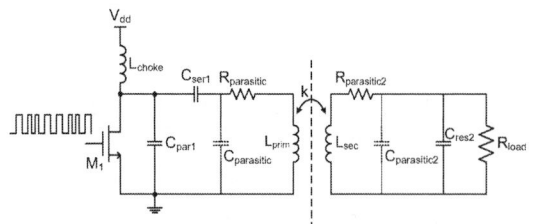

Fig. 2. Class-E type power amplifier with FSK modulation

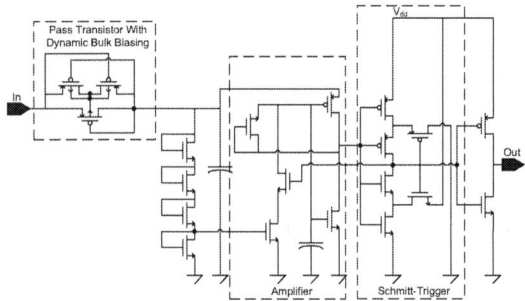

Fig. 3. ASK demodulator for downlink communication

pass through this switch. This decreases the overall efficiency, which makes it a practical but an inefficient solution depending on the application.

III. ASK DEMODULATOR

The ASK demodulator at the implant side is chosen by looking to the power budget and datarate limits. In addition to these, since remote powering AC signal carries the downlink data, and the amplitude of this signal increases or decreases with the misalignment or movement of the secondary side, the designed demodulator should have self-sampling property. The design in [3] seemed to be a suitable solution. The modified (addition of the Schmitt-trigger) and fabricated version of this design is seen in figure 3. This design also played a role in [4]. The input of the demodulator is an envelope detector with diode connected transistors and filtering capacitance. The envelope feeds the nonlinear amplifier to create a rail-to-rail swing. The output of this amplifier feeds a Schmitt-trigger whose output is also used as a feedback for the amplifier. The voltage supply of the Schmitt-trigger and the output buffer is chosen as the regulated supply, whereas the voltage supply of the input stage is created from the input itself.

With these modifications the modulator can demodulate up to 1 Mb/sec, and down to 20 kb/sec. The power consumption of the demodulator is less than 15 μW, which is much lower than the power consumption of the sensors which are going to be used in the final SimoS application. As mentioned above, it can demodulate the modulated carrier independent of the amplitude of the input. For proper operation of this demodulator, the signal amplitude should be higher than the threshold of two NMOS and V_{dsat} of two NMOS transistors.

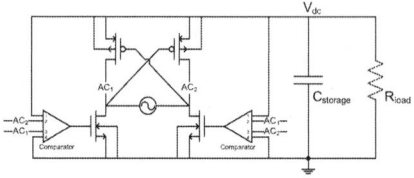

Fig. 4. Rectifier

There is no V_{inH} or V_{inL}, which means that only the MI is important. This is due to its self comparison property. This allows to use the remote powering signal which varies with the movement of the secondary side or the change of the load of the PA. The fabricated demodulator can reach a minimum MI of 2% which is fine for this application.

The lowest MI that can be achieved depends on the demodulator performance, whereas the upper limit depends on the rectifier and regulator chain. If the MI is too high and the data rate and the storage capacitance is too low, the regulated output tends to decrease and follow the output voltage of the rectifier. This creates operational problems at the system. To avoid this, the above mentioned limitations should not be discarded. The easiest solution could be increasing the value of the storage capacitor if there is available place for off-chip capacitor. The only drawback is the long charge-up time that depends on the turn on resistance of the rectifier. On the other hand, if the capacitance is bigger, the ripple at the output of the rectifier is less. However these are not concerns for this demodulator, since it can demodulate signals with small MI.

IV. RECTIFIER AND REGULATOR

The supply voltage is chosen as 1.8V. To create this voltage after the inductive link, a rectifier whose operation was mentioned in [5] is necessary. The modified and fabricated version of this design was published also in [6]. The necessary input peak voltage amplitude should be one V_{dsat} higher than the desired output voltage of the rectifier at the steady state. Since the rectifier is a full-wave rectifier, the AC input voltage elevates on a DC value which increases with the output voltage of the rectifier. At steady state the lowest peak value of the input is one V_{dsat} less than the GND level. The schematic of this rectifier is seen in figure 4.

The regulator is chosen from commercially available components. Minimum required voltage to operate is 2.0V which can be provided by the rectifier, even the downlink modulation is being occurred. For the highest possible efficiency one of the most critical issue to decide is when to switch on and off. The comparators used for this purpose in this rectifier to decide when to pass the current through the switches are shown in figure 5.

V. FSK TRANSMITTER

For uplink communication, a different carrier frequency is generated. The fabricated FSK transmitter is a version of [7]. This transmitter indeed is a low power frequency multiplier.

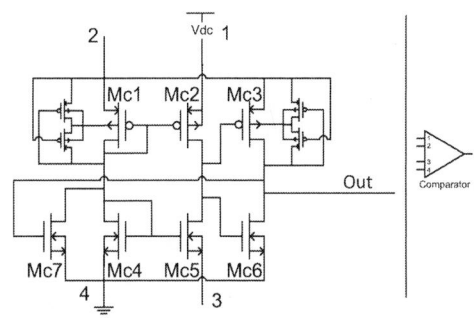

Fig. 5. Comparator used in the rectifier

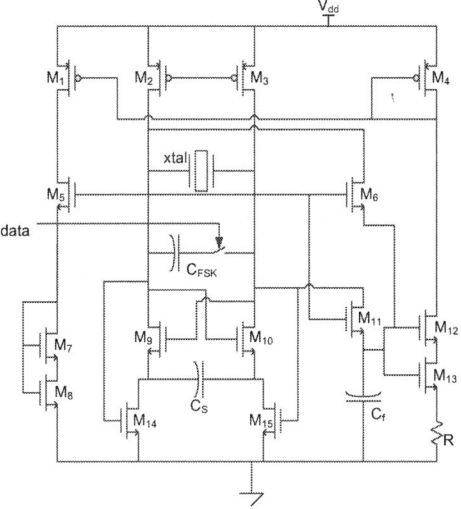

Fig. 6. Crystal oscillator with FSK modulation

The fabricated transmitter consumes less than 250 μW while delivering 25μW to a 50Ω antenna.

A. Crystal oscillator

The basic topology of crystal oscillator is the low power differential oscillator of [8]. The reference frequency is chosen as 48 MHz, which is then multiplied by 9 to obtain 432 MHz carrier. Modulation capability is added to this basic topology by changing the parallel capacitance of the crystal seen in figure 6, the output frequency is pulled. The capacitance should be chosen according to the desired frequency pulling. However higher the frequency pulling is higher the power consumption. So the frequency difference is chosen as 4.5 kHz. By doing so, the crystal oscillator continues to work in either position of the switch which changes the parallel capacitance value. The differential outputs are amplified separately and rail-to-rail signals achieved. Either one of these signals feed the injection locked oscillator by using a multiplexer controlled by the data signal. The power consumption of the fabricated crystal oscillator simulated as 80 μW.

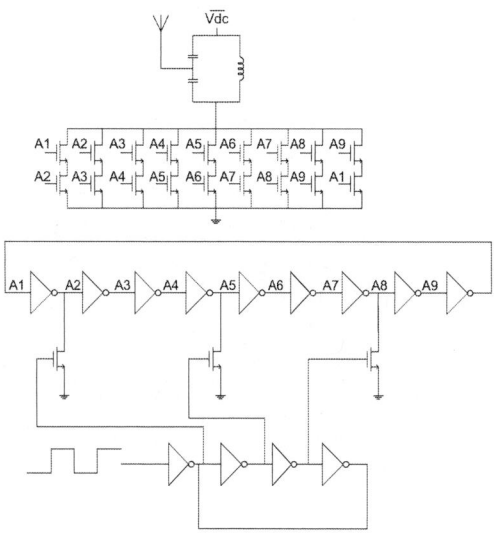

Fig. 7. Injection locked oscillators

B. Injection Locked Oscillator

As it is mentioned, the generated reference frequency is multiplied by an injection locked oscillator. As it is shown in figure 7 by using two ring oscillators and a reference frequency generator a low power transmitter was designed[7]. This design is a version of it. Whereas the frequency deviation at 48 MHz is 4.5 kHz, the frequency deviation after multiplying by 9 becomes 40 kHz. The power consumption of the transmitter with the crystal oscillator is measured as less than 250 μW while radiated signal is demodulated properly despite the harmonics of the PA.

VI. FSK RECEIVER

For the receiver, the ADF7020 series from Analog Devices is used. The flexibility of the carrier frequency adjustment and different demodulation capabilities made it a reasonable choice. The receiver configured to demodulate an FSK modulated signal at the desired carrier frequency. One additional good thing of this receiver is that it can demodulate the signal even the bandwidth of the data is higher than the frequency deviation of the modulator. This can be done because the properties of the modulated signal can be defined by programming the receiver. By this technique, an FSK modulated signal (frequency deviation of 40 kHz), whose data rate can go up to 172 kb/sec, can be demodulated by the receiver.

VII. MEASUREMENT RESULTS

All mentioned blocks were combined together to have remotely powered secondary node communicating with its controller. The test set-up shown in figure 8 consists of the mentioned elements. All the circuit blocks except the regulator and the programmable receiver were fabricated by using 0.18μm CMOS process. Although the ultimate aim is

Paper T3A4   *PRIME 2013, Villach, Austria*

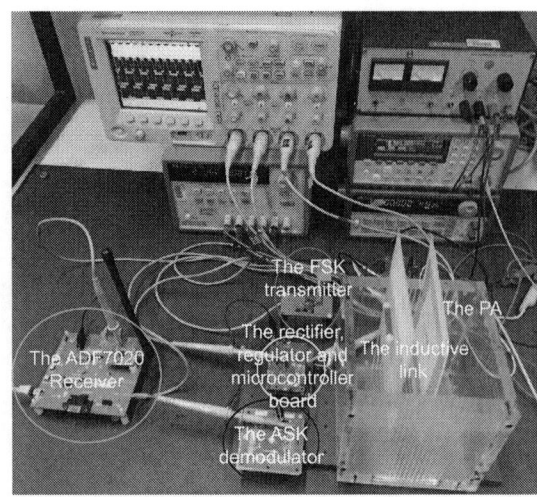

Fig. 8.   Measurement setup

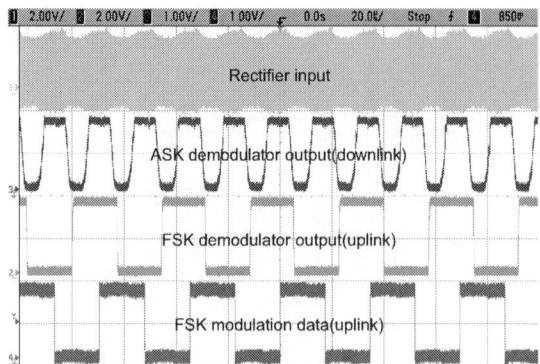

Fig. 9.   Demodulated uplink and downlink data waveforms

full integration, at this level the circuits were not connected internally. These blocks are indicated with their names on the figure. The PA is controlled via a signal source who is providing the square wave for PA switch control. The signal output of this source is FSK modulated, which created ASK modulated signal at the output of PA as aforementioned. The normal operation frequency was 13.56 MHz, whereas the hopping frequency was chosen as 13.46 MHz. By doing so the achieved MI of the ASK modulated signal is 8%. The remote powering signal is received by the secondary node via the inductive link, rectified, and regulated to provide the power to the implant microcontroller. After the power-up, modulated signal is demodulated and fed the implant microcontroller. The microcntroller takes the data, and replies via the FSK transmitter. Some selected waveforms are seen in figure 9. The modulated remote powering signal has peak-to-peak amplitude of 2.4V, with a MI of 8% for easy visualization. It was seen that the signal whose MI reaches up to 2% can be correctly demodulated. The BER tests are not performed yet, but it is clear that the BER performance would decrease in this case. The datarate for downlink was chosen as 114 kbit/s to be consistent with the uplink communication because of practical reasons. Thus the uplink datarate was determined as 57 kb/s which the commercial receiver is working better because of its reference frequency. The demodulated uplink signal is also shown in the same figure, which is the output of the commercial receiver.

VIII. CONCLUSION

In this paper, full-duplex communication realization with remote powering for SimoS project is presented. The remote powering is achieved by using an inductive link with a power amplifier driving that. The induced voltage at the secondary side is rectified, and regulated to obtain the DC supply for the sensor electronics and the microcontroller. For the

downlink communication, the control signal of the PA switch is modulated with an FSK signal, and at the primary and the secondary coils it is seen as ASK modulation. The modulated signal is demodulated by a low power ASK demodulator. The achieved rail-to-rail data signal feeds the MSP430 family microcontroller. It is also shown that depending on the commands, the stored data is transmitted via the second channel which is chosen as 432 MHz. The second carrier is generated in two steps, first the reference frequency is created by using a crystal oscillator and by multiplying this reference frequency, the carrier frequency is obtained. The FSK modulation is achieved by changing the parallel capacitance of the crystal. A frequency deviation of 40 kHz is achieved. It was seen that this deviation is enough even for the cases that the bandwidth of the data is higher than the frequency deviation with the help of a programmable receiver. All the circuit blocks except the regulator and the programmable receiver were fabricated by using $0.18\mu m$ CMOS process and tested.

ACKNOWLEDGEMENT

The SimoS project is a Nano-Tera project and supported by Swiss National Funding (SNF).

REFERENCES

[1] Arami, A. et al.,"Instrumented Knee Prosthesis for Force and Kinematics Measurements," *IEEE Trans. Automation Science and Engineering*, vol.PP, no.99, pp.1-10, 2013
[2] N. Sokal and A. Sokal, Class E-A new class of high-efficiency tuned single-ended switching power amplifiers, *IEEE Journal of Solid-State Circuits*, vol. 10, pp. 168176, June 1975.
[3] C.-C. Wang et al., "Self-Sampled All-MOS ASK Demodulator for Lower ISM Band Applications," *IEEE Trans. Circuits and Systems II: Express Briefs*, , vol.57, no.4, pp.265-269, April 2010
[4] G. Yilmaz et al., Wireless Data and Power Transmission Aiming Intracranial Epilepsy Monitoring. *SPIE Conference*, to be published.
[5] Y.-H. Lam et al., Integrated Low-Loss CMOS Active Rectifier for Wirelessly Powered Devices, *IEEE Trans. Circuits and Systems II: Express Briefs*, vol. 53, no. 12, pp. 13781382, Dec. 2006.
[6] O. Atasoy and C. Dehollain, Remote powering realization for smart orthopedic implants, *NEWCAS*, Montreal, 2012, pp. 521524.
[7] J. Pandey and B. Otis, A Sub-100 uW MICS/ISM Band Transmitter Based on Injection-Locking and Frequency Multiplication, *IEEE Journal of Solid-State Circuits*, vol. 46, no. 5, pp. 10491058, May 2011.
[8] D. Ruffieux, A high-stability, ultra-low-power quartz differential oscillator circuit for demanding radio applications, *ESSCIRC*,2002, pp. 8588, 2002.

978-1-4673-4579-8/13 $31.00 © 2013 IEEE

PRIME 2013, Villach, Austria　　　　　　*Session T1B – Transmitters and Receivers*

Efficiency Enhancement of Burst Mode Transmitters by RF Energy Recovery

David Seebacher, Wolfgang Bösch
Graz University of Technology
Email: {david.seebacher, wolfgang.boesch}@tugraz.at

Peter Singerl, Christian Schuberth
Infineon Technologies Austria
Email: {peter.singerl, christian.schuberth}@infineon.com

Abstract—**Modern communication standards have high peak to average power ratios (PAPR) to achieve high spectral efficiency, trading of power amplifier (PA) efficiency. In this paper a method to directly reuse the RF energy reflected from carrier bursting PAs to improve the efficiency will be introduced and compared to conventional operation with and without DC energy recovery. Efficiency enhancement utilizing DC energy recovery is limited by the rectifier and PA efficiency. Therefore a method to reuse the RF energy directly, which does not suffer from these limitations is presented. It will be shown that it is possible to maintain high efficiency in back off with this method.**

I. Introduction

Modern communication standards such as Long Term Evolution (LTE) inherently show highly fluctuating envelope signals with high PAPR in order to efficiently use the available spectrum. But the PA efficiency is severely degraded for these type of signals, as the PA is operated far from maximum output power and hence maximum efficiency for most of the time. This results in a rather poor average PA efficiency. Therefore several techniques have been developed to increase the efficiency at back-off operation. Carrier bursting [1]–[3] is one of them and is based on operating the PA either at full output power or completely switched off, meaning that the PA is always operated at its maximum efficiency. In Fig. 1 such a carrier bursting system is shown [2], [4], [5]. The signal generation for the highly efficient operation is done by coding the amplitude a(t) by means of PWM and converting it up by the phase modulated carrier c(t). The load is connected to the PA by means of a circulator and a band pass filter. Only frequencies around the carrier pass the filter and the modulation sidebands resulting from the PWM coding are reflected and dissipated in the sidebands termination resistor R_{SB}.

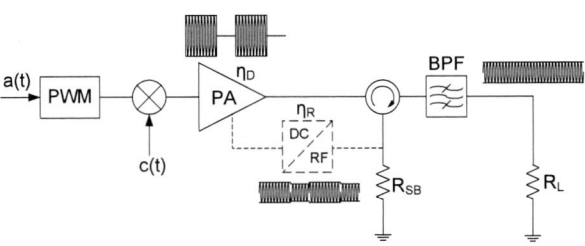

Fig. 1.　Block diagram of a conventional burst mode transmitter.

In Fig. 2 the drain current and drain voltage of a Class-B PA

[6] in burst mode operation with a duty cycle of 40 %, a supply voltage of 20 V and a load resistance of 50 Ω are plotted. The carrier, with a frequency of 1 GHz, is PWM modulated with a modulation frequency of 50 MHz. It can be seen that the drain voltage is at maximum swing during the first period (t$_{ON}$) and zero during the second period (t$_{OFF}$). The drain current is modulated accordingly and hence the amplifier is always efficient. For the simulations presented in this paper a Class-B PA with a parallel resonator, placed in parallel to the DC decoupled transistor drain, is used [6]. It provides an open circuit for the carrier and the modulation sidebands and short circuits all the carrier harmonics. The output filter is implemented as a high-Q parallel resonator.

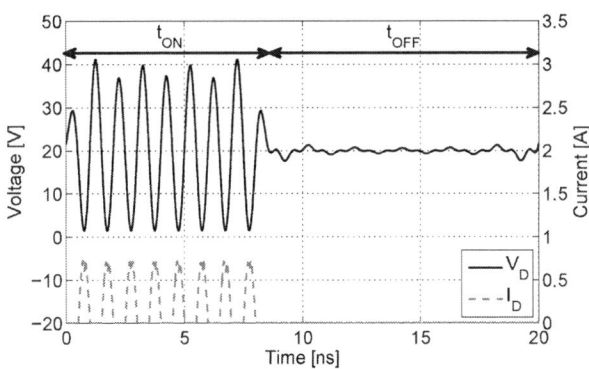

Fig. 2.　Drain voltage and current for a duty cycle of 40 % and conventional burst mode PA operation.

In Fig. 3 the voltage across R_{SB} for a duty cycle of 40 % is given. It can be seen that the voltage at the sidebands termination resistor is the difference between the (DC decoupled) drain voltage and the average output voltage. In this case the average output voltage is 8 V due to the 20 V supply voltage and the 40 % duty cycle. This means that the voltage amplitude is 12 V during t$_{ON}$ and -8V during t$_{OFF}$. This leads to a significant energy dissipation and resulting in poor overall efficiency and effectively the same efficiency as a Class-B PA [1]–[3]. The corresponding efficiency curve η_B is given in Fig. 8.

II. DC Energy Recovery

One method to reuse this energy is DC-energy recovery, as covered in [4], [5], [7]–[10]. Instead of dissipating the energy in the resistor R_{SB} the signal is rectified and fed back to the PA

978-1-4673-4579-8/13 $31.00 © 2013 IEEE　　　　221

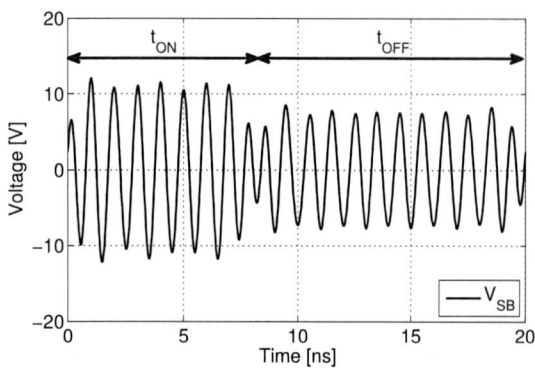

Fig. 3. Voltage across sidebands termination resistor R_{SB} for a duty cycle of 40%.

supply, as indicated in Fig. 1. With this the improved efficiency is given by

$$\eta_{DC} = \frac{V_L \, \eta_D}{1 - (1 - V) \, \eta_R \, \eta_D}, \qquad (1)$$

where V_L is the normalized load voltage, η_D denotes the drain efficiency of the PA and η_R is the rectifier efficiency. In Fig. 8 the efficiency η_{DC} of a Class-B PA with DC energy recovery applied is plotted. Although a rectifier efficiency η_R of 100% is considered the efficiency enhancement in backoff is limited by the drain efficiency η_D of roughly 71%. This results in additional losses that cannot be recovered and limits the efficiency enhancement in back off. Thus for high efficiency in back off a highly efficient PA in combination with a rectifier operating highly efficient over a wide power range is required.

RF energy recovery on the other hand aims to mitigate these problems by directly reusing the energy on RF level and will be introduced and discussed in the following section.

III. RF ENERGY RECOVERY

Instead of rectifying the RF energy and feeding it back to the DC supply, the RF energy can be directly reused, mitigating the rectifier and its losses. To reuse the energy it is required to reflect the signal from the sidebands termination resistor R_{SB} in the correct phase to the PA in order to have constructive interference, as it is shown in Fig. 4. When the switch S_1 is open and S_2 is closed the termination is 50 Ω and no energy is reflected. But by closing switch S_1 the termination is a short circuit and by leaving both switches open an open circuit is present. With this configuration, based on ideal switches, it is possible to either terminate the port of the isolator or to reflect the signal in the desired phase (0° or 180°). Three different types of operation are possible, energy recovery during t_{OFF}, t_{ON} and full energy recovery. In the following subsections energy recovery during t_{OFF} and t_{ON} and finally full energy recovery during the whole period will be introduced.

A. Reuse During OFF Period

One possibility to improve efficiency is to recover the energy during t_{OFF}. During the ON period the circulator port is terminated with 50 Ω, but during t_{OFF} the signal is reflected.

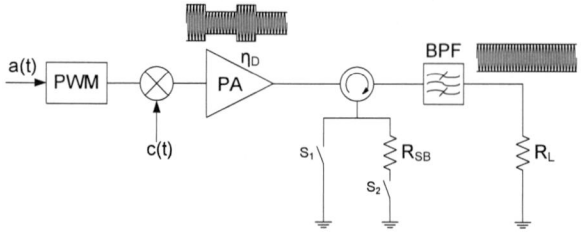

Fig. 4. RF energy recovery block diagram.

The correct phase to achieve constructive interference of the drain voltage depends on the used type of band pass filter. A parallel resonator, as considered in this simulations, provides a short circuit for the modulation sidebands and therefore an open circuit is required to reflect the signal with the correct phase for constructive interference. The series resonator on the other hand provides an open circuit for the modulation sidebands and thus a short circuit at the sideband resistor would be required.

The resulting drain current and voltage for a duty cycle of 40% and energy recovery during t_{OFF} are plotted in Fig. 5. It can be seen that also during the OFF period a drain voltage is present. The amplitude of the drain voltage corresponds to the output voltage, as the signal is fully reflected from the sideband termination and no energy is lost during this time. The drain voltage and current during t_{ON} are the same as for the conventional operation presented in Fig. 2, meaning the PA effectively sees the load resistor during this time.

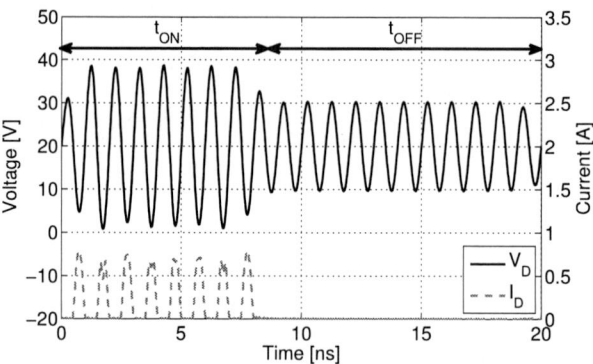

Fig. 5. Drain voltage and current for RF energy recovery during t_{OFF} for a duty cycle of 40%.

The presence of the drain voltage also during the OFF period causes a nonlinear dependence of the output voltage on the duty cycle $d = t_{ON}/(t_{ON} + t_{OFF})$ and results in

$$V_{L_{OFF}} = \frac{2 \, d}{1 + d}. \qquad (2)$$

The simulated output voltage dependence curve OFF is plotted in Fig. 9. It corresponds well to the theoretically calculated output voltage defined in (2). As can bee seen in Fig. 9 the resulting output voltage $V_{L_{OFF}}$ is larger for low duty cycles compared to conventional operation $V_{L^{NO}}$ or DC energy recovery $V_{L^{DC}}$. Also the efficiency is larger than for conventional operation and can be calculated by

$$\eta_{OFF} = V_L \left(2 - V_L\right) \eta_D, \tag{3}$$

where V_L denotes the normalized load voltage. The efficiency enhancement is caused by the fact that no energy is dissipated in the sideband termination resistor R_{SB} during the off period anymore. In Fig. 8 the corresponding simulated efficiency curve η_{OFF} is plotted. For low output voltages the efficiency enhancement is limited, but for higher output voltages the efficiency can be significantly improved and it even outperforms DC energy recovery η_{DC}. It shall be noted that all simulations for efficiency and load voltage are performed with a modulation frequency of 5 MHz in order to reduce the influence of band limitation and aliasing effects [11], whereas for time domain plots a modulation frequency of 50 MHz is used for illustrative purposes.

B. Reuse During ON Period

To recover the energy during t_{ON}, the energy from the sidebands termination resistor is reflected back to the PA during the first period, and terminated during t_{OFF}. As for the energy recovery during t_{OFF} the required reflection depends on the used type of band pass filter. For a parallel filter a short circuit is required during t_{ON}, whereas a series filter requires an open circuit. The signal is reflected in such a way, that constructive interference for the drain voltage and destructive interference for the drain current occurs, as can be seen in Fig. 6. This means that the equivalent load resistance for the PA is increased, which can be seen by the lower drain current in comparison to conventional operation (Fig. 2) during t_{ON}.

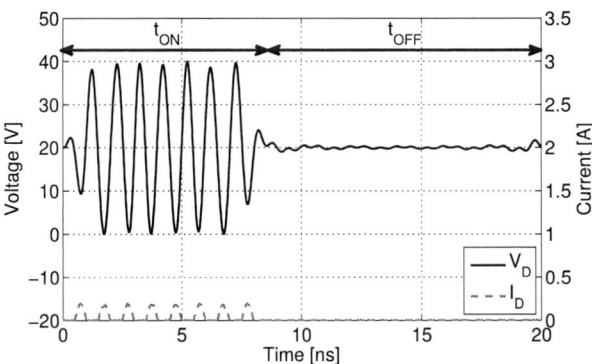

Fig. 6. Drain voltage and current for RF energy recovery during t_{ON} for a duty cycle of 40 %.

Additionally to the increased equivalent load resistance no energy is dissipated in the sideband's resistor during t_{ON}, which leads to a significant efficiency improvement. The resulting efficiency in dependence of the normalized load voltage V_L is given by

$$\eta_{ON} = \frac{1 + V_L}{2} \eta_D. \tag{4}$$

The simulated efficiency curve η_{ON} is depicted in Fig. 8. A significant efficiency improvement is possible, especially for

low output voltages. The efficiency deviates from the theoretical limit for very low output voltages due to the finite rise and fall times of the pulses caused by bandwidth constraints.

As well as for the energy recovery during t_{OFF} the output, respectively load voltage does not depend linearly on the duty cycle d anymore and is given by

$$V_{L_{ON}} = \frac{d}{2 - d}. \tag{5}$$

The corresponding control characteristic resulting from simulation can be seen in trace $V_{L_{ON}}$ in Fig. 9. For low duty cycles the output voltage is smaller than for conventional operation and increases rapidly to its maximum value towards the maximum duty cycle.

C. Full Reuse

Up to now only reusing the energy either during the ON or OFF period was considered. Both methods have the drawback of nonlinear output voltage control characteristic and still lose energy in the sidebands termination resistor, which lowers efficiency. By combining both methods the energy in the sidebands can be fully reused and efficiency can be enhanced. This means that recovery takes place during the whole cycle, it is only required to reflect the signal from the sideband's resistor R_{SB} in the correct phase according to the period. This means that the signal envelope at the sidebands termination resistor (Fig. 3) is rectified and fed back to the drain of the PA. Constructive interference for the drain voltage during t_{ON} and t_{OFF} occurs. The resulting drain voltage and current waveforms for a duty cycle of 40 % can be seen in Fig. 7.

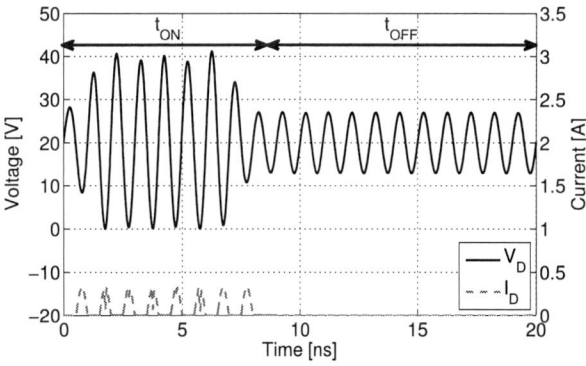

Fig. 7. Drain voltage and current for full RF energy recovery for a duty cycle of 40 %.

The drain current during t_{ON} is lower than for conventional operation, meaning that the equivalent load resistance for the PA also increases for this type of operation. During t_{OFF} the load voltage is present at the drain of the PA. As no energy at all is dissipated in the sidebands termination resistor anymore, the efficiency η_{FULL} is simply the drain efficiency η_D. The corresponding simulated efficiency curve η_{FULL} is given in Fig. 8. For lower output voltages the efficiency starts to slightly increase, as the influence of the knee voltage is reduced for higher load resistances, and decreases for very low output voltages. This decrease is also present for the energy recovery

Paper T1B1   *PRIME 2013, Villach, Austria*

during t_{ON}, due to the influence of the signal transitions on the drain efficiency.

For a practical system implementation a very compact solution, minimizing all delays, is desired. A delay in the system will result in a different phase of the reflected signal and impacts constructive interference, which can be overcome by providing the correct reflection coefficients for the sideband termination. In case of a delay there will also be a short overlap of both recovery phases. This can be solved by either introducing a short dead time between both recovery phases or by performing partial energy recovery and selecting the most efficient mode according to the duty cycle.

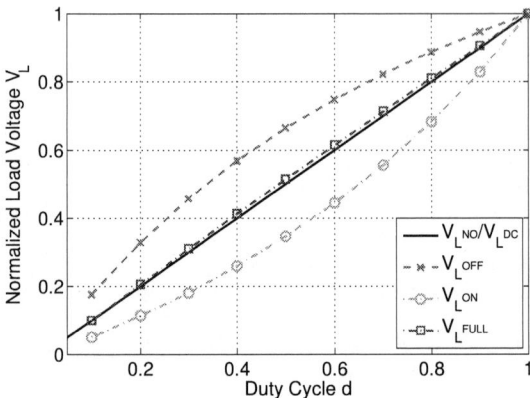

Fig. 9. Normalized load voltage V_L versus duty cycle d for Class-B PA $V_{L_{NO}}$ with DC energy recovery $V_{L_{DC}}$, partial RF energy recovery $V_{L_{ON}}$, $V_{L_{OFF}}$ and full RF energy recovery $V_{L_{FULL}}$.

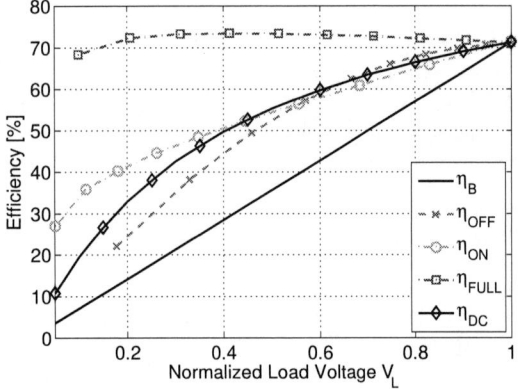

Fig. 8. Efficiency of a Class-B PA η_B with different energy recovery techniques applied, like DC energy recovery η_{DC}, partial RF energy recovery η_{ON}, η_{OFF} and full RF energy recovery η_{FULL}.

In contrast to the nonlinear dependence of the output voltage on the duty cycle for partial energy recovery, depends the output voltage linearly on the duty cycle for full energy recovery. This can be seen in the simulation results in Fig. 9 in trace $V_{L_{FULL}}$. The slight increase in comparison to conventional operation is due to the increase of the equivalent load resistance presented to the PA during t_{ON}, resulting in reduced influence of the knee voltage, respectively on-resistance of the transistor, for the same drive level.

IV. CONCLUSION

In this paper it is shown that by directly recovering the energy on RF level instead of converting it back to DC a significant efficiency improvement is possible. To enable the recovery a termination capable of varying its reflection coefficient according to the duty cycle is required. In the most simple case this can be a switch. By reflecting the signal from the sidebands with the correct phase, according to the duty cycle, constructive interference of the drain voltage can be achieved. This interference increases the effective load resistance presented to the transistor and performs a kind of load modulation. This load modulation in combination with mitigation of energy dissipation in the isolator enables highly efficient operation. It has been shown that RF energy recovery is capable of outperforming DC energy recovery and also results in a reduced circuit complexity due to the mitigation of the rectifier and the required DC/DC converter.

REFERENCES

[1] T. Blocher and P. Singerl, "Coding Efficiency for Different Switched-Mode RF Transmitter Architectures," in *Proc. 52nd IEEE Int. Midwest Symp. Circuits and Systems MWSCAS '09*, 2009, pp. 276–279.

[2] C. Schuberth, P. Singerl, M. Gadringer, H. Arthaber, A. Wiesbauer, and G. Magerl, "Highly Efficient Switched-Mode Transmitter Using a Current Mode Class-D RF Amplifier," *International Journal of RF and Microwave Computer-Aided Engineering*, vol. 20, no. 4, pp. 446–457, 2010.

[3] J. Young-Sang and N. Sangwook, "A Discrete-Amplitude Pulse Width Modulation for a High-Efficiency Linear Power Amplifier," *ETRI Journal*, vol. 33, no. 5, pp. 679–688, Oct. 2011.

[4] Y.-S. Jeon, H.-S. Yang, and S. Nam, "A Power Re-Use Technique for Improved Efficiency of Pulsed Oscillating Amplifiers," *IEEE Microw. Wireless Compon. Lett.*, vol. 16, no. 10, pp. 567–569, 2006.

[5] S. Ali and T. Johnson, "A New High Efficiency RF Switch-Mode Power Amplifier Architecture for Pulse Encoded Signals," in *Wireless and Microwave Technology Conference (WAMICON), 2012 IEEE 13th Annual*, April 2012.

[6] S. Cripps, *RF Power Amplifiers for Wireless Communications*. Artech House, 2006.

[7] S. N. Ali and T. E. Johnson, "RF Switch-Mode Power Amplifier with an Integrated Diplexer for Signal Reconstruction and Energy Recovery," in *Microwave Symposium Digest (MTT), 2012 IEEE MTT-S International*, June 2012.

[8] R. Langridge, T. Thornton, P. Asbeck, and L. Larson, "A Power Re-Use Technique for Improved Efficiency of Outphasing Microwave Power Amplifiers," *Microwave Theory and Techniques, IEEE Transactions on*, vol. 47, no. 8, pp. 1467–1470, Aug. 1999.

[9] X. Zhang, L. Larson, P. Asbeck, and R. Langridge, "Analysis of Power Recycling Techniques for RF and Microwave Outphasing Power Amplifiers," *Circuits and Systems II: Analog and Digital Signal Processing, IEEE Transactions on*, vol. 49, no. 5, pp. 312–320, May. 2002.

[10] P. Godoy, D. Perreault, and J. Dawson, "Outphasing Energy Recovery Amplifier With Resistance Compression for Improved Efficiency," *Microwave Theory and Techniques, IEEE Transactions on*, vol. 57, no. 12, pp. 2895–2906, Dec. 2009.

[11] S. Chi, C. Vogel, and P. Singerl, "The Frequency Spectrum of Polar Modulated PWM Signals and the Image Problem," in *Electronics, Circuits, and Systems (ICECS), 2010 17th IEEE International Conference on*, Dec. 2010, pp. 679–682.

A 23mW 4.5/8 GHz IR-UWB Transmitter in 65nm TSMC CMOS Technology

A. Donno[1,2], S. D'Amico[1,2], M. De Matteis[3], A. Baschirotto[3]

[1]Department of Innovation Engineering, University of Salento, Italy

[2] INFN-Lecce,

[3]Department of Physics, University of Milano Bicocca, Italy

Abstract— **This paper presents a low power transmitter for Impulse-Radio Ultra-Wideband (IR-UWB) applications. It generate short duration bi-phase modulated UWB pulses with a center frequency of 4.5 / 8 GHz according to the selected channel. A simplified transmitter architecture enabling low power consumption has been adopted. The key circuit is a phase shifter used to obtain positive and negative pulses. Generated pulses comply with requirements of the IEEE 802.15.4a standard. The transmitter is designed in 65nm CMOS technology. Simulations results show that the transmitter consumes 23 mW peak power from a 1.2V supply at 8 GHz of work frequency.**

Index Terms— **CMOS, impulse radio, transmitter, ultra-wideband.**

I. Introduction

Radio transceivers relying on Impulse Radio UWB signals show a strong potential for low data rate communications at an ultra low power consumption. They are for instance proposed by the IEEE 802.15.4a to support low data rates, low power and low complexity short-range radio communications.

Power consumption is the key issues of such portable applications. Most of the power consumption in analog front-end of IR-UWB transceiver lies on the transmitter side. In this paper, a IEEE 802.15.4a standard compliant, low power transmitter is presented. The transmitter consumes only 23mW peak power, from 1.2V supply. It enables operation in the two mandatory channels included in the IEEE 802.15.4a standard, in high band (7.9872 GHz carrier frequency), and low band (4.4928 GHz carrier frequency) [1], [5].

The proposed transmitter architecture consists of a phase shifter and a power amplifier (PA) driving an external antenna. This simplified architecture guarantees low power consumption. The paper is organized as following. Section II describe the transmitter architecture. Phase shifter circuit design is shown in Section III. Power Amplifier design is shown in Section IV. Section V presents simulations results. Finally Section VI draws conclusions.

II. Transmitter Architecture

A IR-UWB transmitter based on IEEE 802.15.4a can transmit positive or negative pulses. In order for a IR-UWB transmitter

to be compliant with the standard 802.15.4a transmitted pulse must meet two requirements. The pulse is constrained by the shape of its cross correlation function with a standard reference pulse, which is a root raised cosine pulse with roll-off factor of $\beta = 0.6$. The main lobe of the magnitude of the normalized cross correlation function should be greater or equal to 0.8 for a duration of at least 0.5 ns (default pulses of 2 ns) and any side lobe shall not be greater than 0.3. Secondly, the transmitted spectrum must fit within a specific mask (fig.1) defined such that the spectrum shall be less than -10 dBr (dB relative to the maximum spectral density of the signal) for $0.65/T_p < |f - f_c| < 0.8/T_p$, and -18 dBr for $|f - f_c| > 0.8/T_p$, with T_p being the pulse duration.

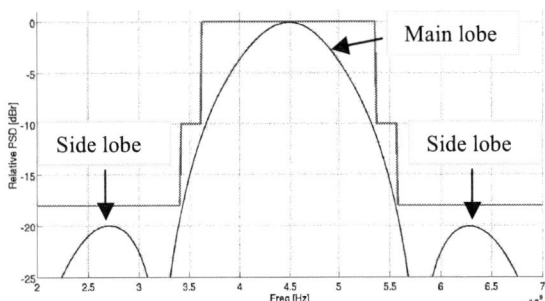

Fig. 1. Mask required to be fit by the IEEE 802.15.4a standard, and an example of spectrum of a compliant pulse.

Furthermore there are regulatory requirements defined for compliance with FCC, European and Japanese rules each with their own UWB emission limits.

Fig. 2 shows typical IR-UWB transmitter architecture [2]. It includes a pulse generator, an up-conversion mixer and a power amplifier. Fig. 3 shows the block diagram of the proposed transmitter. The architecture consists of a phase shifter and a power amplifier that drive the antenna.

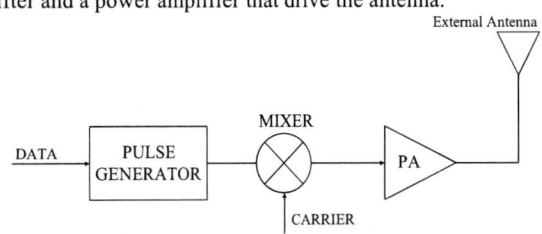

Fig. 2. Typical IR-UWB transmitter architecture.

Paper T1B2　　　　　　　　　　　　　　　　　　　　　　　　*PRIME 2013, Villach, Austria*

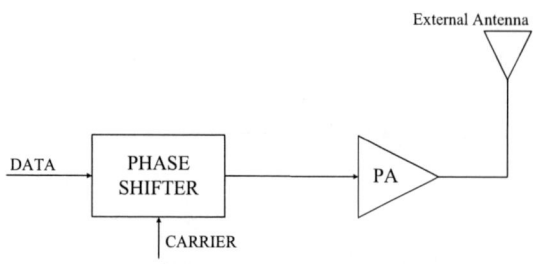

Fig. 3. Proposed transmitter architecture.

As for the typical IR-UWB transmitter, a power amplifier driving the external antenna is included. However, the proposed architecture includes a phase shifter instead of an up-conversion mixer and a pulse generator. Pulses are obtained by allowing the carrier to drive the power amplifier for a time interval equal to the pulse duration (2ns). Acting on the phase of the carrier it is possible to generate positive or negative pulses according to digital data. These operation are performed by the phase shifter. In the following paragraph detailed descriptions of the phase shifter and the power amplifier are reported.

III. PHASE SHIFTER CIRCUIT DESCRIPTION

Phase shifter is the key circuit block of the proposed transmitter. The phase shifter has in input the carrier and a 2 bits (b1 and b2) data stream generated by the baseband processor. If b1 is high, then the carrier is transmitted to the power amplifier, otherwise it is stopped. The b1 data bit is high for the time duration of the pulse. The b2 data bit set the phase of the carrier. If b2 is zero, than the carrier phase is inverted, otherwise it is not. The proposed phase shifter has an all digital architecture. Fig. 4 depicts the schematic of the phase shifter.

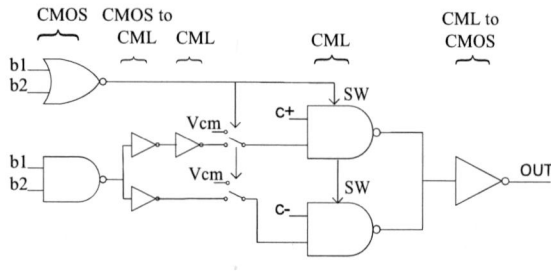

Fig. 4. Circuit implementation of the of phase shifter.

Table I. Truth table of phase shifter.

b1	b2	OUT
1	1	carrier
0	0	no carrier
1	0	carrier with inverted phase

Labels c+ and c- represents the differential carrier which is a 4.5 or 8 GHz 200 mV-pp wave sine. Truth table of phase shifter is reported in table I. Nand gate with the carrier at their

inputs have been implemented in current-mode logic (CML). The choice of CML has be determined considering by its features of s high-speed capabilities as well as its inherent differential operation, which results in lower random and deterministic jitter as compared to the standard digital single-ended CMOS [3]. Fig. 5 shows the adopted NAND CML and inverter schematics. When the carrier has not to be transmitted CML NAND gates are disabled (acting on sw control bit of fig. 5), while their input are set to Vcm (600mV), in order to speed up the gates power up. Inverter circuit convert the CML input signal into CMOS.

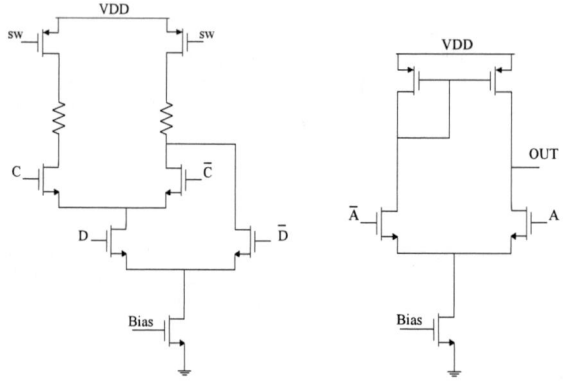

Fig. 5. a) NAND CML schematic, b) Inverter schematic.

The phase shifter has been opportunely designed in order make the transmitted pulse spectrum to fit the specific mask required by the IEEE 802.15.4a standard. In fact, in frequency domain, the main lobe width of the transmitted pulse mainly depends on the pulse duration, which is set by b1 data bit. However, the side lobes amplitude is inversely proportional to the rise and fall times of the pulse envelopment [4], which are determined by the phase shifter. The phase shifter has a dominant pole at the output node, where a large parasitic capacitance ($C_{o,PS}$) mainly due to the input parasitic capacitance of the PA ($C_{in,PA}$ in fig. 6). In order to allows the output spectrum to fit the specific mask required by the IEEE 802.15.4a standard, the design condition is:

$$T_{\text{rise time}} = T_{\text{fall time}} = r_{out} \cdot C_{o,PS} < 0.35 \cdot T_p \qquad (1)$$

where r_{out} is the output resistance of the output inverter, and T_p is the pulse duration (2ns) time.

IV. POWER AMPLIFIER DESIGN

Architecture of PA relies on a two-stage amplifier in order to achieve optimum output power and gain while maintaining a wide bandwidth. The proposed two-stage PA is shown in Fig. 6. This PA employs a cascode topology on first stage while second stage is a common source topology [6], [7] AC coupled through C4 [11], [12], [13].

978-1-4673-4579-8/13 $31.00 © 2013 IEEE　　　　　226

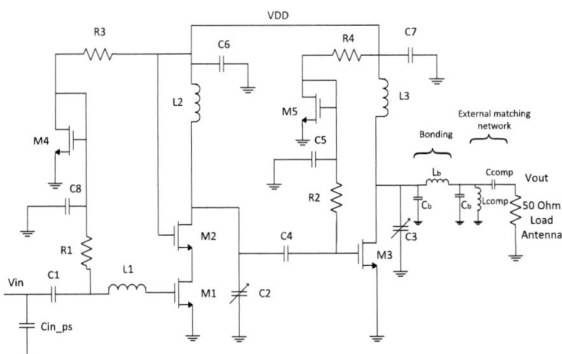

Fig. 6. Schematic of the proposed two-stage UWB PA.

L_1 gives linearity and stability enhancements. A cascode architecture in first stage is used to increase the signal amplitude. The L_2 and L_3 inductors are placed as shunt peaking inductor resonating with C_2 and C_3 capacitances, respectively. A large value of peaking inductance L_2 is used as RF choke to the DC supply and to compensate the power consumption. A common source second stage is added to further improve the gain of the PA. The size of $M3$ is optimized to match low power consumption and output gain. C_2 and C_3 are implemented as 3 bits digitally programmable capacitive arrays. They are programmed to adjust the resonance frequency according to the selected channel at 4.5 GHz or 8 GHz carrier frequency. An external matching network is needed to match output impedance to 50 Ω antenna.

V. SIMULATION RESULTS

The proposed transmitter is designed using 65 nm CMOS technology with a supply voltage of 1.2 V. Fig. 7 shows the output pulse at 4.5 GHz carrier frequency.

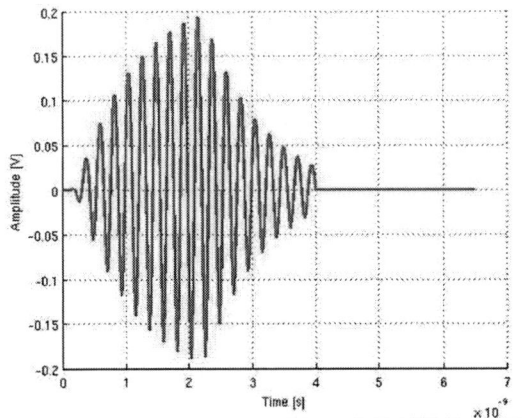

Fig. 7. Output pulse at 4.5 GHz carrier frequency.

The pulse amplitude is 400mV$_{pp}$, Fig. 8 shows the spectrum of the transmitted pulse at 4.5GHz carrier frequency.

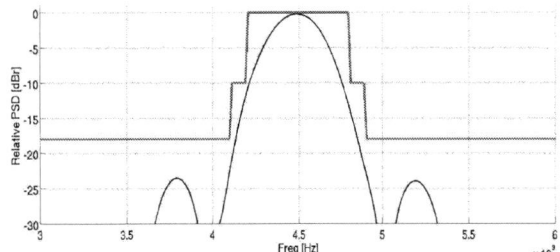

Fig. 8. Spectrum of the pulse out simulation at 4.5 GHz.

As it is evident from Fig. 8 the transmitted pulse fits the IEEE 802.15.4a mask. Fig. 9 shows the simulations results at 8 GHz carrier frequency.

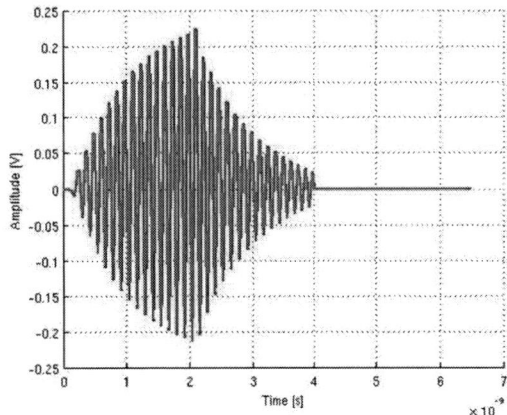

Fig. 9. Output pulse at 8 GHz carrier frequency.

At 8GHz carrier frequency the transmitted pulse has 450 mV$_{pp}$ amplitude.

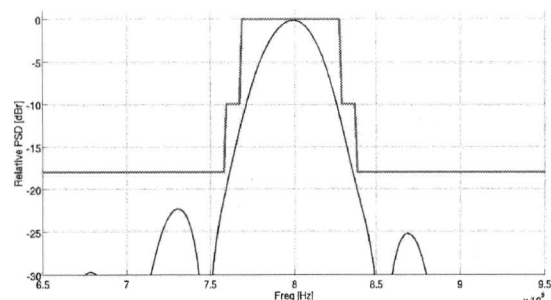

Fig. 10. Spectrum of the pulse out simulation at 8 GHz.

As it is evident from Fig. 10, the spectrum of the transmitted pulse at 8GHz carrier frequency, fits the mask required by the IEEE 802.15.4a standard. For both pulses at 4.5 GHz and 8 GHz, the shape constrain of their cross correlation function with a standard reference pulse required by IEEE 802.15.4a standard has been verified.

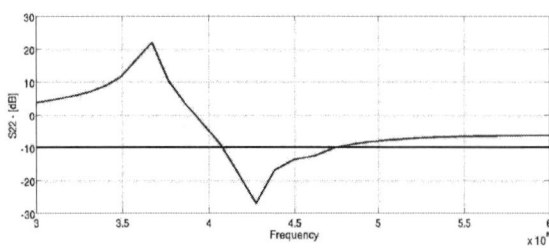

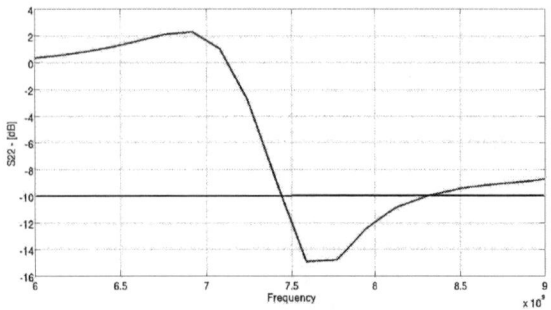

Fig. 11. S_{22} scattering parameters at a) 4.5 GHz b) 8 GHz work frequency.

Fig. 11a and b show the S_{22} scattering parameters of PA at a 4.5 GHz and 8 GHz carrier frequency, respectively. As it is evident from Fig. 11a and b the output matching is guaranteed in the overall channel bandwidth. The peak power consumption of the proposed transmitter is 23mW. Fig. 12 shows layout of the transmitter. It occupies 0.16mm² die area. Table II summarizes the performance of the proposed transmitter, with a comparison to previously published UWB transmitters. The proposed transmitter results compared well in terms of peak power consumption.

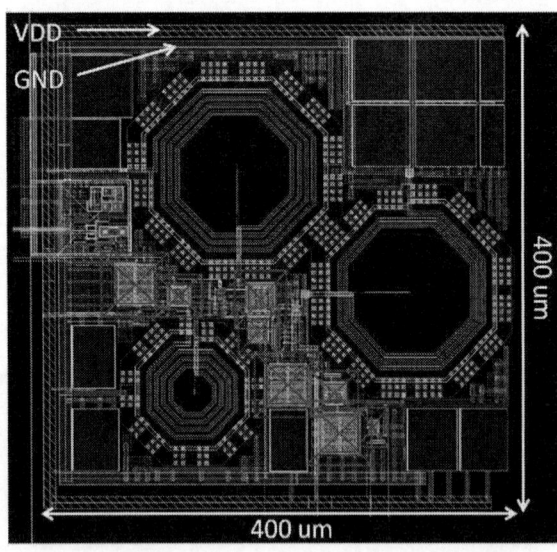

Fig. 12. Transmitter layout.

TABLE II. PERFORMANCES SUMMARY AND COMPARISON

Refs.	Carrier frequency (GHz)	Peak to peak Output voltage (mV)	Peak power consumption (mW)	Vdd (V)	Tech.
[8]	4	640	29.7	2.2	CMOS 0.18 um
[9]	15	200	80	1.2	CMOS 90 nm
[10]	3 - 5	2000	66.6	1.8	CMOS 0.18 m
This work	**4.5 - 8**	**450**	**23**	**1.2**	**CMOS 65 nm**

VI. CONCLUSION

A 65nm CMOS IEEE 802.15.4a standard compliant IR-UWB transmitter has been presented in this work. The transmitter generated bi-phase modulated signal with pulses of 2ns duration at 4.5/8 GHz center frequencies, according to the selected channel. The peak power consumption is 23 mW. The proposed transmitter results well compared in term of power consumption with respect to the state of the art transmitters.

REFERENCES

[1] 'Final Rule of the Federal Communications Commission, 47 CFR Part 15, Sec. 503', FCC, Federal Register, vol. 67, no. 95, May 2002.

[2] M. Ghavami, Lachlan Michael, Ryuji Kohno, "Ultra Wideband Signals and Systems in Communication Engineering", John Wiley & Sons, 2007.

[3] Kalantari, Nader, "All-CMOS High-Speed CML Gates with Active Shunt-Peaking", Circuits and Systems, 2007. ISCAS 2007. IEEE International Symposium on, 27-30 May 2007,pp.2554 - 2557.

[4] Pflug, Hans W., "UWB Pulse Shaping for IEEE 802.15.4a", Microwave Conference, 2008. EuMC 2008. 38th European, 27-31 Oct. 2008, 713-716.

[5] X. Wang et al., "A high-band IR-UWB chipset for real-time duty-cycled communication and localization systems", Solid State Circuits Conference (A-SSCC), IEEE Asian, pp. 381-384, 2011.

[6] S. D'Amico, M. De Blasi, M. De Matteis, A. Baschirotto, "A 255 MHz Programmable Gain Amplifier and Low-Pass Filter for Ultra Low Power Impulse-Radio UWB Receivers" Circuits and Systems I: Regular Papers, IEEE Transactions on, Vol.59, no. 2,. pp. 337-345,2012.

[7] Wong, Sew-Kin, "High efficiency CMOS power amplifier for 3 to 5GHz ultra-wideband (UWB) application", Consumer Electronics, IEEE Transactions on,vol.55,pp. 1546-1550, August 2009.

[8] Norimatsu, Takayasu, "A novel UWB impulse-radio transmitter with all-digitally-controlled pulse generator", Solid-State Circuits Conference, 2005. ESSCIRC 2005. Proceedings of the 31st European, 12-16 Sept. 2005, pp. 267- 270.

[9] Li, Jun, "Reconfigurable, spectrally efficient, high data rate IR-UWB transmitter design using a Δ–Σ PLL driven ILO and a 7-tap FIR filter", VLSI Design, Automation and Test (VLSI-DAT), 2011 International Symposium on, 25-28 April 2011, pp. 1 - 4.

[10] Diao, Shengxi Xi, "3–5GHz IR-UWB timed array transmitter in 0.18μm CMOS", Solid-State Circuits Conference, 2009. A-SSCC 2009. IEEE Asian, 16-18 Nov. 2009, pp. 365 – 368.

[11] Baschirotto, A.; Cocciolo, G.; De Matteis, M.; Giachero, A.; Gotti, C.; Maino, M.; Pessina, G. "A fast and low noise charge sensitive preamplifier in 90 nm CMOS technology". Journal of Instrumentation. Volume 7, January 2012. pp.1-8.

[12] De Matteis, M.; D'Amico, S. ; Costantini, A. ; Pezzotta, A. ; Baschirotto, A."A 1.25mW 3rd-order Active-Gm-RC 250MHz-bandwidth analog filter based on power-stability optimization" Electronics Circuits and Systems (ICECS), 2012 19th IEEE International Conference on Publication Year: 2012 , Page(s): 260 - 263 h.

[13] Costantini, A. ; Pezzotta, A. ; Baschirotto, A. ; De Matteis, M.; D'Amico, S. ; Murtas, F. ; Gorini, G. "A CMOS 0.13um low power front-end for GEM detectors". Electronics Circuits and Systems (ICECS), 2012 19th IEEE International, Page(s): 193 – 196.

System Design for an Experimental Cognitive Transceiver

Arun Ashok, Iyappan Subbiah, and Stefan Heinen
Chair of Integrated Analog Circuits and RF Systems
RWTH Aachen, 52062, Germany
Email:ias@rwth-aachen.de

Abstract—**The aim of this paper is to show a detailed system design and bring out the challenges for cognitive transceivers in UHF band. An experimental cognitive transceiver is proposed based on commercially available components. The complete system aspects of a filterless implementation are discussed. Along with the intermodulation test for LTE specified by the 3GPP, the intermodulation of Tx leakage with the blocker arising in an FDD scenario is also dealt with. It is shown that even with highly linear RF components, the receiver cannot sustain Tx leakage. The blocker performance and the channel filtering requirements for a SAW-less transceiver are also shown along with an AGC algorithm for a steady SNR. With a brief system analysis, the effect of transmitter noise leaking into the Rx band and its impact on the sensitivity are discussed. The experimental setup has the provision to test the impact of full-FDD, half-FDD and TDD modes on the RF performance in a cognitive environment.**

I. INTRODUCTION

The recent release Rel-11 by the 3GPP [1] shows the crowded scenario of the existing spectrum for telecommunications. While on one hand we have a crowded spectrum, on the other hand local regulations have lead to inefficient spectrum utilization in the UHF bands which can otherwise be wisely used for long range communications owing to their excellent propagation characteristics. Such a situation has given rise to spectrum holes in the DVB-T bands. As a result cognitive radio (CR) architectures have gained interest recently. A CR is essentially a broadband transceiver which can sense sprectrum holes and register itself as a secondary user. Such a system not only has to satisfy the test conditions of the secondary user, but also has to respect the specifications of the primary user. Here we demonstrate the system design of an experimental cognitive transceiver for the DVB-T2 and LTE bands (470 MHz to 890 MHz). The front end of such a setup is frequency agile and can be tuned to any desired band. While a wideband front end is relatively easier to design, it brings out several challenges to the radio engineer such as frequency selectivity, linearity to tolerate the in-band and out-of-band blockers and low noise figure to satisfy the sensitivity requirements. Besides, the transmitter leakage and noise arising in an FDD scenario can completely mask the received signal in a filterless front end. The aim of our setup is to develop a highly linear receiver from off-the-shelf components and provide insights and challenges for state-of-the-art transceivers.

II. SYSTEM DESIGN

Among the various possible architectures, direct down-conversion is the preferred choice for the multiband LTE transceiver owing to its reduced complexity. While the RF front

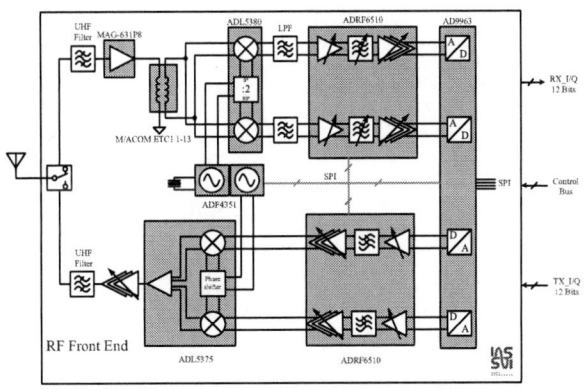

Fig. 1: Experimental Cognitive Transceiver

end provides a minimum gain for sensitivity, most of the gain can be provided at the baseband relaxing the RF front end design. In order to support multiband operation in a cost effective way, a common front end is devised rather than separate Rx/Tx chains for every band of operation. The performance of such a SAW-less architecture is constrained by the linearity of the components. Hence for our cognitive transceiver setup, highly linear commercial components are chosen to ensure high dynamic range. The proposed architecture is shown in Fig. 1 and the specifications of the components are listed in Table I. The CR prototype is designed for operation in LTE bandwidths of 1.4 MHz, 5 MHz, 10 MHz and 20 MHz. While the RF side is same, VGA at the baseband section provides the necessary gain and bandwidth settings for the band of interest. The gain of the VGA can be varied over 50 dB with the bandwidth programmable from 1 MHz to 30 MHz in steps of 1 MHz. The RF front end can receive the DVB-T and LTE bands from 470 MHz to 890 MHz. The gain of the LNA is a compromise between the noise figure of the receiver and its linearity (IIP3). LNA with 17 dB is hence selected here. The differential signal coming from Balun is downconverted and fed into the VGA which provides most of the gain and bandwidth selection. In order to improve the frequency selectivity of receiver, a passive filter is proposed between the Mixer and the VGA. The filtered analog information is digitized by the ADC. Since some of the blocker power reaches ADC unfiltered, the sampling rate is taken as twice the frequency of the significant blocker. Although the experimental transceiver has to support all LTE bandwidths, for the ease of design, the system is evaluated for 5 MHz at first which can be later extended for all other bandwidths. The sensitivity (REFsens) of the receiver can be

TABLE I: Components of the Experimental CR setup

Component	Gain	IIP3	NF
LNA	17 dB	4 dBm	0.60 dB
IQDownconverter	7 dB	30 dBm	11 dB
VGA	40 dBV	4 dBVrms	17.78 nV/sqrt(Hz)

obtained from the noise figure and the SNR as in (1). An SNR of 7 dB is assumed here. The cascaded noise figure can be found from the equation (2). Since the RF and baseband are in power and voltage domain respectively, the noise voltages of the VGA and the ADC has to be brought before the IQMixer (which is 50Ω terminated) where they are converted into power domain. This noise information in $\frac{nV}{\sqrt{Hz}}$ depends on the gain and bandwidth settings of the VGA. The noise voltage of the ADC is a function of the fullscale rms voltage, number of bits and the signal bandwidth as in (3). Finally the IIP3 which quantifies the linearity is calculated using (4).

$$REFsens = 10log(kT) + 10log(BW) + NF_{rxr} + SNR \quad (1)$$

$$F = F_1 + \sum_{i=1}^{N} \frac{F_i - 1}{\prod_{j=1}^{i-1} Gain_j} \quad (2)$$

$$NV_{ADC} = \frac{V_{fsrms}}{SNDR * \sqrt{Fbw}} \quad (3)$$

$$\frac{1}{IIP3} = \frac{1}{IIP3_1} + \sum_{i=1}^{N} \frac{\prod_{j=1}^{i-1} A_j}{IIP3_i} \quad (4)$$

The complete level plan is shown in Table II. The noise figure, IIP3 and the REFsens of the experimental CR is obtained as 8.8 dB, -8.3 dBm and -103 dBm respectively. The sensitivity is 6 dB better than the 5 MHz LTE reference sensitivity of -97 dBm. With separate antennas for Tx and Rx, the CR can be operated in FDD, TDD and half-FDD modes and the impact of various duplex modes on RF performance can be evaluated.

III. INTERMODULATION TEST

The 3GPP Rel-11 [1] prescribes the intermodulation test criteria for LTE systems. For the 5 MHz band, the system is excited with two tones of -46 dBm at a frequency spacing of 10 MHz and 20 MHz. The desired signal is placed 6 dB higher than the REFsens. The inherent nonlinearity will cause the intermodulation (IM) of the two tones and generate IM noise which falls over the received signal band. The output IM3 noise (OIM3) in dBVrms is calculated as in (5). Fig. 2 shows the OIM3 at the output of each stage of receiver.

$$P_{OIM3} = 3P_{out} - 2P_{OIP3} \quad (5)$$

As one can see from Fig. 2 the IM3 noise floor is at 50 dBc below the desired signal and is insensitive to the intermodulation (test case). But in a filterless implementation effect of Tx leakage has to be considered as well. A typical duplexer can provide an isolation of 40 dB resulting in -20 dBm of Tx power at the Rx front end when the Tx transmits at 20 dBm. Such a Tx leakage at a duplex distance of 30 MHz (for the concerned

TABLE II: Complete Level Plan

	LNA	IQ-Mixer	VGA	ADC
Power gain(dB)	17			
Voltage Gain(dBV)	17	7	40	
Noise Figure (dB)	0,6	11,		
Noise Factor	1,15	12,6		
Noise Density(dBV/√Hz)			-155	-155
Noise voltage (nV/√Hz)	0,35	3,1	17,78	18,6
IIP3 dBm	4,00	30		
dBVrms	-9,01	17	4	50
Cascaded Noise				
Noise voltage (nV/√Hz)		16,2	17,8	18,6
Noise Factor		324,2		
Noise Figure(dB)	8,8	25,1		
Cascade IIP3				
dBVrms		-4	3	50
dBm	-8,3	9		

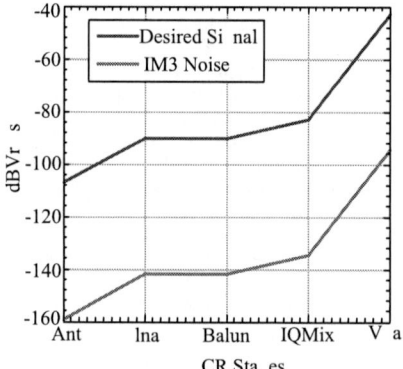

Fig. 2: IM3 Test for CR

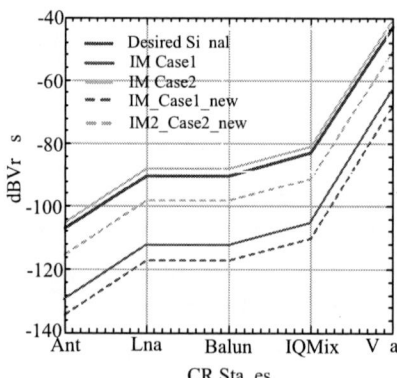

Fig. 3: Intermodulation of Tx leakage with Blockers

LTE bands here) can intermodulate with the blockers present at 15 MHz (case 1) or 60 MHz (case 2). The OIM3 in these cases where the tones are not in equal amplitude can be calculated as in (6) and are shown in Fig. 3

$$v_{OIM3} = \frac{v_{int1}^2 v_{int2} Av}{v_{IIP3}^2} \quad (6)$$

As shown, in both cases the Tx IM noise completely masks the desired signal. [2] shows that minimum of 26 dBm IIP3 is required for a SAW-less CR working in FDD. As with each dB of Tx leakage attenuation, IIP3 improves by same amount, performance can be improved with reduced Tx power and improving Rx/Tx isolation with separate antennas and incorporating low pass filters (LPF) before VGA. With an assumed Tx/Rx isolation of 40 dB, 5 dB Tx power backoff and 3rd order LPF, the IM distance increases by 24.8 dBc and 7 dBc respectively for the two cases. While usage of high order filter increases the implementation complexity, Tx power backoff will result in loss of coverage. Hence only limited FDD operation is possible. But the availabilty of bandwidth and better propagation conditions in a cognitive scenario can outweigh this loss of performance.

IV. BLOCKER TEST AND PHASE NOISE

Blocking characteristics indicate the measure of receiver's ability to receive a wanted signal in the presence of unwanted interferers in the neighbouring channels. These neighbouring channels are specified at predetermined offsets from the assigned channel frequency. The blocking specifications are usually categorized into two subsets, the in-band blocking (InB) and the out-of-band blocking (OoB). The in-band blockers are closeby the desired frequency and they influence the selectivity of the receiver, whereas the out-of-band blockers are far-away and decide the linearity of the receiver. Table III represents

TABLE III: Blocking requirements for 1.4 MHz LTE channel

Freq.Offset [MHz]	Blocker Level [dBm]
2.8	-56
4.2	-44
15	-44
60	-30
85	-15

the blocking levels for 1.4 MHz LTE channel. Since any practical LTE transceiver also needs to support the standard legacy systems, the radio has to be compatible to standards like GSM and UMTS as well. Additionally, unlike GSM or UMTS, LTE in itself requires operation in a multitude of channel bandwidths. Consequently, a fully compatible LTE transceiver that provides good selectivity for a channel bandwidth of 20 MHz by filtering out the unwanted frequencies cannot provide a good selectivity for a channel bandwidth of 1.4 MHz without altering its channel filtering. In our transceiver, this functionality is provided by the VGA which has a programmable bandwidth. For a SAW-less receiver, there is no external filtering of the blockers and the attenuation has to be provided by the on-chip filtering in each of the system blocks. Since the farthest out-of-band blocker is situated at 90 MHz offset from the desired channel and system being wideband, all the blockers proceed through the system blocks maintaining their relative power levels to that of the desired signal till they are filtered in the VGA. Thus, from Fig. 4 it can be observed that before the VGA, the blocker levels are high with respect to the desired signal. Also it is known that for a successful demodulation the necessary SNR for the desired signal has to be maintained at the input of the ADC. Alternatively, some of the blocker filtering can be postponed to the digital domain. In such a strategy, some of the blockers can be allowed to be higher than the desired signal at the ADC. However, for such an approach to be effective, the Nyquist-bandwidth of

the ADC should also accomodate the farthest blocker that is still above the ADC noise floor.

As an estimate for the total order of filtering that would be required for suppressing the blocker levels, the calculations are done with the level plan such that the blocker levels at the ADC are suppressed at least to the noise floor of the receiver, whereby the noise floor would not be raised by more than 3 dB. It is inferred that the highest filter order is demanded for the narrowest channel bandwidth case of 1.4 MHz. Here, a filter order of 6 is needed to suppress the closest in-band blocker. For other channel bandwidths of 3 MHz, 5 MHz and 10 MHz, a 5th-order filter is required. The filter-order is much relaxed for the wide channel bandwidths of 15 MHz and 20 MHz. In all these calculations, the corner frequency of the filter is determined such that within the baseband of each radio channel bandwidths, the attenuation in magnitude is no worse than 0.1 dB. In our transceiver, the VGA has an on-chip 6th order butterworth filter and Fig. 4 shows the blocker and signal levels for the 5 MHz channel bandwidth case.

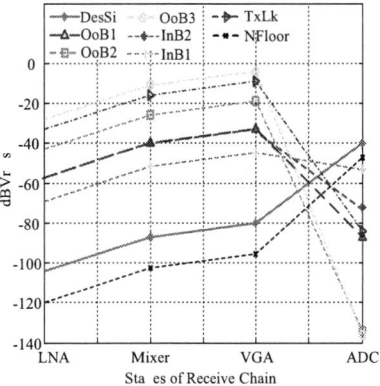

Fig. 4: Blocker and Signal levels along the Receiver

Phase noise requirements dictate the necessary spectral purity that is expected of the local oscillator. The phase noise can reciprocally mix with the blockers and fall into the Rx band. The allowed phase noise levels at each oscillator frequency offset can be deduced from

$$PN_{\Delta f} = P_{desired} - P_{\Delta fblocker} - 10log(BW) - SNR \quad (7)$$

Accordingly, the phase noise requirements for a 1.4 MHz channel requiring a maximum SNR of 22.6 dB [3] are laid out in Table IV.

TABLE IV: Phase Noise for the narrowband LTE channel

Freq.Offset [MHz]	Phase Noise [dBc/Hz]
2.8	-134.26
4.2	-146.26
15	-146.26
60	-160.26
85	-175.26

V. AUTOMATIC GAIN CONTROL

As shown in Table III, the Rx has to sustain the blockers as high as -15 dBm. As the dynamic range required is 102 dB [2]

Paper T1B3   *PRIME 2013, Villach, Austria*

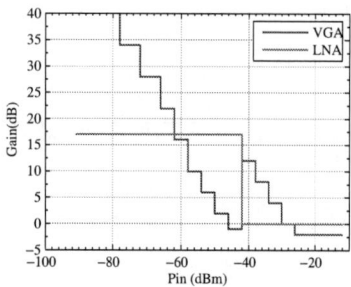

Fig. 5: Automatic Gain Control

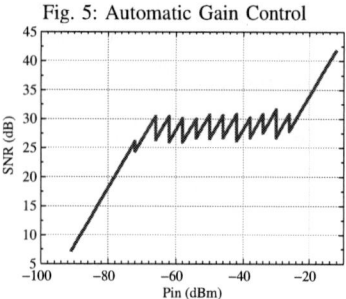

Fig. 6: SNR at ADC with Input power

which is beyond state-of-the-art ADCs, automatic gain control (AGC) is needed to reduce the dynamic range requirement of the ADC. The AGC we propose here not only acheives that, but also prevents the saturation of the front end maintaining a constant SNR for better user performance. The input power can vary from the REFsens of -97 dBm to -25 dBm. But taking into account PAPR of OFDM downlink signal as well as the Tx leakage the maximum input for the LTE reaches -12 dBm [3]. In our experimental setup when the VGA provides 50 dB of gain control with 1 dB step, LNA has 1 step gain control. The received signal strength indication is assumed to be provided by the baseband and gain settings are then modified based on a look up table. The power of the adjacent channel which is at -31.5 dBc is also considered here as the input. The gain control and the corresponding SNR are shown in Fig. 5 and Fig. 6 respectively. From [3] and the SNR plot our experimental setup can support QPSK for powers from REFsens till -90 dBm (SNR<6 dB), 16QAM for powers greater than -90 dBm (6 dB<SNR<20 dB), and 64QAM for powers greater than -83 dBm (SNR>20 dB). For a steady performance we maintain a minimum guarenteed SNR with input power once SNR threshold is reached. With our proposed AGC control an SNR of 25 dB is maintained most of input power range providing a steady performance at the user end.

VI. TRANSMITTER AND SPECTRUM

For Tx design, the transmitted power, the in-band performance in terms of Error Vector Magnitude (EVM), out-of-band (OoB) emissions, far out spurious emissions and excess noise in the receive band are of importance. Tx power according to 3GPP is 24 dBm ± 2 dB. The OoB emissions are specified by Spectrum Emission Mask (SEM) and Adjacent Channel Leakage power Ratio (ACLR). The power amplifier (PA) was chosen such that it satisfies the power and spectrum mask requirements of the 3GPP. Another problem in the FDD scenario apart from the Tx leakage is the excess noise

falling into the receive band. This is aggravated by the fact that the LTE bands under consideration (12, 13, 14 and 17) have smallest duplex distance of 30 MHz. For the 5 MHz bandwidth, noise levels in CR stages are shown in Table V. The output noise level is weighted by the gain of succeeding stage and propagated in the direction of transmission. The noise and gain of the respective stages are evaluated at the 30 MHz offset. The broadband PA amplifies the noise with its gain of 18 dB resulting in -101 dBm/Hz at the Rx input reducing its sensitivity. In order to make Rx insensitive to the noise, the Tx noise contribution should be 10 dB less than the thermal noise floor. This implies, all the noise contributions have to be attentuated with an interstage filter. A passive LPF after VGA is needed hence. But duplex distance being small, the attenuation possible with reasonable complexity is limited. Reasonable attenuation through an LPF, increased Tx/Rx isolation with separate antennas and the Tx backoff can reduce the Tx noise contribution for our homodyne setup. As the spectrum mask specification for the CR scenario being in research phase, the aim here is to make the transmitter compliant to the LTE specifications. A field test with the experimental transceiver will assist to understand and set limits on the out of channel emissions and system requirements on the CR.

TABLE V: Tx Noise Calculation

		DAC		VGA	50R	IQMix		PA	
Rx Band Gain	dB							18	
	dBV			0		-3,5			
O/p Noise	dBm/Hz	-147		-		-150			
	dBV/$\sqrt{Hz}$			-130		-163			
Cascaded Noise	dBm/Hz		-147	-			-119	-101	
	dBV/$\sqrt{Hz}$		-		-129	-132		-	
Tx Noise in Rx Band	dBm/Hz							**-101**	
Thermal Noise	dBm/Hz							-174	

VII. CONCLUSION

A complete system design for an experimental cognitive transceiver has been carried out. Along with the basic test cases for intermodulation and blockers, effects of Tx leakage and Tx noise on Rx performance are also shown. It is emphasized that a frequency agile cognitive radio can operate only in limited FDD mode; conversely for maximum coverage half-FDD or TDD has to be chosen. The proposed transceiver can operate under different modulation schemes for various input powers and can maintain a steady SNR for most of the power levels. Some insights are also given into the phase noise requirements and Tx system design in this paper.

REFERENCES

[1] "ETSI, LTE; Evolved Universal Terrestrial Radio Access (E-UTRA) User Equipment (UE) radio transmission and reception,3GPP TS 36.101 version 11.2.0 Release 11 ."

[2] S. Heinen and R. Wunderlich, "High dynamic range RF frontends from multiband multistandard to Cognitive Radio," in *Semiconductor Conference Dresden (SCD)*, pp. 1–8, sept. 2011.

[3] S. Sesia, I. Toufik, and M. Baker, *LTE, The UMTS Long Term Evolution: From Theory to Practice*. Wiley Publishing, 2009.

978-1-4673-4579-8/13 $31.00 © 2013 IEEE

High Speed Interface for Digital Centric Transmitters

Bastian Mohr*, Jan Henning Mueller*, Ye Zhang*, Richard Leys[†], Sven Schenk[†],
Ulrich Bruening[†] and Stefan Heinen*
*Chair of Integrated Analog Circuits and RF Systems
RWTH Aachen University, D-52056 Aachen, Germany
[†]Computer Architecture Group
University of Heidelberg, D-68131 Mannheim, Germany
Email: bastian.mohr@ias.rwth-aachen.de

Abstract—This paper presents a high-speed serial PLL-less interface suitable for usage in mobile transmitters. The interface uses current mode signaling to reduce both ground bouncing and the crosstalk impact on the mobile frontend. A digital controlled delay line is employed to adjust the sampling point of the high-speed serial clock. The data is 8b/10b encoded for word recovery and signaling of configuration packets. The interface is self-initializing and distinguishes between signal and configuration data. It consists of three lanes from the FPGA to the ASIC and one lane in backward direction to debug the internal ASIC signals. The interface is able to transfer up to 1.6 Gbit/s per lane. It consumes 3 mA from a 1.2 V supply.

I. INTRODUCTION

A modern communication system-on-chip (SoC) has to cover a wide range of mobile and wireless standards, such as IEEE 802.11 (WLAN), WCDMA based UMTS, next generation LTE or legacy standards such as GSM. An interface for the communication between the baseband processor and the radio frequency IC (RFIC) has to provide high data rates for standards such as WLAN and LTE. On the other hand, the crosstalk impact on the frontend needs to be very small, as for example UMTS minimum output power is at -50 dBm. The interface does not only act as an aggressor but also as a victim in the system as it suffers from crosstalk introduced by the transmitter.

This paper is organized as follows: In Section II, the requirements are discussed and several implementation types of the interface are compared. In Section III, the interface is explained on a system level with detailed description of the initialization procedure. Section IV is focused on the circuit implementation of the driver cells. Measurement results are presented in Section VI.

II. INTERFACES REQUIREMENTS

There are several options to implement a high-speed serial interface for baseband to RFIC communication, each with its own advantages and drawbacks. First the minimum requirements are elaborated. The interface should use low voltage signaling (LVDS), not only because it is faster, but also it enables lower signal levels reducing the crosstalk impact on the radio frontend and minimize the ground bouncing [1]. The

highest data rate the RFIC supports is 30.72 MHz, which is full bandwidth LTE. To ensure high linearity at the output of the transmitter frontend both inphase- (I) and quadrature-path (Q) are represented by 12 bit. Thus, the minimum data rate R of the interface is

$$R_{min} = 30.72\,\text{MHz} \cdot 12\,\text{bit} \cdot 2 = 737.38\,\text{Mbit/s}. \quad (1)$$

The interface should transfer I and Q signals as well as reconfigure the RFIC. Hence 8b/10b encoding [2] is used to indicate start and stop of the data and control packages respectively. It is also used for word clock recovery and state information during the interface initialization procedure.

The most pin saving way is to transfer I and Q on the same lane, using a PLL to recover the bit clock. In this approach only two pins are necessary (for the positive and negative polarity of the differential signal), but the interface needs a dedicated PLL and the overall data rate is limited. Using a separate lane for each I and Q enables more bandwidth for the signals and allows for bandwidth increasing nonlinear digital preprocessing at the baseband processor e.g. predistortion or Cartesian-to-polar conversion.

For the presented implementation, an additional lane for the serial clock was added, rendering the clock recovery PLL unnecessary. While this approach is pin consuming, it offers the highest flexibility for testing purposes and saves on-chip power consumption. Without clock recovery it is crucial to match the delays between the data and clock paths to ensure an optimum sampling of the data and to avoid unwanted delay between the I and Q path. With one additional lane in reverse direction for initialization and debugging the total pin count of the interface is 8.

III. SYSTEM DESCRIPTION

The block diagram of the interface is depicted in Fig. 1. The interface is suitable for any transmitter, conventional design and digital centric as presented in [3]. The baseband data could be generated on a baseband processor, i.e. a baseband IC or an FPGA. In the presented work an FPGA is used for baseband signal generation and preprocessing of the data.

978-1-4673-4579-8/13 $31.00 © 2013 IEEE

Paper T2B1

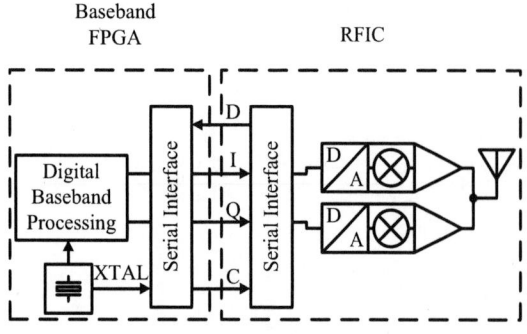

Fig. 1. Interface block diagram.

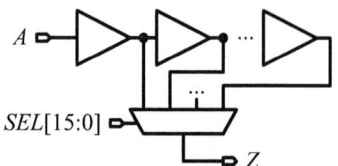

Fig. 2. Block diagram of the delay chain.

Inphase and quadrature data (I and Q) are transferred to the RFIC on separate lanes. As the interface does not contain a phas-locked-loop (PLL), the serial bit clock (C) is also transferred to the RFIC. From the RFIC to the FPGA also one debug path (D) is available. Each lane is designed to transmit data with a rate of 1.6 Gbit/s, which results in an overall data rate of 3.2 Gbit/s from baseband IC to the RFIC and 1.6 Gbit/s in the opposite direction. The 12 bit of each data word are filled up with 4 additional bits which can be used to send additional information (such as frame number or power control for example). After 8b/10b encoding every data word consists of 20 bit, thus the line rate R and baseband sampling frequency f_{BB} are related as given in (2). Typical baseband frequencies are 61.44 MHz ($2 \cdot f_{LTE}$) and 80 MHz ($4 \cdot f_{WLAN}$)

.

$$R = 20 \text{ bit} \cdot f_{BB}. \tag{2}$$

The interface is capable of transferring either signal data or configuration information. In case of configuration I and D lane transfer the register addresses and contents to and from the RFIC respectively.

On the FPGA, the interfaced is realized using Xilinx GTX transceivers supporting data rates up to 6.6 Gbit/s, 8b/10b encoding, comma alignment and dual data rate (DDR) transmission. The data is two times oversampled, this allows for adjusting serial clock delay by half a cycle with respect to the data. DDR transmission results in a serial clock period of

$$T_{clk} = \left[\frac{1}{2} \cdot R \cdot \frac{1}{1\,\text{bit}} \right]^{-1} = 1.25 \text{ ns}. \tag{3}$$

A. Bit Clock Alignment

Each lane from the FPGA to the RFIC is equipped with an on-chip delay line as depicted in Fig. 2. The output of all 16 delay elements is connected to the input of a multiplexer. A one-hot signal controls the multiplexer to select the delayed signal. This ensures that I and Q path can be delayed separately if any path delays are between them. It is also possible to delay the clock (C) to achieve a negative path delay. The delay should be adjustable by a quarter clock cycle which

ranges from 250 ps to 312.5 ps, depending on the baseband data rate. The step size should be around 50 ps to ensure a high resolution in the step sizes.

B. Word Clock Alignment

When transferring serial data it is important to detect the boundaries of the data words. A straight forward approach would be an additional world clock, but in this case path delay will corrupt the data. The presented interface uses 8b/10b encoding [2] to allow correct detection of word boundaries. Another advantage of this encoding is the possibility to detect errors and to use AC coupling, as the data is DC balanced. As for each 8 bit two encoding bits are necessary, 8b/10b encoding reduces the effective data rate to

$$R_{eff} = R \cdot \frac{8}{10} = 1.28 \text{ Gbit/s}. \tag{4}$$

C. Initialization of the Interface

After start-up the FPGA first sends the serial clock to the RFIC. The RFIC responds with this clock on the debug lane. If the FPGA receives this clock, an initialization pattern is sent to the RFIC. The interface logic then starts to delay both I and Q path and compares the resulting bit sequence with the known pattern. In this way it is possible to estimate the clock alignment. The value of the minimum and the maximum delay where no bit errors occur is saved. The delay is adjusted to select the arithmetic mean of both values which guarantees a wide open eye.

After the bit clock is aligned, the RFIC sends a special character (so called komma character in [2]) to the FPGA to start the next initialization step. The FPGA then starts sending another komma character which the RFIC tries to detect. As long as the komma character is not detected, the deserialized data is barrel shifted by one more bit as long as the data is correctly aligned. Afterwards the RFIC indicates the FPGA to start the transmission of data.

D. Data and Control Transmission

As the data transferred to the RFIC should not be limited to a certain burst length the data package can be arbitrary long. It is started by a *DATA_SOP* (start of data package) character. If the FPGA intents to reconfigure the RFIC an end of package character *DATA_EOP* is sent. Afterward control words, consisting of address and data are transmitted, enclosed by *CTRL_SOP* and *CTRL_EOP* characters.

978-1-4673-4579-8/13 $31.00 © 2013 IEEE

PRIME 2013, Villach, Austria   *Session T2B – Digital Systems*

E. Receiving data from the RFIC

In general, the interface is symmetrical and can transfer data in both directions which makes it suitable to be used in a transceiver. In the tested transmitter design [3] the backward line is mainly used for initialization and debugging. It is possible to read out all the registers on the RFIC and also to receive data processed on the RFIC to check for issues in the digital parts. The data is 8b/10b encoded using the same komma characters as on the FPGA, serialized with the received bit clock and sent back to the FPGA.

IV. CIRCUIT IMPLEMENTATION

The interface was implemented using current mode logic (CML), which offers some advantages over conventional CMOS drivers. Due to its differential structure it is less sensitive to electromagnetic coupling [4] and increases the maximum operating frequency compared to static CMOS logic [5]. Another advantage is the continuously flowing current from supply to ground, reducing inductive voltage drop, which is crucial for sensitive analog circuits as spurious emissions could leak into the output spectrum of the transmitter [1].

A. Deserializer

Fig. 3 shows the components of the deserializer which are identical for all three receiving paths (I, Q, C). The converter CML2CMOS converts the differential CML signal into single-ended CMOS. Its schematic is depicted in Fig. 4.

Fig. 3. Deserializer.

The CML2CMOS converter consists of a $100\,\Omega$ terminated CML buffer, which also acts as a differential to single-ended converter. The signal is buffered and converted to a full scale CMOS signal using two inverters.

The delay line is implemented using non-inverting CMOS buffers. Table I shows the delay of one buffer from extracted corners simulation. For DDR transmission as required for the interface delays for the rising ($t_{d,LH}$) and falling ($t_{d,HL}$) edge must match closely. The delay covers in fast case 16 ps to 256 ps which is almost one quarter period of an 800 MHz clock (312.5 ps). In slow corner the delay is adjustable in steps of 29 ps which is fine enough to adjust the bit clock in steps of $0.023 \cdot T_{clk}$.

The serial-to-parallel converter is fully described in hardware description language (HDL) and generated by hardware synthesis.

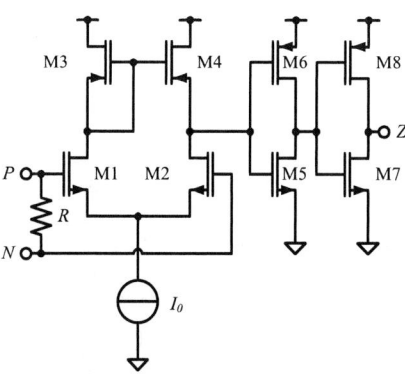

Fig. 4. Schematic of CML2CMOS converter.

TABLE I
DELAY ELEMENTS CHARACTERISTICS.

	Slow Corner	Typical Corner	Fast Corner
$t_{d,LH}$	28 ps	21 ps	16 ps
$t_{d,HL}$	29 ps	22 ps	16 ps

B. Serializer

The data is serialized by a HDL described serializer. The given technology offers negative edge triggered registers so a DDR serializer can be fully generated by hardware synthesis. The single ended serial data is converted to CML afterward using the circuit depicted in Fig. 5 [6]. The output ports P and

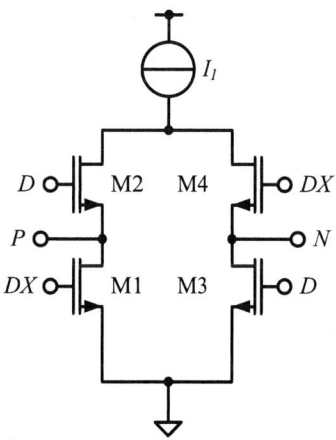

Fig. 5. Schematic of CMOS2CML converter.

N are connected to a $100\,\Omega$ resistance at the FPGA receiver. When D has high value the inverted input DX is low. In this case, current flows from P over N to ground. When D is low current flows from N to P, inverting the output at the receiver. The current I_1, is always constant as is the current through the ground node.

978-1-4673-4579-8/13 $31.00 © 2013 IEEE   235

V. MEASUREMENT RESULTS

The interface was implemented in an RFDAC based transmitter architecture using a 65 nm low power process [3]. Fig. 6 shows a micrograph of the RFIC. The whole interface uses an area of 740 μm x 35 μm, including supply caps.

Fig. 6. ASIC micrograph.

Running at 80 MHz baseband frequency, the CML drivers and the delay chain consume about 3 mA from a 1.2 V source. The synthesized serializer and deserializer use about 5 mA from a 1.2 V source.

Figure 7 shows the output of the serializer (D), while transmitting data to the FPGA at a baseband rate of 80 MHz at supply voltage of 1.2 V. The DC level is around 700 mV. The

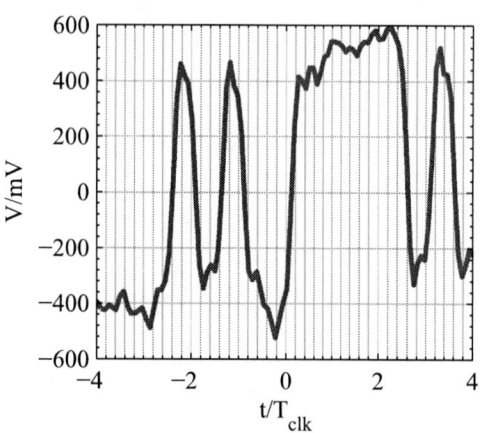

Fig. 7. Serialized data at 800 MHz serial clock.

x-axis was normalized to the clock period (1.25 ns). High and low levels are almost equivalent. Also rise and fall times are close together with 0.23 and 0.24 clock cycles approximately (taking 350 mV as full scale level).

The interface is designed for 800 MHz. However, if the supply voltage of the synthesized serializer is increased to 1.3 V it is also possible to transmit data with up to 2 Gbit/s

per channel at cost of higher power dissipation in the digital building block.

Fig. 8 shows the output spectrum of interface at 80 MHz baseband data rate measured at the transmitter output port. The largest peak is at 1.12 GHz at a level of −69 dBm. This

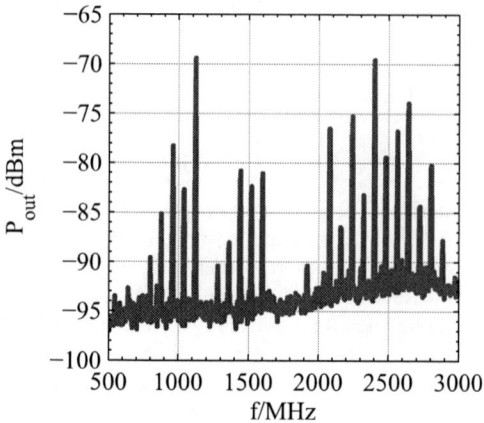

Fig. 8. Output spectrum at 800 MHz serial clock.

is 20 dB below the transmission level for UMTS which is one of the mobile standards with the lowest output levels.

VI. CONCLUSION

This paper presents an interface for mobile transmitters adjusted for the test environment. The interface is able to transfer data from the baseband processing FPGA to the RFIC with a data rate of 1.6 Gbit/s per lane. Employing 8b/10b encoding results in an total data rate for I and Q of 2 · 1.28 Gbit/s. The interface initializes automatically and provides a backward lane for debug purposes. The analog blocks of the interface consume 3 mA from an 1.2 V supply. Increasing the driver strength in the digital domain makes the interface suitable for even higher datarates up to 2 Gbit/s.

REFERENCES

[1] K. Chabrak, F. Bachmann, G. Hueber, K. Seemann, L. Maurer, Z. Boos, and R. Weigel, "Design of a high speed digital interface for multi-standard mobile transceiver rfics in 0.13 mu;m cmos," in *Microwave Conference, 2005 European*, vol. 3, oct. 2005, p. 4 pp.

[2] A. X. Widmer and P. A. Franaszek, "A dc-balanced, partitioned-block, 8b/10b transmission code," *IBM Journal of Research and Development*, vol. 27, no. 5, pp. 440–451, Sept. 1983.

[3] B. Mohr, N. Zimmermann, B. T. Thiel, J. H. Mueller, Y. Wang, Y. Zhang, F. Lemke, R. Leys, S. Schenk, U. Bruening, R. Negra, and S. Heinen, "An rfdac based reconfigurable multistandard transmitter in 65 nm cmos," in *Radio Frequency Integrated Circuits Symposium (RFIC), 2012 IEEE*, june 2012, pp. 109 –112.

[4] P. Heydari and R. Mohanavelu, "Design of ultrahigh-speed low-voltage cmos cml buffers and latches," *Very Large Scale Integration (VLSI) Systems, IEEE Transactions on*, vol. 12, no. 10, pp. 1081 –1093, oct. 2004.

[5] J. M. Rabaey, *Digital Integrated Circuits: A Design Perspective*, 2nd ed. Prentice Hall, 2002.

[6] *LVDS Owner's Manual*, National Semiconductor Std., 2008.

Design of a Reconfigurable Multi-Core Architecture for Streaming Applications with a Case study on Performance Evaluation of FIR-Filters

Leyla S. Ghazanfari, Roberto Airoldi, Jari Nurmi and Tapani Ahonen
Department of Electronics and Communications Engineering
Tampere University of Technology, Tampere, Finland
Email:*name.surname*@tut.fi

Abstract—This paper presents a reconfigurable multi-core architecture which is composed of nine nodes connected to each other in a mesh topology network. The central node hosts a general purpose processor and serves as the system controller and the surrounding nodes are equipped with coarse grain reconfigurable arrays that provide hardware flexibility and parallelism. This architecture is mainly targeted for streaming applications. As a case study, performance and hardware usage for implementations of FIR filters with different number of taps are presented. Each implementation was prototyped on an Altera FPGA and to ensure the correctness of the system, RTL simulations were carried out. The performance results were compared with the same applications implemented on a single-core processor. The timing analyses show speed-ups up to 3.6x.

Keywords: Reconfigurable architecture, Multi-processor system on chip, FIR filter

I. INTRODUCTION

Multiprocessor Systems-on-Chip (MPSoC) play an essential role in digital embedded systems. Recent applications in modern systems require complex multiprocessor designs to reach Real-Time deadline. Furthermore these systems should meet other critical constraints such as power consumption and low area. MPSoC seem to be the solution for many applications such as networking and multimedia that require such complex systems [1]. In addition to MPSoC, the consideration of Coarse Grain Reconfigurable Arrays (CGRA) in embedded systems has been growing in recent years. Researchers exploit the use of CGRAs in their systems to minimize power while maintaining high performance, efficiency and flexibility [2].

SmartCell [3] and Floating-point Reconfigurable Array (FloRA) [4] are two examples of the CGRA architectures in the domain of multimedia and digital signal processing applications [2]. SmartCell is composed of three major units that are: the cell, the reconfigurable interconnect fabric and the high-speed I/O. In a typical SmartCell each cell is composed of four processors, control and data memories. For different applications the data flow can be dynamically reconfigured. In addition number of processing elements (PE) for every application and their functionality can be changed in real time. Although SmartCell is unable to perform floating point operations FloRA is able to perform both integer and floating point operations. This platform contains two dimensional array of PEs that are paired together while performing floating point operations. It also supports speculative execution to handle control intensive kernels. PE operands are assigned from data memory. This memory is filled through a bus connection.

The platform presented here, is a multi-core enriched with CGRAs. The cores are connected to each other through a Network-on-Chip (NoC) utilizing the system with the option of broadcasting while streaming the data to the nodes buffers. CGRAs have been embedded in the nodes that surround the central core to provide system with high performance parallelism and programmable hardware through reconfiguration. These CGRAs are also able to perform floating point operations. In order to evaluate the performance several Finite Impulse Response (FIR) filters are applied to the platform and their speedup and memory usage is compared with same applications mapped to a single General Purpose Processor (GPP), similar to the central core of the reconfigurable multi-core.

The paper is organized as follows: In the next section we will present an overview of our platform and explain its components. In section three the implementation of the filters and how they are mapped on to our reconfigurable multi-core and single-core platforms are described. Afterwards we go through analyzing the performance and memory usage of the filter and compare the results between the two platforms. Finally we conclude in the fourth section.

II. OVERVIEW OF PROPOSED ARCHITECTURE

The idea behind this architecture was presented in [5]. In this section we give an overview of our platform.

The platform shown in Figure 1.b is reconfigurable multi-core which is composed of nine cores connected to each other via a NoC [6]. The central node acts as the main controller of the system. This node hosts a COFFEE processor which is a general purpose processor [7]. The central controller runs a C code which is mapped on to the platform.

The network is the interface between the central node and the other reconfigurable nodes. The current implementation of the network provides 16 different addressing paths [8]. In the current platform, we needed to activate the broadcast option of the network, therefore the access paths to each node were reduced from two to one and an extra path was used to enable the network broadcast option.

The eight cores that are located around the central cores are the reconfigurable nodes. Figure 2 shows the internal

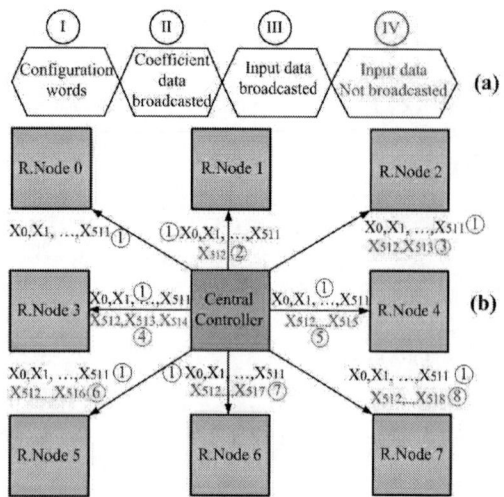

Fig. 1. (a) Four steps of data flow at system level. (b) When the number of filter taps is 512, at first the 512 data are broadcasted to all nodes, afterwards one sample is sent to R.Node 1, two samples to R.node 2 and finally seven samples to R.Node7. This is done to fill the buffer with inputs that are shifted.

architecture of a single reconfigurable node. Each reconfigurable node is composed of seven main sections, they are: the Network Interface (NI), Configuration memory (Conf.MEM), buffer, Reconfigurable Processing Element (RPE) plus three individual controllers that belong to the latter three elements.

NI is the interface between the network and the Reconfigurable node. Every package that enters the node is composed of data, address and some other control signals. The data of the entry package enters the Conf.MEM if it is a configuration data

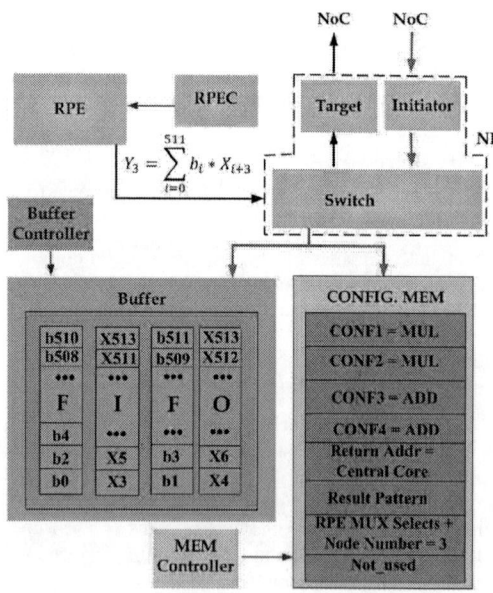

Fig. 2. Inside the reconfigurable node. The data shows the flow in R.Node number 3 when applying FIR filter with 512 taps application

or the Buffer if it is a data to be processed. The Conf.MEM is composed of eight registers; they each have their enable signal which is activated by the Conf.MEM Controller. The address of the entry package is sent to the Conf.MEM Controller and if that is an address of a configuration data the corresponding enable signal is activated and the data of the package enters the correct register inside the Conf.MEM. The data that are stored in this memory are mainly used for configuring the RPE. The address of the node to which the RPE result is being sent to is also stored in this memory.

Buffer is another component of the reconfigurable node. It is composed of four First In First Out (FIFO) blocks that can read and write at different clock cycles [9]. The data of the input packages that should be processed for the application (e.g: filter input signals and coefficients) are stored in the buffer. Data is written to the buffer one by one with on every clock cycle and when the buffer is full the Buffer Controller activates the FIFOs read signals and data is read from the buffer four at a time. The read clock cycle is four times the length of the write clock cycle. The data are sent from the buffer to the RPE. Figure 4 shows the structure of the RPE element.

The RPE component contains four coarse-grained reconfigurable cells [10]. These cells have a 2x2 structure in which the data can flow in a pipeline fashion. RPE receives four data at every clock cycle and after the process is done it has four outputs ready to be sent out to the network. The RPE controller (RPEC) selects the desired result(s). This controller is also responsible for generating the destination address and a write enable signal for every result that is sent back to the network.

III. IMPLEMENTATION OF FIR-FILTER

The applications implemented, are Finite Impulse Response (FIR) filters. FIR filters is a filter whose impulse response is of finite duration that is because it settles to zero in finite time. For a discrete-time FIR filter, the output is a finite weighted sum of the current and previous values of the input. Equation 1 describes the operation.

$$Y_n = \sum_{i=0}^{N} X_{n-i} \cdot b_i \qquad (1)$$

In this equation: $Y[n]$ represents the outputs, $X[n]$ is the input samples ,every b_i is one of the filter coefficients and N is the order of the filter; an Nth-order filter has $(N+1)$ terms on the right-hand side. $X[n-i]$ in these terms are commonly called taps based on the structure of a tapped delay line that in

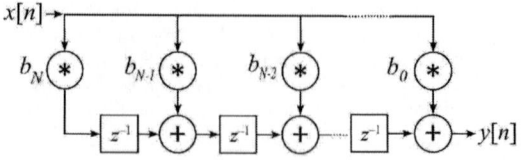

Fig. 3. Transpose-form of FIR filter structure

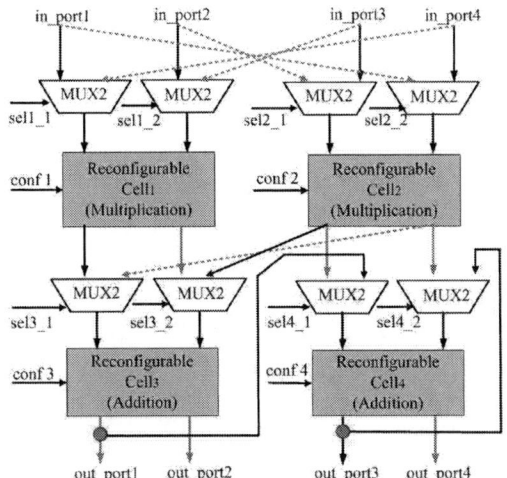

Fig. 4. RPE is configured to process the data. The black lines are the configured paths while the gray lines show the other possibilities.

many implementations or block diagrams provides the delayed inputs to the multiplication operations [11].

To implement FIR filter we used the transposed form of the filters structure. This structure is shown in Figure 3. This implementation is applied for FIR filters with ten different numbers of taps. The performance of each filter is evaluated both on the reconfigurable multi-core and single-core platform.

For the single core (COFFEE) platform the sum of products of the input signals and the coefficients are generated. This is written in a C code. The code for each filter is mapped to the COFFEE core.

To map the application to our platform the data are sent to the reconfigurable nodes in four main steps. This is illustrated in Figure 1.a. In step one the configuration data are sent to each node. These words are stored in the CONF.MEM. Afterwards the coefficients are broadcasted, meaning that they are sent to the nodes at the same time. The coefficients are located in the first and third FIFO of the buffer and these two FIFOs are filled up row by row. In the third step the input data are broadcasted. These values are stored in the second and fourth FIFO of the buffer. In this case also the data fill up the FIFOs row by row. However, considering that this is an FIR filter application, the input data should be shifted one place in every node. Hence when the input data are received by the nodes the buffer controller will allow all of them to enter the buffer in $node_0$. The first input data will not enter the buffer in $node_1$, the first two data will not enter the buffer in $node_2$ and this is the procedure until the last node in which the first seven data are not located in the buffer. In the fourth step the top empty spaces of the FIFOs dedicated to the input data are filled. This time the data are not broadcasted but sent to each node individually. Figure 1.b shows steps three and four in more detail. This figure shows an example of a FIR filter with 512 taps. The order in which the input samples are sent to the nodes are shown in the figure with a circled number. It is visible that first X input signals from X_0 to X_{511} are broadcasted at the same time, and afterwards the other X samples are sent to fill

the buffer. Figure 2 is an example of $node_3$ in a FIR filter with 512 taps. The main difference between the architectures of filters with different number of taps are the length of their buffer FIFOs. The FIFOs length is half the number of taps.

After the data are sent for processing, every node's buffer is full. At this time the data are streamed out from the buffer to the RPE. The configuration operations for the four processing elements of the RPE are read from the Config.MEM. Figure 4 shows an example of a configured RPE. As it is shown for this application the first row of RPE cells are configured to multiplication and second row to addition. The buffer sends two pairs of coefficients and input data to the RPE in the first cycle they are multiplied in $cell_1$ and $cell_2$ in the next cycle the results of the first row are added in $cell_3$ and on the next cycle $cell_4$ adds up the current result of $cell_3$ with its own output. On the first round $cell_4$ is adding $cell_3$ result with zero but on the next rounds it accumulates all the addition results from $cell_3$. The counter inside the RPEC realizes when the last accumulation is processed in $cell_4$. At this point the select signal which is the RPEC output chooses the accumulated result among the four RPE outputs to be sent out to the network. At the same time RPEC activates the write signal and generates the destination address. The destination address is the address of the central node. The result data and the destination address along with the write signal are packed and sent to the network. Each reconfigurable node generates one result. Therefore in total we produce eight results for every FIR filter implementation.

IV. RESULTS AND ANALYSIS

The reconfigurable multi-core were simulated for every FIR Filter and synthesized on target Stratix IV FPGA. TableI

TABLE I. FPGA SYNTHESIS RESULTS OF PROPOSED ARCHITECTURE FOR AN FIR FILTER WITH 512 NUMBER OF TAPS

Parts	No. of Registers	No. of Logic Elements	No. of DSP
Total System	20965	47812	144
Central Core	5506	5743	16
NoC	5065	6233	0
Reconfigurable Node	1291	4473	16
RPE	283	3545	16
Buffer	288	262	0
Config.MEM	131	15	0
NI	272	284	0

TABLE II. BLOCK MEMORY BITS

| FIR Filter Taps | Block Memory Number of Bits | | | Compared to Single Core |
	Single Reconfigurable Node	Central Node	Total	
4	768	3145728	3151872	1,00295313
8	1024	3145728	3153920	1,00260417
16	1536	3145728	3158016	1,00390625
32	2506	3145728	3166208	1,00651042
64	4608	3145728	3182592	1,01171875
128	8704	3145728	3215360	1,02213542
256	16896	3145728	3280896	1,04296875
512	33280	3145728	3411968	1,08463542
1024	66048	3145728	3674112	1,16796875
2048	131584	3145728	4198400	1,33463542

TABLE III.   PROCESS CYCLES AND SPEED UP

FIR Filter	Single core No. of clks	Reconfigurable Multi-Core No. of clks		Process	Total
Taps	Process	Send	Process	Speedup	Speedup
4	439	179	22 (117)	19,9545	2,18408
8	759	283	26 (145)	29,1923	2,45631
16	1399	444	41 (161)	34,1219	2,88454
32	2679	763	74 (193)	36,2027	3,20072
64	5239	1403	138 (257)	37,9638	3,39974
128	10359	2683	263 (382)	39,3878	3,51629
256	20599	5243	582 (641)	35,3935	3,53631
512	41080	10363	1034 (1153)	39,7292	3,60446
1024	82040	20602	2060 (2177)	39,8252	3,62016
2048	163959	41083	4106 (4225)	39,9316	3,62829

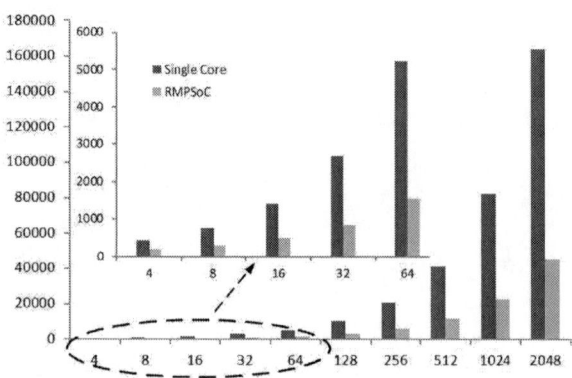

Fig. 5.   Comparing the total number of clock cycles consumed for FIR filter application in reconfigurable multi-core and single-core platforms

presents the synthesis results of the proposed RMPSoC when the buffer is large enough for implementing FIR filter with 512 number of taps. Table II shows the amount of memory bits used for FIR filters with different number of taps. The main portion of the memory is consumed by the central general processor. Depending on the number of taps the memory is increased from 0.002% to 0.33% compared to a single core. According to the table we can also observe that the total memory bits of each system is equal to the number of bits of the central node plus eight times the number of bits of a single reconfigurable node of the same system.

$$Total = 8 * R.Node + CentralNode(MemoryBits) \quad (2)$$

Table III shows the comparison between the number of cycles that it took for the filters to be processed on a single-core and the number of cycles it took for filters to send data to the reconfigurable nodes and process them in proposed platform. There are two sets of numbers under the process column of reconfigurable multi-core. The larger number inside the brackets are the number of cycles that the actual process takes but since the process can start before the communication finishes there will be some clock cycles of overlaps between them. Therefore the numbers outside the brackets show the number of process clock cycles without any overlap from the communication step. As shown the improvement for the process time is very significant and in the filters with larger number of taps we can gain up to 39.9x speedup in process but the main bottleneck that keeps us from having such a gain is the communication cycles. Figure 5 shows the total clock cycles that it takes for FIR filters with different number of taps to perform. The darker bars represent the application on the single-core while the lighter bars present it on reconfigurable multi-core platform.

V. CONCLUSION

We presented a reconfigurable multi-core platform to take advantage of two levels of hardware parallelism, one at network level and the other inside the reconfigurable nodes in the level of CGRAs. The speedup effect of this platform was compared with a single GPP core while mapping same FIR filter applications. The results show that we can gain up to 3.6x speed-up with filters with higher number of taps and the main bottleneck for not obtaining higher gain is

our limitation in the network communication. In addition to speedup results, memory usage for different filters was also presented to provide the reader with the hardware costs of such accelerations. Further work will be finding solutions for reducing the communication time in the network and increase the speedup even more.

ACKNOWLEDGMENT

This work was supported by CRAFTERS project (#295371) in ARTEMIS programme, DEFT project in the Academy of Finland (#258506) and GETA.

REFERENCES

[1] T. Dorta, J. Jiménez, J. L. Martín, U. Bidarte, , and A. Astarloa, "Reconfigurable multiprocessor systems: A review," *International Journal of Reconfigurable Computing*, 2010.

[2] V. Tehre and R. Kshirsagar, "Survey on coarse grained reconfigurable architectures," *International Journal of Computer Applications (0975 888)*, vol. Volume 48 No.16, June 2012.

[3] C. Liang, "Smartcell: An energy efficient reconfigurable architecture for stream processing," Ph.D. dissertation, WPI, 2009.

[4] D. Lee, M. Jo, K. Han, and K. Choi, "Flora: coarse-grained reconfigurable architecture with floating-point operation capability," in *International Conference on Field-Programmable Technology*, 2009.

[5] L. S. Ghazanfari, R. Airoldi, J. Nurmi, and T. Ahonen, "Reconfigurable multi-processor architecture for streaming applications," in *FPL*, 2012.

[6] J. Nurmi, "Silicon Café: a Heterogeneous Multi-Processor Platform based on Coffee (RISC Core)," in *8th International Forum on Application-Specific Multi-Processor SoC*, 2008.

[7] J. Kylliäinen, T. Ahonen, and J. Nurmi, *Processor Design: System-on-Chip Computing for ASICs and FPGAs*. Springer, 2007, ch. General-purpose embedded processor cores - the COFFEE RISC example, ch5, pp. 83–100.

[8] T. Ahonen and J. Nurmi, "Hierarchically Heterogeneous Network-on-Chip," in *Proc. International Conference on "Computer as a Tool" EUROCON*, 9–12 Sept. 2007, pp. 2580–2586.

[9] X. Wang, T. Ahonen, and J. Nurmi, "A synthesizable rtl design of asynchronous fifo," in *Proc. Int System-on-Chip Symp*, 2004, pp. 123–128.

[10] F. Garzia, W. Hussain, and J. Nurmi, "Crema: A coarse-grain reconfigurable array with mapping adaptiveness," in *Proc. Int. Conf. Field Programmable Logic and Applications FPL 2009*, 2009, pp. 708–712.

[11] "Finite impulse response digital filters practical approaches, by the technical staff of: Signal processing group inc. 561 e. elliot road, no.171,chandler, arizona 85225. 2003."

On the calculation of Hall factors for the characterization of electronic devices

V. Uhnevionak[1,*], A. Burenkov[1], P. Pichler[1,2]

[1]Fraunhofer Institute for Integrated Systems and Device Technology, Erlangen, Germany
[2]Chair of Electron Devices, University of Erlangen-Nuremberg, Erlangen, Germany
[*]viktoryia.uhnevionak@iisb.fraunhofer.de

Abstract—**In this work, a new method for the calculation of Hall factors is described. It is based on the interdependence with mobility components via the respective relaxation (scattering) times. The new method allows an accurate determination of mobility and carrier sheet concentration from Hall-effect measurements and can not only be applied to homogeneously doped substrates but also at the interfaces of electronic devices such as field-effect transistors. To demonstrate the general applicability of the method, we use it to predict the dependence of the Hall factor on dopant concentration in silicon and compare it with measured Hall factors reported in the literature.**

Keywords—*Hall factor, Hall measurements, relaxation (scattering) time, mobility, bulk silicon.*

I. INTRODUCTION

Hall-effect measurements in combination with sheet resistivity measurements are well established to determine sheet concentration and mobility of charge carriers independently in semiconductors and metals. However, from the combination of the measured values one obtains an effective Hall mobility μ_H rather than the drift mobility μ of the majority charge carriers. The ratio of the two is called Hall factor or scattering factor and needs to be known a-priori for an accurate interpretation. It is common practice to assume that the Hall factor equals unity. From theory it is known that this is correct only in strong magnetic fields for which the condition $\mu B \gg 1$ holds. For a magnetic field of $B = 1$ T, as an example, the mobility should be significantly higher than 10.000 cm²/Vs. For lower fields and non-degenerate semiconductors, the Hall factor will depend on the particular mechanisms by which the charge carriers are scattered and this may depend in turn on material and measurement condition. A typical example where the assumption of a Hall factor equal to unity is not justified is the inversion channel of a MOSFET because of the rather low channel mobility. In consequence, when the Hall factor is ignored, the sheet concentration and mobility values extracted from Hall-effect measurements may be wrong by some 10% which is considerable for such devices.

In literature, various approaches have been presented to calculate transport properties and Hall factors. They are typically based on Monte Carlo methods or solutions of the Boltzmann transport equation taking the full band structure into account [1,2]. Despite being computationally demanding, they still need comparison with experiments and cannot be easily applied to general conditions in macroscopic devices. In this work we present a new method for the calculation of Hall

factors. It is based on the fact that both Hall factor and mobility depend on the mechanisms by which the charge carriers are actually scattered. The links between the various mobility components and the Hall factor are the respective relaxation (scattering) times. A particular advantage of our method is that it can be easily applied for different materials under different measurement conditions. To show the applicability of the method, we calculated the Hall factors in bulk silicon as a function of doping concentration and compared them with the values reported in literature.

II. METHOD OF THE CALCULATION

The Hall factor r_{H0} is defined as the ratio of the Hall mobility to drift mobility and can be expressed in the relaxation time approximation through certain mean values of relaxation times:

$$r_{H0} = \frac{\mu_H}{\mu} = \frac{<\tau^2>}{<\tau>^2} \qquad (1)$$

Therein, the parameter τ denotes the relaxation time, i.e. the mean free flight time between carrier collisions. In general, it will depend on the energy of the charge carriers. The squared brackets denote an averaging over the energy of the charge carriers. The inverse of the relaxation time $1/\tau$ is the average number of scattering events per unit time. For many scattering processes, the energy dependence of the relaxation time can be expressed as a simple power law:

$$\tau = aE^{-s} \qquad (2)$$

where a and s are constants whose values depend on the involved types of scattering mechanisms. For example, for semiconductors with acoustic phonon or ionized impurity scattering, τ is proportional to $E^{-1/2}$ ($s = 1/2$) or to $E^{3/2}$ ($s = -3/2$), respectively [3].

The values of the averages $<\tau>$ and $<\tau^2>$ can be calculated analytically as long as the relaxation time can be expressed in the form of a power law. They take the form

$$<\tau> = \frac{a(kT)^{-s}\,\Gamma(\tfrac{5}{2}-s)}{\Gamma(\tfrac{5}{2})}, \qquad (3)$$

$$<\tau^2> = \frac{a^2(kT)^{-2s}\,\Gamma\!\left(\tfrac{5}{2}-2s\right)}{\Gamma\!\left(\tfrac{5}{2}\right)} \qquad (4)$$

978-1-4673-4579-8/13 $31.00 © 2013 IEEE

with k standing for Boltzmann's constant, T temperature, and $\Gamma(n)$ the gamma function defined as:

$$\Gamma(n) \equiv \int_0^\infty x^{n-1} e^{-x} dx \qquad (5)$$

In accordance with (1), it follows from the expressions above that the Hall factor becomes $r_{H0} = 1.18$ for acoustic phonon scattering and $r_{H0} = 1.93$ for ionized impurity scattering. It was also suggested by Erginsoy [4] that the relaxation time for scattering at neutral impurities is independent of the energy of the charge carriers ($s = 0$). For such a mechanism dominating at extremely low temperature or for heavily doped semiconductors, the Hall factor becomes $r_{H0} = 1$. In summary, depending on the actual scattering mechanism, values of the Hall factor r_{H0} may vary between 1 and 1.93.

In a real electronic device, several charge scattering mechanisms are usually involved in parallel. In order to calculate the Hall factor from (1), we need to determine the values of $<\tau>$ and $<\tau^2>$ for such conditions. This can be done by accounting these several scattering mechanisms when averaging relaxation times over energy. As suggested by Iwata [5], the various scattering mechanisms can be combined in the form

$$<\tau> = \frac{\int_0^\infty \frac{x^{3/2} e^{-x} dx}{\Sigma_i \tau_i^{-1}}}{\int_0^\infty x^{3/2} e^{-x} dx} \qquad (6)$$

$$<\tau^2> = \frac{\int_0^\infty \frac{x^{3/2} e^{-x} dx}{[\Sigma_i \tau_i^{-1}]^2}}{\int_0^\infty x^{3/2} e^{-x} dx} \qquad (7)$$

where x stands for ε/kT with ε denoting the kinetic energy of an electron and the index i the particular scattering mechanism.

To apply (6) and (7) to certain experimental conditions, we have to know the relaxation times for each of the scattering mechanisms and their dependence on the energy of the charge carriers under investigation. For example, for measurements in a MOS structure, only scattering of the prevalent charge carriers in the inversion layers should be considered. A peculiarity of Hall-effect measurements in MOS structures is scattering at interface defects which is largely negligible for measurements on bulk semiconductor samples. This difference in the measurement conditions can manifest itself in different Hall-factor values. The new method for the calculation of Hall factors presented in this work allows us to master the complexity of the necessary calculations. As indicated above, it takes advantage of the interdependence of the scattering time with mobility. In general, the electron mobility in a semiconductor can be expressed as:

$$\mu_e = \frac{e <\tau>}{m_e^*} \qquad (8)$$

where $<\tau>$ is the mean relaxation time that may contain the contributions from several scattering processes, e denotes

elementary charge, and m_e^* stands for the effective electron mass. It is further assumed that the scattering events are independent of each other so that neither of the τ_i is affected by other scattering process. The contributions of the individual scattering times τ_i to the global scattering time τ can then be combined in the form

$$\frac{1}{\tau} = \sum_i \frac{1}{\tau_i}. \qquad (9)$$

Since the electron mobility is proportional to τ, (9) can be rewritten in terms of the mobility components μ_i associated with the various scattering mechanisms. The result,

$$\frac{1}{\mu} = \sum_i \frac{1}{\mu_i}, \qquad (10)$$

is known as the Matthiessen's rule. From (8) we see that relaxation time and electron mobility differ only by a constant factor of e/m_e^*. Having the possibility to calculate specific mobility components μ_i, the components of scattering time τ_i associated with these mechanisms can be easily found. If a power-law dependence of the relaxation time on the electron energy can be assumed, the unknown constant a in (2) can be calculated from (3) and (8). Given a, the energy dependence of the relaxation time for ith scattering component can be derived as:

$$\tau_i(E) = \frac{\mu_i m_e^* \Gamma\left(\frac{5}{2}\right) E^{-s_i}}{e (kT)^{-s_i} \Gamma\left(\frac{5}{2} - s_i\right)} \qquad (11)$$

It should be noted that these formulas are valid for non-degenerate semiconductors with a single-valley spherical energy band. For multi-valley semiconductors, such as Si, SiC and Ge, the expression for the Hall factor has to be modified. For this purpose, the Hall factor given by (1) for the single-valley model is multiplied by a constant a_0 that accounts for the anisotropy of the effective electron mass. Thus, the Hall factor for the case of a multiple conduction valley semiconductor can be expressed in the form [6]

$$r_H = r_{H0} a_0 \qquad (12)$$

where r_{H0} is the scattering factor given by (1) for single-valley semiconductors assuming isotropic model, and a_0 is known as the "Hall mass factor" or anisotropy factor of the Hall-effect. The latter is given by [6]

$$a_0 = \frac{3K(K + 2)}{(2K + 1)^2} \qquad (13)$$

with K being the ratio of longitudinal and transverse effective masses of the electrons. For Si, SiC and Ge, the Hall mass factor a_0 takes values of 0.87, 0.98 and 0.785, respectively.

Using (6) – (11), and assuming the concurrent action of different scattering mechanisms, the Hall factor can be easily determined and corrected with the anisotropy model given in (12).

978-1-4673-4579-8/13 $31.00 © 2013 IEEE

III. HALL FACTOR IN BULK SILICON

To check the feasibility of our new method for the calculation of Hall factors, we applied it to the calculation of the Hall factor in bulk silicon, which was extensively investigated in the literature. Specifically, we investigated the dependence of the Hall factor in n-type silicon at room temperature on the doping concentration. For this purpose, we performed simulations of electronic transport in phosphorus-doped silicon samples for the range of phosphorus concentration from 1×10^{17} cm^{-3} to 2×10^{20} cm^{-3}. Given bulk silicon, we assumed that the mobility is determined by Coulomb scattering on ionized impurities and by bulk phonon scattering. To model the mobility component due to bulk phonon scattering, a constant mobility of $\mu_{ph} = 1417$ cm^2/Vs at $T = 300$ K was adopted from the work of Lombardi et al. [7]. The mobility component due to charged impurity scattering was calculated using the Arora model as implemented in TCAD Sentaurus. It can be described by [8]

$$\mu_{dop} = \mu_{min} + \frac{\mu_d}{1 + (\frac{N_D}{N_0})^{A^*}} \qquad (14)$$

with N_D standing for the donor concentration and μ_{min}, μ_d, N_0 and $A^* > 0$ for some given parameters of the model. To predict the dependence of the Hall factor on the dopant concentration, we started with an analysis of (14). It is clear from the functional dependence that the doping-related component will strictly decrease with an increasing doping concentration. This is shown also in Fig. 1. It can be interpreted as an increase of the number of scattering centers in the bulk with an increasing dopant concentration, leading further to a monotonic increase of Coulomb scattering. Keeping in mind that the Hall factor is $r_{H0} = 1.18$ for the limiting case of exclusive acoustic phonon scattering and $r_{H0} = 1.93$ for the exclusive case of ionized impurity scattering, one expects also an increase of the Hall factor with doping concentration. On the other hand, several groups found experimentally that the Hall factor in n-type silicon increases with donor concentrations but only until a doping level of about 3×10^{18} cm^{-3} [9-13]. With a further increase of doping concentration, the Hall factor decreases and, for doping concentrations exceeding 10^{20} cm^{-3}, the Hall factor approaches the theoretical value of $r_H = 0.87$ [11]. This value of r_H results from $r_{H0} = 1$ in (12) with the anisotropy of the electron mass in silicon taken into consideration. It is important here to point out that a value of $r_{H0} = 1$ indicates that the semiconductor has already become degenerate. Vice versa, the reduction of the Hall factor can be taken as an indication that the importance of scattering at charged impurities decreases at sufficiently high doping levels. The apparent contradiction between the Hall factor found experimentally and predicted by our model can be reconciled by noting that the effect of the screening due to the presence of majority charge carriers has not been accounted for. This kind of screening is expected at very high doping levels at which doping acts not only as a source of Coulomb scattering at ionized impurities but simultaneously reduces it due to the increase in the concentration of majority carriers.

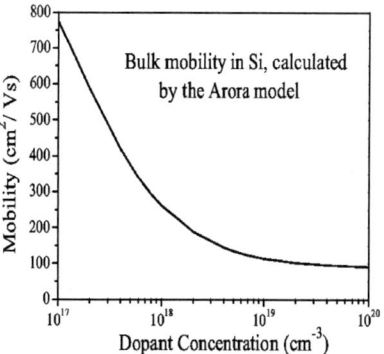

Fig. 1. Bulk mobility in silicon as a function of phosphorus concentration.

To account for the screening of ionized impurities by majority carriers, we have to adapt our model accordingly. The suggested modification for the degenerate case is based on an analogy between screened ionized impurities and neutral impurities. At sufficiently low dopant concentration, the distances between the ionized impurities are sufficiently large so that the free electrons can be considered to exist separately from them. As the impurity concentration and with it the concentration of free electrons increases, electrons will spend on their paths more and more time in the vicinity of ionized impurities. For very high dopant concentrations, impurities will find themselves surrounded by alternating electrons which compensate their charge. In terms of the power-law energy dependence of the relaxation times, the power index s in (2) should change accordingly from the value of -3/2 typical for ionized impurity scattering to the value of 0 that is valid for the scattering on the neutral centers. Evidence from the measurements of Hall factors indicates that this decrease happens at a doping level exceeding about 3×10^{18} cm^{-3} which corresponds closely to the Mott transition [14]. To include the effect in our Hall-factor calculation, based on the reasoning above, we model the predicted dependence of the power index s on doping concentration N_i empirically in the form

$$s = 1.5 \times \left(\frac{1}{1 + \left(\frac{3\times10^{19}}{N_i}\right)^{1.2}} - 1 \right) \qquad (15)$$

A graphical representation of this relationship can be found in Fig. 2. In accordance with (15), the power index s changes from -3/2 to 0 when dopant concentration grows from zero to infinity. At low dopant densities, when scattering at ionized centers is dominant, s has its minimum equal to -3/2. With increasing screening, the importance of Coulomb scattering will decrease, which is reflected in the increase of the constant s. Finally, at high doping levels around 2×10^{20} cm^{-3}, the scattering is similar to that at neutral impurities, which is reflected in the tendency of s towards the value of 0.

Paper T3B1   *PRIME 2013, Villach, Austria*

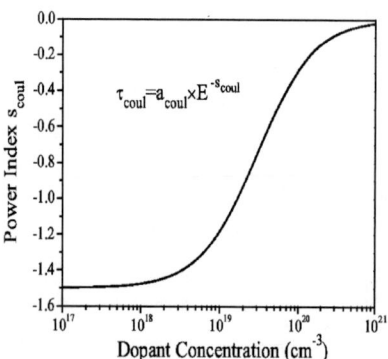

Fig. 2. The dependence of the power index *s* for Coulomb scattering on dopant density including screening.

In accordance with the method presented in the previous section, taking the mobility degradation due to phonon and Coulomb scattering in the phosphorus-doped silicon bulk material and introducing the effect of screening at high doping levels discussed above into account, the Hall factor in the bulk Si was calculated as a function of dopant concentrations. The results of this calculation are shown in Fig. 3 in comparison with the measurements reported in the literature for Si [9-13].

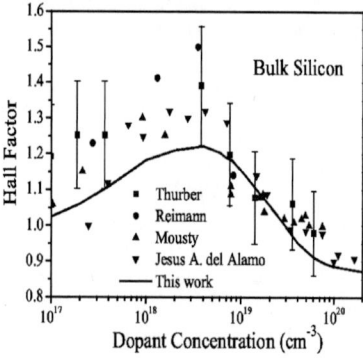

Fig. 3. The electron Hall factor against phosphorus doping concentration in silicon.

In accordance with the predictions of our method, the Hall factor in bulk silicon increases up to a doping level of 4×10^{18} cm^{-3}. When the phosphorus concentration increases further, the screening of the ionized impurities by the majority charge carriers sets in and leads to a continuous reduction of r_H towards a value of 0.87 when the dopant density reaches a value of 2×10^{20} cm^{-3}.

For a correct interpretation of the results, we also present in Fig. 3 typical error bars of these measurements which are about 12% and independent of doping level. It should be noted our calculated values are within the experimental error of nearly all experimental data points. The excellent agreement between calculations and experimental results confirms the validity of our new method for the calculation of Hall factors and allows us to apply this method further to electronic

devices for which the additional complication by the interface imposes conventional approaches.

V. SUMMARY

In this work a new method for the calculation of Hall factors has been presented which is based on its interdependence with mobility via the relaxation (scattering) time. The method was validated by applying it to the dopant dependence in bulk silicon and comparing the calculated values to the experiments available. The simplicity of the method suggested allows to apply it further to the interpretation of Hall-effect measurements on electronic devices like MOSFETs where the presence of the interface to the gate oxide poses an additional severe complication.

ACKNOWLEDGMENT

This work has been carried out in the framework of the project MobiSiC (Mobility engineering for SiC devices) and supported by the Program Inter Carnot Fraunhofer (PICF 2010) by BMBF (Grant 01SF0804) and ANR.

REFERENCES

[1] J. E. Dijkstra and W. Th. Wenckebach, "Calculation of the Hall scattering factor using a Monte Carlo technique", Appl. Phys. Lett., vol. 70, no. 18, pp. 2428-2430, 1997.

[2] G. Ng, D. Vasileska, and D. K. Schroder, "Calculation of the electron Hall mobility and Hall scattering factor in 6H-SiC," J. Appl. Phys., vol. 106, pp. 053719-1-053719-6, 2009.

[3] S.M. Sze, Physics of Semiconductor Devices. New York: Wiley, 1981.

[4] C. Erginsoy, "Neutral impurity scattering in semiconductors," Phys. Rev., vol. 79, pp. 1013-1014, 1950.

[5] H. Iwata, K.M. Itoh, "Donor and acceptor concentration dependence of the electron Hall mobility and the Hall scattering factor in n-type 4H-and 6H-SiC," J. Appl. Phys., vol.89, no. 11, pp. 6228-6234, 2011.

[6] S. S. Li, Semiconductor Physical Electronics. New York: Plenum Press, 1993.

[7] C. Lombardi, S. Manzini, A. Saporito, and M. Vanzi, "A physically based mobility model for numerical simulation of nonplanar devices," IEEE Trans. Computer-Aided Design of Integrated Circuits and Systems, vol. 7, no. 11, pp. 1164-1171, 1988.

[8] Synopsis, Inc.,Sentaurus Device User Guide. Mountain View, CA: 2012.

[9] F. Mousty, P. Ostoja, and L. Passari, "Relationship between resistivity and phosphorus concentration in silicon," J. Appl. Phys., vol. 45, no. 10, pp. 4576-4580, 1974.

[10] J. A. del Alamo and R. M. Swanson, "Measurements of Hall scattering factor in phosphorus-doped silicon," J. Appl. Phys., vol. 57, no. 6, pp. 2314-2317, 1985.

[11] R.S. Popovic, Hall Effect Devices.UK: IOP Publishing Ltd, 2004.

[12] P.L. Reimann, A.K. Walton, "Hall scattering constant in n-type Si;" phys. stat. sol. b, vol. 48, pp. 161-164, 1971.

[13] W.R. Thurber, "A comparison of measurement techniques for determining phosphorus densities in semiconductor Silicon," J. Electron. Mater. vol. 9, no. 3, pp. 551-560, 1980.

[14] P. P. Altermatt, A. Schenk, and G. Heiser, "A simulation model for the density of states and for incomplete ionization in crystalline silicon," J. Appl. Phys., vol. 100, pp.113714-1-113714-10, 2006.

Analysis and Modeling of Minority Carrier Injection in Deep–Trench Based BCD Technologies

Michael Kollmitzer
Infineon Technologies Austria AG
Siemensstrasse 2, 9500 Villach, Austria
Email: michael.kollmitzer@infineon.com

Markus Olbrich
Institute of Microelectronic Systems
Leibniz Universität Hannover, Germany
Email: markus.olbrich@ims.uni-hannover.de

Erich Barke
Institute of Microelectronic Systems
Leibniz Universität Hannover, Germany
Email: eb@ims.uni-hannover.de

Abstract—This paper proposes a methodology for circuit simulation of parasitic effects caused by minority carrier injection into the substrate of a deep-trench based BCD technology. An equivalent circuit is used containing pre-calculated macro models for the injecting diode, the substrate of the chip and the sensitive diode. The macro models are generated by means of TCAD simulations which determine the carrier density distribution in the substrate. The carrier density in the substrate at the sensitive pn-junction is directly related to the parasitic current of the device. The results of the simulations are verified by test chip measurements.

I. Introduction

In modern Smart Power Integrated Circuits (ICs), many different devices are integrated monolithically on the same substrate including digital, analog and high-voltage power transistors in order to reduce overall system cost. A continuous increase in complexity can be seen as additional functions and protective circuits – like open-load or short-to-ground detection, and especially thermal protection – are implemented on the same chip. Reduction of technological feature size decreases the distance between sensitive devices and power stages thus increasing susceptibility to parasitic coupling effects.

As stated in [1], Smart Power technologies can be classified by their isolation techniques, which are self, junction and dielectric isolation. The parasitic coupling effects discussed in this paper are based on a Bipolar-CMOS-DMOS (BCD) technology utilizing dielectric isolation technique which is realized by a poly-silicon filled deep-trench as illustrated in Figure 1. The p-doped poly-silicon within the deep-trench is used as low-ohmic connection to the p-doped substrate. The deep-trench separates the sidewalls of the n-doped pockets of epitaxial wells from each other. The substrate and the epi-well are separated by a highly n-doped buried layer which ensures a low-ohmic drain connection for vertical DMOS structures. During normal operation, the diode D_{sub} in Figure 1 is reverse biased and thus isolates the n-pockets from the common p-substrate. However, there are certain modes of operation during which this diode becomes forward biased.

In an H-bridge configuration driving inductive loads – a typical application implemented in Smart Power ICs – a situation occurs during switching events in which the voltage of the drain terminal becomes negative with respect to the ground potential (see Figure 2). At first, transistors $HS1$ and $LS2$ are turned on driving the load and the switches $LS1$ and $HS2$ are turned off. At some point in time, all four power DMOS are turned off and the body diodes of DMOS $LS1$ and $HS2$ act as flyback diodes. In this case, the voltage at node V_1 becomes negative and both, the body diode D_{body} and the substrate diode D_{sub} (see Figure 1) become forward biased which causes injection of electrons into the common substrate of the chip. These electrons can diffuse over large distances [2] and disturb other devices in their normal operation. As stated in [3], even parasitic currents of as low as 1 μA can influence the behavior of sensitive circuits.

Even today, there are no conventional computer aided design (CAD) tools available which consider the effect of minority carrier injection. A full chip technology CAD (TCAD) simulation approach for substrate current safe designs was investigated by [4] and showed good results but the disadvantage is the required simulation time. Due to the npn-configuration (or pnp for an n-doped substrate) of the parasitic device, modeling approaches based on bipolar transistors are common [1]. However, the arbitrary geometry of the base of the transistor significantly complicates definition of device parameters. [5] introduces an extracted transistor model for circuit simulation but effects caused by electric fields in the substrate cannot be taken into account. With this limitation, verification of active protection measures is not possible. A different approach is introduced by [6]. The parasitic coupling effect is modeled by a network of special resistors which account for majority and minority carriers. These resistors are connected to the junction-isolation diodes of the devices and substrate-coupled currents can be simulated even for high temperatures.

In this paper, pre-calculated macro models are generated by means of TCAD simulation to account for the parasitic coupling effect caused by minority carrier injection into the substrate of an IC. The effect is modeled by an equivalent circuit containing the injector diode, the substrate model and the sensor diode. Each of the models provides information about the carrier densities and the electric potential at its boundaries which, in turn, is used as input to the following model. The entire system describes the relation between injected current and the current at the sensitive node. The simulation results gained by this approach are in good agreement with measurements from a test chip.

978-1-4673-4579-8/13 $31.00 © 2013 IEEE

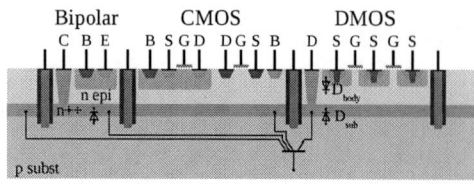

Fig. 1. Simplified cross-section of a dielectric isolation BCD technology

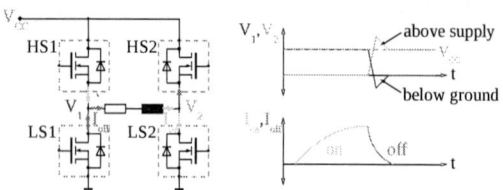

Fig. 2. H-bridge with voltages during a switching event

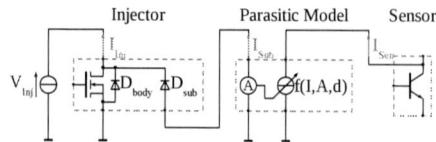

Fig. 3. Circuit simulator model

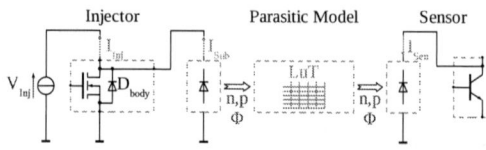

Fig. 4. Macro model setup

II. SIMULATION METHODOLOGY

The proposed methodology addresses the issue of determining the parasitic effects caused by substrate-coupled currents during the design phase of a chip. While the focus at an early stage of the design is to identify all sensitive structures and to gain information about the magnitude of interference, it is necessary to verify that implemented counteractive measures – if any – provide sufficient protection. Distance between the injector and the sensitive device is one of the simplest forms of protection, however, it comes at a high price – area of the chip. Especially in multi-channel power ICs, there can be multiple active injectors and sufficient distance to all of them may not be achieved. Another common layout-based approach to reduce the parasitic effect is to implement guard rings which suppress minority carriers at certain regions (see [1], [2], [7]–[11]). The effectiveness of such structures need to be determined as well.

A. Circuit Simulator Model

A well established method to estimate parasitic substrate currents – also at an early stage in the development phase – is shown in Figure 3. An injected current I_{Inj} is divided between two forward-biased diodes (D_{body} and D_{sub}). The current I_{Sub} is measured at the substrate terminal of the power device and is the input to a current-controlled current source which is connected to the epi-well terminal of the sensitive device and thus generates the parasitic current I_{Sen}. The current source is governed by

$$I_{Sen} = I_{Sub} \cdot A_{Sen} \cdot \alpha \cdot e^{-\beta \cdot d}, \quad (1)$$

where A_{Sen} is the area of the sensitive device, d is the distance between injector and sensor, and the fitting parameters α and β are based on empirical data. Since only limited information about the layout is available at this phase of the design, the application of the model is also limited. It is valid for an evenly distributed network of low-ohmic substrate

connections and epitaxial wells. It cannot handle the influence of intermediary structures between injector and sensor. Thus, a more comprehensive model is required.

B. Pre-calculated Macro Model

The circuit simulator setup for the pre-calculated macro model (Figure 4) is deduced from the previously discussed model but includes significant improvements concerning the calculation of the parasitic currents. Instead of recognizing parasitic bipolar transistors, parasitic substrate currents are modeled by a diode representing the pn-junction of the injector, another diode for the pn-junction of the sensitive epi-well and a pre-calculated macro model containing a look-up table (LuT) to emulate the behavior of the substrate. The methodology is similar to [6], but instead of creating a mesh of special resistors which accounts for the carrier densities, this approach uses a single connection for each injector-sensor pair, thus reducing the overall size of the simulation network. Since only few devices are sensitive to substrate currents in their normal operation, the number of injector-sensor pairs can be reduced even more.

As shown in Figure 4, the substrate diode D_{sub} – which was part of the injector's equivalent circuit – is replaced by a diode model which determines the carrier densities n and p, and the electric potential Φ at the border to the substrate. The look-up table of the parasitic model contains the relationship between the injected carrier densities and the carrier densities at the position of the sensitive device. The key parameters to identify the parasitic coupling behavior are the distance between the injector and the sensor, the epi-well geometries of the two diodes, and any structure in close proximity which influences the carriers in the substrate. The carrier density distribution at the pn-junction of the sensitive device is used to simulate the parasitic current I_{Sen}.

The content of the macro models is determined by TCAD simulations as illustrated in Figure 5. The substrate diodes D_{sub} of the injector and the sensor are confined by deep-trenches on all four sides and therefore can be simulated in 1D under the assumption that the doping profile is laterally

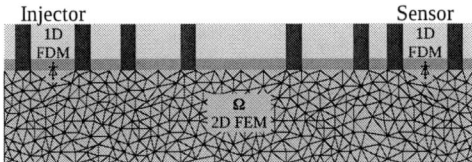

Fig. 5. TCAD simulator setup

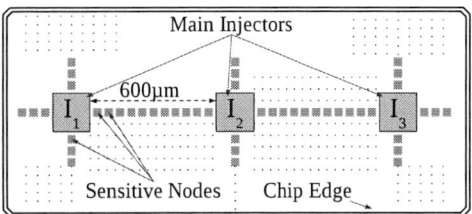

Fig. 6. Sketch of the test chip layout

homogeneous. The solution of the injecting diode provides one of the boundary conditions $\partial\Omega$ for the simulation domain Ω which represents a cross-section of the chip's substrate. Instead of using a 1D finite difference method (FDM) like for D_{sub}, the substrate TCAD simulation is performed by 2D finite element method (FEM). The top edge of Ω is located at the lower edge of the deep-trench structures and is connected to the bottom edge of the injectors, sensor and the substrate connections. The system simulation is set up by defining the boundary conditions $\partial\Omega$ which are obtained from the solutions of the diode models and the deep-trench connections. All three TCAD models – the injector, the sensor and the substrate – are based on drift-diffusion Equations 2 to 6 for semiconductors [12] and are implemented in Matlab® to ensure fast simulation setup for different geometrical configurations. Additionally, the boundary conditions for the carrier densities and the electric potential can be chosen freely to account for active guard ring configurations as well.

$$-\nabla^2\Phi = \frac{q}{\epsilon_{Si}} \cdot \left(p - n + N_D^+ - N_A^-\right) \qquad (2)$$

$$\vec{J}_n = q \cdot \mu_n \cdot n \cdot \vec{E} + q \cdot D_n \cdot \nabla n \qquad (3)$$

$$\vec{J}_p = q \cdot \mu_p \cdot p \cdot \vec{E} - q \cdot D_p \cdot \nabla p \qquad (4)$$

$$\frac{\partial n}{\partial t} - \frac{1}{q} \cdot \nabla \cdot \vec{J}_n = G_n - R_n \qquad (5)$$

$$\frac{\partial p}{\partial t} + \frac{1}{q} \cdot \nabla \cdot \vec{J}_p = G_p - R_p \qquad (6)$$

The parameter extraction for the system simulation setup is performed by means of layout verification software. It recognizes each injector-sensor pair and determines critical structures that influence the parasitic coupling behavior. The key elements include substrate connections, proximity of floating epi-wells, distance between injector and sensor and also the geometry of the epi-wells.

III. TEST CIRCUIT

The layout of the test chip is as close as possible to a productive chip with regard to the arrangement of the epi-wells but still symmetrical enough to gain universally valid results. It is implemented in a deep submicron BCD technology for automotive applications and contains three major injector structures (DMOS devices) which are capable of injecting up to 1 A directly into the substrate. In order to measure the influence of injected substrate currents at different locations the entire area of the chip is covered by separated epi-wells

which act as sensitive nodes, having either a high-ohmic or low-ohmic connection to the buried layer. A simplified illustration of the test chip setup is shown in Figure 6. To study the influence of the device area, sensors have different sizes ranging from 600 μm^2 to 65000 μm^2. Between the epi-wells, a mesh of deep-trenches serves as ground network and substrate connection. All in all, the test chip contains more than 130 pads which are connected to the segments of the ground network and to the sensitive nodes. Additionally, two of the main injectors have configurable epi-wells and substrate connections placed adjacent to them to analyze the effects of layout-based counteractive measures. With this setup, a wide variety of measurements can be performed to reproduce the parasitic coupling behavior. This includes superposition of multiple active injectors, analysis of ground network connectivity influence, floating wells and epi-well electric potential analysis.

IV. MEASUREMENT

As shown in Figure 7, the basic measurement setup consists of a stabilized current source which is connected to the drain terminal of the DMOS. The entire current is injected into the substrate since all other terminals of the power device are kept floating. This situation, however, does not occur on a productive chip since the current I_{Inj} would be divided by the parallel configuration of the diodes D_{sub} and D_{body}. This setup was chosen to reduce the influence of the body diode on the measurements of the parasitic coupling effect. All measurements were performed at room temperature with floating backside (dielectric glue) of the test chip. The results in Figure 8 show the parasitic current I_{Sen} as a function over the distance and the injected current I_{Sub}. The current at the sensitive device is almost linearly dependent on the injected current and shows an exponential decline over the distance. This simple relationship is used for the circuit simulator model shown in Figure 3.

Especially in multi-channel power devices several injectors may be active at the same time. As long as the injectors do not influence each other by a voltage shift of the ground potential, the effect at any sensitive node is equal to the sum of parasitic currents of each injector (see Figure 9).

V. SIMULATIONS

The results shown in Figure 10 are TCAD simulations run in Matlab® for the full system setup (see Figure 5).

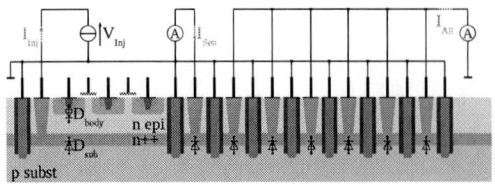

Fig. 7. Basic measurement setup

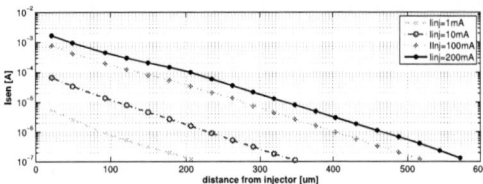

Fig. 8. Measurements of current dependency over distance

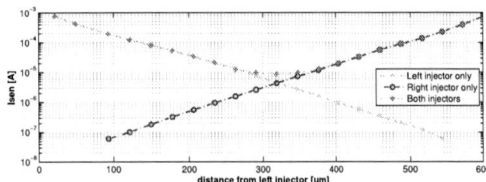

Fig. 9. Measurement of injector superposition

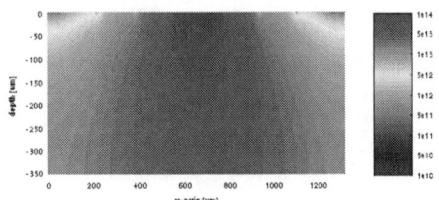

Fig. 10. Carrier density profile during injection

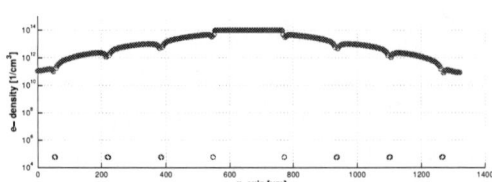

Fig. 11. Carrier density at the top edge of the substrate

TABLE I
PARASITIC CURRENT COMPARISON

Distance $[\mu m]$	$I_{Sen}[\mu A]$	
	Simulation	Measurement
20	108.13	99.39
230	2.25	1.35
400	0.162	0.036

a substrate model for each injector-sensor pair. These devices contain pre-calculated macro models determined by TCAD system simulations. The results agree with the measurements obtained from a test chip. Future work includes improvement of the diode and substrate model regarding high-carrier injection, temperature behavior and effects caused by electric fields.

The carrier density information for the pre-calculated macro models is determined by analyzing data on the top boundary of the simulation domain Ω as illustrated in Figure 11. The simulation shows promising results for low-level injection but there are still problems for high-level injection and electric fields. Additionally, the carrier density gradient along the top boundary does not yet match the physical behavior because the boundary conditions $\partial\Omega$ do not accurately account for extracted carriers at the sensitive nodes. This can be seen by the increasing mismatch for larger distances from the injector (see Table I).

VI. CONCLUSION

In this paper a new methodology is presented to determine the effects of minority carrier injection during circuit simulation. It introduces a combination of two diodes – representing the pn-junction of the injector and the sensitive element – and

REFERENCES

[1] B. Murari, F. Bertotti, G. Vignola, *Smart Power ICs,* 2nd Edition, Springer Verlag, Berlin, 2002.
[2] R.J. Widlar, *Controlling Substrate Currents in Junction-Isolated IC's,* in IEEE Journal of Solid-State Circuits, Vol. 26, No. 8, pp. 1090-1097, August 1991.
[3] W. Horn, H. Zitta, *A Robust Smart Power Bandgap Reference Circuit for Use in an Automotive Environment,* in IEEE Journal of Solid-State Circuits, Vol. 37, No. 7, pp. 949-952, July 2002.
[4] M. Schenkel, *Substrate Current Effects in Smart Power ICs,* PhD Thesis, ETH Zürich, 2002, ISBN 3-89649-848-7.
[5] J. Oehmen, M. Olbrich, L. Hedrich, E. Barke, *Modeling Lateral Parasitic Transistors in Smart Power ICs,* IEEE Trans. on Device and Materials Reliability, Vol. 6, No. 3, September 2006, pp. 408-420.
[6] F. Lo Conte, J.-M. Sallese, M. Kayal, *Circuit Level Modeling Methodology of Parasitic Substrate Current Injection from a High-Voltage H-bridge at High Temperature,* in IEEE Transactions on Power Electronics, Vol.26, No.10, pp. 2788-2793, Oct. 2011.
[7] O. Gonnard, G. Charitat, P. Lance, M. Suquet, M. Bafleur, J.P. Laine, A. Peyre-Lavigne, *Multi-ring Active Analogic Protection for Minority Carrier Injection Suppression in Smart Power Technology,* in Proc. of the Int. Symposium on Power Semiconductors and Devices, pp. 351-354, 2001.
[8] O. Gonnard, G. Charitat, P. Lance, E. Stefanov, M. Suquet, M. Bafleur, N. Mauran, A. Peyre-Lavigne, *Substrate Current Protection in Smart Power IC's,* in Proc. of the Int. Symposium on Power Semiconductors and Devices, pp. 169-172, 2000.
[9] G. de Cremoux, E. Dubois, S. Bardy, J. Lebailly, *Simulations and Measurements of Cross-Talk Phenomena in BiCMOS Technology for Hard Disk Drives,* in IEEE Int. Electron Devices Meeting, pp. 481-484, 1996.
[10] R. Peppiette, *A New Protection Technique for Ground Recirculation Parasitics in Monolithic Power I.C.s,* Sanken Technical Report, Vol. 26, No. 1, 1994, pp. 91-97.
[11] M. Feldtkeller, *Integrierte Schaltungsanordnung zur Vermeidung von Minoritätsträgerinjektion,* Patentschrift DE 4209523C1, Deutsches Patentamt.
[12] A. Jüngel, *Transport Equations for Semiconductors,* Lecture Notes in Physics, Vol. 773, 2009.

S-parameter Models for transient simulation in Verilog-A

Tobias Maier
Robert Bosch GmbH
Reutlingen, Germany
Email: tobias.maier2@de.bosch.com

Dirk Droste
Robert Bosch GmbH
Reutlingen, Germany
Email: dirk.droste@de.bosch.com

Michael Siegel
IMS, KIT
Karlsruhe, Germany
Email: michael.siegel@kit.edu

Abstract—**This work presents a new method to model analog modules for transient simulations with their S-parameters and DC behavior. The model can be used for verification of analog circuits in a system simulation where the modeling is necessary to speed up transient simulations in analog domain for systems like Sigma Delta modulators (SDM) or DC/DC converters. The model includes circuitry to separate DC from AC in transient simulations to test various distortions on system level, like the power supply rejection. It is functional and shows high speed up for certain simulations.**

I. Introduction

In recent years switched-capacitor (sc) analog integrated circuits became larger with respect to functionality and complexity. Therefore, it becomes more important to verify these circuits by exhaustive simulations before tape-out. A first system verification is done by abstract models which contain only the most important block features. These models are often created in discrete time step environments, e.g. in matlab or in a continuous simulation with a hardware description language like Verilog [1][2] and represent good behavior for the signal paths.

After implementation, the transistor circuit is verified on its own and also within the system. Therefore, transient and AC analysis must be executed. The AC analysis is interesting in two points, first the signal path and second the error paths as needed for the Power Supply Rejection (PSR) analysis. If an AC analysis is not possible because of changing operation points, a long transient analysis must be done to see error effects, especially if a detailed fast Fourier transform (FFT) is intended to be applied. This simulation is also necessary for multiple distortion frequencies to estimate the frequency response of the system to the distortion. To verify sc circuits, a periodic steady state simulation helps to analyze AC behavior if the circuit only needs to settle and has periodic states afterwards.

For a load change at a DC-DC sc regulator a transient simulation is needed as it is needed for Sigma Delta Modulators (SDM), which can only have periodic states with DC input and hence would create unwanted tones in the output spectrum. Even the periodic time depends on the DC input level of the SDM and can be very long. The transient simulation of systems, like sc regulators or SDMs on transistor level, takes huge real time for only very small simulation time. To verify the Noise Transfer Function (NTF) of an SDM, it is necessary to create a frequency spectrum of the bitstream output. If this is done by simple models, only the mathematical calculated NTF would be given. However, to see the non-ideal behavior of the elements within a circuit on the signal, transistor level simulations are needed. Especially, if power supply distortions are taken into concern, models often fail to represent them.

To overcome the problem of fast but inaccurate models, many ideas were elaborated. One is to use event driven models [3] where an event is created when the input changes by a certain value. This avoids a calculation at every time step and so decreases the simulation time. This is also stated in [4]. The paper also emphasizes, that it is important to create pin accurate models that can be exchanged by the transistor level blocks. This is also important in [5], where a good approach for verifying is presented, but the analog blocks are already very abstract, e.g. a SDM would be described in one block with its functionality and not with the blocks contained in the system (e.g. amplifiers and switched caps). For a functional verification of a circuit this model might be enough but for a distortion analysis it is not sufficient.

There also exists some work to case dependent improvements in simulation time [6] where the digital domain is connected to analog or vice versa. Transfer gates or transfer transistors are replaced by ideal components here and e.g. a digital to analog converter (DAC) is created by ideal voltage or current sources. This approach is useful for functional verification but not for a distortion analysis.

Another approach to improve simulation speed is to reduce the model order [7], i.e. to reduce the size of the Matrix which must be solved at every time step. This can be a very useful approach, but attention must be paid to not remove too much information from the model. Also the computation time for the algorithm must be taken into account as with huge matrices it can increase significantly.

The idea presented in this paper also wants to implement pin compatible models with a DC, AC and transient behavior. The data for the models is gained by simulation of each block alone and on system level. The data consist of the DC behavior of the model and the AC behavior represented by the S-parameters data. This data is post-processed and then transferred into a model that has decreased transient simulation time. After a short recap of S-parameters in Section II, Section III shows how the data was gained and post-processed while Section IV shows how the model was implemented. In Section V the results are presented and the last section shows an outlook on further improvements.

II. Review of Scattering Parameters

S-parameters describe the behavior of a linear electrical network. In contrast to other parameters like the Z- or Y-parameters, the S-parameters don't need open or short conditions to be measured. A matched load is used instead, which often is $50\,\Omega$. The scattering variables a_i and b_i of all ports of an n-port are calculated in reference to the matched load Z_{Bi} or the internal resistance of the source at port i, Z_i [8].

Figure 1 shows the general circuitry at a port for calculating the waves a_i and b_i. The incident power wave a_i and reflected

Paper T3B3

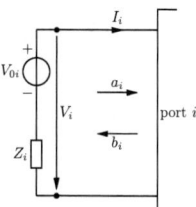

Fig. 1.   Current, Voltages and Waves at port i [8]

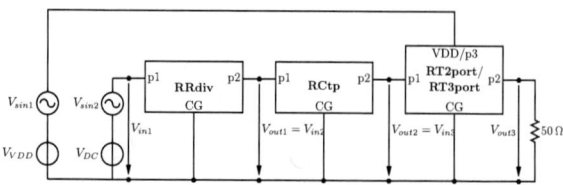

Fig. 2.   Testsystem

power wave b_i are shown in (1) and (2), respectively. From (1) it becomes clear that if no source is switched on at port i then $a_i = 0$.

$$a_i = \frac{V_i + Z_i I_i}{2\sqrt{Z_i}} = \frac{V_{0i}}{2\sqrt{Z_i}} \tag{1}$$

$$b_i = \frac{V_i - Z_i I_i}{2\sqrt{Z_i}} \tag{2}$$

The relation between a_i and b_j of port i and j of a n-port is the Sji parameter. (3) shows that the calculation of b_j is the superposition of the $S_{ij}a_i$ products. The equations also reveal the meaning of S-parameters, which show how a power wave inserted at port j is transmitted to port i, or reflected to itself in case of the S_{ii} parameters.

$$b_j = \sum_{i=1}^{n} S_{ij}a_i \forall j \in \{1..n\} \tag{3}$$

According to (3) the S-parameters S_{ij} are calculated by capturing all waves $b_j, j \in \{1 \ldots n\}$ while only one wave a_i is nonzero, for all a_i waves. The frequency data of the S-Matrix most often is saved in the touchstone format. If two ports are connected together, the reflected wave of the one port is the transmitted wave of the other and vice versa [8].

III.   Data capturing and post-processing

A. Capture DC data

For capturing the DC data of all blocks in a system (e.g. block RRdiv and RCtp in Fig.2), the system is driven to its DC operation point. This can be done by a DC analysis or by a transient analysis whose final value is the DC operation point (dcOp)[1]. Then all the input and output currents and voltages (see arrows in Fig.2) of the blocks are saved for later gain of S-parameters' data and for the creation of the Verilog-A model.

B. Capture SP data

To separate the DC bias from the AC stimulation during S-parameters simulation, a BiasTee (BT) is used to bias the block with the DC values already simulated (see III-A). As a node can be either biased by a voltage source or a current source for correct setup, the right choice for the source must be made. The input current is thereby defined by the input voltage and by the output bias (current or voltage) as in the dcOp case only resistivity behavior counts for voltage and current calculation. This why a voltage source at the input and a voltage source at the output also determine the input current. Thus, the input current and voltage as well as the output voltage are defined. By the output voltage the output current is set and so all DC properties are well specified and

---
[1]Of course it is also possible to take any transient step as dcOp

biasing is finished. This biasing rule not only counts for 2-port blocks, but also for blocks with 3, 4, or even more ports. In case where RCtp is the device under test, the corresponding testbench bias voltages V_{p1DC} and V_{p2DC} would be V_{in2} and V_{out2} of Fig.2, respectively.

The S-data is then gained by a SP analysis with spectre and a port-element at each AC pin of the BTs. Afterwards, the data is saved for later post-processing.

C. Post-processing

The S-parameters' data gained from SP simulation was post processed and all the single S-functions (s11, s12,...) were analyzed for poles and zeros. The description of the S-parameters by gain, poles and zeros is essential for the model creation, as the S-parameters can be described by a Laplace function in the model instead of a vector of frequency-amplitude tuples as it is the case in usual nport elements. The Laplace function in Verilog-A is laplace_zp($expr$, ζ,ρ), where $expr$ is the Signal, ζ is the zeros vector and ρ is the vector containing the poles. The resulting transfer function is given in (4). Fitting the data to that format was done by some algorithms in Matlab.

$$H(s) = \frac{\prod_{k=0}^{M-1} \left(1 - \frac{s}{\zeta_k^r + j\zeta_k^i}\right)}{\prod_{k=0}^{N-1} \left(1 - \frac{s}{\rho_k^r + j\rho_k^i}\right)} \tag{4}$$

IV.   Model implementation

The implementation of the model is demonstrated in the following on the module RCtp, which represents an RC lowpass. Parts of the implementation are listed in listing 1. As the model is implemented in Verilog-A, the module (line 1) has the same pins as the schematic with the same direction (line 2) and the property electrical (line 3).

A. DC part implementation

The splitting from transient signal into DC and AC part, where the DC part is the static part and the AC part is the varying part, is done by the circuit shown in Fig.3. The measured DC currents and voltages at the inputs and outputs of the device under test are created within the model by DC voltage and current sources. The splitting is necessary as the AC functions would multiply the DC values with the DC gain of the AC function and so wrong DC voltages would be created. E.g. an operational amplifier with an DC gain of some tens dB would also multiply the DC signal with that gain and show a magnified voltage and not the real DC voltage of the amplifier at the output. To implement the splitting, a new AC node must be created within the model for each port (line 4). The I_{dc1} current source in Fig.3 is implemented by a current definition from the port to ground (line 9), where the value of I_{dc1} is implemented as a parameter. The V_{dc} source is

978-1-4673-4579-8/13 $31.00 © 2013 IEEE

PRIME 2013, Villach, Austria   *Session T3B – Modeling*

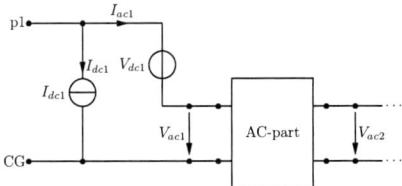

Fig. 3.   Splitting circuit

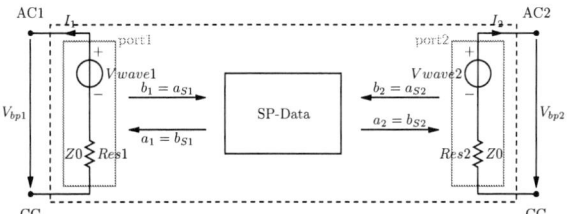

Fig. 4.   AC part

defined by a voltage from the port's positive node to the new node AC1 (line 10). The splitting must be made for every port of the model.

For block RCtp in Fig.2, the voltage V_{dc1} would be V_{in2} and the current I_{dc1} would be equal to I_{in2}.

B. AC / SP part implementation

A port exists of a voltage source and an impedance, e.g. port1[2] in Fig.4. In the Verilog-A model the voltage over this port (V_{bp1}) can be obtained by taking the voltage (line 12, 1st summand) of a defined branch from node AC1 to CG (line 5). For ease of implementation and readability of code, a new node p1Res (line 4) is inserted between the voltage source and the resistor of Fig.4. The voltage source V_{wave1} is defined by setting up a voltage (line 13-17) between AC1 and the new node (line 6). For Res1 in Fig.4 also a branch is defined (line 7). The Voltage over the resistor is set up by Kirchhoff's law (line 11), where the current I(Res1) has the opposite direction of the current I_1 in Fig.4.

The wave b_1 going out from port1 is according to (2) defined by $b_1 = \frac{V_{bp1} - Z0\,I_1}{2\sqrt{Z0}}$. $V_{bp1} - Z0\,I_1$ is calculated with a temporary voltage (line 4,11). The changed sign of the second summand (line 11) comes from the previously mentioned opposite current direction of I(Res1) and I_1.

As port1 and the SP-Data block in Fig.4 are connected directly, the outgoing wave of port1 is the incoming wave a_{S1} of SP-Data's port1 (not shown in Fig.4).The same counts for the outgoing wave b_{S1} of SP-Data's port1, which is also the incoming wave of port1 a_1. If we now look at the definition (1) of the incoming wave at port1, we see that $V_1 + Z0\,I_1 = V_{01}$ and $V_{01} = a_1\,2\sqrt{Z0}$. Substituting a_1 by the outgoing wave of the SP-Data block to port1 ($\sum_{j=1}^{n} S_{1j}\,b_j$) and inserting b_j we get (5), which corresponds to $Vwave1$ in Fig.4. We also see that $2\sqrt{Z0}$ cancels out of $Vwave1$, which makes calculation much easier.

$$V_{0i} = \sum_{j=1}^{n} S_{ij}\,(V_j - Z0\,I_j) \qquad (5)$$

The calculation of $Vwave1$ is done by the laplace_zp [9] implementation of the frequency behavior of the S-parameters (line 13,16). As described in III-C the measured data is post-processed and then inserted by the poles, zeros and gain representation, where the gain is represented by k_{ij}. j is the number of the incoming port, i the one of the outgoing port.

As there can be more than one zero or pole and each zero/pole exists of an imaginary and real part, the zeros and poles are numbered by the postfixes $n_z r$, $n_z i$, $n_p r$ and $n_p i$, where n_z is the number of the zero of a specific S_{ij} and n_p

---

[2]for ease of reading port1 is regarded in the following, but all calculations and definitions also count for port2 in Fig.4 and of course for all ports in a general n-port

the number of the pole, respectively. r or i specify the real or imaginary part of poles/zeros. So the parameter name used for laplace calculation is in the form of $xi_{ij}\_n_z r$ and $xi_{ij}\_n_z i$ for the zeros of S_{ij} (e.g. parameter xi11_1r = ..; for S_{11} first zero's real part) and $rho_{ij}\_n_p r$ and $rho_{ij}\_n_p i$ for the poles. The sum is broken up into multiple commands for automated implementation and better readability (12-16).

C. Connection of DC and AC part

The connection of DC and AC part is done in the model by the branch bp1. The advantage of the implementation here is that the model reacts to both, a current and a voltage change. In Fig.3 the voltage over the branch is called V_{ac1}, but in Fig.4 the same voltage has the more adequate name V_{bp1}.

Listing 1.   Verilog-A extract of RCtp

```
    module RCtp(CG, p1, p2);
 2    inout CG, p1, p2;
      electrical CG, p1, p2;
 4    electrical AC1, p1Res, Vtemp1;
      branch (AC1,CG) bp1;
 6    branch (AC1,p1Res) Vwave1;
      branch (p1Res,CG) Res1;
 8    ...
      I(p1,CG) <+ I1_dc;
10    V(p1,AC1) <+ V1_dc;
      V(Res1) <+ Z0*I(Res1);
12    V(Vtemp1) <+ V(bp1)+ V(Res1);
      V(Vwave1) <+ k11 * laplace_zp(V(Vtemp1),
14        {xi11_1r,xi11_1i},
          {rho11_1r,rho11_1i}); // S11
16    V(Vwave1) <+ k12 * laplace_zp(V(Vtemp2),,
          {rho11_1r,rho11_1i}); // S12
```

V. RESULTS

Multiple test circuits were implemented to verify the models functionality step by step. First, a resistive voltage devider, second, a RC lowpass and then an active structure with 2 transistors as 2-port and 3-port. Also the testsystem (Fig.2) was used in the 2- and 3-port configuration for verification. At last, a real circuit of a bandgap with 4 major ports was implemented and tested.

Corresponding to the described method of getting DC data in the previous sections, the DC data of the testsystem was received by dcOp simulation. The SP data was simulated for each subcircuit with a corresponding SP-testbench. Each model's S-parameters show a very good fit to the schematic one's and also the transient simulations shows very good results. The testsystem's simulation results with models matched those of the the schematic one's very well.

For the bandgap again the DC and S-parameters data of the subsystem were captured, post-processed and then implemented. The match of the S-parameters again was very

---

978-1-4673-4579-8/13 $31.00 © 2013 IEEE   251

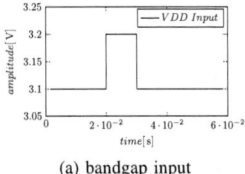
(a) bandgap input

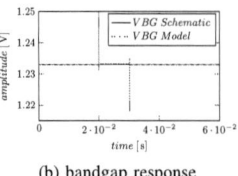
(b) bandgap response

Fig. 5. Simulation results of bandgap

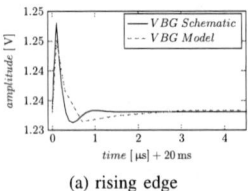
(a) rising edge

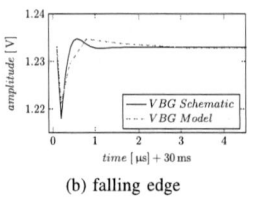
(b) falling edge

Fig. 6. Detailed view of spike on edges of VDD

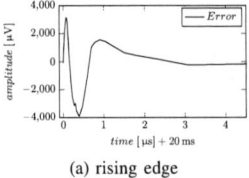
(a) rising edge

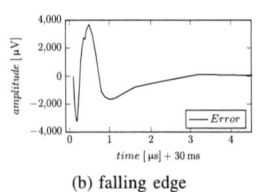
(b) falling edge

Fig. 7. Error view of spike on edges of VDD

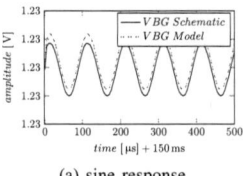
(a) sine response

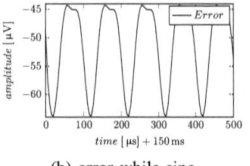
(b) error while sine

Fig. 8. VBG output response to a 100mV sine at VDD

TABLE I.   SCHEMATIC TO MODEL COMPARISON OF TRANSIENT
SIMULATION OF BANDGAP

	model	schematic
time in sec	2.8202	11.6787
steps solved	10108	10192

needed to create the model. As the system testbench exists, only little time has to be spent to capture DC values. The creation of the S-parameters testbench is a straight forward procedure as well as the capturing of the S-parameters. The post-processing is done mostly automatically. The check of S-parameters and model by simulation is again straight forward. The time needed for model creation is depending on the model size and occurring difficulties in implementation. But this time is gained back when multiple long-time simulations have to be run, where million of steps have to be calculated.

VI.   CONCLUSION

In this paper, the new method shows excellent results with transient and AC/SP simulations. The reduction in simulation time is most effective with huge circuits as here the overhead introduced through the DC/AC splitting and the S-parameters block is compensated by the high reduction of elements and nodes. As the purpose of this models differs from that of others, like described in the introduction, a performance comparison between them and the one proposed in this work is not relevant. Further investigations must be made for usage with circuits like amplifiers which often change their operation point as the S-parameters are changed depending on the operation point.

REFERENCES

[1] G. Coram, "How to (and how not to) write a compact model in verilog-a," in *Behavioral Modeling and Simulation Conference, 2004. BMAS 2004. Proceedings of the 2004 IEEE International*, 2004, pp. 97–106.

[2] J. Chen, "A modeling methodology for verifying functionality of a wireless chip," in *Behavioral Modeling and Simulation Workshop, 2009. BMAS 2009. IEEE*, 2009, pp. 96–101.

[3] S. Joeres, H.-W. Groh, and S. Heinen, "Event driven analog modeling of rf frontends," in *Behavioral Modeling and Simulation Workshop, 2007. BMAS 2007. IEEE International*, 2007, pp. 46–51.

[4] Y. Wang, S. Joeres, R. Wunderlich, and S. Heinen, "Issues on view switching for rf soc verification," in *Behavioral Modeling and Simulation Workshop, 2008. BMAS 2008. IEEE International*, Sept. 2008, pp. 72–77.

[5] K. Kundert and H. Chang, "Verifying all of an soc-analog circuitry included," *Solid-State Circuits Magazine, IEEE*, vol. 1, no. 4, pp. 26–32, 2009.

[6] T. Sheffler, "Design of a switch-level analog model for verilog," in *Behavioral Modeling and Simulation Workshop, 2008. BMAS 2008. IEEE International*, 2008, pp. 118–123.

[7] Y. Shi, L. He, and C. J. R. Shi, "Scalable symbolic model order reduction," in *Behavioral Modeling and Simulation Workshop, 2008. BMAS 2008. IEEE International*, 2008, pp. 112–117.

[8] D.-I. W. Klein, *Mehrtortheorie (3. Auflage)*. Akademie-Verlag Berlin, 1976.

[9] I. Cadence Design Systems, *Cadence Verilog-A Language Reference*, product version 7.2 ed., Cadence Design Systems, Inc., Cadence Design Systems, Inc. (Cadence), 2655 Seely Ave., San Jose, CA 95134, USA., December 2009.

good and therefore only the transient results, which are of more interest regarding the topic of this work, are shown.

In Fig.5a the supply voltage input (VDD) of the bandgap is shown. It has a $100\,\text{mV}$ jump up at $20\,\text{ms}$ and down at $30\,\text{mV}$. From $150\,\text{ms}$ it also has a sine distortion with $100\,\text{mV}$ amplitude with a frequency of $10\,\text{kHz}$ (not shown in Fig.5a). The response of the bandgap to that input voltage is shown in Fig.5b. There are spikes at the VDD jumps. Fig.6a and Fig.6b show a detailed view of the VBG spikes and it can further be seen that schematic and model match very well. The DC error is $49.5\,\mu\text{V}$ at normal DC input and $61.8\,\mu\text{V}$ during the increased jump voltage, while the maximum dynamic error is about $4\,\text{mV}$ at both jumps. The jumps are still very well created by the model. Also the damped amplitude of the sine distortion matches very well, as illustrated in Fig.8a. The peak to peak error of the amplitude of the damped sine is $19.5\,\mu\text{V}$.

The models were developed to decrease simulation time while only having a very small error. So, next to the already shown error discussion, the simulation time should be discussed. As the simulation time depends on the number of equations which must be solved at every step, the number of steps which were solved is also listed in the results Table I. From that table it can clearly be seen, that the improvement in time, with the same number of calculated steps, even with small analog circuits like a bandgap. For circuits like Rt3port, the time spent is slightly higher for the model as there the adoption overhead increases the number of nodes which must be solved. However, for real implemented circuits, the model works, as shown with the bandgap circuit.

The improvement in simulation time is bought by the time,

Extended model for platinum diffusion in silicon

Elie Badr[1,*], Peter Pichler[1,2]
[1]Fraunhofer IISB, Erlangen, Germany
[2]Chair of Electron Devices, University of Erlangen-Nuremberg, Erlangen, Germany
[*]Elie.Badr@iisb.fraunhofer.de

Gerhard Schmidt
Infineon Technologies Austria AG
Villach, Austria
Gerhard.Schmidt@infineon.com

Abstract—Close to the wafer surfaces, our platinum diffusion experiments were found to be at gross discrepancy with the predictions of well-established diffusion models. These differences are associated with the ramping-down of the temperature at the end of the diffusion processes. To obtain a consistent model able to explain the experiments reported previously in the literature together with our experiments, energy barriers had to be included for the various reactions rate. For the Frank-Turnbull, kick-out and bulk recombination reactions, barrier heights of 0.55, 0.16, and 0.57 eV were determined, respectively. The newly established model is able to reproduce platinum diffusion for a considerably wider range of experimental conditions than models before.

Keywords: platinum diffusion, silicon, DLTS.

I. INTRODUCTION

Platinum diffusion into silicon is used in industry for life-time control in silicon power devices. The acceptor level of substitutional platinum located at $Ec - 0.243$ eV enables minority-carrier lifetime reduction at low leakage currents [1]. For a thorough understanding and an optimization of the processing conditions, an accurate description of the profiles after a series of thermal treatments by numerical simulation is required. A well-established model for platinum diffusion at low temperatures has been presented by Jacob et al. [2]. However, the purpose of their work was to characterize vacancy profiles in silicon and, in their experiments, the samples were rapidly unloaded after annealing. Ramping rates as they are required in industrial processes were not considered. When applying the model to industrial processes, we found that the model of Jacob et al. is able to describe the platinum profiles in the bulk of our samples well. Within about 10 microns from the surface, however, gross differences between simulations and experiments were observed. To explain them quantitatively, we had to take energy barriers for both platinum-point-defects reactions and bulk recombination into account. In this article we describe a series of dedicated experiments in which the ramping rate after thermal annealing was varied to allow a determination of the energy barriers.

II. EXPERIMENTAL PROCEDURE

Our experiments were performed with n-type (100)-oriented phosphorus-doped dislocation-free FZ-grown silicon substrates. The wafers used were polished on both sides and had thicknesses of either 390 or 508 μm and resistivities of 2.1

and 5 Ω cm, respectively. After an oxidation step to eliminate grown-in vacancies, platinum was evaporated on one side of the wafers and driven in at 832 °C for two hours. At the end of this process, the wafers were cooled down with rates of 4.6, 2.3, 1.2 and 0.6 K/min. The depth profiles of substitutional platinum in the bulk were obtained from deep level transient spectroscopy (DLTS) measurements on beveled samples. For beveling, the wafers were cut into pieces of approximately 1 cm², polished at an oblique angle and chemically etched (2 parts 50% HF, 1 part 50% HNO₃ and 2 parts 70% CH₃COOH). Close to the surface, depth profiling could be performed by changing the DLTS pulse and reverse voltages. For this, the samples used were only HF dipped to remove the oxide layers. Prior to the evaporation of gold to form the Schottky contacts required for DLTS, the samples were annealed at 450 °C for one hour in order to dissociate the Pt-H complexes that were formed by the wet chemical etching treatments [3]. The ohmic contacts at the back-side were made by scratching an Al-Ga alloy onto the respective side of the samples.

III. THEORETICAL MODEL

Platinum diffuses in silicon via the Frank-Turnbull and kick-out mechanisms. These mechanisms postulate the quasi-chemical reactions

$$Pt_i + V \underset{K_{FT\leftarrow}}{\overset{K_{FT\rightarrow}}{\rightleftharpoons}} Pt_S \qquad (1)$$

$$Pt_i \underset{K_{KO\leftarrow}}{\overset{K_{KO\rightarrow}}{\rightleftharpoons}} Pt_S + I \qquad (2)$$

between interstitial platinum Pt_i, substitutional platinum Pt_s, self-interstitials I, and vacancies V. In addition, bulk recombination

$$I + V \underset{K_{IV\leftarrow}}{\overset{K_{IV\rightarrow}}{\rightleftharpoons}} 0 \qquad (3)$$

has to be taken into account with O standing for the undisturbed lattice. $K_{FT\rightarrow}$, $K_{FT\leftarrow}$, $K_{KO\rightarrow}$, $K_{KO\leftarrow}$, $K_{IV\rightarrow}$, and $K_{IV\leftarrow}$ are the reaction constants in forward and backward direction,

978-1-4673-4579-8/13 $31.00 © 2013 IEEE

respectively. From these reactions, a system of partial differential equations for the substitutional platinum, interstitial platinum, self-interstitials and vacancies can be deduced directly [4]:

$$\frac{\partial C_{Pts}}{\partial t} = K_{FT\to}C_{Pti}C_V - K_{FT\gets}C_{Pts} + K_{KO\to}C_{Pti} \\ - K_{KO\gets}C_{Pts}C_I \tag{4}$$

$$\frac{\partial C_{Pti}}{\partial t} = \text{div}(D_{Pti}\,\text{grad}\,C_{Pti}) - K_{FT\to}C_{Pti}C_V + K_{FT\gets}C_{Pts} \\ - K_{KO\to}C_{Pti} + K_{KO\gets}C_{Pts}C_I \tag{5}$$

$$\frac{\partial C_I}{\partial t} = \text{div}(D_I\,\text{grad}\,C_I) + K_{KO\to}C_{Pti} - K_{KO\gets}C_{Pts}C_I \\ - K_{IV\to}C_I C_V + K_{IV\gets} \tag{6}$$

$$\frac{\partial C_V}{\partial t} = \text{div}(D_V\,\text{grad}\,C_V) - K_{FT\to}C_{Pti}C_V + K_{FT\gets}C_{Pts} \\ - K_{IV\to}C_I C_V + K_{IV\gets} \tag{7}$$

Therein, the symbols C and D stand for concentrations and diffusion coefficients, respectively. Substitutional platinum is considered imobile. Assuming the reactions to be diffusion-limited, the reaction constants can be expressed as

$$K_{FT\to} = 4.\pi.a_{FT}.(D_{Pt_i} + D_V).\exp\left(\frac{-E_{FT}}{kT}\right) \tag{8}$$

$$K_{KO\to} = 4.\pi.a_{KO}.D_{Pt_i}.\exp\left(\frac{-E_{KO}}{kT}\right) \tag{9}$$

$$K_{IV\to} = 4.\pi.a_{IV}.(D_I + D_V).\exp\left(\frac{-E_{IV}}{kT}\right) \tag{10}$$

with a_{FT}, a_{KO} and a_{IV} denoting the respective capture radii. E_{FT}, E_{KO} and E_{IV} represent the barrier energies of the reactions. A similar approach has been used by Ghaderi et al. [5] to simulate the diffusion of gold in silicon.

In thermal equilibrium, the backward reaction constant can be expressed by the forward reaction constant and the equilibrium concentrations (indicated by a superscript eq).

$$K_{FT\gets} = K_{FT\to}\frac{C_{Pti}^{eq}C_V^{eq}}{C_{Pts}^{eq}} \tag{11}$$

$$K_{KO\gets} = K_{KO\to}\frac{C_{Pti}^{eq}}{C_{Pts}^{eq}C_I^{eq}} \tag{12}$$

$$K_{IV\gets} = K_{IV\to}C_I^{eq}C_V^{eq} \tag{13}$$

The coupled differential equations were integrated using the general purpose solver PROMIS [6]. As boundary conditions

for self-interstitials and vacancies at the wafer surfaces, Dirichlet boundary conditions with the respective equilibrium concentrations were assumed. For interstitial platinum, Dirichlet boundary conditions were used below the PtSi layer, and Neumann boundary condition at the other side.

IV. RESULTS AND DISCUSSIONS

For calibration of the model parameters, we took the experiments with the longest diffusion times from the work of Jacob et al. [2] as well as our own experimental results. The model parameters were then modified by an optimization program in order to minimize the differences between the experimental results and the simulations. The parameters obtained are given in Table 1.

Table 1: Parameters for simulation of platinum diffusion.

Parameter	Value
a_{FT}	5.11 nm
a_{KO}	4.83 nm
a_{IV}	5.58 nm
E_{FT}	0.55 eV
E_{KO}	0.16 eV
E_{IV}	0.57 eV
C_{Pts}^{eq}	$9.96\times10^{29} \times \exp(-3.32\ eV\,/\,kT)\ cm^{-3}$
C_{Pti}^{eq}	$8.69\times10^{19} \times \exp(-1.87\ eV\,/\,kT)\ cm^{-3}$
C_I^{eq}	$3.8\times10^{30} \times \exp(-4.62\ eV\,/\,kT)\ cm^{-3}$
C_V^{eq}	$1.66\times10^{21} \times \exp(-1.99\ eV\,/\,K.T)\ cm^{-3}$
D_{Pti}	$21.2 \times \exp(-1.1\ eV\,/\,kT)\ cm^2/s$
D_I	$1.5\times10^{-4} \times \exp(-0.42\ eV\,/\,kT)\ cm^2/s$
D_V	$13.15 \times \exp(-2.03\ eV\,/\,kT)\ cm^2/s$

Capture radii are usually assumed to be on the order of the distance between two substitutional silicon atoms (0.25 nm). The somewhat larger values obtained are largely consistent with this estimate. The values of the energies barrier found are of the same order as those reported by Ghaderi et al. [5] from short-time diffusion of gold. Since metal diffusion is known to depend largely on the products of diffusion coefficients and equilibrium concentrations, only limited significance can be given to the individual prefactors and energy barriers. Other parameters will be discussed in the context of literature knowledge below.

Figure 1 shows the depth profile of substitutional platinum resulting from diffusion at 832 °C with a cooling rate of 4.6 K/min. The simulation based on the work of Jacob et al. shows a very steep decrease of platinum at the surfaces which is not observed experimentally. On the contrary, the simulation based on our model fits the experimental data throughout the complete depth of the sample.

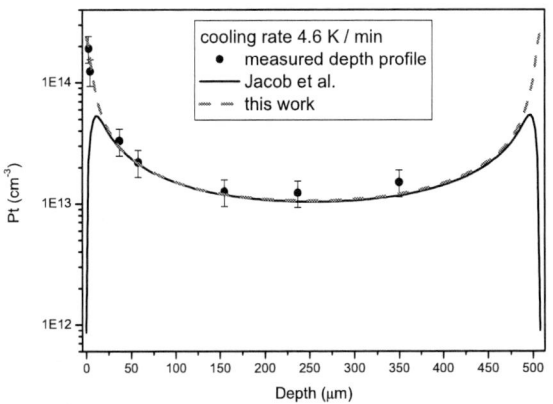

Figure 1: Depth profile of substitutional Pt obtained from DLTS measurements (scatter points). The solid line represents the simulation of the profile based on the model of Jacob et al., the dashed line is based on this work.

In Figure 2, depth profiles measured close to the surface of samples processed with different cooling rates are compared to simulations. During ramping-down, the equilibrium concentration of platinum reduces markedly the lower the temperature. Thus, since the formation of a PtSi phase in the silicon is unlikely because of kinetic reasons, out-diffusion of platinum to the surface is expected. However, the model of Jacob et al., which assumes diffusion-limited reactions, overestimates this effect clearly. Moreover, the profiles shown are the result of stopping the ramping at a temperature of 700 °C. Continuing the ramping to lower temperatures would lead to still lower platinum concentrations at the surfaces.

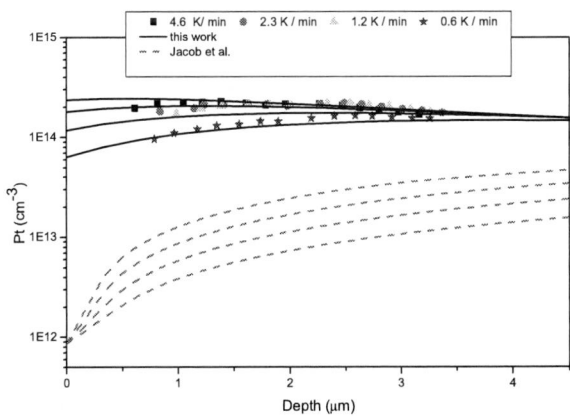

Figure 2: Substitutional Pt depth profiles from the samples' surface for 4.6, 2.3, 1.2 and 0.6 K/min cooling rate with the comparison between simulations based on this work and the work of Jacob et al.

Since the diffusion kinetics of the platinum atoms and its temperature dependence is largely established already by the in-diffusion experiments, it is necessary to assume that the platinum atoms become "frozen in" at their substitutional sites during ramping-down. This motivates the barriers assumed in this work for the reactions rendering the substitutional platinum atoms mobile. Simulations with our calibrated model accordingly give a much better description of the platinum

concentrations near the surface. A closer look at the measured profiles is shown in Figure 3. The shapes of the profiles indicate indeed the expected out-diffusion of substitutional platinum. However, it is kinetically limited and even for the slowest ramp of 0.6 K/min, the sheet concentration of out-diffused platinum is limited and significantly smaller than predicted by the model of Jacob et al. As far as we know, this is the first report of out-diffusion of platinum during ramping-down after a thermal process.

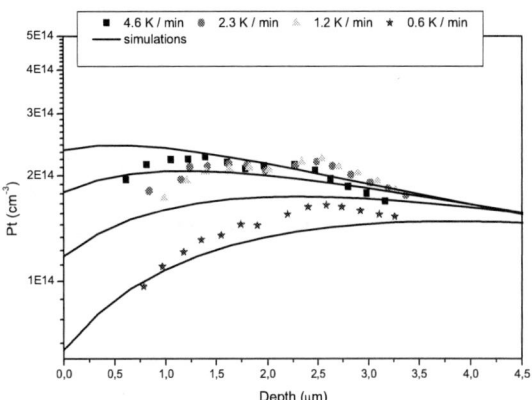

Figure 3: Comparison between the measured results and subsequent simulations.

As shown in Figure 4, our simulations are also in good agreement with the experimental data of Jacob et al. [2] for platinum diffusion at 730 °C for 240 minutes and at 780 °C for 20, 60 and 240 minutes. This shows that the suggested model is suitable to describe platinum diffusion in the temperature range from 730 to 832 °C. It now remains to compare the parameters obtained with literature values.

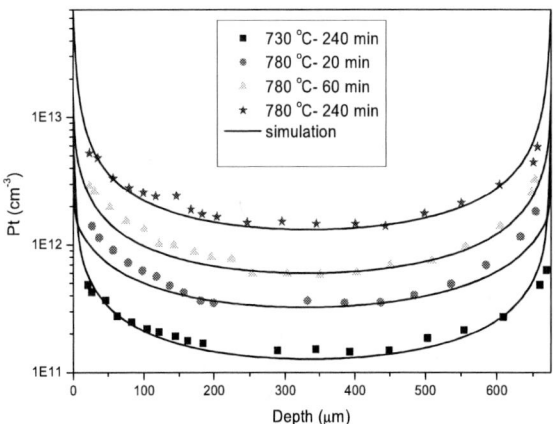

Figure 4: Comparison between experimental data obtained from [2] (scatter points) and the simulation based on this work (lines).

In the literature, there are no direct measurements for substitutional platinum c_{Pts}^{eq} below 800 °C. The values used are usually deduced from fitting diffusion models. Figure 5

Paper T3B4   *PRIME 2013, Villach, Austria*

shows a comparison of the values found in our work to the work of Zimmermann and Ryssel [7].

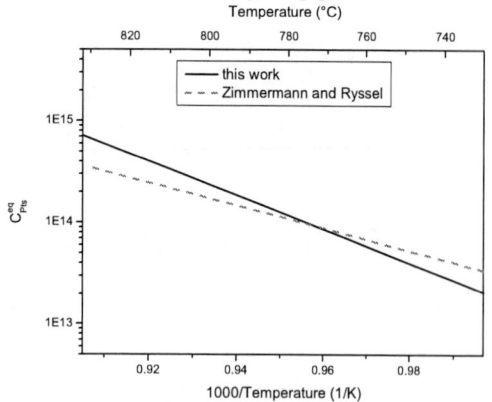

Figure 5: Comparison between c_{Pts}^{eq} of this work and the work of Zimmermann and Ryssel.

Lerch et al. [8] deduced an Arrhenius expression for the product $c_{Pti}^{eq} \cdot D_{Pti}$ for the temperature range from 950 to 1200 °C from the investigation of platinum diffusion into dislocated silicon. An extrapolation of his expression to the temperature range from 730 to 832 °C in Figure 6 shows an excellent agreement with our values.

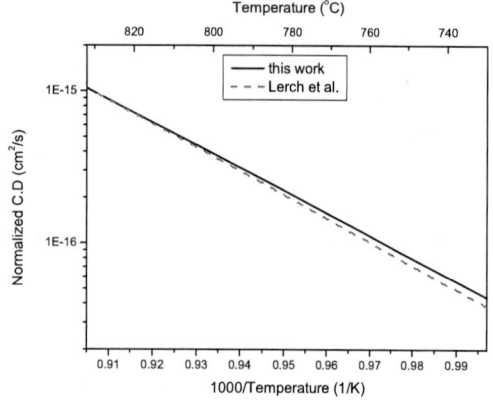

Figure 6: Comparison between the product $c_{Ptt}^{eq} \cdot D_{Ptt}$ of this work and the work of Lerch et al.

The tracer diffusion coefficient D_T of silicon can be expressed in the form

$$D_T = f^I \cdot D_I \cdot \frac{C_I^{eq}}{C_{Si}} + f^V \cdot D_V \cdot \frac{C_V^{eq}}{C_{Si}} \qquad (14)$$

where f^I and f^V are the tracer correlation coefficients which take values of 0.7273 and 0.5, respectively [9, 10]. In Section 2.3 of his book, Pichler [4] summarized the experiments on tracer diffusion reported in the literature. Figure 7 shows a comparison with the values obtained in our work. It can be seen that the values of our unbiased optimization are, by a

factor of about two, larger and within the experimental error of most tracer diffusion experiments.

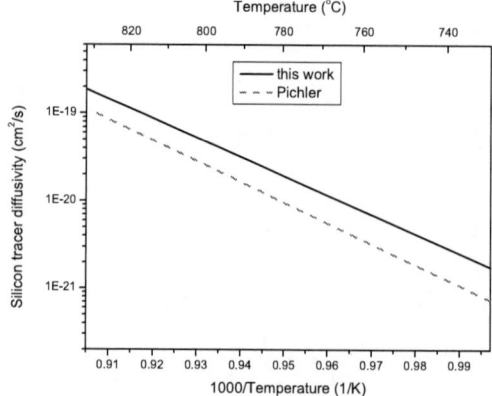

Figure 7: Comparison between the silicon tracer diffusivity of this work and the expression obtained from Pichler.

V. CONCLUSION

The influence of the cooling rate on platinum profiles has been investigated. The experimental results indicated out-diffusion of platinum but at a significantly smaller rate than predicted by existing diffusion models. Including reaction barriers enabled reproducing previously published experiments as well as ours.

REFERENCES

[1] J. Vobecky and P. Hazdra, "High-power P-i-N diode with the local lifetime control based on the proximity gettering of platinum," *IEEE Electron Device Letters*, vol. 23, pp. 392-394, 2002.

[2] M. Jacob, P. Pichler, H. Ryssel, and R. Falster, "Determination of vacancy concentrations in the bulk of silicon wafers by platinum diffusion experiments," *J. Appl. Phys*, vol. 82, pp. 182-191, 1997.

[3] J.-U. Sachse, E. Ö. Sveinbjörnsson, W. Jost, and J. Weber, "Electrical properties of platinum-hydrogen complexes in silicon," *Phys. Rev. B*, vol. 55, pp. 16176-16185, 1997.

[4] P. Pichler, Intrinsic Piont Defects, Impurities, and Their Diffusion in Silicon, Wien-New York: Springer, 2004.

[5] K. Ghaderi, G. Hobler, M. Budil, L. Mader, and H. J. Schulze, "Determination of silicon point defect parameters and reaction barrier energies from gold diffusion experiments," *J. Appl. Phys.*, vol. 77, pp. 1320-1322, 1995.

[6] P. Pichler, W. Jüngling, S. Selberherr, E. Guerrero, and H. W. Pötzl, "Simulation of Critical IC-Fabrication Steps," *IEEE Trans. Electron Devices*, vol. 32, pp. 1940-1953, 1985.

[7] H. Zimmermann and H. Ryssel, "The Modeling of Platinum Diffusion in Silicon under Non-equilibrium Conditions," *J. Electrochem. Soc.*, vol. 139, pp. 256-262, 1992.

[8] W. Lerch, N. A. Stolwijk, H. Mehrer and C. Poisson, "Diffusion of platinum into dislocated and non-dislocated silicon," *Semicond. Sci. Technol.*, vol. 10, pp. 1257-1263, 1995.

[9] K. Compaan and Y. Haven, "Correlation Factors for Diffusion in Solids," *Trans. Faraday Soc.*, vol. 54, pp. 1498-1508, 1958.

[10] M. Yoshida, "Diffusion of Group V Impurity in Silicon," *Jpn. J. Appl. Phys.*, vol. 10, pp. 702-713, 1971.

978-1-4673-4579-8/13 $31.00 © 2013 IEEE

A SiGe Wideband VCO and Divider MMIC with Low Gain Variation for Multi-band Systems at 2.4 and 5.8 GHz

Niko Joram, Robert Wolf, and Frank Ellinger
Chair for Circuit Design and Network Theory
Technische Universität Dresden, 01062 Dresden, Germany

Abstract—A wideband LC voltage controlled oscillator (VCO) is presented in this work. Measurements show a wide relative tuning range of 37.6% from 4.57 GHz to 6.69 GHz with a tuning voltage from 0 to 3 V. Using an integrated divide-by-2 circuit, both the 2.4 GHz and 5.8 GHz ISM bands can be covered. A key feature is the low gain variation, which stays below 12.5 % up to 6 GHz and below 33.3 % for the whole tuning range. At 5.725 GHz, the phase noise at 1 MHz offset was measured to -105 dBc/Hz. The VCO core consumes 4 mA of current from a 3 V supply voltage. It is implemented using an IBM 180 nm SiGe BiCMOS process and the core area amounts to 0.5 mm×0.4 mm.

Index Terms—wideband voltage controlled oscillators, chirp radar, MMICs, SiGe BiCMOS

I. INTRODUCTION

The trend in modern radio frequency (RF) frontends goes towards integration of several standards into one chip to reduce size and cost. For communications systems, versatility is improved. But multi-band frontends can also be useful for radar systems, for example based on the frequency modulated continuous wave (FMCW) approach. In such systems, this can lead to improved robustness and accuracy by gaining multiple uncorrelated ranging data [1]. However, there are several design issues involved, including the shape of the tuning curve, when generating chirps for FMCW. A linear tuning characteristic with low variations is important to reach stable and predictable behaviour of the phase locked loop (PLL) in the system over the whole chirp bandwidth, ensuring a good chirp linearity and therefore high accuracy of the radar.

Usually, switched capacitor arrays are used in wideband VCOs [2]–[4] to enhance the tuning range, but the wiring and switches required reduce the quality factor of the varactor and take up space. There are approaches using multiple tuned elements [5], leading to a more complex circuit, control and design. Also, tunable active inductor VCOs allow a very wide relative tuning range in the order of 100 % [6], [7]. Unfortunately, all those architectures usually show a quite large gain variation over the tuning range.

The VCO presented in this paper uses a differential hyperabrupt varactor to provide continuous frequency tuning without using switches, achieving a wide tuning range while keeping the chip size and complexity to a minimum. It can be used in both ISM bands at 2.4 GHz and 5.8 GHz by just switching the included dividy-by-2 frequency divider. Because

of the low gain variation, applications include PLL-stabilized FMCW radar systems.

II. DESIGN

A. Gain Variation

The gain of a VCO is the slope of its frequency over tuning voltage characteristic, which is

$$K_{\text{VCO}} = \frac{\partial f(V_{\text{tune}})}{\partial V_{\text{tune}}}. \tag{1}$$

The relative gain variation $\varepsilon_{K_{\text{VCO}}}$ can then be defined as the deviation of the actual gain characteristic from an ideal constant gain characteristic. Commonly, it is calculated as [3]

$$\varepsilon_{K_{\text{VCO}}} = \frac{K_{\text{VCO,max}} - K_{\text{VCO,min}}}{K_{\text{VCO,max}} + K_{\text{VCO,min}}}. \tag{2}$$

Keeping gain variations of the VCO small helps maintaining the PLL dynamics over the whole system bandwidth. Thereby, the transient response of the loop can be optimized. In FMCW radar systems, where a frequency chirp with greatest possible bandwidth needs to be generated, chirp nonlinearities degrade spatial resolution.

The resonant frequency range of the oscillator can be estimated using

$$f(V_{\text{tune}}) = \frac{1}{2\pi\sqrt{L(C_{\text{var}}(V_{\text{tune}}) + C_{\text{par}})}} \tag{3}$$

with $C_{\text{var}}(V_{\text{tune}})$ being the voltage-dependent varactor capacitance and C_{par} the equivalent parasitic capacitance from the wiring, transistors and output load. The dependency of the oscillation frequency from the varactor capacitance as outlined in (3) is nonlinear, therefore a matching $C-V$ characteristic of the used varactor is important to minimize the gain variation. A model for estimating the junction capacitance of a reverse-biased hyperabrupt varactor is given as follows [8]

$$C_{\text{var}}(V_{\text{tune}}) = \frac{C_{j0}}{\left(1 - \frac{V_{\text{tune}}}{V_j}\right)^m}. \tag{4}$$

The parameters of the equation can be matched to the hyperabrupt varactor, which is available in the employed SiGe technology, with $C_{j0} \approx 1.69$ pF, $V_j \approx 2.5$ V and $m \approx 1.5$. A value of $m > 0.5$ indicates a hyperabrupt junction. Inserting (4) in (3) yields an estimate for the tuning characteristic of the

Paper W1A1

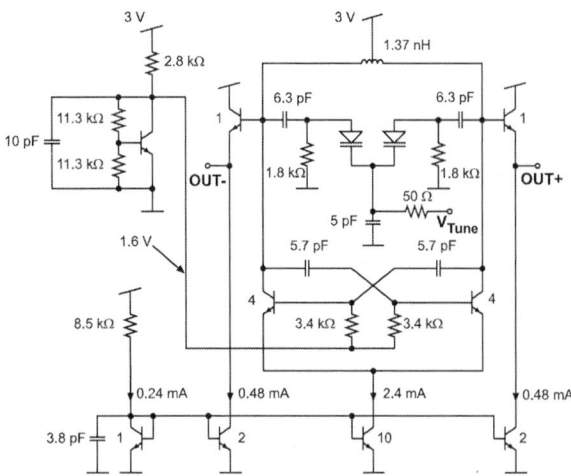

Fig. 1. VCO core circuit diagram (BJT emitter areas $0.25\,\mu m \times 2.5\mu m$, multiplier given)

VCO, which is depicted along with the corresponding gain in Fig. 5. The results are discussed in the following sections.

B. VCO Core

The VCO core uses a commonly known differential cross-coupled topology with npn bipolar junction transistors (BJTs). The schematic and component values are shown in Fig. 1. The oscillator tank consists of an inductor with center tap and differential varactor diodes, which form a parallel resonant circuit. The center tap of the symmetric inductor is connected to the supply voltage and is also used to bias the collector of the cross-coupled BJT pair. The varactor diodes are decoupled by large RF blocking capacitors of 6.3 pF from the collector nodes, because they need to be reverse-biased. Therefore, their anodes are biased to ground using a high-ohmic resistor and the cathodes are connected together to form the input for the tuning voltage, including a RC low pass filter. The varactors consist of an array of 16 diodes, each having 2 anodes. It allows for a large capacitance change for a given voltage change. The relative capacitance change of the used varactor is about 3.3:1 (or 1.69 pF to 520 fF absolute) single-ended for a voltage change from 0 to 3 V.

The quality factor Q of a varactor or inductor can be expressed as the ratio of its reactance X to its resistance R.

$$Q = \frac{X}{R} \tag{5}$$

For the used varactor, a differential Q of about 15 to 20 was simulated at 5 GHz, depending on the tuning voltage. Since the varactor array is quite large and therefore involves increased wiring effort, the quality factor is slightly lower than in narrowband VCOs, where 20 to 30 is common. On the other hand, the differential Q of the employed inductor was simulated to be 17 at 5 GHz, which is a good value.

The effective inductance L for calculating the oscillation frequency is half of the used differential inductance of 1.37 nH.

Simulations showed that $C_{\mathrm{par}} \approx 250\,\mathrm{fF}$, resulting from the wiring and the parasitics of the cross-coupled pair and output buffer transistors. The calculation using the hyperabrupt varactor model from the previous subsection then yields a tuning range from 4.37 GHz to 6.94 GHz with a maximum gain variation of 16 % below 6 GHz and 27 % for the whole bandwidth, which is a promising result and indicates that the varactor is applicable for wideband linear VCOs.

To decouple the tank from the output stage and frequency divider, small common-collector buffers are employed, which add only little parasitic capacitance to the critical collector nodes. The bases of the cross-coupled pair are biased via resistors by a V_{BE} multiplier. Simulations showed that this biasing scheme is advantageous to a simple resistor divider in terms of phase noise.

The tail current sources for the core and output buffers are biased resistively rather than using a transistor current source. Low frequency flicker noise of such a current source would translate into phase noise in the oscillator core, degrading the performance at offset frequencies in the kHz range. To face this problem, the transistor widths of the current source could be made very large. However, in order to save area, a small bias resistor provides superior noise performance.

C. Frequency Divider

The employed frequency divider consists of two cascaded latches in emitter-coupled logic (ECL) architecture. Those latches form a delay flip-flop, where the inverted data output is fed back to the data input, thus halving the applied clock frequency. To allow switching the frequency divider, the buffered oscillator output and the divider output both are open-collector and operate on a common resistive load. Therefore, the divider can be activated and deactivated by switching the tail current sources of the involved differential amplifiers. The divider can also be switched off completely, saving its contribution of 26 mW to the total power budget.

III. MEASUREMENT

The VCO was manufactured in an IBM 180 nm SiGe BiCMOS technology with a maximum transit frequency of 60 GHz. Fig. 2 shows a chip micrograph of the VCO integrated into a transceiver chip together with the frequency divider by 2 and a totem-pole output stage [9].

All measurements were done with the transceiver chip bonded onto a PCB. The chip is connected to an external PLL to be able to control the frequency of the VCO. The spectrum of the VCO output at the frequency of 6.7 GHz with 3 V tuning voltage is shown in Fig. 3. This spectrum snapshot was created with the PLL turned off.

The phase noise measurement was done in the application system and using the PLL, which has 150 kHz loop bandwidth. The result is presented in Fig. 4. Because of the large gain, the oscillation frequency is very suceptible to noise and interference, making phase noise measurements of the free-running oscillator with fixed control voltage inaccurate. Instead, the PLL filters out the interference below 150 kHz

PRIME 2013, Villach, Austria   *Session W1A – High-Frequency Analog Circuits*

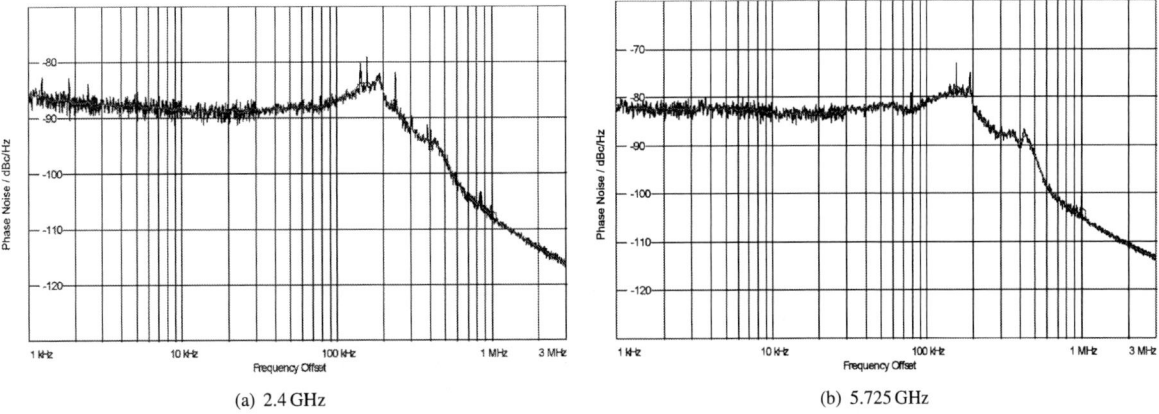

(a) 2.4 GHz   (b) 5.725 GHz

Fig. 4.   Measured phase noise plot of the VCO within a PLL with 150 kHz loop bandwidth at 2.4 GHz and 5.725 GHz oscillation frequency, showing -108 dBc/Hz and -105 dBc/Hz, respectively, at 1 MHz offset

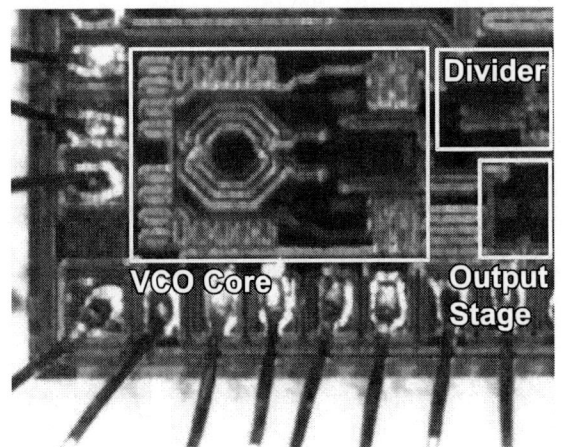

Fig. 2.   Chip micrograph, VCO core area 0.5×0.4 mm², divider area 0.3×0.2 mm²

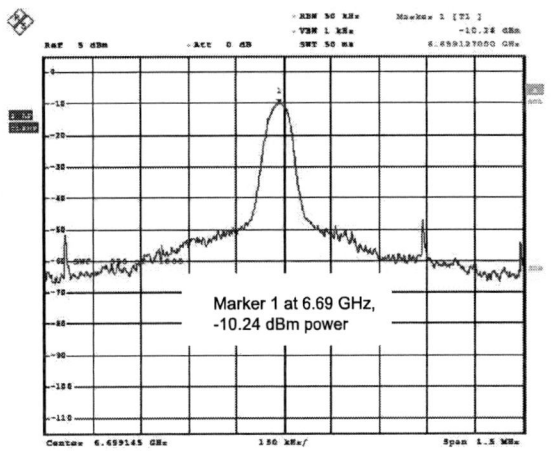

Fig. 3.   Measured spectrum of the free-running VCO

to reach a stable frequency output and the measurement then was done using a Rhode&Schwarz FSU-67 spectrum analyzer with phase noise measurement software option. The phase noise at 1 MHz offset from the carrier, which lies out of band of the PLL, amounts to -108 dBc/Hz at 2.4 GHz or -105 dBc/Hz at 5.725 GHz oscillation frequency. Phase noise at offset frequencies below the PLL bandwidth is mainly determined by the other components of the PLL rather than the VCO.

The gain of the oscillator at low tuning voltages is approximately 900 MHz/V. The gain variation amounts to only 12.5 % from 4.57 GHz up to 6 GHz, which is the frequency range for use in the ISM bands, and 33.3 % for the whole tuning range up to 6.7 GHz. The calculated, measured and simulated tuning curves are shown in Fig. 5 along with the gain. It can be seen, that the model calculation gives a good estimate of the oscillation frequency and gain variations to be expected. At low oscillation frequencies up to 6 GHz, the simulated

and measured values match well. At higher frequencies, the simulated curve is more flat than the measured one, showing 200 MHz less tuning range. This may be due to the fact, that the used standard parasitic extraction does not include parasitic inductances. Since the hyperabrupt varactor is specified up to 6 V of reverse bias voltage, the frequency could be increased further, e.g. to 7 GHz at 4 V. However, the supply voltage of the system is specified to be 3 V and an additional voltage source would then be needed to supply the PLL charge pump. Besides, the tuning curve becomes flat in the region above 3 V, increasing the gain variation.

Table I compares key parameters of current wideband VCOs with the oscillator presented in this paper. There are only some isolated publications of wideband VCOs covering the 4-6 GHz range, mostly using multiple tuned elements or tuned inductors [6], [7]. Gain variations of those types can be in the order of 50 % and above with tuning curves shaped after exponential or tanh functions. Furthermore, phase noise

978-1-4673-4579-8/13 $31.00 © 2013 IEEE   259

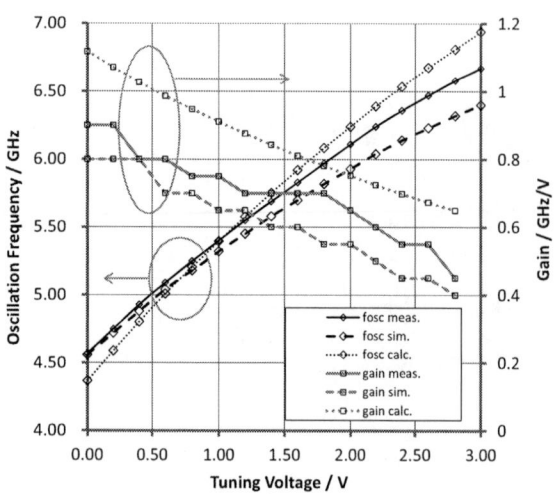

Fig. 5. Calculated, simulated and measured VCO tuning characteristic and gain

TABLE I
COMPARISON WITH STATE-OF-THE-ART WIDEBAND VCOS

Reference	[3]	[6]	[7]	This work
Technology	65 nm	SiGe	180 nm	SiGe
Tuning Range / GHz	1.51-3.06 (67.8 %)	1.7-5.2 (101.4 %)	3.8-7.4 (64.3 %)	4.57-6.69 (37.6 %) 2.29-3.35 with divider
Phase Noise (1 MHz offset) / dBc/Hz	-115 to -120	-94 to -103	-75 to -92	-108 to -105
Gain Variation	49.5 %	-	-	33.3 % 12.5 % below 6 GHz
Core area / mm²	0.39	0.5	-	0.26 with divider
Supply power / mW	1	3.7	29	12 VCO 26 divider

performance seems to be degraded by those approaches. Of the referenced publications, the gain variation of the presented VCO is among the lowest with respect to oscillation frequency and tuning range.

IV. CONCLUSION

This work presented a VCO with large tuning range and low gain variation compared to other wideband VCOs. It can be tuned continuously without switching capacitor banks from 4.6 GHz to 6.7 GHz. The low gain variation within the ISM bands allows for effective and stable PLL designs covering large bandwidths. It is suitable for operating in both, the 2.4 GHz and 5.8 GHz ISM bands by only employing a divide-by-2 circuit. The oscillator can be integrated into multi-band FMCW radar or communications systems.

ACKNOWLEDGMENT

The research leading to these results has received funding from the European Community's Seventh Framework Programme (FP7/2007-2013) under grant agreement n°242411 (E-SPONDER) and by the Federal Ministry of Education and Research (BMBF) in the excellence cluster Cool Silicon, project Cool Broadcast Repeater.

REFERENCES

[1] N. Joram, B. Al-Qudsi, J. Wagner, A. Strobel, and F. Ellinger, "Design of a multi-band fmcw radar module," in *10th Workshop on Positioning, Navigation and Communication 2013 (WPNC'13)*, march 2013, p. on CD.

[2] K. Fu, Z. Cheng, J. Li, Y. Zhou, and X. Zhou, "Design of wideband lc vco with small kvco fluctuation for rfid synthesizer application," in *11th IEEE International Conference on Communication Technology, 2008. ICCT 2008.*, nov. 2008, pp. 286 –289.

[3] Y. Weng, H. Gao, and L. Sun, "A 1.5-3.0ghz wideband vco with low gain variation," in *Electrical Design of Advanced Packaging and Systems Symposium (EDAPS), 2011 IEEE*, dec. 2011, pp. 1–4.

[4] Z. Mingzhu, "A low phase noise wideband vco in 65nm rf cmos for low power applications," in *2012 International Conference on Microwave and Millimeter Wave Technology (ICMMT)*, vol. 1, may 2012, pp. 1–4.

[5] M. Tsuru, K. Kawakami, K. Tajima, K. Miyamoto, M. Nakane, K. Itoh, M. Miyazaki, and Y. Isota, "A triple-tuned ultra-wideband vco," *IEEE Transactions on Microwave Theory and Techniques*, vol. 56, no. 2, pp. 346–354, feb. 2008.

[6] R. Mukhopadhyay, Y. Park, S. Yoon, C.-H. Lee, S. Nuttinck, J. Cressler, and J. Laskar, "Active-inductor-based low-power broadband harmonic vco in sige technology for wideband and multi-standard applications," in *2005 IEEE MTT-S International Microwave Symposium Digest*, june 2005, pp. 1349–1352.

[7] M. Mehrabian, A. Nabavi, and N. Rashidi, "A 4-7ghz ultra wideband vco with tunable active inductor," in *IEEE International Conference on Ultra-Wideband, 2008. ICUWB 2008.*, vol. 2, sept. 2008, pp. 21–24.

[8] V. I. Cojocaru and T. J. Brazil, "A large-signal equivalent circuit model for hyperabrupt p-n junction varactor diodes," in *22nd European Microwave Conference, 1992.*, vol. 2, sept. 1992, pp. 1115–1121.

[9] M. Wickert, R. Wolf, and F. Ellinger, "Analysis of totem-pole drivers in sige for rf and wideband applications," *International Journal of Microwave and Wireless Technologies*, vol. 4, no. 2, pp. 245–255, april 2012.

PRIME 2013, Villach, Austria *Session W1A – High-Frequency Analog Circuits*

A CMOS-28nm 880-MHz 4th-Order Low-Pass Active-RC Filter for 60 GHz Transceivers

A. Pezzotta[1], M. De Matteis[1,2], S. D'Amico[2] and A. Baschirotto[1]

[1]Dept. of Physics, University of Milano-Bicocca, Milano, Italy; [2]Dept. of Innovation Engineering, University of Salento, Lecce, Italy

a.pezzotta1@campus.unimib.it, marcello.dematteis, andrea.baschirotto@unimib.it

Abstract— **In this paper a 4th order low-pass continuous time analog filter in CMOS 28nm technology is presented. The filter complies with the specifications of the 60GHz next-generation transceivers. A novel circuital topology is presented suitable to perform the low-pass filtering of the in-band signal, while the in-band thermal noise is high-pass filtered, improving Signal-To-Noise-Ratio. 880MHz -3dB bandwidth is obtained by a prototype of the filter simulated in 28nm CMOS technology. The overall power consumption is 4.8mW from a single 1 V supply voltage. 51dB@THD>40dBc is the SNR at 0dB gain.**

Keywords— Analog filter, Active RC, Wireless, CMOS 28nm, Low-Power, Low-Voltage

I. INTRODUCTION

Nowadays, despite digital signal processing has achieved very high performance, many functions are still implemented into analog domain, such as bandwidth selection, anti-aliasing filtering for the following A-to-D-Converter and out-of-band interferes rejection[1]. In base-band chain of telecommunications transceivers closed-loop Active-RC filters are a winning choice for continuous-time filter design, due to their ability to achieve better linearity and high frequency response accuracy performance, with respect to g_m-C filters[3]. On the other hand, power and thermal noise is critical, and this aspect is further stressed in 60GHz transceivers where baseband chain bandwidth is about 1GHz, very much larger than other typical telecommunication standards (like GSM, UMTS, WLAN, UWB) [1].

Due to the larger bandwidth, at constant in-band IRN-PSD (Input Referred Noise Power Spectral Density), in-band integrated noise power tends to increases, degrading Signal-To-Noise-Ratio. This scenario forces analog designers to explore innovative circuital solutions, in order to avoid SNR degradations. Moreover, the larger bandwidth imposes high integration scale, so that typically CMOS 65nm and down nodes are adopted due to their larger transition frequency. Furthermore, technological scaling-down leads to lower supply voltage and lower transistor intrinsic gain (defined as $g_m \cdot r_{ds}$). The effect is that SNR degradation (due to the larger in-band integrated noise power) cannot be recovered by processing a larger output voltage signal.

In this scenario, the analog filter here presented represents an evolution of an existent 11MHz-bandwidth topology, suitable to reduce the in-band integrated noise power[4]. This design allows to reduce thermal noise contributions to only three components. The idea is to improve the circuit presented in [4], enabling larger bandwidth (880MHz), and increasing integration scale, by implementing a prototype of the filter in CMOS 28nm technology.

A prototype of this filter has been designed with 880MHz –3dB-bandwidth. The simulation results shows that this

topology guarantees very interesting results even adapting specifications to different and challenging standards, such as 60 GHz communications, maintaining a global improvement comparing with the state-of-the-art,[5]-[8].

The paper has been organized as follows. Section II introduces the main noise/linearity/mask specifications, explains the filter transfer function calculus, describing the most important design aspects. Simulation results are shown in Section III, while at the end conclusions are drawn.

II. 4TH ORDER FILTER TRANSFER FUNCTION

A. Specifications.

Starting from system level considerations[2] the analog filter has to comply with the requirements shown in Table 1.

At least 50 dB-SNR is required, while the output signal swing must perform 40dBc of Total-Harmonic-Distortion.

Four poles are required, at 0dB dc-gain. The main transfer function parameters (poles frequency and quality factor) are then reported in Table 1.

Parameter	Value	
Transfer Function	(Low-Pass) 4th Order	
Pass-Band Gain - G	0 dB	
f@-3 dB	880MHz	
In-band IRN PSD	< 15 nV/√Hz	
SNR@(THD ≥ 40 dBc)	> 50 dB	
Parameter	**Poles Pair 1**	**Poles Pair 2**
Poles Frequency - ω_0	1 GHz	1 GHz
Quality Factor - Q	$Q_1=1.31$	$Q_2=0.50$
Real pole Frequency - ω_0	1 GHz	

Table 1: Filter Global Requirements

B. Filter Transfer Function.

The Multipath topology is based on the 4th Order Butterworth low-pass filter transfer function, shown in (1).

This guarantees a flat in-band response, through the dimensioning of quality factors Q_1 and Q_2.

$$H(s) = \cfrac{1}{\left(\cfrac{s^2}{\omega_0^2} + \cfrac{s}{\omega_0 \cdot Q_1} + 1\right) \cdot \left(\cfrac{s^2}{\omega_0^2} + \cfrac{s}{\omega_0 \cdot Q_2} + 1\right)} \quad (1)$$

As regards the implementation, cascading two Active-RC cells (one cell for each complex poles pair), has been demonstrated to be not convenient regarding noise and power consumption [4], because at 0dB dc-gain every cell contributes to the overall thermal noise power. For 60 GHz communications standard[2], the maximum allowable noise is about 15nV/√Hz, resulting in very low resistors (<1 kΩ). In these conditions, larger power has to be paid for (low) resistive load driving, and Opamp input stage g_m (for noise power reduction).

978-1-4673-4579-8/13 $31.00 © 2013 IEEE   261

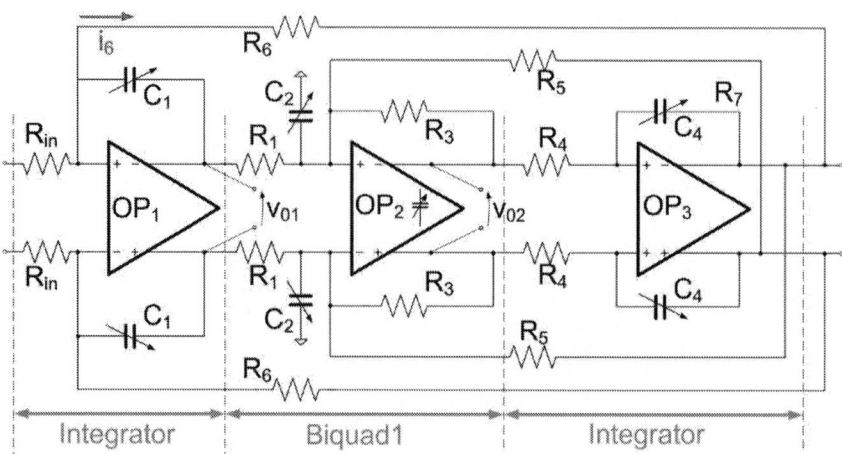

Figure 1: Global Filter Schematic

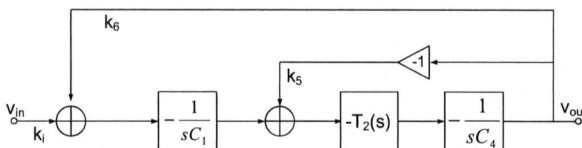

Figure 2: Filter Functional Scheme

The presented design approach aims to demonstrate that the multipath filter can represent a valid alternative with high-frequency standards, together with more scaled CMOS technology issues. Figure 2 shows the functional scheme of the multipath filter. The three main blocks consists in two integrators and one biquadratic cell.

Two feedbacks are then needed to synthesize a global 4th order low-pass transfer function. The schematic implementation of the multipath filter is illustrated in Figure 1. The integrators are designed by using a simple single-Opamp topology, and the biquadratic cell is implemented in Active-G_m-RC configuration. That allows to synthesize a complex poles pair adding only one Opamp, with respect to the most popular Tow-Thomas biquad[1].

The filter transfer function can then be expressed as a function of the design parameters: R-C values, and OP_2 unity gain bandwidth (ω_u, since in Active-G_m-RC cell the Opamp is used as integrator). The resulting transfer function is shown in (2), where a_i parameters are given by (3).

$$H(s) = -\frac{R_6}{R_i} \cdot \frac{1}{(a_4 \cdot s^4 + a_3 \cdot s^3 + a_2 \cdot s^2 + a_1 \cdot s + a_0)} \quad (2)$$

$$\begin{cases} a_4 = \dfrac{C_1 C_2 C_4 R_1 R_4 R_6}{\omega_u} \\[2mm] a_3 = \dfrac{C_1 C_4 R_4 R_6}{R_3 R_5 \omega_u} \cdot (R_1 R_3 + R_1 R_5 + R_3 R_5) \\[2mm] a_2 = \dfrac{C_1 C_4 R_1 R_4 R_6}{R_3} \quad , \quad a_1 = \dfrac{C_1 R_1 R_6}{R_5}, \; a_0 = 1 \end{cases} \quad (3)$$

The ω_0-Q_1-Q_2-G parameters are available in Table 1. Note that the more interesting aspect of this circuit is that the only thermal noise sources affecting the global filter noise power spectral density are related to R_i, R_6 and OP_1.

In fact, at low frequency the C_1 capacitor features very high impedance, so that in first approximation no feedback is

present between the OP_1 output and its inverting. So that if the signal is not coming from v_{in}, i_6 current is always zero.

Obviously, increasing frequency, the C_1 impedance decreases and the inside-loop noise contributions becomes higher. However, the in-band noise is strongly attenuated.

Neglecting the flicker noise contribution, the filter input noise power spectral density (IRN_{Filter}) is given by (4).

$$IRN_{Filter}^2 = 8kT \cdot R_i + 8kT \cdot R_6 \cdot (1/G)^2 + IRN_{OP1}^2 \cdot (1 + 1/G)^2 \quad (4)$$

C.   R-C and Filter Sizing.

In order to implement the multipath 4th order filter, R-C values and OP_2 unity gain bandwidth (ω_u) have to be set. A proper Matlab script has been used for the calculus. The idea is to set the system level inputs considering three different set of input data.

Noise. With reference to (4), once the IRN_{OP1} has been estimated, the overall filter in-band IRN and dc-gain G (available in Table 1) determine the R_i and R_6 values.

Area. The condition about C_1/C_4 ratio allows to manage the reduction of the implementation area of the circuit.

Transfer Function. The ideal transfer function in (1) is totally defined by the ω-Q values in Table 1. Imposing the equivalence between (1) and (2), four equations are set, one for each a_i term in (3) (except for a_0), shown in (5).

Linearity. The last data-set is imposed by linearity considerations. Linearity performance are strongly related with the filter signal swing at the OP_1 and OP_2 output nodes (v_{01}/v_{in} and v_{02}/v_{in} vs. frequency are plotted in Figure 4).

In fact, at low frequency, due to the virtual ground principle, the input stage of every Opamp in the filter is expected to face very low signal swing, with the consequence that linearity is dominated by the Opamps output stage.

In order to comply with the THD requirement, system-level simulations show that in case of maximum filter output swing, the signal at the OP_1 output (v_{01}/v_{in}) have to be maintained -3dB lower than the input signal. If v_{01}/v_{in} overcomes this limit, the OP_1 output stage distortion becomes critical, breaking the minimum 40dBc-THD requirement.

From a dc analysis, the v_{01} voltage at the OP_1 output (vs. v_{in}) is given by (5), where G=1, is the filter dc-gain (Table 1). Its value imposes the R_1/R_5 ratio, as reported in Table 2. By setting it as shown in (5), the signal swing at the OP_1 output node is totally controlled and the required THD can be satisfied.

978-1-4673-4579-8/13 $31.00 © 2013 IEEE

At low frequency the OP_2 output nodes are connected to the OP_3 virtual ground. At dc, in fact, no current is flowing through R_4, leading the v_{02} node to be a virtual ground too.

In order to complete linearity considerations, the design has to be analysed around the poles frequency (880 MHz). At this frequency is clear that the virtual ground principle is weaker, due to the limited Opamp bandwidth, and at the same time C_4 capacitor impedance becomes comparable with the other filter impedances, so that v_{02} swing is not negligible.

For v_{02}/v_{in} minimization, the minimum Unity-Gain Bandwidth for OP_3 is about 3GHz.

Parameter	Value	Parameter	Value
R_{in}	406 Ω	R_6	406 Ω
R_1	1.02 kΩ	C_1	1.77 pF
R_3	721 Ω	C_3	0.2 pF
R_4	653 Ω	C_4	1.77 pF
R_5	1.44 kΩ	ω_u	2π·2.48 Grad/s

Table 2: Matlab Script Output Results

$$
\begin{cases}
R_{in} = \dfrac{IRN_{Filter}^2 - (1+1/G)^2 \cdot IRN_{OP1}^2}{8kT \cdot (1+1/G)} \\[2mm]
R_6 = G \cdot R_{in}, \quad \dfrac{1}{\omega_0^4} = a_4 \\[2mm]
\left(\dfrac{1}{Q_1} + \dfrac{1}{Q_2}\right) \cdot \dfrac{1}{\omega_0^3} = a_3 \\[2mm]
\left(2 + \dfrac{1}{Q_1 Q_2}\right) \cdot \dfrac{1}{\omega_0^2} = a_2 \\[2mm]
\left(\dfrac{1}{Q_1} + \dfrac{1}{Q_2}\right) \cdot \dfrac{1}{\omega_0} = a_1 \\[2mm]
R_1/R_5 = \dfrac{\sqrt{2}}{2}; \quad C_1 = 10 \cdot C_4; \quad R_5/R_3 = 2
\end{cases}
\tag{5}
$$

III. SIMULATION RESULTS

A prototype of this filter has been designed in CMOS 28 nm technology. The Opamps are of class-A Miller-Compensated type (Figure 3), while their main performance is available in Table 3.

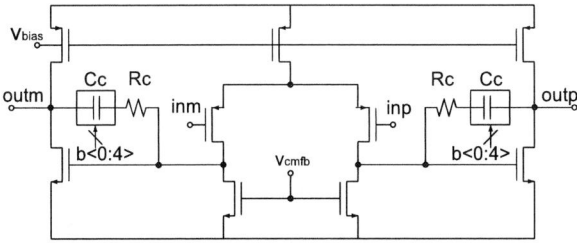

Figure 3: Opamps Structure

Parameter	OP_1	OP_2	OP_3
DC-Gain	43 dB	45 dB	43 dB
Unity-Gain BW	3.1 GHz	2.48 GHz	3.1 GHz
IRN	6.4 nV/√Hz	6.5 nV/√Hz	6.4 nV/√Hz
Current Cons.	1.6 mA	1.7 mA	1.6 mA

Table 3: Opamps Performance

Frequency Response. The nominal frequency response is plotted in Figure 4, with v_{01}/v_{in} and v_{02}/v_{in} frequency response. Note that these results confirm what concerns linearity issues, in terms that at low frequencies v_{01}/v_{in} is -3dB lower than v_{out}/v_{in}, as discussed in Section II. As a further validation, in Figure 5, v_{01}, v_{02} and v_{out} nodes signals vs. time are depicted in the case of low (2 MHz) and high frequency (800 MHz).

Notice that in case of 2MHz input signal, due to the virtual ground principle v_{02} is quite negligible comparing with v_{01} and v_{out}.

The frequency response features 880MHz -3dB-bandwidth, and it is externally tuned using variable capacitors[10]-[12], as it is the compensation capacitance included in OP_2, directly involved into the transfer function calculus.

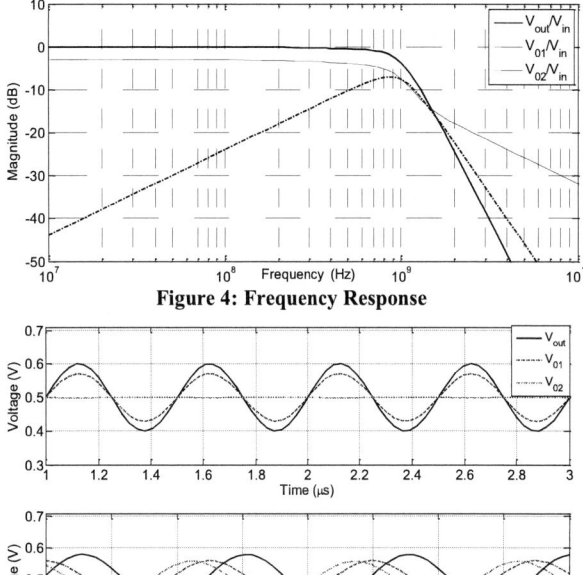

Figure 4: Frequency Response

Figure 5: Opamps Output Signal (2 MHz & 800 MHz)

Noise Power Spectral Density. Figure 6 and Figure 7 show the noise transfer function of some components, described in Section II. Note that the R_{in} thermal noise contribution is constant over the entire bandwidth (up to 880MHz), as in (4) results for R_6 and OP_1. This fact is due to the presence of C_1 capacitance, so the remaining noise sources are high-pass filtered. Although this fact, some peaking in noise transfer function comes out around the poles frequency (see overall noise PSD at the filter output, in Figure 8), due to the other resistors and OP_2-OP_3 (some examples are shown in Figure 7. This peaking leads to a small noise increasing with respect to (4). The overall in-band output integrated noise results to be 480μV$_{rms}$. The effective ORN calculus, considering also the peaking effect (resulting 14.8 nV/√Hz) is a little bit higher than the ORN predicted (13 nV/√Hz). This demonstrates the in-band noise reduction rejects all thermal noise contributions included inside the R_i-R_6 loop, at the cost of a slight noise increasing, estimated to be 12% (from 13 to 14.8nV√Hz).

Linearity. The Input-IP3 (3rd-Order Harmonic Intercept Point) and 1dBcP (1dB Compression Point) simulation results are available respectively in Figure 9 and Figure 10. In detail, the IIP3 is about 7.32dBm and has been evaluated using two input tones at 2 MHz and 3 MHz. 1dBcP is obtained with a 0.306mV$_{0-pk}$ differential input signal.

IV. CONCLUSIONS

The filter here presented exploits an interesting circuital topology suitable to synthesize four poles by using a single compact cell and high-pass filtering the in-band noise. Table 4 resumes the most important filter performance. The Figure-of-Merit (6) used in [5]-[8] has been used for comparison, where PW is the total power consumption, BW is the f@3dB bandwidth, N the number of poles.

Parameter	Value
DC-Gain	0 dB
f@-3dB	878 MHz
Supply Voltage	1 V
CMOS Technology	28 nm
Power Consumption	4.9 mW
Output Integrated Noise – (100KHz-400MHz)	480 μVrms
IRN Spectral Density@200MHz	13 nV/√Hz
IIP3 - $v_{in}=v_{in1}+v_{in2}$ - v_{in1}@2MHz, v_{in2}@3MHz	7.32 dBm
1dB Compression Point	0.306 mV$_{0-pk}$
SNR@THD=40dBc	51 dB

Table 4: Overall Filter Performance

This work (based on simulation results) compares favorably with other filters present in literature. Figure 11 shows the FoM.

$$FoM = \frac{PW}{8 \cdot k \cdot T \cdot BW \cdot SNR \cdot N} \quad (6)$$

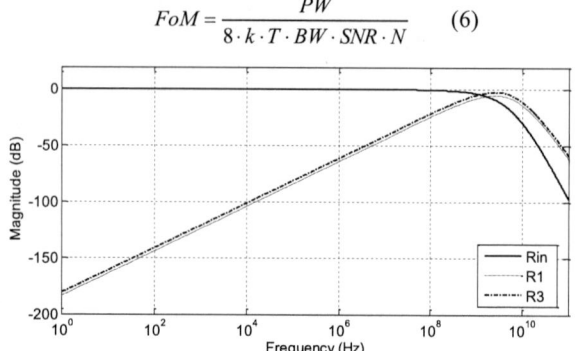

Figure 6: R$_{in}$ R$_1$ R$_3$ Output Noise Transfer Function

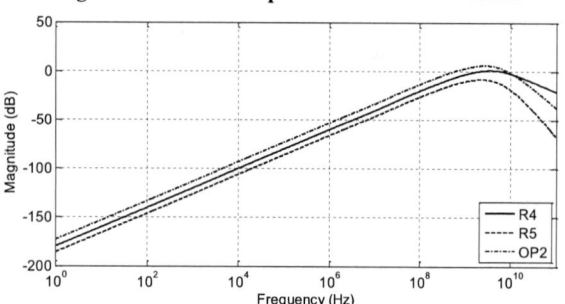

Figure 7: R$_4$ R$_5$ OP$_2$ Output Noise Transfer Function

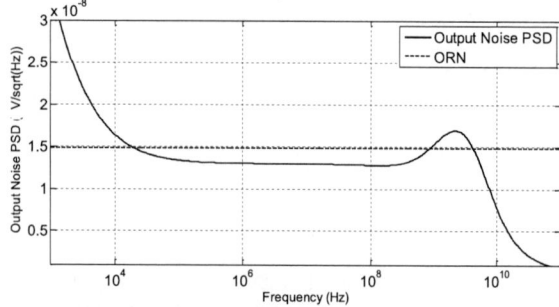

Figure 8: Output Noise Power Spectral Density

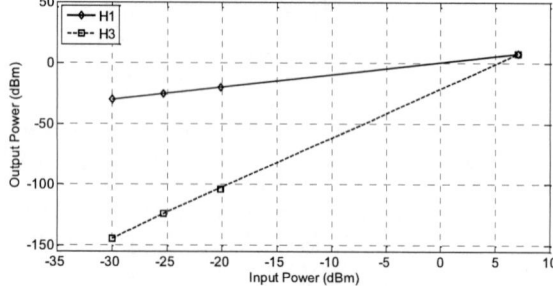

Figure 9: IP3 Simulation Results

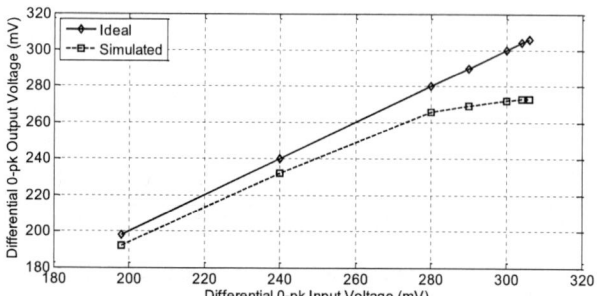

Figure 10: 1dB Compression Point Simulation Results

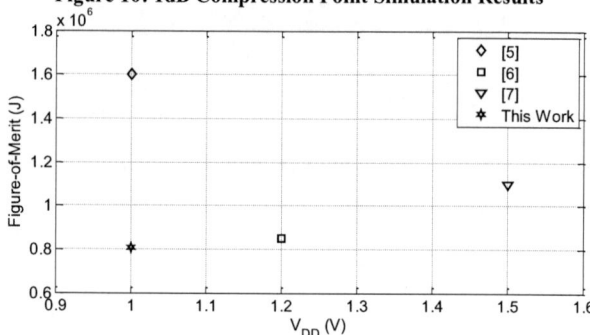

Figure 11: Figure-of-Merit

REFERENCES

[1] Baschirotto et al. "Advances on analog filters for telecommunications". *Advanced Signal Processing, Circuits, and System Design Techniques for Communications. 2006 IEEE International Symposium on Circuits and Systems*, , pp. 131-138.

[2] J. Borremans, et al. "A digitally-controlled compact 57-66GHz receiver front-end for phased-arrays in 45nm digital CMOS", *Proceedings of ISSCC*, pp. 492-493, February 2009.

[3] D'Amico et al. "A 6.4 mW 4.9 nV/√Hz, 24 dBm IIP3 VGA for a multi-standard (WLAN, UMTS, and Bluetooth) receiver", *Analog Integrated Circuits and Signal Processing*, 2009, Vol. 61, Issue 1, pp 1-7

[4] De Matteis, M. et al. "4th-Order 84dB-DR CMOS-90nm low-pass filter for WLAN receivers". *Circuits and Systems (ISCAS), 2011 IEEE International Symposium on*. Page(s): 1644 - 1647R.

[5] A.Vasilopoulos et al. "A Low-Power Wideband Reconfigurable Integrated Active-RC Filter With 73 dB SFDR" *JSSC* vol.41 No.9.Sept.06

[6] S. Kousai et al. "A 19.7 MHz, Fifth-Order Active-RC Chebyshev LPF for Draft IEEE 802.11n With Automatic..". *JSSC* vol.42, No.11, 2006.

[7] M. De Matteis, et al." A 550mV 8dBm IIP3 4th Order Analog Base Band Filter for WLAN Receivers", *33rd ESSCIRC*, 2007.

[8] De Matteis, M.; et al. "A programmable active-RC complex filter for wireless communications". *Electronics, Circuits, and Systems, 2009. ICECS 2009. 16th IEEE International Conference on*, pp. 187- 190.

[9] Pirola, A.; Liscidini, A.; Castello, R. "Current-Mode, WCDMA Channel Filter With In-Band Noise Shaping". *Solid-State Circuits, IEEE Journal of*. Volume: 45 , Issue: 9, 2010 , Page(s): 1770 – 178.

[10] D'Amico et al. "A 255 MHz Programmable Gain Amplifier and Low-Pass Filter for Ultra Low Power Impulse-Radio UWB Receivers". *Circuits and Systems I: Regular Papers, IEEE Transactions on*. 2012. Vol.59 n. 2. pp. 337-345.

[11] Carniti, P. et al. "CLARO-CMOS, a very low power ASIC for fast Photon counting with pixellated photodetectors". *JINST* Vol. 7, Nov. 2012. pp: 1-24.

[12] Costantini, A.; Pezzotta, A.; Baschirotto, A.; De Matteis, M.; D'Amico, S.; Murtas, F.; Gorini, G., "A CMOS 0.13μm low power front-end for GEM detectors," *Electronics, Circuits and Systems (ICECS), 2012 19th IEEE International Conference on*, pp.193,196, 9-12 Dec. 2012

An Inductorless 34 Gbit/s Half-Rate
4:1 Multiplexer in 0.25-μm SiGe Technology

Mahdi Khafaji*, Corrado Carta*, Elena Sobotta*, Daniel Micusik[†] and Frank Ellinger*
*Chair for Circuit Design and Network Theory,
Technische Universität Dresden, 01062 Dresden, Germany
[†]IHP Microelectronics, Im Technologiepark 25, 15236 Frankfurt (Oder), Germany

Abstract—This paper demonstrates a 34 Gbit/s 4:1 multiplexer with 200 mW power dissipation in a 0.25-μm SiGe process. A multiphase clock architecture at half-rate clock is chosen to reduce the power dissipation. In addition, stability issues caused by the capacitively loaded clock driver are studied and a solution is proposed with a suitable compensation circuit. Measurement results are presented as well. The output signals yield open eye diagrams at the highest available test rate of 34 Gbit/s, while presenting rise and fall times below 12 ps.

I. INTRODUCTION

A parallel-to-serial data multiplexer (MUX) is one of the key components in optical communication systems as well as measurement instrumentation. Open literature offers many examples of silicon implementations of such component capable of operation at rates up to 30-44 Gbps [1]– [4]. While CMOS designs often provide lower power dissipation than their SiGe HBT counterparts, they need several inductors to work at the desired rate. In addition to larger area, this requires electromagnetic simulations of each specific inductor structure. On the other hand, SiGe-based designs are inherently suited for stacked transistor implementations of current-mode logic (CML) gates, thanks to their high breakdown voltages and small signal amplitudes required for switching. This eases an integration with high swing laser drivers in SiGe technology. The disadvantage of using such SiGe technology is the power required to drive other gates, as the loading effect of the following logic gate is higher than that of CMOS. In this work, a 0.25-μm SiGe process is utilized to realize a 4:1 MUX operating at 32 Gbit/s. This bit rate has over 10% margin over the specification for recent 100 Gbit/s optical communication systems, where due to high spectral efficiency modulation, the electronics need to operate at 28 Gbit/s. This paper presents the design and characterization of an inductorless 34 Gbit/s multiplexer in 0.25-μm SiGe technology. Design techniques at both system and circuit level are employed to achieve a power dissipation in the order of CMOS implementations.

II. MULTIPHASE MUX

The tree architecture, shown in Fig. 1, is suitable for implementation of MUXs operating at very high speed. However, it incorporates five latches in front of every 2:1 MUX, which makes the design of low power clock signal drivers challenging. The latches are utilized to generate a time shift between the input data and the applied clock signal. This, in turn, decreases the output glitches caused by synchronous data and clock changes at the input of the MUX. One possible method of omitting the aforementioned latches is to generate a

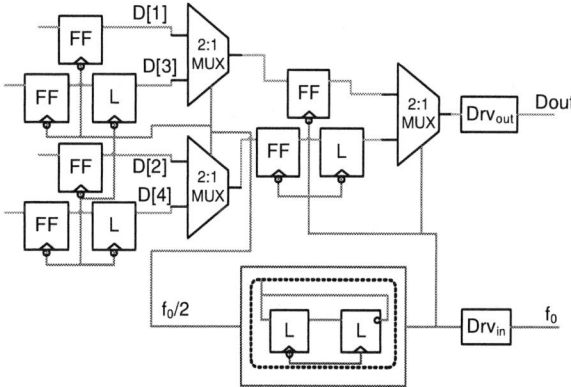

Fig. 1: Conventional 4:1 MUX tree architecture.

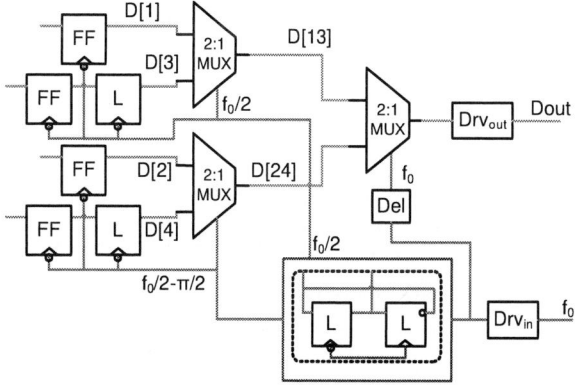

Fig. 2: Multiphase 4:1 MUX architecture.

multiphase clock and apply it to different MUXes, as shown in Fig. 2. In this case, instead of delaying the input data, the clock timing is adjusted such that synchronous change of the data inputs and clock signal is avoided [5]. Since the clock divider circuit is often realized by two fed-back latches, the required clock phases for this architecture are easily available. In [5], to improve the quality of the output signal, a full-rate clock has been applied. In this work, a multiphase clock architecture with a half-rate clock signal has been used. This allows to benefit from the lowered power gates at half the clock speed.

TABLE I: A comparison of power dissipation in multiphase and conventional implementations

	Multiphase 4:1	Conventional 4:1	Multiphase 8:1	Conventional 8:1
Clock Divider	$2 \cdot VI$	$2 \cdot VI$	$4 \cdot VI$	$3 \cdot VI$
MUX 2:1	$(1+2 \times 0.5) \cdot VI$[a]	$(1+2 \times 0.5) \cdot VI$	$(1+2 \times 0.5 + 4 \times 0.25) \cdot VI$	$(1+2 \times 0.5 + 4 \times 0.25) \cdot VI$
Latches	$(1+10 \times 0.5) \cdot VI$	$(5+10 \times 0.5) \cdot VI$	$(3+20 \times 0.25) \cdot VI$	$(5+10 \times 0.5 + 20 \times 0.25) \cdot VI$
Clock Driver	$(1+12 \times 0.5)/3 \cdot VI$[b]	$(6+12 \times 0.5)/3 \cdot VI$	$(24 \times 0.25)/3 \cdot VI$	$(6+12 \times 0.5 + 24 \times 0.25)/3 \cdot VI$
Sum	**12.3·VI**	**18·VI**	**17·VI**	**27·VI**

[a] The power of a gate at half the speed is assumed to be halved as well.
[b] The power of the clock driver is assumed to be 1/3 of the driven gates.

It is important to notice that in the multiphase architecture the clock phase margin is smaller and hence the timing of the clock applied to 2:1 MUXes should be adjusted properly. This can be achieved by adding some delay gates in the MUX clock signal path, as shown in Fig. 2. Using CML logic is beneficial in this case, as the gate latency is fairly robust to temperature and supply variations.

Adding the delay gates increases the power dissipation. On the other hand, having several different paths for the clock signal eases the layout of the clock distribution network. The multiphase architecture can be extended to higher MUX ratios. The power saved in 4:1 and 8:1 MUXes is compared to a conventional design in Table I. The comparison shows that the power required for latches is almost halved and so is the clock driver loading. This leads to more than 30% and 35% reduction in the dissipated power for 4:1 and 8:1 MUXes, respectively. The timing diagram of the 4:1 MUX is also shown in Fig. 3.

III. CIRCUIT DESIGN

Emitter-coupled logic (ECL) is normally used to realize GHz-range logic gates. In this logic, between every two gates at least one emitter-follower (EF) pair is utilized. The current required for the EFs is normally in the same order of the logic gate current; for some specific load conditions, it could be even more than that of the logic gate itself. Additionally, using two subsequent EFs for improved isolation can cause stability issues. In this work, the current of the EFs was carefully adjusted for low power functionality. The block diagram is shown in Fig. 2. The latches are based on a conventional implementation and are not discussed here.

A. Input Stage

At the clock and data inputs, a shunt 50 Ω resistor at the base of an EF provides the required broadband matching. The output of the EFs is connected directly to the latches for the data input. For the clock, an amplification stage is added. The purpose is to decrease the sensitivity to the amplitude of the input clock signal and generate a differential clock with well-defined amplitude levels from a single-ended input. The amplification also helps to decrease the rise and fall time of the input clock, whereas an external source generates a sinusoidal signal above 10 GHz. This can provide a faster switching for the following stages. Simulations showed that one Cherry-Hooper (CH) amplifier is sufficient to meet the bandwidth and rise time constraints for the target frequency of operation. The circuit, together with the input stage is shown in Fig. 4. The output of the CH amplifier is connected to two EFs and then to the dividing latches. Also the same signal is connected to the delaying block. The delay is implemented as a XOR logic, to provide a delay matched to one latch latency [5].

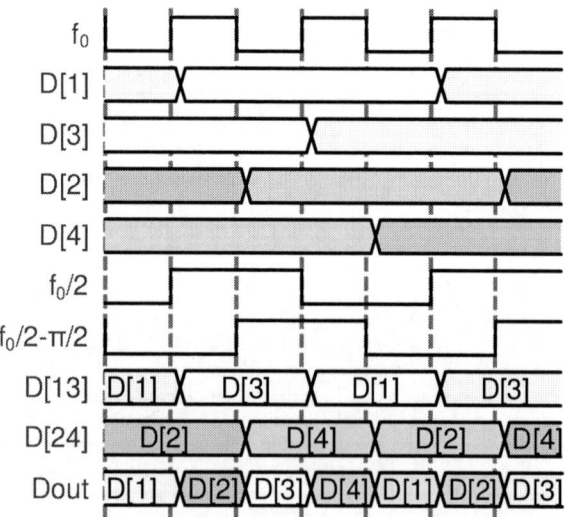

Fig. 3: Time diagram of multiphase 4:1 MUX.

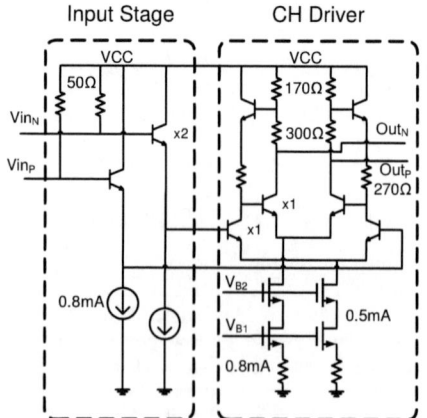

Fig. 4: Input stage and clock driver.

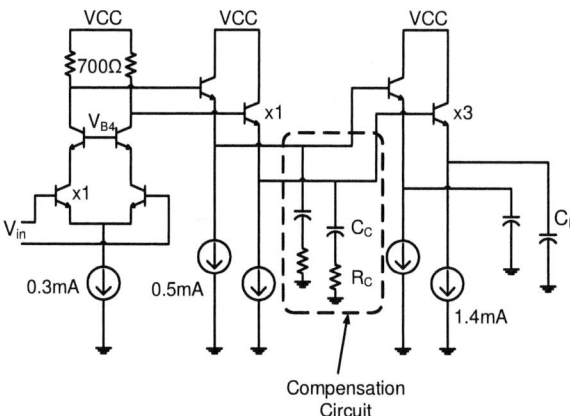

Fig. 5: 8 GHz clock driver.

B. Clock Driver

Once the 16 GHz clock is divided by two, the output should be fed to the input latches and the half speed 2:1 MUX. Hence, a clock driver is needed. After a differential amplifier, which serves as driver, the signal output level is close to the supply voltage. However, it has to be shifted by at least two EFs for appropriate functionality at the input of latches. Adding two EFs, provides very good isolation from the loading capacitance. This helps reducing the current of the amplifier. In our case, as illustrated in Fig. 5, only 0.3 mA was required for the cascode amplifier. On the other hand, two EF pairs with capacitive load, require much higher current and can influence the stability of the circuit. The load (C_L), is the input capacitance of other latches in addition to parasitic wiring capacitances. The value is approximated to be nearly 350 fF. To switch a bipolar differential pair, a voltage of five times the thermal voltage (V_T) is required. If we consider a quarter of the clock period as switching time, minimum current providing the slew rate is obtained as follows:

$$I_{min} = \frac{C_L \times (5V_T)}{\Delta t} = \frac{350fF \times 125mV}{1/4 \times (8GHz)^{-1}} = 1.4\,mA. \quad (1)$$

Because of current density limitations, a device three times the minimum transistor size was required. The current of the first EF is then chosen to be one-third and so minimum size transistor was used. Still, the output signal exhibits more than 20% overshoot and strong ringing. The reason is related to impedance transformation of EFs. Here, a strong capacitive load is translated to a negative resistance between the first and second EF [6]. This node is a low ohmic node because of the low impedance at the output of the first EF, which can easily make the circuit unstable. One approach to overcome this issue is the adding of a pole at this node, which can be implemented as a series RC network. The values of the compensation network are [6]

$$R_c = \frac{(C_\pi + C_L)^2}{g_m C_\pi C_L}, \quad \text{and} \quad C_c = \frac{C_\pi C_L}{C_\pi + C_L}; \quad (2)$$

in which g_m and C_π are the transconductance and base-emitter capacitance of the transistors in the second EF. The calculated

values in this design were C_c =40 fF and R_c =185 Ω. Simulations showed that, to have the best performance, the values should be C_c =38 fF and R_c =240 Ω, which is close to the calculated ones.

C. 2:1 MUX and Output Driver

Fig. 6 shows the circuits of the 32 Gbit/s 2:1 MUX and the output driver. The standard ECL topology was used for the MUX implementation. Minimum size transistors are biased close to their peak transit frequency point. A pair of EFs isolate the MUX circuit from the output driver. The output driver has to provide a signal amplitude higher than 150 mV at the external 50 Ω load. Basically by increasing the value of the internal load resistance, lower current can produce the same swing at the output, while tolerating some reflections. In this case, the internal resistance was not completely matched to the external one. A power reduction of more than 10% was achieved at the output driver by utilizing 65 Ω load resistors. To improve the signal opening at the output of the MUX, a CH amplifier stage was chosen [7]. Here, one stage could adequately improve the rise/fall times of the MUX and provide the required amplitude to drive the external load.

IV. MEASUREMENT AND COMPARISON

The circuit was fabricated in the 0.25-μm SiGe technology of IHP with 180 GHz transit frequency. The process also offers five metallization layers, in which the two topmost ones are thick to ease the implementation of microwave passive components. The die photograph is shown in Fig. 7. The chip dimensions are 1.1 mm×0.9 mm with total area of 1 mm². All the single-ended data inputs are on the same side (south), as well as each of the differential clocks and outputs. The external clock signal can be single-ended as well as differential. The north side is used for power supply, bias and other DC control voltages. The chip works from a single 3.3 V supply and requires 60 mA of supply current .

The characterization was carried out on wafer. An SHF 12100B bit pattern generator provides the high speed input bit stream. A commercial 1:4 demultiplexer then generates the 4 data inputs. These signals are fed to the MUX. The length differences on the four data paths were compensated by means

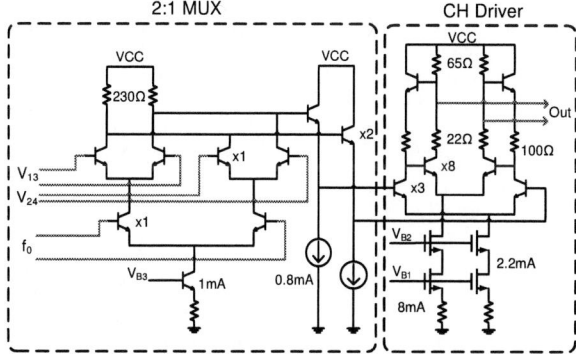

Fig. 6: 32 Gbit/s 2:1 MUX and CH output driver.

978-1-4673-4579-8/13 $31.00 © 2013 IEEE

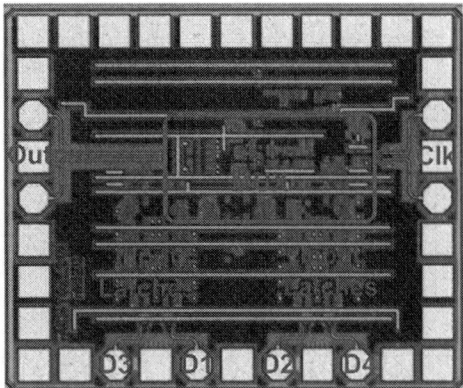

Fig. 7: Die photograph.

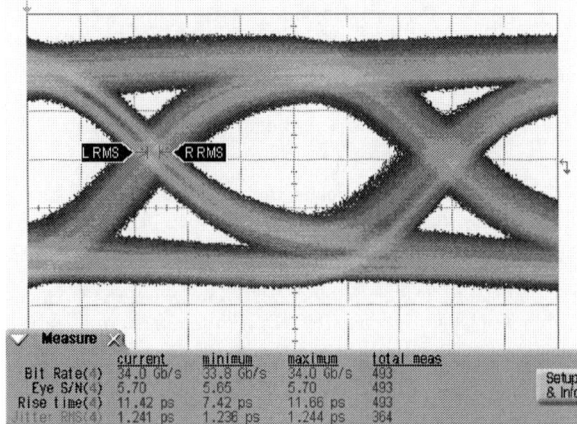

Fig. 8: Measured eye diagram of the single-ended output, the input is a PRBS 2^9-1 signal.

of mechanically-tunable phase shifters. An AC-coupled single-ended clock was used for the clock input. One of outputs is connected to a sampling scope, and the other one is connected to the SHF12100A error analyzer. Error-free functionality was observed for a PRBS 2^9-1 input stream at 34 Gbit/s. Fig. 8 shows the measured single-ended eye diagram for the same input pattern at 34Gbit/s. The low frequency probe used on the north side of the chip was longer than the chip length. This did not allow the high frequency probes to properly contact the chip, and thus the ground pins of the clock probe did not contact the on-chip ground pads. This, to some extent, affected the output signal quality and prevented operation at slightly higher frequencies. A comparison with state-of-the-art low power 4:1 MUXes is presented in Table II.

V. CONCLUSION

A 34 Gbit/s 4:1 MUX is designed in SiGe technology. It is shown that by using a multiphase architecture and appropriately adjusting the EF currents in ECL logic gates, the power dissipation can be decreased. Using half-rate clock

TABLE II: Comparison with other 4:1 MUX designs

Reference	[1]	[8]	[9]	[10]	This Work
Clock Freq. (GHz)	44.6	20	25	15	17
Output Data rate (GSps)	44.6	40	50	30	34
Power Diss. (mW)	635	220[a]	1475	70[b]	200
Swing (mV)	500	560	400	120	160
Technology	CMOS	CMOS	SiGe	CMOS	SiGe
Process Node (nm)	65	90	180	90	250

[a] The clock driver power is not included.
[b] The power of input latches is not included.

eliminates the very high speed gates, thereby significantly reducing the power consumption. The total power in this design was 200mW. A rise/fall time below 12ps was measured. The stability issues of two consequentive EFs with capacitive load, which are typically used for clock drivers, are discussed and a proper compensation circuit was designed.

ACKNOWLEDGMENT

This work has been partly supported by the German Research Foundation (DFG) in the Collaborative Research Center 912 Highly Adaptive Energy-Efficient Computing. The authors would like to thank IHP for providing the measurement equipments. M. Khafaji thanks Abo-Al-Fadhl for his help.

REFERENCES

[1] S. Kaeriyama et al., "A 40 Gb/s multi-data-rate CMOS transmitter and receiver chipset with SFI-5 interface for optical transmission systems," IEEE J. Solid-State Circuits, vol. 44, no. 12, pp. 3568–3579, Dec. 2009.

[2] H. Tao et al., "40-43-Gb/s OC-768 16:1 MUX/CMU chipset with SFI-5 compliance," IEEE J. Solid-State Circuits, vol. 38, no. 12, pp. 2169–2180, Dec. 2003.

[3] K. Kanda et al., "A single-40 Gb/s dual-20 Gb/s serializer IC with SFI-5.2 interface in 65 nm CMOS," IEEE J. Solid-State Circuits, vol. 44, no. 12, pp. 3580–3589, Dec. 2009.

[4] N. Nedovic et al., "A 3 Watt 39.8–44.6 Gb/s dual-mode SFI5.2 SerDes chip set in 65 nm CMOS," IEEE J. Solid-State Circuits, vol. 45, no. 10, pp. 2016–2029, Oct. 2010.

[5] T. Suzuki et al., "A 50-Gbit/s 450-mW full-rate 4:1 multiplexer with multiphase clock architecture in 0.13-μm InP HEMT technology," IEEE J. Solid-State Circuits, vol. 42, no. 3, pp. 637–646, Mar. 2007.

[6] P. Starič and E. Margan, Wideband Amplifiers. Dordrecht, The Netherlands: Springer, 2007.

[7] C. Holdenried, J. Haslett, and M. Lynch, "Analysis and design of HBT cherry-hooper amplifiers with emitter-follower feedback for optical communications," IEEE J. Solid-State Circuits, vol. 39, no. 11, pp. 1959–1967, Nov. 2004.

[8] H. Wang and J. Lee, "A 40-Gb/s transmitter with 4:1 MUX and subharmonically injection-locked CMU in 90-nm CMOS technology," in Proc. Symp. VLSI Circuits, Japan, Jun. 2009, pp. 48–49.

[9] M. Meghelli, A. Rylyakov, and L. Shan, "50-Gb/s SiGe BiCMOS 4:1 multiplexer and 1:4 demultiplexer for serial communication systems," IEEE J. Solid-State Circuits, vol. 37, no. 12, pp. 1790–1794, Dec. 2002.

[10] D. Kehrer and H.-D. Wohlmuth, "A 30-Gb/s 70-mW one-stage 4:1 multiplexer in 0.13-μm CMOS," IEEE J. Solid-State Circuits, vol. 39, no. 7, pp. 1140–1147, Jul. 2004.

A Nano-power Power Management IC for Piezoelectric Energy Harvesting Applications

M. Dini, M. Filippi, M. Tartagni, A. Romani

Department of Electrical, Electronic and Information Engineering "Guglielmo Marconi" (DEI)
University of Bologna
Via Venezia 52, 47521 Cesena, Italy
mdini@arces.unibo.it

Abstract—This paper describes a power management IC for piezoelectric energy harvesting which integrates an active AC-DC converter with residual charge inversion together with a smart self-supply architecture to speed up the start-up phase and increase harvesting effectiveness. Due to randomness of available power and to its typical intrinsic limitation to tens of μW the IC has been carefully designed to reduce quiescent current down to 150 nA while the efficiency is estimated to be at least 81%. Moreover this allows the system to be fully autonomous even with very low input energy. The IC has been designed in a 0.32 μm BCD technology from STMicroelectrics and its area is 4.6 mm².

Keywords— energy harvesting; piezoelectric transducer; nano-power circuits; synchronous charge extraction; energy storage.

I. INTRODUCTION

Energy harvesting has been widely explored in the last years with the purpose of developing a new class of autonomous systems, supplied by environmental energy. Vibrational energy is widely diffused, and piezoelectric transducers allow to achieve high power density during energy conversion [1]. Conversion schemes based on synchronous electrical charge extraction (SECE) [2-4] have been extensively studied (Fig. 1(a)) and have proved to be more efficient than passive interfaces (e.g. based on diode rectifiers). In many foreseen applications of autonomous systems it is crucial to store high amounts of energy, for example by using supercapacitors in order to compensate for fluctuations of harvested energy while sustaining load power requirements. Moreover, storing large amount of energy on small capacitances leads to voltage levels which are not compatible with new CMOS technologies. However, for self-powered systems, a passive (i.e. not actively controlled) conduction path, as could be a simple diode, is required to charge C_{ST} up to the minimum necessary voltage for active operations. In case of supercapacitors this phase can be very long, reducing the overall effectiveness of the energy harvester.

In this work a new highly efficient IC for piezoelectric energy harvesting and power management is presented. The IC is fully autonomous and presents intrinsic nano-power consumption. The proposed architecture, which introduces residual charge inversion and a two-way storage topology, implements optimized power management policies.

This research was funded by the European Community's FP7 2007-2013 under grant agreement Nanofunction no. 257375 and by the ENIAC Joint Undertaking under grant agreement END no. 120214.

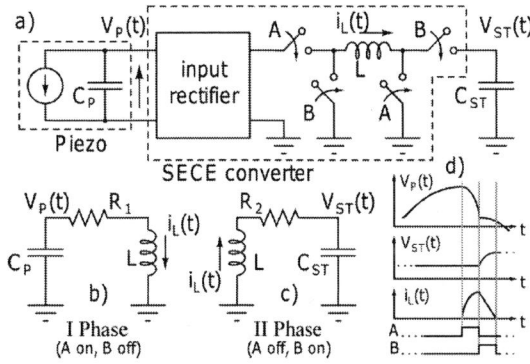

Fig. 1. (a) block diagram of a SECE converter, (b) equivalent circuit during first phase of SECE, (c) equivalent circuit during second phase of SECE, (d) qualitative waveforms under ideal conditions.

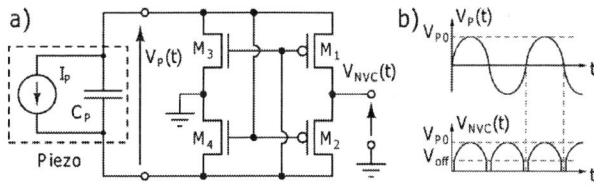

Fig. 2. (a) NVC schematic with piezoelectric transducer connected, (b) qualitative waveforms of NVC input and output.

II. INPUT STAGE

As the typical output voltage $V_P(t)$ of a piezoelectric transducer (PZ) is an AC signal, a rectifier stage is required. The SECE conversion scheme extracts energy from C_P when it reaches a local maximum. Thus, a negative voltage converter (NVC, Fig. 2 (a)), which converts the negative half-waves into positive ones but does not force a current direction, can be used instead of a classical diode full bridge. An advantage of this choice is that the NVC does not have voltage drop due to diode threshold and behaves like a resistor (the MOSFETs in Fig. 2 (a) work in triode region). In order to operate correctly, the NVC needs a minimum input voltage V_{off} (as in Fig. 2 (b)) equal to the highest threshold voltage between the N and P channel MOSFET. Similarly, the first phase of SECE conversion (Fig. 1 (b), signal A=on, B=off) cannot extract the whole energy on C_P as the NVC stops conducting before

978-1-4673-4579-8/13 $31.00 © 2013 IEEE   269

reaching C_P=0 C. Moreover, the residual charge Q_R=$V_{off}C_P$ on C_P prevents $V_P(t)$ from reaching the maximum possible absolute value after the next peak-to-peak elongation since the current I_P is integrated on C_P starting from V_{off} (i.e. Q_R) and not from $-V_{off}$ (i.e. $-Q_R$) as in Fig. 3 (a) (starting from $-V_{off}$ is advantageous because the sign of I_P in a semi-period is opposite with respect to the previous one, therefore $|V_P(t)|$ can reach $|V_{P0}+2V_{off}|$ instead of $|V_{P0}|$). An inversion of residual charge (RCI) is proposed, in a way similar to the SSHI principle in [2], by exploiting a resonant circuit (Fig. 3 (b) and (d)) to change the sign of the residual charge Q_R after an energy extraction cycle has been executed, in order to introduce an advantageous offset in $V_P(t)$ for the next elongation as shown in Fig. 3 (a). The parallel resonant circuit (Fig. 3 (d)) is composed of C_P and L_2 plus the resistance R_X of the switch and of L_2. Equation (1) shows the expression of the current $i_{L2}(t)$ considering t=0 s at the beginning of the RCI phase and where

$$a=R_X/2L_2, \ \omega_{02} = 1 \big/ \sqrt{L_2 C_P} \ , \ \omega_2 = \sqrt{\omega_{02}^2 - a^2} \ .$$

$$i_{L2}(t) = \frac{Q_R}{\omega_2 L_2 C_P} e^{-a2t} \sin(\omega_2 t) \qquad (1)$$

The duration t_{RCI} of the RCI phase (signal X=on in Fig. 3 (b)) is half of the oscillation period: at time t_{RCI}=π/ω_2 $i_{L2}(t_{RCI})$=0 A, Q_R has been inverted and the switch can be safely opened. Integrating (1) and considering the initial conditions ($V_P(0)$=Q_R/C_P) the PZ voltage after the RCI phase can be expressed as

$$V_P(t_{RCI}) = \frac{1}{C_P} \left(Q_R - \int_0^{t_{RCI}} i_{L2}(t)dt \right) = -\frac{Q_R}{C_P} e^{\frac{a\pi}{\omega}} . \qquad (2)$$

The increase of performance of the converter then depends on Q_R (i.e. the threshold voltage of the NVC), the precision of timing and the parameter a. RCI provides a significant increase to the maximum voltage reached by $V_P(t)$ and thus to the available energy. Fig. 4 shows the theoretical effect of RCI on the available energy, obtained from analytical calculations on energy relations and numerical simulations to evaluate the effect of realistic component values. RCI increases its relative effectiveness when the maximum of V_P is close to V_{off} (about

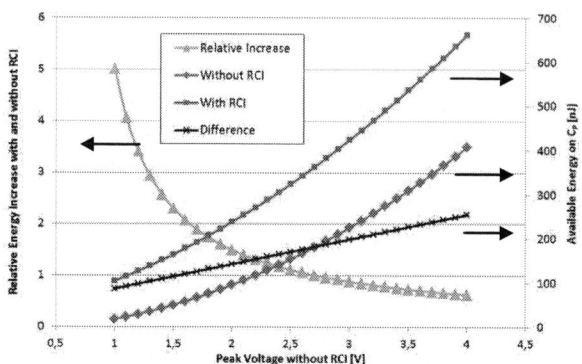

Fig. 4. Energy on C_P with RCI (red, square) and without RCI (blue, diamond), difference (absolute energy gain) between them (black, cross) and relative increase of energy (green, triangle). Arrows specify the correct axis for each curve. Curves are obtain from analytical calculations and numerical simulations with C_P=52 nF, L_2=10000 µH and R_X=20 Ω, V_{off}=580 mV.

580 mV in this design), i.e. in case of weak vibrations, the absolute energy gain is directly proportional to the maximum of V_P.

III. IC ARCHITECTURE

Fig. 5 depicts the architecture of the system, designed in a 0.32 µm BCD technology from STMicroelectronics. As shown (gray boxes), the converter can be divided in sub-systems with specific tasks. The required external components are the inductors L_1 and L_2, the capacitors C_{DD} and C_{ST}, and the resistor shown. A single inductor could be theoretically used, as it would still be used with a very low duty cycle. In order to speed up the start-up time, an additional capacitor C_{DD} has been added (two-way storage) and is responsible for supplying the power conversion circuits. The use of C_{DD} allows the active operation of the converter even if C_{ST} is fully discharged. The converter is fully operative when V_{DD}>1.4 V. Until that moment C_{DD} is passively charged through a conductive path that is cut off by an under-voltage lock-out (UVLO) circuit. A second UVLO (Voltage Monitor block in Fig. 6), with a nominal threshold of 2.5 V and an hysteresis windows of several hundreds of mV, decides if the energy harvested by the SECE circuit has to be directed towards C_{DD} or C_{ST}, by controlling two active rectifiers. This policy allows to charge

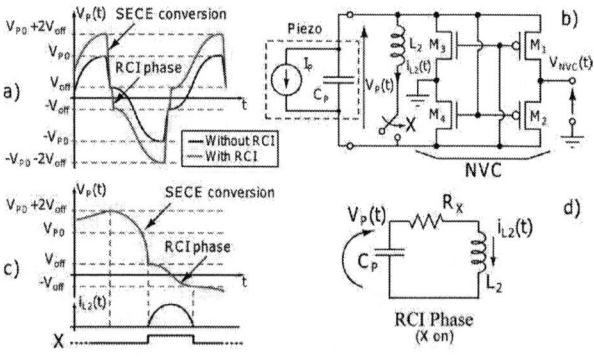

Fig. 3. (a) Qualitative waveforms of $V_P(t)$ with (red) and without (black) RCI, (b) NVC with RCI circuit, (c) qualitative waveforms of SECE extraction cycle and RCI phase, (d) equivalent circuit during RCI phase.

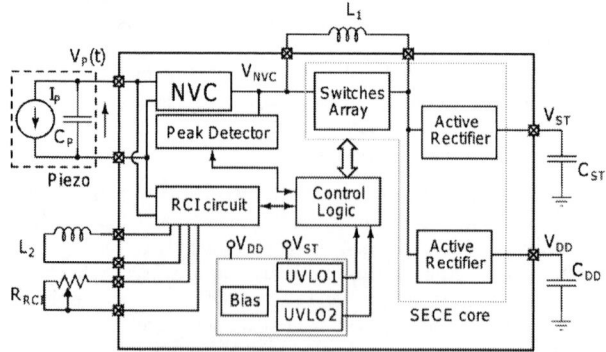

Fig. 5. block diagram of the designed converter.

C_{DD} with priority when it is getting too discharged to supply the converter.

The piezoelectric transducer (PZ) is connected to a NVC which converts the negative half-wave to a positive one, without forcing the current direction. As shown in Fig. 6, an analog peak detector monitors V_{NVC} determines the correct timing for starting each conversion cycle. However, the PZ voltage can exceed the supply voltage of the converter and, for this reason, the peak detector core circuitry is supplied by the higher voltage between V_{NVC} and V_{DD}. This is accomplished with the circuit formed by M_{14} and M_{15} in Fig. 6. In case V_{DD} and V_{VNC} are almost equal, conduction occurs through the body diodes of $M_{14,15}$ and V_{HH} is thus lower of about a pMOS threshold voltage. However, a diode-connected pMOS M_{13}, which has a lower threshold voltage, also feeds the V_{HH} node from V_{DD}, with a limited amount of current (10 times I_{bias}), in order to reduce the voltage drop on V_{HH}. In order to be compliant with the common mode input voltage range V_{CMi} of comparators and in order to avoid malfunctions on supply switching, the peak detector is fed by a down-shifted version of V_{NVC} ($M_{5...7}$ in Fig. 6). Circuit simulations show an increase of available input energy of at least 42% (with respect to the same converter without RCI and with V_{P0}=3.7 V) due to RCI, which is performed by connecting the PZ to L_2 for half of the oscillation period of the C_P-L_2 resonant circuit. The duration t_{RCI} of RCI is constant and in first approximation depends on C_P and L_2. In this implementation an external trimmer is used to set the proper timing (R_{RCI} in Fig. 5). RCI is started together with the second phase of SECE (Fig. 9).

The SECE core integrates the switches for connecting L_1 to the two active rectifiers, for charging alternatively C_{DD} and C_{ST}, and the analog circuitry to detect the end of the first and second phases of SECE (ZCV and ZCC signals, respectively activated at zero voltage/current on the inductor) as shown in Fig. 7. The

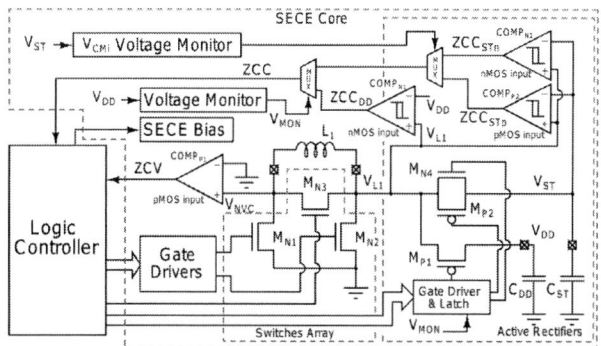

Fig. 7. SECE core schematic. Acronyms ZCV (Zero-crossing voltage) and ZCC (Zero-crossing current) represent the stop signals of Phases I and II of SECE conversion.

controller is implemented with asynchronous logic in order to reduce power consumption due to clock switching and distribution. Moreover, it communicates to the other blocks (RCI, PZ circuitry) the start and stop commands of each phase. The designed active rectifiers are slightly different as shown in in Fig. 7: the one responsible for charging C_{DD} has only a pMOS switch, since the voltage on C_{DD} is always in the V_{CMi} range of a n-type differential couple; the other one has a nMOS/pMOS switch and two comparators with different V_{CMi} and the correct output is chosen by a voltage monitor (V_{CMi} Voltage Monitor block in Fig. 7). In order to minimize circuit absorption the comparators in Fig. 7 are kept shut down except during conversion phases. M_{N3} has the function of creating a conductive path in parallel to L_1 to slowly dissipate the residual energy and so avoid ringing. The same principle is implemented in the RCI circuit for L_2.

The circuit is fully-autonomous and the quiescent current is 150 nA, leading to 400 nW at 2.7 V on C_{DD}. For the simulations of the IC, performed with Eldo® from Mentor Graphics®, the values in Table I have been used and simulation waveforms are shown in Fig. 8 and Fig. 9. The net efficiency, including circuit consumption, of the SECE converter is at least 81% during active operation of the IC. The

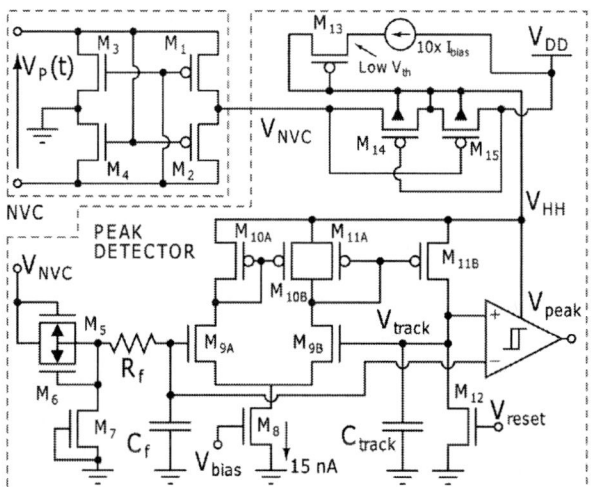

Fig. 6. schematic of PZ-specific circuitry: M_1-M_4 NVC, M_5-M_7 and R_f, C_f create a scaled and filtered version of V_{NVC} to feed the peak detector, M_8-M_{12} and the comparator form the peak detector, M_{13}-M_{15} and the current generator realize the circuit to obtain the higher supply (V_{HH}) between V_{DD} and V_{NVC}.

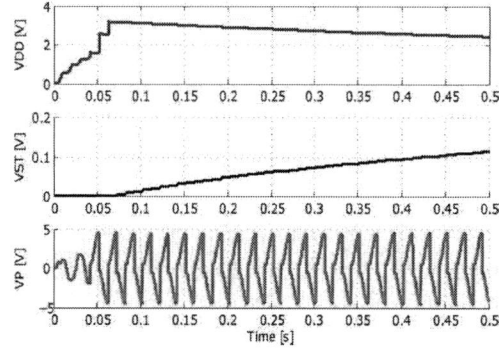

Fig. 8. Simulation waveforms of start-up showing the two-way storage (V_{DD}, the blue curve on top, is quickly charged before V_{ST}, black curve). Until about 40 ms the IC operates in passive mode, as can be seen in the V_P curve (in red, on bottom); instead in the active operation mode V_P is chopped.

Paper W2A1   *PRIME 2013, Villach, Austria*

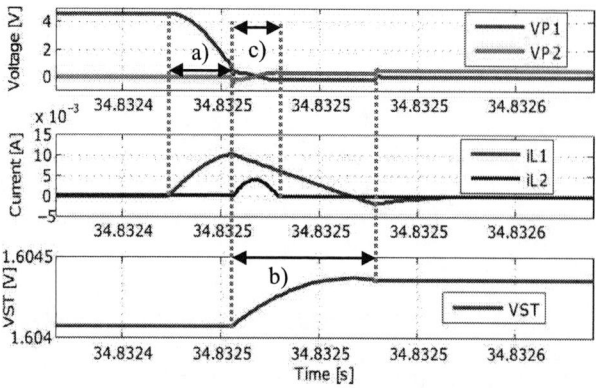

Fig. 9. simulation waveforms showing a SECE conversion and an RCI cycle; a) I phase of SECE conversion, b) II phase of SECE conversion, c) RCI cycle.

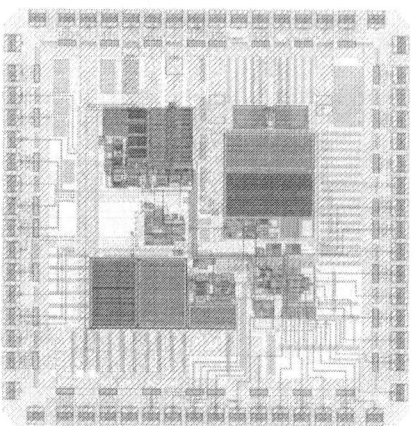

Fig. 11. layout of the designed IC.

effectiveness of the designed two-way storage is show in Fig. 10: simulations, performed with the same storage capacitor C_{ST}=1 mF and same input vibrations, show that the presence of the two-way storage improve performance significantly in the initial start-up with respect to the same converter without the presence of C_{DD} and its associated active rectifier. The small value of C_{DD}, comparable with C_P, allows the converter to start almost immediately in the active operation mode, increasing effectiveness of energy extraction (blue waveform in Fig. 10) with respect to the passive operation mode (red waveform in Fig. 10). Supply voltage V_{DD} is kept in the 2.2-3 V range (the UVLO circuit has a smaller hysteresis window but, due to the value of C_{DD} comparable with C_P, it is not possible to have a close control on the upper limit of V_{DD} as each SECE conversion contributes with an unpredictable energy packet which can exceed the nominal required energy). Fig. 11 shows the final layout of the designed IC. Chip dimensions, which were not specifically optimized, are 2142 µm for each side and total area is slightly less than 4.6 mm² and the active area used by the converter is 0.92 mm².

TABLE I. COMPONENT VALUE USED IN SIMULATIONS.

Name	Value	Name	Value
L_1	10 mH	C_{DD}	220 nF
L_2	10 mH	C_{ST}	1 mF
R_{RCI}	13.95 MΩ	C_P	52 nF

IV. CONCLUSIONS

A power management IC for piezoelectric energy harvesting has been designed in a 0.32 µm BCD technology from STMicroelectronics. It integrates an actively controlled AC-DC converter with RCI to enhance energy extraction of over 42% and a smart supply scheme to speed up the start-up phase to increase efficiency in charging a supercapacitor. Due to its quiescent consumption, as low as 150 nA, the IC is fully autonomous and can achieve a positive energy budget even with very low and irregular vibrations.

REFERENCES

[1] S. Roundy, et al., "A study of low level vibrations as a power source for wireless sensor nodes," *Computer Communications*, vol. 26, no. 11, pp. 1131-1144, Jul. 2003

[2] E. Lefeuvre, A. Badel, C. Richard, L. Petit, D. Guyomar, A comparison between several vibration-powered piezoelectric generators for standalone systems, Sensors and Actuators A: Physical, Volume 126, Issue 2, 14 February 2006, Pages 405-416

[3] A. Romani, C. Tamburini, R. P. Paganelli, A. Golfarelli, R. Codeluppi, E. Sangiorgi, M. Tartagni, *"Dynamic Switching Conversion for Piezoelectric Energy Harvesting Systems"*, IEEE Sensors 2008, pp. 689-692

[4] Dallago, E.; Miatton, D.; Venchi, G.; Bottarel, V.; Frattini, G.; Ricotti, G.; Schipani, M., "Electronic interface for Piezoelectric Energy Scavenging System," Solid-State Circuits Conference, 2008. ESSCIRC 2008. 34th European , vol., no., pp.402,405, 15-19 Sept. 2008

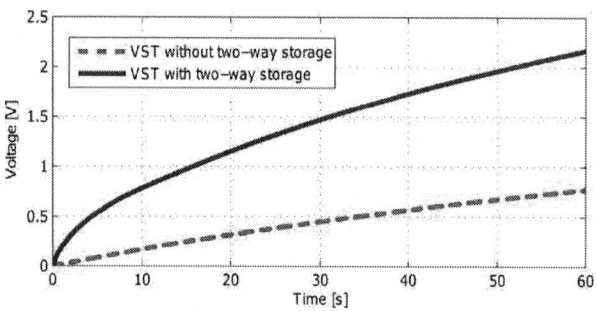

Fig. 10. Simulation waweforms showing the comparison of the same designed SECE converter with and without the two-way storage. The presence of C_{DD} speeds up the start-up time of the converter (blue waveform) and increase havesting efficiency with respect to the same converter without C_{DD} and its associated active rectifier. For both simulations C_{ST}=1 mF and same input vibrations are used.

Portable Battery Charging circuits for Enhanced Magnetic Resonance Wireless Power Transfer (WPT) System

Hyun Jin Choi, Eun Hye Ahn, Sung Yeul Park and Jun Rim Choi
School of Electronics Engineering
Kyungpook National University
Daegu, Republic of Korea, 702-701
Email: hjchoi21@gmail.com

Abstract—In recent years there has been significant development in the field of consumer electronics and biomedical applications based on wireless power transfer technology. In this paper, we propose efficiency improvement method for magnetic resonance Wireless Power Transfer (WPT) at a particular frequency within ISM band, applying controlled Impedance Matching technique. The impedance matching network guides the system to maintain the resonance condition when distance between Transmitter (Tx) and Receiver (Rx) is varied, which results in high power transfer efficiency. Measured results with proposed scheme showed an average of 7.21% improvement in power efficiency for distance between 10cm to 100cm. In addition an on-chip high efficient power recovery scheme is designed and fabricated in Dongbu HiTek's (DBHs) 0.35μm BCD technology based on LOCOS 0.35μm CMOS process for portable battery recharging applications. A maximum overall efficiency be 70.2% was determined for the entire system. This work shows the potential of automatic adjustments of parameters, such as impedance, in order to maintain resonance frequency to attain the maximum power efficiency and also to make the device portable.

Keywords—Magnetic resonance, WPT, impedance matching, power recovery, system on chip (SoC).

I. INTRODUCTION

The coupled magnetic resonance WPT system proposed in [1], [2], [3], improved power efficiency and transmission range compared to the inductive coupled systems, was a potential breakthrough for mid-range energy transfer. In this paper we propose a method to improve delivered power efficiency in coupled magnetic resonance WPT system by controlled impedance matching technique. The miniaturized resonator structure was analyzed through structural simulation and prototype production for experimental verification of the proposed technique. The associated Q-factor of the resonator enables to analyze the transfer efficiency of the WPT system. Revision process of the Q-factor guides the proper matching of the impedance to attain high power transfer efficiency, which was successfully verified in our design. In addition an integrated circuit was designed with the concept of reducing power loss in the rectifier and Low Drop Out (LDO) voltage regulator for effective use of received power. The design was implemented to effectively improve the total power efficiency of WPT and power recovery. The system for designed for the purpose of smart phone battery charging application.

II. SYSTEM BLOCK

The system block diagram of WPT and Power Recovery scheme is shown in Fig. 1. The controlled impedance matching network aids the efficiency improvement of the WPT system by maintaining resonance condition through proper matching of impedance. The Integrated Chip (IC) recovery system effectively converts the received power to safely charge portable batteries.

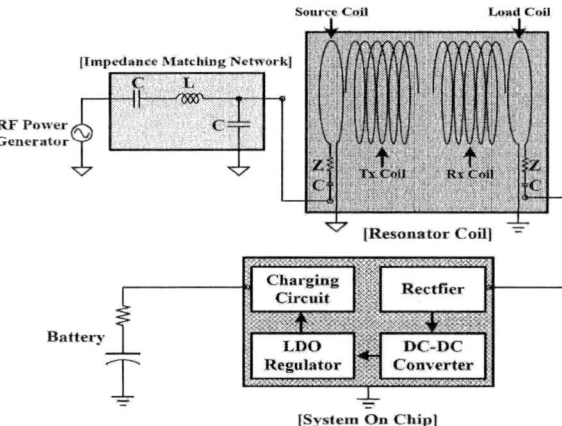

Fig. 1. Block diagram of the WPT and Power Recovery System.

A. Resonator structure design

The resonator structure for the experiment was designed and verified with High Frequency Structure Simulator (HFSS) tool, which involved the interpretation of electromagnetic field in various environment. We incorporated helical structure in our design process and optimized the size to attain maximum power transfer efficiency, for a distance varying from 10cm to 100cm. We employed copper EMG cable of diameter 20cm, 5 turns, 5mm thickness and 5mm pitch for the design purpose. The designed resonator structure and magnetic field generated in HFSS is shown in Fig. 2.

The simulated power transfer efficiency curve for varying distances is depicted in Fig. 3. The graph shows maximum efficiency at 20cm where the Q-factor is determined to be the highest, that allows the system to resonate and transfer maximum power. When Tx and Rx are $10cm$ apart the reflected power from the Rx end reduces the transfer efficiency of the system. The graph shows gradual decrease from $20cm$ to $30cm$, which shows the optimum distance of power transfer for the designed helical structure. Beyond $30cm$ efficiency curve shows a steep descent, due to the variation in Q-factor that results from poor coupling, increased distance between Tx & Rx and, environmental factors. In this scenario coupling between the coils can be increased by proper impedance matching of the system, which we realize for obtaining higher power transfer efficiency.

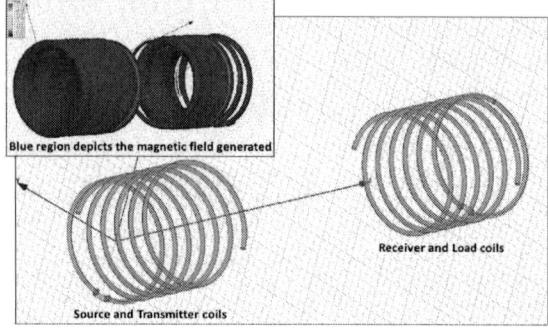

Fig. 2. Resonator structure simulation using HFSS.

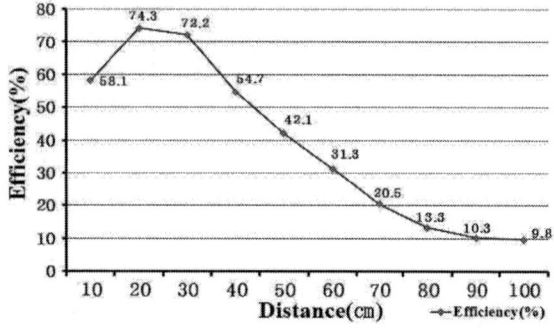

Fig. 3. Transfer efficiency curve for varying distance (HFSS simulation).

B. Analysis of Controlled Impedance Matched WPT

Magnetic resonant coupling is based on generating LC resonance and transmitting the power with electromagnetic coupling. Increasing the air gap between the Tx and Rx coils will weaken the the coupling, thus reducing the coupling coefficient. The variation in coupling coefficient result in impedance mismatch which further results in resonance frequency splitting and reduced power transfer efficiency [2], [3]. The resonating coil can be represented as a combination of capacitance and inductance [4].

Resonance phenomenon in WPT system can be maintained either by varying the optimum resonance frequency or by proper impedance . The latter method is apt for maintaining resonance at preferred frequencies and more economical as it requires simple and less expensive hardware. The impedance

matching technique usually involves inserting a matching network such as LC network to minimize the reflection to the power source thus improving power efficiency. We employ this technique in our proposed scheme. Initially a perfectly matched resonant system is projected, which resonates at 13.56 MHz, delivering maximum power to the load. We then vary distance between Tx and Rx which would change the coupling coefficient between coils and resulting in impedance mismatch and dual resonant frequency. We bring back the system to resonating state, for the new distance, by impedance matching with the help of variable capacitors within the tuning network as in Fig. 2. The auto tuning of impedance can be brought about by including a directional coupler and Micro Controller Unit (MCU) at transmitter side. The directional coupler and MCU sense the reflected power and adjust the impedance of the circuit until the reflected power gets minimized. The minimum power reflected under ideal operating conditions results only when coils are resonating at the supply frequency. The final equivalent circuit of WPT system with impedance matching network is shown in Fig. 4. The steady-state admittance of the circuit is calculated as (1)

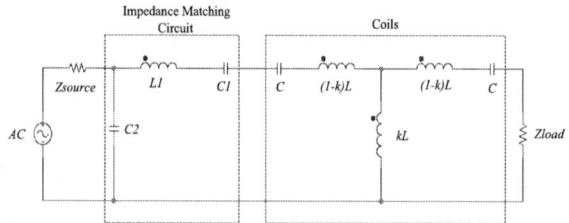

Fig. 4. Equivalent circuit with impedance matching Network.

$$Y = R + \frac{(\frac{A}{B} + j(\omega((1-k)L + L1) - \frac{1}{\omega}(\frac{C*C1}{C+C1}))) * \frac{-j}{\omega c2}}{\frac{A}{B} + j(\omega((1-k)L + L1) - \frac{1}{\omega}(\frac{1}{C} + \frac{1}{C1} + \frac{1}{C2}))} \quad (1)$$

$$A = (R + j(\omega(1-k)L - \frac{1}{\omega C})) * j\omega kL, B = R + j(\omega - \frac{1}{\omega C}) \quad (2)$$

where k is the coupling coefficient and R is the source impedance which is purely resistive. The imaginary part of (1) is equated to 0 to obtain resonance frequency, which is now dependent on the C1, C2, and L1. The designed structure has been simulated using the equivalent circuit in ADS tool to verify the working of proposed impedance matching technique.

C. Power Recovering Topology

Most of the power recovery schemes with high power efficiency, were based on low voltage recovery circuits, with intention to increase the total efficiency of the system. To overcome the flaws, in the less power efficient high voltage prior designs employing discrete components, full on chip high voltage design is recommended [5]. Our case study involved designing a scheme for recharging a portable battery with specifications $3.7V$, $6.11Wh$, and $1680mAh$.

978-1-4673-4579-8/13 $31.00 © 2013 IEEE

1) Semi-Active High Voltage Rectifier and DC-DC Boost Converter : We propose a new semi-active rectifier in HV-MOS technology based on fractional and adaptive threshold cancellation technique. The bias voltage of HM4 and HM3 is maintained between 85% and 65% of threshold voltage in the conduction and non-conduction phases, reducing the leakage currents for obtaining high PCE.

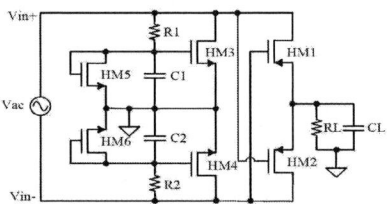

Fig. 5. Proposed high voltage rectifier Scheme.

An on chip boost converter has been proposed and designed to boost the voltages below $6V$, the control circuitry designed was inspired from [6], the proposed design was capable of boosting the voltages as low as $3V$ to $6V$.

2) Temperature Independent Voltage and Current Reference Generator: We designed a voltage reference and bias current generator independent of supply voltage and temperature variations based on subthreshold MOSFET technique. We exploited the features in [7] & [8], to realize our design in HV environment ranging to $40V$.

3) LDO Voltage Regulator and Charging circuit : We designed a LDO voltage regulator based on cascode compensation technique that offers unconditional stability and a wider range of output capacitance [9], [10]. The wide range of capacitances provide the circuit to be employed to charge batteries with various configurations. The proposed LDO is shown in Fig. 6

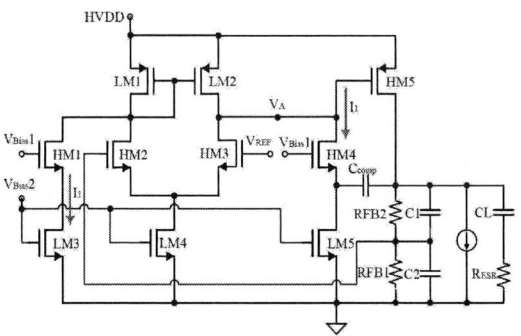

Fig. 6. LDO voltage regulator

The charging circuit developed was inspired from model in [11]. To ensure safe charging of the battery we developed the control circuit to follow the typical charging profile during the charging, i.e, constant current (CC) followed by constant voltage (CV) charging. Die microphotograph of the fabricated system on chip is shown in Fig. 7

Fig. 7. Die microphotograph of the integrated chip, size $5mm$ x $5mm$

III. RESULTS

The proposed scheme was fabricated for real time experiments. The coils were constructed using copper. An RF power generator with negligible internal loses, supplied 2W RF power at $13.56MHz$ frequency. A maximum Power transfer efficiency of 74.3% was determined when the coils were $20cm$ apart. The measured resonating frequency varied within $\pm20KHz$ to the estimated resonant frequency. The Fig. 8 depicts the the measured voltage and current at the Rx end after proper impedance matching is established. The frequency is swept from 0 to $15MHz$ and we observe that the system resonates approximately at $13.56MHz$. The voltage and current harmonics that exists at other frequencies are due to the Q-factor variations, that deprives the system to attain the expected power transfer efficiency.

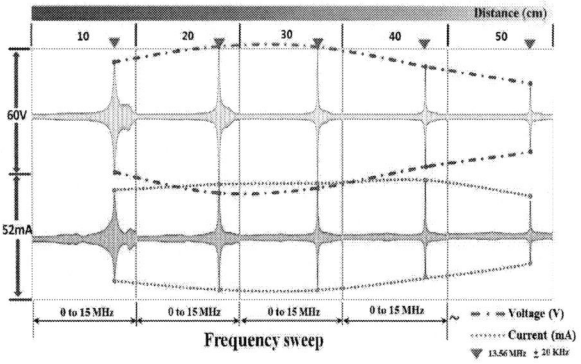

Fig. 8. Measured current and voltage at the load coil.

Fig. 9 shows the efficiency of the WPT system before & after impedance matching, and the improvement in efficiency for varying distances.

Measured results imply an average of 7.21% improvement in power efficiency with the proposed scheme when distance between Tx and Rx is varied from $10cm$ to $100cm$.

When on-chip circuits were analyzed, the rectifier obtained maximum PCE of 93.5% and voltage conversion efficiency of 87.8% with $5K\Omega$ load resistance when the supply voltage was $50V$ peak to peak. The DC-DC converter boosts voltages between $2.7V$ and $6V$ to approximately $7.3V$ and determined

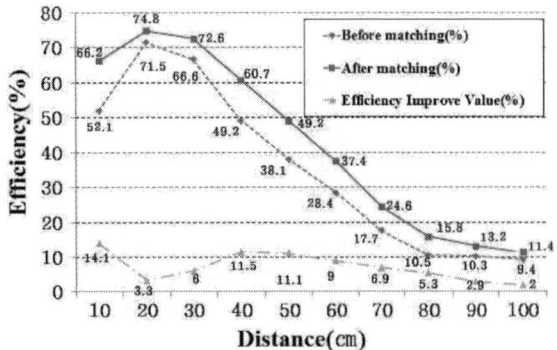

Fig. 9.   Power transfer efficiency before and after impedance matching.

power efficiency was 47% when the input was $4V$. The reference voltage generator, generates two bias voltages $4V$ and $1.5V$. The temperature sensitivity of reference voltage was determined to be $8mVp - p$ and supply voltage sensitivity was $6mV/V$. The results obtained for the LDO regulator are depicted in table 1. The final experimental setup is shown in Fig. 10. The figure includes the RF power generator, the controlled impedance matching network and the resonator. The charging of the smart phone using the fabricated circuit is also depicted in the figure.

Fig. 10.   Experimental setup.

TABLE I.   PERFORMANCE SUMMARY OF THE LDO-REGULATOR

Parameter	Value	Conditions
Output Voltage	$8V \pm 90mV$	Vin=8 to $40V$; Iout=0.5 to $200mA$
Dropout voltage	$180mV$ Mean	Iout=$1mA$
Line Regulation	$12mV$	Vin=8 to $40V$; Iout=$1mA$
Load Regulation	$22mV$	Vin=$8V$; Iout=1 to $200mA$

The designed charging control circuit, ensures safe and reliable battery charging. The transition from CC to CV was observed to gradual and immediate. The current dropped from $180mA$ at $3.68V$ to $40mA$ at $3.7V$, ensuring the safety of the battery. The excess $40mA$ current was sourced out by the external protection circuit preventing overcharge of battery. The final experimental setup of the is shown in Fig. 10. The

maximum overall efficiency of the system based on measurement results was determined to be 70.2% approximately.

IV. CONCLUSION

In this paper, we proposed an enhanced magnetic resonance WPT system for restricted frequency band and a high voltage power recovery scheme for charging portable battery. Our main objective was to improve the WPT for particular frequency was achieved by employing impedance matching technique and also to implement a fully integrated power recovery scheme for safe charging of batteries. We reported the design on high voltage IC fabricated in DBHs $0.35\mu m$ BCD technology based on LOCOS $0.35\mu m$ CMOS process. The power transfer efficiency can be further increased if the material with very low resistance is used for the design of resonator. This work has great contributions towards consumer & medical electronics other applications.

ACKNOWLEDGMENT

This work was supported by Ministry of Knowledge Economy(MKE) and IDEC Platform center(IPC) and Industrial Strategic Technology Development Program funded by the Ministry of Knowledge Economy (MKE, Korea)(10039173, Design of Core SoCs for Multimedia Mobiles).

REFERENCES

[1] A. Kurs, A. Karalis, R. Moffatt, J. D. Joannopoulos, P. Fisher, and M. Soljačić, "Wireless power transfer via strongly coupled magnetic resonances," *Science*, vol. 317, no. 5834, pp. 83–86, July 2007.

[2] A. Karalis, J. D. Joannopoulos, and M. Soljačić, "Efficient wireless non-radiative mid-range energy transfer," *Ann.Phys.*, vol. 323, no. 1, pp. 34–48, January 2008.

[3] M. K. Kazimierczuk and D. Czarkowski, *Resonant Power Converters*, 2nd ed.   New York, NY: IEEE Press and John Wiley & Sons, 2011, no. 978-0-470-90538-8.

[4] T. Imura, H. Okabe, and Y. Hori, "Basic experimental study on helical antennas of wireless power transfer for electric vehicles by using magnetic resonant couplings," in *IEEE Vehicle Power and Propulsion Conference*, September 2009, pp. 936–940.

[5] F. Mounaïm and M. Sawan, "Integrated high-voltage inductive power and data-recovery front end dedicated to implantable devices," *IEEE Transactions on Bio-Medical Circuits and Systems*, vol. 5, no. 3, pp. 283–291, June 2011.

[6] C. S. Lee, E. J. Kim, M. Gendensuren, N. S. Kim, and K. Y. Na, "High performance current-mode dc-dc boost converter in bicmos integrated circuits," *Transactions on Electrical and Electronic Materials*, vol. 12, no. 6, pp. 262–266, Dec 2011.

[7] P. Huang, H. Lin, and Y. T. Lin, "A simple subthreshold cmos voltage reference circuit with channel- length modulation compensation," *IEEE Transactions on Circuits and Systems II: Express Briefs*, vol. 53, no. 9, pp. 882–885, September 2006.

[8] H. Ballan, M. Declercq, and F. Krummenacher, "Design and optimization of high voltage analog and digital circuits built in a standard 5v cmos technology," in *IEEE Custom Integrated circuits Conference*, 1994, pp. 574–577.

[9] J. B. Wiser and R. Reed, "Current source frequency compensation for a cmos amplifier," United States Patent Patent 4,484,148, November 20, 1984.

[10] G. A. Rincon-Mora and A. Gabriel, *Analog IC Design With Low-dropout Regulators*.   McGraw-Hill, 2009.

[11] P. Li, J. C. Principe, and R. Bashirullah, "A wireless power interface for rechargeable battery operated neural recording implants," in *Proceedings of the 28th IEEE EMBS Annual International Conference*, Aug 2006, pp. 6253–6256.

Monolithic Power Management Front End with High Voltage Dense Energy Storage for Wireless Powering

Hans Meyvaert and Michiel Steyaert
Katholieke Universiteit Leuven
Dept. Elektrotechniek, afd. ESAT-MICAS
Heverlee, Belgium
{hans.meyvaert, michiel.steyaert}@esat.kuleuven.be

Arne Crouwels and Stijn Indevuyst
{arne.crouwels, stijn.indevuyst}@gmail.com

Abstract—**A monolithic power management system is proposed, enabling an energy storage density improvement of more than an order of magnitude with respect to the current state of the art. This is made possible by increasing the storage voltage V_{bat} towards 10V. A proof of concept prototype was designed in a 90nm bulk CMOS technology with an integrated 1.8mm dipole antenna using the surrounding energy available in the 5.8GHz ISM band. In order to achieve the large voltage conversion under aim of less than 100mV to 10V, a two stage approach is found necessary as the input impedance mismatch of a single stage passive voltage multiplier with the antenna would render the solution infeasible. First, the antenna voltage is passively rectified and multiplied to an intermediate system supply voltage V_{PMU} of 1.2V, aftwer which this voltage is pumped up to the storage voltage level V_{bat} of 10V by an active DC-DC converter. This, for a minimal system startup voltage of 53mV.**

Keywords—*Wireless powering, on-chip antenna, AC-DC voltage multiplier, capacitive DC-DC converter, high voltage energy storage.*

I. INTRODUCTION

Together with the increasing demand for radio frequency identification (RFID) and other stand-alone wireless systems such as in sensor networks, there is a growing need for custom power management units (PMUs) that enable proper operation of the load circuits for any given incident energy. The energy budget of wireless systems is typically low and the necessity to also wirelessly transfer data only complicates the use of this budget. It is therefore key to be able to store sufficient energy to sustain circuit operation until the end of an action, even when the incoming power level is low or zero.

This work investigates the issue of dense energy storage to enable improved performance in low cost, small size wireless powered circuits. Instead of storing the incident energy on a capacitor at less than 2V, this work proposes to use a much higher voltage to increase the stored energy by more than an order of magnitude. The storage density gained by such an approach can be valorised by having an increased energy budget for an identically sized system. Alternatively, the same energy budget can now be achieved with a relaxed specification for the storage capacitor. This can allow a previously external capacitor to be replaced by an integrated capacitor on chip, substantially decreasing the cost and size of the end solution.

Previously reported systems tackle the energy budget problem either using external components such as antennas to

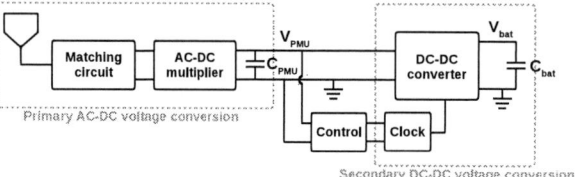

Fig. 1. Block diagram of the proposed energy storage system.

boost the capture of incident power [1] and external storage capacitors to achieve a sufficient energy storage [2] or end up with a limited energy budget, but achieve full integration such as [3]. The integrated storage of [3] consists of a 6nF capacitor charged to 1.8V, which can be discharged down to 1.2V while a linear regulator provides provides a regulated supply voltage. If 10V instead of 1.8V was stored, then the useful energy would be increased by more than a factor 50. It is clear that even with step-down conversion efficiency loss, which is not yet taken into account, that there are substantial improvements to be achieved by going to higher storage voltages.

This paper is organised as follows. Section II introduces the proposed system architecture. The primary voltage conversion step from the incidental RF waves into a workable DC system supply voltage is discussed in Section III followed by Section IV, which continues with the secondary voltage conversion towards the final storage voltage. Both integration of these building blocks into an operational system and consequent simulation results are presented in Section V. Concluding remarks finish this work in Section VI.

II. SYSTEM ARCHITECURE

This work targets to increase the energy storage density in wireless powered systems by increasing the voltage to which the storage capacitor C_{bat} is charged, as shown in the proposed system architecture of Fig. 1. Since the stored energy E_{bat} on a capacitor C_{bat} is proportional to square of the storage voltage V_{bat} (Eq. 1), this proves to be a very effective technique. However, not all energy in C_{bat} can be utilized to power a circuit as V_{bat} must remain higher than the nominal voltage V_{nom}, required by the circuit to operate. The usable stored energy E_{usable} is then given by Eq. 2 and Fig. 2 demonstrates this concept.

$$E_{bat} = \frac{C_{bat}V_{bat}^2}{2} \tag{1}$$

978-1-4673-4579-8/13 $31.00 © 2013 IEEE

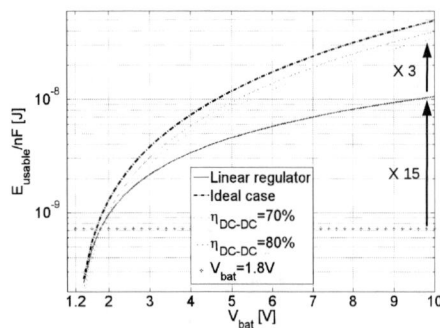

Fig. 2. Concept of the system: representation of additionally stored energy (per nF of C_{bat}) as function of V_{bat} for the ideal case of Eq. 2 (slash-dotted line) and other non ideal example cases.

$$E_{usable} = \frac{C_{bat}(V_{bat}^2 - V_{nom}^2)}{2} \qquad (2)$$

Fig. 2 plots the ideal usable stored energy, of Eq. 2, per nF of storage capacitance C_{bat} as function of V_{bat}. More realistic results are also provided by the non ideal examples, incorporating the voltage conversion efficiency necessary before this energy can be used by the load. It can be seen in Fig. 2 that the energy budget increases to 15 times its original value if only simple linear regulation for step-down is considered. This factor can still be increased another 3 times with a DC-DC step-down converter at an efficiency of 70%, which is moderate in comparison to the similar voltage conversion ratio of [4].

The proposed architecture of Fig. 1 consists of two main voltage conversion blocks and a control block to ensure proper operation. The primary voltage conversion rectifies and steps up the voltage available at the antenna terminals to a system supply voltage V_{PMU}, which is used as a starting point for the second active step-up voltage conversion. Due to the wireless origin of the incident energy, very low power levels in the order of nA's can be expected and a major challenge in this design will be to minimize the overhead consumption of the PMU. Therefore, the first voltage conversion is implemented as a passive Greinacher multiplier. For the secondary step-up conversion, a Dickson DC-DC converter [5] is selected because of its robustness, low control complexity and resulting low power operation.

The target input frequency range is 5.8GHz to facilitate full integration of the antenna. For impedance matching to be feasible, the capacitive part of in voltage multiplier's input impedance is limited. This means that only a limited amount of multiplication cells can be cascaded, unable to generate the target V_{bat} voltage of 10V. Therefore a two stage voltage conversion approach is used offering the benefit of both relaxing the specification of first conversion stage as well as a higher conversion efficiency in the second converter due to the larger voltage increase per stage.

To cope for input power fluctuations, the system can automatically suspend the operation of the Dickson converter. While V_{PMU} is passively generated and requires no attention, the Dickson converter can be disconnected by the control circuitry when needed. If V_{PMU} has reached upper threshold value V_H, the DC-DC converter is switched in and pumps up

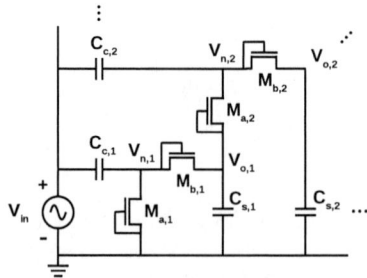

Fig. 3. The AC-DC voltage multiplier.

charge until intermediate buffer capacitor C_{PMU} reaches the lower threshold V_L. As such, the Dickson converter is only active when V_{PMU} is between V_L and V_H. A minimum AC voltage of 53mV is required out of the antenna in order for V_{PMU} to reach V_H and start up the system.

III. PASSIVE AC-DC VOLTAGE MULTICATION

The primary voltage conversion of this design consists of an on-chip antenna, an AC-DC multiplier and a matching circuit. The incident EM waves are converted into a DC voltage V_{PMU} of 1.2V, used as the global supply voltage of the PMU.

A. Antenna

Due to a die size limitation typical in monolithic integrated circuits, the implemented antenna is an electrically small dipole antenna. The open circuit voltage V_{ant} as function of the distance d, dipole antenna length l_{ant}, $\eta_{air} \approx \eta_{vacuum} = 377 \ \Omega$ and power of the transmitter P_t [2], is given in Eq. 3.

$$V_{ant} = \frac{l_{ant}}{2d}\sqrt{\frac{\eta_{air}P_t}{2\pi}} \qquad (3)$$

For an antenna length of 1.8mm and a transmitted equivalent isotropically radiated power of 4W, the voltage V_{ant} is equal to 100mV at a distance of 13.9cm in open air. Integrating the antenna in the top layer of the metal stack reduces this distance by 37 % to 8.8cm due to reflection on the dielectric layers.

B. AC-DC multiplier

The topology of the voltage multiplier is given in Fig. 3. It is built up by a cascade of capacitive coupled rectifier stages, each increasing the voltage. In order to generate a target V_{PMU} in the range of 1.2V from an input voltage of 100mV, the number of cascaded stages and the size of its components need to be optimized. Also an output current $i_{AC-DC,out}$ of 50nA is assumed in the optimalization of the AC-DC converter as this is an acceptable power budget to efficiently operate the rest of the system. An evolutionary algorithm is used to optimize this set of design variables, resulting in a V_{PMU} voltage of 1.22V at an $i_{AC-DC,out}$ of 50nA. The transistors are all equally sized with a W/L of 7.2/0.08μm. Each capacitor $C_{c,x}$ is equal to 4pF and each $C_{s,x}$ is equal to 300fF. In total 23 stages are cascaded and a power efficiency of 18% is reached.

PRIME 2013, Villach, Austria
Session W2A – Power Management

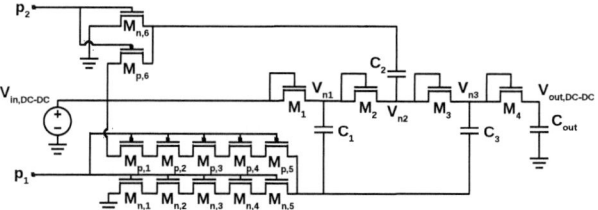

Fig. 4. A three stage Dickson DC-DC converter with diode-connected transistors M_1 to M_4 and two clock buffers.

C. Matching circuit

A matching circuit is designed to boost the input voltage of the AC-DC multiplier. Because of the highly capacitive antenna impedance and the capacitive input impedance of the multiplier, the matching must be inductive. Due to the low quality factor and large area overhead [2] for inductors at the selected frequency of operation, practical inductance values are limited to 2.6nH in the used CMOS technology. This is too low to make a suitable power matching but voltage matching can still be performed with an integrated inductor of 1.6nH, placed in between the terminals of the antenna. This enables an input voltage again of around 100mV for the voltage multiplier at a powering distance of 8.8cm.

IV. ACTIVE STEP-UP DC-DC VOLTAGE CONVERTER

The second voltage conversion consists of a DC-DC converter and a low power clock generator to step up V_{PMU} to the level of the storage voltage V_{bat}. A Dickson DC-DC converter topology can perform DC-DC step-up with a minimum of necessary control inputs making it the perfect match given the design requirements. This converter can efficiently reach high voltages, only needing a complementary drive. The topology is shown in Fig. 4 for a three stage implementation. M_{1-4} are diode connected transistors and even though this causes threshold voltage V_{th} drops [6] as given by Eq. 4, the benefit of this switches being passive outweighs the otherwise necessary power consumption in level shifting and gate drive.

$$V_{out,N} = (N+1)V_{in} - (N+1)V_{th} \qquad (4)$$

Control signals p_1 and p_2 are generated by a low power clock generator, implemented by a ring oscillator topology running at a frequency f_{osc} of around 20kHz. Multiple techniques are applied to suppress the power consumption in the ring oscillator. Stacked diode transistors reduce the voltage swing [7] in the 3-stage ring oscillator core as can be seen in Fig. 5. In the subsequent regeneration inverters leakage is lowered through implementing stacked LLHVT devices [8] and a non minimal gate length L decreases the short circuit current [9]. The resulting clock generator consumes 14nA at 1.2V to create a clock signal of 19kHz.

Fig. 4 shows the clock buffers and 3 stages of a Dickson converter. In the implemented design, 9 Dickson stages are necessary to generate a V_{bat} close to 10V. An output impedance minimization approach, as described in [10], is used to achieve the optimal switch $W_{switch,i}$ and flying capacitor C_i converter parameters. The diode connected transistors are sized with a W/L of 0.12/0.18 µm and flying capacitors are each 6.25pF.

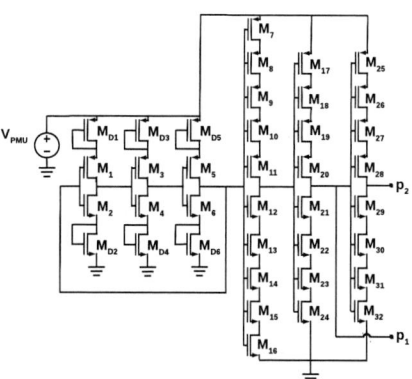

Fig. 5. The low power clock generator with regenerated output signals p_1 and p_2.

When supplying 4nA, an output voltage of 8.3V is generated from a 1.2V input at an efficiency of 75% and 60% respectively considering only the converter core and the converter core together with the clock generator. 10V is achieved when the output is open circuit.

V. SYSTEM INTEGRATION AND SIMULATION

A. System control

A robust design is realized due to the voltage detector [2], [3] control block that automatically turns on the secondary voltage conversion, only when the boundary conditions of correct and efficient operation are met. This is necessary because the Dickson converter, unlike the passive voltage multiplier, depends on the active clock drive to function. Two voltage thresholds $V_H = 1.24$V and $V_L = 1.08$V are implemented in the voltage detector to control the activity of the Dickson converter. When V_{PMU} reaches the upper threshold value V_H due to the passive voltage multiplier, the Dickson converter will transfer charge from C_{PMU} to C_{bat}. When V_{PMU} has discharged to lower threshold V_L, it is again disconnected in order to allow V_{PMU} to recharge. A robust system is achieved due to the fact the both voltage converters can retain their stored voltages even when they are not driven. This is due to the use of transistor diodes instead of active switches, which can lead to large reverse leakage if driven improperly.

Fig. 6 shows the control circuit and consists of a voltage reference, a voltage divider and comparator. The reference voltage [11] is composed of 11 stages to generate a 500mV reference. Each transistor operates in weak inversion to reduce the power consumption. As long as $V_{div} < V_{ref}$, the comparator output equals zero. When the input voltage reaches V_H, V_{div} equals V_{ref}. The output voltage of the comparator will then be equal to the input voltage of the detector and the voltage division will be adapted in such a way that, if the detector input drops to V_L, V_{div} is again equal to V_{ref}. The bias currents in the detector are generated by a transistor with a V_{GS} of zero to limit the power consumption further. The control current consumption is limited to 3.8nA. Since V_H determines the activity of the Dickson converter, it also determines the total system startup voltage to be 53mV at the input of the voltage multiplier, translating in a maximum powering distance of 16.7 cm, given an isotropically radiated power of 4W according Eq. 3.

978-1-4673-4579-8/13 $31.00 © 2013 IEEE

Paper W2A3   *PRIME 2013, Villach, Austria*

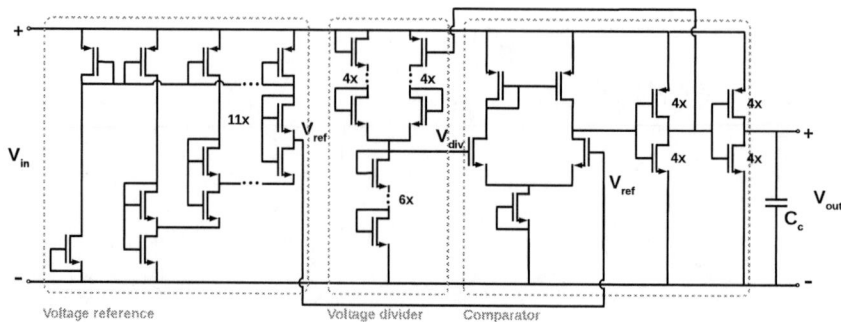

Fig. 6.   The voltage detector control block.

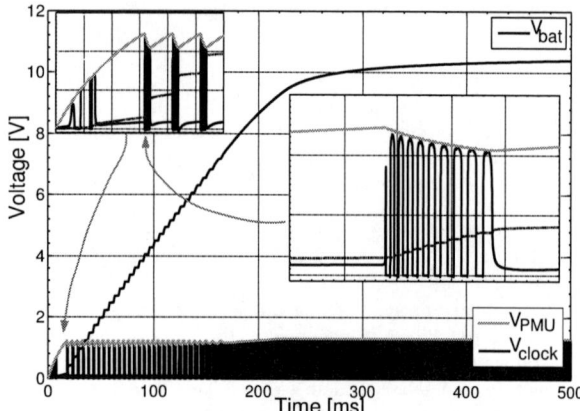

Fig. 7.   Simulation of the storage system at a powering distance of 8.8cm. If the output of the AC-DC multiplier $V_{PMU.}$ is higher than the threshold value V_H, the 19kHz clock generator V_{clock} is activated. If $V_{mult.}$ is lower than the threshold value V_L, the clock is disabled. V_{bat} is the target output storage voltage.

B. System simulation

Fig. 7 shows a simulation of the total system with its most important waveforms. At startup, the Dickson converter's capacitors are uncharged and a higher than 50nA current is drawn. As this can not be delivered continuously by the voltage multiplier, the voltage detector will control the clock generator to operate the DC-DC converter in short intervals, keeping V_{PMU} within V_L and the V_H threshold. A more detailed representation of this is given in zoom boxes of Fig 7. The DC-DC converter's input current is decreased after startup at 180ms and the clock can then operate continuously.

VI. CONCLUSION

A monolithic power management system is presented, enabling an energy storage density improvement of more than an order of magnitude. This is made possible by increasing the storage voltage towards 10V. To demonstrate this concept, a prototype was designed in a 90nm bulk CMOS technology with an integrated 1.8mm dipole antenna using the surrounding energy available in the 5.8GHz ISM band. In a first voltage conversion step, the antenna voltage is passively rectified and multiplied to an intermediate system supply voltage V_{PMU} of 1.2V. Subsequently this voltage is actively stepped up further to

the final storage voltage level V_{bat} of 10V. A low power voltage detector controls the system and ensures correct operation. By means of shifting a part of the total voltage conversion to a second active converter, the specifications on the passive voltage multiplier are kept feasible and a minimum startup voltage of 53mV is necessary for the output to be fully charged. The AC-DC multiplier and the DC-DC converter reach an efficiency of 18 % and 60 % respectively, controlled by a voltage detector circuit consuming 3.8nA and a 19kHz clock generator with a consumption of 14nA. The complete system enables dense energy storage at 10V up to a powering distance of 16.7cm, which is more than twice as large compared to previously reported [3] fully integrated systems at 5.8 GHz.

REFERENCES

[1] T. Le, K. Mayaram, and T. Fiez, "Efficient far-field radio frequency energy harvesting for passively powered sensor networks," *IEEE Journal of Solid-State Circuits*, vol. 43, no. 5, pp. 1287–1302, 2008.

[2] C. De Roover and M. Steyaert, "Energy Supply and ULP Detection Circuits for an RFID Localization System in 130 nm CMOS," *Solid-State Circuits, IEEE Journal of*, vol. 45, no. 7, pp. 1273 –1285, july 2010.

[3] S. Radiom, M. Baghaei-Nejad, K. Aghdam, G. Vandenbosch, L.-R. Zheng, and G. Gielen, "Far-Field On-Chip Antennas Monolithically Integrated in a Wireless-Powered 5.8-GHz Downlink/UWB Uplink RFID Tag in 0.18-μm Standard CMOS," *Solid-State Circuits, IEEE Journal of*, vol. 45, no. 9, pp. 1746 –1758, sept. 2010.

[4] V. Ng and S. Sanders, "A high-efficiency wide-input-voltage range switched capacitor point-of-load DC-DC converter," *IEEE Transactions on Power Electronics*, vol. 28, no. 9, pp. 4335–4341, 2013.

[5] J. F. Dickson, "On-Chip High-Voltage Generation in NMOS Integrated Circuits Using an Improved Voltage Multiplier Technique," *IEEE Journal of Solid-State Circuits*, vol. 11, no. 3, pp. 374–378, 1976.

[6] L. Pylarinos, "Charge Pumps: An Overview," Ph.D. dissertation, University of Toronto.

[7] M. Azarmehr, R. Rashidzadeh, and M. Ahmadi, "Low-power oscillator for passive radio frequency identification transponders," *Circuits, Devices Systems, IET*, vol. 6, no. 2, pp. 79 –84, march 2012.

[8] K. Roy, S. Mukhopadhyay, and H. Mahmoodi-Meimand, "Leakage current mechanisms and leakage reduction techniques in deep-submicrometer CMOS circuits," *Proceedings of the IEEE*, vol. 91, no. 2, pp. 305 – 327, feb 2003.

[9] A. Morgenshtein, "Short-circuit power reduction by using high-threshold transistors," *Journal of Low Power Electronics and Applications*, vol. 2, no. 1, pp. 69–78, 2012.

[10] M. D. Seeman and S. R. Sanders, "Analysis and Optimization of Switched-Capacitor DC-DC Converters," *IEEE Transactions on Power Electronics*, vol. 23, no. 2, pp. 841–851, 2008.

[11] E. Vittoz and O. Neyroud, "A low-voltage CMOS bandgap reference," *Solid-State Circuits, IEEE Journal of*, vol. 14, no. 3, pp. 573 –579, june 1979.

978-1-4673-4579-8/13 $31.00 © 2013 IEEE

Application of Bayesian Networks to Predict SMART Power Semiconductor Lifetime

Kathrin Plankensteiner*[†], Olivia Bluder*, Jürgen Pilz[†]

*KAI Kompetenzzentrum Automobil- und Industrie-Elektronik GmbH, Villach, Austria
[†]Institute of Statistics, Alpen-Adria-University of Klagenfurt, Austria
Email: kathrin.plankensteiner@k-ai.at

Abstract—In this paper Bayesian networks are used to model semiconductor lifetime data from a cyclic stress test system. The data of interest is a mixture of log-normal distributions, representing different failure mechanisms and moreover, the data is censored. To understand the complex lifetime behavior, interactions between test settings, geometric designs, material properties and physical parameters of the semiconductor device are modeled by a Bayesian network. For the network's structure and parameter learning statistical toolboxes in MATLAB have been extended and applied. Due to censored observations MCMC simulations are necessary to determine the posterior distribution.

For model selection the ARD algorithm and goodness of fit criteria such as marginal likelihoods, Bayes factors, posterior predictive density distributions and SSEPs are used.

The results indicate that the application of Bayesian networks to semiconductor reliability provides useful information about the interactions between covariates and serves as a reliable alternative to currently applied methods.

I. INTRODUCTION

In automotive industry, end-of-life tests are necessary to verify that semiconductor products operate reliably. Since resources are limited, accelerated stress tests [8] in combination with statistical models are commonly applied to predict the lifetime of devices. Previous investigations showed that the lifetime follows a mixture of two log-normal distributions, representing two different dominant physical failure mechanisms. Due to the complex behavior of the data, common acceleration models, e.g. Arrhenius, failed. The application of a Bayesian Mixtures-of-Experts extended Coffin-Manson approach [1] led to a satisfying interpolation quality, but still lacks accuracy in case of extrapolation. This inaccuracy is assumed to be caused by the fact that the model does not include parameters reflecting interactions between different geometric designs or material properties of the semiconductor device [19]. To compensate for this weakness, a Bayesian network including these parameters for each component is proposed.

II. DATA & CHALLENGES

In this paper censored lifetime data obtained under different electrical and thermal stress conditions are analyzed. The accelerated repetitive stress is produced by a cycle stress test system [8]. To stress the semiconductor devices, a combination of five test settings has to be defined by the operator: the ambient temperature (T_{amb}), the stress pulse shape (triangle or rectangle), the induced peak current (I), the pulse length (t_p) and the repetition time (t_{rep}). The clamping voltage (V) is defined by the device itself.

During the test, the system counts the number of applied stress cycles. If the Device Under Test (DUT) fails during the stress test, the exact lifetime, measured in cycles to failure (CTF), is logged by the test system. If the DUT did not fail by the end of the test, its status is survivor (SURV) and the total number of applied stress cycles is logged for this DUT. The lifetime for this DUT is therefore censored.

Previous investigations [1], [19] showed that the logarithmic transformed CTF (logCTF) follows a mixture of two normal distributions. Fig. 1, a normal probability plot with a logarithmic x-axis, shows the outcome of three tests. For each test 16 DUTs are stressed, illustrated by blue triangles (a high-stress test), green circles (a moderate-stress test) and red squares (a low-stress test). As expected, increasing the stress level leads to a shorter lifetime on average, cf. 50% quantiles. Taking a closer look at the lifetime of DUTs tested with the moderate-stress level, a double distribution with CTF $< 10^6$ for the first component and CTF $> 10^6$ for the second component is clearly visible. The other two test results show single component distributions. For the high-stress test, all DUTs fail within the first 10^4 cycles, whereas three DUTs of the low-stress test survive. For this data, highlighted by the black rectangle, the test system logs the total number of applied stress pulses at the end of this test. Physical failure inspections indicate that DUTs surviving less than $\sim 10^6$ stress cycles show a different failure mechanism than DUTs failing after 10^6 stress cycles [2].

Due to limited resources, it is neither possible to test nor to inspect all devices. Thus, reliable forecasts for the lifetime are needed. Previous investigations point out that the currently applied Mixtures-of-Experts (MoE) extended Coffin-Manson model [2] is insufficient for extrapolation because it does not include physical parameters representing the geometric design of the DUT nor any electro-thermal or thermo-mechanical effects caused by repetitive stress [19]. To compensate for this weakness, the geometric design of the DUT needs to be considered. Taking this into account, further covariates such as the current density (J), the power (P), the

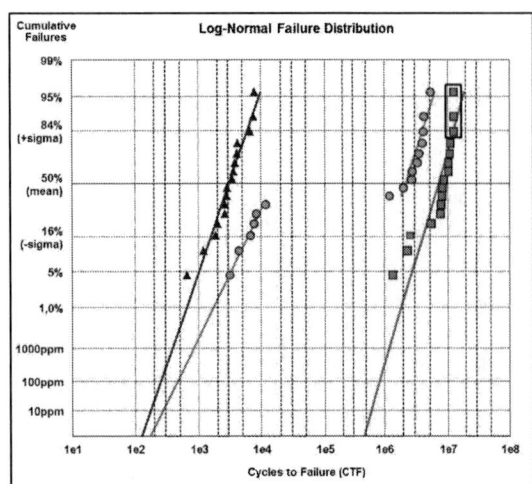

Fig. 1. Semiconductor lifetime probability plots from three stress tests.

cumulative energy (E) and the energy per area (E_{area}) are gained. For the previous model the case temperature (T_{case}) of the DUT was approximated analytically. For these tests measured temperatures are available which are used for the Bayesian network model.

Caused by electric power dissipation the DUTs heat up and cool down during each stress pulse. The induced temperature rise (ΔT) can either be simulated with the Finite-Element-Method (FEM) [14], [20] or it can be approximated analytically [19]. In the following, simulated temperature rises are denoted by ΔT_{FEM} and analytically derived approximations are denoted by ΔT_{SOA}. Given these temperature rises, the corresponding peak temperatures (T_{peak}) as well as the stress level (L_{SOA}) are defined by

$$
\begin{aligned}
T_{peak-FEM} &= T_{case} + \Delta T_{FEM} \\
T_{peak-SOA} &= T_{case} + \Delta T_{SOA} \\
L_{SOA} &= \frac{\Delta T_{SOA}}{T_{dest} - T_{case}}
\end{aligned}
$$

where T_{dest} denotes the device specific destruction temperature. This is the temperature where the device fails at the first stress pulse (i.e. the SOA limit).

Since thermo-mechanical effects are also significant for device failure, three analytical models are applied to gain information about the equivalent stress (s) [22], the shearing stress (τ) and the peeling stress (ρ) [3]. In this work all thermo-mechanical stress values are based on simulated temperatures, ΔT_{FEM} and $T_{peak-FEM}$, on known geometric designs and material properties of the DUTs.

Due to the availability of lifetime data from five different designs, another covariate, the device category (*cat*), is introduced. Since T_{amb} and pulse shape did not vary in the investigated tests, they are neglected for modeling. Altogether

18 covariates are available for the Bayesian network to model the lifetime of semiconductor devices. Since logCTF is a mixture of two normally distributed components representing two different failure mechanisms, the dataset is divided into two subsets. The first and second subset contain 169 and 867 datapoints, respectively, tested under 65 different stress conditions. Both subsets include censored data. The mean lifetime of the first subset (3.5 logCTF) and the mean lifetime of the second subset (7.5 logCTF) differ significantly. On average, the variance of the first subset is higher than the variance of the second one.

III. BAYESIAN NETWORK & APPLICATION

A Bayesian network (BN), also called belief network, is a directed acyclic graph (DAG) and belongs to the family of probabilistic graphical models. It is defined as a pair $G = (V, E)$, where V is the set of vertices and E is the set of directed edges. The set of vertices may represent random variables in the Bayesian sense, observable quantities, latent variables, unknown parameters or hypotheses. The edges represent conditional dependencies between the variables. If an edge from node A to node B is in the graph, we say, that A influences B. A is then called a parent of B and B is called a child of A. Since the graph is defined as acyclic, no random variable is allowed to interact with itself. Therefore, a BN represents a joint probability model of given variables including the conditional independence statement, that each variable is independent of its nondescendents given the state of its parents. This Markovian property can be used for efficient parameter estimation and is the main benefit of this kind of models. Based on this property, the joint probability distribution $P(X_1, X_2, \ldots, X_n)$ can be factorized [5], [13], [15], [21] to

$$
P(X_1, X_2, \ldots, X_n) = \prod_{1 \leq i \leq n} P(X_i | pa(X_i)) \qquad (1)
$$

where $V = \{X_1, X_2, \ldots, X_n\}$ and $pa(X_i)$ denotes the set of parents of node X_i. For this representation corresponding conditional probability distributions (CPDs), $P(X_i | pa(X_i))$, have to be defined.

The application of BNs consists of two parts: the qualitative part (structure learning) and the quantitative part (parameter learning). For these purposes the BNT toolbox [18] and the extended MoE toolbox [1], [7] for MATLAB are used. Since the available BNT toolbox does neither provide Bayesian inference for continuous data nor regression, it had to be extended for an efficient data handling.

For modeling all nodes are continuous except the node corresponding to the categorical variable *cat*, which is discrete. For the corresponding CPDs root and gaussian nodes are used. Root nodes represent fixed input variables, whereas gaussian nodes represent random variables that are influenced by other factors within the model. If Y is a gaussian node and X is the set of parents of Y, the CPD of Y is defined by [17]

$$Y|X = x \sim \mathcal{N}(\beta_0 + x^t\beta, \sigma^2). \qquad (2)$$

To define the model, the following assumptions are made:

- root nodes (no parents allowed): (1) cat, (2) V, (3) I, (4) t_{rep} and (5) t_p
- gaussian nodes (parents and children allowed): (6) J, (7) P, (8) E, (9) E_{area}, (10) T_{case}, (11) ΔT_{SOA}, (12) $T_{peak-SOA}$, (13) ΔT_{FEM}, (14) $T_{peak-FEM}$, (15) L_{SOA}, (16) s, (17) τ and (18) ρ
- gaussian nodes (no children allowed): logCTF.

Equation 2 defines the BN as hierarchical model with a linear regression in each submodel. Since physical relationships between covariates are effects that stack multiplicatively, but the defined BN requires additive models, all continuous covariates are logarithmic transformed and standardized. Due to the assumption that the model might be too complex for the given amount of data, five different approaches are investigated: A BN including

(N1) all 18 covariates
(N2) 16 covariates (ΔT_{FEM} and $T_{peak-FEM}$ excluded)
(N3) 16 covariates (ΔT_{SOA} and $T_{peak-SOA}$ excluded)
(N4) a significant subset of covariates
(N5) a significant subset of covariates and prior information on edges.

For the selection of a significant subset of covariates, different methods are known, e.g. principal component analysis (PCA) [12] or forward selection/ backward elimination [11]. Since the automatic relevance determination (ARD) [23] is the most common one for structure discovery in BNs, a Bayesian concept of ARD is applied for the data of interest. Fig. 2 shows the results. The labels on the x-axis refer to numbered covariates above and the bars demonstrate the corresponding posterior means of the weights. If the absolute value of the posterior mean, also called Bayes point [10], [23] is large, the covariate is assumed to be significant. Since the dataset consists of five different designs, this evaluation is done component and device specific, e.g. for device A the covariates t_{rep}, E, E_{area} and L_{SOA} are significant for the mean lifetime of both components. The covariate T_{case} seems to be only significant for the mean lifetime of the first component. Further, it is conspicuous that the electro-thermal parameter L_{SOA} and the thermo-mechanical parameter s have high weights especially for the second component. This confirms the hypothesis of two different dominant failure mechanisms representing the two components, because each component (failure mechanism) depends on another combination of covariates (influences). ΔT_{SOA}, $T_{peak-SOA}$, ΔT_{FEM} and $T_{peak-FEM}$ are not significant for both components, but they are indirectly included by L_{SOA} and s.

Knowing the active area of the DUT, the covariates I and J as well as E and E_{area} are directly proportional to each other.

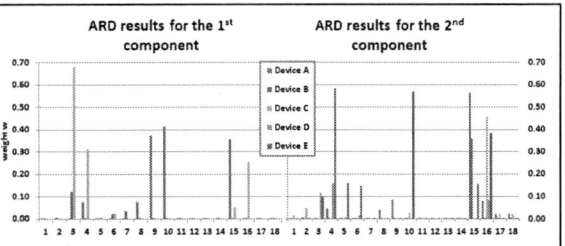

Fig. 2. Device specific ARD results for the first and second component. The application of different sets of covariates for each component is proposed.

Since cat is in the model, these covariates are equivalent and thus it is sufficient to use only one of them. With this assumption the set of covariates can be reduced to cat, t_{rep}, t_p, J, E_{area}, T_{case}, L_{SOA} and s. Moreover, due to known physical relationships between the variables prior information on edges is used additionally.

To develop a BN, structure and parameter learning are performed in MATLAB. For the structure learning a Metropolis-Hastings algorithm [6] with the Bayes Factor (BF) as scoring function is used. To determine the score of the graphs, the Bayesian Information Criterion (BIC) as approximation for the marginal likelihood is applied [4]. For this purpose, the posterior density distributions of the model parameters are needed which can be sampled with a combination of data augmentation and the Slice-within-Gibbs sampler [1]. Due to the vagueness of prior information about the model parameters, flat normal priors for means and hierarchical inverse-Gamma priors for variances are applied

$$\begin{aligned}
\beta_0 &\sim \mathcal{N}(0, 10) \\
\beta &\sim \mathcal{N}(0_d, 10 \cdot I_d) \\
\sigma^2 &\sim \mathcal{G}^{-1}(c_0, C_0) \\
C_0 &\sim \mathcal{G}(g_0, G_0) \qquad (3)
\end{aligned}$$

where d is the cardinality of X and I_d is the unit matrix of dimension d. For c_0, C_0, g_0, and G_0 the values suggested by Frühwirth-Schnatter [6] are used: $c_0 = 2.5$, $g_0 = 0.5$ and $G_0 = 0.5 \cdot s_y^2 \cdot g_0 \cdot (c_0 - 1)^{-1}$, where s_y^2 is defined to be the sample variance.

For the structure and parameter learning a sample size of 1000 with a burn-in of 500 is used. The goodness of fit is evaluated with Draper's approximation for the marginal likelihood [4]. With this criterion the BN with all available information (N1) gives the best fit for the first and second component. The mean lifetime of the first component is modeled depending on P, E_{area}, ΔT_{SOA} and ΔT_{FEM}. The mean lifetime of the second component is modeled by t_{rep}, L_{SOA}, ρ. The summary statistics for the posterior density distributions of the model parameters is given in Table I. The posterior densities of the model parameters show narrow Highest Posterior Density (HPD) regions and therefore indicate a good fit.

TABLE I
SUMMARY STATISTICS OF POSTERIOR DENSITY DISTRIBUTIONS

First Component			Second Component		
	Q50	95% HPD		Q50	95% HPD
Intercept	3.4	[1.9; 4.9]	Intercept	12.0	[11.7; 12.4]
P	2.3	[1.0; 3.6]	t_{rep}	1.15	[0.8; 1.5]
E_{area}	2.1	[0.3; 3.9]	L_{SOA}	−5.2	[−5.6; −4.7]
ΔT_{SOA}	−4.4	[−5.8; −2.8]	ρ	−1.9	[−2.2; −1.6]
ΔT_{FEM}	1.3	[0.2; 2.3]			
σ^2	0.4	[0.3; 0.4]	σ^2	0.3	[0.3; 0.4]

TABLE II
LIST OF MEANSSEPS

model	Device A	Device B	Device C	Device D	Device E
# tests	33	3	5	4	6
MoE	6.08	3.88	0.89	2.44	0.52
BN	3.83	4.12	0.96	5.26	0.22

IV. PREDICTION RESULTS

To evaluate the prediction quality, cross-validations using posterior predictive distributions [16] and the sum of squared errors of predictions (SSEPs) [9] are applied. Since the MoE model was developed based on a subset of data, the same subset is used to learn the BNs. With this a direct comparison between the predictive power of the two different approaches is provided. For the lifetime prediction the posterior predictive distribution for each component is sampled independently and mixed by estimated mixture weights. The mixture weights are modeled by a cumulative Beta distribution function being dependent on t_p and L_{SOA} [19]. The meanSSEPs are compared with results gained by the currently applied MoE model (see Table II). Since it is infeasible to determine SSEPs for tests with no fails, they are neglected for this evaluation. Thus, the number of tests is reduced to 51. The results show that the predicted outcomes of BN and MoE model are comparable.

V. CONCLUSION

In this paper different Bayesian networks were proposed to model mixed distributed semiconductor lifetime data. For the model 18 possible covariates were available. For complexity reduction, the ARD algorithm as well as prior information about edges were applied. Based on the best structure, the posterior density distributions of the model parameters were simulated using extended MATLAB toolboxes. The posterior densities showed small variations and indicated therefore a good fit.

The evaluation of the prediction quality was performed with cross-validations using posterior predictive distributions. The outcomes were compared with the predictions made by the MoE model. The results showed that the application of a BN provides useful information about interactions of covariates and that it represents a reliable alternative to currently applied methods.

ACKNOWLEDGMENT

The authors would like to thank Roland Sleik and Michael Ebner for the measurement support, as well as Michael Glavanovics, Michael Nelhiebel and Christoph Schreiber for valuable discussions on the topic.

This work was jointly funded by the Austrian Research Promotion Agency (FFG, Project No. 831163) and the Carinthian Economic Promotion Fund (KWF, contract KWF-1521|22741|34186).

REFERENCES

[1] O. Bluder, *Prediction of Smart Power device lifetime based on Bayesian modeling*, PhD thesis Alpen-Adria-University of Klagenfurt, 2011.
[2] O. Bluder, J. Pilz, M. Glavanovics and K. Plankensteiner, *A Bayesian Mixture Coffin-Manson Approach to Predict Semiconductor Lifetime*, SMTDA 2012: Stochastic Modeling Techniques and Data Analysis, 2012.
[3] W. T. Chen and C. W. Nelson, *Thermal Stress in Bonded Joints*, IBM Journal of Research and Development, 1979, Vol.23(2), pp. 179-188.
[4] D. M. Chickering and D. Heckerman, *Efficient approximations for the marginal likelihood of Bayesian networks with hidden variables*, Machine Learning, 1997, Vol.29, pp. 181-212.
[5] D. M. Chickering, *Optimal Structure Identificatoin with Greedy Search*, Journal of Machine Learning Research, 2002, Vol.3, pp. 507-554.
[6] S. Frühwirth-Schnatter, *Finite mixture and Markov switching models*, Springer Series in Statistics, New York, 2006.
[7] S. Frühwirth-Schnatter, *Manual: MATLAB package* bayesf *Version 2.0*, http://www.jku.at, 2008.
[8] M. Glavanovics, H. Köck, V. Kosel and T. Smorodin, *A new cycle test system emulating inductive switching waveforms*, Proceedings of the 12th European Conference on Power Electronics and Application, 2007, pp. 1-9.
[9] L. Held, *Methoden der statistischen Inferenz: Likelihood und Bayes*, Springer Heidelberg, 2008.
[10] R. Herbrich, T. Graepel and C. Campbell, *Bayes point machine*, Journal of Machine Learning Research, 2001, pp. 245-279.
[11] R. Hocking, *A Biometrics Invited Paper: The Analysis and Selection of Variables in Linear Regression.*, Biometrics, 1976, Vol.32, pp. 1-49.
[12] I.T. Jolliffe, *Principal Component Analysis*, Springer Series in Statistics, New York, 2002.
[13] J. Pearl, *Probabilistic Reasoning in Intelligent Systems*, Morgen Kaufmann, San Mateo, California, 1998.
[14] V. Kosel, R. Sleik and M. Glavanovics, *Transient Non-linear Thermal FEM Simulation of Smart Power Switches and Verifcation by Measurements*, Proceedings of THERMINIC 2007-THERMal INvestigations of ICs and Systems, 2007, pp. 110-114.
[15] S. Hojsgaard, D. Edwards and S. Lauritzen, *Graphical Models with R*, Springer, New York Dordrecht Heidelberg London, 2012.
[16] S. M. Lockwood and M. J. Schervish, *MCMC strategies for computing Bayesian predictive densities for censored multivariate data*, Journal of Computational and Graphical Statistics, 2005, Vol.14(2), pp. 395-414.
[17] K. P. Murphy, *An introduction to graphical models*, Technical Report of UBC, 2001.
[18] K. P. Murphy, *Bayes Net Toolbox*, https://code.google.com/p/bnt, 2007.
[19] K. Plankensteiner, *Application of Bayesian Models to Predict Smart Power Switch Lifetime*, Master thesis, Alpen-Adria-University of Klagenfurt, 2011.
[20] M. Riccio, A. Irace, G. Breglio, P. Spirito, V. Kosel, M. Glavanovics and A. Satka, *Thermal simulation and ultrafast IR temperature mapping of a Smart Power Switch for automotive applications*, ISPSD 2009 - 21st International Symposium on Power Semiconductor Devices & IC's 2009, pp. 200-203.
[21] F. Ruggeri, F. Faltin and R. Kenett, *Bayesian Networks*, Encyclopedia of Statistics in Quality and Reliability, UK, Wiley and Sons, 2007.
[22] S. Suresh, *Fatigue of Materials*, Cambridge solid state science series, Cambridge University Press, 1998.
[23] Y. Qi, T. P. Minka, R. W. Picard and Z. Ghahramani, *Predictive Automatic Relevance Determination by Expectation Propagation*, Proceedings of the 21st International Conference on Machine Learning, 2004, pp. 671-678.

Optimal design of experiments for semiconductor lifetime data

Anja Zernig[*†], Olivia Bluder[*], Gunter Spöck[†]

[*]KAI Kompetenzzentrum Automobil- und Industrie-Elektronik GmbH, Villach, Austria
[†]Institute for Statistics, Alpen-Adria-University of Klagenfurt, Austria
Email: anja.zernig@k-ai.at

Abstract—**Performing experiments is necessary to find influences of different factors on the measured output. In semiconductor industry experiments are mainly performed following predefined specifications and guidelines, given by experts for the device under test (DUT). The statistical method design of experiments (DoE) provides an objective solution to the question: which experiments have to be performed to get the most information concerning main influencing factors and interaction between factors. In practical usage classical DoE often reach their limits, especially when resources for experiments are meagre. A remedy is given by optimal DoEs. They are more flexible and offer the possibility to optimize e.g. the prediction accuracy on a pre-defined area, where performing measurements is difficult. For this purpose the IV-optimality criterion is used in this paper. On the basis of already performed experiments, an exchange algorithm proposed by Spöck and Pilz [2] was used to select 3 further desired experiments. After their performance they were evaluated and, as expected, an improvement in the mean squared error of prediction (MSEP) was observed.**

I. Introduction

In semiconductor industry reliable devices are essential. To test their reliability, lifetime stress tests on the devices are common practice. Getting results within a reasonable amount of time, accelerated stress conditions with e.g. increased voltage and faster cycling frequencies (in the range of milliseconds), compared to use conditions are performed. This results in a reasonable test time and therewith implies lower costs and saved resources. The downside of accelerated stress tests is that the lifetime at use conditions needs to be extrapolated based on statistical models. The quality of the prediction is subject to the quality of the model, whereas the model itself depends on the quality of the data.

The aim of DoE is to get the most informative measurements for the statistical model on the basis of a minimum numbers of experiments.

To guarantee high quality devices, lifetime tests under well defined accelerated test conditions are performed. A modified Coffin-Manson model is used to extrapolate the lifetime measured in Cycles to Failure (CTF) to real life behaviour. In general, life tests are performed on a set of test conditions, selected according to specifications or based on a classical DoE. A more advanced and flexible method to select the most appropriate sample of test conditions is optimal DoE, which provides an optimal set of test conditions considering the underlying model, possible restrictions on the combination of test conditions and already performed tests. The resulting set is called a design.

Summarized, the procedure of optimal DoE suggests the most informative test conditions for the given experimental situation. A test condition in terms of DoE is a combination of different design factors levels, which have to be measurable. In this paper a comparison between the classical and the optimal approach is given in Section II. The theory behind optimal DoE and two optimality criteria are quoted in Section III followed by the definition of the workspace in Section IV, which has to be done to apply the exchange algorithm, based on Spöck and Pilz [2], in Section V. Finally, in Section VI, the selected test conditions generated with this algorithm are evaluated regarding their contribution to the improvement in prediction accuracy.

II. Advantages of optimal designs compared to classical DoEs

Classical DoE often requires a fixed number of experiments at fixed levels to be able to cover the effects of a specified linear model. This means that the classical approach is limited to strict rules. Optimal DoE instead is more flexible, especially in conjunction with non-standard design regions [6]. Such a region e.g. is one, where not every factor level combination can be performed, mostly edges of the design region, see Figure 1. This means handling diverse restrictions on the design region.

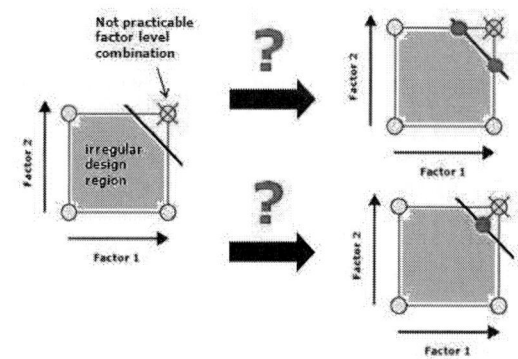

Fig. 1.   Possible intuitive solutions for non-standard design regions [4]

In contrast to the classical approach of DoE, where the number of experiments grow very fast with the number of factors and levels, the advantage of optimal DoE is the freely selectable number of experiments, only restricted by the underlying model and its number of estimated parameters. In practice, an upper limitation for the number of experiments is given by constraints like available test resources, execution time and other cost factors.

The classical approach prefers a small number of factors and corresponding levels. Commonly not more than three factor levels (low, middle and high level) are used. For linear models even a reduction to low and high factor levels is appropriate, see Figure 2(a). One advantage of the optimal approach is the freely selectable number of factors with individual and also freely selectable number of levels, see Figure 2(b). With the further advantage of specifying the underlying model, optimal DoE is indispensable.

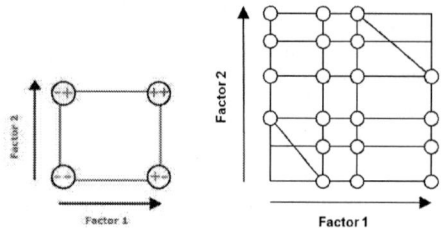

(a) Classical DoE factor levels   (b) Optimal DoE factor levels

Fig. 2. Possible factor levels for classical vs. optimal DoE. (a) classical full factorial design for two factors with two levels, '-' (low) and '+' (high). (b) restricted 2-dimensional design area with 4 levels for factor 1 and 6 levels for factor 2

III. Optimal Experimental Design Theory

Optimal experimental design theory distinguishes between continuous and exact design problems, where an exact design is a special case of a continuous design and required in practical applications. The difference lies in the weighting factors of the design points. More concrete, an experimental design ξ_n with n support points can be written as a set [3]

$$\xi_n = \left\{ \begin{array}{cccc} x_1 & x_2 & \dots & x_n \\ w_1 & w_2 & \dots & w_n \end{array} \right\} \tag{1}$$

where $x_i = (x_{i1}, \dots, x_{ik})$ is the vector of the i-th experiment. If the weights are handled as arbitrary probability measures, we call the design ξ continuous. If these weights are random integer multiples of $\frac{1}{n}$, it represents an exact n-point design $\xi_n \in \Xi_n$ with a discrete probability measure. More detailed, the weights can be written as relative frequencies

$$w_i = \frac{n_i}{n} \text{ with } \sum_{i=1}^{k} n_i = n, n_i \in \mathbb{N} \text{ and } x_i \neq x_j. \tag{2}$$

The solution of an exact optimal experimental design problem can be written as

$$\xi_n^* = \arg \min_{\xi_n \in \Xi_n} \Psi(\xi_n) \tag{3}$$

where ξ_n^* is the optimal design out of all possible designs $\xi_n \in \Xi_n$ regarding an optimality criterion. The design functional Ψ is defined by the chosen optimality criterion and is used to judge the quality of the design.

A. Optimality criteria

Optimality criteria can be roughly divided into two different groups of interest:

(i)   increasing the parameter estimation quality or

(ii)   increasing the prediction accuracy.

Both interests are achieved by minimizing variances.

Optimality criteria regarding parameter estimation quality
To minimize the variance in parameter estimation, the most common criterion is the D-optimality, where D stands for determinant. Its aim is to minimize the general variance in parameter estimation by reducing the volume of the confidence ellipsoid. The smaller the confidence region the more accurate the parameter estimation. The D-optimality criterion can therefore be written as

$$\Psi_D(\xi_n) = \left| M(\xi_n)^{-1} \right| \to \min_{\xi_n \in \Xi_n}, \tag{4}$$

where $M(\xi_n) = \frac{1}{n} X_n^t X_n$ denotes the information matrix, containing the n-points of the design ξ_n.

Optimality criteria regarding prediction accuracy
To increase prediction accuracy the IV-optimality criterion is introduced. Its aim is to minimize the variance in the prediction area $\mathcal{X}_P$. For the prediction of a single point $x \in \mathcal{X}_P$ the definition of the variance can be given as

$$var(x, \xi_n) = f(x)^t M(\xi_n)^{-1} f(x) \quad \text{with } x \in \mathcal{X}_P, \tag{5}$$

where $M(\xi_n)$ is again the information matrix. Considering now l points in the prediction area, they can be summarized in the matrix U with

$$U = \frac{1}{l} \sum_{i=1}^{l} f(x_i) f(x_i)^t P(dx_i) \quad \text{with } x_i \in \mathcal{X}_P, \tag{6}$$

where $P(dx_i)$ is a weighting factor for each point of the prediction area. The resulting IV design functional for minimizing the prediction variance over the prediction area $\mathcal{X}_P$ can be written as

$$\Psi_{IV}(\xi_n) = tr(U M(\xi_n)^{-1}) \to \min_{\xi_n \in \Xi_n}. \tag{7}$$

Figure 3 shows the transition from a theoretical design to a calculable quantity, which is used to evaluate the goodness of a design, regarding its optimality criterion.

IV. Preliminary Work for the Construction of an Optimal DoE

As already mentioned at the beginning, optimal DoE need an underlying model, related to the optimal choice of experiments. Further, the design region where it is possible to make experiments, has to be defined. With the intention of increasing prediction accuracy the desired prediction area has to be selected as well. Using a Bayesian approach, also already performed measurements can be taken into account. Summarized, these definitions create the workspace for the optimal DoE and will be explained more detailed in the following subsections.

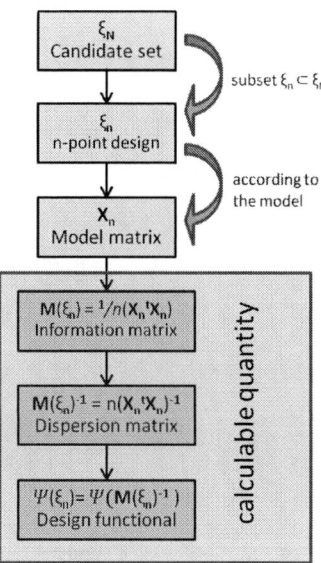

Fig. 3. Connection between a theoretical design and a calculable quantity [4]

A. Underlying model

Previous research of Bluder [1] and Plankensteiner [5] suggest a modification of the well-known physical Coffin-Manson model (Eq. 8) as the best fit between stressed devices and the resulting CTF.

$$E(CTF) = A \cdot t_p^{-\alpha} \cdot \Delta T^{-\beta} \cdot e^{\frac{\Delta H}{k \cdot T_{peak}}} \quad (8)$$

where k is the Boltzmann constant and A denotes a scaling factor. The three design factors for the DoE are therefore the pulse length t_p, the temperature rise ΔT and the peak temperature T_{peak}.

B. Design region $\mathcal{X}_E$

The design region $\mathcal{X}_E$, also known as experimental region, is characterized through measurable experiments. For the three design factors in Eq. 8 the ranges are denoted in Table I.

TABLE I.   DESIGN FACTOR RANGES

Factor	Range
$t_p [\mu s]$	{200, 500, 1500, 3500}
$\Delta T [^\circ C]$	[200; 250]
$T_{peak} [^\circ C]$	[350; 400]

The possible levels of the design factors depend also on the test equipment and technology parameters. Consequently, the design region is limited to a set of performable points, which form the candidate set ξ_N, see pink points in Figure 4.

C. Already performed experiments

For the investigated semiconductor technology, $m = 16$ already performed experiments (see blue points in Figure 4), produced due to specifications or standard tests, can be included. Optimal DoE considers this available knowledge in the selection of further points from the candidate set.

D. Prediction area $\mathcal{X}_P$

Contrary to the design region, the prediction area $\mathcal{X}_P$ is not necessarily measurable, this means that $\mathcal{X}_P \subseteq \mathcal{X}_E$ or $\mathcal{X}_P \bigcap \mathcal{X}_E = \emptyset$. With optimal DoE it is possible to optimize the prediction accuracy for a specified area without taking measurements there, although measurements in the prediction area are desirable to improve the quality of the results significantly.

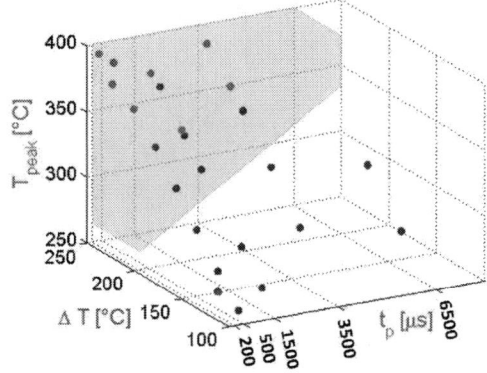

Fig. 4. Work space: candidate set ξ_N (pink points), the desired prediction area $\mathcal{X}_P$ (green) and the already performed experiments (blue points) [4]

V. CONSTRUCTION OF AN OPTIMAL DoE USING AN EXCHANGE ALGORITHM

After the definition of the workspace, an exchange algorithm, used to generate optimal designs, can be applied. Common exchange algorithms [7] start with an initial n-point design out of the candidate set $\xi_n \subset \xi_N$ and try to improve the initial design by exchanging points with the remaining points of the candidate set $\xi_N \setminus \xi_n$. After each exchange step (exchanges one or more points simultaneously) the design functional regarding the chosen optimality criterion is calculated. The algorithm stops if further exchange steps do not lead to an improvement, this means that the best combination of n points is found. In contrast to this, Spöck and Pilz [2] suggest to iterate an addition-deletion procedure right at the beginning until an n-point design is collected. A guideline for this procedure is given in Figure 5. Starting with m already available performed experiments, $s = 2$ further points are added regarding the IV-optimality criterion. Afterwards $j = 1$ point is optimally deleted. This procedure is iterated k times, representing the number of desired additional experiments.

VI. RESULTS AND CONCLUSION

Based on $m = 16$ already performed experiments and available resources for further $k = 3$ test runs, an optimal DoE was performed. The 3 selected optimal design points (DPs) are visualized as red stars in Figure 6.

These 3 experiments were performed and the improvement in prediction quality was evaluated through the mean squared error of prediction (MSEP). Compared to the model based on the already performed experiments (basic), the improvement by adding 3 optimal chosen DPs is displayed in Figure 7.

Paper W3A2

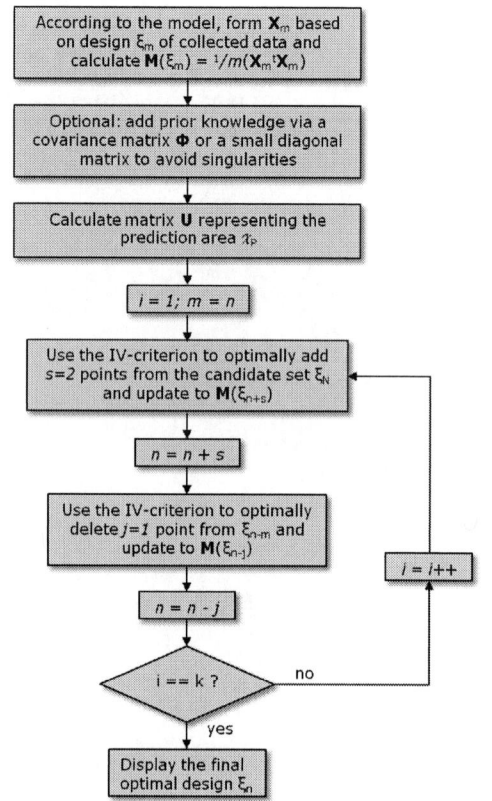

Fig. 5. Flowchart of the exchange algorithm proposed by Spöck and Pilz [2], [4]

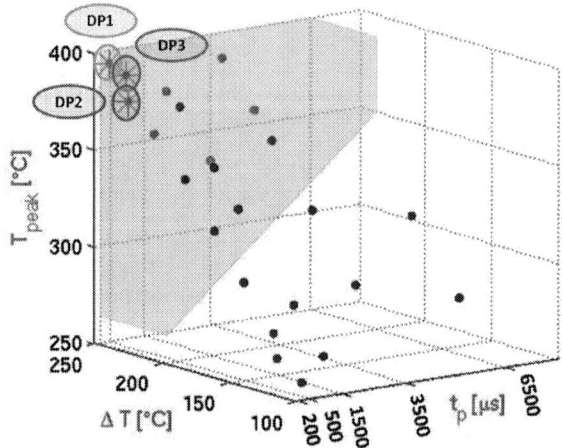

Fig. 6. The whole work space for the DoE with the 3 optimally chosen design points (red stars) [4]

Due to the addition-deletion procedure, there is no ranking between DP1, DP2 and DP3. All 3 DPs have to be performed to guarantee an optimal result. As expected, an improvement

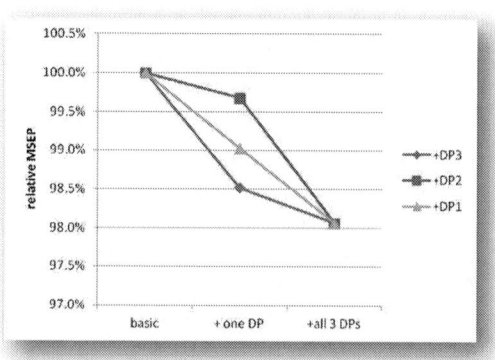

Fig. 7. Development of MSEP when adding sequentially three optimal design points to a basic ξ_{16} design. As expected, the MSEP decreases when adding the optimal design points [4]

in prediction accuracy of $\sim$2% under the IV-optimality criterion was observed, using the MSEP as a quality measure.

Summarized, the advantages of optimal designs are:

- the freely selectable number of test runs on non-standard design regions which implies a reduction of resources and therewith costs,

- the possibility to improve the prediction accuracy instead of the parameter estimation and

- already performed experiments or such, which have to be performed due to specifications (whether they are optimal or not) can be included as prior knowledge.

ACKNOWLEDGMENT

The authors would like to thank Michael Ebner (Infineon) for the measurement support, as well as Michael Glavanovics (KAI) and Andreas Spitzer (Infineon) for valuable discussions on the topic.

This work was jointly funded by the Austrian Research Promotion Agency (FFG, Project No. 831163) and the Carinthian Economic Promotion Fund (KWF, contract KWF-1521—22741—34186).

REFERENCES

[1] O. Bluder, *Prediction of Smart Power device lifetime based on Bayesian modeling*, PhD thesis, Alpen-Adria-University of Klagenfurt, 2011.

[2] G. Spoeck and J. Pilz, *Spatial sampling design and covariance-robust minimax prediction based on convex design ideas*, Stochastic Environmental Research and Risk Assessment 24, 2010.

[3] A. C. Atkinson and A. N. Donev, *Optimum Experimental Design*, Oxford University Press, 1992.

[4] A. Zernig, *Optimal Design of Experiments for Semiconductor Data following a Mixtures-of-Experts Model*, Master thesis, Alpen-Adria-University of Klagenfurt, 2013.

[5] K. Plankensteiner, *Application of Bayesian Models to Predict Smart Power Switch Lifetime*, Master thesis, Alpen-Adria-University of Klagenfurt, 2011.

[6] D. C. Montgomery, *Design and Analysis of Experiments*, 7th ed., John Wiley & Sons Asia, 2009.

[7] V. Fedorov, *Theory of optimal experiments*, Academic Press, New York, 1972.

A Metric Driven Verification and Validation Approach for Smart Power Devices

Oleksandr Melnychenko
Institute of Computer Engineering
Vienna University of Technology
Vienna, Austria
Email: oleksandr.melnychenko@tuwien.ac.at

Hans-Peter Kreuter
Infineon Technologies Austria AG
Villach, Austria
Email: hanspeter.kreuter@infineon.com

Abstract—We propose a method for the pre-silicon metric driven verification and post-silicon validation of automotive Smart Power devices. The method is based on extending the UVM architecture with hardware drivers and monitors. We present the implementation of the hardware test bench components and the required software interfaces. The method is finally demonstrated by verifying a serial interface of a high-side power switch showing promising results.

I. INTRODUCTION

In automotive industry electromechanical relays, fuses and discrete circuits have been replaced with Smart Power switches (SPSs). These devices are built using power MOSFET transistors together with CMOS logic circuitry for integrated control [1]. Additionally to their switching capability SPSs provide protective and monitoring functions. They protect the power transistor and the load from over voltage, over current, loss of ground and implement thermal shutdown. Moreover, various features are provided like pulse-width modulation (PWM) generators, analogue current feedback, status feedback, failsafe and low quiescent current consumption in standby mode. In fact, simple transistor switches evolved to complex Systems-on-Chip (SOCs) [2].

It is a general trend in all areas of microelectronics that the complexity of products is increasing due to the high degree of miniaturization. The high level of integration causes the verification process to consume more than 50% of the total development time [3]. The verification process is important to ensure that the circuit meets the specification requirements. Design bugs in a delivered IC will definitely have a negative impact on the quality of the product itself. On the other hand, the short time-to-market in the competitive automotive world calls for fast verification and development processes.

Verification is normally accomplished before the silicon is available (pre-silicon) using models and simulations tools, and after IC fabrication (post-silicon) in laboratory environment and with Automated Test Equipment (ATE) [3]. Quite often hardware prototyping is involved in pre-silicon verification. It provides more information about the design functionality and speeds up the development process due to short feedback loops between design and verification [4].

Testing of digital circuits by simulation is a well-developed area in micro-electronics design. Powerful simulation tools from different vendors are available like QuestaSim from Mentor Graphics or INCISIVE from Cadence. A virtual verification environment, that is executed on the simulation platform, can be built using hardware verification languages (HVLs), e.g. SystemVerilog [5]. HVLs define various powerful non-synthesizable constructs that help to build the test bench. To leverage the power of HVLs and introduce better standardization a Unified Verification Methodology (UVM) has been recently developed [6], currently as a library on top of SystemVerilog. UVM test benches are focused on coverage driven functional verification—a state-of-the-art technique in pre-silicon testing [6]. In addition high level of test bench reconfigurability and re-usability can be easily achieved.

Post-silicon testing process is not so formalized and automated. A lot of manual interactions are required, e.g. test scenarios are created by lab engineers according to the device specification. The tests are usually of pass/fail type. This situation results in a low level of test bench reuse and poor coverage metrics. Close to the tapeout of an IC all most important test cases, coverage goals and verification plans are already defined and achieved by simulations. However this pre-silicon test benches are fairly reused in post-silicon testing [2].

Recent improvements in this direction have been achieved. The approach proposed in [3] introduces a test system independent verification plan and formalisation of SPSs test process by means of simple test configuration language. Unfortunately, this method is only applicable for chip parameter measurement, but not for functional testing. The idea of introducing coverage driven verification techniques into laboratory test flow is promoted in [2]. It brings a lot of advantages, however a very limited SystemVerilog/UVM subset can be ported to a laboratory test equipment. In [4] different pre- and post-silicon test benches are generated from the same meta-language description. The overall verification process is metric driven and controlled by common verification plan. This method was developed for pure digital circuits verification, but it is hardly applicable for mixed-signal designs like SPSs.

The method presented in this work is intended to overcome limitations of the currently known approaches. The key idea is to integrate hardware simulations into PC simulation flow instead of adopting PC simulation techniques for hardware test flow. The practical way to do this is to let the SystemVerilog/UVM test bench access the hardware units with implemented Design Under Test (DUT). With this approach the full power of coverage driven verification becomes available for laboratory testing. Moreover, a UVM random stability property

978-1-4673-4579-8/13 $31.00 © 2013 IEEE

Paper W3A3   *PRIME 2013, Villach, Austria*

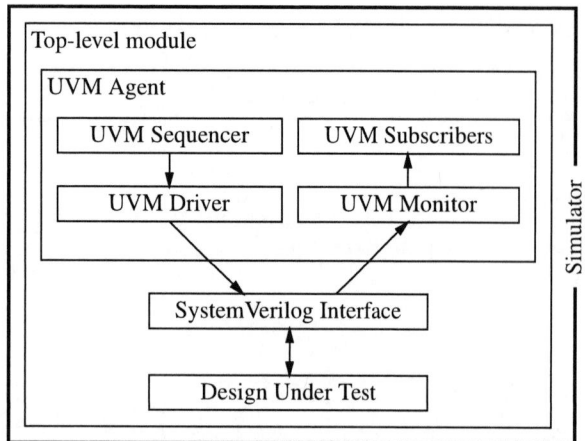

Fig. 1.   Simple UVM test bench

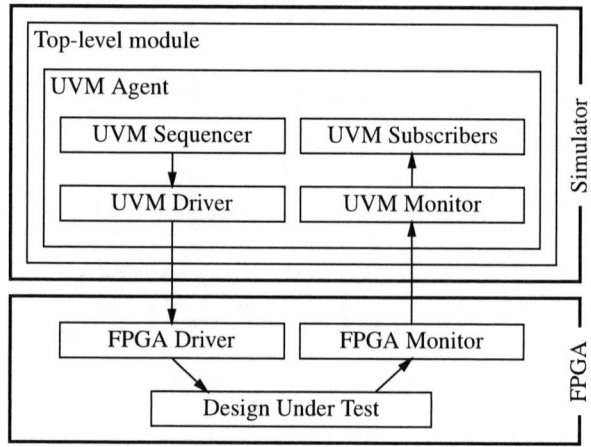

Fig. 2.   Proposed test bench

gives a possibility to run exactly the same test sequences on simulation and measurement platforms [7].

II.   UVM AGENT EXTENDED WITH HARDWARE INTERFACE

A. Traditional UVM Agent

A simple transaction-level test bench is made of the following components:

- A sequencer to create streams of transactions;
- A driver to apply these transactions to the DUT;
- A monitor to collect the activities of the DUT;
- Subscribers to perform coverage and checking.

In order to reuse this quadruple it is recommended to group this components and form a so called Agent as shown in the Fig. 1.

B. Concept Overview

The authors propose to extend the UVM agent with hardware drivers and monitors to reuse the Agent for laboratory testing or validation of hardware prototypes. Therefore the bit-level activities of the driver and monitor are implemented on an FPGA (Fig. 2). The FPGA module must provide an API to control the hardware components. Consequently the hardware drivers and the monitors can be connected to the UVM test bench with software interfaces written in C and SystemVerilog.

Fast and straightforward implementation of the method is only possible with the proper choice of hardware and software.

C. UVM Driver and Monitor—FPGA Implementation

The DUT, the hardware driver and monitor may be placed on a PCI-bus compatible FPGA module. The LabVIEW development environment provides VHDL code integration into the LabVIEW block diagram. Moreover the FPGA design can be accessed with a C API. Consequently the VHDL prototype

of the DUT, the driver and the monitor can be implemented with VHDL and connected on the LabVIEW block diagram as shown in Fig. 3.

Referring to Fig. 3 "SPOC_TB" is an Intellectual Property (IP) integration block for VHDL code that represents the DUT model together with the hardware driver and monitor. "REQ" and "RES" blocks are FIFOs for the communication with the PC. This FIFOs enable a communication link between the FPGA module and the PC over the PCI bus. FIFO is the most suitable interface in the presented method since it ensures that no transactions are lost and helps to improve simulation speed. Required execution logic is provided by loops and sequential frames. Finally, C API functions should be generated out of this LabVIEW structure to access the FPGA module.

D. UVM Driver and Monitor—Host Implementation

SystemVerilog provides the Direct Programming Interface (DPI) to call code written in a foreign languages like C. The proposed method replaces the bit-level activities in the UVM test bench with DPI function calls to access the hardware FIFO interfaces (Fig. 4). As mentioned before, the FIFOs can be accessed with the generated C API. These functions ("rio_wr_req" and "rio_rd_res") are called inside "run_phase" tasks of the driver and the monitor performing reading and writing of the complete transactions.

There are some DPI functions left out in Fig. 4 that are required for the FPGA initialization and finalization. These functions must be called before the test starts and after the results are available.

E. DPI Functions

The DPI code developed for the presented method consists of C language and SystemVerilog layers. The functions "rio_wr_req", "rio_rd_res" and other functions for FPGA control mentioned before are implemented in the C layer (Fig. 5), while SystemVerilog layer contains their prototypes declared as imported functions (Fig. 6).

978-1-4673-4579-8/13 $31.00 © 2013 IEEE   290

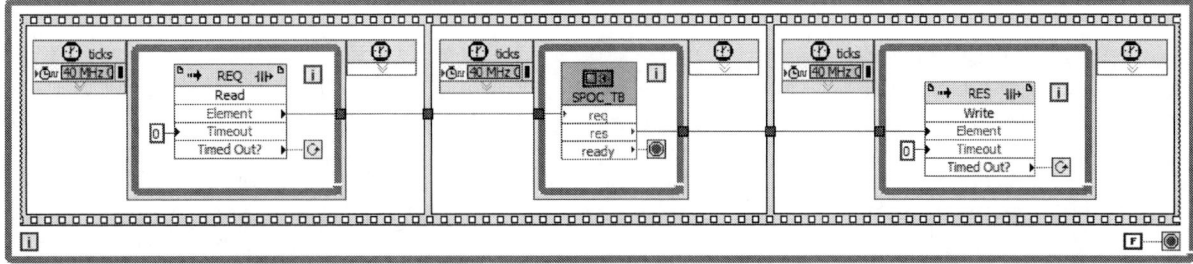

Fig. 3. LabView block diagram for FPGA prototyping platform

```
class any_driver extends uvm_driver #(any_transaction);
   ...
   task run_phase(uvm_phase phase);
      forever begin
         seq_item_port. get (req);
         rio_wr_req(req);
         #1us;
      end
   endtask
endclass


class any_monitor extends uvm_monitor;
   ...
   task run_phase(uvm_phase phase);
      ...
      forever begin
         res = rio_rd_res() ;
         post_process (res );
         #1us;
      end
   endtask
endclass
```

Fig. 4. UVM driver and monitor

```
int rio_laststatus ;

void rio_checknewstatus (int status ) {
   if ( NiFpga_IsError(NiFpga_MergeStatus(
      &rio_laststatus , status )))
      rio_triggeralert (rio_laststatus ); }

void rio_wr_req(uint8_t req) {
   uint8_t buffer [1] = {req};
   rio_checknewstatus(
      NiFpga_WriteFifoU8(
         ..., buffer , ..., NiFpga_InfiniteTimeout , ...) ); }

uint16_t rio_rd_res() {
   uint16_t buffer [1];
   rio_checknewstatus(
      NiFpga_ReadFifoU16(
         ..., buffer , ..., NiFpga_InfiniteTimeout , ...) );
   return buffer [0]; }
```

Fig. 5. DPI functions, C layer

```
typedef byte        unsigned u8;
typedef shortint    unsigned u16;
event rio_alert ;

export "DPI−C" function rio_triggeralert ;
function void rio_triggeralert (input int status );
   $display ("Error in VI, error code %d", status );
   −> rio_alert;
endfunction

import "DPI−C" context function void rio_wr_req(u8 req);
import "DPI−C" context function u16 rio_rd_res() ;
```

Fig. 6. DPI functions, SystemVerilog layer

The DPI functions to access and control the FPGA module, call the C API functions generated by the LabVIEW development environment. The data transfer functions "rio_wr_req" and "rio_rd_res" set infinite timeout when accessing FPGA FIFOs to ensure that none of the transactions are lost.

In order to decrease the amount of data transferred through the DPI interface and to simplify the usage of the DPI functions the FPGA session ID and the error status variable are declared in the DPI C layer. Additionally the error handling is encapsulated in the function "rio_checknewstatus" (Fig. 5). In case of an FPGA API interface error the function "rio_checknewstatus" calls the "rio_triggeralert" function and passes the error code. The UVM verification engineer can implement any FPGA error handling action by overwriting the "rio_triggeralert" body without changing the DPI C code. In the presented example an error message is printed to the console and an event "rio_alert" is raised (Fig. 6).

III. EXPERIMENTAL RESULTS

A. Test Bench

The method presented in this work is used for the functional verification of SPSs developed for automotive applications. Digital behavioural VHDL models of these mixed-signal devices are used to perform the verification tasks and to enable FPGA prototyping. The control, configuration and status registers of the DUT can be accessed with a Serial Peripheral Interface (SPI), which is a simple 4-wire bidirectional hardware interface.

In the test bench the UVM sequencer generates random SPI frames and sends them to extended UVM driver that

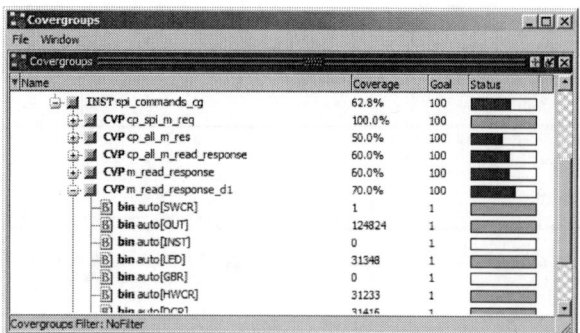

Fig. 7. An example of coverage results

performs the bit-level switching activity and drives the pins of the DUT on the FPGA. Simultaneously the activity of the pins is monitored and packed into SPI response frames by the extended UVM monitor. The obtained SPI transactions are analyzed with standard UVM components like comparators, coverage collectors, loggers etc. shown in the Fig. 1 as UVM subscribers.

B. System Configuration

A PC workstation, equipped with a processor Intel Core i7-3770 @3.40 GHz and 16 GB RAM running OS Windows 7 Enterprise SP1 64-bit, has been prepared to perform experimental research on the proposed method. The hardware platform was represented by NI device PCI-7831R, installed as a PCI extension card in the PC workstation. This PCI card incorporates a Virtex2 FPGA, 8 DACs, 8 ADCs and 96 digital IOs. Questa Sim 10.0e 32-bit and LabVIEW 2012 32-bit software versions were used to build the test bench and run the simulation.

C. Results

Simulation with an unconstrained random test sequence of length $N = 10^6$ transactions was performed both with the traditional and proposed test bench. A basic coverage model for SPI requests and responses was defined in the UVM test bench to observe the full potential of the simulation process. Fig. 7 presents coverage results of the simulation. They are identical both in the traditional and proposed test bench.

Fig. 7 clearly shows that in order to meet 100% coverage more agents, constraints and tests cases should be created. Obviously a more detailed coverage model and coverage goals should be defined. However this is not covered here but will be addressed in a future work.

Table I compares the simulation speed of the the traditional and proposed test benches. The test time t was measured with a simple DPI wrapper for the system time function. The average simulation throughput p was computed as $p = N/t$. It could be shown, that the throughput of the extended test bench is more than twice as high as for the traditional one. The effect is achieved because driver and monitor bit-level activity does not produce a computational load on the PC, when FPGA co-simulation is performed.

TABLE I. COMPARISON BETWEEN SIMULATION SPEED

Test bench	Test time, s	Throughput, trans./s
Traditional UVM	480	$2.1 \cdot 10^3$
Proposed UVM / hardware	224	$4.5 \cdot 10^3$

IV. SUMMARY AND OUTLOOK

A new approach for automotive semiconductor device verification and validation was presented in this work. The major advantages with respect to the currently used methods is the combination of SystemVerilog/UVM constructs with embedded hardware (e.g. FPGA), in order to perform testing. With the proposed method there is no need to develop a new test bench for hardware prototyping platform or multi-environment verification plan. A metric driven test process setup for hardware prototyping platform is easy and effective with the new method. The proposed method can be easily reused for post-silicon validation and to improve the simulation speed.

Obviously, in case of prototyping the complexity of the simulated device is limited by the embedded hardware characteristics. However, if the DUT is available in silicon, it can be validated regardless its complexity. In this case only hardware drivers and monitors should be implemented, while the DUT should be connected directly.

While the example demonstrated here may be simple, it nevertheless proves the concept of heterogeneous UVM Agents. Future work will focus on mixed-signal metric driven verification and extending the test bench with multiple agents.

ACKNOWLEDGMENT

This work was jointly funded by the Federal Ministry of Economics and Labor of the Republic of Austria and the Carinthian Economic Promotion Fund (KWF).

REFERENCES

[1] B. Murari, F. Bertotti, and G. A. Vignola, *Smart Power ICs—Technologies and Applications*. Springer Verlag, 1995.

[2] T. Nirmaier, V. Meyer zu Bexten, M. Tristl, M. Harrant, M. Kunze, M. Rafaila, J. Lau, and G. Pelz, "Measuring and improving the robustness of automotive smart power microelectronics," in *Design, Automation & Test in Europe Conference & Exhibition*, 2012, pp. 872–873.

[3] A. Pirker-Frühauf, "New methods and concepts to speed-up the verification process of integrated circuits," Ph.D. dissertation, Technische Universität Wien, 2009.

[4] A. Adir, S. Copty, S. Landa, A. Nahir, G. Shurek, A. Ziv, C. Meissner, and J. Schumann, "A unified methodology for pre-silicon verification and post-silicon validation," in *Design, Automation & Test in Europe Conference & Exhibition*, 2011, pp. 1–6.

[5] J. Bromley and M. Smith, "A user's experience with SystemVerilog," in *SNUG Europe*, 2004.

[6] S. Rosenberg and K. A. Meade, *A Practical Guide to Adopting the Universal Verification Methodology (UVM)*, 1st ed. Cadence Design Systems, 2010.

[7] A. Efody, "UVM random stability. Don't leave it to chance," in *Design & Verification Conference & Exhibition, DVCon*, Feb. 2012.

Modeling of an electrostatically actuated microelectromechanical (MEMS) speaker system

David Tumpold*, Manfred Kaltenbacher
Institute of Mechanics and Mechatronics
University of Technology,
Vienna, Austria
*david.tumpold@tuwien.ac.at

Christoph Glacer, Alfons Dehé, Mohsin Nawaz
Infineon Technologies AG
Munich, Germany

Abstract—The market for tablets, laptops and mobile devices is increasing rapidly. Device housings get thinner and energy efficiency plays a major role for battery-powered devices. Microelectromechanical (MEMS) loudspeakers, fabricated in complementary metal oxide semiconductor (CMOS) compatible technology merge energy efficient driving technology with cost economical fabrication processes. Fabricating these devices is a elaborating and expensively task. Therefore, the need of computer models, capable of precisely simulating the multi-field interactions is strongly increasing. We use a system of coupled partial differential equations (PDEs) describing the interaction between the electrostatic, mechanical and acoustical field and apply finite element method FEM to solve them. Additionally, we fully take nonlinear effects like large deformations or stress stiffening effects into account. Mortar FEM is used, to efficiently handle the coupling between mechanical and acoustical field. In combination with special boundary conditions, like perfectly matched layers (PML) truncated propagation regions can be applied in the model. We will present simulations of a MEMS speaker system based on a single sided driving mechanism starting at the electric potential applied on the two electrodes and resulting in the generated sound pressure level (SPL).

Keywords; MEMS, finite element modeling, speaker, nonlinerities, electrostatic force, mechanics, acoustics, sound pressure level

I. INTRODUCTION

In this paper we present a micro speaker consisting of two circular shaped electrodes with a constant gap between them. The top electrode has a thin insulation coating on it and represents the back plate. The membrane, which is located below the back plate is mechanically softer compared to the insulated back plate, hence it will start to bend while the back plate is stiff and can be assumed as fixed. The back plate is perforated and clamped with a spider suspension fixing to avoid squeeze film damping and keep its mechanical strength.

Investigations on maximum sound pressure level (SPL) possible and optimization methods are computer aided with finite element models (FEM). First of all, the microelectromechanical transducer can be split up into its physical fields, where the corresponding partial differential equations (PDE) are coupled to describe the working principle according to its physical behavior. The resulting interaction between the electrostatic, mechanical and acoustic field is depicted in Fig. 1. Since we assume no back coupling from the acoustics onto the mechanical field [1], the acoustic calculation is done in a separate model and using the mechanical displacement as input data. Furthermore, this enables us an additional interface for model verification. To minimize the computational time and amount of memory the complex three dimensional mechanical model was reduced into a two dimensional axis symmetric model. Model parameters for the perforated back plate in axis rotational domains and characterization of fabrication processes was determined in prior work by [2] on the MEMS microphone.

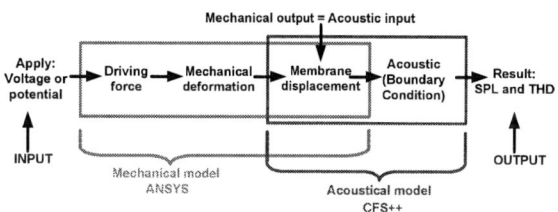

Fig. 1. Major parts for modeling the MEMS micro speaker. Starting on the left side with an applied voltage resulting in a sound pressure level on the right side of the modeling chain.

In the upcoming section the modeled electrostatic force coupling to the mechanical model is presented, where ANSYS is used as tool. Continuing the model chain, the acoustic part is presented, which is computed by the FEM solver CFS++. Open domain computations are realized on a truncated propagation region surrounded with PML elements. The final output is the sound pressure level of the electrostatically actuated, mechanically oscillating membrane.

II. MODELING METHODS

A. Electrostatic mechanical model

Electrostatic force was modeled with TRANS126 elements, which have a single degree of freedom in translation and for electric voltage. As displayed in Fig. 2 a.), the electro-mechanical transducer element is located between both conducting electrodes. The membrane and the back plate. The back plate features an insulation layer with a relative permittivity of $\varepsilon_r = 7.5$ and a very high tensile pre stress compared to the poly silicon. The electro mechanical transducer (EMT) input data is a polynomial fit of a capacitance versus gap function as described in [3]. Hence the gap between both electrodes is modeled with capacitances in series, as can be seen in Fig. 2 b.). The capacitance in the insulation layer is fixed and specifies the maximum capacitance of the system, if the membrane is mechanically in contact with the back plate and the gap is closed. The gap versus capacitance relation is depicted in Fig. 2 c.) where Gap_{max} is related to minimum capacitance and Gap_{min} is related to maximum capacitance.

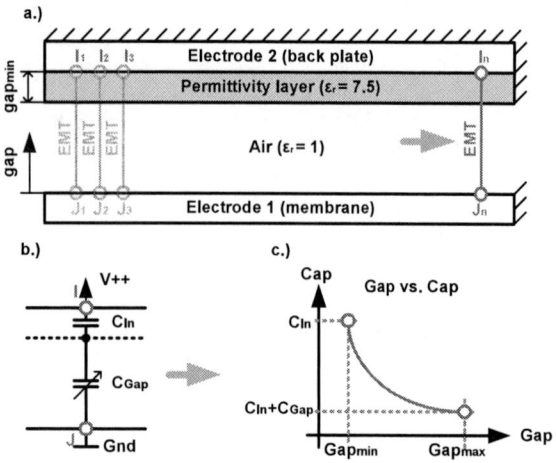

Fig. 2. a.) TRANS126 element implementation between nodes I and J results in an b.) analogue circuit diagram and corresponding gap versus capacitance function in c.).

C_{In} is computed with

$$C_{max} = C_{In} = \frac{\varepsilon_1 A}{G_{min}} = \frac{\varepsilon_r \varepsilon_0 A}{d_1} \qquad (1)$$

and C_{Gap} with

$$C_{Gap} = \frac{\varepsilon_2 A}{G(y)} = \frac{\varepsilon_0 A}{d_2 - y}, \qquad (2)$$

where A is the axis rotational surface of the layer of one element (results in a ring shaped area), d_1 the thickness of insulation layer, d_2 the thickness of initial gap and ε_r the relative permittivity for the insulation layer. Furthermore the minimum capacitance at Gap_{max} can be computed by adding the air capacitance C_{Gap} in series to insulation capacitance C_{In}

$$C_{min} = \frac{C_{In} C_{Gap}}{C_{In} + C_{Gap}} = \frac{\varepsilon_1 \varepsilon_2 A}{\varepsilon_1 d_2 + \varepsilon_2 d_1 - \varepsilon_1 y} \qquad (3)$$

As a result the correct capacitance to gap relation according to Fig. 2 c.) can be computed with

$$C(y) = \frac{\lambda}{G_{min} - G_0} \ or \ \frac{\lambda}{G_{max} - G_0} \qquad (4)$$

where

$$G_0 = \frac{C_{max} G_{min} - C_{min} G_{max}}{C_{max} - C_{min}} \qquad (5)$$

and

$$\lambda = C_{max}(G_{min} - G_0). \qquad (6)$$

Since numerical derivations of polynomial functions are simpler to implement in comparison to reciprocal functions like the capacitance to gap relation (4), the implementation in ANSYS has to be done with polynomial coefficients [4]. Hence the coefficients are computed with

$$\mathbf{A} \cdot \mathbf{x} = \mathbf{b} \qquad (7)$$

Solving (7) results in the polynomial coefficients representing (4). Each electro-mechanical transducer element along the radius owns an individual capacitance to gap relation, which must be computed and set separately. The electrostatic force is computed by the virtual work principle resulting for our case in [3]

$$F_e = \frac{\partial W_e}{\partial y} = \frac{1}{2} \frac{\partial C}{\partial y} U^2, \qquad (8)$$

with U the applied electric voltage. We found out that the capacitance to gap relation can not be fitted with the polynomial function for every setup. High relative permittivity layers, small gaps or bad chosen sampling points show characteristics, where the polynomial fitted function starts to oscillate. These oscillations can lead to alternating signs of the electrostatic attractive force, caused by the gradient of the capacitance derivation along the gap (8). This must be taken into consideration, when modeling with TRANS126 elements and checked carefully before solving.

Geometric nonlinearities like large deformation and stress stiffening are fully included in the model, since stress stiffening effects play a major role in modeling various layers of different silicon types. Material properties were taken from [2]. Computational background information about structural nonlinear effects can be found in [1, 3] for FEM and in [5] for structural mechanics.

B. Mechanical-acoustical model

The acoustic field (according to Fig. 1) was computed with FEM with CFS++ used as tool. As described in [1] weak coupling between mechanical and acoustic fields can be asumed if the pressure force of the fluid on the structure is negligible. The most important thing at the acoustic model is the free field or open domain condition. To prevent reflections of waves leaving the propagation region, special boundary conditions are required to be able to model truncated propagation regions with open domain characteristics. Absorbing boundary conditions (ABC) or perfectly matched layers (PML) have been developed for this purpose. ABC of first order absorbs only the normal component of the impinging wave front. But leads to reflections for other angles of incident. Grote [6] shows investigations and improvements of higher order ABCs. Previous tests [7] have shown, that PML works better for our computational domains.

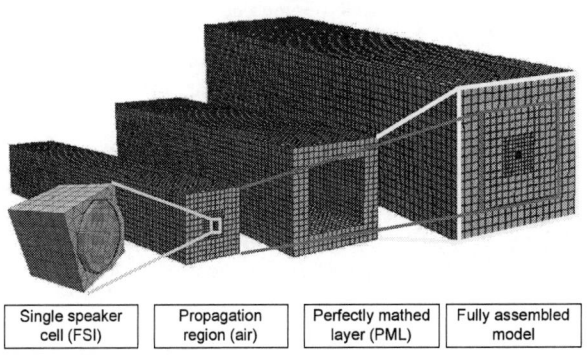

| Single speaker cell (FSI) | Propagation region (air) | Perfectly mathed layer (PML) | Fully assembled model |

Fig. 3. Mechanical-acoustic model with propagation region glued with Mortar FEM and PML to model open domain characteristics.

As depicted in Fig. 3, the acoustic model was built up with different modules (assembled with various regions). These modules are put together with Mortar FEM. The smallest part represents the speaker cell with the fluid structure interface between mechanical model and structural model. The speaker cell block is surrounded by four shells of propagation region (characteristics of air). In this ambient air region the discretization size increases from $\lambda/20$ to $\lambda/10$ in four steps, to reduce the total amount of elements. With the help of Mortar FEM, nonconforming grids are used [8], reducing the total amount of elements and avoiding distorted elements near the mechanical-acoustical interface. Further information on PML can be found in [1, 9, 10] and for Mortar FEM in [8].

III. RESULTS

Measured data for verifying the simulated results were provided by Infineon Technologies AG Munich.

A. Electrostatic mechanical model verification

Membrane displacement was determined optically with a laser scanning vibrometer and measured electrically with a capacitive voltage divider as discussed in [7, 11] in more detail.

Fig. 4 displays the membrane displacement versus voltage applied. On the left side the center point displacement of the membrane is shown, where the snap-in point is computed between 10.9 V and 11 V. The snap-in occurs, when the electrostatic force and the mechanical pull back force of the membrane are unequal and the system falls into an unstable mode. The membrane is rapidly pulled towards the insulation layer of the back plate. From the acoustically point of view, snap-in results in a highly distorted sound, because of the high positive and negative accelerations and therefore from the sound pressure level point of view, the amplitude and therefore the loudness increases. On the right side of Fig. 4 a slice view of the membrane is depicted, where the nonlinear voltage to displacement relation and the snap-in can be seen very well. The snap-in voltage was determined by measurement at 13 V ±2 V for arbitrary chosen speaker cells of various lots.

We found out that the derivation between simulation and measurement concerning the snap-in point is caused by fabrication tolerances in gap distances, as this parameter is linked quadratic within the electrostatic force. Furthermore we have introduced a second deviation from the experiments due to our axis symmetric model. Thereby the perforated back plate with its spider suspension fixing was assumed as solid body. This results in an effective lower surface and higher electric fringe field rates, hence a lower effective capacitance. As a result the electrostatic force is stronger in the model compared with the device. That effect was investigated on effective capacitance computation for model simplifications for the microphone in [2] before and confirms our results. Considering these two facts the snap-in point is determined accurately within the simulation.

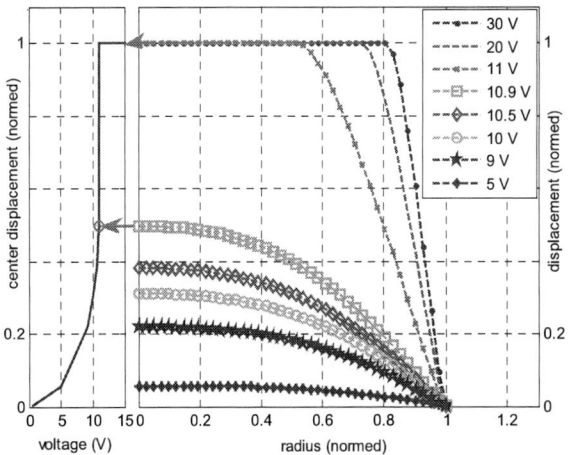

Fig. 4. Static membrane displacement under applied voltage(right) and center displacement versus applied voltage with snap-in (left)

B. Mechanical-acousitc model verification

Since a single speaker cell operating in non snap-in mode, shows a very low sound pressure level and the signal to noise ratio is in bad condition, a small array with four by four (16 cells) was used. This results in a theoretical increase of +24 dB$_{SPL}$ (+6 dB$_{SPL}$ each time the active speaker area is

Paper W4A1   *PRIME 2013, Villach, Austria*

doubled). Open domain was realized with perfectly matched layers, since investigations (see [7]) show that PML are well suited for arbitrary impinging wave fronts.

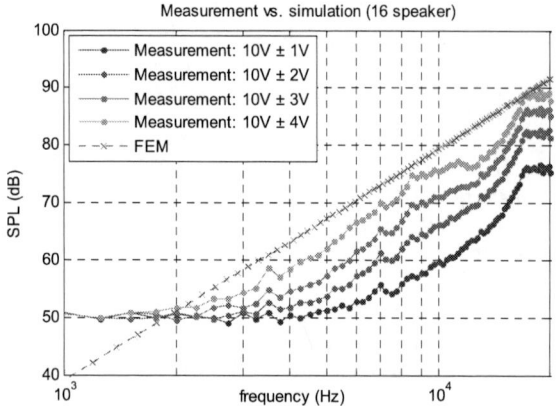

Fig. 5. Non snap-in moving membrane characteristics with bias voltage versus sound pressure level measurement compared to FEM result.

For this measurement the speaker was driven in non snap-in mode, with a bias voltage of 10 V and an audio signal level (peak to peak) of 1 V to 4 V. We observe for increasing the voltage, the SPL increases as well, which is a result of the membrane stroke level. The membrane stroke level was determined as described before (Fig. 4) with the mechanical model, but assumed as piston movement with the center point displacement as stroke level. As a result the expected SPL is higher than the measured and represents the maximum possible SPL. More accurate SPL can be achieved by modeling the correct deformed displacement of the membrane, which leads to a finer grid and results in a tradeoff between computational time and accuracy. Particularly for smaller diameter to displacement ratios this effect must be taken into consideration.

IV. SUMMARY AND OUTLOOK

Bearing in mind that the actual loudspeaker system is an inverse operating microphone and was primary not designed to produce high sound pressure levels, the resulting sound pressure level is acceptable. However, because of its physical dimension the MEMS speaker system is not able to generate low frequency sound. Maximum sound pressure level depends on active surface, stroke level and frequency. In our case, the stroke level and the surface are limited by fabrication technology. Frequency is limited by human auditory between 20 Hz and 20 kHz. A possibility to increase the SPL by increasing the active surface was demonstrated by the array setup. These points open up the opportunity to change the driving principle from single sided actuator to a double sided push-pull or pull-pull system, since the gap distance and the non snap-in operation mode increase. We have considered all necessary nonlinearities and a method in ANSYS to model the nonlinear electrostatic force for simple geometries. The electrostatic-mechanical model was verified with optical and

electrical measurements. In the acoustic model, we presented an ansatz for open domain modeling and truncated propagation regions. Additionally, the mechanical-acoustic coupling uses Mortar FEM to reduce the total amount of elements and avoid distorted elements near mechanical-acoustic interface.

ACKNOWLEDGMENT

The author would like to thank Infineon Technologies AG Munich (Germany) and Villach (Austria) for great teamwork and making available fabricated speaker devices and measurement results to evaluate the modeling process.

REFERENCES

[1] M. Kaltenbacher, *Numerical Simulation of Mechatronic Sensors and Actuators*, 2nd ed. Berlin Springer 2007.

[2] M. Füldner, "Modellierung und Herstellung kapazitiver Mikrofone in BiCMOS-Technologie," PhD, Universität Erlangen-Nürnberg, München, 2004.

[3] ANSYS, "Theory Reference for the Mechanical APDL and Mechanical Applications," vol. 12.1, ed. Canonsburg,PA, 2009, p. 1228.

[4] ANSYS, "Element Reference," vol. 12.0, ed. Canonsburg,PA, 2009, p. 1690.

[5] R. C. Hibbeler, *Technische Mechanik 2. Festigkeitslehre* vol. 5. Louisiana: Pearson Studium, 2005.

[6] M. J. Grote and J. B. Keller, "Nonreflecting boundary conditions for time-dependent scattering," *J. Comput. Phys.*, vol. 127, pp. 52-65, 1996.

[7] D. Tumpold, "MEMS based speaker system," Master Thesis, Applied Mechatronics, Alpen Adria Universität, Klagenfurt, 2011.

[8] S. Triebenbacher, *et al.*, "Applications of the Mortar Finite Element Method in Vibroacoustics and Flow Induced Noise Computations," *Acta Acustica united with Acustica*, vol. 96, pp. 536-553, 2010.

[9] Andreas Hüppe and M. Kaltenbacher, "Spectral Finite Elements for Computational Aeroacoustics using Acoustic Perturbation Equation," *Journal of Computational Acoustics*, vol. 20, p. 13, 2012.

[10] Barbara Kaltenbacher, *et al.*, "A modified and stable version of a perfectly matched layer technique for the 3-d second order wave equation in time domain with an application to aeroacoustics," *Journal of Computational Physics*, vol. 235, pp. 407-422, 15 Februrary 2013 2013.

[11] C. Glacer, "Reversible akustische Wandler in MEMS Technologie," Dipl.-Ing., Theoretische Elektrotechnik und Mikroelektronik (ITEM), University Bremen, Bremen, 2011.

978-1-4673-4579-8/13 $31.00 © 2013 IEEE

A promising technology of Schottky diode based on 4H-SiC for high temperature application

Razvan Pascu, Florea Craciunoiu, Mihaela Kusko
Laboratory of Nanobiotechnologies
National Institute for Research and Development in
Microtechnologies – IMT
Bucharest, Romania
razvan.pascu@imt.ro

Abstract – **A Schottky diode technology based on silicon carbide (SiC) with ramp oxide termination is presented. The improvement of the Schottky, respectively ohmic contact is conditioned by a rapid thermal processing to form Ni-silicide. The Schottky diodes (SD) have been electrical characterized in forward bias in range of temperature 300-573 K to verify their capacity to operate like temperature sensors. The results show an improvement of the electrical parameters with increasing of the temperature.**

Keywords – *SiC; Schottky diode; ramp oxide termination; Schottky Barrier Height; high temperature; temperature sensor.*

I. INTRODUCTION

The devices based on silicon carbide (SiC), a wide band-gap semiconductor with low intrinsic carrier concentration, gain an increased scientific interest due to their properties [1]. Besides the fact that SiC is a material with high thermal conductivity, high saturated drift velocity and high breakdown electric field, it has a native oxide layer which represents an advantage for developing further processing technologies [2]. Thus, high-power, high-frequency and high-temperature applications are requirements where SiC successfully performs.

In this paper the fabrication technology of the temperature sensor based on metal-semiconductor SiC Schottky diode (SD) is presented. An advantage of this contact is the absence of the minority charge carriers leading to a fast switching. For a better operation at high temperatures the requirements are: low leakage current and a large Schottky Barrier Height (SBH). Different metals like Pd, Au, Ni and Pt have been tested in order to obtain a high SBH, with the best results obtained for Ni [3].

Moreover, the Ni work function is $\Phi_{Ni} = 5.15eV$ and a theoretical value of the SBH is done by the difference between the metal work function and the semiconductor electron affinity. A value of the electron affinity is in range $X_{SiC}=3.3-3.7eV$ [4]. Making the difference between Ni metal work function and a value from the range mentioned, we expect the SBH to be in range 1.45-1.85eV. The low electrical resistivity at room temperature and thermodynamic stability [5] are also important properties that make Ni favourite and our work aims to propose a novel design and fabrication technology for diodes which improve the reported characteristics.

Therefore, for a better efficiency and an operating regime close to the theoretical limits, a structure with oxide ramp termination has been designed. The field plate technique is one of the simplest configurations, consisting in metal overlapping an oxide layer intended to smooth the field line at the contact periphery. However, a better solution was demonstrated to be the usage of a three step field plate, where an improvement in electrical performance was achieved. Going further, we can say that the multiple step field plate assures a proper operation of the device. Consequently, the ramp oxide termination, where infinite steps can be considered, might be the right solution for this application. The efficiency and the performance of the device increases with the ramp angle decrease [6].

The SD have been electrical characterized at different temperature in the range of 300-573K, obtaining important information about the electrical parameters: ideality factor (n), SBH, leakage current (I_s) and series resistance (R_s). The SD based on SiC operates in normal parameters up to 573K.

II. TECHNOLOGY

In this section, the technology of the SD based on SiC having a ramp oxide termination is briefly presented:

- the starting substrate was n^+/n^- 4H-SiC (0001) wafer, 7.93° off Si face epi-layer and 0.02Ωcm resistivity, purchased from Cree Inc. The donor concentration was 10^{18} cm^{-3} in n^+-section and respectively $2.23 \cdot 10^{15}$ cm^{-3} in the lightly doped n^- epi-layer. The latter is the active layer to form the Schottky contact.
- prior the oxidation, the wafers were degreased in piranha and 10% HF for 15s at room temperature.
- two silicon dioxide (SiO_2) layers have been deposited by Low Pressure Chemical Vapour Deposition (LPCVD) from tetraethoxysilane (TEOS) at 780°C. The first layer of SiO_2 has an initial thickness of 700nm, but after an annealing in dry oxygen at 1100°C, the oxide thickness contracted at 670nm. Apart from the thickness decrease, the oxide became more compact. The second layer has been deposited in normal conditions at 780°C with a thickness of 950nm.

■ a sandwich formed by Ni(150nm) / Cr(10nm) was deposited on the back-side for the ohmic contact. The Cr layer is used in order to prevent the Ni oxidation, therefore the two layers, Ni respectively Cr, are deposited in the same cycle. In order to improve the ohmic contact, a rapid thermal annealing in Ar atmosphere at 800°C for 2 minutes has been used.

■ a photolithographic process to open the contact window in oxide was performed. Due to the difference between the densities of the two layers of SiO_2, after the etching process in NH_4F/CH_3-COOH (180/200ml), a ramp oxide termination has been achieved, a SEM image of the experimental structure cross-section being presented in figure 1:

Fig. 1 SEM image of the ramp oxide termination

The measured value of the first angle ramp was 4.5°, which is a good performance, having in mind that it has been demonstrated that a ramp angle lower than 5° assures an uniform distribution of the current density and a value of the breakdown voltage close to the theoretical values [7].

■ in order to obtain an optimal SBH, before to metal deposition, the oxide remained on SiC must be completely removed, since any residue left on wafer lead to a decrease of the SBH [8].

■ after the opening of the windows in oxide, the Ni/4H-SiC Schottky contact has been obtained by deposition of Ni(150nm) / Cr(10nm) bilayers by e-beam evaporation.

■ to improve the Schottky contact, a rapid thermal annealing in Ar atmosphere, at 600°C for 2 minutes is preferred [9]. After that, the nickel silicide is formed at interface, which has two important roles: the first is to eliminate the inhomogeneities from metal-semiconductor interface and the second is to embed the oxide residue.

A schematized cross-section of the fabricated Schottky diode is presented in figure 2:

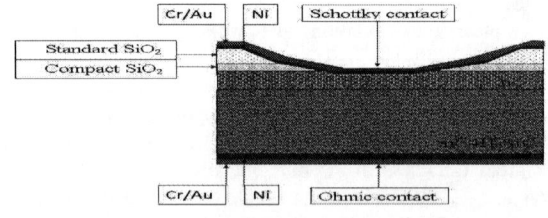

Fig.2 Schematic cross section of metal/4H- SiC Schottky diodes.

Several diodes have been packaged in TO39 capsules for electrical tests. A gold layer (150 nm) was deposited on both sides of the samples using chromium adhesion layer and patterned by photolithography. The cathode of the chip has been connected to capsule by sticking with silver paste, whereas the wire bonding technology has been used for the anode.

III. I-V-T BEHAVIOR

The electrical behavior of the SD based on SiC at different temperature has a crucial role taking account that SD operating like temperature sensor. The forward voltage variation with temperature is an important parameter for such sensor. Thus, the SDs have been tested with PARSTAT 2273 system using a temperature controller which enabled us to make measurements in the temperature range 300–600 K. The test structures have been measured in forward biased in range of 0-2.5V, with a step of 25mV at 12 different temperatures in the range of 300-573K.

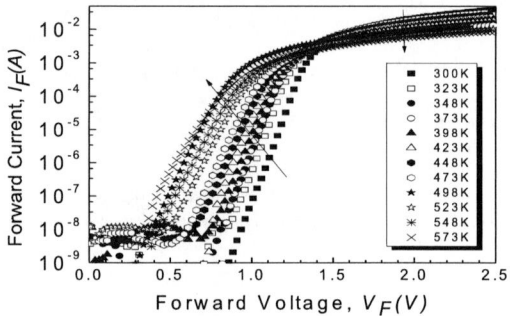

Fig.3 I-V-T Characteristics of the Schottky diode

From forward characteristics at different temperatures, the electrical parameters of the SD can be extracted. According to the thermionic emission law [10], the current-voltage relation is:

$$I = I_s[exp(qV/nkT)-1] \qquad (1)$$

$$I_s = A^*A_jT^2exp[-q\Phi_B/kT] \qquad (2)$$

where : A^* is the Richardson's constant (A^*=146A/cm^{2}K^2), A_j is the active area of the diode ($A_j = 3.14*10^{-4}$cm^2) and Φ_B represent the Schottky Barrier Height (SBH).

Different regimes can be observed for the data presented in Fig.3:

- for the bias voltage lower than 3kT/q, V < 3kT/q ≈ 0.75V the variation of the forward current with the forward voltage is linear, having a resistive behavior, where I ~ I_s

- when V > 3kT/q, then the exponential expression from (1) is dominant, and in this case, the forward current is:

$$I=I_sexp(qV/nkT) \qquad (3)$$

This fact can be observed in Fig 3. The exponential characteristics on several decade are visible on the whole range of temperature, and their fitting allows calculation of three electrical parameters for SD: n, Φ_B and I_s.

PRIME 2013, Villach, Austria   *Session W4A – Advanced Devices*

- at high bias voltage, the characteristic is again linear, interfering series resistance limitation. R_s has been calculated from a linear fit at high bias voltage (> 1.5V) on the I-V characteristic. The reverse slope represents the series resistance of the SD.

The temperature dependent values of various parameters determined from the forward bias $I-V$ characteristics of values for the tested SD are listed in table 1:

Table 1 The parameters of the Schottky diode based on SiC

T(K)	n	I_S(A)	SBH (V)	Rs (Ω)
300	1.295	6.627E-21	1.416	28.76
323	1.21	2.54E-20	1.492	33.6247
348	1.16	2.27E-19	1.546	40.7166
373	1.067	4.5E-19	1.6399	49.091
398	1.094	1.85382	1.626	58.823
423	1.04	6.768E-17	1.686	68.917
448	1.07	1.595E-15	1.668	81.168
473	1.048	8.94E-15	1.6954	95.1474
498	1.02	4.02E-14	1.725	109.53
523	1.032	2.849E-13	1.7277	128.534
548	1.023	1.46E-12	1.737	146.198
573	1.157	6.911E-11	1.63	164.47

As can be seen, their values are in good agreement with those reported in literature, the proposed SD operating without problems up to 573K.

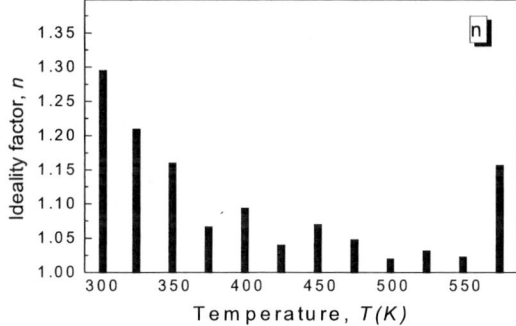

Fig. 4 Ideality factor versus temperature

The ideality factor has a constant value with temperature, being close to his ideal value, 1. An improvement is observed with the temperature raising, thus the values are constants with temperature.

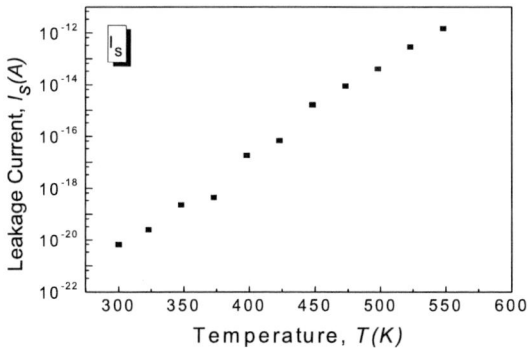

Fig. 5 Leakage Current versus Temperature

As we can observe in Fig.5, the leakage current exponentially increases with temperature, but it remains at a low value (order to picoamperes at T=573K) which proves that the Schottky diodes operating very well at high temperature.

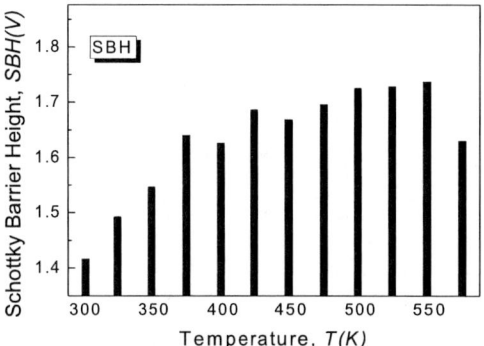

Fig. 6 Schottky Barrier Height versus temperature

At room temperature, SBH has a 1.416V value, but with temperature raising, the SBH is improving, getting to a constant value with temperature. The maximum SBH is 1.737V at T=548K. These values are enclosed in the analytical calculated range, the experimental results being in a good agreement with the analytical results.

Obtaining a small leakage current and a high SBH, we can say that this diode acts successfully at high temperature.

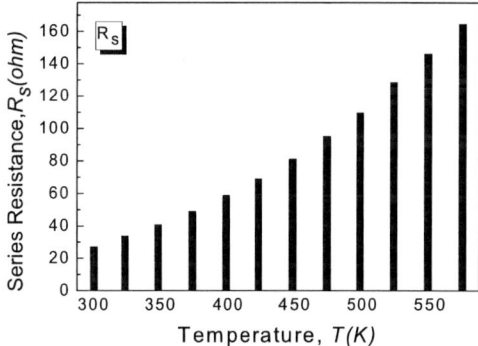

Fig.7 Series resistance versus temperature

At room temperature, the series resistance has a R_s=28.76 Ω value, this increasing up to R_s=164.47 Ω at T=573K. Follow with attention the parameters value, we observe that an annealing effect is present, which can be assigned to the barrier inhomogenities reducing within the contact area [10]. More precisely, if the temperature measurements get higher, then, the electrical parameters are improving.

When operating as a temperature sensor, the SD is kept at a constant forward current value. In this case, the voltage variation with temperature is done by a thermionic emission expression derived: [11]

$$V_F(T) = n\Phi_B - [n\Phi_B - V_F(T_0)]T/T_0 \qquad (4)$$

where : T_0 is a reference temperature

978-1-4673-4579-8/13 $31.00 © 2013 IEEE

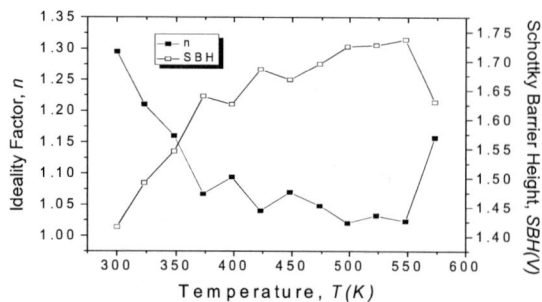

Fig.8 Temperature dependence of the experimental Schottky barrier height and experimental ideality factor

Because the $n\Phi_B$ product is approximately constant with temperature, (see Fig. 8), the forward voltage decreases linear with temperature.

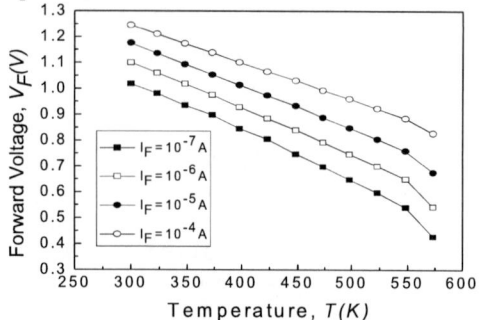

Fig. 9 Forward voltage as a function of temperature measured at several constant currents

Fig.9 represents the experimental demonstration that the forward voltage decreases linear in the whole range of temperature, respecting (4). Therefore, a linear decrease of the forward voltage up to 548K is observed, becoming out of the linear trajectory at 573K.

An important parameter of the temperature sensor is sensitivity(S). This is determined by a linear fit on the curves from Fig.9. The slope represents the sensitivity value.

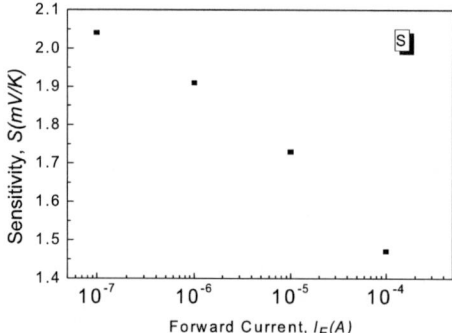

Fig.10 Sensor sensitivity versus forward current

At a constant current, $I_F=10^{-7}A$, the sensor sensitivity is S=2.04mV/K. This value decreases with current increasing, having a S=1.47mV/K value at $I_F=10^{-4}A$.

CONCLUSIONS

Schottky diodes based on SiC have been fabricated for temperature sensors. A novel technology based on a ramp oxide termination which assures an optimal operating is presented. Moreover, it has been shown that additional improvement of the Schottky, respectively ohmic contact is conditioned by a post-metallization rapid thermal annealing. The SD electrical tests demonstrate that they successfully operate in the whole range of the temperature 300-573K, having values of the electrical parameters in a good agreement with the value from literature. The forward voltage linearly decreases with temperature increase at different constant currents. The sensitivity of the sensor is in range of S = [1.47-2.04] mV/K and it decreases with current increasing.

ACKNOWLEDGMENT

The authors gratefully acknowledge the support of the Romanian Ministry of Education and Research through the contract no. 21/2012 (PN II–PT-PCCA-2011-3.2-0487).

REFERENCES

[1] Stephen E. Saddow, Anant Agarwal, "Advanced in silicon carbide processing and applications", Semiconductor Materials and Devices Series, Artech House, Inc., 2004

[2] P. Drabek, "Using of the modern semiconductor devices based on the SiC", Advances in Electrical and Electronic Enginnering, 2008, pp 106-109.

[3] T.N. Oder, T.L. Sung, M. Barlow, J.R. Williams, A.C. Ahyi, T. Isaacs-Smith, "Improved Ni Scottky contacts on n-type 4H-SiC using thermal processing", Journal of Electronic Materials, Vol. 38, No. 6, 2009, pp. 772-777.

[4] F. Roccaforte, F. La Via, V. Raineri, P. Musumeci, L. Calcagno, G.G. Condorelli, "Highly reproductible ideal SiC Schottky rectifiers: effects of surface preparation and thermal annealing on the Ni/6H-SiC barrier height", Applied Physics A – Materials Science & Processing, Vol. 77, 2003, pp. 827-833..

[5] Sanjeev K. Gupta, A. Azam, J. Akhtar, "Improved electrical parameters of vacuum annealed Ni/4H-SiC (0001) Schottky barrier diode", Physica B, Vol. 406, 2011, pp. 3030-3035.

[6] Gheorghe Brezeanu, "High performance power diodes", Proceedings of the Romanian Academy, Series A, Vol. 8, number 3, 2007, pp. 000–000

[7] R. Pascu, F. Draghici, M. Badila, F. Craciunoiu, G. Brezeanu, A. Dinescu, I. Rusu, "High temperature sensor based on SiC Schottky diodes with undoped oxide ramp termination", International Conference Semiconductor, Vol. 2, 2011, pp. 379-382.

[8] F. Roccaforte, F. La Via, V. Raineri, "High reproducible ideal SiC Schottky rectifiers by controlling surface preparation and thermal treatments", Solid state device research conference, 2002, pp. 543-546.

[9] D.J. Morrison, N.G. Wright, A.B. Horsfall, C.M.Johnson, A.G. O'Neil, A.P. Knights, K.P. Hilton, M.J. Uren, "Effect of post-implantation anneal on the electrical characteristics of Ni 4H-SiC Schottky barrier diodes terminated using self-aligned argon ion implantation", Solid-State Electronics, Vol. 44, 2000, pp. 1879-1885.

[10] Vik Saxena, Jian Nong (Jim) Su, Andrew J. Steckl, "High voltage Ni- and Pt-SiC Schottky diodes utilizing metal field plate termination", IEEE Transactions on Electron Devices Vol. 46, 1999, pp. 456-463.

[11] R. Pascu, F. Craciunoiu, M. Kusko, F. Draghici, A. Dinescu, M. Danila, "The effect of the post-metallization annealing of Ni/n-type 4H-SiC Schottky contact", International Semiconductor Conference, Vol.2, 2012, pp. 457-460.

Comparative analysis of methods for computing pole angle offsets in magnetic pole wheels using ABS sensor

Muhammad Adnan
Alpen-Adria Universität
Klagenfurt, Austria.
Adnan.Muhammad@ctr.at

Hammerschmidt Dirk
Infineon Technologies AG
Villach, Austria.
Dirk.Hammerschmidth@infineon.com

Abstract—**Magnetic pole wheels, that are typically used in vehicles to measure speed, have inherent manufacturing intolerance resulting in variable pole length in each pole wheel. Moreover, due their use in harsh and prone to dust environment, magnetic characteristics of each pole change over a period of time. This gives rise to change in pole pitch reported by the ABS sensors. The target here is to accurately measure pole angles by analyzing the ABS sensor signal. This paper describes two methods to achieve optimum pole angle or offset calculation by Synchronous Averaging and Kalman Filter accompanied by a novel Pole Skipping technique to compensate for the speed variations. The computed pole angles can be used to calculate the speed signal more accurately. Besides that, application of pole offset correction will remove the regular pattern from the speed signal data, which prevails in its spectrum suppressing important information related to tire vibrations.**

I. Introduction

Output of an ABS sensor is a square wave whose frequency is proportional to the rotational speed of a pole wheel. Zero crossings of this square wave are used to calculate the time it takes for a single pole to pass across the sensor. Detailed explanation of pole wheel in conjunction with the ABS sensor can be found in [1]. The basic principal of an angle calculation from one revolution is described in [2] and [3]. If a pole wheel has N poles and it takes $t_1, t_2, ..., t_n$ time in seconds for each pole passing across the sensor, then the average angular speed of pole wheel for this revolution can be calculated by

$$\omega_{rev} = \frac{2\pi}{\sum_{i=1}^{N} t_i} \qquad (1)$$

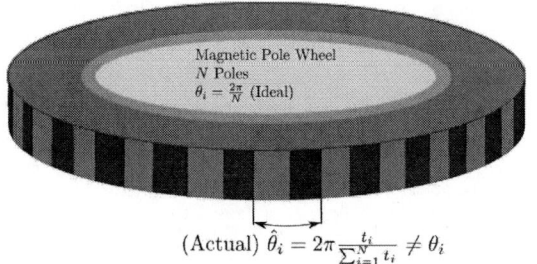

(Actual) $\hat{\theta}_i = 2\pi \frac{t_i}{\sum_{i=1}^{N} t_i} \neq \theta_i$

Fig. 1: Ideal and Actual pole angles

Let $\theta_i = \frac{2\pi}{N}$ is the ideal pole angle of each pole, supposing that all poles have equal length. If $\hat{\theta}_i$ is the pole angle of the nth pole calculated from the average speed over one revolution, then

$$\hat{\theta}_i = w_{rev} t_i \qquad (2)$$

Now the offset is calculated by

$$\delta_i = \theta_i - \hat{\theta}_i \qquad (3)$$

where

$$\sum_{i=1}^{N} \delta_i = 0 \ provided \ that \ \sum_{i=1}^{N} \hat{\theta}_i = 2\pi \qquad (4)$$

Testbench measurements have shown that even at constant speed of pole wheel in a lab environment, recorded time stamps have some noise parameters which arise because of clock jitter and vibrations coming from neighboring machine peripherals or speed regulators. This gives rise to a variable computed angle of each pole for each revolution. A common technique to remove noise elements and calculate the optimum pole angle or offset is synchronous averaging [4]. Another method to compute angle offset of pole angles is by Kalman filter, which inherently removes noise by taking into account the variance of the recorded values. We shall analyze both techniques in the following.

II. Synchronous Averaging

Let $\omega_1, \omega_2, ..., \omega_M$ be the angular speed calculated for M revolutions. From this we can calculate the pole angles by using equation (2), which can be written in matrix form as

$$\hat{\theta} = \begin{pmatrix} \theta_{11} & \theta_{12} & \dots & \theta_{1M} \\ \theta_{21} & \theta_{22} & \dots & \theta_{2M} \\ \vdots & \vdots & \ddots & \vdots \\ \theta_{N1} & \theta_{N2} & \dots & \theta_{NM} \end{pmatrix} \qquad (5)$$

Where N is the number of Poles and M is the total number of revolutions. Average of each angle is given by

$$\hat{\theta}_i = \frac{1}{M} \sum_{n=1}^{M} \theta_{i,n} \qquad (6)$$

The target here is to find the number of revolutions M, which are sufficient for the optimum angle calculation. For this reason, a recursive averaging algorithm can be used to compute successive approximations of the pole angles. After an arbitrary number of revolution m, we will have the value of angle averaged over all the previous revolutions. This technique is more efficient in terms of memory requirements.

$$\hat{\theta}_i = \frac{m-1}{m}\hat{\theta}_i^{m-1} + \frac{1}{m}\theta_i^m \qquad (7)$$

Where $\hat{\theta}_i$ is the average value of angle for pole i , after m revolutions. θ_i^m is the pole angle calculated for the mth revolution and $\hat{\theta}_i^{m-1}$ is the previous average for $m-1$ revolutions. To achieve more effective averaging, we can average the angular speed over all the revolutions and then compute and average the pole angles. This in essence shall eliminate the rapid changes in angular speed one step before the angle calculations and hence shall be more effective in calculating optimum pole angle.

$$\hat{\omega}_i^m = \frac{1}{m}\sum_{i=1}^m \omega_i \qquad (8)$$

Where $\hat{\omega}_i^m$ is the speed averaged up to m revolutions, which is then used to compute pole angles for the mth revolution. This can only be used for the case when the angular speed is constant with small variations from revolution to revolution. $\hat{\omega}_i^m$ is same for all poles for one particular revolution.

III. Kalman Filter

Kalman Filter can estimate the pole angle offset by utilizing the variance of pole angle calculated from each revolution. This variance is proportional to the variance of the recorded time stamps for one particular pole. Following system of equations reflect a simplest kalman filter neglecting some system dependant variables in the original filter [5], which are not appropriate to our problem. Equations (9), (10) and (11) are Measurement update equations.

$$\begin{cases} K_p = \sigma^2(\sigma_p^2 + \sigma_m^2)^{-1} & (9) \\ \hat{\varphi}_p' = \hat{\varphi}_p + K_p(\delta_p - \hat{\varphi}_p) & (10) \\ \hat{\sigma}_p^2 = \sigma_p^2(1 - K_p) & (11) \end{cases}$$

While Equations (12) and (13) are time update eqations for Kalman filter

$$\begin{cases} \sigma_p^2 = \hat{\sigma}_p^2 + Q_p & (12) \\ \hat{\varphi}_p = \hat{\varphi}_p' & (13) \end{cases}$$

Where K_p is the Kalman Gain, σ_m^2 is the measurement variance, which is considered to be same for all poles for simplicity, σ_p^2 is the old variance of Pole Offset, $\hat{\sigma}_p'$ is the new estimated variance of Pole p , $\hat{\sigma}_p$ is the Old estimated variance of pole angle offset, δ_p is the calculate pole offset from one revolution and Q_p is the noise factor. $\hat{\varphi}_p$ is the old estimate of the pole angle offset and $\hat{\varphi}_p'$ is the new estimate of the pole angle offset which is also the output of the Kalman filter.

The benefit of Kalman Filter arises from the fact that the filter can be tuned to achieve optimal results using filter parameters like noise factor etc. Also, the convergence of pole angle offset is faster with Kalman filter as compared to Synchronous averaging.

IV. VARIABLE SPEED COMPENSATION BY POLE SKIPPING

Although above mentioned techniques work well when the speed is constant but with an accelerating or decelerating pole wheel, the computed pole angle values start fluctuating as long the acceleration or deceleration continues and then they stabilize again once the speed is constant. In this case offset of each pole shall have an additional parameter of acceleration.

$$\delta_p = \delta_p' + p\alpha_m \qquad (14)$$

Where δ_p is the pole angle offset, p is the pole number $(p = 1, 2, ..., N)$, δ_p' is the pole offset when the speed is constant and α_m is the acceleration factor for m^{th} revolution. α_m is considered to be constant for all poles for one particular revolution.

This drawback can be eliminated by a Pole Skipping technique, where one pole is skipped after each revolution to compensate for the speed change during one revolution. The final result can then be smoothed out by a running average of length N followed by an $N \times 1$ Decimation filter.

V. Simulation and Results

For the testing of aforementioned techniques, pulse data has been recorded from an ABS sensor with a $25\,nsec$ clock. This high frequency clock is sufficient to sample the pulse signal from ABS sensor whose pulse frequency usually remain sub $1\,kHz$. Fig. 2 shows the overall simulation flow graph.

Fig. 3 shows the histogram plot of first four poles which shows the distribution of clock pulse counts for a constant speed. It indicates that the measurements for each pole are following near Guassian distribution. Fig. 4 shows pole angle offsets variations corresponding to time slots in Fig. 3.

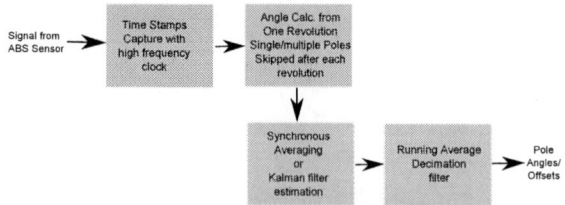

Fig. 2: Simulation flowgraph

Fig. 5 shows the angles offsets when averaged recursively over the number of revolutions. It shows that the calculated angle values converges in about 500 revolutions for a constant angular speed.

For the Kalman filter computation, we use the variance information from the recorded data. We estimate our initial offsets values to be zero for all poles. Fig. 6 shows the pole angle offset convergence with Kalman Filter.

PRIME 2013, Villach, Austria   *Session W4A – Advanced Devices*

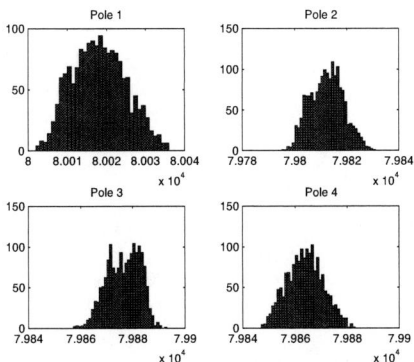

Fig. 3: Histogram plot of clock pulse counts for first 4 poles (clock period = 25 *nsec*. $N = 32$. Clock pulse count for 1677 revolutions. 800 *rpm*)

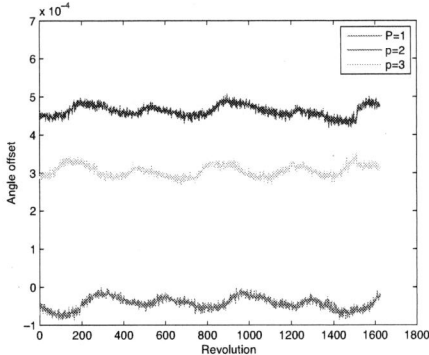

Fig. 4: Pole Angles Calculated for first 3 poles. ($N = 32$ Poles).

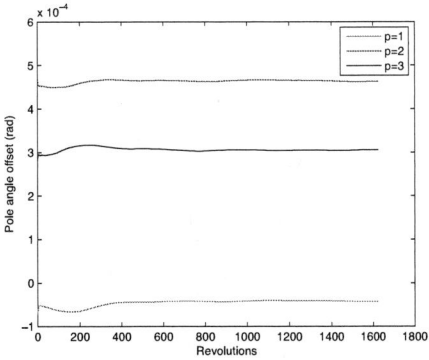

Fig. 5: Recursive Synchronous averaging over number of revolutions ($N = 32$). Horizontal axis shows the number of revolutions over which the averaging has been done.

Speed change has been simulated by multiplying the recorded time stamps with an increasing or decreasing ramp (Fig. 7). Fig. 8 and Fig. 9 show the result of an additional running average filter of length N after Synchronous averaging and Kalaman filter respectively. This can be cascaded with a decimation filter of N:1 to avoid unnecessary calculations and

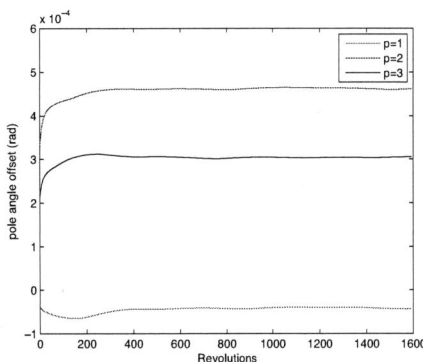

Fig. 6: Pole angle offsets relative to zero. Calculated with Kalman filter

storage.

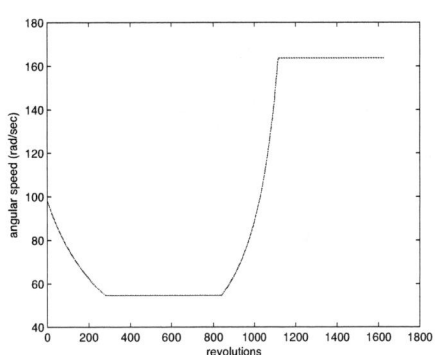

Fig. 7: Acceleration and decceleration simulation

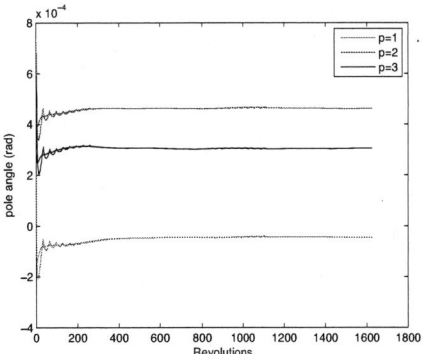

Fig. 8: Offset calculations by Synchronous Recursive Averaging with Pole Skipping for variable speed. Result of a subsequent running average filter of length N is superimposed

VI. IMPLEMENTATION ASPECTS

The main computationally expansive block in the Fig. 2, is the Synchronous Averaging or Kalman Filter block. This

978-1-4673-4579-8/13 $31.00 © 2013 IEEE   303

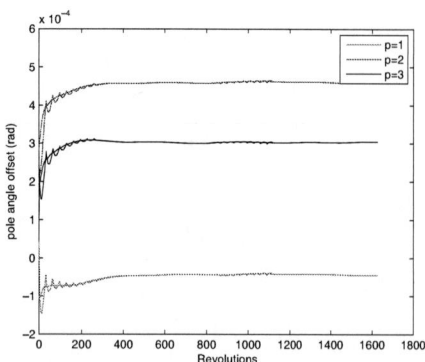

Fig. 9: Result of Kalman filter with Pole Skipping. Output of a subsequent running average filter of length N is superimposed.

tasks can be implemented either in parallel (for all poles) in an FPGA or in sequential scheme, in an FPGA or micro controller. We assume that pole angle offsets are pre-calculated from each revolution of the pole wheel. The basic idea for this implementation, where we have the pole angle offsets in milliradians, is to scale up the pre-computed angle offsets by a suitable multiplier (depending on number of poles), to minimize fixed point saturations. This can be achieved by a shift operation when scaling parameter is a power of 2. Equation (7), is the main equation for computing pole angle offsets with Synchronous Averaging. This equation requires two divisions, two multiplications and one addition per iteration for one pole.

Apparently, Kalman filter is computationally more expansive as compared to Synchronous Averaging algorithm as it utilizes equation (9) through equation (13). Nevertheless, these equations can be reduced by recognizing that each subsequent computation of Kalman Gain and Variance updates, equations (9), (11), (12), does not incorporate the new measurements. So these can be computed off line and stored in memory to use in real-time targets. The rest we are left with one measurement update equation (10) and one time update equation (13). Hence Kalman implementation is relatively simple as compared to Synchronous Averaging, provided that we have accurate knowledge of system noise parameters as well as enough memory to store Gain and Variance values for all the iterations until steady state has been obtained.

VII. CONCLUSIONS

Pole angle offsets in a pole wheel, whether because of manufacturing intolerance or wear and tear, can be calculated accurately by utilizing Recursive Synchronous Averaging or Kalman Filter. Analyses have shown that both techniques produce similar results. Implementation of Kalman filter is more complex as compared to synchronous averaging when gain and variance values are not pre computed. Moreover Kalman Filter produces slightly faster convergence of offsets than Synchronous Averaging and the filter can be tuned to one particular setup. False calculation of pole angles due to variable speed can be compensated by Pole Skipping technique. Moreover, a running average filter shall eliminate the variations produced due to the Pole Skipping.

VIII. ACKNOWLEDGMENTS

Measurement data for this analysis has been obtained from Infineon Technology AG Villach for the research on new version of Magnetic ABS Sensors in collaboration with Carinthia Tech Research AG and Alpen Adria Universität Klagenfurt.

REFERENCES

[1] R. Schwarz, "Increasing signal accuracy of automotive wheel-speed sensors by online learning," in *Proc. IEEE American Control Conference, 1997*, Fuel Utility Syst., Goodrich Corp., Vergennes, VT, USA.Nelles, Oliver, Scheerer, Peter, Isermann and Rolf, Jun. 1997, pp. 1131–1135 vol.2.

[2] M. D. Fredrik Gustafsson and N. Persson, "Tire pressure estimation," Swedish Patent 7 240 542, Jul. 10, 2007.

[3] N. P. F. Gustafsson, "Event based sampling with application to vibration anaylysis in pneumatic tires," in *Proc. IEEE in Acoustics, Speech, and Signal Processing, 2001*, Dept. of Electr. Eng., Linkoping Univ., Sweden. Gustafsson, Fredrik, 2001, pp. 3885 – 3888 vol.6.

[4] D. Hochmann and M. Sadok, "Theory of synchronous averaging," in *Proc. IEEE Aerospace Conference Proceedings*, Inst. of Autom. Control, Darmstadt Univ. of Technol., Germany, Mar. 2004, pp. 3636–3653 Vol. 6.

[5] G. Bishop and G. Welch. (2013, Mar.) An introduction to the kalman filter. [Online]. Available: http://www.cs.unc.edu/ welch/kalman/

PRIME 2013, Villach, Austria

Session W1B – RF Techniques 2

A Wideband Planar Microstrip to Coplanar Stripline Transition (Balun) at 35 GHz

Jan Dirk Leufker, Axel Strobel, Corrado Carta, Frank Ellinger
Chair for Circuit Design and Network Theory
Technische Universität Dresden, EUI, IEE
01062 Dresden, Germany

Abstract—**A novel approach for wideband planar baluns is presented, which enables the realization with just two instead of three transmission lines. The analysis of the principle of operation and a suitable design procedure are described. Measurement results of test structures are compared to EM simulations. Two variants of this balun were designed for a center frequency of 35 GHz. Measurements show an absolute 3 dB-back-to-back insertion loss bandwidth of 16.7 GHz and 10.1 GHz with minimum losses of 0.38 dB and 0.71 dB respectively. The extrapolated absolute 1 dB insertion loss bandwidths of the single baluns are 14.5 GHz and 14.4 GHz respectively.**

Keywords—balun; transition; wideband; millimeter wave; microstrip; coplanar stripline;

I. Introduction

In the design of electronic circuits and systems for the most diverse applications and frequencies, it is often necessary to interface single ended components to differential ones. This is particularly common for chip-on-board systems operating in microwave and millimeter wave bands. In frequent scenarios the high frequency frontend of wireless communication systems consists of differential building blocks, while the input and output signals must be fed to single ended antennas or coaxial connectors. The transformation of the signal from single ended to differential or vice versa is performed by a dedicated component – a balun - which can be realized with active or fully passive topologies. At higher frequencies the balun is more often realized with passive structures on a printed circuit board (PCB). Depending on the available number of PCB layers these structures can be couplers ([1],[2],[3]) or microstrip structures ([4],[5],[6],[7]).

This paper presents the design and characterization of a novel microstrip balun. Section II describes the circuit topology and basic circuit operation. General design guidelines are provided in section III. Section IV describes the implementation details of the balun, while in section V measurement results are presented and compared with simulations. Finally section VI gives a conclusion of this work including an outlook to further work based on this design principle.

II. Architecture and Theory

Fig. 1 a) shows the schematic of a common balun based on microstrips ([5],[6],[7]). Pin 1 and 2 act as differential port with differential port impedance Z_d and common mode port impedance Z_{cm} while pin 3 is the single ended port with port impedance Z_s. To achieve $180\,^\circ$ phase difference between pin 1 and 2 the difference in length of lines TL_1 and TL_2 must be half of the wavelength λ_c at the design frequency f_c. In typical designs the characteristic impedance $Z_{1/2}$ of lines TL_1 and TL_2 is set to half of the port impedance Z_d. If Z_s isn't equal to $Z_d/2$, which is almost always the case, an additional $\lambda_c/4$ transformer with characteristic impedance $Z_3 = \sqrt{Z_d/2 \cdot Z_s}$ is necessary for matching the connecting point of TL_1 and TL_2 to Z_s. This introduces an additional discontinuity, which increases the modeling complexity.

In the proposed design lines TL_4 and TL_5 are set to provide a $180\,^\circ$ phase difference between pin 1 and 2 as well as impedance matching between single ended and differential port. In order to simultaneously meet both conditions the lengths of TL_4 and TL_5 have to be

$$l_{TL_4} = (2n-1)\cdot \lambda_c / 4 \qquad (1)$$

and

$$l_{TL_5} = (2n+1)\cdot \lambda_c / 4 \qquad (2)$$

respectively $(n = 1,2,3,...)$. Furthermore the characteristic impedance can be calculated as follows:

$$Z_{4/5} = \sqrt{Z_d/2 \cdot Z_s \cdot 2} = \sqrt{Z_d \cdot Z_s} \qquad (3).$$

If the conditions in (1)-(3) are satisfied, no extra $\lambda_c/4$ transformer is necessary and the balun schematic can be simplified as shown in Fig. 1 b). Additionally the impedance $Z_{4/5}$ is higher than $Z_{1/2}$ and this may simplify the design of the balun at higher frequencies thanks to the smaller widths of lines TL_4 and TL_5. For some applications a drawback of this

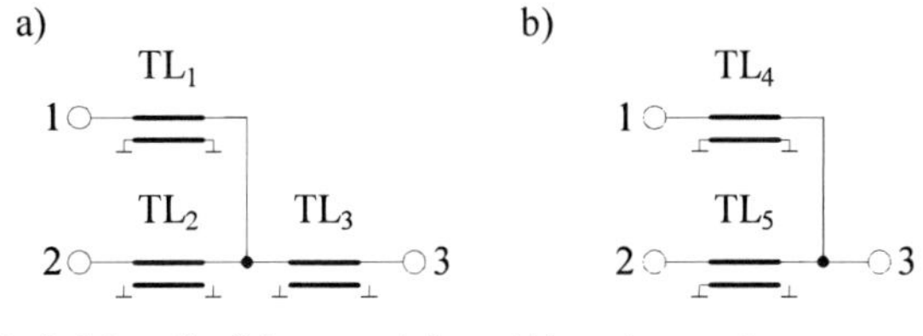

a)   b)

Fig. 1: Schematic of a) common balun and b) novel approach

978-1-4673-4579-8/13 $31.00 © 2013 IEEE   305

approach can be a smaller insertion loss bandwidth compared to the standard balun, because lines TL_4 and TL_5 are simultaneously acting as transformers and $180°$ phase shifters, resulting in a higher quality factor of the impedance matching.

III. DESIGN PROCEDURE

The design procedure of this novel balun architecture is very similar to the description provided in [6]:

1. All dimension parameters of the lines are determined. In this first step simple microstrip and stripline models can be used to estimate the geometry of the lines as well as performance parameters, such as bandwidth and losses.

2. All microstrip and stripline structures are separately modeled and simulated within an EM simulator.

3. If suitable software is available, optimizations are run at this point to investigate - e.g. - the impact of slightly different line widths or lengths on bandwidth and loss of the balun.

4. The whole balun structure is modeled and simulated within the EM simulator. Fig. 6 shows one example of structure modeled in the simulator. $90°$ bends should be realized with proper chamfering like described in [8] to reduce further parasitics. The lengths of TL_4 and TL_5 are independent parameters for this simulation. If needed, the width of these two lines may also be varied in a small range around the calculated one. Hereby losses in each branch can be equalized.

5. If the process tolerances of the used technology are available, etch compensation is to be applied to the final layout.

In general, parameter optimizations targeting individual performance figures require shorter execution times, because the parameter set varies over a smaller domain. The optimization of the proposed balun design benefits further from this effect, since the $\lambda_c/4$ transformer does not need to be modeled and adjusted. Furthermore the interfaces of the balun are directly the ones defined by the port impedances Z_s and Z_d and no further discontinuities are added.

IV. IMPLEMENTATION

The mixed mode scattering parameters of a balun can be calculated with (4) corresponding to the pin configuration in Fig. 1. $\underline{S}_{d13}$ and $\underline{S}_{3d1}$ can be described as insertion loss of a balun whereas $\underline{S}_{dd11}$ and $\underline{S}_{33}$ are the most interesting input reflection parameters. As discussed in section II the 1 dB insertion loss bandwidth of the proposed balun design is smaller than in the typical balun design. Fig. 2 shows simulations of baluns with ideal components. The absolute 1 dB insertion loss bandwidth is reduced by 8.6 % and 33 %

for $n = 1$ and $n = 2$ respectively compared to the typical balun approach. However the losses of actual components and additional discontinuities in the typical approach should attenuate this effect. On the other hand this reduction of bandwidth can relax the requirements on the antenna filter.

In open literature the experimental characterization of baluns is often reported as indirect measurement: typically two baluns are connected at their differential ports in at least two different combinations to ease the requirements on the measurement setup as well as on de-embedding structures. These combinations are referred to as unbalanced and balanced [5] and shown in Fig. 3. The overall loss of this test structure is referred to as back-to-back insertion loss. It is a key performance parameter useful to compare different balun circuits. As it can be seen in Fig. 4 the unbalanced test structure shows more clearly the circuit bandwidth, so it is to be preferred as test structure.

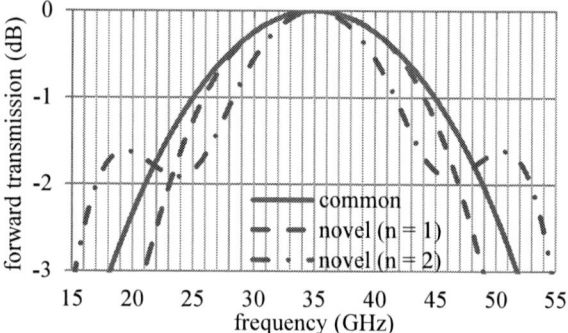

Fig. 2: Simulations of ideal balun topologies a) and b)

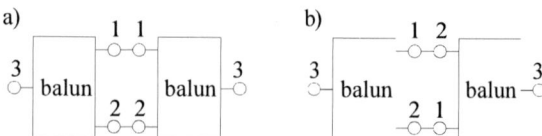

Fig. 3: Block diagram of a) unbalanced and b) balanced test structure

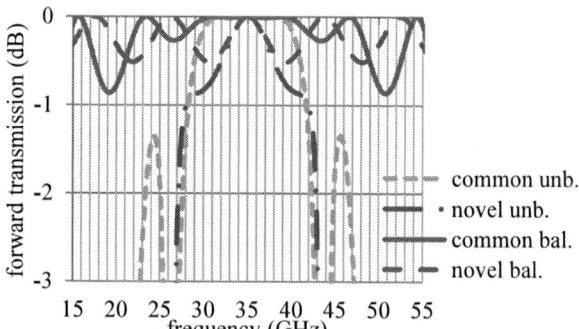

Fig. 4: Simulations of ideal balanced and unbalanced test structures ($n = 2$)

$$\underline{S}_{MM} = \begin{pmatrix} \underline{S}_{dd11} & \underline{S}_{dc11} & \underline{S}_{d13} \\ \underline{S}_{cd11} & \underline{S}_{cc11} & \underline{S}_{c13} \\ \underline{S}_{3d1} & \underline{S}_{3c1} & \underline{S}_{33} \end{pmatrix} = 0.5 \cdot \begin{pmatrix} \underline{S}_{11} - \underline{S}_{21} - \underline{S}_{12} + \underline{S}_{22} & \underline{S}_{11} - \underline{S}_{21} + \underline{S}_{12} - \underline{S}_{22} & \sqrt{2} \cdot (\underline{S}_{13} - \underline{S}_{23}) \\ \underline{S}_{11} + \underline{S}_{21} - \underline{S}_{12} - \underline{S}_{22} & \underline{S}_{11} + \underline{S}_{21} + \underline{S}_{12} + \underline{S}_{22} & \sqrt{2} \cdot (\underline{S}_{13} + \underline{S}_{23}) \\ \sqrt{2} \cdot (\underline{S}_{31} - \underline{S}_{32}) & \sqrt{2} \cdot (\underline{S}_{31} + \underline{S}_{32}) & \underline{S}_{33} \end{pmatrix} \quad (4)$$

Since the presented balun design will be used in a 35 GHz local positioning system and scaled for a 60 GHz wireless communication system, the Rogers RO3003 was chosen as substrate material for its low dissipation factor tan δ = 0.0013 at high frequencies. Other material parameters relevant to this implementation are: dielectric constant $\varepsilon_r = 3.0$, substrate thickness $h = 250$ μm, copper layer thickness $t = 35$ μm, distance to side walls $d = 1$ mm and minimum spacing between polygons/lines $s = 150$ μm. In this design these values lead to a width $w = 320$ μm for $TL_{4/5}$ with $Z_{4/5} = 70.7$ Ω and $w = 600$ μm and $w = 370$ μm for microstrip and stripline feed lines with $Z_s = 50$ Ω and $Z_d = 100$ Ω, respectively (cf. Fig. 5). Although a minimum substrate thickness of $h = 250$ μm was chosen, for higher frequencies f_c the width w of a 35 Ω line already exceeds $\lambda_c/4$. This would make the design of a $\lambda_c/4$ transformer very challenging and lead to the proposed approach.

Two versions of this balun were fabricated with $n = 1$ and $n = 2$ as defined in (1) and (2), both with center frequency $f_c = 35$ GHz. Each structure was also simulated with Sonnet EM simulator ([9]). 3D views of both Sonnet balun models are shown in Fig. 6. The simulation box sizes are 3.59 x 5.52 mm² and 6.48 x 6.0 mm²; feed lines with length of 1 mm each and side walls are also included at each port to reduce the impact of the simulation box on the simulation results. Total length of all test structures was set to 50 mm. They were always connected on both sides through a custom designed V connector fixture.

V. MEASUREMENT RESULTS

All measurements were run using an Anritsu 37397D network analyzer with bandwidth of 65 GHz. Figs.7-8 show the back-to-back forward transmission of the unbalanced measurement structure as described in [5] while Figs. 9-10 show the input reflection of these same structures. Fig. 11 depicts balun structures under test (left) as well as the measurement setup (right). Balanced test structures were also fabricated and successfully characterized, but as described in section IV their performance is of lesser interest than their unbalanced counterpart. Therefore no corresponding plots are included here.

The measured minimum back-to-back insertion losses of the two structures are 0.38 dB and 0.71 dB. Both values are lower than in [10]. The absolute measured 1 dB-back-to-back insertion loss bandwidth is 5.7 GHz (28.3...34 GHz, $n = 1$) and 4.4 GHz (31.4...35.8 GHz, $n = 2$) while absolute measured 3 dB-back-to-back insertion loss bandwidths are 16.7 GHz (27.5...43.7 GHz) and 10.1 GHz (29.9...40.0 GHz). Within corresponding ranges of 10.4 GHz (27.2...37.6 GHz) and 7.1 GHz (29.7...36.8 GHz) the input reflection is below -10 dB. Measurement results of all test structures are fitting well to their simulations. When integrated in the target system, balun version with $n = 1$ should show a 1 dB conversion loss bandwidth (from single ended port to differential one) which would cover the whole Ka-band (26.5...40 GHz) although this was no design goal. The EM simulations show good agreement with the ideal behavior presented in Fig. 12. The notches in the curves of the EM simulations at 32 GHz, 34 GHz and 42 GHz are related to

simulation box sizes and should not be seen in the real behavior of the circuits.

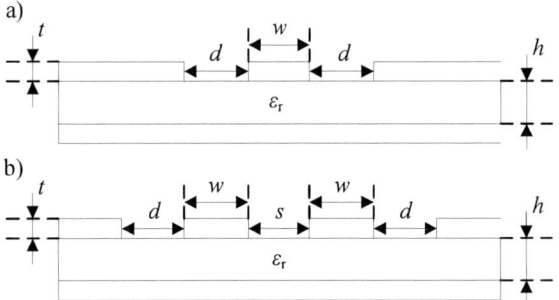

Fig. 5: Cross section of a) microstrip and b) stripline

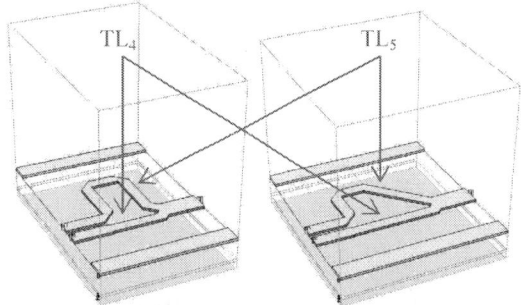

Fig. 6: 3D views of both Sonnet balun models

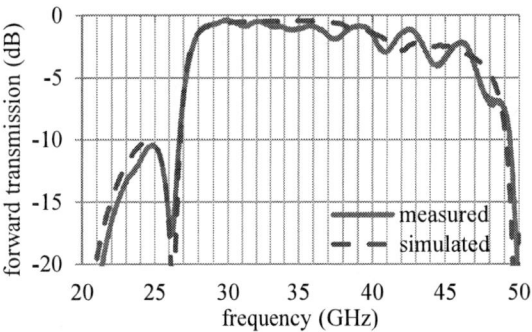

Fig. 7: Back-to-back forward transmission ($n = 1$)

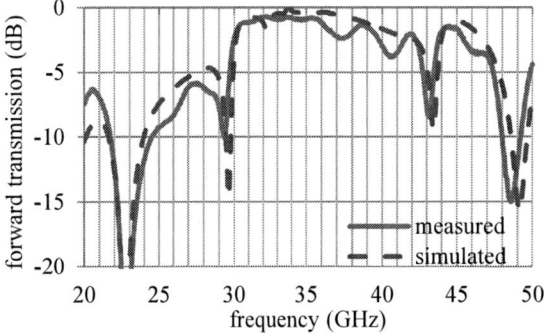

Fig. 8: Back-to-back forward transmission ($n = 2$)

Paper W1B1

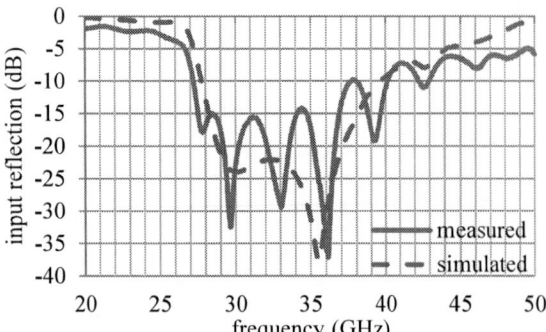

Fig. 9: Input reflection ($n = 1$)

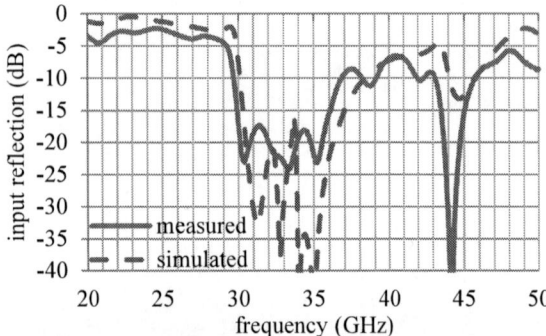

Fig. 10: Input reflection ($n = 2$)

Fig. 11: Balun structures under test (left) and measurement setup (right)

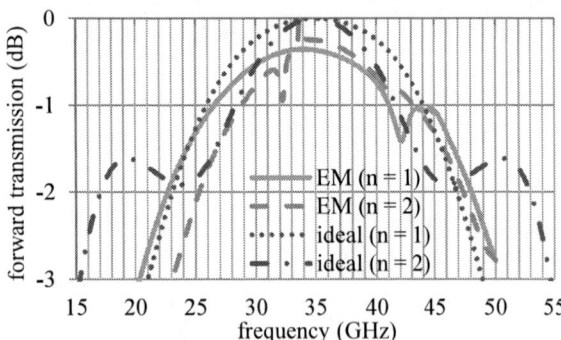

Fig. 12: Simulated forward transmission of single baluns

VI. CONCLUSION

A novel balun approach has been presented, which enables the realization with just two transmission lines, compared to common approaches requiring three transmission lines. Hence, compared to the state of the art, the design is simpler. The design approach, implementation and characterization of a 35 GHz balun have been presented. Measurements of two back-to-back baluns show that this approach can offer a loss of less than 1 dB over the whole Ka-band. This performance compares very well with works available in open literature ([10], [3]).

This approach is successfully implemented in a 35 GHz local positioning system ([11]). It is also well suited for adoption at other frequencies and implementations in different technologies: a 60 GHz design of a scaled version in the same technology is in progress and a 180 GHz integrated mixer on a 130 nm SiGe-BiCMOS process was designed where the balun was necessary to ease the interfacing to the measurement instrumentation.

ACKNOWLEDGMENT

This work has been partly supported by the Staatsministerium für Wissenschaft und Kunst (SMWK) within the frame of the research project "Cool Wireless Audio" and by the German Research Foundation (DFG) in the Collaborative Research Center 912 "Highly Adaptive Energy-Efficient Computing" and within the project "Novel Techniques, Theories and Circuits for Locatable mm-Wave RFID Tags" (grant EL 506/1-1).

REFERENCES

[1] J. Qi, C. Domier and N.C. Luhmann, "A wideband low loss planar microstrip-to-CPS balun", *Asia-Pacific Microwave Conference Proceedings (APMC)*, pp. 1205-1207, 4-7 Dec. 2012.

[2] M. Vahdani and X. Begaud, "Sinuous antenna fed by a microstrip to CPS balun", *3rd European Conference on Antennas and Propagation. EuCAP*, pp. 1622-1626, 23-27 March 2009.

[3] K. Young-Gon, W. Dong-Sik, K. Kang Wook and C. Young-Ki, "A New Ultra-wideband Microstrip-to-CPS Transition", *IEEE/MTT-S International Microwave Symposium*, pp. 1563-1566, 3-8 June 2007.

[4] H. Kyu Hwan, B. Lacroix, J. Papapolymerou and M. Swaminathan, "New microstrip-to-CPS transition for millimeter-wave application", *IEEE 61st Electronic Components and Technology Conference (ECTC)*, pp. 1052-1057, May 31 2011-June 3 2011.

[5] Q. Yongxi and T. Itoh, "A broadband uniplanar microstrip-to-CPS transition", *Asia-Pacific Microwave Conference Proceedings. APMC*, pp. 609-612 vol.602, 2-5 Dec 1997.

[6] L. Duixian and S. Reynolds, "A systematic design of microstrip-to-CPS transition for 60-GHz package applications", *IEEE International Symposium on Antennas and Propagation (APSURSI)*, pp. 504-507, 3-8 July 2011.

[7] N.I. Dib, R.N. Simons and L.P.B. Katehi, "New uniplanar transitions for circuit and antenna applications", *IEEE Transactions on Microwave Theory and Techniques*, vol. 43, pp. 2868-2873, 1995.

[8] I. Bahl: "Lumped Elements for RF and Microwave Circuits" (Artech House, 2002)

[9] http://www.sonnetsoftware.com/

[10] L. Jin, Y. Qiancheng and X. Jun, "A full ka-band microstrip-to-waveguide transition using tapered CPS probe and V-antenna", *The 7th German Microwave Conference (GeMiC)*, pp. 1-4, 12-14 March 2012.

[11] A. Strobel, C. Carlowitz, R. Wolf, F. Ellinger and M. Vossiek, "A mm-wave low power active backscatter tag for FMCW radar systems", *IEEE Transactions on Microwave Theory and Techniques*, in press.

PRIME 2013, Villach, Austria — *Session W1B – RF Techniques 2*

Cascode class-E power amplifier in 180/350 nm CMOS for EER system

Ivan A. Rumyancev, Alexander S. Korotkov
Integrated Electronics Dept.
St. Petersburg State Polytechnical University
St. Petersburg, Russia
i.a.rumyancev@gmail.com, korotkov@rphf.spbstu.ru

Johann Hauer
Integrated Circuits and Systems Dept.
Fraunhofer Institute for Integrated Circuits
Erlangen, Germany
johann.hauer@iis.fraunhofer.de

Abstract— **Cascode class-E power amplifier in 180/350 nm CMOS for envelope elimination and restoration system is presented. Designed amplifier achieves 21 dBm output power and 44% PAE at 1900 MHz with 3 V supply voltage. Charging acceleration technique is used for efficiency enhancement. ESD protection problem is discussed.**

Keywords— *power amplifier, class-E, cascode, envelope elimination and restoration, charging acceleration technique, ESD protection*

I. INTRODUCTION

Modern wireless communication protocols, such as WCDMA, WiMAX, LTE use non-constant envelope modulated signals with large peak-to-average ratios and require high linearity. Envelope elimination and restoration (EER) is a widespread technique [1, 2, 3] for amplification of non-constant envelope modulated signals. Functional scheme of this system is shown on Fig. 1. It consists of a limiter, an envelope detector, a high frequency switch mode power amplifier (RF) and an envelope amplifier (EA). The input signal is divided into phase- and amplitude modulated components. The phase modulated component switches RF amplifier. The amplitude modulated component is amplified by EA and modulates supply voltage of RF amplifier.

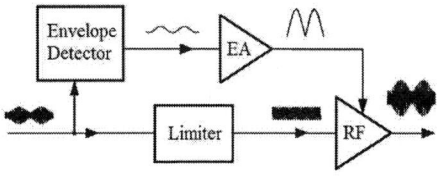

Fig. 1. Envelope elimination and restoration system

The main switch mode power amplifier classes are D, E, F. Class-E is the most perspective class of operations for realization of CMOS RF power amplifier because of its circuit simplicity and high practical efficiency that can be as high as 65 % [4]. Despite that design of a class-E power amplifier is well studied, ESD protection of a circuit is rarely discussed. This work is devoted to the design of the CMOS RF class-E power amplifier based on the cascode structure with different thickness of the gate oxide of the transistors combined with charging acceleration technique and ESD protection. The paper

is organized as follows. Section II describes common principals of operations of conventional and cascode class-E power amplifiers. The main ways to increase the efficiency of the cascode circuit is also discussed. Sections III describes designed power amplifier schematic and simulation results. Section IV presents short conclusion.

II. CLASS-E POWER AMPLIFIER

A. Conventional Circuit

Conventional class-E power amplifier (see Fig. 2) consists of the transistor T, which acts as a switch, the shunt capacitor C_p, the series resonator LC, the choke inductance L_{ch}, and the load impedance $Z_L = R_L + jX_L$ [5].

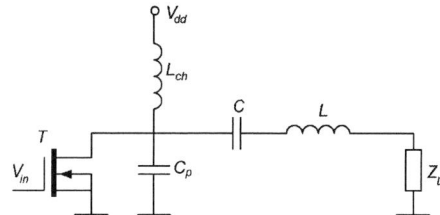

Fig. 2. Conventional class-E power amplifier circuit

In the ideal case there are two states (see Fig. 3) of the switch in class-E operation: closed and opened. While the switch is closed the current through it is high, but the voltage on it is zero. While the switch is opened there is a charging/discharging process of the shunt capacitor, so the voltage on the switch is high, but the current through it is tended to zero. In both cases dissipated power is tended to zero.

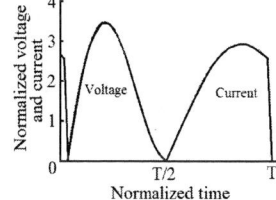

Fig. 3. Switch voltage and current in ideal class-E operation

The main reason of efficiency degradation is a parasitic channel resistance r_{ds} of the transistor, which can be approximated as:

978-1-4673-4579-8/13 $31.00 © 2013 IEEE   309

$$r_{ds}(\omega t) = \frac{L}{\mu C_{ox} W (V_{gs}(\omega t) - V_{th})},$$

where μ – charge-carrier effective mobility C_{ox} – gate oxide capacitance per unit area, W – transistor width, L – transistor length, V_{gs} – gate-source voltage, V_{th} – threshold voltage. Thus, to increase efficiency the transistor width should be increased. Unfortunate, this causes increasing of the parasitic capacitances. Maximum value of the output parasitic capacitance is limited by calculated value of C_p. Normally capacitor C_p is replaced by the parasitic capacitance of the transistor. The main disadvantage of the class-E amplifier is a high peak voltage on the transistor because of charging/discharging process of the capacitor C_p. This peak voltage can be as high as $5V_{dd}$ depending on output power level and duty cycle. Usually V_{pk} is about $3.5V_{dd}$. High peak voltage can cause gate oxide breakdown of the thin oxide transistor which is implemented in CMOS technology. To solve this problem cascode circuit is used widely [6, 7, 8].

B. Cascode Circuit

In the cascode class-E power amplifier (see Fig. 4) the thin oxide transistor T_1 stack in series with the thick oxide transistor T_2. The transistor T_1 is switched by the input signal. The transistor T_2 protects circuit from the gate oxide breakdown. Capacitor C_p is the same shunt capacitor as in the conventional circuit.

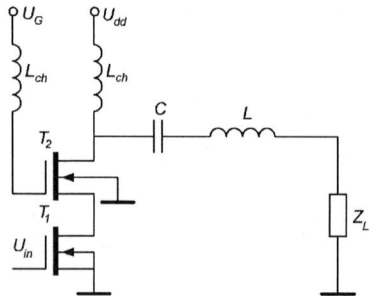

Fig. 4 – Cascode class-E power amplifier

Simplified equivalent circuit of the cascode class-E amplifier is shown in Fig. 5. Capacitors C_{p1} and C_{p2} are an equivalent to summarized parasitic capacitances of the transistors. Sizes of the transistors are chosen so that the total output capacitance equal to the calculated value C_p.

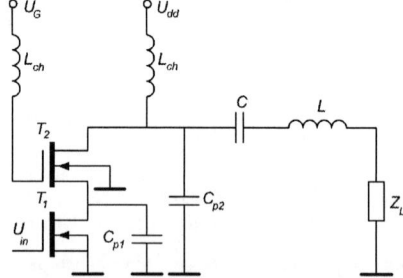

Fig. 5 – Simplified equivalent circuit Cascode class-E power amplifier

The principle of operation of the cascode circuit is the same as of the conventional circuit. However in the cascode circuit transistors cannot switch simultaneously because of the parasitic capacitance C_{p1} causing additional losses. To explain the mechanism of losses in the cascode amplifier a simplified equivalent circuit can be used. There are four states of the circuit for a cycle [6]:

1. Both transistors are in the triode region. In this state losses are the same as in the conventional class-E amplifier and are caused by the parasitic channel resistance.

2. The transistor T_1 operates in the cut-off region and the transistor T_2 is still closed. When the transistor T_1 is closing, charging process of the capacitance C_{p1} begins and the gate-source voltage V_{gs_T2} of the transistor T_2 is decreasing. To make the transistor T_2 operates in the cut-off region the following condition must be met:

$$V_{gs\_T2} \leq V_{th\_T2}.$$

With increasing of the source potential of the transistor T_2 the channel resistance r_{ds} is increasing as well. This causes decreasing of the current i_{ds} which is charging the capacitance C_{p1}. Despite that the current i_{ds} is small, the voltage V_{ds} is large, so the dissipated power is large too. As shown in [8] this switching losses can be up to 5 times higher than the losses due to the parasitic resistances.

3. Both transistors operate in the cut-off region. There is no current through the transistors therefore no power dissipation on the parasitic resistances.

4. The transistor T_1 operates in the cut-off region and the transistor T_2 is still closed. In this state losses are caused by the transistor parasitic resistance similar to the state 2. Nevertheless these losses are much smaller because the voltage on the transistor T_2 is low.

C. Cascode Circuit Efficiency Enhansment

There are two ways to reduce the effect of the parasitic capacitance C_{p1}. First is adding an inductance with blocking capacitor between the drain and the source of the transistor T_1 [8].The inductor resonates with the parasitic capacitance C_{p1} at operating frequency. Second (see Fig. 6) is adding the capacitor C_{ca} between the drain and the source of the transistor T_2 [6].

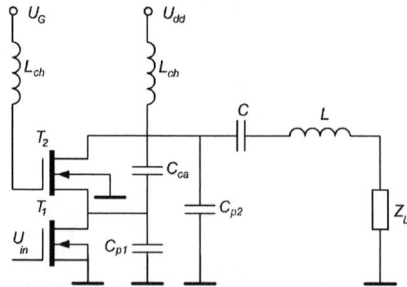

Fig. 6 – Cascode class-E power amplifier simplified equivalent circuit

When the transistor T_1 is opening, the current through the transistor T_2 is high enough. But with growing of the resistance r_{ds} of the transistor T_2 the current through is decreasing. This cause decelerating of charging of the parasitic capacitance C_{p1}.

The capacitor C_{ca} provides an additional current path for charging the parasitic capacitance. Since the capacitor C_{ca} is a part of the output capacitance of the circuit, its value should be taken into account when choosing the size of the transistors to meet the expression:

$$C_p = C_{p1} \| C_{ca} + C_{p2}.$$

In [8] the possibility of improving of this approach is also shown. It is the illumination of the additional capacitor and connection of the bulk of the transistor T_2 to its source. In such configuration the parasitic capacitance C_{db} of the transistor T_2 acts as a charging acceleration capacitor. Another benefit of this improvement is reducing of the parasitic capacitances. Therefore width of the transistor can be increased. However for this approach deep N-well process is needed.

Thus the second way seems the most effective because it is frequency independent, the capacitor require less area and there is no need in the additional technological process.

III. DESIGN AND SIMULATION

Schematic of the designed cascode power amplifier is shown on Fig. 7. The UMC CMOS models have been used for the design. The input signal from goes through the matching circuit $C_1C_2L_1$ and switches the 180 nm transistor T_1. The output signal goes through the filter L_3C_5, the matching network L_3C_6 and gets to the load resistor R_L. The transistor T_1 gate is DC grounded through the inductor L_2. The capacitor C_3 resonates with the inductor L_2 at the operating frequency. The gate of the 350 nm transistor T_2 is AC grounded by the capacitor C_7 to eliminate a feedback through the parasitic capacitance C_{gd} of the transistor T_2. The capacitor C_4 is charging acceleration capacitor. The inductances L_4 and L_5 are chokes. Because of a low quality factor of a planar inductor it was decided to make the inductance L_3 by a bond wire. Both transistors have the same width about 1.7 mm.

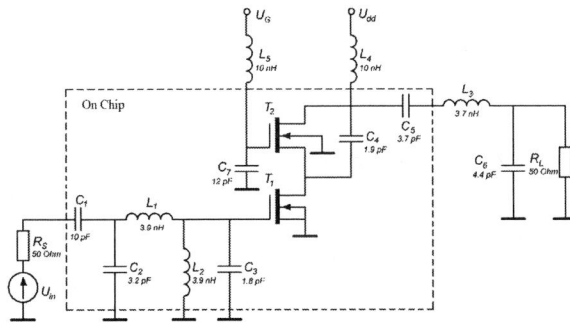

Fig. 7 – Designed cascode class-E power amplifier schematic

Design of microelectronic circuits with protection against electrostatic discharge (ESD) is an important task. In the discussed amplifier the gate of the 180 nm transistor T_1 is ESD protected by the input network. Simulation results of the human-body model test for 1 kV and 1.5 kV input are shown on Fig. 8. The Faraday library cells are used for protection of other chip pads. Due to the high voltage at the drain of the 350 nm transistor T_2 the use of the Faraday library cells is not possible. Therefore, to improve the reliability the drain of the 350 nm transistor is connected to its gate through 500 Ohms

resistor. For a similar reason, 10 kOhm resistor is added at the output of the chip. These resistors caused 8 % efficiency degradation.

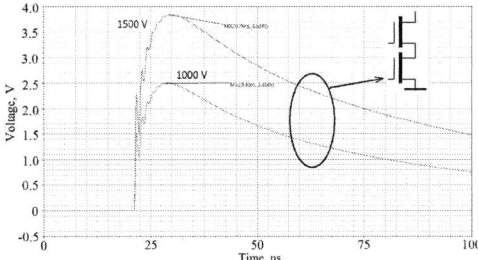

Fig. 8 – Gate voltage of the transistor T_1 with 1 kV and 1.5 kV inputs

The chip layout is shown on Fig. 9. The chip area is about 1.8 sq. mm with ESD protection cells and about 1.1 sq. mm without it.

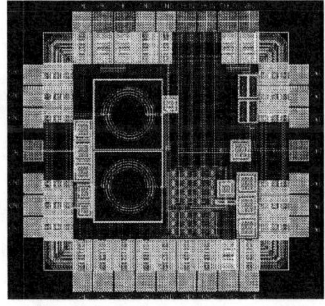

Fig. 9 – Layout of the designed cascode class-E power amplifier

Designed power amplifier was simulated in Cadence Virtuoso v. 5.141. Frequency dependences of the output power and the power added efficiency after compensation of the layout parasitics are shown on Fig. 10 and Fig. 11 respectively.

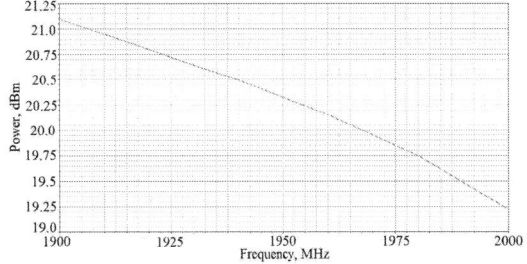

Fig. 10 – Output power vs Frequency (P_{in} = 10 dBm)

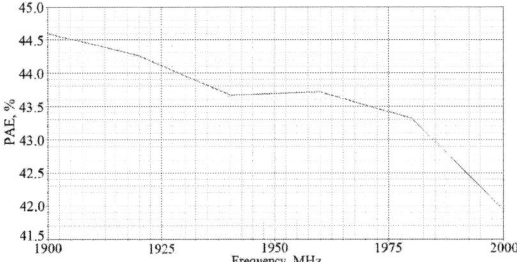

Fig. 11 – PAE vs Frequency

Simulation of the designed power amplifier in the EER system with WCDMA signal has been done. It was suggested that the envelope detector, the envelope amplifier and the limiter (see Fig. 1) are ideal. Say other words, nonlinear distortions and delays of these elements are not taken into account. The input signal has been generated and divided into phase- and amplitude modulated components in Matlab. These components have been used during simulation in Cadence software. Normalized spectrums of the designed power amplifier input and output signals are shown on Fig. 12. The input signal is shown by gray color. The output signal is shown by black color. Taking into account possible nonlinear distortion caused by real elements of EER system, a linearization circuit will be needed.

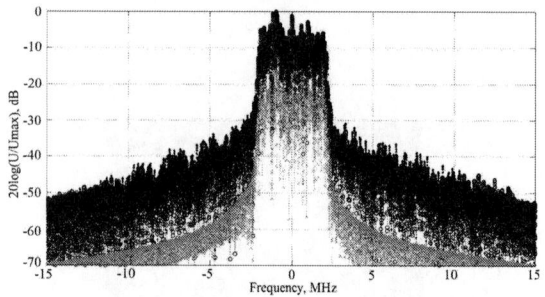

Fig. 12 – Normalized spectrums of the input and output signals

A comparison of the design power amplifier with other class-E power amplifiers is summarized in table 1.

TABLE I. POWER AMPLIFIERS PERFORMANCE COMPARISON

Parameter	[8]	[9]	[7]	[10]	[6]	This work
Process, nm	280	180	65	350	180	180
Freq., MHz	1700	950	2000	900	2000	1900
V_{dd}, V	2.5	1.8	5	2.2	3.3	3
Power, dBm	23	23.5	30	26.8	30.7	21
PAE, %	67	41	60	43	46	44
Area, sq. mm	0.9	–	0.29	–	1.72	1.1
Year	2006	2009	2009	2010	2010	2013

Despite the reduction in efficiency due to additional resistors designed amplifier shows promised characteristics and has ESD protection elements. PAE also can be increased by using better technology process, e.g. 65 nm as in [7], with lower parasitic resistances and capacitances. In contrast to [8], where high PAE has been achieved due to realization of the input and output inductances as bond wires, in this work there is only one bond wire inductance. Thus, the approach proposed in [8] will improve the efficiency manufacturing of a several

bond wires with exact values is a difficult task. The charging acceleration technique will give better results if it is combined with deep N-well process [6].

IV. CONCLUSIONS

Simulation results of the cascode class-E power amplifier in 180/350 nm CMOS for EER amplifier is presented. Designed amplifier achieves 21 dBm output power and 44% PAE at 1900 MHz with 3 V supply voltage. Charging acceleration technique is used for efficiency enhancement. Circuit inputs are ESD protected. As shown by simulations, implementation of the amplifier in EER system will need a linearization circuit to decrease nonlinear distortions.

AKNOWLEDGMENT

The work has been funded by Federal Target Program "Scientific and pedagogical human resources of innovation Russia for 2009-2013 years", Russian Federation. Authors thank the Program administration for the valuable support.

REFERENCES

[1] F. Wang, D. Kimball, J. Popp, A. Yang, D. Y. C. Lie, P. Asbeck, L. Larson, "Wideband envelope elimination and restoration power amplifier with high efficiency wideband envelope amplifier for WLAN 802.11g applications," IEEE MTT-S Int. Microwave Symp. Digest, 12-17 June 2005, pp. 645-648.

[2] M. Vasic, O. Garcia, J.A. Oliver, P. Alou, D. Diaz, J.A. Cobos, A. Gimeno, J.M. Pardo, C. Benavente, F.J. Ortega, "High efficiency power amplifier for high frequency radio transmitters," Proc. Twenty-Fifth Annual IEEE Applied Power Electronics Conference and Exposition, 21-25 Feb. 2010, pp. 729-736.

[3] D. Milosevic, J. van der Tang, A. van Roermund, "On the feasibility of application of class E RF power amplifiers in UMTS," Proc. Int. Symp. Circuits and Systems, 25-28 May 2003, vol.1, pp. 149-152.

[4] R. Zhang, M. Acar, M. P. van der Heijden, M. Apostolidou, L. C. N. de Vreede, D. M. W. Leenaerts, "A 550 – 1050MHz +30dBm class-E power amplifier in 65nm CMOS," Proc. Radio Frequency Integrated Circuits Symp, 5-7 June 2011, pp. 1-4.

[5] S. Cripps "RF power amplifiers for wireless communications," Artech House, 1999.

[6] O. Lee, J. Han, K. H. An, D. H. Lee, K.-S. Lee, S. Hong, C.-H. Lee, "A charging acceleration technique for highly efficient cascode class-E CMOS power amplifiers," IEEE J. Solid-State Circuits, vol. 45, no. 10, Oct. 2010, pp. 2184-2197.

[7] M. Apostolidou, M.P. Heijden, D.M.W. Leenaerts, J. Sonsky, A. Heringa, I. Volokhine, "A 65 nm CMOS 30 dBm class-E RF power amplifier with 60% PAE and 40% PAE at 16 dB back-off," IEEE J. Solid-State Circuits, vol. 44, no. 5, May 2009, pp. 1372-1379.

[8] A. Mazzanti, L. Larcher, R. Brama, F. Svelto, "Analysis of reliability and power efficiency in cascode class-E PAs," IEEE J. Solid-State Circuits, vol. 41, no. 5, May 2006, pp. 1222-1229.

[9] S. Datta, H. Saha, "A 950-MHz fully differential class-E power amplifier in 0.18µm CMOS for wireless communications," Proc. Int. Conf. Emerging Trends in Electronic and Photonic Devices & Systems, 22-24 Dec. 2009, pp. 88-91.

[10] H. R. Khan, Q. Wahab, J. Fritzin, A. Alvandpour, "A 900 MHz 26.8 dBm differential class-E CMOS power amplifier," Proc. German Microwave Conf., 15-17 March 2010, pp. 276-279.

PRIME 2013, Villach, Austria   *Session W1B – RF Techniques 2*

A SiGe LTE Power Amplifier with Capacitive Tuning for Size-Reduction of Biasing Inductor

Jinshu Zhao, Robert Wolf, Frank Ellinger, Senior Member, IEEE

Chair for Circuit Design and Network Theory, Technische Universität Dresden
01062 Dresden, Germany
jinshu.zhao@mailbox.tu-dresden.de

Abstract—This paper demonstrates a fully integrated power amplifier for long term evolution (LTE) applications. The power amplifier is adopting a tuning capacitor in parallel with the choke inductor, which reduces the DC power consumption of the inductor and decreases the chip size. The total power added efficiency (PAE) is enhanced as a result. The proposed power amplifier is manufactured using a 0.25 μm SiGe hetero-junction bipolar transistor (HBT) process. For a LTE signal with a centre frequency of 2.1 GHz, the PA delivers an output power of 21.3 dBm with the power added efficiency (PAE) of 21% and power gain of 19.5 dB at the 1 dB compression point. For the LTE downlink quadrature amplitude modulation (QAM64) signal at 2.11 GHz with 20 MHz channel bandwidth, the measured ACLR is below -30 dBc for the output power of 18 dBm with PAE above 15%. The error vector magnitude (EVM) specifications are met for LTE 20 MHz bandwidth QPSK modulation scheme, and complied with 1 dB power back-off for QAM16 signal and at 2 dB power back-off for QAM64 modulation scheme. The measurement results verify that the proposed method is promising for the application of the fully integrated PA.

Keywords— Power amplifier (PA), power added efficiency (PAE), long term evolution (LTE), hetero-junction bipolar transistors (HBT), choke inductor, adjacent channel leakage power ratio (ACLR), error vector magnitude (EVM).

I. INTRODUCTION

3GPP LTE (Long Term Evolution) is one of the promising 3G wireless standards, which enables higher data rates and lower latency. LTE adopts Single-Carrier-Frequency-Division Multiple Access (SC-FDMA) for the uplink and Orthogonal Frequency-Division Multiple Access (OFDMA) on downlink [1], [2]. Considering the linearity requirement of LTE, the transmitter must have good linearity performance, which induces high linearity requirement for power amplifier. Besides, the power-added-efficiency for power amplifier is highly concerned. Compared to CMOS technology, the SiGe BiCMOS technology has very attractive advantage for power amplifier performance, which is higher efficiency and output power for single stage of PA. Moreover, SiGe BiCMOS technology provides a possibility of integrating RF components and digital parts on a single chip. However, one of the big challenges for fully integrated PAs is the very low quality factor of the coils, which induces high loss and hence low power efficiency.

This work is supported by the CoolBaseStations Project, and sponsored by the Federal Ministry of Education and Research (BMBF).

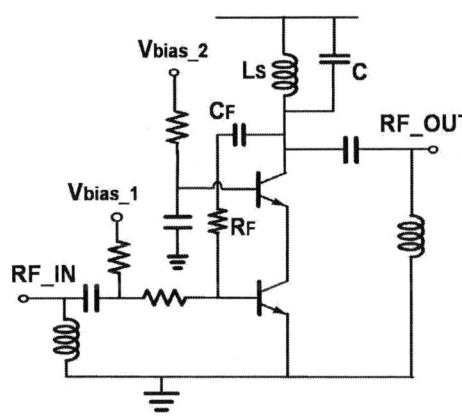

Fig. 1 Schematic of the proposed power amplifier

In this paper, a fully integrated SiGe HBT class A power amplifier is demonstrated. To diminish the influence of the high loss of the on-chip inductors, this PA adopts a tuning capacitor in parallel with the choke inductor, which reduces the PA's DC power consumption. The related theory is analysed in this paper. The measurement demonstrates promising results of this method.

II. POWER AMPLIFIER DESIGN

A. PA configuration

The PA schematic is shown in Fig. 1. The cascode PA with RC negative feedback can obtain broad bandwidth with flat gain and good input and output matching [3], [4]. Since the inductors are integrated on-chip, the inductor losses are much higher than the discrete solutions. Therefore, we add a capacitor in parallel with the choke inductor to reduce the value of the required inductance without degenerating the performance of the entire PA. In this case the quality factor of the LC network is increased. As a result, the operating bandwidth is reduced.

B. Choke inductor with tuning capacitor

The choke inductor compensates the parasitic capacitor of the bipolar transistors. For the linear PA, the choke inductor is on the path of very large DC current. The resistance of the inductor plays an important role in the DC power consumption.

978-1-4673-4579-8/13 $31.00 © 2013 IEEE   313

Paper W1B3

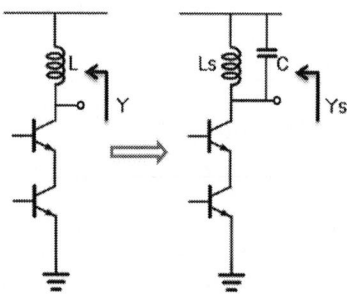

Fig. 2 Cascode PA adopting tuning capacitor structure

For example, if the DC current of the PA is 200 mA, and the choke inductor has an inner resistance of 2 Ω. Then the DC power consumption of the inductor will be 80 mW, which is 12% for the PA with 600 mW power consumption. The on-chip inductors are important components in the fully integrated RF systems. However, the integrated coils have much lower quality factor compared with their discrete counterparts [5], [6]. Although one can vary a little bit the size of the coil by optimizing the wiring space and the metal width, the optimized spiral size of the coil has already been determined by the process as long as the inductance value is decided. In general the higher the inductance value, the larger the resistance it contains. Hence the power loss generated by the inductors is unavoidable. To increase the equivalent inductance with relatively small coil size, the structure of an inductor and a capacitor connected in parallel is considered, as shown in Fig. 2.

Assuming that the inductor resistance is in linear proportional to the inductance,

$$R = qL \tag{1}$$

q is the proportional ratio.

At the centre frequency f_c, the admittance of the choke inductor is

$$Y = \frac{1}{R + j\omega_c L} = \frac{1}{L}\frac{q - j\omega_c}{q^2 + \omega_c^2} \tag{2}$$

For the LC network depicted in Fig. 2, the admittance can be expressed as

$$Y_S = j\omega_c C + \frac{1}{R_S + j\omega_c L_S} = j\omega_c C + \frac{1}{L_S}\frac{q - j\omega_c}{q^2 + \omega_c^2} \tag{3}$$

Comparing (2) and (3), if $\text{Im}(Y) = \text{Im}(Y_S)$, deriving from the upper equations,

$$C = \frac{1}{q^2 + \omega_c^2}(\frac{1}{L_S} - \frac{1}{L}) \tag{4}$$

Then the combination of the inductor L_S and capacitor C can be considered as an equivalent inductor with the inductance L. In this case the input and output matching of the PA are not influenced. By choosing L_S and C properly, the choke inductor can be smaller than the original value without introducing performance degeneration.

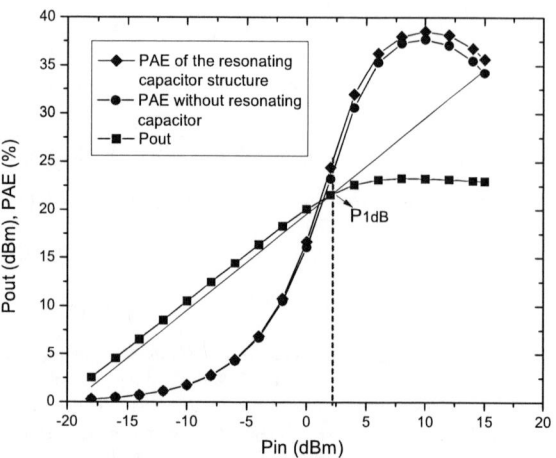

Fig. 3 Comparison of the power added efficiency and the output power (post simulation results)

For class A PA, the DC power consumption equals the product of the average DC current and voltage, which can be written as

$$P_{DC} = V_{DD}I_{DC} \tag{5}$$

The efficiency of the PA can be derived as,

$$\eta = \frac{P_{RF}}{P_{DC}} = \frac{P_{RF,T} - P_{Loss,AC}}{P_{DC,T} + P_{Loss,DC}} \tag{6}$$

Where $P_{RF,T} = \eta_T P_{DC,T}$ is the RF power of the transistor array, $P_{Loss,AC}$ is the RF loss of the LC network, $P_{DC,T}$ is the DC power consumption of the transistor array, and $P_{Loss,DC}$ is the DC loss caused by the inductor resistance.

Then (6) can be expressed as

$$\eta = \frac{\eta_T(V_{sat} + \hat{V}_s)I_{op} - \frac{\hat{V}_s^2}{2}\frac{q}{q^2 + \omega_c^2}\frac{1}{L_S}}{(V_{sat} + \hat{V}_s)I_{op} + I_{op}^2\frac{q}{k}L_S} \tag{7}$$

In which V_{sat} is the saturation voltage, and $\hat{V}_s$ is the RF swing of the transistor array. The factor k takes the skin effect into account.

Substituting $V_{sat} + \hat{V}_s$ with V_{op}, then to differentiate (7), we can get

$$\frac{\partial \eta}{\partial L_S} = \frac{\frac{qI_{op}\hat{V}_s^2V_{op}}{2(q^2 + \omega_c^2)}\frac{1}{L_S^2} + \frac{q^2I_{op}^2\hat{V}_s^2}{k(q^2 + \omega_c^2)}\frac{1}{L_S} - \eta_T\frac{q}{k}V_{op}I_{op}^3}{(V_{op}I_{op} + I_{op}^2\frac{q}{k}L_S)^2} \tag{8}$$

When the numerator of (8) equals to 0, the efficiency reaches its maximum value. Hence the optimized value of L_S for maximum efficiency can be expressed as

$$L_S = \frac{q\hat{V}_s^2 + \hat{V}_s\sqrt{q^2\hat{V}_s^2 + 2\eta_T V_{op}^2 k(q^2 + \omega_c^2)}}{2\eta_T V_{op} I_{op}(q^2 + \omega_c^2)} \qquad (9)$$

The self-resonance frequency of the LC network shown in Fig. 2 is

$$f_0 = \frac{1}{2\pi\sqrt{L_S C}} \qquad (10)$$

Above the self-resonance frequency, the impedance becomes capacitive. Hence the $\sqrt{L_S C}$ value should be small enough to keep f_0 higher than the PA's operation frequency.

C. Simulation results

To prove the idea stated in subsection A and B, a class A PA has been designed based on the tuning capacitor theory. According to simulation, the inductance L without tuning capacitor should be 1.8 nH. Adopting (9), which is derived in subsection B, we can calculate that the optimized inductance L_s is around 1 nH. Then the tuning capacitor value C can be obtained by (4). The value C and L_s should comply with the relationship stated in (10). After simulating and optimizing with the calculated data, we choose the solution of a 1.1 nH inductor and a 1.8 pF capacitor connected in parallel, which reduces the coil resistance by approximate 30%. As a result, the coil area is also reduced by 20-30%.

In Fig. 3 the post layout simulation results of the original and proposed PA are demonstrated for comparison. The output power of the tuning capacitor PA is kept the same as the original PA. In the meanwhile, the power added efficiency of the proposed PA is enhanced by 3% at the compression point. Therefore, the simulation results prove the efficiency enhancement of the proposed tuning capacitor method. Moreover, since the LC network and the PA load form a band pass filter, the higher order harmonics are filtered, which improve the PA's linearity.

III. MEASUREMENT RESULTS

The cascode PA with tuning capacitor was fabricated in a 0.25 μm SiGe HBT technology. The chip size is 1×1 mm². The chip photo of the PA is demonstrated in Fig. 4.

The S parameter measurement results are demonstrated in Fig. 5. The S_{11} reaches -21.7 dB and S_{22} is -4.9 dB at the frequency of 2.1 GHz, which indicate sufficient network matching.

The power performance results are shown in Fig. 6, which shows that the output power at the 1 dB compression point is 21.3 dBm and the power added efficiency reaches 21% for the compression point at frequency of 2.1 GHz. The power gain of 19.5 dB is sufficient for the LTE application.

For testing the linearity performance of the fabricated power amplifier, a LTE downlink quadrature amplitude modulation (QAM64) signal at 2.11 GHz with 20 MHz channel bandwidth is used as the input signal. Fig. 7 shows the spectral mask at the frequency of 2.11 GHz. No predistortion was applied to the PA.

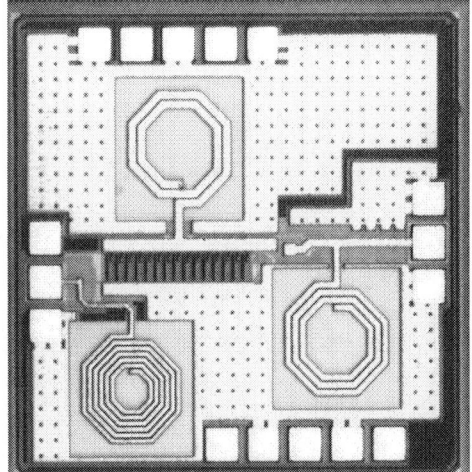

Fig. 4 Chip photo of the tuning capacitor PA

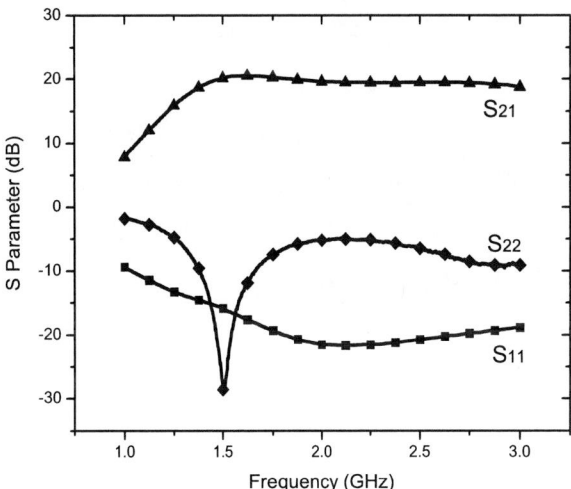

Fig. 5 Measured S parameter

The adjacent channel leakage power ratio (ACLR) is defined as the ratio of the power at the adjacent channel to the in-channel power. The ACLR of Evolved Universal Terrestrial Ratio Access (E-UTRA) at the lower and the upper adjacent channels were measured at the offset frequencies. The measured ACLR versus the output power for QAM64 modulation scheme is shown in Fig. 8. The ACLR is lower than -30 dBc for the output power of 18 dBm, leading to the PAE above 15% (with -3 dB back-off).

The error vector magnitude (EVM) requirements for LTE are 17.5% for quadrature phase-shift keying (QPSK), 12.5% for quadrature amplitude modulation 16 (QAM16), and 8% for quadrature amplitude modulation 64 (QAM64). The measured EVM results versus the output power of the PA are plotted in Fig. 9. The EVM specifications are always met for QPSK modulation scheme, and complied at -1 dB back-off for QAM16 signal and at -2 dB back-off for QAM64 modulation scheme.

Paper W1B3   *PRIME 2013, Villach, Austria*

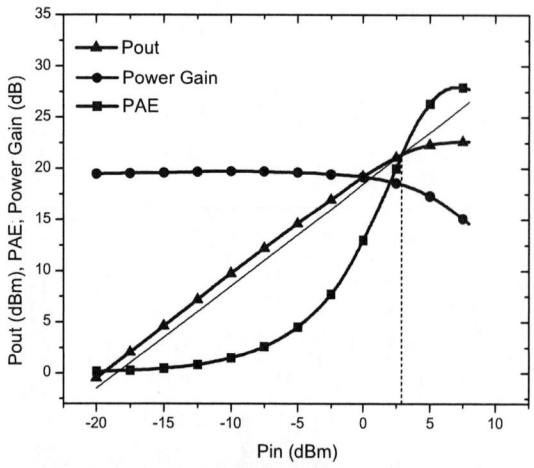

Fig. 6 Measured power characteristics of the fabricated PA at 2.1 GHz

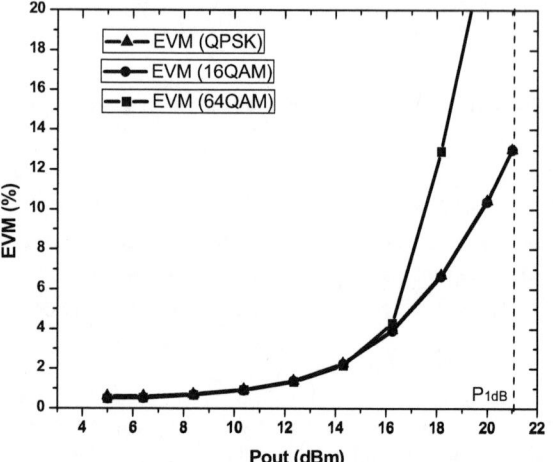

Fig.9 Measured EVM with LTE standard (Measuring frequency: 2.11 GHz; Bandwidth: 20 MHz; Modulation scheme QAM64, QAM16, and QPSK)

The measurement results demonstrate a fully integrated PA with a good linearity performance and a relatively high PAE, which show the potential of the idea of including tuning capacitor for the fully integrated power amplifier design.

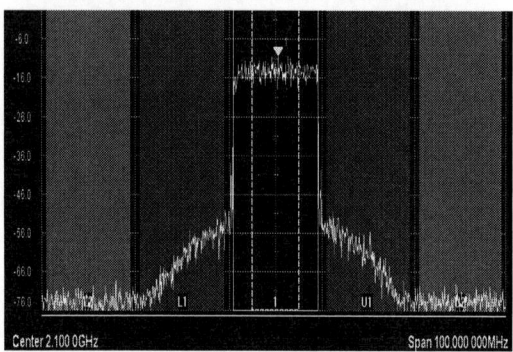

Fig. 7 Measured spectrum mask with LTE 20 MHz signals at 7 dBm power back-off

IV. CONCLUSIONS

In this paper, we demonstrate a fully on-chip power amplifier adopting tuning capacitor. The tuning capacitor structure reduces the choke inductor size, which decrease the DC power consumption as well. The proposed PA delivers an output power of 18 dBm with a PAE above 15%, satisfying the ACLR and standard EVM requirements for the QPSK, QAM16 and QAM64 modulation scheme for 2.11 GHz LTE downlink signal with 20 MHz bandwidth. The measured results prove that the tuning capacitor method enhances the power efficiency without decreasing the linearity performance.

REFERENCES

[1] 3GPP technical specification, TS 36.101 V8.9.0 (2010-03)
[2] B. A. Bjerke, "LTE-advanced and the Evolution of LTE Deployments", *Wireless Communications, IEEE*, vol. 18(5), pp. 4-5, 2011
[3] S. C. Cripps, RF Power Amplifiers for Wireless Communications. Norwood, MA: Artech House, 1999
[4] J. Tajima, "Monolithic Low Power Amplifiers with RC Parallel Feedback", *IEEE Trans. Microwave Theory Tech.*, vol. MTT-32, pp. 542-545, 1984
[5] C. P. Yue, and S. S. Wong, "Design Strategy of On-chip Inductors for Highly Integrated RF Systems," *Proc. IEEE Design Automation Conference*, pp. 982-987, 1999
[6] T. D. Stetzler, I. G. Post, J. H. Havens, and M. Koyama, "A 2.7-4.5 V Single Chip GSM Transceiver RF Integrated Circuit," *IEEE Journal of Solid-state Circuits*, vol. 30, pp. 1421-1429, Dec. 1995

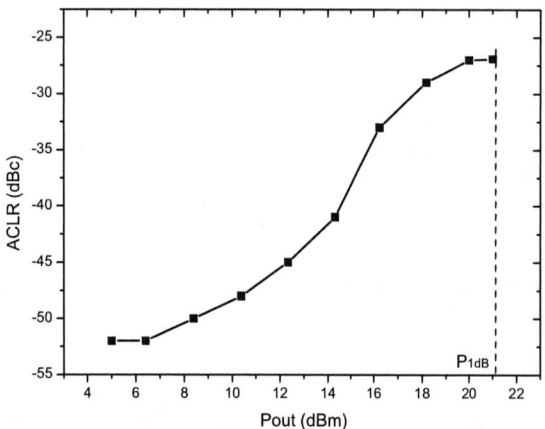

Fig. 8 Measured ACLR with LTE standard (Bandwidth: 20 MHz; Modulation scheme QAM64)

PRIME 2013, Villach, Austria

Session W2B – Sensors

An Ultra-Thin CMOS In-Plane Stress Sensor

Yigit U. Mahsereci, Nicoleta Wacker, Harald Richter, and Joachim N. Burghartz
Institut für Mikroelektronik Stuttgart, IMS CHIPS,
Allmandring 30a, 70569 Stuttgart, Germany
{mahsereci|wacker|richter|burghartz}@ims-chips.de

Abstract—This paper presents a CMOS stress sensor for uniaxial, in-plane flex measurement. The sensor is to be fabricated as an ultra-thin chip in order to adapt to surface shape. The sensor generates 2×10-bit data from piezo-resistive elements, via on chip stress insensitive readout circuitry. Unlike the directional selectivity of the strain gauges, the uniaxial in-plane stress vector can be measured in any direction by this sensor. In addition, the MOSFET stress macro model used for the sensor flex simulations is presented. It is used for the sensor flex simulations. It supports DC, AC, and transient analysis with any analog simulator.

Keywords—flexible electronics, piezo-resistive effect, stress sensor, ultra-thin chip, stress simulation.

I. INTRODUCTION

Flexible electronics, based on hybrid system-in-foil technology, is expected to exhibit rapid market growth. Ultra-thin silicon (Si) chips are key components for this technology [1]. By using the physical properties of the ultra-thin chips, smart and highly sensitive stress sensors can be developed. Besides enabling new applications, the flexible chips include design challenges such as variable strain in silicon. Strain changes the electrical behavior of devices, leading to performance degradation or even failure. Therefore, flexible chips require an extra design effort in order to account for the effects of variable strain in Si.

In this paper, we present a flexible complementary metal oxide semiconductor (CMOS) stress sensor for measuring 2-D stress components on a flexible surface. The sensor provides 2×10-bit data, from which the angle and the magnitude of the local in-plane stress vector are calculated. It also compensates the temperature effect and the out of plane stress components; hence the relevant stress data is not corrupted. The readout circuit is insensitive to uniaxial stress, hence the sensor maintains precision under variable stress. Special transistor models are required for evaluating the circuit performance under variable stress. An extended version of the stress dependent metal-oxide-semiconductor field effect transistor (MOSFET) macro model [2] is also introduced in this paper. The macro model enables the simulation of any stress induced change in the electrical behavior of a MOSFET, at any DC operating point. Furthermore, the model can be used in any analog simulator. The presented model is used in the design of the stress sensor.

II. STRESS-DEPENDENT MOSFET MACRO MODEL

In this section, we present the extended version of the stress-dependent macro model for analog simulation, as introduced in [2]. First its functionality is explained and then, the equivalent MOSFET circuit is introduced. The functionality of the macro model is than verified in two cases.

A. MOSFET Macro Model Structure

The macro model comprises three functional modules. The first module calculates the local stress in MOSFET $|\sigma_{local}|$ based on the magnitude of the applied uniaxial stress $|\sigma|$ and the angle that defines the stress application direction φ, as shown in Fig. 1. The local stress is given by

$$|\sigma_{local}| = \sqrt{\sigma_{11}^2 + \sigma_{12}^2} \tag{1}$$

where $|\sigma_{11}| \cong |\sigma| \cdot cos\,\varphi$ and $|\sigma_{12}| \cong |\sigma| \cdot sin\,\varphi$ are the stress components along the reference axes x_1 and x_2.

The second functional module calculates the drain current (I_{DS}) correction factor PS. The experimental results show that applied uniaxial stress mainly affects the effective carrier mobility [2], [3]. This variation can be expressed as

$$U0(\sigma_{local}) = U0(0) \cdot (1 + \Pi(\theta, \varphi, \Pi_{11}, \Pi_{12}, \Pi_{44}) \cdot |\sigma_{local}|) \tag{2}$$

where U0 is the BSIM3v3 low-field mobility parameter, and θ specifies the drain current flow direction (Fig. 1). We denote $PS = \Pi \cdot |\sigma_{local}|$. Π is the piezo-resistive coefficient, whose general relation for the (001) Si-plane [2] is

$$\Pi(\theta, \varphi, \Pi_{11}, \Pi_{12}, \Pi_{44}) = \Pi_{11} \cdot (cos^2\,\theta \cdot cos^2\,\varphi + sin^2\,\theta \cdot sin^2\,\varphi)$$
$$+ \Pi_{12} \cdot (cos^2\,\theta \cdot sin^2\,\varphi + sin^2\,\theta \cdot cos^2\,\varphi)$$
$$+ 2 \cdot \Pi_{44} \cdot sin\,\theta \cdot cos\,\theta \cdot sin\,\varphi \cdot cos\,\varphi \tag{3}$$

$(\Pi_{11}, \Pi_{12}, \Pi_{44})$ are the three fundamental piezo-resistive coefficients of the (001) Si-plane. Their values, which depend on the applied electric field and on the doping density [4], are summarized in Table I for two CMOS technologies.

The third functional module of the MOSFET macro model corrects the drain current for the stress-induced effects and also

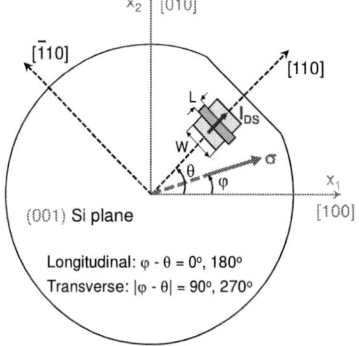

Fig. 1. The drain current I_{DS} flow direction defined by θ and the stress application direction defined by φ in (001) silicon (Si) plane. Both angles are measured with respect to [100] Si-direction.

978-1-4673-4579-8/13 $31.00 © 2013 IEEE   317

TABLE I. MEASURED[a] AND ESTIMATED[b] PIEZO-RESISTIVE COEFFICIENTS OF BULK MOSFETS FABRICATED IN 0.8 AND 0.5 μm TECHNOLOGIES ($\times 10^{-12}$ Pa)

Piezo-resistive Coefficient	0.8 μm tech.		0.5 μm tech.	
	NMOS	PMOS	NMOS	PMOS
$\Pi_{11} + \Pi_{12}{}^a$	650($\pm$26)	$-$180($\pm$16)	494($\pm$12)	15($\pm$15)
$\Pi_{44}{}^{a,b}$	310($\pm$13)	$-$1060($\pm$95)	140($\pm$2)	$-$863($\pm$18)
$\Pi_{11}{}^b$	420	101	650	240
$\Pi_{12}{}^b$	260	$-$280	$-$150	$-$260

accounts for the I_{DS} variation with U0, at different applied voltages. The drain current variation with stress is obtained by expanding I_{DS} in Taylor series, and using (1) and (2).

$$
\begin{aligned}
I_{DS}(U0(\sigma_{local}),V,T) &= I_{DS}(U0(0),V,T) \\
&+ \frac{I_{DS}((U0(0)+U0(0)\cdot d_0),V,T) - I_{DS}(U0(0),V,T)}{d_0} \cdot PS
\end{aligned}
\tag{4}
$$

where V represent the voltages and T the temperature. d_0 is a finite difference corresponding to $\sim$ 30 % of the maximum observed change of I_{ID} due to stress.

B. MOSFET Equivalent Circuit

Fig. 2 shows the extended MOSFET circuit, containing a sub-circuit type representation of the transistor, built based on (4). I_{DS} is changed with a current source **BIstress**, placed in parallel to the transistor's channel. The term $I_{DS}(U0(0) + U0(0)\cdot d_0), V, T$ of (4) is represented by the nominally strained transistor M_2, whereas the term $I_{DS}(U0(0), V, T)$ corresponds to the unstrained transistor M_1. The stress components $|\sigma_{11}|$ and $|\sigma_{12}|$ are translated into the voltages $V(S_{x1})$ and $V(S_{x2})$ based on the convention that 1 MPa corresponds to 1 V.

C. MOSFET Macro Model Verification

Two examples of simulated strained bulk NMOSFETs, performed using the estimated piezo-resistive coefficients given

in Table I, are presented. The results correspond to different transistor geometries, in both linear and saturation regimes. The first example presents the measured and simulated results for strained transistors $(\theta, \varphi) = (45^0, 45^0)$ produced in a 0.8 μm CMOS process. The simulated results reproduce the measurements, as shown in Fig. 3. The second example shows a less common case, defined by $(\theta - \varphi) = 45^0$. Also in this case, the simulated results, for bulk NMOSFETs fabricated in a 0.5 μm CMOS technology, reproduce the measurements, as shown in Fig. 4. The macro model was also verified successfully in case of strained PMOSFETs.

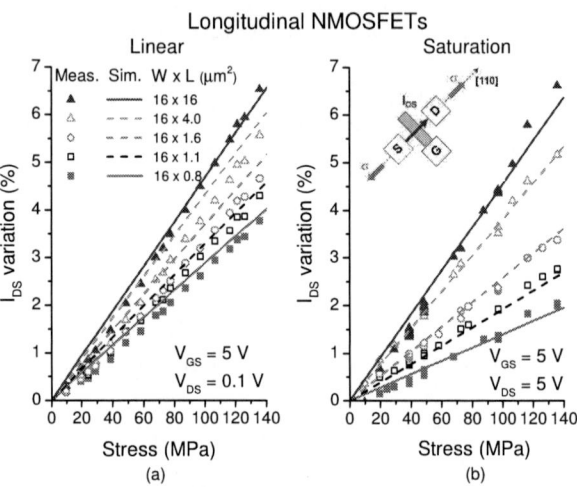

Fig. 3. Measured (dots) and simulated (lines) results for NMOS devices of different geometries in (a) linear and (b) saturation regimes. Transistors were produced in a 0.8 μm CMOS process. ($|\sigma_{11}| = |\sigma_{12}| \in [0, 100\,]MPa$).

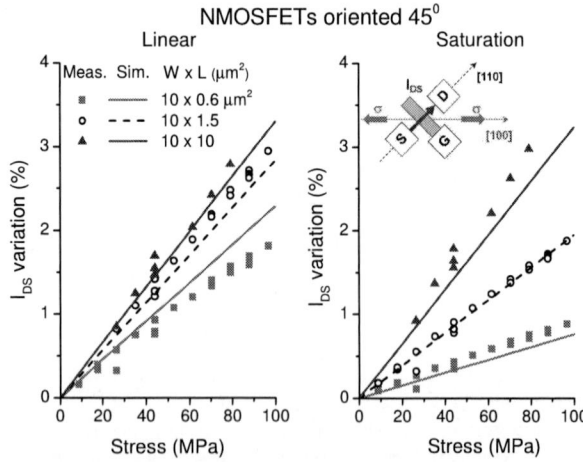

Fig. 4. Measured (dots) and simulated (lines) results for NMOS devices of different geometries in (a) linear and (b) saturation regimes. Transistors were produced in a 0.5 μm CMOS process. ($\theta - \varphi = 45^0$)

III. DESIGN OF THE FLEXIBLE STRESS SENSOR

In this section, we present the design of the proposed in-plane stress sensor. Fig. 5 illustrates the sensor system.

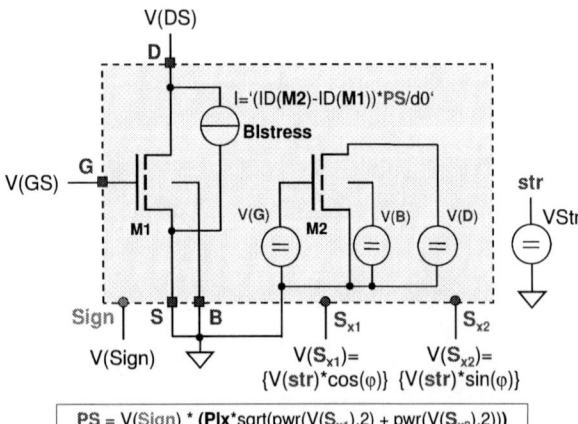

Fig. 2. NMOS device equivalent circuit for simulation of variable uniaxial stress effects. $V(Sign)(\pm 1$ V) establishes the sign of PS, allowing to switch between tensile and compressive stress. PI$_x$ is the general piezo-resistive coefficient (3). The voltages $V(S_{x1})$ and $V(S_{x2})$ correspond to the stress components $|\sigma_{11}|$ and $|\sigma_{12}|$. $d_0 = 0.02$, represents a nominal 2% change of U0.

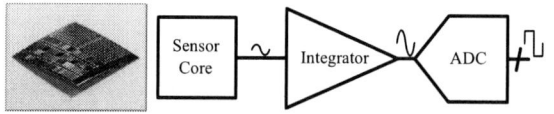

Fig. 5. Stress sensor system diagram. The circuitry is exposed to strain due to flexibility.

A. Sensor Core

The stress sensing elements are MOSFETs, in which the primarily stress affected parameter is mobility. Secondary effects, such as threshold variation or geometric changes, are either intrinsically negligible or managed by proper biasing of devices [2]. The mobility change results in current variations, as given in (4). The currents of orthogonal transistor pairs are subtracted in order to compensate the temperature dependency and out-of-plane stress component. The resulting relations are

$$\theta(0^0, 90^0): \frac{\Delta I_{DS[100]} - \Delta I_{DS[010]}}{I_{DS[100],[010]}} = -(\Pi_{11} - \Pi_{12}) \cdot cos2\varphi \cdot |\sigma| \tag{5}$$

$$\theta(45^0, 135^0): \frac{\Delta I_{DS[110]} - \Delta I_{DS[\bar{1}10]}}{I_{DS[110],[\bar{1}10]}} = -\Pi_{44} \cdot sin2\varphi \cdot |\sigma| \tag{6}$$

where $I_{DS[100],[010]}$ and $I_{DS[110],[\bar{1}10]}$ are the matched drain currents of orthogonal transistor pairs. The subtraction is only valid if the correlation of the canceled terms is maintained by compact layout. Given the thickness of the chip, significant temperature gradients may occur over a large area. Fig. 6 shows the circuit diagram of the sensor core. Two types of cores are designed for optimizing the responsivity of the sensor to normal and shear stress components. The $[110]-[\bar{1}10]$ and $[100]-[010]$ pairs are constructed with PMOS and NMOS devices respectively. M_3-M_8 forms the current mirror. The drain current of M_1 is mirrored to M_2. The resulting current from the subtraction is fed to the readout at I_{diff} node. A divided copy of the drain current of M_1 is also available as I_{DS}, required to calculate the stress in (5) and (6). I_{DS} is affected by the temperature fluctuations, causing gain error. A compensation technique, such as chip stacking [3], can be used. In the current design, this effect is not considered. Three-stage cascode current mirror enables precise subtraction with low offset, yet leaves enough voltage headroom for M_1-M_2 to operate in the saturation region due to the 5 V supply [5].

The size of M_3-M_8 is selected to minimize V_{DS} swing as I_{DS} changes for M_1. The voltage stability of the I_{diff} node is maintained by the integrator in the readout circuit. By keeping the node voltages of the orthogonal transistor pair stable, I_{diff} is linearly dependent on the induced stress.

B. Readout

The readout chain consists of an integrator, a track-and-hold stage, and an ADC, as shown in Fig. 7. The integrator topology has a number of advantages for this design. First, it serves as a low pass filter, and decreases the noise. Second, the voltage of the I_{diff} node is constant due to the feedback. Finally and most importantly, the feedback loop only accommodates a capacitor, for which the effect of stress is negligible. The feedback behavior remains the same under varying stress, unlike the resistive feedback structures. A two stage Miller op-amp is used for the integrator. It is designed to remain within operation limits up to 200 MPa, which may cause a mobility change in range of ±15 % for both NMOS and PMOS transistors, depending on the direction of the stress vector, and device. The gain and the phase margin are explicitly affected by the varying mobility, therefore the phase margin is pushed above 60^0 at the cost of higher power consumption and the minimum gain is set to be 80 dB, to minimize the voltage drift at the input node under stress. The bandwidth of the op-amp is set to be large enough so that input node voltage is also kept constant during integration. Since the input node is connected to drain of M_2 of the sensor core, the linearity of the sensor is maintained under both variable stress and readout operation. The V_{ref} voltage is externally provided. It is to be $V_{DD}/2$ for symmetric signal range and low-noise for overall system noise performance.

The track-and-hold circuit is intrinsically not affected by stress. The size of the capacitor is selected considering the op-amp load and the kickback of the ADC. Complementary switches with charge injection cancelation are used. A 10-bit successive approximation register ADC from the IMS CHIPS intellectual property library is used for digitization. Its stress insensitivity is arranged by design and checked by simulation.

C. Digital Controller

The management of the analog blocks and external communication is done by the controller. The VHDL-based design is realized using Si-compiling tools. The system performance figures, such as sensor bandwidth, noise, and sensitivity, can be optimized by configuring the analog timing. The controller provides two synchronous serial interfaces for timing configurations and data output. The stress mainly affects the speed performance of the digital cells. The circuit is synthesized for worst case specs and layout techniques are used for stress management.

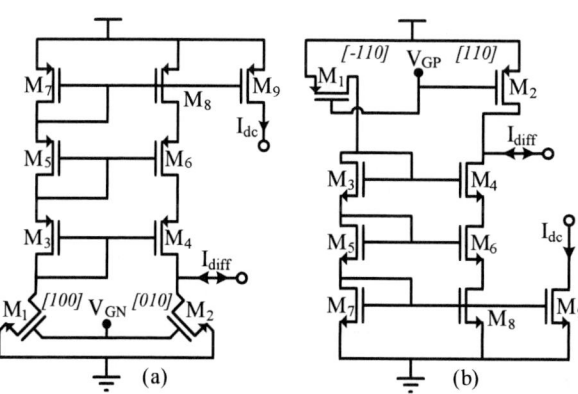

Fig. 6. Stress sensing blocks which employ (a) NMOS and (b) PMOS orthogonal transistor pairs.

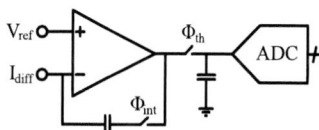

Fig. 7. Circuit diagram of the readout chain. It consists of an integrator, a track-and-hold stage, and an ADC.

D. Layout Considerations

All circuits except stress sensor cores are constructed with uniformly oriented transistors. Fig. 8 illustrates the orthogonal and uniformly oriented transistors. Differential branches in the analog and digital circuitry that benefit from uniform alignment as stress effects are nulled out. Furthermore, global orientation results in common mobility for all devices of the same kind in worst cases and simplifies the design considerations as well as the simulation effort.

IV. SIMULATION RESULTS

The models explained in Section II are used for the simulation. The estimated piezo-resistive coefficients given in Table I, are used for the variable direction uniaxial stress simulation. The simulation flow is shown in Fig. 9. The stress information is fed to the model in circuit simulation, and the sensor output is analyzed for performance evaluation. Therefore, total introduced error is considered. Fig. 10 shows the input and output stress and angle sweep simulation results. The relatively large deviation of the sensor's raw output is due to the offset errors in circuitry, and it is fixed by an offset calibration in the calculation phase. According to the simulation results, a sensitivity of 1.2 MPa for magnitude and 0.8^0 for angle are expected. Also, simulated noise floor is half LSB, therefore a true 10-bit performance is expected.

V. CONCLUSION

The sensor compensates the temperature effect, filters the out-of-plane stress components, and generates 2×10-bit digital data using the on-chip readout. A simulated linearity of 5 % for the vector magnitude, and 1 % for the angle are provided. The sensor can be used for uniaxial stress measurements in any in-plane direction, and for stress magnitudes up to $\sim$ 200 MPa. The sensor is to be tested with static and dynamic bending setups after fabrication. The stress dependent MOSFET macro model can be used in any analog simulator without introducing

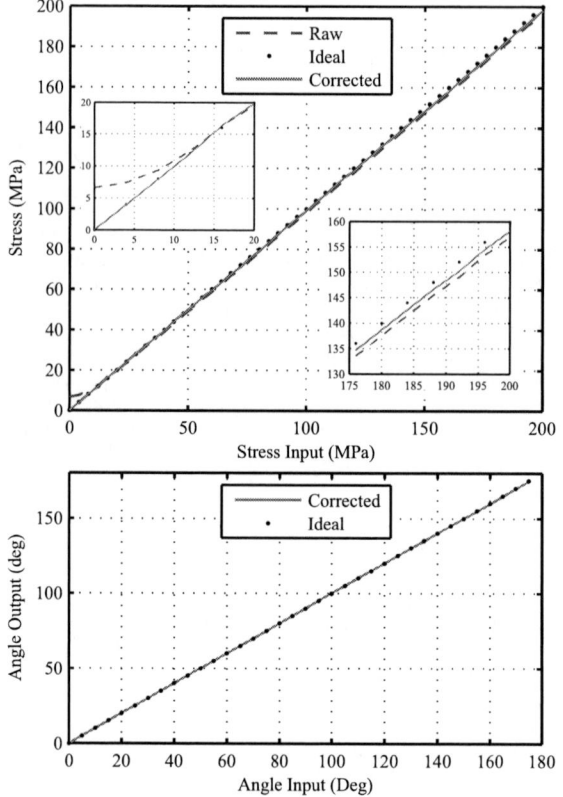

Fig. 10. Magnitude and angle sweep results. Linearity is enhanced by offset correction. Nonlinearities of 5% for magnitude and 1% for angle are calculated after correction.

new model code. Moreover, it is transferable to other CMOS technologies [2].

ACKNOWLEDGMENT

The authors would like to thank Horst Rempp, Stefan Endler, Cor Scherjon, and IMS CHIPS staff for sharing their experience and support.

REFERENCES

[1] J. N. Burghartz, W. Appel, C. Harendt, H. Rempp, H. Richter, and M. Zimmermann, "Ultra-thin chips and related applications, a new paradigm in silicon technology," in *ESSCIRC, 2009. ESSCIRC'09. Proceedings of.* IEEE, 2009, pp. 28–35.

[2] N. Wacker, H. Richter, M.-U. Hassan, H. Rempp, and J. N. Burghartz, "Compact modeling of CMOS transistors under variable uniaxial stress," *SSE*, vol. 57, no. 1, pp. 52–60, 2011.

[3] S. Endler, "Bending-stress management by stacking of ultrathin integrated circuit chips," Ph.D. dissertation, University of Stuttgart, 2012.

[4] M. Chu, T. Nishida, X. Lv, N. Mohta, and S. E. Thompson, "Comparison between high-field piezoresistance coefficients of Si metal-oxide-semiconductor field-effect transistors and bulk Si under uniaxial and biaxial stress," *J. Appl. Ph.*, vol. 103, no. 11, pp. 113 704–113 704, 2008.

[5] Y. Chen, R. C. Jaeger, and J. C. Suhling, "CMOS sensor arrays for high resolution die stress mapping in packaged integrated circuits," *IEEE Sensors Journal*, vol. 13, no. 6, pp. 2066–2076, June 2013.

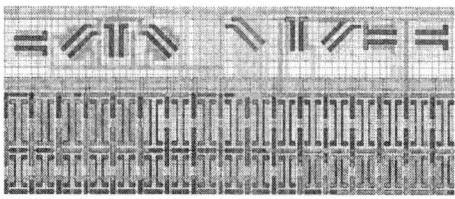

Fig. 8. A section of the sensor layout. All transistors, except the orthogonal pairs, have the same orientation. The final sensor size is 4.8×4.8 mm^2.

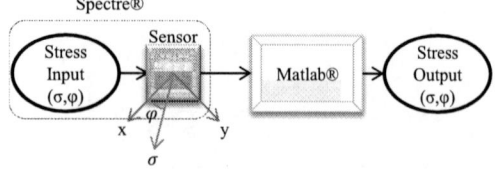

Fig. 9. Flow diagram for simulation. Stress magnitude and angle are supplied to the sensor in circuit simulator, and then sensor output is analyzed for performance figures.

A Low-Power Sensor Interface Circuit for Remotely Powered Implants

Xiao Liu and Catherine Dehollain

Radio Frequency Integrated Circuit Group (RFIC)
Ecole Polytechnique Fédérale de Lausanne (EPFL)
CH-1015 Lausanne, Switzerland
E-mail: xiao.liu@epfl.ch, catherine.dehollain@epfl.ch

Abstract—This paper proposes an interface circuit for wirelessly powered implants for mice temperature monitoring. The circuit contains the read-out circuit and a data converter. By using a novel approach to cancel the non-linear effects of thermistor sensors, the circuit achieves 27dB higher SFDR. The system is designed in UMC CMOS 0.18μm technology, which has 200KHz sampling rate, 8-bit resolution, and total power-consumption of 29μW. The successive approximation analog-to-digital convertor (SAR ADC) is fabricated. The measurement result shows an Effective Number of Bits (ENOB) of 7.

I. INTRODUCTION

With the advance of RFID technology, a continuously increasing number of remotely powered devices, which wirelessly power the sensor and transmit its output, are used in bio-medical applications [1]. To monitor bio-signals, a remotely powered sensor node includes the RF circuits, which transmit data and receive power, and the interface circuits which receive analog data from sensor and convert it to digital codes before being transmitted. Compared with traditional battery-powered devices which require replacing battery, remotely powered devices are capable of consistent power supply and thus fit the requirements of long-term implantable or wearable devices.

The interface circuits for such device have two major challenges, the limitation of power budget and compatibility to handle different bio-signals. Because of the limited power efficiency of wireless power transmission, the power consumption is usually kept at micro-watt level. Also, in order to adapt the vital bio-signals for disease identification or scientific research, such as EEG, heartbeats, blood pressure, and body temperature [2], the circuits needs to realize a moderate range of bandwidth and amplitude for various signals [3].

This paper focuses on a temperature monitoring application, proposing a low-power 8-bit SAR ADC, which operates at 200KHz sampling frequency with reduced digital-to-analog convertor (DAC) capacitors and no amplifiers.

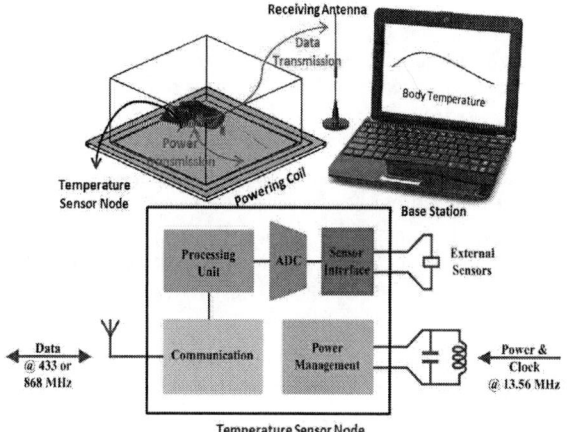

Figure 1. System overview

II. SYSTEM OVERVIEW

The main aim of this system is to monitor the temperature change in a particular tissue inside a mouse without interference on its normal activity. Attaching cables or replacing batteries would lead to inflammation, which can influence the temperature. Moreover, the use of a cable will affect the movements of a live animal. Therefore, it is necessary to develop a wireless powered sensor node.

The system is composed by two main parts, the interface blocks for sensors, and RF circuits for power and data transmission. An example of such system is shown in [4]. Fig.1 shows the system overview of proposed application. The sensor node is implanted inside the mouse. A coil is placed under the cage, generating an electro-magnetic field, through which the implant absorbs power. At the same time, an antenna is placed near the cage, which receives the signal transmitted back from the implant. In our system, the power signal is at 13.56MHz and the communication signal is at 868MHz. Total available power for the interfacing circuits is 100μW.

978-1-4673-4579-8/13 $31.00 © 2013 IEEE

As shown in Fig. 2, the temperature monitoring circuits are composed by read-out circuits followed by a SAR ADC, where R_x is the sensor resistance, R_{ref} is the reference resistance, and $V_{bias1,2}$ provides power to the resistive bridge.

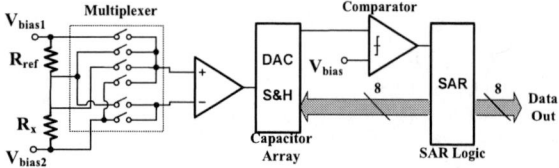

Figure 2.   Block diagram of interface circuits

A four-wire measurement topology is used [5], where a multiplexer switches the input signal between the sensor output and reference voltage. The input could also be set to short-circuit to measure the offset voltage of interface circuits.

An amplifier is put after the read-out circuits, in order to match the dynamic ranges of sensor output with that one of the ADC. And also it acts like a low-pass filter, which cancels the out-band noises.

In this application, a SAR ADC is used because of its low power consumption. The power scales with clock frequency, and thus could adapt different signal frequencies without wasting power.

After converting the signal into digital domain, it would be processed by a transmitter and sent back to base station.

A. Sensor Non-linearity Compensation

The 100K6MCD1 thermistor probe from Measurement Specialties Company is used as temperature sensor for its high sensitivity and relatively small volume. However, the thermistor sensor has a high non-linear characteristic [6]. Approximately, the relationship between sensor resistance and temperature can be expressed by

$$\frac{1}{T} = C_1 + C_2 \ln(R_x) + C_3 \ln(R_x)^3 \qquad (1)$$

where T is the temperature, $C_{1,2,3}$ are constant coefficients, and R_x is the thermistor resistance. The function is plotted in Fig.3. For this application, the temperature inside the mouse varies from 24C° to 48C° range with a resolution of 0.1C°, which equals to 8-bits accuracy. The sensor's response is highly non-linear in this range, which has a spurious free dynamic range (SFDR) of only 24dB. To achieve desired resolution, calibration is needed.

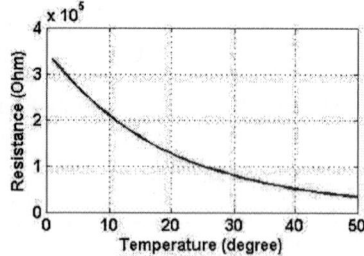

Figure 3.   Non-linear characteristics of the thermistor

Conventionally, the signal is calibrated after analog to digital conversion. The sensor's non-linearity is maintained during the conversion. Afterwards, the temperature signal is recovered according to the characteristic function of sensor.

The interface circuits would add new components into the sensor's non-linearity. This effect may change the characteristic function, and degrade the accuracy of recovered temperature signal. Besides, this method requires the interface circuits to identify the minimum step size. As shown in Fig.3, at 24C°, the step size is 505Ω per 0.1C°, and it decreases to 154Ω at 48C°. Accordingly, the interface circuits need a step size of 154Ω over 68.3kΩ range, which requires at least 9-bits resolution.

When an extra 1-bit resolution is added, it leads to doubled capacitor size, 12% lower conversion rate, more power consumptions and increased complexity of the interface circuits.

Since the power consumption is critical in wireless applications, a simple technique is used to avoid the extra resolution. The idea is to cancel partially the sensor non-linearity with a non-linear resistor bridge, which compensates the signal before conversion. Therefore, the ADC does not need extra resolutions or calibration afterwards.

The voltage response of resistive bridge in Fig.2 is expressed as

$$V_x = \frac{R_x}{R_x + R_{ref}} \cdot (V_{bias1} - V_{bias2}) \qquad (2)$$

where V_x is the output signal of the bridge. As the function shows, the denominator is affected by R_x, which leads to non-linear relationship from R_x to V_x. This effect usually leads to a decrease in SFDR. However, for a thermistor probe, this effect helps to compensate its own non-linearity. By choosing the right reference resistor, the SFDR of sensor signal can be increased to the required accuracy.

As shown in Fig.4, with a 24-48C° sinus temperature signal applied as input, the SFDR of resistive bridge output V_x is plotted versus R_{ref}. By biasing the sensor with 48kΩ reference, the SFDR of the signal is actually increased to 51dB after compensation, which already satisfies the required 8-bits accuracy.

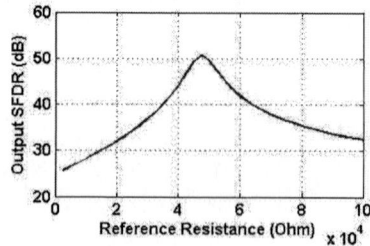

Figure 4.   SFDR after compensated by different reference resistor

By using this technique, the signal can be processed with an 8-bit interface. Without necessity for extra resolutions, the capacitor size, power consumption, and conversion rates are all optimized.

B. Pre-amplifier

As shown in Fig.5, a folded-cascode amplifier is used as the pre-amplifier. This topology provides a large dynamic range, higher gain, and relatively small power consumption. $V_{b1,b4}$ are generated by current mirror, and $V_{b2,b3}$ are DC bias voltages.

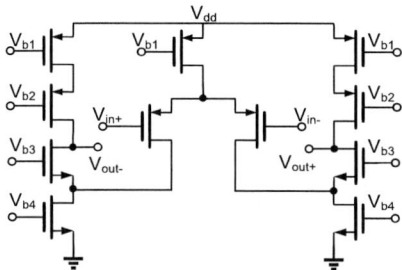

Figure 5.　Schematic of the pre-amplifier

C. SAR ADC structure

The SAR ADC is composed by 3 main blocks, including a DAC, a comparator and SAR logic circuits. The conversion is done by comparing the sampled signal with a series of different voltage levels, which are combined with a fraction of the total dynamic range V_{ref}. The comparison starts from ½ V_{ref}, and gradually scales down to the required accuracy. Conversion result is stored as a set of digital codes D_{1-k}.

The temperature signal lies in very low frequency, however, some margin is left to make the circuit adaptable for chopping technique, which leads to a minimum 5KHz input frequency. The sampling frequency is 200KHz, which could over-sample the input signal by factor up to 20. To fully cover the temperature range, an 8-bit resolution is needed.

1) DAC capacitor array

An array of capacitors is implemented for both sampling capacitor and DAC. As shown in Fig.6, it is composed of two capacitor arrays, where C_u stands for the unit capacitance. A cascade capacitor is placed between the arrays, which results in a smaller total capacitance [7].

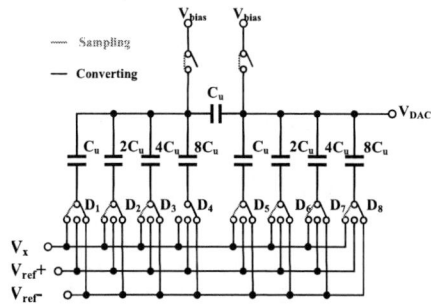

Figure 6.　Schematic of DAC capacitor array

The power consumption of the ADC is mainly composed by switching of digital circuits, and the charging/discharging of the DAC array [8]. Compared with traditional binary capacitor array, the capacitance is reduced from 51pf to 6.2pf,

which leads to a lower power consumption and faster settling speed.

2) Comparator

A dynamic comparator is used because it has no idle current consumption.

As shown in Fig.7, the input signals $V_{in\pm}$ are feed to a differential n-MOS pair $MN_{1,2}$, and by cross-connected p-MOS transistors, a positive feed-back loop is formed as the load.

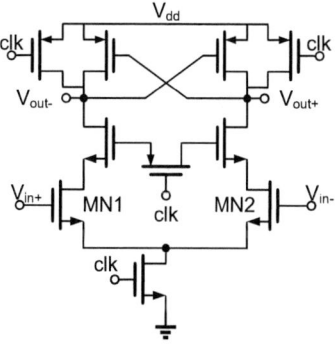

Figure 7.　Schematic of the comparator

3) Conversion Timing and SAR Logic

The ADC receives a high frequency clock signal from the RF circuits, from which it generates two internal clocks. Aside from a main clock, another clock signal is generated with a ¼ cycle delay. With these two clock signals, the ADC operation is divided into several phases, which is shown in Fig.8.

The input signal is sampled every 10 clock cycles. During the following 8 clock cycles, the ADC decides the 8 bits from MSB to the LSB. For each cycle, the first quarter is used for DAC to generate and settle to the required voltage. The two quarters afterwards are used for comparator settling. When the comparison is done, the result is restored and the comparator is reset. After all bits are decided, the ADC outputs a valid code.

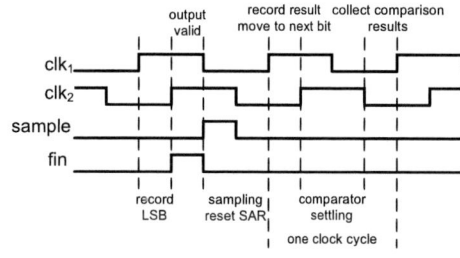

Figure 8.　Timing of ADC conversion

III. SIMULATION AND MEASUREMENT RESULTS

The interface circuit is designed in 0.18um CMOS technology. The circuits are validated by Cadence simulation, and one prototype ADC is fabricated.

The read-out circuit is as shown in fig.2, where R_{ref} is chosen as 48kΩ to suppress sensor non-linearity. In order to reduce the power-consumption, voltage between V_{bias1} and V_{bias2} equals to 900mV, which leads to 5.4μW power at 24C°.

The pre-amplifier is designed and currently under fabrication. The pre-amplifier has 55dB gain, 80° phase margin and 34MHz gain-bandwidth. The maximum gain covers 5KHz band. The power-consumption is 3.6μW. The AC-simulation results are shown in Fig.9.

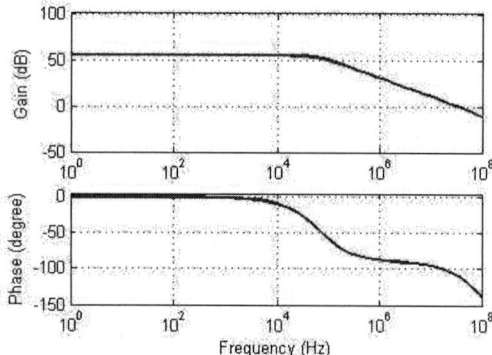

Figure 9. AC simulation results for pre-amplifier

An 8-bits SAR ADC is fabricated, the layout and fabricated chip is shown in Fig.10.

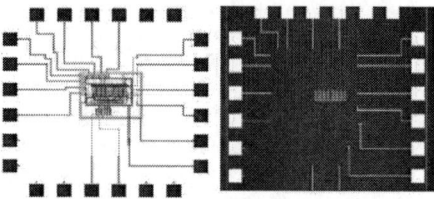

Figure 10. 8-bit SAR ADC layout and chip view

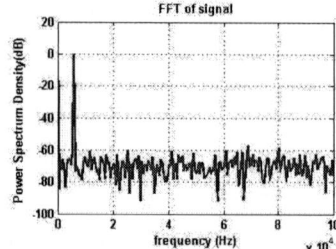

Figure 11. ADC output spectrum by measurement

To measure its accuracy, the ADC's response to a full-range 5KHz input sine wave, from 450mV to 1.35V, is recorded. By analyzing the power spectrum density, the SNR and ENOB are calculated.

The fabricated ADC operates at a sampling rate of 200KHz. The spectrum of output signal is shown in Fig.11. The fabricated ADC achieves a SNR of 43.9dB, which leads to an ENOB of 7.0. The total power consumption is 20μW.

The measurement result of fabricated ADC is listed in Table I.

TABLE I.   MEASUREMENT RESULTS OF FABRICATED ADC

Process	0.18 μm CMOS	Resolution	8 bits
Supply	1.8 V	Power Consumption	20 μW
Sampling Rate	200 KHz	SNDR	43.9 dB
ENOB	7.0 bits	Core Area	303×405 μm²

The total power consumption of the system is 29μW, including the read-out circuit, pre-amplifier, and measured power-consumption of the ADC. It is within the power-budget.

IV. CONCLUSION

This paper proposed a low-power inter-face circuit for wirelessly powered temperature monitoring. The system uses a resistive bridge to cancel the non-linearity of thermistor, which increases the SFDR by 27dB for this application. The read-out circuits and ADC is validated with Cadence simulation and measurement results. The performance of the proposed interface is adequate for the temperature monitoring application.

ACKNOWLEDGMENT

The research leading to this result has received funding from the Swiss National Funding (SNF) NANO-TERA initiative for the PlaCiTUS project.

REFERENCES

[1] I. Korhonen, J. Parkka, and M. Van Gils, "Health monitoring in the home of the future," IEEE Engineering in Medicine and Biology Magazine, vol.22, no.3, pp. 66-73, 2003.

[2] B. Hoit, S. Kiatchoosakun, J. Restivo, D. Kirkpatrick, K. Olszens, H. Shao, Y. Pao, and J. Nadeau, "Naturally Occurring Variation in Cardiovascular Traits among Inbred Mouse Strains," Genomics, vol.79, no.5, pp.679–685, May 2002.

[3] Z. Xiaodan, X. Xiaoyuan, Y. Libin, and L. Yong, "A 1-V 450-nW Fully Integrated Programmable Biomedical Sensor Interface Chip," IEEE Journal of Solid-State Circuits, vol.44, no.4, pp. 1067-1077, April 2009.

[4] C. Peng, N. Chaimanonart, W.H. Ko, and D.J. Young, "A Wireless and Batteryless 10-Bit Implantable Blood Pressure Sensing Microsystem With Adaptive RF Powering for Real-Time Laboratory Mice Monitoring," IEEE Journal of Solid-State Circuits, vol.44, no.12, pp.3631-3644, December 2009.

[5] L. Xiujun, and G.C.M. Meijer, "A High-Performance Interface for Platinum Temperature Sensors with Long-Cable," 9th International Conference on Solid-State and Integrated-Circuit Technology, pp.1709-1712, October 2008.

[6] B.W. Mangum, "Triple Point of Succinonitrile and Its Use in the Calibration of Thermistor Thermometers," Review of Scientific Instruments, vol.54, no.12, pp.1687-1692, December 1983.

[7] F. Maloberti, Data Convertors, Springer Int., 2007, pp.107–110.

[8] P. Kamalinejad, S. Mirabbasi, and V.C.M. Leung, "An Ultra-Low-Power SAR ADC with An Area-Efficient DAC Architecture," IEEE International Symposium on Circuits and Systems, pp.13-16, May 2011

PRIME 2013, Villach, Austria

Session W2B – Sensors

Low-Noise Low-Offset Current-Mode Hall Sensors

Hadi Heidari, Umberto Gatti, Edoardo Bonizzoni, and Franco Maloberti

Dipartimento di Ingegneria Industriale e dell'Informazione

University of Pavia, Pavia, Italy

E-mails: hadi.heidari@unipv.it, gattiu@alice.it, edoardo.bonizzoni@unipv.it, franco.maloberti@unipv.it

Abstract—The performances of a current-mode Hall sensor featuring output current signals are discussed. The current-mode approach is analyzed by applying for first time to our best knowledge the spinning current technique to Hall plate working in current-mode to eliminate offset and $1/f$ noise. Among different geometries that have been studied and simulated using COMSOL Multiphysics™, cross-shaped model displayed the lowest noise and residual offset and the best sensitivity. The COMSOL results determined a behavioral model implemented in Verilog-A for simulations in the Cadence environment. Simulations results achieved in COMSOL and in Cadence environment show the potentiality, thus demonstrating the effectiveness of the approach, for a possible use of the device with remarkable performances.

I. Introduction

Hall effect sensors have been used for various applications such as current sensing, position detecting, electronic compass and contactless switching within automotive and industrial electronics, which are fully compatible with commercial integrated circuit technologies of CMOS process, [1-3].

Signal to noise ratio (SNR) and offset are important features in Hall sensors performance evaluation. Several techniques have been developed in order to improve these characteristics. Most of them, [1-4], are based on conventional voltage-mode Hall transducers. When used in the voltage-mode, Hall sensors convert the magnetic field to be measured into an output voltage, as shown in Fig. 1(a). For many years voltage-mode Hall sensors have been absolutely dominant in most of the applications.

This paper uses a cross-shaped Hall plate in the current-mode in order to have current and not voltage as an output signal. The proposed scheme is illustrated in Fig. 1(b). A bias current is injected into two terminals of the Hall plate (A and B in this figure) and, in presence of a magnetic field, two differential output currents are available at the other two terminals. The current-mode Hall sensor has advantages when compared to its voltage-mode counterpart such as less number of terminals, easier ultimate miniaturization of the device, less parasitic effects in its high-frequency operation, no variation of the terminal potentials and so no influence of the parasitic capacitances, [1].

This paper considers several current-mode Hall effect plates modeled and evaluated with respect to noise, offset and sensitivity using 2D COMSOL Multiphysics™, [5]. The cross-shaped model emerged as the optimum plate to fit the lowest noise and residual offset and the best sensitivity. The symmetrical cross-shaped Hall plate is widely used because of its high sensitivity and immunity to alignment tolerances

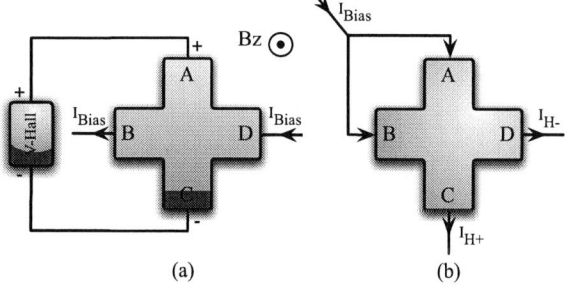

Fig. 1. (a) Hall plate operating in the voltage-mode. (b) Proposed Hall plate working in the current-mode.

resulting from the fabrication process. Generally, the output offset strongly limits the DC resolution of the sensors. The use of the crossed-shaped Hall plate and of the proposed current-mode approach enables to first time at our best knowledge the use of current spinning technique, [6], in current-mode Hall sensors to reduce offset and $1/f$ noise.

This paper also analyzes the influence of the current injection and current output signals on the Hall effect sensors performances, including noise and offset, with the aid of COMSOL Multiphysics simulations. Moreover, an accurate 8-resistor network model for the cross-shaped Hall plate is described in Verilog-A and tested in a Cadence environment, [7]. Simulation results obtained in COMSOL and in Cadence environment show excellent system potentiality.

The paper is organized as follows: in the next Section the novel current-mode technique is discussed. In Section III, simulations in 2D COMSOL Multiphysics environment of the Hall plate with and without mismatch are presented. Section IV presents and discusses the model of Hall plate implemented in Verilog-A. Comparison of simulations results achieved in COMSOL and Cadence environments are also given. Section V draws some conclusions.

II. Methodology

Typically, Hall plates are used in voltage-mode. This means that the magnetic field to be measured is converted into an output voltage. The idea behind a current-mode Hall sensor is to have current and not voltage as an output signal.

The physical structure of the proposed current-mode Hall sensor is exactly the same of modern devices, with the

978-1-4673-4579-8/13 $31.00 © 2013 IEEE

same possibility of compensating for the offset caused by mismatch (current spinning). The difference is in the driving and extracting the signal.

Consider the device of Fig. 1(a); the sensor bias current I_{Bias} flows from one arm to the other in front (the symmetrical structure is for current spinning). A magnetic field B_z gives rise to the Hall voltage across the two orthogonal arms. Here, it is commonly accepted that the Hall voltage can be calculated as

$$V_H = S_I I_{Bias} B_z \qquad (1)$$

where I_{Bias} is the sensor bias current, B_z the perpendicular magnetic field and S_I the current-related sensitivity. The latter is given by

$$S_I = G \frac{r_H}{qnt} \qquad (2)$$

where r_H is the Hall factor, q the elementary carrier charge, n the carriers density, t the thickness of the sensor, and G a geometrical correction factor having a value in the [0, 1] interval, depending on the dimensions of the sensor.

The connection realized in the scheme of Fig. 1(b) injects the current laterally in two consecutive arms and a magnetic field causes an unbalancement of the two output currents. A difference of these output currents could be represented by an equivalent current source of a Hall current, I_{Hall}. The current-mode Hall sensor principle has been already described in [3], where it has been found that the Hall current is

$$I_{Hall} = \mu_H \frac{w}{l} I_{Bias} B_z \qquad (3)$$

Here, μ_H is the Hall mobility of majority carriers, I_{Bias} is the total bias current, B_z is a normal magnetic field and $\frac{w}{l}$ is the width-to-length ratio of the plate.

This mode of operation has been extensively studied to estimate the benefit of the current-mode approach. The output currents (I_{H+}, I_{H-}) are calculated as:

$$I_{H+} = \frac{I_{Bias}}{2} + \frac{I_{Hall}}{2} \qquad (4)$$

$$I_{H-} = \frac{I_{Bias}}{2} - \frac{I_{Hall}}{2} \qquad (5)$$

The Hall current (I_{Hall}) is also proportional to the external magnetic field (B_z), biasing current of the Hall plate (I_{Bias}) and magnetic resistance coefficient (β). This current, for the cross-shaped Hall plate, can be expressed as:

$$I_{Hall} = \frac{\beta B_z I_{Bias}}{1 - (\beta B_z)^2} \qquad (6)$$

In (6), β is magnetic resistance coefficient in presence of a magnetic field and is calculated as:

$$\beta = \frac{R_{(B_z)} - R_{(B_z=0)}}{R_{(B_z=0)} B_z} \qquad (7)$$

where $R_{(B_z=0)}$ and $R_{(B_z)}$ define the Hall plate resistance in absence and presence of an external magnetic field, respectively.

TABLE I
MODEL PARAMETERS

Symbol	Value	Parameter
q [C]	$-1.602e - 19$	Electron Charge
n [cm^{-3}]	$-2.6e16$	Doping
μ [cm^2/Vsec]	1200	Mobility
sigma0 [S/m]	-q*n*μ	Silicon Conductivity
R_h [m^3/C]	$-1/(q*n)$	Hall Coefficient
B_z [mT]	$0 - 20$	Magnetic Field
V_0 [V]	0.02	Applied Voltage
t_si [m]	$0.3e - 6$	Silicon Tickness
I_0 [μA]	12	Input Current

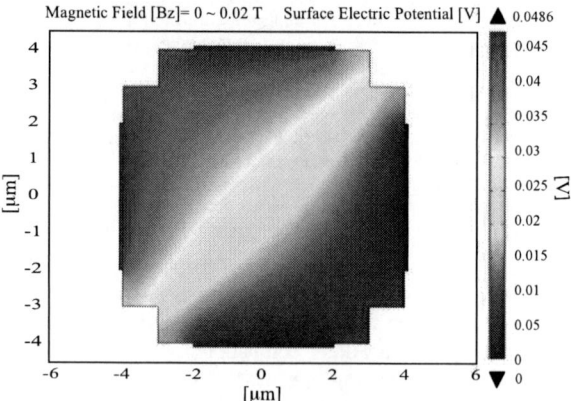

Fig. 2. COMSOL simulation.

III. SIMULATION MODEL

The current-mode technique applies to a two-dimensional model of the Hall plate simulated in COMSOL Multiphysics, [5], with parameters summarized in Table I. Fig. 2 shows the model geometry of the Hall plate: the maximum width and length is 8 μm. The figure also shows the surface electrical distribution when a magnetic field of 20 mT is applied. The simulation uses the nominal bias current of 12 μA injected in terminals A and B. The model geometry of the Hall plate has been simulated both without any mismatch and with mismatch. Results are summarized in the following sub-sections.

A. The ideal Hall plate (without mismatch)

Consider again the scheme of Fig. 1(b). Fig. 3 shows the simulated input and output currents of the Hall plate with no-mismatch. The solid line shows the input currents of A and B (equal to 6 μA) terminals while the dash-dotted and the dotted lines represent the currents at C and D terminals. These output currents increase and decrease, respectively, by changing the magnetic field within the 0 to 20 mT range. Table II summarizes the simulation results plotted in Fig. 3. The maximum differential output current is 13.3 nA for a magnetic field equal to 20 mT. These current levels are pretty

PRIME 2013, Villach, Austria

Session W2B – Sensors

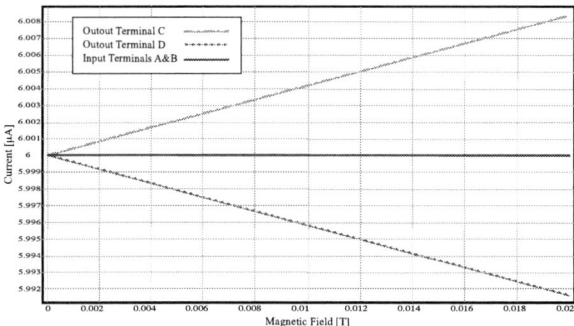

Fig. 3. Simulated input and output currents of the current-mode Hall plate.

TABLE II
SIMULATION RESULTS

Magnetic Field	Current[μA]		
B_z [T]	A+B	C	D
0	12	6	6
0.001	12	6.00033	5.99967
0.002	12	6.00067	5.99933
0.003	12	6.001	5.999
0.004	12	6.00133	5.99867
0.005	12	6.00166	5.99834
0.006	12	6.002	5.998
0.007	12	6.00233	5.99767
0.008	12	6.00266	5.99734
0.009	12	6.00299	5.99701
0.01	12	6.00333	5.99667
0.011	12	6.00366	5.99634
0.012	12	6.00399	5.99601
0.013	12	6.00432	5.99568
0.014	12	6.00466	5.99534
0.015	12	6.00499	5.99501
0.016	12	6.00532	5.99468
0.017	12	6.00565	5.99435
0.018	12	6.00599	5.99401
0.019	12	6.00632	5.99368
0.02	12	6.00665	5.99335

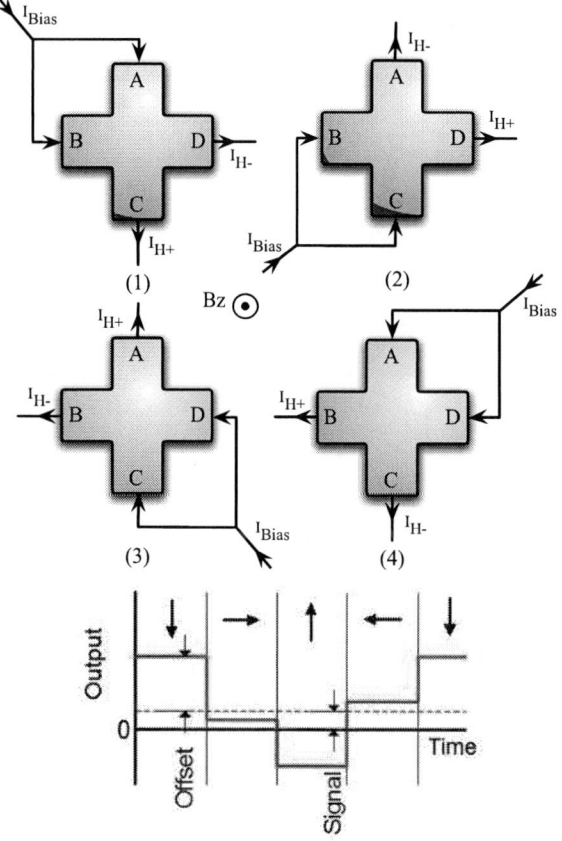

Fig. 4. Spinning current technique and output offset of Hall plate during four states.

small, but the output signal can be increased by integrating the current signal over a given period of time.

B. Mismatched Hall plate and spinning current

Mismatch in Hall plate dimensions due to possible masks misalignment during fabrication limits accuracy and causes offset in the output signal.

As known, the spinning current technique allows strongly reducing the offset of Hall sensors, [6]. The use of the cross-shaped Hall plate and the proposed current-mode approach enables to apply the current spinning technique. The current spinning interchanges periodically the output and supply terminals of the Hall plate so that the input bias current injecting

point is rotated in each state whereas the offset appears at the output terminals. The plate is clocked with four phases and the output currents are summed.

Fig. 4 shows the four states of input and outputs for each 90 ° rotation. From the figure it can be also observed the output offset of Hall plate during the four states. After the fourth phase, it is expected that the average of the offset will be zero.

Fig. 5 shows the average of output currents of the Hall plate after spinning current as simulated in COMSOL. Simulations include a 0.01-μm mismatch in the C terminal of the Hall plate. For the spinning operation there are four phases, summarized in Table III. At the beginning, the bias current is injected into A and B terminals (phase 1). During phase 2, after 90 ° rotation, the bias current is injected into B and C terminals and so on for other phases. The used spinning frequency is 1 MHz.

From Fig. 5 it can be noted that, at the end of complete spinning and average of four phases, the offset for zero magnetic field is eliminated. When increasing the magnetic field from 0 to 20 mT, the positive output (I_{H+}) rises and the

978-1-4673-4579-8/13 $31.00 © 2013 IEEE

327

TABLE III
Four phases for spinning current method

Current	Phase1	Phase2	Phase3	Phase4
I_B	A,B	B,C	C,D	D,A
I_{H-}	D	A	B	C
I_{H+}	C	D	A	B

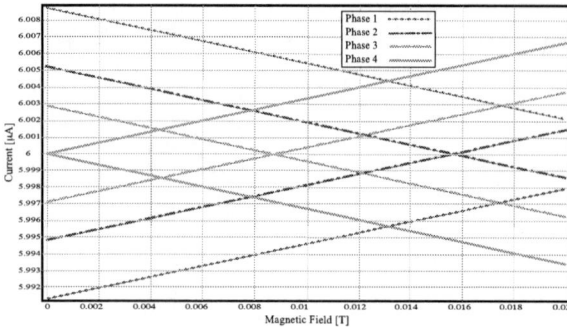

Fig. 5. Simulated average output currents (I_{H+} and I_{H-}) of current-mode Hall sensor plate with mismatch after spinning current method.

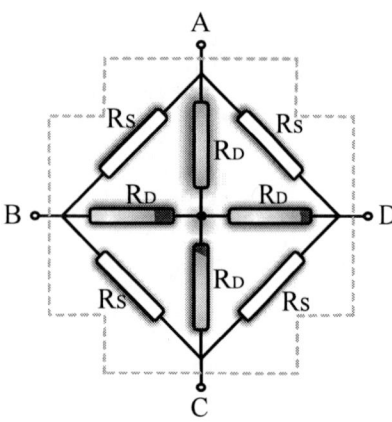

Fig. 6. The model implemented in Verilog-A using eight resistors.

negative output (I_{H-}) decreases.

IV. Verilog-A Model and Simulations Comparison

The Hall plate has been modeled and described using the Verilog-A language so that it can be simulated in the Cadence environment as well, [7].

Fig. 6 shows the Hall plate model. It includes four electrical terminals (A, B, C and D) and eight resistors (four side resistors, R_S, and four diagonal resistors, R_D). The values of these resistors is controlled by three parameters: the external magnetic field, the initial value of resistors, R_0, and the magnetic resistance coefficient, β. The extraction of parameters

from device simulations is as follows. The initial value of

TABLE IV
Comparison between the simulations of the Hall plate in COMSOL and Verilog-A

Magnetic Field	COMSOL		VERILOG-A	
B_z[mT]	$I_D[\mu A]$	$I_C[\mu A]$	$I_D[\mu A]$	$I_C[\mu A]$
0	6	6	6	6
10	6.0033	5.99667	6.003373	5.996627
20	6.00665	5.99335	6.006746	5.993254

resistors, R_0, is obtained in two steps. First step consists of grounding terminals C and D and applying 12 μA to terminals A and B. In the second step, terminals B and D are grounded and the current is applied to terminals A and C. The magnetic resistance coefficient, β, is defined as the average of initial values of resistors, R_0, in presence and absence of the magnetic field.

In order to show the correctness and accuracy of this model, the simulation results have been compared with the ones achieved with the COMSOL model, as summarized in Table IV. I_C and I_D stand for currents of C and D terminals, respectively. The results are in excellent agreement and the difference is less than 0.1%.

The simulations have been performed with 12 μA input bias current (injected in terminals A and B) while the magnetic field has been considered from 0 to 20 mT.

V. Conclusion

The study shows the effectiveness of a current-mode Hall sensor. The cross-shaped model has been simulated at the first using COMSOL Multiphysics and after that it has been modeled with an equivalent circuit using Verilog-A for behavioral simulations in the Cadence environment. Simulation and modeling results revealed that the proposed technique enables superior performance of magnetic field sensitivity, in terms of signal to noise ratio and sensitivity compared to the voltage-mode Hall sensors.

As future work it will be continued the analysis and the design of a complete Hall sensor microsystems, readout circuits, all using a conventional CMOS technology.

References

[1] R. S. Popovic, *Hall Effect Devices*, 2nd ed. Bristol, U.K.: Inst. Physics Publishing, 2004.

[2] M. Pastre, M. Kayal, H. Blanchardl, "A Hall Sensor Analog Front End for Current Measurement with Continuous Gain Calibration", *IEEE International Solid-State Circuits Conference (ISSCC)*, pp. 242-243, 596, February 2005.

[3] A. Ajbl, M. Pastre and M. Kayal, "A Current-Mode Back-End for a Sensor Microsystem", *IEEE New Circuits and Systems Conference (NEWCAS)*, pp. 466-469, June 2011.

[4] G. Boero, D. Memierre, P.-A. Besse, R.S. Popovic, "Micro-Hall devices: performance, technologies and applications", *Sensors and Actuators A*, no. 106, pp. 314-320, 2003.

[5] R. W. Pryor, *Multiphysics Modeling Using COMSOL: A First Principles Approach*. Jones and Bartlett Publishers, 2009.

[6] A. Bilotti, G. Monreal, and R. Vig, "Monolithic Magnetic Hall Sensor Using Dynamic Quadrature Offset Cancellation", *IEEE Journal of Solid-State Circuits*, vol. 32, no. 6, pp. 829-836, June 1997.

[7] K. Kundert and O. Zinke, *The Designer's Guide to Verilog-AMS*. Boston: Kluwert Academic Publishers, 2004.

978-1-4673-4579-8/13 $31.00 © 2013 IEEE

Dynamic Programming Based Grouping Method for RO-PUFs

Giray Kömürcü
National Research Institute of Electronics and Cryptology,
TÜBİTAK, 41470, Kocaeli, TURKEY
Email: girayk@uekae.tubitak.gov.tr

Ali Emre Pusane, Günhan Dündar
Bogazici University, Dept. of Electrical and Electronics Eng.
34342 Bebek, Istanbul, Turkey
Email: {ali.pusane, dundar}@boun.edu.tr

Abstract—Key generation is one of the most promising applications of Physical Unclonable Functions (PUFs), which requires 100% robust bit streams within each circuit and true randomness among a set of circuits. However, due to the noisy nature of PUFs, it is hard to provide stable outputs under changing environmental conditions, such as supply voltage and temperature. In this work, we have adapted Dynamic Programming (DP) to RO-PUFs for the first time in literature, in order to extract maximum entropy with minimum possible resource usage. Next, the robustness of all output bits is guaranteed even in unstable environmental conditions just by measuring a small subset of circuits prior to shipment. Finally, the efficiency of our method is analyzed and validated experimentally with FPGA implementation.

Keywords-PUF, Physical Unclonable Functions, Reliability, Robustness, Ring Oscillator, Dynamic Programming, FPGA.

I. INTRODUCTION

Protection of cryptographic keys is the most important issue in security operations. In conventional systems that have no constant power sources such as smart cards, keys are either transferred once and stored in non-volatile memories, or transferred to the device whenever needed. Secure transfer of keys is problematic in both situations, since it requires implementation of complex protocols to protect the keys against snooping, and non-volatile memory usage is not suitable for resource limited devices.

Physical Unclonable Functions (PUFs) offer promising solutions in the area of key generation and storage. These functions, which have the unique capability of generating chip specific signatures on the fly, were first introduced in 2001 [1]. Among various PUF types presented in the literature such as Arbiter PUFs, SRAM PUFs, Butterfly PUFs, and Glitch PUFs [2]–[7], RO-PUFs are the most convenient type for FPGA implementation [8] and exhibit higher reliability than other PUF types under changing environmental conditions [9]. This work focuses on building a 100% reliable and highly efficient PUF structure based on ROs suitable for cryptographic key generation.

Output generation mechanism of conventional RO-PUFs depends on the pairing approach, which generates one bit output with a pair of ROs [10]. To extract the maximum entropy from the system, frequency ordering of all ROs have to be used, which can generate up to $\lfloor \log_2(N!) \rfloor$ bits by using N ROs [10]. Even though this theoretical upper-bound is not achievable due to noise in the system that is causing

unreliable bits, it is still much higher than the number of bits generated in conventional systems, which is upper-bounded by $N/2$. The frequency ordering approach is used in [11] for output generation, which is called the Longest Increasing Subsequence-Based Grouping Algorithm (LISA) .

In this work, DP algorithm is adapted to RO-PUF output generation for achieving maximum entropy with minimum resources. Secondly, we introduce a parameter called predetermined frequency threshold (f_{thp}) to be used in DP for a 100% reliable PUF. Finally, the effectiveness of the proposed algorithm in terms of entropy generated, computational cost, and area utilization is analyzed. The rest of the paper is organized as follows: In Section 2, background for RO-PUFs is provided and methods of maximizing the entropy are discussed. In Section 3, DP is adapted to RO-PUF output generation for maximum entropy with highest robustness in minimum time. In section 4, experimental validation is performed. Last section concludes the paper.

II. MAXIMIZING ENTROPY AND ROBUSTNESS IN RO-PUF CIRCUITS

RO-PUFs are structures that generate output depending on the frequency differences of identically laid out ROs. For continuous oscillation, an odd number of inverting delay stages are connected serially forming a ring structure. In vast majority of RO-PUFs, one bit output is generated by comparing the frequencies of two ROs [12]–[14]. In such systems, the area of the implementation is usually high due to the low entropy utilization by the structure [11].

To overcome the entropy extraction problem, comparing more than two ROs at a time is required. The first step is grouping of the ROs that will be compared at once. The frequencies of ROs in each group should be adequately seperated from each other, preventing changes in frequency ordering due to noise in the system and temperature or supply voltage fluctuations. For a group of M ROs, $M!$ different orderings that are equally likely may occur. By mapping each different ordering to a bit stream, $\lfloor \log_2(M!) \rfloor$ bits can be generated from each group [10]. In this approach the main problem is to form the largest possible groups from the set of ROs implemented in the circuit.

In the literature, LISA is used to overcome the grouping problem [11]. In LISA, noise in the system is compensated with the f_{th} value and the effect of environmental changes are compensated by measuring each RO in two extreme conditions and using both values in the algorithm. Even though this approach guarantees robustness, it is quite complex, since it requires two measurements for all ROs in all circuits. Also, the computation cost increases since two frequencies are used per RO in the algorithm. In our proposed method, the frequency deviation that may result from temperature and supply voltage changes are compensated by including them in the f_{th} in LISA and this new parameter is called f_{thp}. Therefore, the value of f_{thp} will be higher in DP. With this approach ROs are measured once under normal operating conditions and DP works with only one frequency value per RO.

The key point in this new approach is determining the value of f_{thp}. If the value determined is less than the required amount, bigger groups will be formed, but their RO frequencies will not be far enough from each other. In this case, ordering may change under certain conditions preventing the PUF outputs from being 100% reliable. On the other hand, if the value of f_{thp} is higher than the required amount, smaller but more reliable groups will be formed and the extracted entropy, hence the number of output bits generated, will be lower. In order to determine the optimum f_{thp}, a small subset of circuits is formed and the frequencies of the ROs in this subset is measured at two different temperatures. Normally, all ROs will be slower at higher temperatures, but their frequency change will be different from each other. This difference is the reason for unreliable bits and should be used as minimum f_{thp} in the PUF output generation algorithms. A formal structure of determining the f_{thp} is given in Algorithm 1.

Data:
1. A list of minimum RO frequencies $fmin[n]$.
2. A list of maximum RO frequencies, $fmax[n]$.
Result: f_{thp}
for $i \leftarrow 1$ **to** n **do**
   | $diff(i) = fmax(i) - fmin(i)$;
end
$f_{thp} = max(diff) - min(diff)$;
     Algorithm 1: Determining f_{thp} in pseudo code

With the proposed method, a realistic value of f_{thp} is determined. Since a subset of all circuits is used, a small amount of overhead should be added to this value for guaranteed robustness in all manufactured circuits.

III. Adapting dynamic programming to RO-PUFs

Even though LISA extracts the maximum available entropy from the system with guaranteed robustness, it is very costly in terms of computational power, mainly due to redundant search for ROs to form the optimum groups. By using LISA, it may be hard to achieve low computation times for output generation on devices with limited capability. In addition to this, it requires two measurements for each circuit at two extreme temperatures which complicates and increases the cost of initialization and output generation phases.

To overcome the drawbacks of LISA without decreasing its capability of extracting entropy and achieving high robustness levels, we have adapted DP to this problem. With this approach, the computational complexity of output generation decreases considerably and the requirement to measure each circuit at two extreme temperatures during the initialization phase is avoided. Measuring a small subset of circuits at extreme conditions is enough to determine f_{thp}, which will guarantee reliability with just one measurement for the rest of the circuits.

The inputs to the DP for RO grouping is similar to the inputs of LISA. Each RO frequency is measured during the initialization phase and given as input to the algorithm. In addition to this, f_{thp} is also used by the algorithm for generation of reliable outputs. By using this f_{thp} parameter, it is possible to avoid measuring and working on two different frequencies for each RO, reducing the complexity of the initialization phase and PUF output generation.

DP algorithm has three steps. In the first step, ROs are sorted according to their frequencies in increasing order and $Fsorted[n]$ list is created. In the second step, for each RO, the nearest RO whose frequency is at least f_{thp} higher is found and a linked list is created, $list[n]$. In the third step, groups are formed using the linked list, that satisfy the requirements of maximum entropy and 100% robustness even in unstable environmental conditions. In this step, the algorithm starts from the first position in the list, $list[1]$, and groups RO_1 with the RO that $list[1]$ shows, RO_j. Then, DP continues by jumping to the position in $list$ of last grouped RO, j and group the one that $list[j]$ shows. This continues until last position is reached in $list$. After the first run, the first group is formed and this step is repeated until all ROs are grouped. During the grouping process, if the RO that the list shows is grouped, the nearest ungrouped RO through the end of list is added to the group. DP algorithm is illustrated with a small data set in Figure 1. In this figure, $FreqRO[n]$ represents the frequencies of ROs in MHz. Also, f_{thp} is selected as 1.5 MHz. After the algorithm is applied, 3 distinct groups are formed satisfying the reliable PUF output generation conditions with maximum entropy extraction. The pseudo code of the DP approach is presented in Algorithm 2.

The reduced complexity of the proposed DP approach is a result of avoiding the redundant RO search done by the LISA. For each addition to a group, LISA searches over all ROs that have at least f_{th} higher frequency than the last group member. Once a possible RO is identified, it is added to the group. In the DP, benefiting from the fact that we are operating on a sorted RO frequency list, the first qualifying candidate is chosen. This simplifies the task, since it amounts to choosing the nearest item to the last group member in the sorted RO frequency list. This search is illustrated in Figure 2, where the last group member is assumed to be i. The LISA algorithm searches over the region of the remaining sorted RO list, where there is at least f_{th} frequency difference, whereas the DP approach simply chooses i''. Choosing the first available candidate seems like a suboptimal solution; however, as we

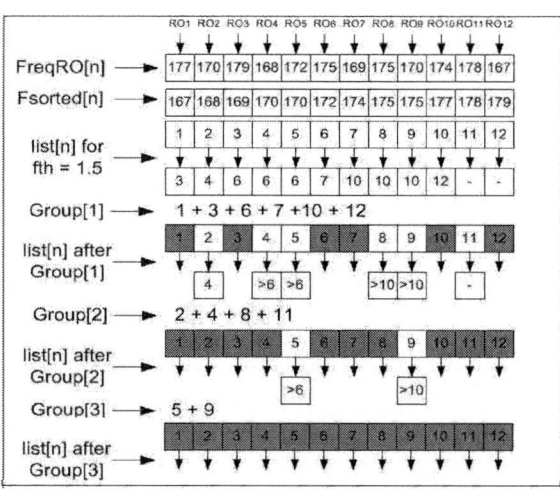

Fig. 1. DP sample execution for 12 elements.

Data:

1. A linked list of ROs with their frequencies measured under normal operating conditions, $FreqRO[n]$.

2. f_{thp} for robustness

Result: Groups of ROs.

Sort $FreqRO[n]$ by frequency in increasing order: $Fsorted[n]$

for $i \leftarrow 1$ to $n-1$ **do**

$\quad$ find the nearest element $Fsorted[j]$ that is $(Fsorted[i] < Fsorted[j]\text{-}f_{thp})$ and link i to j in $list[n]$

end

$i = 1$

while *ungrouped RO exists* **do**

$\quad$ **if** *ROi is ungrouped* **then**

$\quad\quad$ Add ROj to the group of ROi

$\quad\quad$ Jump to $ROj(i = j)$

$\quad$ **end**

$\quad$ **if** *ROi is grouped* **then**

$\quad\quad$ Increment i until ROi is ungrouped

$\quad$ **end**

$\quad$ **if** *i=n and still ungrouped RO exists* **then**

$\quad\quad$ $i = 1$

$\quad$ **end**

end

Algorithm 2: Dynamic Programming in pseudo code

will prove next, this approach is indeed optimal in the sense that it always adds the same group member as LISA, thanks to the sorted nature of the input RO frequency list. We attempt to prove this via proof by contradiction.

Let $\{S_i\}$ denote the largest group that starts at position i (and ends at position n) and g_i denote the size of this group, $1 \leq i \leq n$. Also note that f_i denotes the frequency for the corresponding RO. The LISA algorithm searches over all possible future positions to obtain i' that belongs to the

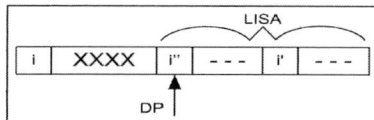

Fig. 2. Search for largest group

TABLE I
MAXIMUM FREQ. DEVIATION OF ROS DUE TO TEMP. CHANGE

Initialization Temperature (°C)	Min./Max. Operation Temp. (°C)	Max. Frequency Deviation (kHz)
20	0	296
20	40	242
20	60	362
20	80	661
20	100	985

largest group, i.e., $i' = \arg\max(g_{i'})$ for all $i' > i$, such that $f_{i'} - f_i > f_{th}$. In this case, we can form the largest group for position i, by simply adding it to this group:

$$g_i = 1 + g_{i'}, \qquad (1)$$
$$S_i = S_{i'} \cup \{i\}. \qquad (2)$$

On the other hand, the DP approach simply looks for the smallest i'' such that the f_{thp} condition is satisfied using $i'' = \arg\min(i'')$ for all $i'' > i$, such that $f_{i''} - f_i > f_{thp}$. The corresponding group can be formed by

$$g_i = 1 + g_{i''}, \qquad (3)$$
$$S_i = S_{i''} \cup \{i\}. \qquad (4)$$

Our claim is that the newly formed group using the DP approach is at least as large as that, formed by the LISA algorithm, i.e., $g_{i''} \geq g_{i'}$. Now, let's assume this is incorrect, i.e., assume that $g_{i''} < g_{i'}$. In this case, we can simply take $S_{i'}$ and replace i' with i'' using $S_{i''} = S_{i'} \setminus \{i'\} \cup \{i''\}$. This is indeed a valid set, since $f_{i''} \leq f_{i'}$ and any group that has the position i' as its lowest value would be still valid if this change is completed. This leads to the fact that we have a group that starts at position i'' and has the same number of elements as that of $S_{i'}$, i.e., $g_{i''} = g_{i'}$. This is a contradiction, since we had started with assuming $g_{i''} < g_{i'}$. Hence, the claim of $g_{i''} \geq g_{i'}$ is valid and the DP can indeed form groups that are at least as large as the ones formed by the LISA.

IV. EXPERIMENTAL ANALYSIS AND VALIDATION

In the system we have set up, 160 ROs are placed on a Xilinx 3S5000 chip and their frequencies are sent to PC via MATLAB. The frequencies of each RO are measured at 6 different temperatures, 0°C, 20°C, 40°C, 60°C, 80°C, 100°C to be able to calculate the related f_{thp} values under different environmental conditions. It is assumed that the initialization of the PUF circuit is done at 20C° and all f_{thp} values are calculated with reference to the frequencies measured at this temperature which are given in Table 1.

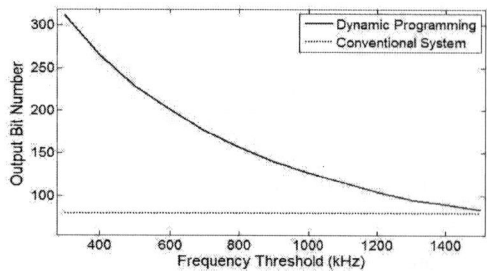

Fig. 3. PUF output bit generation comparison.

TABLE II
AREA REDUCTION WITH DP

f_{thp} (kHz)	Num. of ROs for 80 bit out with Dyn. Prog.	RO num. decrease (%)
600	75	53
1000	105	34
1200	122	23

From this point on, we have analyzed the effectiveness of DP for different values of f_{thp} in a wide range, since different designs may require different operating regimes, hence different f_{thp} values. In the real case, it is the designers responsibility to determine the correct temperature and/or supply voltage for the reference and extreme measurement cases for an effective f_{thp} determination.

In one-by-one comparison based systems N ROs can generate $N/2$ bits without any dependency. In ordering based systems, theoretical upper-bound is $\lfloor \log_2(N!) \rfloor$. By using 160 ROs, 80 bit output can be generated with conventional systems. On the other hand, in ordering based systems, the upper-bound is 1086 bits which is more than 13 times higher. But this is not achievable due to the noise in the system and changing environmental conditions. When f_{thp} is added to compensate these effects, the generated number of output bits decreases. For instance, by using 1000kHz f_{thp}, DP generates 127 bits of output which is significantly higher than the 80 bits of one-by-one comparison systems. Moreover these conventional systems do not guarantee 100% robustness. Number of bits generated by DP with respect to conventional systems by using different f_{thp} values are shown in Figure 3. Since more entropy is extracted from the system with ordering based output generation mechanisms, less RO implementation is enough for the same number of output bits. In Table 2, required number of ROs for DP algorithm is presented for different f_{thp} values. As seen from the table by using 600kHz as f_{thp}, more than 50% area reduction is achieved.

Our another aim was to decrease the computational cost of the output generation in ordering based systems. To analyze the computational cost of mentioned methods, both algorithms are implemented on MATLAB and computation times are measured under the same conditions. As seen in Figure 4, DP has significant advantage over LISA. For instance, for 160 RO implementation, the execution time of LISA is more than 5.5

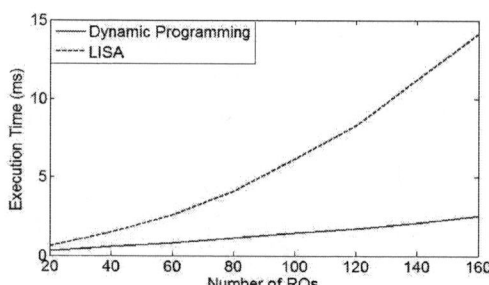

Fig. 4. Execution time of algorithms in MATLAB.

times longer than the execution time of DP.

V. CONCLUSION

We have adapted Dynamic Programming approach to PUF output generation in order to maximize entropy extraction from a certain number of ROs with a minimum computational complexity for the first time in the literature. In addition to this, a method for achieving 100% robustness is proposed. Lastly, the efficiency of proposed methods is analyzed and validated experimentally with FPGA implementation.

REFERENCES

[1] R. S. Pappu, "Physical one-way functions." Ph.D. dissertation, Massachusetts Institute of Technology, 2001.

[2] D. Lim, J. Lee, B. Gasend, G.E.Suh, M. V. Dijk, and S. Devadas, "Extracting secret keys from integrated circuits," *IEEE Transactions on VLSI Systems*, vol. 13, no. 10, pp. 1200–1205, 2005.

[3] B. Gassend, D. Clarke, M. V. Dijk, and S. Devadas, "Delay-based circuit authentication and applications," in *ACM Symposium on Applied Computing*, 2003, pp. 294–301.

[4] B. Gassend, "Physical random functions," M.S. Thesis, Massachusetts Institute of Technology, 2003.

[5] J. Guajardo, S. Kumar, G. Schrijen, and P. Tuyls, "FPGA intrinsic PUFs and their use for IP protection," in *18th Annual Computer Security Applications Conference (CHES)*, vol. 4727, 2007, pp. 63–80.

[6] J. Guajardo, S. Kumar, R. Maes, G. Schrijen, and P. Tuyls, "Extended abstract: The butterfly PUF protecting IP on every FPGA," in *Hardware-Oriented Security and Trust (HOST)*, 2008, pp. 67–70.

[7] D. Suzuki and K. Shimizu, "The glitch PUF: A new delay-PUF architecture exploiting glitch shapes," in *Cryptographic Hardware and Embedded Systems (CHES)*, 2010, pp. 366–382.

[8] A. Maiti and P. Schaumont, "Improved ring oscillator PUF: An FPGA-friendly secure primitive," *Journal of Cryptology*, vol. 24, no. 2, pp. 375–397, 2011.

[9] C. Yin and G. Qu, "Temperature aware cooperative ring oscillator PUF," in *IEEE International Workshop on Hardware Oriented Security and Trust (HOST)*, 2009, pp. 36–42.

[10] G. E. Suh and S. Devadas, "Physical unclonable functions for device authentication and secret key generation," *Design Automation Conference (DAC)*, pp. 9–14, 2007.

[11] C. Yin and G. Qu, "LISA: Maximizing RO-PUF's secret extraction," pp. 100–105, 2010.

[12] A. Maiti and P. Schaumont, "Improving the quality of a physical unclonable function using configurable ring oscillators," *International Conference on Field Programmable Logic and Applications, (FPL)*, pp. 703–707, 2009.

[13] H. Yu, P. Leong, H. Hinkelmann, L. Moller, M. Glesner, and P. Zipf, "Towards a unique FPGA-based identification circuit using process variations," *International Conference on Field Programmable Logic and Applications (FPLA)*, pp. 397–402, 2009.

[14] B. Gassend, D. Clarke, M. van Dijk, and S. Devadas, "Silicon pysical random functions," in *ACM Conference on Computer and Communications Security (CCS)*, 2002, pp. 148–160.

Power-Aware Architectural Exploration of the CORDIC Algorithm

Jennifer Manica, Roberto Passerone and Luca Rizzon

Dipartimento di Ingegneria e Scienza dell'Informazione – Università degli Studi di Trento
Povo di Trento, Italy 38123

Abstract—**Mobile applications require the use of specialized design techniques to reduce power consumption and maximize battery life. New implementation technologies are able to reduce energy at the device level. At the same time, architectural choices can lead to significantly different power consumption profiles for equivalent implementations of the same function. In this paper we focus on the architectural level, and analyze various implementation alternatives for the CORDIC function. Our results provide insight into the trade-offs between area, performance and power consumption, and give designers directions for their architectural choices.**

I. INTRODUCTION

In the last decade, the widespread adoption of battery-powered mobile devices has made power consumption become one of the most critical performance metric in the design of digital electronic systems. In addition to the obvious lifetime improvement, reducing power consumption simplifies problems related to thermal management and battery size, which significantly affect the reliability and cost of a device [1]. At the same time, constraints on area and performance limit the range of optimal implementations, which must be searched through a careful balance of conflicting requirements.

Several factors influence the power consumption of a design. The implementation technology, for instance, determines the overall characteristics of the devices and their power profile. At the design level, techniques such as dynamic voltage and frequency scaling are used to selectively activate and modulate the performance of the computational blocks to adapt to the varying requirements of the applications [2], [1]. Likewise, different implementation alternatives for the core functions of the system exhibit wildly different performance characteristics which are not always easy to predict using intuition and back-of-the-envelope calculations.

In this work, we focus on this last aspect and analyze the performance of the CORDIC algorithm [3], [4], whose flexibility and simplicity make it ideal in diverse applications such as direct frequency synthesis, digital modulation and coding or direct and inverse kinematic computation for robot manipulation [4]. We chose three different implementation architectures, which we have compared on the basis of area, throughput and power consumption. The results, based on an FPGA technology, demonstrate how the specifics of each architecture contribute to the determination of the final performance metrics, and can be used to quickly choose a solution given the system constraints.

A. Related work

We target an FPGA CORDIC implementation, and take as reference the architectures reviewed by Andraka [3]. Other studies have targeted low power implementations of CORDIC. Sarrigeorgidis and Rabaey study methods to reduce the number of iterations and make use of redundant number representations to improve performance [5]. Kim et al. employ hard wired operations and interpolation to achieve the same [6]. Our work is orthogonal, as we wish to study the impact of different architectural choices on the overall performance. Indeed, the above techniques can be applied to all the architectures we consider. Studies similar to ours have been conducted for *other* algorithms in widespread use. Rhee et al. consider the H.264/AVC encoder and use various techniques to trade-off power consumption with compression efficiency [7]. Sherazi et al. focus on digital filters and evaluate several structures in the sub-V_T domain [8]. Roth et al. discuss the trade-off in LDPC decoders [9], while Lin et al. study ways to improve the power consumption of Viterbi decoders [10]. Our results complement the existing literature with a characterization of the CORDIC architectures.

In the following, we first describe the principles of the CORDIC algorithm, and describe the chosen architectural solutions in Section II. Then, Section III discusses the results of the architectural analysis. Section IV concludes the paper outlining directions for future work.

II. CORDIC ALGORITHM AND ARCHITECTURES

The CORDIC algorithm is used in a variety of applications that require the computation of trigonometric and hyperbolic functions. CORDIC, short for Coordinate Rotation Digital Computer, is an iterative algorithm used to efficiently compute vector rotations. In its basic form, the algorithm computes the rotation of a vector (x, y) by an angle φ into a sequence of elementary rotations for angles whose trigonometric tangent is a power of 2. This way, a single elementary rotation can be computed as [3]

$$x_{i+1} = x_i - y_i d_i 2^{-i}, \qquad y_{i+1} = y_i + x_i d_i 2^{-i}$$
$$z_{i+1} = z_i - d_i \tan^{-1}(2^{-i}).$$

Variable z_i, initialized with the desired angle φ, represents the residual rotation and is updated at each iteration with the effective rotation. The coefficient d_i determines the direction of rotation, and is equal to 1 if $z_i \geq 0$, and -1 otherwise. Hence, the algorithm successively approximates the desired rotation with incrementally smaller elementary rotations, until z_i converges to 0. The advantage of this formulation is that it does not need a multiplier, since powers of 2 can be implemented with shifters. It requires a look-up table to compute $\tan^{-1}$, which is nonetheless restricted to a small number of constant values. This algorithm alters the length of the vector while it rotates it by a factor that can be considered constant for a sufficient number of iterations, and will therefore be ignored in this study.

A. CORDIC architectures

The CORDIC iteration can be implemented using several architectures [3]. The iterative nature of the algorithm lends itself to an iterative implementation, as shown in Figure 1. The intermediate values for the vector coordinates are stored

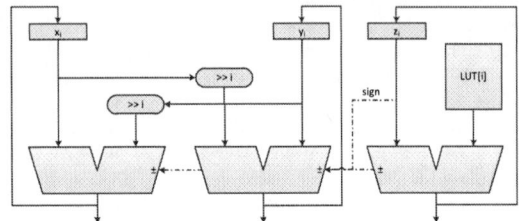

Fig. 1. Iterative CORDIC architecture

in the x_i and y_i registers, while the residual angle in z_i. The shifts are implemented as barrel shifters, shown in the figure as rounded rectangles. A counter, not shown, indexes the iterations and selects the appropriate shift value. The $\tan^{-1}$ operator is implemented as a look-up table (LUT) which stores the required values, also selected by the iteration counter. The three adders execute the main update operation. These adders perform either an addition or a subtraction according to the formula and the sign of z_i.

The values are stored using a fixed point representation, with one bit for the sign (in two-complement), two bits for the integer part, and the remaining bits for the fractional part. All angles are measured in radians. The number of iterations n required to compute the results depends on the total number of bits chosen for the operands. The higher the number of bits, the higher the number of iterations required to exploit the extra representation. In our experiments, we stop the iterations once the $\tan^{-1}$ operator returns a value below the chosen precision (i.e., zero). In particular, we have $n = 6, 9, 13, 18, 21, 29$ iterations for $8, 12, 16, 20, 24$ and 32 bits.

The performance of the iterative architecture depends on the clock frequency f_c and on the number n of iterations. The throughput is equal to $th = f_c/n$, while the latency is $l = n/f_c$. The area increases with the number of bits, because the registers, the shifters, the adder and the LUT become larger. It is instead only marginally affected by the number of iterations, since only the LUT increases in size.

An alternative architecture consists in unrolling the iteration. In this case, one could either construct a *feed-through* architecture, where each adder is connected directly to the following adder, or a *pipeline* architecture, where the stages are separated by registers. The pipeline is shown in Figure 2. The number of stages equals the number of iterations. The advantage of a pipeline is that more than one computation is active at the same time, increasing the throughput which is

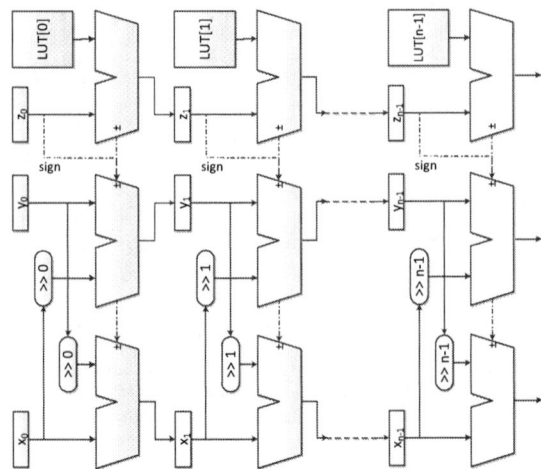

Fig. 2. Pipeline CORDIC architecture

now equal to the clock frequency. The latency, on the other hand, is the same as the iterative solution, since the data has to traverse the entire structure.

The maximum clock frequency of the pipeline architecture is higher than that of the iterative solution. This is because for each stage the amount of shift is known beforehand, so one has simply to re-route wires, instead of using a barrel shifter. Hence, shifting does not incur in any extra delay, and one component is removed from the critical path. For the same reason, the area is not n times as large as that of the iterative architecture, because the shifters are removed.

The feed-through solution does not have pipeline registers. In this case, the clock frequency will be significantly lower, since the critical path now encompasses all the stages of the architecture, but not n times lower, since the delay introduced by the registers is removed. The throughput is equal to the clock frequency, while the latency is equal to the clock period.

For these architectures, the area increases both with the number of bits, as well as with the number of iterations, since every new iteration requires an additional stage.

III. ARCHITECTURE ANALYSIS AND RESULTS

We have developed the above architectures in VHDL, for word sizes ranging from 8 to 32 bits. The code was synthesized for the Altera DE1 Development Board, which employs a 90nm Cyclone II FPGA, using the Quartus tool with default settings. Our architectural analysis is in particular focused on analyzing the power consumption of the different architectures. In this study, we concentrate on dynamic power and neglect the contribution of static power (e.g., leakage) since its contribution is not significant in the chosen FPGA technology. In general, the dynamic power consumption P depends on the number of times that signals switch their state, and is proportional to several factors [1]:

$$P = \alpha \cdot f_{ck} \cdot C_L \cdot V_{DD}^2,$$

where f_{ck} is the clock frequency, V_{DD} is the power supply, C_L is proportional to the total capacitance in the circuit, and α

is a factor that accounts for the switching activity. The various architectures will differ in terms of their size (and thus C_L), of the frequency at which they need to run to achieve a certain throughput, and the activity α in the internal nodes.

Determining the internal node activities may be difficult to do in general. In our study, we have simulated the various architectures driving them with the same test-bench requesting the computation of the trigonometric functions for a series of 32 angles. The results are used by the Quartus tool to precisely determine the internal activity, which is therefore averaged out on thousands of transitions. The tool is then able to provide us with accurate estimates of the power consumption based on the models of the FPGA components. In this study, we have considered the power consumed by the core logic only, and excluded the contribution of the I/Os, which is of comparable size and may therefore obscure the actual dependency of the power from the architectural parameters. For the same reason, it was not possible to compare the results to measurements, since there is no access to the internal circuits. Nevertheless, our estimates were obtained using detailed simulation (after place and route) with the manufacturer provided models.

A. Circuit size

Figure 3(a) shows the dependency of the area of each architectural solution as a function of the word size. The iterative architecture shows a linear area increase with the number of bits, due to the larger size of the registers, adders, the shifters and the look-up table. The size of the look-up table also increases, since the architecture works with more bits.

The size of the pipeline and feed-through architectures also increases with the number of bits in a close to linear fashion. However, the *rate* of increase is higher than that of the iterative architecture, because the algorithm requires a larger number of iterations to converge (and make use of the extra bits), and therefore *more stages*. As expected, the size of the pipeline and feed-through architectures is not n times as large as that of the iterative solution, where n is the number of iterations. This is because, as discussed, the shifters in these architectures can be implemented by simply moving wires, rather than using a barrel shifter. In addition, the iterative architecture requires a small control circuit to keep track of the number of iterations, which is not present in the other two.

In the FPGA implementation, the feed-through architecture requires slightly more area than the pipeline architecture, although it obviously uses fewer components (it does not have intermediate pipeline registers). This result is specific to FPGAs, whose fabric consists of logic blocks that already incorporate a register which is conveniently used by the pipeline implementation, while it is essentially wasted in the feed-through solution. The longer combinational chain also makes it more difficult for the tool to optimize the area, resulting in some overhead. Nonetheless, automatic design tools might be able to reclaim these extra registered for use with the logic required by the surrounding system. This effect would not exist in a standard cell implementation where every function has a dedicated component.

B. Performance

Figure 3(b) shows the maximum throughput achievable by each architecture as a function of the number of bits for

our chosen FPGA. Throughput is computed as the number of computations per unit time. Hence, for the pipeline and feed-through it corresponds to the clock frequency, while for the iterative architecture it must be scaled down to account for the n sequential iterations.

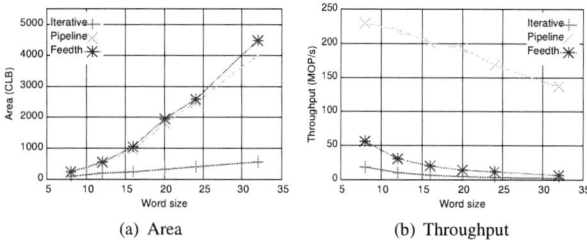

Fig. 3.   CORDIC size (a) and throughput (b) for different architectures

As expected, the pipeline architecture has the largest throughput, since it takes advantage of parallelism. The throughput decreases linearly with increasing precision, because the larger adders incur larger delays. The feed-through architecture performs better than the iterative, since it eliminates the overhead due to intermediate registers, while at the same time eliminates the delay penalty due to the shifters.

Notice that the decrease in maximum performance for the feed-through and the iterative architectures is more than linear with the number of bits. This is because not only do the adders (and the shifters, for the iterative) become slower, but also because the higher precision requires a higher number of iterations for the iterative architecture, and more stages for the feed-through architecture, increasing the length of the critical path. Also the pipeline solution has more stages. However, the pipeline registers break the critical cycles at every stage, making the decrease linear.

C. Power consumption

Figure 4(a) shows the power consumption of each CORDIC architecture as a function of the number of bits, running the computation at its maximum throughput. Power consumption,

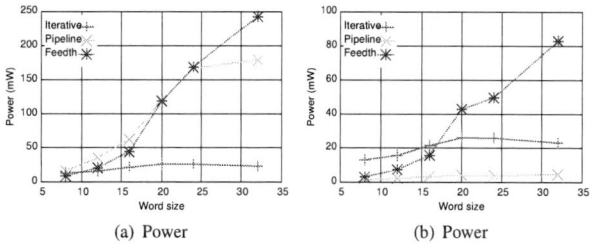

Fig. 4.   CORDIC power consumption at maximum (a) and iterative throughput (b) for different architectures

in general, increases with the number of bits due to the larger size of the circuit. However, higher precision also implies lower throughput, i.e., a lower clock frequency which tends to decrease the power consumption. For this reason, power tends to level off with precision, and in fact eventually decreases in the case of the iterative architecture, and to some extent for the pipeline. The feed-through architecture is not significantly influenced by this effect since clocked registers account for a much lower percentage of the circuit than in the other cases.

A better comparison can be made by running all architectures so that they achieve the same throughput. Figure 4(b) shows how power consumption changes when, for each word size, all architectures are run at the throughput of the iterative solution. We notice how the power consumption of the pipeline architecture is well below that of the iterative solution. This is because the *same* throughput is obtained with a clock frequency which is n *times lower*, while the increase in area and activity factor is not as large. The feed-through version, instead, shows a much more marked increase of power consumption with the increased number of bits. This is due to an increase in the activity factor α. This effect can be explained as follows. The activity factor depends on the amount of switching in the circuit. The adders update their outputs in sequence, starting typically with the least significant bits, and adjusting the higher bits as the carry propagates through the carry chain. The result is that the value at the output is updated several times during the computation. In the feed-through architecture, these updates are immediately propagated to the adder downstream, which compounds its own updates. This produces an elevated amount of switching, which is reflected by the increased power consumption. This cascading effect is instead blocked by the pipeline registers and by the iteration registers in the other two architectures. The graph also shows that this effect is more pronounced as the adders become larger, and as the number of stages increases, so that the feed-through solution quickly becomes the worst performing architecture in terms of power, while it is comparable to the other two for lower precisions.

D. Architectural trade-offs

The data collected in the experiments can be used to construct trade-off graphs where one metric is compared to another. One example is shown in Figure 5, where power consumption and throughput are plotted against each other, on log scale axis, for each of the different architectures and word size. The best solutions are those closer to the left (lower power) and to the top (higher throughput). These plots make the trade-off between the different objectives explicit. In particular, they show when a solution dominates another in more than one metric, making it preferable for the implementation. For instance, the pipelined architecture is to be preferred over the feed-through, especially for large word size, and also over the iterative, when area is not an issue. The plot can be populated with additional data points, not shown here to avoid clutter, where each of the architecture is run at intermediate speeds to achieve different trade-offs, to construct a Pareto front. The data is also useful for higher level design exploration algorithms to help them choose the best implementation.

IV. CONCLUSIONS

We have presented the architectural analysis of various CORDIC architectures. The data collected through accurate simulation allows a designer to choose the architecture with the best performance in terms of area, throughput and power, relative to the system requirements. There are several directions for improving this study. One may consider the effect of lowering

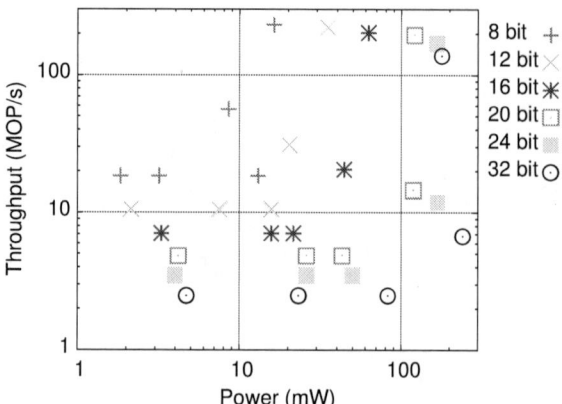

Fig. 5. Power vs. throughput trade-off

the supply voltage when running at clock frequencies below the maximum (voltage scaling is not possible in our targeted FPGA). From an architectural point of view, one may consider intermediate solutions where, for instance, the iteration is only partially unrolled into a pipeline to strike the desired trade-off between area and performance. In an orthogonal direction, one may consider different implementations for the components used in the architectures, such as different kinds of adders whose carry propagation has a large influence on the feed-through architecture. These considerations are part of our future work.

REFERENCES

[1] J. Rabaey, *Low Power Design Essentials*. New York, NY, USA: Springer, 2009.

[2] P. J. M. Havinga and G. J. M. Smit, "Design techniques for low-power systems," *J. Syst. Archit.*, vol. 46, no. 1, pp. 1–21, January 2000.

[3] R. Andraka, "A survey of CORDIC algorithms for FPGA based computers," in *Sixth International Symposium on Field Programmable Gate Arrays*, Monterey, CA, February 22-24, 1998.

[4] P. K. Meher, J. Valls, T.-B. Juang, K. Sridharan, and K. Maharatna, "50 years of CORDIC: Algorithms, architectures, and applications," *Trans. Cir. Sys. Part I*, vol. 56, no. 9, pp. 1893–1907, September 2009.

[5] K. Sarrigeorgidis and J. Rabaey, "Ultra low power CORDIC processor for wireless communication algorithms," *Journal of VLSI Signal Processing 38*, pp. 115–130, 2004.

[6] Y. B. Kim, Y.-B. Kim, and J. T. Doyle, "A low power CMOS CORDIC processor design for wireless telecommunication," in *50th Midwest Symp. on Circuits and Systems*, Montreal, Québec, August 5-8, 2007.

[7] H. Kim, C. E. Rhee, J.-S. Kim, S. Kim, and H.-J. Lee, "Power-aware design with various low-power algorithms for an H.264/AVC encoder," in *IEEE International Symposium on Circuits and Systems (ISCAS11)*, Rio de Janerio, Brazil, May 15-18, 2011, pp. 571–574.

[8] S. Sherazi, J. Rodrigues, O. Akgun, H. Sjöland, and P. Nilsson, "Ultra low energy vs throughput design exploration of 65 nm sub-VT CMOS digital filters," in *NORCHIP*, Tampere, Finland, November 15-16, 2010.

[9] C. Roth, A. Cevrero, C. Studer, Y. Leblebici, and A. Burg, "Area, throughput, and energy-efficiency trade-offs in the VLSI implementation of LDPC decoders," in *IEEE International Symposium on Circuits and Systems (ISCAS11)*, Rio de Janerio, Brazil, May 15-18, 2011.

[10] Y.-M. Lin, W.-C. Liu, L.-Y. Chang, C.-Y. Lien, P.-Y. Chen, and S.-C. Chen, "A low-power IP design of Viterbi decoder with dynamic threshold setting," in *IEEE International Symposium on Circuits and Systems (ISCAS10)*, Paris, France, May 30-June 2, 2010, pp. 585–588.

Zoom FFT for Precise Spectrum Calculation in FMCW Radar using FPGA

Belal Al-Qudsi, Niko Joram, Axel Strobel and Frank Ellinger

Chair for Circuit Design and Network Theory,
Technische Universität Dresden, D-01062 Dresden, Germany

Abstract—The zoom FFT (ZFFT) has been utilized in various digital signal processing systems, it is used when a fine spectral resolution is needed within a small portion of a signal's overall frequency range. It has been implemented in different fields and for different applications. This paper presents an implementation method of the ZFFT approach to estimate the spectral peak in the FMCW radar using a field programmable gate arrays (FPGA). It has been utilized to decrease the calculations complexity, hence, reduction the power consumption. The approach has been realized and a complexity reduction factor of up to 8 has been gained and tested on an FMCW radar processing system.

Index Terms—*Discrete Fourier transforms, chirp radar, radar position measurement, field programmable gate arrays.*

I. INTRODUCTION

Power consumption is a key factor in many radar applications that need to be mobile. Lots of research has been carried out to improve the power efficiency of the radio frequency (RF) front-end of the radar system, whereas most radar systems needs to accommodate a processing unit which considered as the main power consumer unit already in many of them. The FMCW radar is an example of a radar system that needs to include a processing unit to perform complex calculations for getting a high resolution spectral estimation of its resultant signal.

The European project "E-SPONDER" [1] is intended to support the first responder in the crisis situations where the system needs to estimate the position for a mobile node using the range information in an indoor scenario. A hardware part of an FMCW radar unit has already been developed [2] to perform the task.

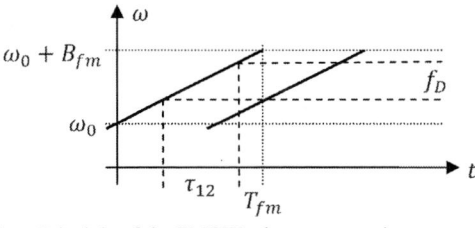

Fig. 1. Principle of the FMCW where a transmitter generates a linear frequency chirp with bandwidth B_{fm}, starting frequency ω_o and duration T_{fm} and its delayed received frequency chirp [2]

The principle of FMCW radar [3] is illustrated in Fig.1. The ranging measurement is established basically by multiplying a transmitted frequency chirp signal with its reflected version to produce the so called "beat signal". The reflector is considered to be passive in a traditional FMCW radar system [3], whereas the system that has been used in this research is relying on an active reflector for its

synchronization, which adds an additional uncertain delay to the time of flight. Hence, the beat frequency is usually not as confined as in the passive FMCW radar. Therefore, a larger frequency band has to be searched to detect the line-of-sight (LOS) beat signal.

A traditional FFT algorithm was implemented to estimate the frequency value of the "beat signal", which is expected to appear in a wide frequency band in the case of the active reflector FMCW radar. The overall resolution of the ranging metric is directly affected by the resolution of the FFT algorithm which is limited by the amount of resources of the FPGA chip used to perform the task.

In this paper, an optimization method for using the ZFFT approach to detect the spectral peak of the FMCW radar is being discussed and implemented to reduce the calculation complexity and increase the peak estimator resolution. Although this method is perfectly matched to the LOS scenario where the peak of the spectrum is the only interesting frequency point, it might be also applied for the non-line-of-sight (NLOS) algorithms which rely on the spectrum peak detection.

The rest of this paper is divided into two parts: Section II is the discussion of the theoretical background behind the ZFFT, and section III is presenting the implementation methodology and its testing result.

II. PRINCIPLE OF ZOOM FFT APPROACH

ZFFT is a well-known technique used to apply the traditional FFT algorithm to a relatively small frequency bandwidth within the spectrum [4], as a kind of focusing technique. The ZFFT is a perfect method for spectrum analysis within a small portion of the frequency spectrum.

The theory behind the ZFFT is based on the traditional Fourier transform algorithm. Consider N points of the digital sequence $x(n)$, where $X(k)$ is the DFT of the sequence.

$$X(k) = \sum_{n=0}^{N-1} x(n)W_N^{kn} \qquad 0 \le k \le (N-1) \qquad (1)$$

where $\qquad W_N^{kn} = e^{-j2\pi kn/N} \quad 0 \le k \le (N-1) \qquad (2)$

N is the number of samples to be transformed, n is the sample index of the time domain signal and k is the frequency index of the frequency domain signal which refers to frequency step values. The resolution of the resultant spectrum from equation (2) is limited by the sampling rate fs and the number of calculated samples.

978-1-4673-4579-8/13 $31.00 © 2013 IEEE

$$\Delta f = \frac{fs}{N} \qquad (3)$$

In order to improve the frequency resolution, the number of sample points could be increased or the sampling rate could be decreased as well. Increasing the number of samples can directly improve the spectral resolution but it also dramatically increases the complexity of the FFT, which is discussed in [4]. One can easily prove that the number of complex multiplications required to perform the FFT algorithm is $(\frac{N}{2}\log N)$, where N is the number of FFT samples. Back to equation (3), another factor which may increase the spectral resolution, is the sampling rate. Decreasing the sampling rate will increase the resolution as well. However, the disadvantage of down sampling is basically in its resultant bandwidth. The bandwidth will be divided by the down sampling ratio, which is equivalent to the zooming factor D, and the high frequencies will be lost.

A two-way time of arrival (TW-TOA) protocol was used to perform the ranging measurement between two asynchronous FMCW radar units [2]. The TW-TOA calculation requires an estimation of the time shift between the received and the local chirp signal in each radar unit. To estimate that, each of the radar unit generates its own beat signal by mixing the received signal with a local asynchronous chirp signal leading to a large ambiguity in the frequency of the resultant beat signal.

Using the traditional FFT algorithm, a wide spectral range needs to be calculated. The idea of using the ZFFT in FMCW radar is to move the interesting part of the spectrum to a lower frequency band by mixing it with a local oscillator, and then down sample the sequence. Fig. 2 shows a brief explanation of the zooming method which will be used in this paper.

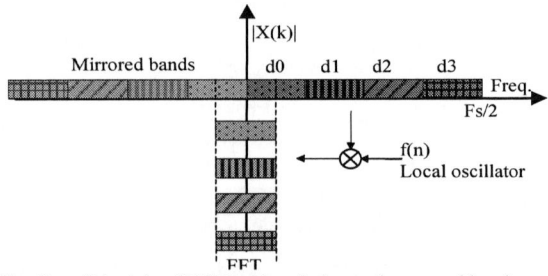

Fig. 2.　Principle of ZFFT, d is refering to the specral band

The number of complex multiplications required to perform N points ZFFT will be reduced dramatically. Due to the decimation process, the number of calculated FFT input samples will be divided by the zooming factor (D) and the number of complex multiplications will be reduced to $(\frac{N}{2D}\log\frac{N}{D})$.

Before going through the implementation of ZFFT, to get an idea about the calculation gain expected by implementing the ZFFT, the complexity ratio was simulated for different zooming factors. The complexity ratio is defined as the ratio of the number of multiplications required by the conventional FFT to those required by the ZFFT. Fig.3 shows the simulation result for the complexity

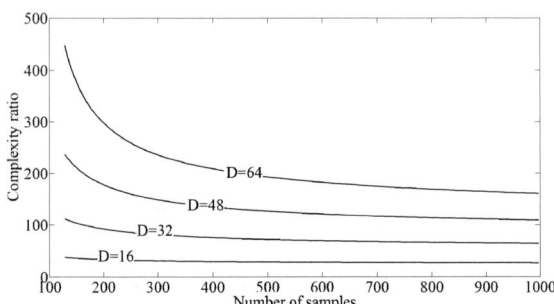

Fig. 3.　Complexity ratio for different zooming factor (D) with respect to the number of samples

ratio for different zooming factor. Moreover, because of the complex mixing of the signal, the resulting signal has no redundant mirrored frequency in its frequency spectrum. A complex mixer would move the bandwidth of interest to the center of the spectrum, and this will increase the efficiency of the ZFFT by producing a non-mirrored signal spectrum, decreasing the redundant information.

Although the complexity is decreasing with high zooming factors, another complexity will be added due to the requirement of a sharp narrow low pass filter; this will be discussed in the filter subsection of this paper.

III. SIMULATION AND IMPLEMENTATION

A. Overview

The method described in Fig. 4 was realized in an FPGA chip, namely, the Spartan 6 chip from Xilinx, using a special tool called "System generator" [5].

As shown briefly in Fig. 4, the design consists of the following blocks:

   a.　Preprocessing unit.
   b.　Direct digital synthesizer (DDS) and complex multiplier.
   c.　Low pass digital filters.
   d.　Down sampler and FFT block.

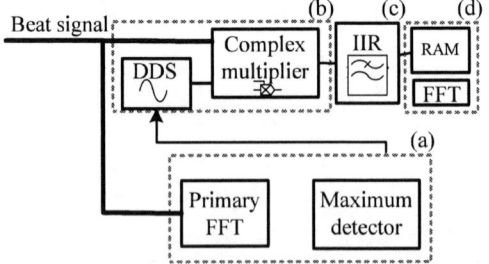

Fig. 4.　Block diagram representation for the implanted ZFFT

B. Preprocessing unit

To perform a ZFFT with a zooming factor of 8 (The zooming factor is limited by the low pass filter complexity and the available FPGA resources), a primary low resolution FFT is required to divide the spectrum bandwidth into 4 sub-bands. A 64 sample FFT was chosen

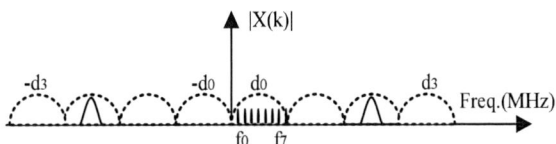

Fig. 5. Frequency spectrum patitioning by the primary FFT, where d and f are refering to the specral band and frequency bin index respectively

to produce 32 frequency bins. As shown in Fig. 5, each 8 frequency bins are grouped to form a sub-band. The resultant power is being calculated by a maximum detector for each sub-band to produce a metric for the next step to choose the correct local oscillation frequency out of the DDS.

The maximum detector search process is basically storing the index of the region where the maximum power is showing up. This index is translated to a phase value to feed the DDS with the proper phase to produce the correct local frequency and centralize the sub-band around the zero line of the spectrum.

C. Direct Digital Synthesizer (DDS) and complex multiplier

The complex mixer is multiplying the input signal $x(n)$ with a complex local signal $(e^{-j2\pi f_c n/f_s})$ which can be expressed as two orthogonal signals $(\cos(2\pi f_c n/f_s) - j\sin(2\pi f_c n/f_s))$.

$$y(n) = x(n).e^{-j2\pi f_c n/f_s} \qquad (4)$$

A separate DDS for real and imaginary part (The orthogonal signal pair has been generated using two DDS blocks to perform the complex multiplication) was used to generate an adaptable local oscillator by controlling its phase increment using the output of the control block. All of the complex arithmetic processes, such as multiplication blocks, were implemented using the DSP slices of the FPGA chip.

D. Low pass Digital filters

Due to the complex multiplication, high and low frequency signals will be generated. A low pass digital filter is required to eliminate the high frequency part of the resultant signal.

Generally, there are two direct possibilities to build a digital filter, the finite impulse response (FIR) and infinite impulse response (IIR). Both of them were compared for the requirements and implemented using the "System generator" tool to compare their performance and estimate the required resources for each of them. Some of the comparison points between both FIR and IIR filter with similar magnitude response are recorded in Table. 1. The current peak detector is intended to detect a peak which is related to the range between two nodes, phase change is negligible. The IIR filter was chosen for simplicity in implementation, with respect to the number of multipliers. Fig. 6 shows the frequency response simulation result for both of them.

TABLE I
COMPARISON BETWEEN THE SELECTED FIR AND IIR FILTER

	FIR	**IIR**
Order	50	6
FPGA resources	51 multiplier 50 adder	12 multiplier 12 adder
Phase linearity	linear	non linear

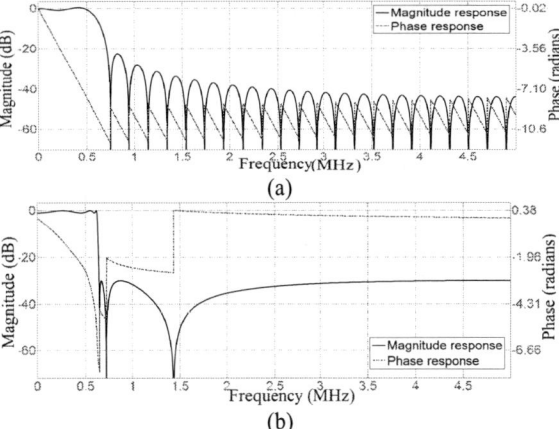

Fig. 6. Magnitude and frequency response of (a) FIR and (b) IIR filter

Another approach using cascaded integrator-comb (CIC) was discussed in [6] as a method to increase the efficiency of the ZFFT.

E. Down sampler and FFT block

The down sampler is intended to relax the calculation complexity of the FFT block. It determines the zoom factor for the whole design. Down sampling factor of 8 is performed after the low pass filter.

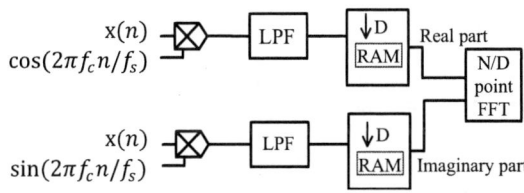

Fig. 7. Realization of decimator using a block RAM, D represents the zooming factor

Instead of direct down sampling the filtered signal, the down sampler (decimator) is implemented by saving the down sampled data in a block RAM, and the samples are being read by the FFT block using the same clock of the original input signal. The whole unit is running on a common clock. This saving method enhances the calculation time dramatically.

F. Verification:

First of all, the ZFFT was simulated using MATLAB. By using a multi tone signal with a narrow bandwidth, the effect of ZFFT has been illustrated. Using two frequency spectrums of the multi tone signal (0.998, 1 and 1.002) MHz respectively one with 10000 points (zero padding was used) conventional FFT and the other using 10000 points ZFFT, the sampling frequency is considered to be 100 MHz for both cases. Simulation shows that the spectrum's details are increased using the ZFFT approach while using the same number of samples as in the conventional approach.

The verification of functionality was a primary step before going through the implementation procedure. The second step is the implementation of the ZFFT in the FPGA chip. The old FMCW radar processing unit was utilizing 64K samples FFT block to detect the spectrum peak. An FFT block of 8K samples has been chosen to replace the 64K FFT; this replacement requires a zooming factor of 8. It has been implemented using the "System generator" [5] tool. Fig. 8 shows a simplified block diagram for its implementation.

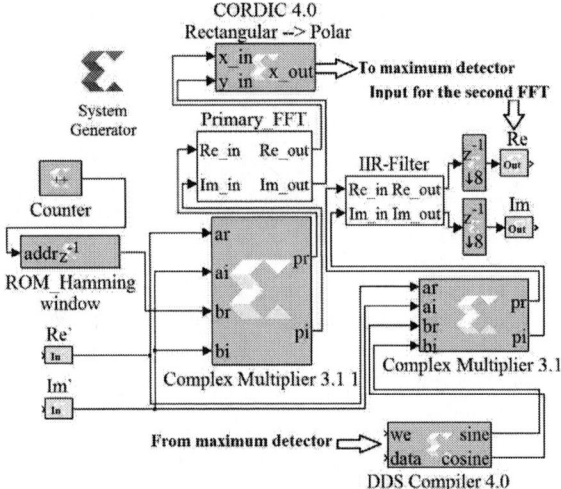

Fig. 8. Simplified block diagram for zoom FFT implementation in System generator, for more detail about each block please refer to [5]

In order to visualize the effect of ZFFT with a zooming factor of 8 a dual tone signal has been saved in a ROM and used as an input signal. The signal has two peaks with a frequency difference of around 1 kHz and 10 MHz sampling frequency. It was used as an input signal for three different FFT blocks, namely, 64K-FFT, 8K-FFT and 64K-ZFFT. Fig.9 compares the resultant frequency spectrum for all of the three possibilities.

One can easily note that the implemented 8K-ZFFT has a frequency resolution equivalent to the 64K-FFT, and the normal 8K-FFT was not able to detect the two peaks. This approach has been integrated with a processing unit for the FMCW radar to detect the spectral peak of its beat signal.

IV. CONCLUSION

This paper presented an implementation of the ZFFT algorithm for the asynchronous FMCW radar system to reduce the complexity in its spectral calculation. It was realized in an FPGA using "System generator" tool from

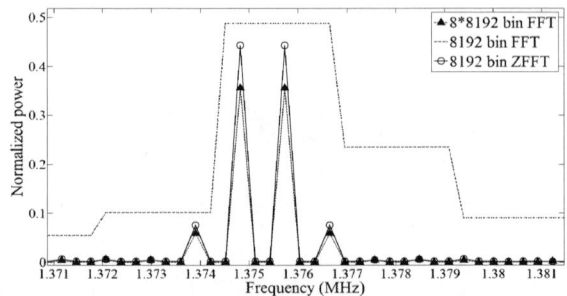

Fig. 9. Comparison between three possible FFT implementation methods using "System generator" from Xilinx

Xilinx. The design has been integrated to replace a 64K traditional FFT by a much less complex 8K samples ZFFT block which decreases the complexity of calculating the FFT by a factor of 8, it can be easily modified to serve different requirements for other systems. The ZFFT can provide a valuable improvement with respect to FPGA resources optimization and power consumption accordingly.

ACKNOWLEDGEMENT

The research leading to these results has received funding from the European Community's Seventh Framework Program (FP7/2007-2013) under grant agreement n°242411 (E-SPONDER).

REFERENCES

[1] D. Vassiliadis, A. Garbi, G. Calarco, M. Casoni, A. Paganelli, R. Morera, C. M. Chen, and M. Wodczak, "Wireless networks at the service of effective first response work: The E-SPONDER vision," *5th IEEE International symposium on Wireless Pervasive Computing (ISWPC),* pp. 210-214, May 2010.

[2] N. Joram, J. Wagner, A. Strobel and F. Ellinger, "5.8 GHz demonstration system for evaluation of FMCW ranging," *9th Workshop on Positioning Navigation and Communication (WPNC),* pp. 137-141, 15-16 March, 2012.

[3] N.C. Currie and C.E. Brown, Principles and Applications of Millimeter- Wave Radar, Norwood: Artech House, 1987.

[4] Richard G. Lyons, "The zoom FFT," in *Understanding Digital Signal Processing,* Pearson Education, 2010.

[5] M. Ownby and W.H Mahmoud, "A design methodology for implementing DSP with Xilinx® System Generator for Matlab®," *System Theory, 2003. Proceedings of the 35th Southeastern Symposium,* pp. 404- 408, March 2003.

[6] Dong Pei, Shuo Yang, Hongwu Yang, Quanzhou Wang and Manman Li, "High efficient and real-time realization of Zoom FFT based on FPGA," *Computer Application and System Modeling (ICCASM) International Conference,* vol. 2, pp. 669-673, Oct. 2010.

A JTAG based 3D DfT architecture using automatic die detection

Yassine Fkih[1,2], Pascal Vivet[1]
[1]CEA-Leti, MINATEC Campus,
F38054, Grenoble, France
Email : first_name.last_name@cea.fr

Bruno Rouzeyre[2], Marie-lise Flottes[2], Giorgio Di Natale[2]
[2]LIRMM – Université Montpellier II/CNRS
Montpellier, France
Email: first_name.last_name@lirmm.fr

Abstract- **3D stacked integrated circuits based on Through Silicon Vias (TSV) are promising with their high performances and small form factor. However, these circuits present many test issues including the test at all 3D manufacturing levels: pre, mid, and post bond levels. In this paper we propose an automatic die-detection mechanism able to detect the presence of upper and lower dies, and its integration within a JTAG based 3D test architecture. 3D die-detectors permit the optimization of the overall 3D test architecture: making it usable at all stacking levels without requiring a configuration step. This paper presents also synthesis results of the proposed test architecture with die-detectors.**

Key words: 3D IC, DfT, JTAG IEEE 1149.1 std, die detectors

I. INTRODUCTION

The stacking process of integrated circuits using TSVs (Through Silicon Via) is a promising technology that keeps the development of the integration more than Moore's law, where TSVs enable to tightly integrate various dies in a 3D fashion. Regarding applications, 3D stacking allows a wide range of new SoC applications, such as heterogeneous stacking (Digital, Memory, RF, Mems); Interposers for multi-chip connection are becoming similar to a silicon board. The first envisaged 3D applications are mainly the WideIO DRAM 3D memory interface for high throughput and low power memory-on-logic stacking [1]. Nevertheless, 3D integrated circuits present many test challenges including the test at different levels of the 3D fabrication process: pre-, mid-, and post- bond tests. Pre-bond test targets the individual dies at wafer level, mid-bond test targets the test of partially assembled 3D stacks, whereas finally post-bond test targets the final 3D circuit. It is generally admitted that a 3D test flow [2] should involve test procedures at all stacking levels of the 3D components. In this paper, we explore a 3D test architecture based on automatic die-detection mechanism; we give advantages of using die-detectors in 3D DfT (Design for testability) context and some synthesis results. This paper is organized as follows. In section II, we introduce the state of the art of Design For Test of 3D integrated circuits, in section III we introduce a 3D test architecture using die-detectors, in section IV we show the 3D test at all test levels, in section V we give the 3D DfT implementation and in section VI we give conclusions and future work.

The work of Y. Fkih is supported in part by the project Master3D

II. STATE OF THE ART OF 3D DFT

Many DfT architectures have been proposed for testing 3D integrated circuits. The first papers addressed pre-bond test of 3D processors using "scan islands" and so called layer test controller (LTC) [3], scan chain optimization approaches [4], and other test issues like test cost optimization [5]. More recent works propose "die level wrappers" based either on IEEE 1500 [6] or IEEE 1149.1 [7] test standards that allow 3D test at all levels: pre, mid, and post bond. In any case, the test architecture includes three features: use of dedicated probe pads for non-bottom dies to perform pre-bond die testing, use of "TestElevators" to drive test signals up and down during post-bond test, and use of a hierarchical WIR chain to configure test interconnects. These features satisfy 3D circuits testing requirements but can be enhanced especially by avoiding the configuration time of the hierarchical WIR, in particular for mid-bond and post-bond tests.

3D DfT architecture can be based either on the two existing standards, IEEE 1149.1 or IEEE 1500, proposing difference design and test tradeoffs. The main difference between these two test standards is the number of used pins: 4 (and 1 optional) for 1149.1 and 8 (and 1 optional) for 1500.

In 3D context, this is an advantage for 1149.1 since the constraint of using less I/Os (either TSVs or pads) is a major constraint for 3D circuits. It is notable that IEEE 1500 wrapper can be controlled using an IEEE 1149.1 Test Access Port, and a dedicated "Test Controller" [8], which can be used in the case of a 3D integrated circuit, where stacked dies have heterogeneous test architectures: some dies wrapped by 1500 and other dies wrapped by 1149.1. The 1149.1 TAP controller is used to control 1500 wrappers while the opposite is not possible.

Parallel testing is better supported by 1500 standard, but can also be done in 1149.1 standard [9]: it is sufficient to use boundary-scan cells with parallel inputs and outputs, and a more developed parallel test scheme can be done as in [10].

In the remainder of this paper we will consider a 3D test architecture based on IEEE 1149.1 (JTAG) standard to illustrate the integration of the die-detection mechanism within a 3D test architecture, but it can be used within any other test standard such as IEEE 1500, IEEE 1149.7, or IEEE P1687.

Design For Testablity of 3D integrated circuits should allow test at all 3D fabrication levels. On the other hand decreasing test time and area overhead of the DfT are very important constraints, which impose reusing of DfT logic as much as possible for pre, mid, and post-bond levels: in pre-bond, test is performed through ordinary pads, and in post-bond, access is done from adjacent dies (for non bottom die) through TSVs. In order to use the same DfT in different stacking levels, multiplexing logic is required to separate TSV path from pad path.

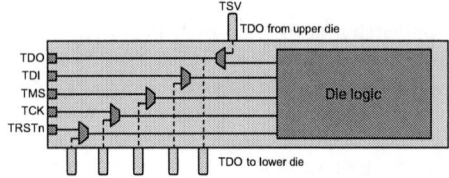

Figure 1. Single die with dedicated JTAG interface

This is illustrated in figure 1, where JTAG inputs are multiplexed between pads and TSVs from bottom and top dies. An obvious solution for this issue is to use multiplexers. In [7], these multiplexers are controlled by the WIR (the Wrapper Instruction Register of the IEEE 1500 std), which means that a step of configuration through 1500 WIR or the 1149.1 IR is necessary before running any test. To avoid this additional test configuration step, we introduce an automatic die detection mechanism which controls the JTAG path within the 3D stack.

III. 3D TEST ARCHITECTURE USING DIE DETECTORS

A. Automatic die detection mechanism

The principle of the automatic die detection mechanism is to use combinational logic to detect the presence of an adjacent die (bottom or top). Each die drives a logic '1' to the upper and lower dies as shown in figure 2, and two die detectors using a pull-down resistance allow the capture of the presence of a top and/or bottom dies: either 'Z' when not connected or '1' when connected.

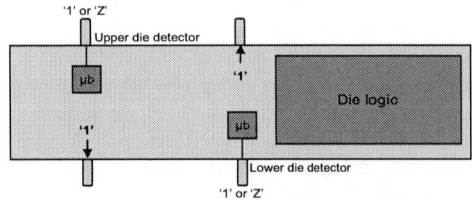

Figure 2. Single die with die-detector mechanism

The automatic die detection is implemented using a specific cell called micro-buffer, containing a buffer to drive the 3D interconnect (TSV and micro-bump) and a pull-down resistance. When the combinational cell is not connected, high impedance 'Z' on its input is detected, the cell drives a '0' logic on its output.

With two die detectors, the position of the current die in the stack is perfectly determined: bottom, middle, or top position. In case of many middle dies in the stack, i.e a 3D circuit with more than 3 dies, their configuration is the same no matter the position in the stack. Truth table of the detection mechanism is given in table 1.

Lower die detector		Upper die detector		3D stack
Input	Output	Input	Output	die position
'Z'	'0'	'Z'	'0'	Bottom (pre-bond)
'1'	'1'	'Z'	'0'	Top
'Z'	'0'	'1'	'1'	Bottom (post-bond)
'1'	'1'	'1'	'1'	Middle

Table 1: Die position definition using die-detectors

B. JTAG based 3D test architecture

The proposed automatic die-detection mechanism is used to define a JTAG based 3D DfT architecture, as presented in figure 3. The die logic is tested through a JTAG port which can be accessed either from pads (for pre-bond test) or from TSVs (for mid, and post-bond tests). The 5 JTAG port signals, TCK, TMS, TDI and TDO, and optionally TRSTn [9], need to be multiplexed according to the test level. The die-detection mechanism is used to define the position of the die in the stack: bottom, middle, or top. The JTAG signals multiplexers are controlled using the two die detectors.

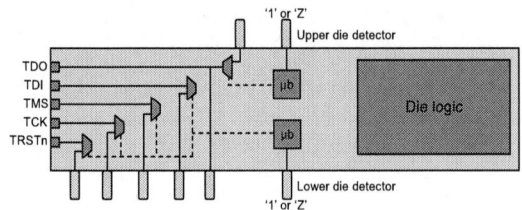

Figure 3. A single die with 2 die detectors

The JTAG multiplexers are controlled by the die-detectors according to four possibilities detailed in table 2.

Die position	Control signals from micro-buffers	
	Lower	Upper
Bottom (pre-bond)	'0' Select pad path	'0' Select TDO of current die
Bottom (post-bond)	'0' Select pad path	'1' Select TDO of upper die
Middle	'1' Select TSV path	'1' Select TDO of upper die
Top	'1' Select TSV path	'0' Select TDO of current die

Table 2: Configuration of multiplexers using die detectors

The lower die detector controls the multiplexer of JTAG inputs: TDI, TMS, TCK, and TRSTn by generating logic '0' to select input from pads and '1' to select inputs from TSVs. The upper die-detector controls multiplexers of the JTAG output TDO by generating logic '0' to select TDO from current die, and '1' to select TDO from upper die. Next section details practical implementation in a 3D test architecture at all 3D testing levels.

IV. 3D TEST STACK TEST AT ALL TEST LEVELS

A. 3D Test Architecture at pre-bond level

Pre-bond test is done at wafer level for each die: this allows to qualify dies to be known as good (KGD). We assume that dedicated test pads are added for the JTAG port for pre-bond test purpose like in [6].

In this situation, die-detectors, which are not shown in the figure 4 for clarity purpose, configure multiplexers of JTAG inputs to be driven from pads, and multiplexer of JTAG output to be driven from current die.

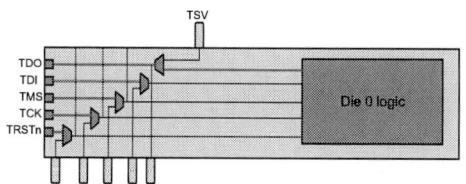

Figure 4. Pre-bond test using die-detectors

The output of the TDO multiplexer is always driven to the output pad and to the lower die through the TSV. The output TDO from die logic is always driven to upper die through TSV as shown in figure 4.

With this test architecture, access to the die is ensured, internal die tests can be performed by shifting test patterns using the serial JTAG input TDI, capturing test results and shift them out through the JTAG output TDO. Parallel test access can also be performed, but this will require additional pads and multiplexing logic.

B. 3D Test Architecture at mid-bond level

Mid-bond test targets partial 3D stacks, where die and interconnects are tested. KGD in pre-bond test phase could be corrupted due to the 3D stacking process (thinning, bonding), that's why DfT architecture should enable this testing level.

An example of 3D partial stack of only 2 dies is considered, where TDO of the bottom die is driven up to the top die to connect its TDI, and TDO of the top die is selected to be driven to the TDO output pad as shown in figure 5.

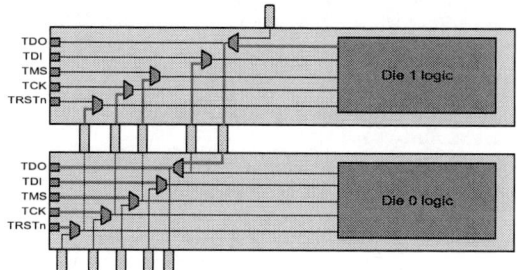

Figure 5. Mid-bond test using die-detectors

Lower die-detectors of the bottom die configure multiplexers to drive JTAG inputs from pads, while for the top die a lower die is detected so JTAG inputs are selected from the bottom die through TSVs.

Upper-die detector of the bottom die configures multiplexer of TDO to be driven from upper die through TSV, while for the top die it does not detect an upper die so it drives TDO of current die out through TSV.

Since, die-detectors and multiplexing logic are fully combinational, this configuration is available for any number of partialy stacked 3D circuit. It is also available for a stack of the same die stacked multiple times. Mid-bond test allow to have Known Good Stack(KGS), which can increase the final 3D stack yield.

C. 3D Test Architecture at post-bond level

Post-bond test targets final 3D stack, where all dies are bonded. An example of a 3D circuit with only 3 dies is considered, the TDI-TDO serial chain is formed as shown in figure 6.

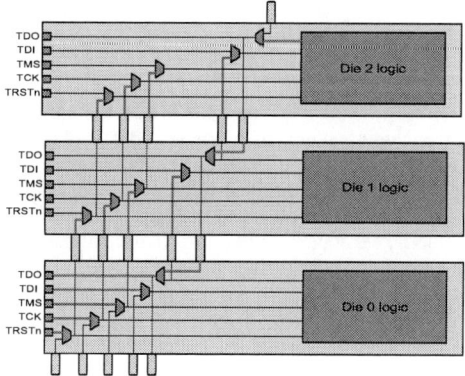

Figure 6. A 3D circuit with 3 dies including JTAG interfaces

The upper die-detector of top die configures the multiplexer of the JTAG output TDO to be driven from the top die to the bottom die through TSVs; so that the TDI-TDO chain is formed serially from down to top.

The 3D circuit can be then compared to a printed circuit board with 3 integrated circuits serially chained, where all JTAG instructions can be executed in sequence: internal test of individual dies, external test of interconnections: TSVs which are vertical interconnections between stacked dies, and pads which connect the 3D circuit to external components.

V. 3D DFT IMPLEMENTATION

A. 3D DfT architetcure design details

In figure 7, we give the details of the proposed 3D test architecture including JTAG infrastructure (TAP controller, instruction decoder, and data registers including the boundary-scan register), upper and lower die-detectors and multiplexing logic.

Mutiplexers colored in pink are controlled by the lower die-detector, and multiplexer colored in green is controlled by the upper die-detector. TSVs at the top of the die are driven to the upper die, and TSVs at the bottom are received from a lower die. Test architecture can be applied to any die no matter its position in the stack.

As in [7], pads are added to perform parallel test in pre-bond level. Lower die-detector is used to multiplex TSV and pad paths of die inputs. Pre-bond test of individual dies can be performed through pads without requiring a configuration step with the JTAG logic.

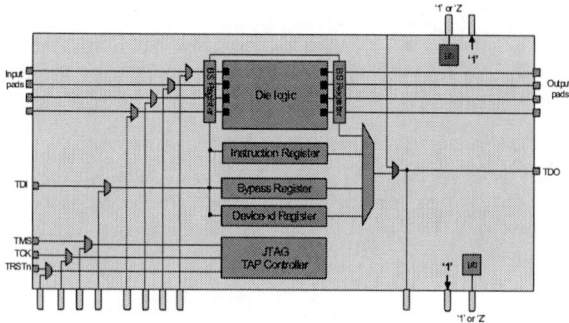

Figure 7. JTAG based *3D test architecture with die-detectors*

It can be noted that it is possible to use de-multiplexers for outputs: including functional outputs and JTAG output TDO to avoid perturbation of normal mode. In that case de-multiplexers will be controlled by the upper-die detector. In the other case it is sufficient to use adequate buffers to charge the additional capacitance of the pads in normal mode.

B. Synthesis results

The proposed DfT architecture has been implemented in RTL and validated. Logic synthesis has been performed using a STMicroelectronics 65nm standard cell libraries,, including a specific micro-buffer cell. The TSV characteristics are the following: 10μm diameter, and 80μm depth (aspect ratio 1/8).

In term of area overhead, synthesis results of a circuit with 1000 TSVs are shown in table 3. Die-detectors require 2 TSVs plugged at logic '1' to upper and lower dies, and 2 TSVs combined with 2 micro-buffer cells per die. The area cost of die-detectors is rather negligible compared to other DfT components.

The area of boundary-scan register is the highest area, this because of the large number of IOs (1000) and because the generic boundary-scan cells uses 2 flip-flops and 2 multiplexers [9].

Component	Area (μm2)
JTAG control logic	5400
Boundary-scan cells	34600
Die-detectors	7040
Total DfT area	47060

Table 3: DfT components area distribution

TSVs matrix area equals to 1.6 mm2, so the area cost of the test architecture is less than 3% of the TSVs matrix area. To compare with existing DfT architectures, as in [7] two types of multiplexers are used: "m9 and m10", The first type to multiplex pads and TSVs, and the second type to multiplex "turn" and "elevator" modes. These multiplexers are configured by the Wrapper Instruction Register. The execution of this configuration step in a JTAG based 3D test architecture require addtional bits in the instruction register, waiting for shifting and updating time of the instruction op-code in the instruction register through the serial path. This is a constraint of this architecture especially when performing pre-bond tests using I/O pads where there is no need to use the JTAG port. The use of die-detectors within test architecture remove these limitations.

VI. CONCLUSION AND FUTURE WORK

3D test architectures can be based on IEEE 1149.1 or IEEE 1500 test standards or even other test standards, but they have the same purpose of enabling 3D circuits testing at all 3D fabrication levels with minimum of added DfT area.

In this paper we presented an automatic die-detection mechanism using two die-detectors: one for upper-die and one for lower die, implemented using a pull-down resistance. With these 2 die-detectors, the die position in the stack is fully defined, which permits the generation of control signals of 3D JTAG multiplexing logic. Die-detectors are integrated within a JTAG based 3D test architecture: the main idea is to use the lower die-detector to drive JTAG inputs, and the upper die-detector to drive JTAG output TDO. The use of die-detectors solves some limitations of the classical 3D test architecture, especially by avoiding the configuration step of multiplexers through the instruction register.

Future work is to use die-detection mechanism in 3D test architectures based on other test standards.

REFERENCES

[1] WideIO JEDEC standard, see http://www.jedec.org/

[2] E.J. Marinissen and Y. Zorian, "Testing 3D chips containing through-silicon vias", in Proc. ITC, 2009, pp.1-11.

[3] Dean L. Lewis, HsienHsin S. Lee "A Scan Island Based Design Enabling Pre-bond Testability in Die Stacked Microprocessors", ITC 2007

[4] Xiaoxia Wu, Paul Falkenstern, Yuan Xie "Scan Chain Design for Three-dimensional Integrated Circuits (3D ICs)", ICCD 2007

[5] Li Jiang, Lin Huang and Qiang Xu, "Test Architecture Design and Optimization for Three-Dimensional SoCs", DATE'09

[6] E.J Marinissen, J Verbree, M Konijnenburg " A structured and scalable test access architecture for TSV-based 3D stacked ICs", in proceedings of VTS 2010, pp. 269 – 274.

[7] E.J Marinissen, Chun-Chuan Chi, J Verbree, M Konijnenburg "A Standardizable 3D DfT Architecture", 3D-TEST'10

[8] Michael Higgins, Ciaran MacNamee, Brendan Mullane "IEEE 1500 Wrapper Control using an IEEE 1149.1 Test Access Port", ISSC 2008

[9] "IEEE Standard Test Access Port and Boundary-Scan Architecture", IEEE Std 1149.1-2001.

[10] Han Ke, Deng Zhongliang, Huang Jianming "Boundary Scan with Parallel Test Access Mechanism", ICEMI '09

[11] "IEEE Standard Testability Method for Embedded Core-based Integrated Circuits" IEEE Std 1500-2005

PRIME 2013, Villach, Austria

Session W4B – Digital Techniques 2

Simulated Power Analysis attacks on a DDPL crypto-core without routing constraints

Simone Bongiovanni, Giuseppe Scotti, Alessandro Trifiletti

Dipartimento di Ingegneria dell'Informazione, Elettronica e Telecomunicazioni (DIET)
Università degli Studi di Roma "La Sapienza", Rome, Italy
{bongiovanni, scotti, trifiletti}@die.uniroma1.it

Abstract— **Delay-based Dual-rail Pre-charge Logic (DDPL) is a logic style introduced with the aim of hiding power consumption in cryptographic circuits in order to prevent Power Analysis (PA) attacks. Its particular data encoding allows to make the adsorbed current constant for each data input combination, irrespective of capacitive load conditions, which allows to design a PA-resistant circuit without routing constraints. In this work we present a fair comparison between SABL, a well-known state of the art transistor level countermeasure which is sensitive to the capacitive mismatches on the complementary lines and requires a customized routing procedure, and DDPL. After having provided a power model for describing the leakage sources for the above mentioned logics, a simple cryptographic circuit has been designed for both SABL and DDPL, and a CPA attack has been mounted. Simulations results show that when capacitive load unbalances are considered, DDPL strongly outperforms SABL in terms of number of traces required for disclose the secret key.**

Index Terms— **Cryptography, Correlation Power Analysis (CPA), Dual-rail Pre-charge Logic (DPL), Sense Amplifier Based Logic (SABL), Delay-based Dual-rail Pre-charge Logic (DDPL).**

I. INTRODUCTION

A side-channel attack is an attempt to recover confidential data, such as the secret key of a cryptographic algorithm, by exploiting the information leaked by the hardware implementation during the execution of the algorithm [1]. For this reason they represent a critical issue for cryptographic applications where a high level of security is required. Side-channels are strongly related to the existence of a physically observable phenomenon, such as time, power, electromagnetic radiations or noise emitted by the device. In this paper we focus on power consumption which is a frequently considered side-channel in practical attacks. Power analysis attacks exploit the dependence of the dynamic power consumption of the hardware implementation on the switching activity and on the state of internal gates, which are both correlated to the processed data. Many techniques have been introduced to promote power analysis attacks, such as Differential Power Analysis (DPA) [2] and Correlation Power Analysis (CPA) [3], template attacks [4], Mutual Information Analysis (MIA) [5], and more recently Leakage Power Analysis (LPA) [6].

Dual-rail Pre-charge Logic (DPL) styles [7] are transistor level countermeasures which aim at de-correlating power consumption from the processed data by making it constant irrespective to input data statistics. Basically DPLs are new logic families introduced for counteracting power analysis both for FPGA and ASIC applications and are also known as anti-DPA logic styles. Sense Amplifier Based Logic [7] is one of the first full custom DPL styles. Other DPL styles as WDDL [9] and MDPL [10] are based on CMOS-composed standard cells and are also suitable for FPGA. One of the most important leakage factors of DPLs is a strong dependence of the power consumption on the capacitive mismatches of the complementary lines, which is expected to increase with the technology scaling [11]. Several transistor level optimizations and novel routing techniques have been proposed [11] [12] with the aim of solving this issue, but the main drawbacks of these solutions are a more complex design flow and constraints on the routing of the differential lines.

In this paper we review the Dual-Rail Delay-based Pre-charge Logic style [13]. It is a full custom DPL style based on a new data encoding which succeeds in making the power consumption constant irrespective of the capacitive unbalance, relaxing the constraints on the routing of the complementary lines. We execute a fair comparison between DDPL and SABL. A theoretical model of the dynamic power consumption is presented for both, in order to compare in which way an unbalanced load may affect the power consumption. In Section V a simulated Correlation Power Analysis attack is performed against a simple crypto-core, implemented with both SABL and DDPL standard cells.

II. LEAKAGE MODEL IN CMOS

In the static CMOS gates there are three distinct dissipation sources [14]: the leakage currents of the transistors (P_{leak}), the short-circuit currents (P_{sc}), and the dynamic power consumption (P_{dyn}). The latter is particularly relevant from a side-channel point of view since it determines a relationship between the processed data inside the gate and its externally observable consumption. In Fig. 1 the model of power consumption is presented for a CMOS inverter, with indicated the dynamic (dotted arrow) and short circuit currents (point arrow). In Fig. 1a, when a transition from 0 to V_{DD} occurs on the output, capacitance C_L is charged and a peak appears in the pattern of the current adsorbed by the power supply line due to

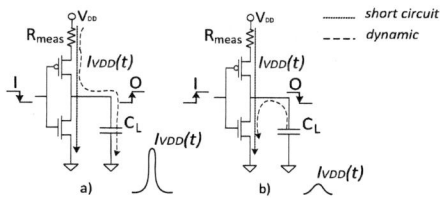

Fig. 1. Model of power consumption in a CMOS inverter.

978-1-4673-4579-8/13 $31.00 © 2013 IEEE

345

the sum of dynamic and short circuit currents. Conversely, when a transition from V_{DD} to 0 occurs as depicted in Fig. 1b, C_L is discharged and the only visible contribution is due to the short circuit current, which can be neglected being much smaller than the dynamic contribution. By generalizing the dynamic power model for the case of a multi ported CMOS logic, suppose that there is a set of N possible configurations of signals at the input. For each data input configuration, the dynamic power consumption is given by the integral of the instantaneous power in a clock cycle, which leads to the well-known formula [14]:

$$P_{dyn}^{(i)} = V_{DD}^2 \, C_L \, f \, P_{0 \to 1}^{(i)} \qquad (1)$$

for i = 1… N. $P_{0 \to 1}^{(i)}$ is the switching activity at the output line for a given input data, and can assume the value 1 if a transition from 0 to V_{DD} occurs or 0 if not. The probability transition factor is given by the frequency of occurrences of a transition 0 → 1, and for N possible data inputs it is:

$$\alpha_t = \frac{\sum_{i=1}^{N} P_{0 \to 1}^{(i)}}{N} \qquad (2)$$

The dynamic power consumption of the gate is the average of the dynamic power for each possible input:

$$P_{dyn} = \frac{1}{N} \sum_{i=1}^{N} V_{DD}^2 \, C_L \, f \, P_{0 \to 1}^{(i)} = V_{DD}^2 \, C_L \, f \, \alpha_t \qquad (3)$$

Equation (3) highlights the dependence of the dynamic power consumption on the probability transition factor, which in turn depends on the switching activity and thus on the data. For instance an inverter has one input, thus N = 2, $\alpha_t = 0.5$.

III. DDPL FUNDAMENTALS

In conventional DPLs (e.g. SABL [8]) data are encoded in the space domain, that is a line is evaluated whereas the other remains stable at the pre-charge value. DDPL introduces a new data encoding based on the time domain: basically the information is encoded according to the order as the lines are charged. A DDPL gate has a fully differential complementary pair composed of an asserted and a not asserted signal, according to which lines is the first to be evaluated (i.e. the asserted signal). The data encoding is characterized by two asynchronous evaluation sub-phases after the rising edge of the clock. Fig. 2 shows the two possible situations for a DDPL data encoding.

During the pre-charge phase the differential lines are set to 0 and, in the evaluation phase, they are both charged to V_{DD} after the clock rising edge. For a logic-1 (Fig. 2a), the first line to be charged is A. Conversely, for a logic-0 (Fig. 2b), the first line to be charged is $\overline{A}$. In [13] some combinatorial and sequential gates have been presented.

IV. DYNAMIC POWER MODELS IN DPLS

The analysis in Section II is extended for DPLs, where the total power consumption is given by the sum of the power consumptions of each differential half circuit. For what concern cryptographic applications, DPLs have the property of making the switching activity of the logic always equal to 1 for each possible input data. Namely at least one output signal makes a transition whereas the other one remains at the pre-charge value [7]. However the mismatch of the output capacitances is a strong limitation in this model.

The architecture of a dynamic cell is reported in Fig. 3 where the real case of a mismatch on the capacitive load is considered. The figure models a DPL one-input logic style (e.g. an inverter). Assuming that the pre-charge is 0, the first column reports the data encoding in the dynamic domain for the evaluated output signals, the columns O_1 and O_2 represent the occurring transition on the respective differential line from the pre-charge to the evaluation value, the other columns indicate the switching factors for each differential line, and the dynamic power consumptions. For a conventional DDPL like SABL, if $C_{L1} \neq C_{L2}$ the overall dynamic power is not constant, highlighting the dependence of the total dynamic power on the value of the single capacitances (see Table I).

A similar model has been extrapolated for DDPL. Unlike SABL, the switching activity for each differential line is always one, which leads to a dynamic power consumption dependent on the sum of the output capacitances and not on their single values. Namely the unbalanced capacitances are both charged and discharged once during a clock cycle. We conclude that the dynamic power consumption for DDPL is constant for each input combination and does not depend on the capacitive load mismatch.

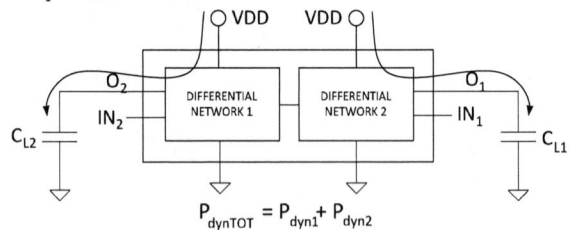

Fig. 3. Architecture of a DPL cell with unbalanced capacitive load.

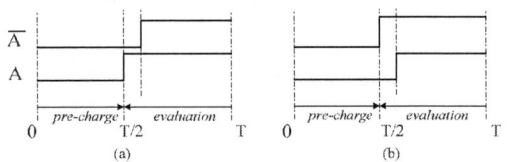

Fig. 2. Time domain data encoding. (a) Logic-1. (b) Logic-0.

Table I. Model of power consumption for SABL and DDPL.

EVALUATION	(O_1,O_2)	O_1	O_2	$P_{0 \to 1}^{O_1}$	$P_{0 \to 1}^{O_2}$	P_{dyn1}	P_{dyn2}	P_{dynTOT}
SABL	(0,1)	$0 \to 0$	$0 \to 1$	0	1	0	$V_{DD}^2 \, C_{L2} \, f$	$V_{DD}^2 \, C_{L2} \, f$
	(1,0)	$0 \to 1$	$1 \to 0$	1	0	$V_{DD}^2 \, C_{L1} \, f$	0	$V_{DD}^2 \, C_{L1} \, f$
DDPL	(0,1)	$0 \to 1$	$0 \to 1$	1	1	$V_{DD}^2 \, C_{L1} \, f$	$V_{DD}^2 \, C_{L2} \, f$	$V_{DD}^2 \, f \, (C_{L1} + C_{L2})$
	(1,0)	$0 \to 1$	$0 \to 1$	1	1	$V_{DD}^2 \, C_{L1} \, f$	$V_{DD}^2 \, C_{L2} \, f$	$V_{DD}^2 \, f \, (C_{L1} + C_{L2})$

PRIME 2013, Villach, Austria *Session W4B – Digital Techniques 2*

V. SIMULATION OF THE CRYPTO-CORE

A case study crypto-core has been attacked with a Correlation Power Analysis (CPA) strategy [3]. It was implemented both with SABL and DDPL standard cells so to compare their CPA resistance. In Fig. 4 the hardware architecture is presented. Basically the core represents a subset of the cryptographic algorithm Serpent [15], in which the 4-bit S-Box S_0 takes as input the XOR between data and key. It was implemented as a 4-bit architecture with two levels of pipeline, and takes three clock cycles for processing a datum.

In Fig. 5 the layout of the DDPL implementation is shown. It was designed with Cadence Virtuoso XL Layout Editor by using the layout views of the BSIM4 low power standard-V_T transistor models available in the STMicroelectronics 65nm-CMOS library. All cells were designed so to have a low area occupation: all transistors were sized with a minimal channel length (65nm) and the aspect ratios of nMOS and pMOS were set to 2 and 4 respectively according to the values used in the library. All DDPL gates in the scheme were cascaded by at most two other DDPL gates in order to require a minimum fan out with no routing congestions. An automatic routing was executed with the Automatic IC Routing Tool of Virtuoso. The design covers a total area of about 0.004 mm².

A post layout simulation allowed to estimate the capacitances at the complementary lines. We loaded the outputs of the gates used in the SABL schematic with the same values so to perform a fair comparison. In the CPA we use the Hamming weight of data at the output of the second stage of registers (i.e. the input of the DDPL-CMOS converters CONV-1) as selection function,. The current adsorption of the entire test chip was measured by simulations in Cadence Spectre environment. A working frequency equal to 100MHz was used for the clock signal and input signals were generated with a rising/falling edge of 20ps. The nominal value 1V was chosen for the power supply which is a typical value for the BSIM4 low power standard-V_T transistors (at the nominal process corner).

The chip was tested by inserting a number of 40.000 randomly generated binary input data with a period of 20ps. In Fig. 6 the pattern of a simulated noisy current trace is shown for the SABL chip, which corresponds exactly to one elaboration. The three clock cycles are clearly visible, and the pre-charge and the evaluation phases (in this chronological order) are also distinguishable. Note that the attack has been performed at the output of the second stage of registers, therefore we expect that some peaks are detectable in the time window between 25ns and 30ns (i.e. the evaluation phase of the third clock cycle). The time resolution was 5 points for clock cycle (i.e. 15 points for elaboration) which corresponds to a sampling frequency of 500MSample/s. Correlation traces for each key guess with 40000 number of queries are shown in Fig. 7. A correlation peak is visible on the last two samples of the trace (black line) which indicates the successful of the attack. Attacks were repeated for the DDPL implementation of the test chip. Fig. 8 and Fig. 9 depict a noisy DDPL current trace and the correlation coefficient for each possible key guess in a CPA with 40000 queries, respectively. Note that the failure of the attack is indicated by the absence of a visible peak in correspondence to the correct key guess (black line).

In order to compare the implementations with an actual security metric we used the minimum number of Measurement To Disclose (MTD) the key [16]. We repeated the attacks for a different number of input queries and calculated the absolute value of the correlation coefficient for each key guess (Fig. 10 and Fig. 11). Results show that SABL implementation is not robust against CPA in presence of capacitive mismatches on the differential lines. On the contrary DDPL helps to enhance the security level of the chip by hiding the power consumption in a crypto test chip irrespective of the geometry of the wires.

In Table II results of the attacks and performances of the chips are summarized. A CPA with 40000 traces was not successful on DDPL, whereas for SABL 1500 traces have been sufficient for disclosing the key. The correlation coefficient for the correct key is also reported for a 40000 input attack.

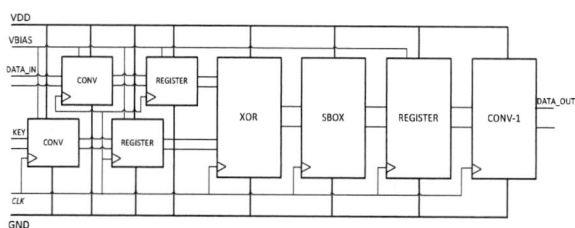

Fig. 4. Hardware architecture of the case study crypto-core

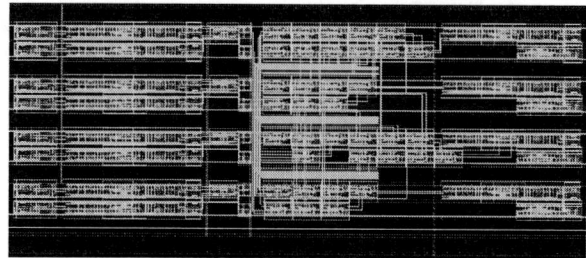

Fig. 5. Layout of the DDPL crypto-core.

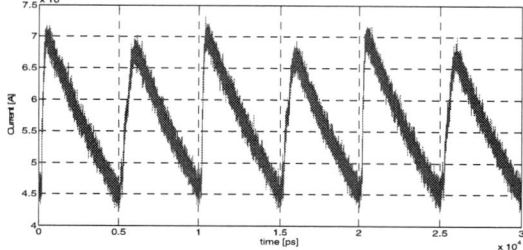

Fig. 6. Simulated current traces of the SABL crypto-core for one elaboration.

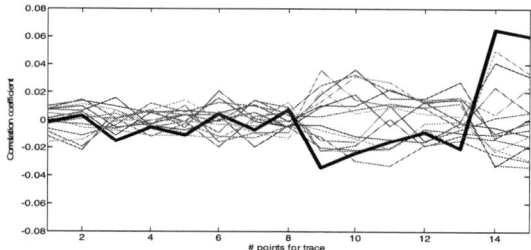

Fig. 7. Correlation coefficients for CPA attacks on the SABL crypto-core.

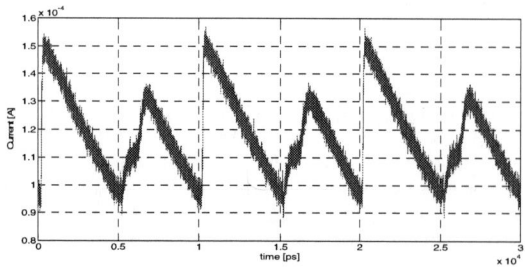

Fig. 8. Simulated current traces of the DDPL crypto-core for one elaboration.

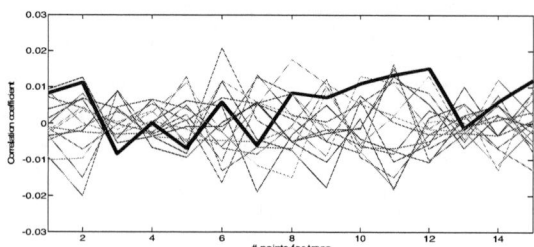

Fig. 9. Correlation coefficients for CPA attacks on the DDPL crypto-core.

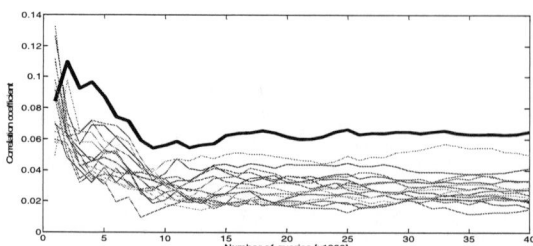

Fig. 10. Success rate of CPA against the SABL implementation.

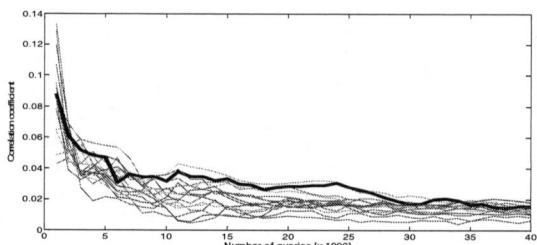

Fig. 11. Success rate of CPA against the DDPL implementation.

Table II. Comparison of performances between the crypto-cores under test

	Area [μm^2]	Average power [μW]	MTD	Correlation coefficient (for q = 40000)
SABL	1703	56.31	~ 1500	0.06
DDPL	2323	117.79	> 40000	0.01

VI. CONCLUSION

In this paper we provided a leakage model for the power consumption of DDPL. DDPL is a delay-based dynamic differential logic which allows to de-correlate the dependence of the power consumption on the unbalanced capacitive loads. We designed a crypto-core by adopting DDPL standard cells and an automatic routing. After having estimated the unbalanced capacitances on each internal complementary line through a post layout analysis, we also designed a SABL implementation of the same crypto-core where the estimated capacitances were inserted so to have a fair comparison when a Correlation Power Analysis attack was mounted. Results demonstrated that a CPA against SABL circuit was successful for a relative few number of input queries (i.e. ~1500), whereas the DDPL design overcomes the latter, being CPA unsuccessful up to 40000 queries. Thus we conclude that DDPL is more resistant on power analysis with respect to SABL, even when internal capacitances mismatches are taken into account, and guarantees an high level of security without requiring ad hoc routing procedures.

REFERENCES

[1] P. C. Kocher. Timing attacks on implementations of Diffie-Hellman, RSA, DSS, and other systems. In Proceedings of Advances in cryptology, CRYPTO '96. Lecture Notes in Computer Science (LNCS), vol. 1109, pp. 104–13, Springer, 1996.

[2] P. C. Kocher, J. Jaffe, and B. Jun. Differential Power Analysis. In Proceedings of Advances in cryptology, CRYPTO '99. LNCS, vol. 1666, pages 388–397. Springer, 1999.

[3] E. Brier, C. Clavier, and F. Olivier. Correlation Power Analysis with a Leakage Model. In Proceedings of CHES 2004, LNCS, vol. 3156. Springer-Verlag, Berlin, 2004, pp. 16-29.

[4] S. Chari, J. Rao, and P. Rohatgi. Template Attacks. In Proceedings of CHES 2002, LNCS, vol 2523, pp. 13-28, USA (CA), August 2002.

[5] B. Gierlichs, L. Batina, P. Tuyls, and B. Preneel. Mutual information analysis. In Proocedings of CHES 2008, LNCS, vol 5154, pp 426-442.

[6] M. Alioto, L. Giancane, G. Scotti, and A. Trifiletti. Leakage Power Analysis Attacks: A Novel Class of Attacks to Nanometer Cryptographic Circuits. IEEE Transactions on Circuits and Systems I, vol. 57, no. 2, pp. 355-367, 2010.

[7] S. Mangard, E. Oswald, and T. Popp. Power Analysis Attacks: Revealing the Secrets of Smart Cards. Springer, 2007.

[8] K. Tiri, M. Akmal, and I. Verbauwhede. A dynamic and differential CMOS logic with signal independent power consumption to withstand differential power analysis on smart cards. In Proocedings of ESSCIRC 2002, pp.403–406.

[9] K. Tiri and I. Verbauwhede. A logic design methodology for a secure DPA resistant ASIC or FPGA implementation. In Proceedings of DATE 2004, pp. 246–251.

[10] T. Popp and S. Mangard. Masked Dual-Rail Pre-charge Logic: DPA Resistance Without Routing Constraints. In *Proceedings of CHES 2005*, LNCS, vol. 3659, pp. 172–186, Springer, 2005.

[11] K. Tiri and I. Verbauwhede. Place and route for secure standard cell design. In Proceedings of CARDIS 2004, pp. 143–158.

[12] S. Guilley, P. Hoogvorst, Y. Mathieu, and R. Pacalet. The "Backend Duplication" Method. In Proceedings of *CHES 2005*, LNCS, vol. 3659, pp. 383–397,Springer, 2005.

[13] M. Bucci, L. Giancane, R. Luzzi, G. Scotti, and A. Trifiletti. Delay-Based Dual-Rail Precharge Logic. IEEE Transactions on Very Large Scale Integration (VLSI) Systems, vol. 19, no. 7, 2011, pp. 1147-1153.

[14] J. M. Rabaey, A. P. Chandrakasan, and B. Nikolic. Digital Integrated Circuits: a Design Perspective. Prentice Hall electronics and VLSI series, II ed., Pearson Education 2003.

[15] R. Anderson, E. Biham, and L. Kundsen. A proposal for the Advanced Encryption Standard. AES proposal, 1998, available at http://www.cl.cam.ac.uk/ftp/users/rja14/serpent.pdf.

[16] K. Tiri, D. Hwang, et al. Prototype IC with WDDL and differential routing - DPA resistance assessment. In Proceedings of CHES 2005, pp. 354-36

An Improved Instruction-Level Energy Model for RISC Microprocessors

Wei Wang and Mark Zwolinski
School of Electronics and Computer Science
University of Southampton
Southampton SO17 1BJ, UK
Email:{ww5g09,mz}@ecs.soton.ac.uk

Abstract—The power and energy consumed by a chip have become primary design constraints for embedded systems and are largely affected by software. However, there is a gap between software and hardware that makes it hard to predict which code consumes the least power before running it. Therefore, it is vital to discover which factors affect a program's energy consumption. In this paper we present an instruction-level power model for a single core, in-order RISC processor architecture. We do not analyze each instruction individually, but we study the average power and running time instead. We find the power in a processor is nearly constant, no matter what instructions are run, but the IO port power is related to the behavior of the program. Furthermore, we provide a model that takes the cache miss rate into consideration.

Keywords—Energy estimation, Energy modeling

I. INTRODUCTION

Power consumption is a crucial aspect of modern digital design, because mobile devices use batteries as their energy sources. However, it is not only the architecture of the processor that can affect the energy consumption but the program as well. This means that a microprocessor system can have different energy consumption when running different programs. Tiwari and Lee [1] state that one of the benefits of an instruction-level analysis is that it provides clues about how to write a effective power savings software applications.

This paper presents a new instruction-level model to estimate the energy consumption of a program based on a single core, in-order RISC processor. One of the contributions of this paper is an investigation into the effect of the cache miss rate and we present an energy model which takes this into consideration. Another contribution is the concept of studying the energy via power and running time rather than studying each instructions individually [2]. Tests show the power of core is nearly a constant regardless of the instruction type. Therefore, in comparison with the other models, the one presented here is simpler but maintains accuracy.

This paper is organized as follows: The background to the work is presented in section 2. Section 3 discusses the basic architecture features of the OpenRISC 1200 processor, the tools used to measure the power and the experimental method. In section 4, the results and analysis of the basic tests are shown. Based on these results, an instruction level energy model is given in section 5. Sections 6 presents the energy estimation and analysis of this model while section 7 compares the results with standard models. Lastly, section 8 summarizes the contributions of this paper.

II. BACKGROUND

A number of models for estimating energy consumption at the instruction level have been proposed. The model in [2] expresses the average energy as:

$$E = \sum_i (B_i \times N_i) + \sum_{i,j} (O_{i,j} \times N_{i,j}) + \sum_k E_k,$$

where E is the total average energy consumed by the program. B is the base cost of instruction which is multiplied by N, the number of this instruction executed in the program. The second term is the additional energy cost due to instruction switching. O is the overhead cost of different instruction sequences and N is the number of occurrences of this sequence. O models the fact that the energy consumed by two different sequentially executed instructions is greater than that of the sum of the basic energy of two individual instructions. This is because more of the circuit state changes when switching between two different instructions compared to that of the same instructions. The last term, E_k, is any additional energy due to cache misses or resource constraints.

This basic model has been used to study the energy usage of processors such as the ARM7TDMI [3], PowerPC 603e [4], and the effect of adjacent instructions is also considered in [5].

When the instruction-level power is analyzed, the result is strongly affected by the operand in some cases [5], [6], and a new model called the data dependency model was created to address this.

$$\begin{aligned} power_{average} &= P_{data} + P_j + P_{i,j} \\ &= K_1 \cdot n_1 + \cdots + K_n \cdot n_n + K_0 + C_{i,j}, \end{aligned}$$

where n_i is the elements which can influence the power consumption, and K_i is their respective weighting. K_0 is the minimum cost for the particular instruction and $C_{i,j}$ is the changing-instruction cost between instructions i and j [6], like the overhead energy discussed above.

However neither of these two models considers a method to analyze cache misses and pipeline stalls. As expected, the results show that the cache miss rate plays an important role in estimating the energy consumption and power usage.

III. EXPERIMENTAL SETUP

A. Target Processor

The OpenRISC 1200 is a 32-bit Harvard architecture scalar RISC processor with a 5 stage integer pipeline and some basic DSP capabilities. It can also support both instruction and data caches with the inclusion of a management memory unit (MMU). OpenRISC 1200 can be configured by the user, for example, to delete or reduce the area of the cache [7].

B. Simulation tools

A standard RTL synthesis tool was used to implement the OpenRISC on a 7 layer metal, 1.05V, low-power high-threshold-voltage 65nm process. A maximum clock speed of 111MHz was achieved. Two files are generated: a netlist and an SDF file (standard delay format) which is used to record the delay of each gate and pin within the netlist.

An industry-standard logic simulator was then used to simulate the different instructions and programs, using the netlist and SDF file to simulate different instructions. After simulation, all of the signal switching information is stored. Synopsys Primetime was used to analyze power for each test. For a CMOS circuit, the power consumption can be divided into two main categories: *dynamic power* when transistors are switching, and *static power* when the transistors remain stable.

IV. RESULTS AND ANALYSIS

The method of testing is to run an infinite loop and the main body of this loop consists of 100 instances of an individual instruction type. The test shows the power consumption of the core and IO individually in 4 cases: the cache can either hit or miss meanwhile the operand switching rate can be either be high or low. In order to prevent the IO ports from sending unnecessary data, the design was modified to filter unnecessary memory access and the processor only sends and receives data from memory when required.

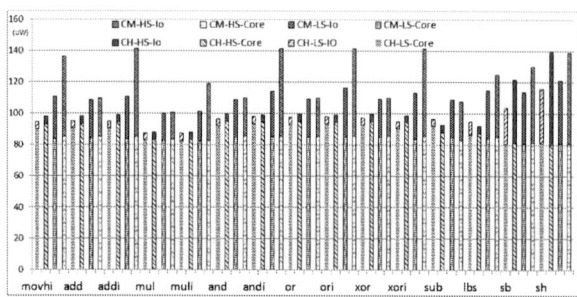

Fig. 1. The Power consumption of OR1200 with 4KB cache

Figure1 shows the power consumption measurement on the OR1200, and in this figure, "CH", "LS","CM","HS", "Core" and "IO" mean cache hit, low switch, cache miss, high switch, the power consumption of core and IO respectively.

For different instructions, the core power consumption is quite similar. For a cache hit, the maximum difference is 15.4% and for a cache miss, the maximum difference is only 6.7%. The reason for this is that nearly all of the instructions need 5 pipeline stages. No matter what the instruction is, the ALU always performs all the different operations and only chooses a specific one as an output. Therefore, logic calculation instructions have similar power consumption. So when analyzing power usage, it is not necessary to consider

the difference between a cache hit or a cache miss and so s constant number can be used to estimate the basic power.

Data switching can affect the core power, but not by very much (4.46%). Because the average number of operand switching bits is about 7 [5], the power found for "low switch" can be used as the standard power consumption of the core in the model. But data switching can affect the IO port power considerably, especially for immediate addressing mode instructions. This is because two consecutive immediate mode operands may have a large Hamming distance. IO ports consume power only when communication happens between the processor and memory. This will only occur following a cache miss, or during execution of load or store instructions. Therefore, our hypothesis is that the power consumption is related to the cache miss rate and Load/Store instruction percentage.

V. INSTRUCTION LEVEL MODELING

Because we are more interested in the energy consumed by a chip than the power, the following section defines an energy model based on instruction level power. The model is described by Equation 1.

$$E = \int_0^T p \times t dt = \bar{P} \times T$$
$$= (\bar{P}_{core} + \bar{P}_{io}) \times (T_{cache\_hit} + T_{cache\_miss}) \quad (1)$$

where $\bar{P}$ is the average power, and T is the total run time. $\bar{P}_{core}$ is the average power of the core, which can be considered a constant, based on the analysis above. $\bar{P}_{io}$ is the average power of the IO ports, which is related to the cache miss, load and store instructions. Hence:

$$\bar{P}_{io} = \bar{P}_{io}(cache\_miss\_rate, ld\_st\_rate), \quad (2)$$

After analysis, the execution time can be divided into two cases: cache hit and cache miss (which leads to a pipeline stall). This is described by the following equations:

$$T_{cache\_hit} = \sum_i N_i \times T_i \quad (3)$$

$$T_{cache\_miss} = R_{cache\_miss} \times \sum_i N_i \times T_{penalty} \quad (4)$$

where T_{cache_hit} and T_{cache_miss} are the times taken for a cache hit and miss respectively. N_i is the number of instructions in each group (MAC, ALU, load and store) and T_i is the timing for each group. R_{cache_miss} is the cache miss rate for the whole program. Based on the analysis above, from equations (1) to (4), we derive equation (5) :

$$E = (\bar{P}_{core} + \bar{P}_{io}(cache\_miss\_rate, ld\_st\_rate)) \times$$
$$(\sum_i N_i \times T_i + R_{cache\_miss} \times \sum_i N_i \times T_{penalty}). \quad (5)$$

VI. ESTIMATION AND ANALYSIS

In order to analyze how the cache and load/store instructions affect the energy usage, we synthesized two different versions of OpenRISC: (1) 512B instruction cache and 512B data cache. (2) 4kB instruction cache and 4kB data cache. The same test programs were run on these two different processors.

978-1-4673-4579-8/13 $31.00 © 2013 IEEE

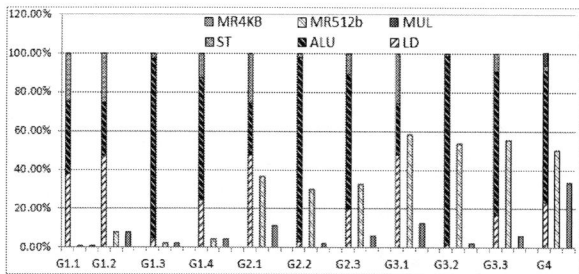

Fig. 2.   The components of each basic test program

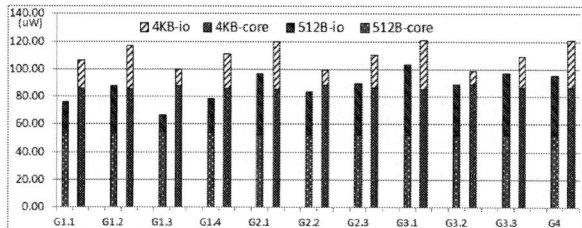

Fig. 3.   The power consumption of each basic test program

Figure 2 presents three columns in each test and the first one shows the components of each test including load instructions, ALU instructions, store and multiply instructions. The second and third columns shows the cache miss rate for 512kB cache and 4kB cache respectively.

The tests are divided into four groups. The main bodies of these eleven basic test programs in Figure 2 are loops with random instruction components. It does not show the *branch* and *nop* components because branch only appear at the end of the program. In the first group, the test loop size is short enough to be stored in the instruction cache completely. In the second group, the loop size is bigger than 512B but smaller than 4kB, which means cache misses can occur. In the third group, the loop size is bigger than in the second group, but smaller than 1kB, which means the cache miss rate is higher than before but cache hits still occur. Group 4 consists of a single program to validate the instruction level energy model works, and is based on matrix multiplication.

Test G1.1 in group 1 is a simple looping C program. Tests G1.2, G2.1 and G3.1, in groups 1, 2 and 3 respectively, have a higher percentage of store instructions compared with other tests in the same group. For tests G1.3, G2.2 and G3.2, the main components of the loop are logic instructions. Tests G1.4, G2.3 and G3.3 have a balance of store and logic instructions. Figure 3 shows the power usage of each test program. From Figures 2 and 3, the following can be observed:

As hypothesized above, the core power of different programs is almost constant, and the maximum difference between test programs is 3.1% and 4.2% for the 512B and 4kB caches respectively. The average power is 52.7μW and 86.7μW, respectively. We use 86.7μW as the estimated power for the core with 4kB cache.

The IO power is related to the proportion of store instructions, as expected. Group 1 shows that the higher the

percentage of store instructions in the program, the more IO power it will consume. The reason for this is that the cache is a one-way direct mapped cache, which means any time the processor transfers data to memory, it will communicate with both cache and memory.

The IO power is also related to the cache miss rate. For the 512B-cache processor in groups 2 and 3 we observe that the lower the cache miss rate, the less IO power is consumed (this can be seen by comparing tests G2.1, G2.2, G2.3 with G3.1, G3.2, G3.3, respectively). Furthermore, for the 4kB processor the cache can store the whole program, which means the processor only experiences cache misses in the first loop iteration, whereas the 512B-cache processor cannot store the whole program. Therefore, it is quite clear that any time a cache miss happens, the 512B-cache processor will communicate with memory via IO ports, and hence will consume more power than the 4kB one.

Based on Figures 2 and 3, we can derive equation (6) by linear regression to describe the IO power, with the cache miss rate and store instruction rate.

$$\bar{P}_{io} = 13.633 + 48.5273 \times P_{ST} + 48.3817 \times P_{miss} - 3.0835 \times P_{ST} \times P_{miss}, \tag{6}$$

where the P_{ST} is the ST instruction percentage of the whole program and P_{miss} is the average cache miss rate.

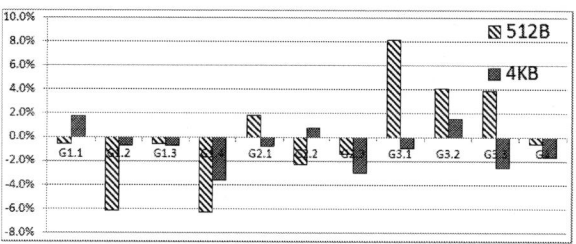

Fig. 4.   Estimation result of basic test

Figure 4 shows the difference between the measured results and the model. Comparing all of the programs in each group shows that the worst case comes from the test G3.1 which is 8.2% for the 512B cache. For most of the other cases, the difference is less than 5% and the minimum difference is only -0.5% and -1.6% for 512B and 4kB caches, respectively.

VII.   COMPARISON

TABLE I.   THE COMPONENTS OF EACH STANDARD BENCHMARK TEST PROGRAM

Name	NOP(%)	BR (%)	LD (%)	ALU (%)	ST (%)	MUL (%)	MR (%)
Fib	14.58	14.58	22.90	31.26	16.67	0.00	1.03
Fir	5.70	5.70	42.75	23.94	18.42	3.50	2.27
Qs	10.35	10.35	40.12	28.71	10.47	0.00	6.58
Tak	13.89	13.89	28.68	22.23	21.31	0.00	0.94
Han	10.95	10.95	26.50	28.13	23.46	0.00	1.13

Five standard benchmarks are shown in Table I, used to test the performance of the model: Fibonacci, FIR filter, Quicksort, Tak and Tower of Hanoi. Columns "NOP" and "BR" shows the nop and branch instruction percentages individually. In this standard test, we consider only the 4KB cache processor.

TABLE II. ESTIMATION RESULTS FOR STANDARD BENCHMARKS

Test	io_m	io_e	io_dif%	P_{core}_m	P_{total}_dif%	Energy dif%
Fib	27.1	22.22	-18.01	86.1	-3.78	-10.58
Fir	26.5	23.66	-10.72	86.1	-1.99	-6.00
Qs	26.1	21.88	-16.17	86.9	-3.91	-11.28
Tak	26.2	24.42	-6.79	85.5	-0.52	-7.06
Han	25	25.56	2.24	85	2.05	-2.69

Table II shows the results of applying our model to standard benchmarks. Columns "io_m" and "io_e" show the simulated and estimated IO power in μW respectively and "io_dif" shows the percentage difference between "io_m" and "io_e". "P_{core}_m" and "P_{total}_dif%" show the measured power consumption of the core and the total power difference between estimation and measurement (including IO and core power). Here, we use 86.7μW as the estimated power consumption of the core. Finally, "Energy dif%" shows the total energy difference between estimation and measurement.

From Table II, we can see that the power estimation is very accurate. The maximum error is -3.91% (quicksort). The IO power has a maximum error of -16.7% (quicksort). The energy estimation is also accurate: the maximum error is -11.28%. Timing errors mostly come from data cache misses. The IO power is fairly stable because the miss rate and ST(%) is low.

TABLE III. COMPARISON WITH PREVIOUS WORK: ENERGY ESTIMATE

	Our method	[3]	[5]	[4]
Fibonacci	-10.58%	—	15.58%	
Fir	-6.00%	-4.05%	—	11.52%
Quicksort	-11.28%	—	11.41%	—

Our method is compared with previously reported results in Table III. The method reported in [3] gives a better estimate because it considers the overhead energy of each instruction pair. In [5], there are two models. For the first model, we have a better performance; the second model is more accurate because it considers factors such as the effect of two adjacent instructions. We have a better result than that in [4].

TABLE IV. INSTRUCTION COMBINATIONS

	model considering overhead
Fibonacci	9+10
Fir	11+13
Quicksort	16+25
General	$4325(93^2/2)$

We do not need to consider the effect of adjacent instructions, thus save a lot of measurement time. Table IV shows how many measurements are needed when considering the effect of adjacent instructions. There are 9, 11 and 16 different instructions used in Fibonacci, FIR filter and Quicksort, respectively. Moreover, there are an additional 10, 13 and 25 instruction pairs in Fibonacci, FIR filter and Quicksort that need to be characterized in the more complex model. However, in our model, the measurement times are proportional to the numbers of instruction types, therefore we need four tests in total. Generally, there are 93 different opcodes in the ISA and 4325 different instruction pairs. Another example which considers the effect of adjacent instructions shows that there are 49 instructions in ISA and 1176 tests are needed for a DSP

56K chip [8]. In our model we consider the energy of the cache miss in terms of a timing penalty. This is because when a cache miss appears, the program will need more time to finish, and will therefore consume more energy. This approach gives an effective method for considering how cache misses affects the energy consumption, which is not considered in [3], [5] or [4].

VIII. CONCLUSIONS

This paper presents an instruction-level energy model for a single core, in-order RISC processor architecture, in which the effect of cache misses is considered. First, the power in the processor does not change much for different operations and operand switch rates, and is thus considered constant. Several tests based on other processors have found the power of the core is fairly constant for different instructions, for example, the StrongARM [9]. Thus, the method may be applied to other RISC processor architectures. Second, the IO port power is related to the percentage of store instructions and the cache miss rate. Using linear regression, an accurate IO port power equation is derived. Instead of analyzing the energy of each instruction individually, we use average power and run time to estimate the total energy. Finally, a timing equation considering the cache miss is also presented. We demonstrate that our model is almost as accurate as those that consider the effect of adjacent instructions, but that the model can be characterized with significantly less effort.

REFERENCES

[1] V. Tiwari and M. Tien-Chien Lee, "Power analysis of a 32-bit embedded microcontroller," in *Proceedings of the ASP-DAC '95/CHDL '95/VLSI '95*, Aug-Sep 1995, pp. 141 –148.

[2] V. Tiwari, S. Malik, and A. Wolfe, "Power analysis of embedded software: a first step towards software power minimization," *Very Large Scale Integration (VLSI) Systems, IEEE Transactions on*, vol. 2, no. 4, pp. 437 –445, Dec. 1994.

[3] N. Kavvadias, P. Neofotistos, S. Nikolaidis, K. Kosmatopoulos, and T. Laopoulos, "Measurements analysis of the software-related power consumption in microprocessors," in *Instrumentation and Measurement Technology Conference. IMTC '03. Proceedings of the 20th IEEE*, vol. 2, May 2003, pp. 981 – 986 vol.2.

[4] O. Acevedo-Patino, M. Jimenez, and A. Cruz-Ayoroa, "Static simulation: A method for power and energy estimation in embedded microprocessors," in *Circuits and Systems (MWSCAS), 53rd IEEE International Midwest Symposium on*, Aug. 2010, pp. 41 –44.

[5] S. Penolazzi, L. Bolognino, and A. Hemani, "Energy and performance model of a sparc leon3 processor," in *Digital System Design, Architectures, Methods and Tools. DSD '09. 12th Euromicro Conference on*, Aug. 2009, pp. 651 –656.

[6] D. Sarta, D. Trifone, and G. Ascia, "A data dependent approach to instruction level power estimation," in *Low-Power Design. Proceedings. IEEE Alessandro Volta Memorial Workshop on*, Mar. 1999, pp. 182 –190.

[7] D. Lampret, "OpenRISC 1200 ip core specification," Jan 2011. [Online]. Available: http://opencores.org/

[8] B. Klass, D. E. Thomas, H. Schmit, and D. F. Nagle, *Modeling inter-instruction energy effects in a digital signal processor*. Citeseer, 1998.

[9] A. Sinha, N. Ickes, and A. Chandrakasan, "Instruction level and operating system profiling for energy exposed software," *Very Large Scale Integration (VLSI) Systems, IEEE Transactions on*, vol. 11, no. 6, pp. 1044 –1057, Dec. 2003.

Author Index

Aaltonen, Lasse, 117
Adnan, Muhammad, 301
Ahmed, Syed Ershad, 141
Ahn, Eun Hye, 273
Ahonen, Tapani, 237
Airoldi, Roberto, 237
Al-Qudsi, Belal, 337
Ashok, Arun, 229
Atasoy, Oguz, 217

Bösch, Wolfgang, 221
Babin, G., 9
Badr, Elie, 253
Barbaro, Massimo, 105
Barke, Erich, 245
Baschirotto, Andrea, 37, 45, 53, 69, 81, 201, 225, 261
Becker, Joachim, 153
Berland, Corinne, 101
Berroth, Manfred, 193
Bianchi, Davide, 61
Bluder, Olivia, 281, 285
Bodano, Emanuele, 105
Bongiovanni, Simone, 345
Bonizzoni, Edoardo, 325
Bruening, Ulrich, 233
Bruschi, Paolo, 65, 197
Bruun, Erik, 57
Burenkov, Alex, 241
Burghartz, Joachim, 41, 317
Butaud, Rémi, 33

Carminati, Marco, 61
Carta, Corrado, 265, 305
Casha, Owen, 113
Cavallo, Domenico, 37
Chekurov, Nikolai, 117
Chen, Wei, 49
Chironi, Vincenzo, 81
Chlis, Ilias, 85
Choi, Hyun Jin, 273
Choi, Jun Rim, 273

Cinti, Alessandro, 129
Congiu, Andrea, 105
Craciunoiu, Florea, 297
Crespi, Lorenzo, 53
Crouwels, Arne, 277
Cuppini, Matteo, 133

D'Amico, Stefano, 81, 201, 225, 261
Dallago, Enrico, 205
De Berti, Claudio, 53
De Blasi, Marco, 69
De Matteis, Marcello, 37, 45, 69, 81, 201, 225, 261
Dehé, Alfons, 125, 293
Dehollain, Catherine, 209, 213, 217, 321
Dekneuvel, Eric, 33
Del Cesta, Francesco, 65, 197
Demir, Alper, 173
Di Natale, Giorgio, 169, 341
Dini, Michele, 269
Donno, Andrea, 201, 225
Draxelmayr, Dieter, 19
Droste, Dirk, 7, 249
Duller, C., 9
Dundar, Günhan, 329

Ellinger, Frank, 93, 189, 257, 265, 305, 313, 337
Estevez, Francisco J., 149

Fanucci, Luca, 121
Ferenci, Damir, 193
Ferrari, Giorgio, 61
Fettweis, Gerhard, 93
Filippi, Matteo, 269
Fkih, Yassine, 341
Flatscher, Martin, 89
Flottes, Marie-Lise, 169, 341
Fornasari, Andrea, 69
Franchi Scarselli, Eleonora, 133
Fulde, Michael, 9

Gaier, Ulrich, 3
Gatt, Edward, 113
Gatti, Umberto, 325

Author Index

Genovese, Mariangela, 137
Gerigk, Janina, 193
Ghazanfari, Leyla, 237
Giotta, Dario, 3
Glösekötter, Peter, 149
Glacer, Christoph, 125, 293
Grözing, Markus, 193
Grech, Ivan, 113
Grimm, Christoph, 165
Gronicz, Jakub, 117
Gruenberger, S., 9
Gruschke, Oliver, 25
Gschier, Tony, 89

Habibovic, H., 9
Halonen, Kari, 77, 97, 117
Hammerschmidt, Dirk, 301
Harrant, Manuel, 165
Hassan, Tarek, 41
Hauer, Johann, 309
Hauptmann, Joerg, 3
Heidari, Hadi, 325
Heinen, Stefan, 229, 233
Herlitschka, Sabine, 1

Indevuyst, Stijn, 277

Jacquemod, Gilles, 33
Joram, Niko, 257, 337
Jouda, Mazin, 25
Jørgensen, Ivan, 57

Kärkkäinen, Mikko, 77, 97
Kömürcü, Giray, 329
Kaltenbacher, Manfred, 293
Kampus, V., 9
Kappert, Holger, 181
Kapucu, Kerem, 213
Khafaji, Mahdi, 265
Khandelwal, Sourabh, 29
Killat, Dirk, 185
Kirscher, Jérôme, 165
Knoblinger, G., 9
Kokozinski, Rainer, 181
Kollmitzer, Michael, 245
Korotkov, Alexander, 309
Korvink, Jan, 25
Krassnitzer, C., 9
Kreuter, Hans-Peter, 289
Kusko, Mihaela, 297
Kuttner, F., 9

Lam, Kai Chi Alex, 177

Lang, Felix, 193
Laur, Rainer, 125
Lazzarini Barnabei, Alessandro, 205
Leggeri, Giuseppina, 37
Leufker, Jan Dirk, 305
Leys, Richard, 233
Li, Fangyan, 33
Liberale, Alessandro, 205
Lindholm, Christian, 19
Liu, Xiao, 321
Longhitano, Aurelio, 65, 197
Lu, Feng, 169

Mahmoudi, Hiwa, 157
Mahmutoglu, A. Gokcen, 173
Mahsereci, Yigit, 317
Maier, Tobias, 249
Malcovati, Piero, 53, 205
Malhotra, Gayatri, 153
Maloberti, Franco, 325
Manica, Jennifer, 333
Marchioro, Alessandro, 45
Melnychenko, Oleksandr, 289
Menicocci, Renato, 145
Merino Panadés, Jose Luis, 213
Meyvaert, Hans, 277
Micusik, Daniel, 265
Mohr, Bastian, 233
Morel, Florent, 73
Mrad, Roberto, 73
Mucci, Claudio, 133
Mueller, Jan Henning, 233

Nagari, Angelo, 73
Napoli, Ettore, 137
Nawaz, Mohsin, 293
Nirmaier, Thomas, 165
Norling, Karl, 19
Nurmi, Jari, 237
Nuszkowski, Heinrich, 93

Olbrich, Markus, 245
Ortmanns, Maurits, 153

Papavassiliou, Christos, 49
Park, Sung Yeul, 273
Parveg, Dristy, 77
Pasca, Mirko, 81
Pascu, Razvan, 297
Pashmineh, Sara, 185
Passerone, Roberto, 333
Pelz, Georg, 165
Pepe, Domenico, 85

Pezzotta, Alessandro, 261
Picciau, Andrea, 105
Pichler, Peter, 241, 253
Pillonnet, Gaël, 73
Pilz, Jürgen, 281
Pipino, Alessandra, 69
Plankensteiner, Kathrin, 281
Ponton, D., 9
Pracný, Peter, 57
Pribyl, Wolfgang, 89
Prou, Nico, 101
Pusane, Ali Emre, 329

Quinn, Patrick, 11

Raič, Dušan, 161
Rebel, Gregor, 149
Ribnikar, Rok, 109
Richter, Harald, 41, 317
Rivière, Fabian, 101
Rizzi, Antonello, 129
Rizzon, Luca, 333
Romani, Aldo, 269
Ronchi, Marco, 37
Rota, Lorenzo, 45
Rouzeyre, Bruno, 169, 341
Rumyancev, Ivan, 309

Sampietro, Marco, 61
Santner, A., 9
Sashank, K. V. S, 141
Scerri, Jeremy, 113
Schenk, Sven, 233
Schmidt, Alexander, 181
Schmidt, Gerhard, 253
Schuberth, Christian, 221
Schulz, Ingo, 149
Schwarzelbach, Oliver, 121
Scotti, Giuseppe, 345
Seebacher, David, 221
Selberherr, Siegfried, 157
Siegel, Michael, 249
Simmarano, Roberto, 65, 197
Singerl, Peter, 221
Singh, Abhishek Kumar, 141

Sisto, Arcangelo, 121
Sobotta, Elena, 265
Spöck, Gunter, 285
Srinivas, M. B., 141
Stadlober, Barbara, 13
Steyaert, Michiel, 277
Strle, Drago, 109, 161
Strobel, Axel, 305, 337
Strobel, Markus, 41
Subbiah, Iyappan, 229
Sverdlov, Viktor, 157

Tartagni, Marco, 269
Thaller, E., 9
Trifiletti, Alessandro, 145, 345
Trifković, Mario, 161
Trotta, Francesco, 145
Tumpold, David, 125, 293

Uhnevionak, Viktoryia, 241
Unterassinger, Hartwig, 89

Vahdati, Ali, 97
Varonen, Mikko, 77, 97
Vergine, Tommaso, 45
Vivet, Pascal, 341
Vollaire, Christian, 73
von Staudt, Martin, 23

Wacker, Nicoleta, 317
Wang, Wei, 349
Wegener, Carsten, 23
Windbacher, Thomas, 157
Wolf, Robert, 93, 189, 257, 313

Xu, Hongcheng, 185

Yilmaz, Gurkan, 209
Ytterdal, Trond, 29

Zernig, Anja, 285
Zhang, Ye, 233
Zhao, Jinshu, 313
Zito, Domenico, 85
Zwolinski, Mark, 177, 349

CURRAN ASSOCIATES INC.
proceedings
.com

9781467345798